HAMMOND'S
COMPREHENSIVE ATLAS
OF THE WORLD

CONTENTS

PART I

NEW YORK
C. S. HAMMOND & COMPANY
1914

Alphabetical List of Countries and States

Indicating the Maps in the Atlas on which they are shown

Hammond's Comprehensive Atlas of the World. Copyright, 1914, by C. S. Hammond & Co., N. Y.

Alphabetical List of Countries and States, (Continued)

INDEX OF PART II

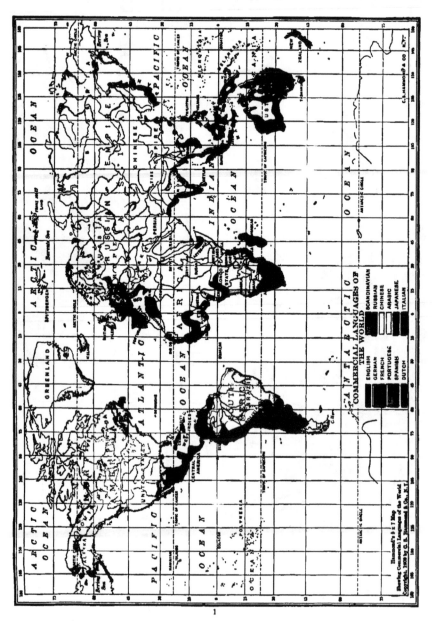

COMMERCIAL LANGUAGES OF
THE WORLD

ENGLISH
GERMAN
FRENCH
PORTUGUESE
SPANISH
DUTCH

SCANDINAVIAN
RUSSIAN
CHINESE
ARABIC
JAPANESE
ITALIAN

Hammond's 8½ Map
Showing Commercial Languages of the World
Copyright 1909 by C. S. Hammond & Co., N.Y.

C. S. HAMMOND & CO., N.Y.

CAM 2

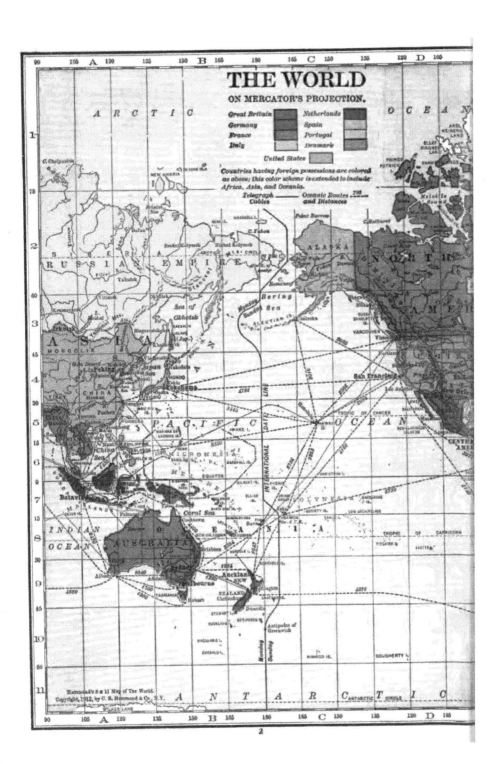

THE WORLD

ON MERCATOR'S PROJECTION.

Great Britain
Germany
France
Italy

Netherlands
Spain
Portugal
Denmark

United States

Countries having foreign possessions are colored as above; this color scheme is extended to include Africa, Asia, and Oceania.

Telegraph Cables — Oceanic Routes and Distances

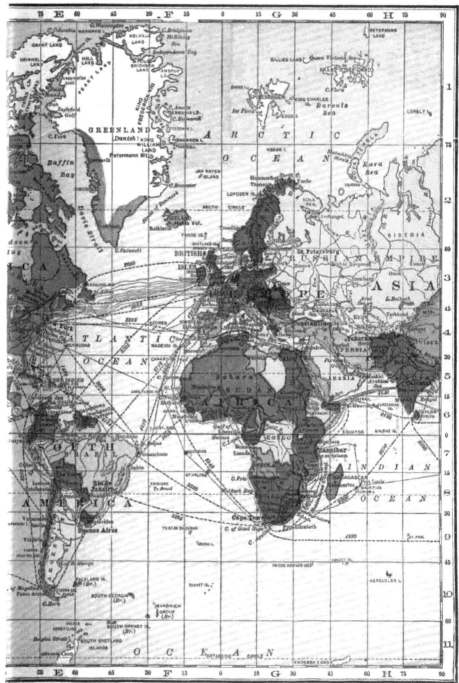

NORTH AMERICA

ENGLISH STATUTE MILES

CARIBBEAN SEA

5

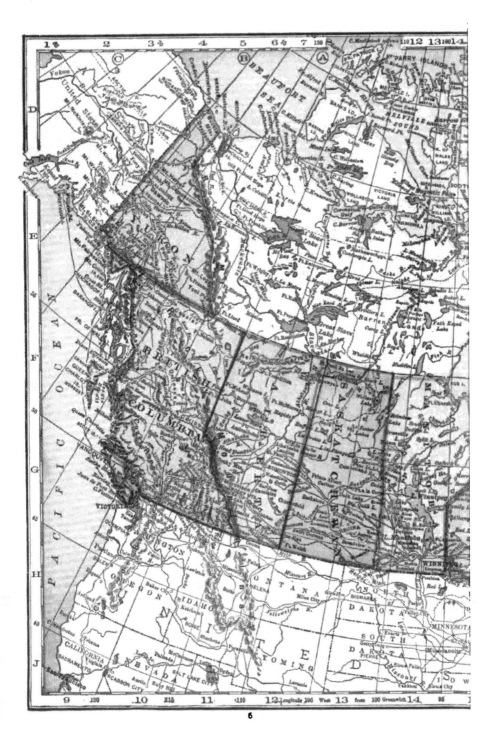

DOMINION OF CANADA
AND NEWFOUNDLAND

ENGLISH STATUTE MILES

Hammond's 8 x 11 map of Dominion of Canada.
Copyright, 1911, by C.S. Hammond & Co., N.Y.

7

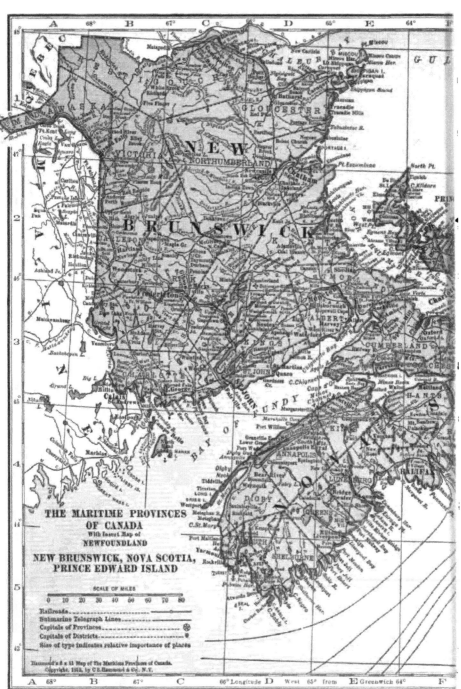

THE MARITIME PROVINCES
OF CANADA
With Insert Map of
NEWFOUNDLAND

NEW BRUNSWICK, NOVA SCOTIA,
PRINCE EDWARD ISLAND

SCALE OF MILES
0   10   20   30   40   50   60   70   80

Railroads
Submarine Telegraph Lines
Capitals of Provinces
Capitals of Districts
Size of type indicates relative importance of places

Hammond's 8 x 11 Map of The Maritime Provinces of Canada.
Copyright, 1913, by C.S. Hammond & Co., N.Y.

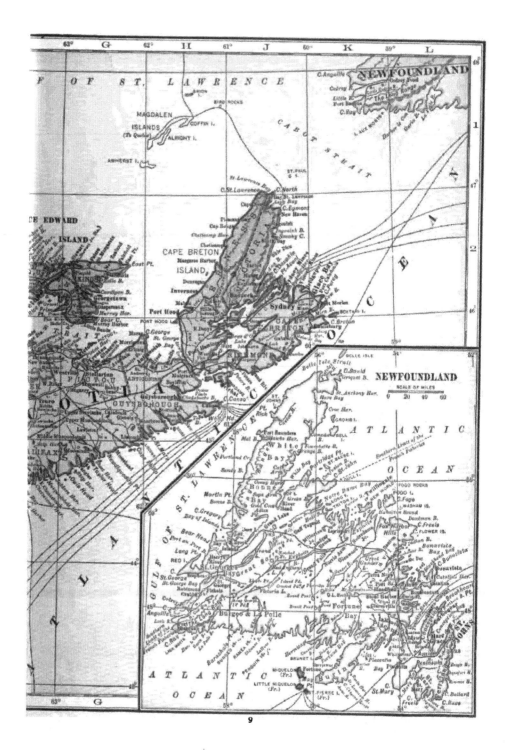

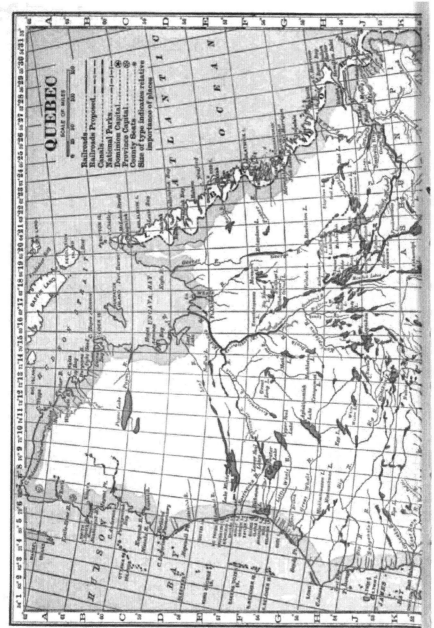

QUEBEC

SCALE OF MILE

Railroads
Railroads Proposed
Canals
National Parks
Dominion Capital
Province Capital
County Seats
Size of type indicates relative
importance of places

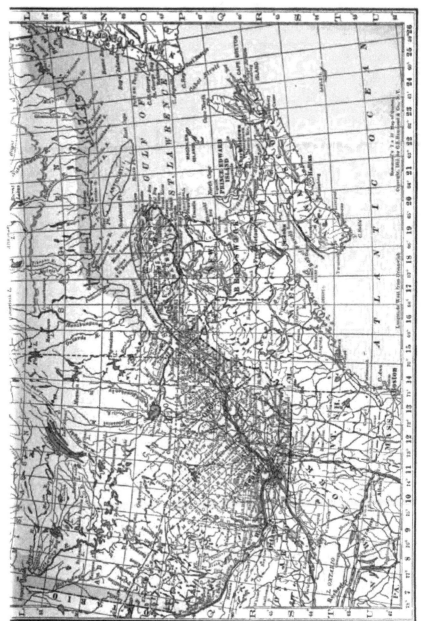

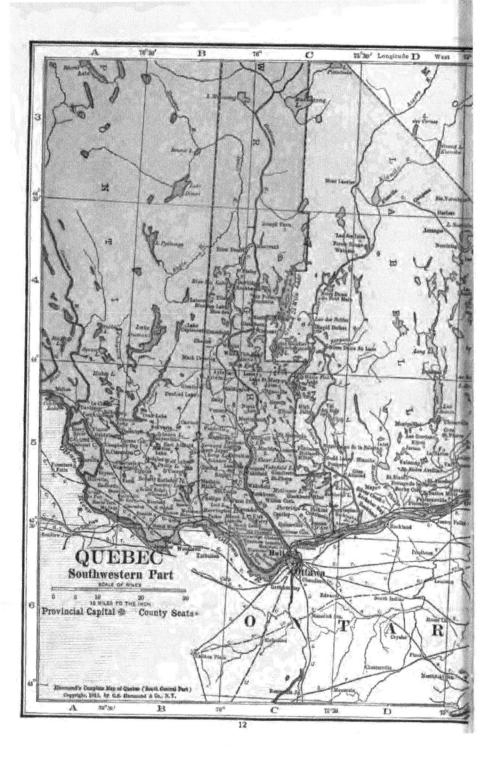

QUEBEC
Southwestern Part

SCALE OF MILES

Provincial Capital ⊕   County Seats

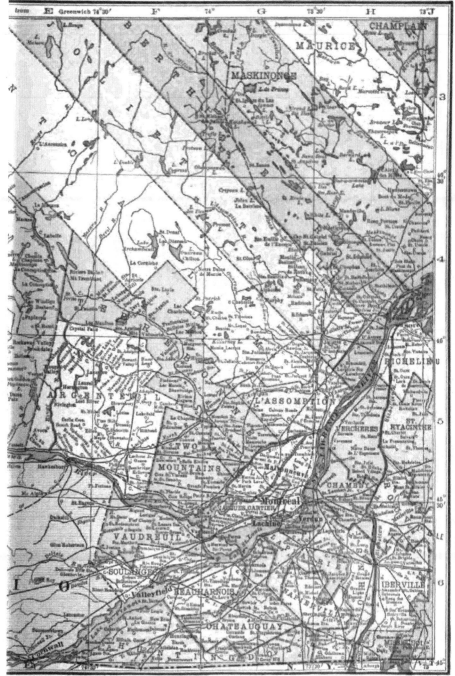

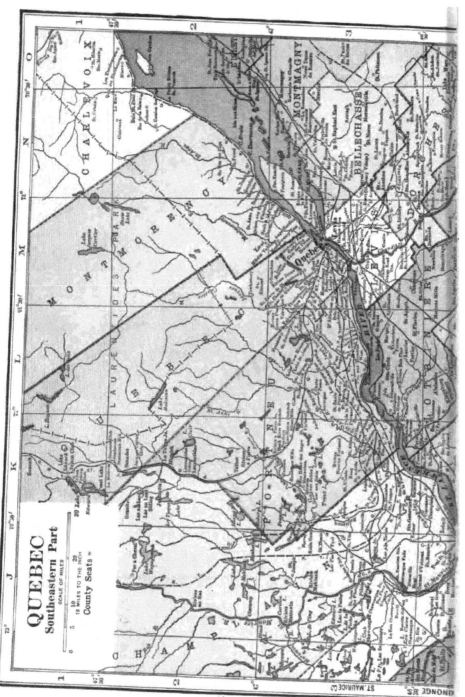

QUEBEC
Southeastern Part
SCALE OF MILES
18 MILES TO THE INCH
County Seats ⊙

CHARLEVOIX

MONTMORENCY

LAURENTIDES PARK

MONTMORENCY

BELLECHASSE

DORCHESTER

Quebec

PORTNEUF

BEAUCE

ST. MAURICE CO.

GRANGE RS.

CHAMPLAIN

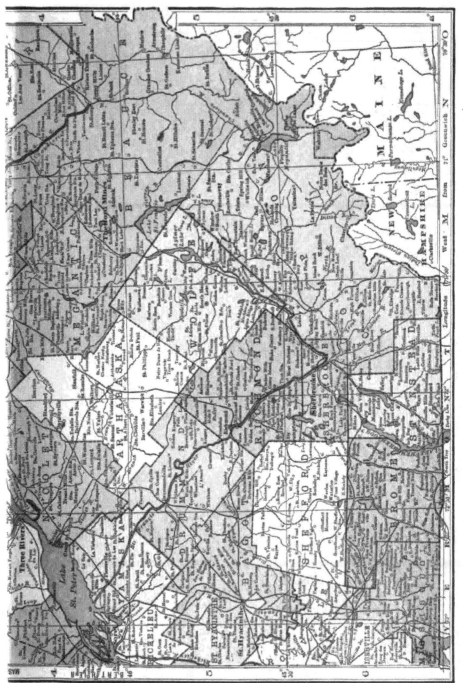

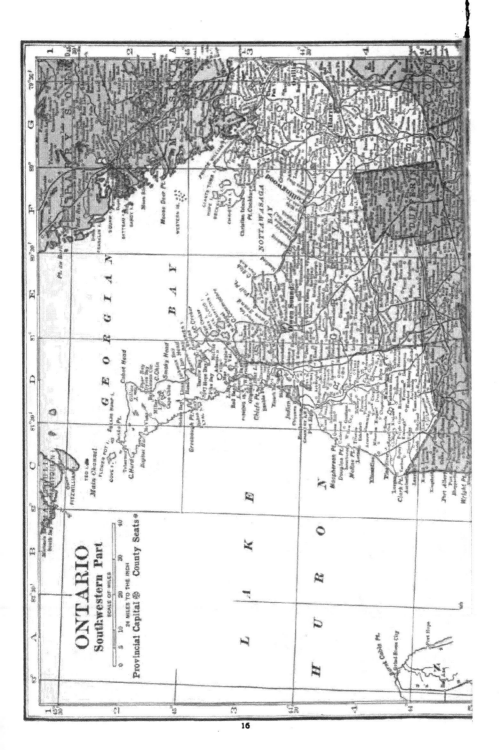

ONTARIO
Southwestern Part

SCALE OF MILES

24 MILES TO THE INCH

Provincial Capital ⊛   County Seats⊛

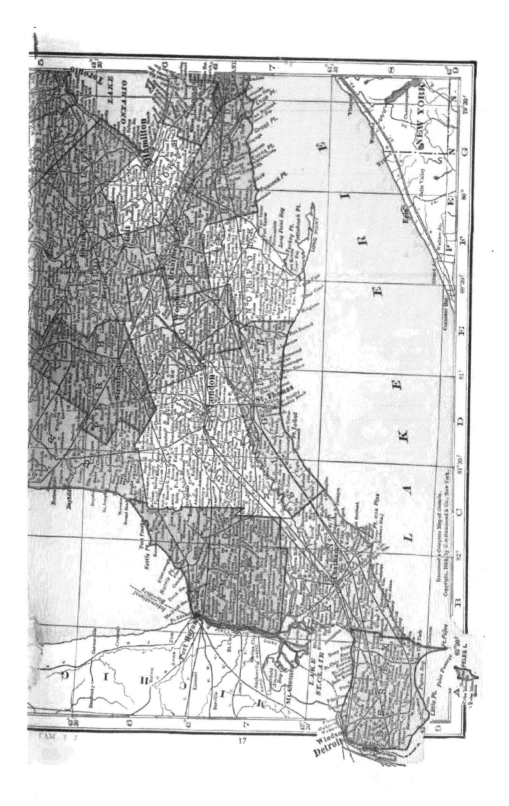

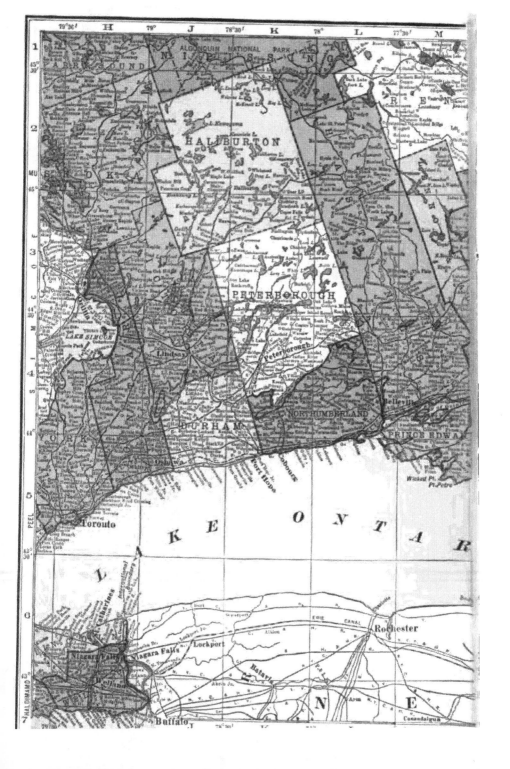

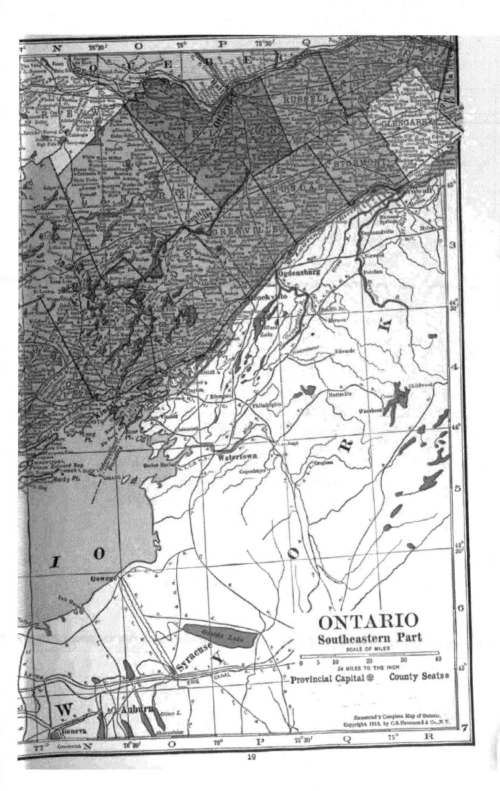

ONTARIO
Southeastern Part
SCALE OF MILES

0   5   10   20   30   40
24 MILES TO THE INCH

Provincial Capital ⊛   County Seats ●

Hammond's Complete Map of Ontario.
Copyright 1915, by C.S. Hammond & Co., N.Y.

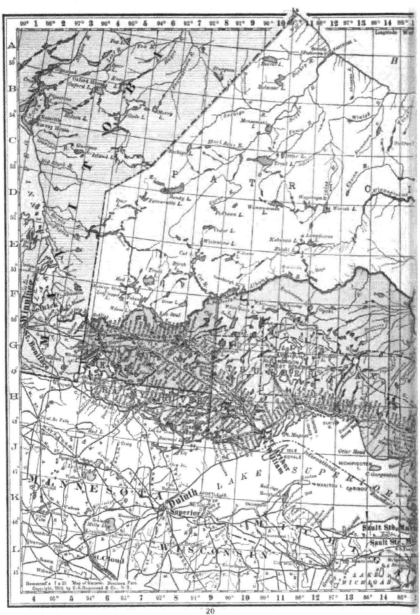

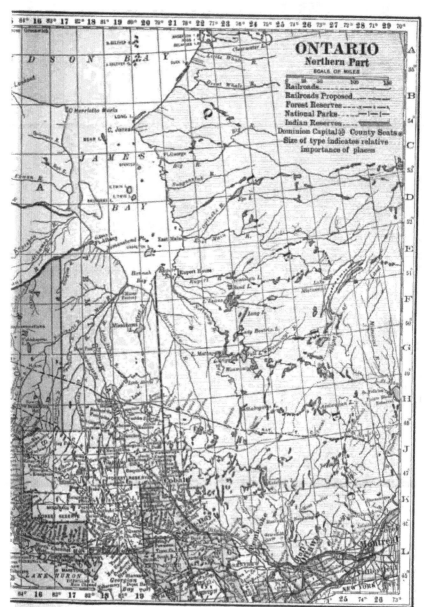

ONTARIO
Northern Part
SCALE OF MILES

Railroads
Railroads Proposed
Forest Reserves
National Parks
Indian Reserves
Dominion Capital County Seats
Size of type indicates relative
importance of places

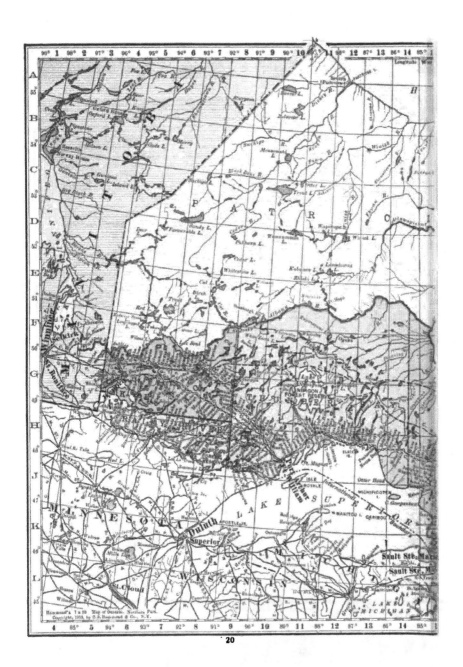

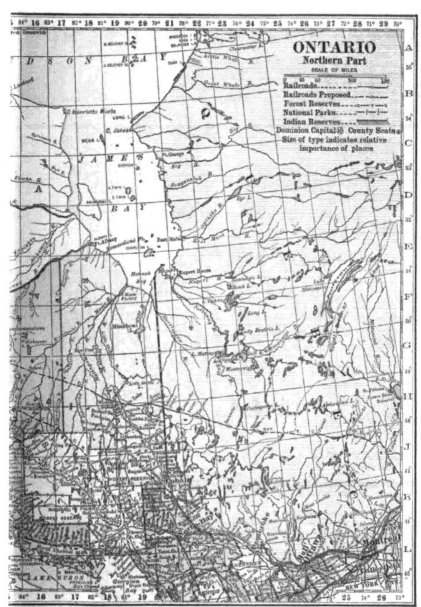

ONTARIO
Northern Part
SCALE OF MILES

Railroads.........................
Railroads Proposed...............
Forest Reserves..................
National Parks...................
Indian Reserves..................
Dominion Capital⊛ County Seats●
Size of type indicates relative
importance of places

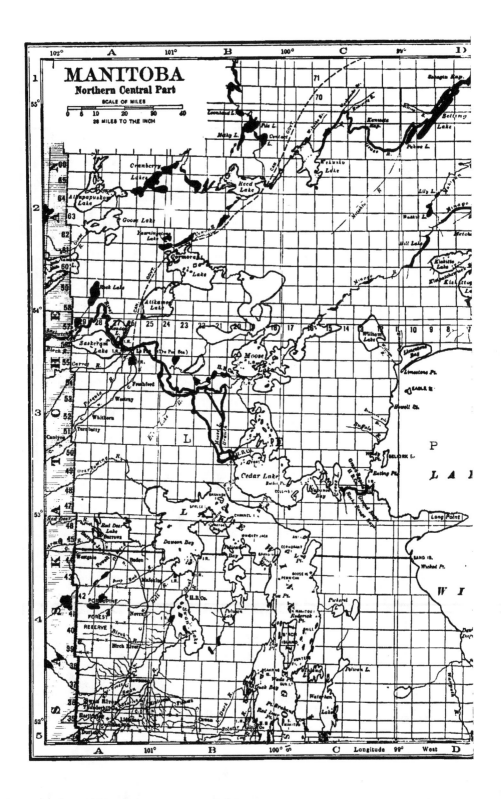

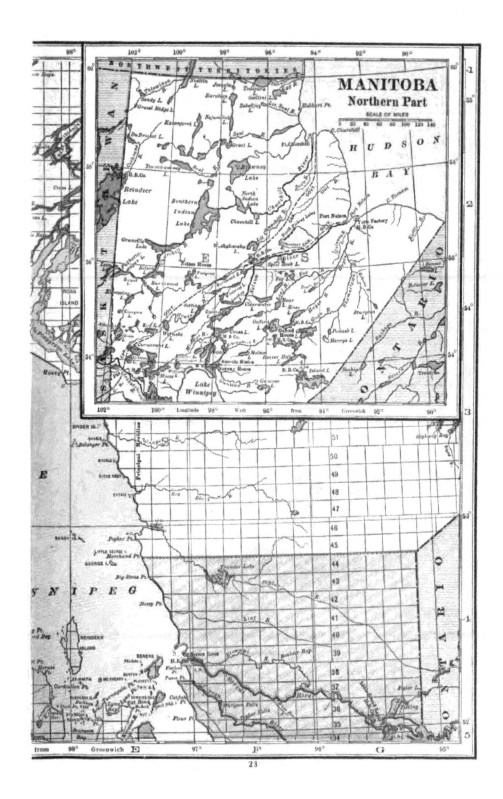

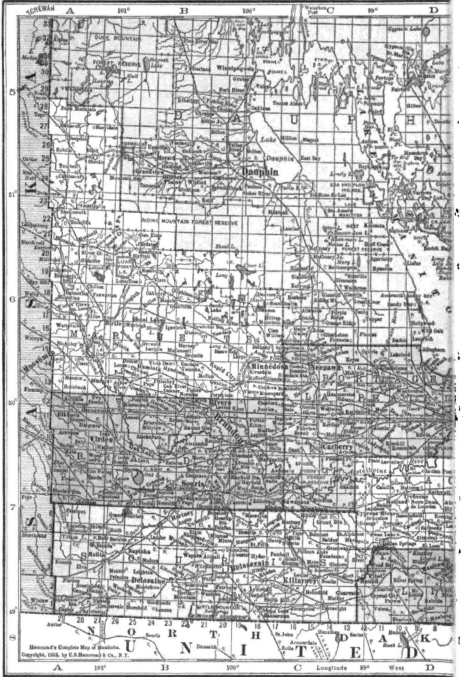

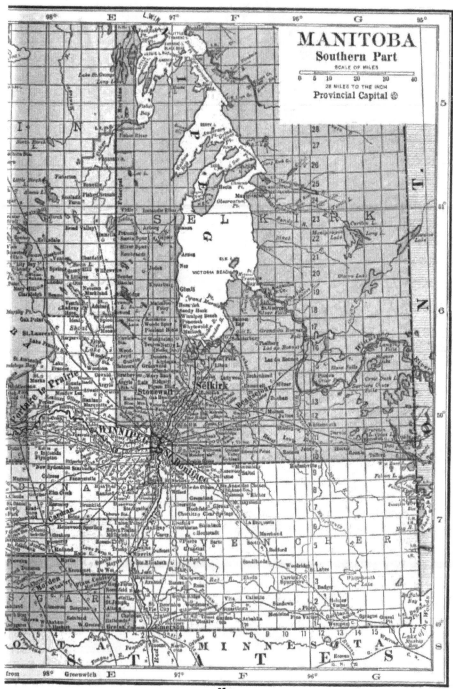

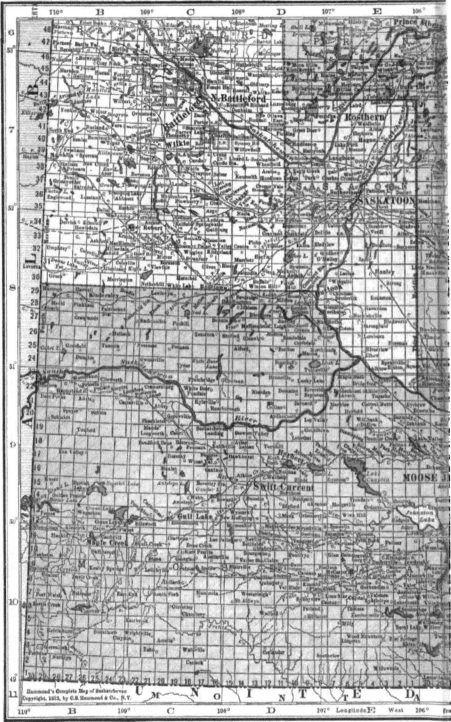

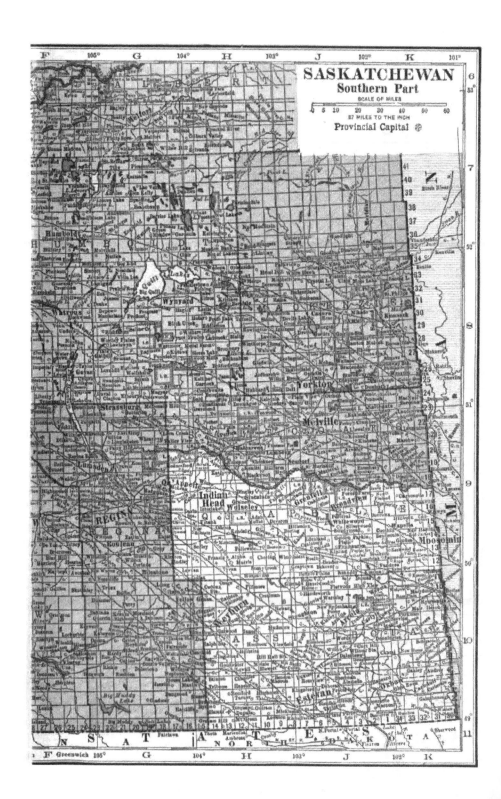

SASKATCHEWAN
Southern Part

SCALE OF MILES

87 MILES TO THE INCH

Provincial Capital ⊕

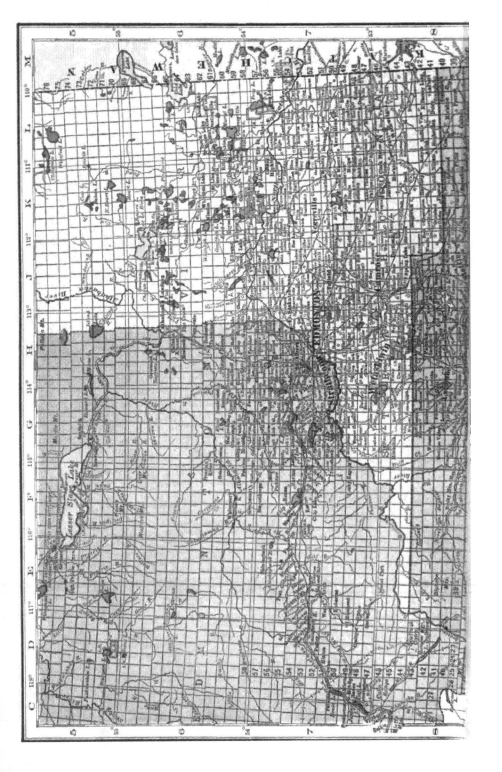

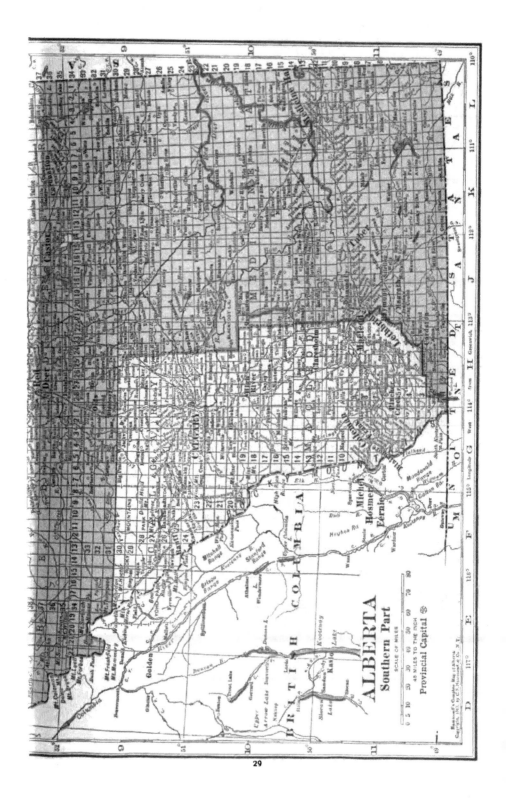

ALBERTA
Southern Part

SCALE OF MILES

Provincial Capital ⊛

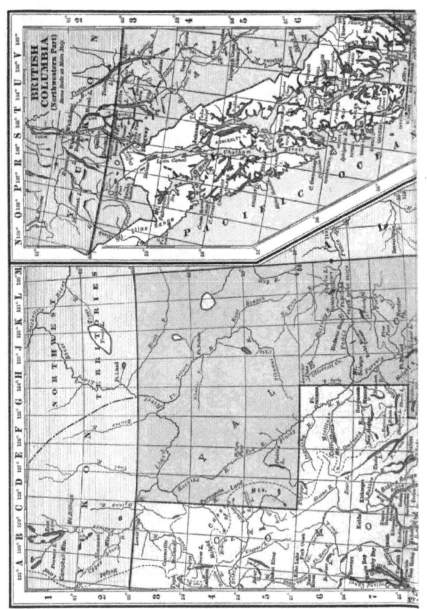

BRITISH
COLUMBIA
(Northwestern Part)

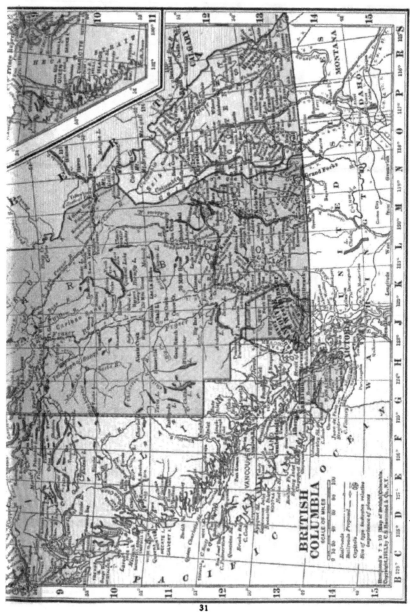

BRITISH COLUMBIA
SCALE OF MILES

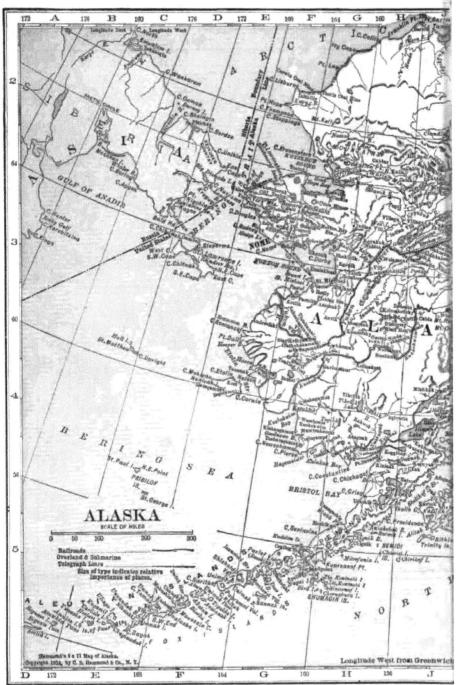

ALASKA

SCALE OF MILES

0   50   100   200   300

Railroads
Overland & Submarine
Telegraph Lines
Size of type indicates relative
importance of places.

Hammond's 8 x 11 Map of Alaska.
Copyright 1914, by C. S. Hammond & Co., N. Y.

Longitude West from Greenwich

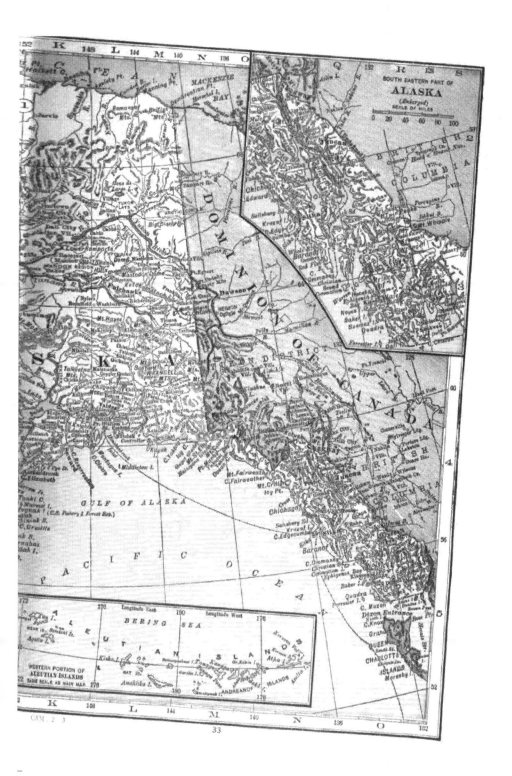

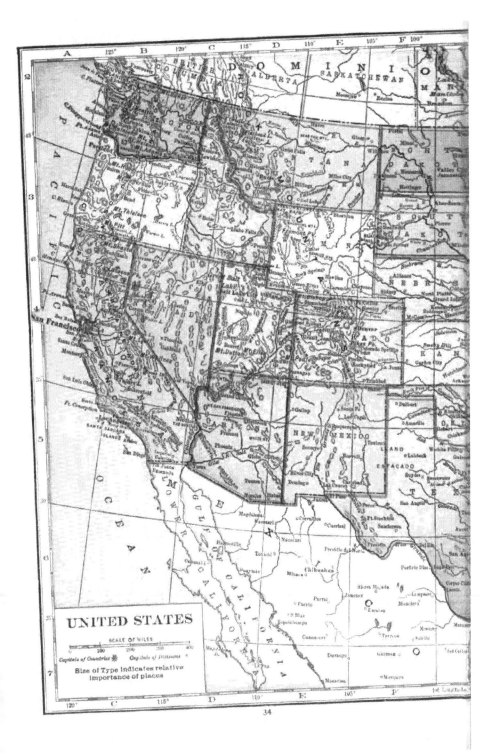

UNITED STATES

SCALE OF MILES

Capitals of Countries ✸   Capitals of Divisions ✳

Size of Type indicates relative
importance of places

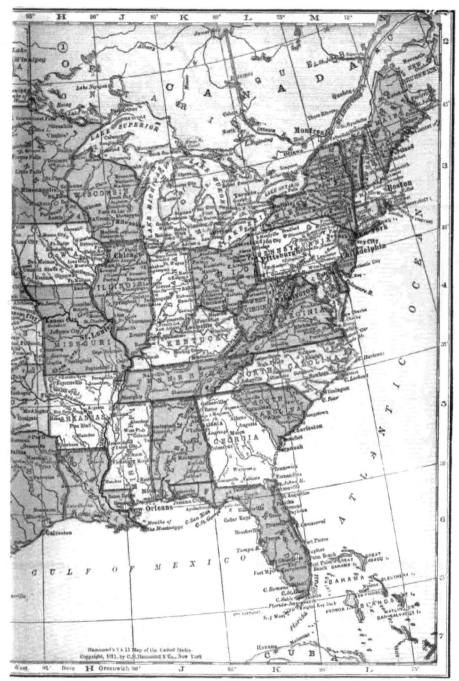

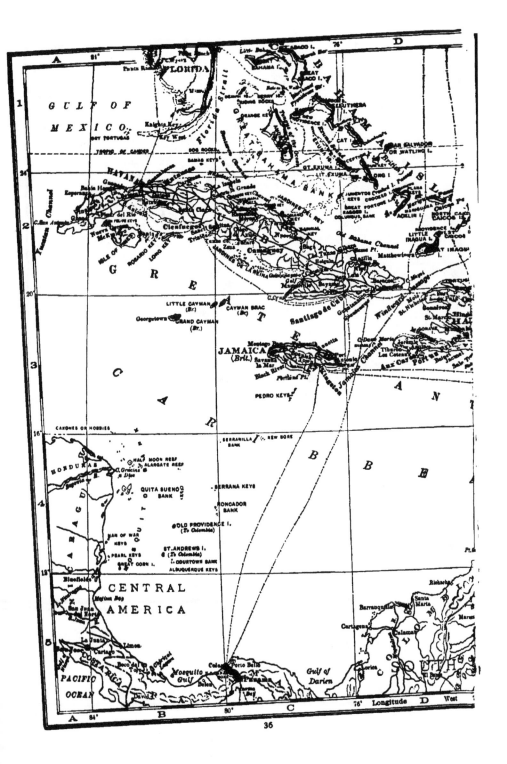

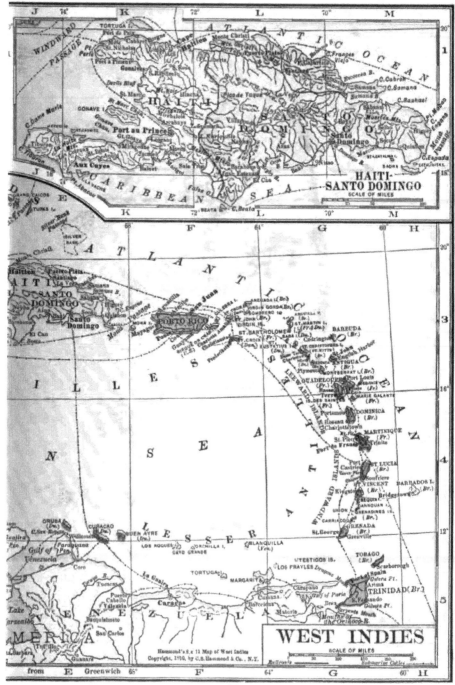

HAITI-
SANTO DOMINGO
SCALE OF MILES

WEST INDIES
SCALE OF MILES

Hammond's 8 x 11 Map of West Indies
Copyright, 1910, by C.S. Hammond & Co., N.Y.

37

CUBA

Scale of Miles

0 10 20 30 40 50   75   100

Explanation:

Capital of Cuba thus: **HAVANA** ✸
Capitals of Province thus: Matanzas ⊙
Railroads: ....

Hammond's 8 x 1" Map of Cuba.

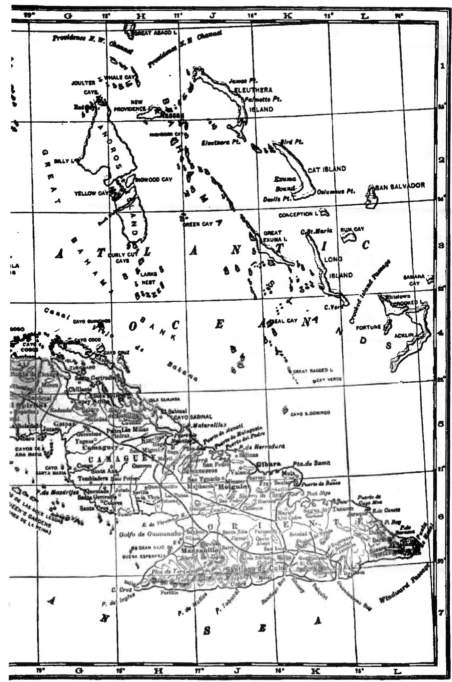

PORTO R[ICO]

SCALE OF MILES

0   5   10

Railroads
Improved Roads
Size of type indicates relative imp[ortance]

Hammond's 8 x 11 Map of Porto Rico
Copyright, 1914, by C. S. Hammond & Co., N.Y.

ISLANDS WEST OF
PORTO RICO
SAME SCALE

Longitude   C   West from   66°30'   Gre[enwich]

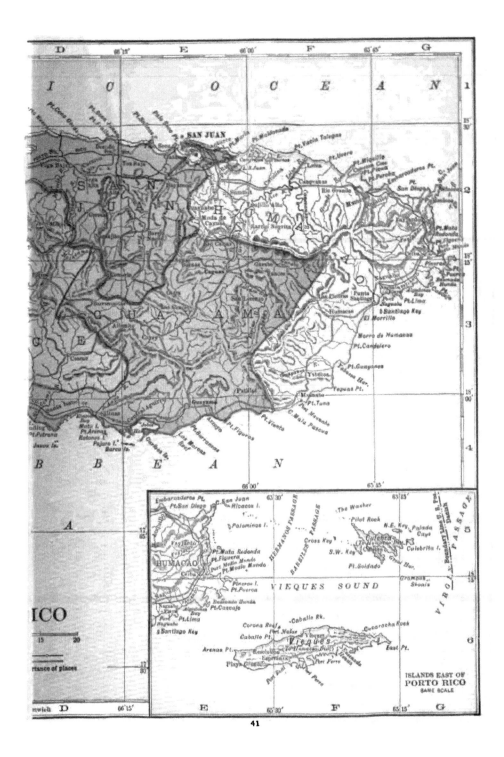

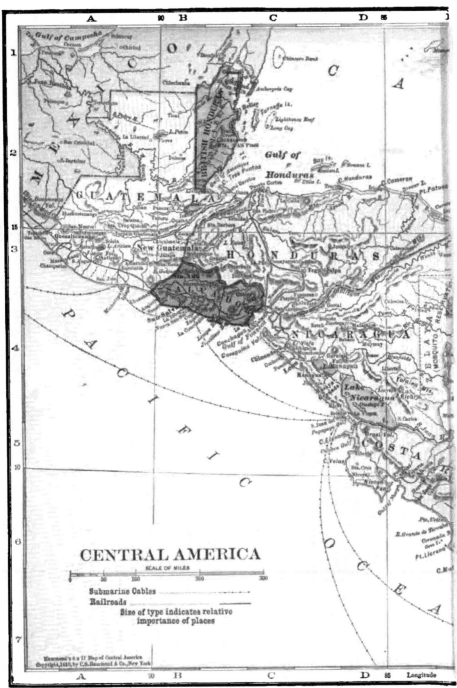

CENTRAL AMERICA

SCALE OF MILES

50   100   200   300

Submarine Cables

Railroads

Size of type indicates relative
importance of places

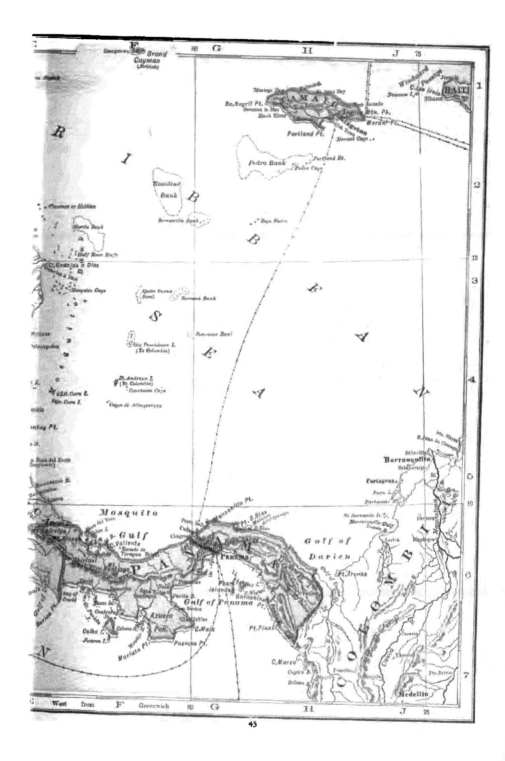

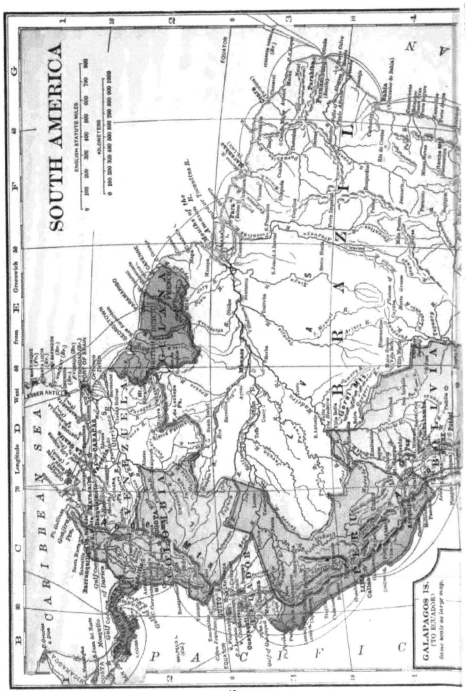

SOUTH AMERICA

ENGLISH STATUTE MILES

0   100   200   300   400   500   600   700   800

KILOMETERS

0   100   200   300   400   500   600   700   800   900   1000

GALAPAGOS IS.
(TO ECUADOR)
Same scale as large map.

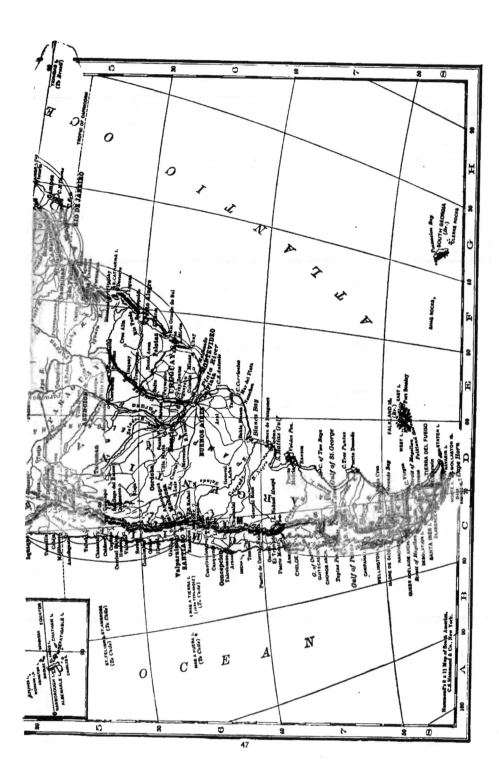

SOUTH AMERICA
(NORTHERN PART)

SCALE OF MILES

0   50 100   200   300   400   500

Railroads ....

Submarine Cables ....

Size of type indicates relative
importance of places

Reproduced 8 x 11 Map of South America (Northern Part)
Copyright 1910 by C.S. Hammond & Co.

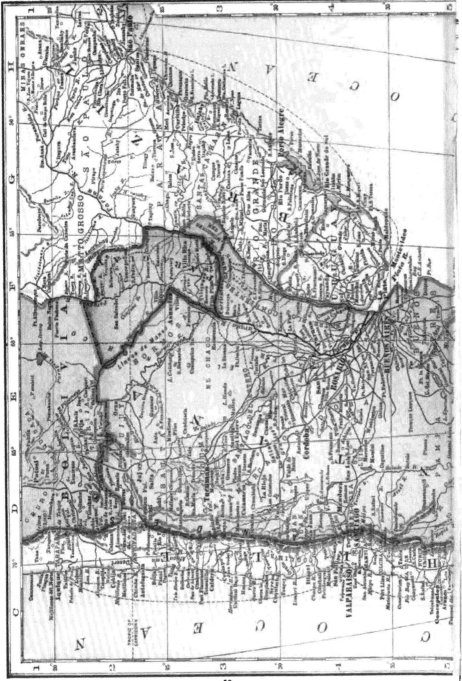

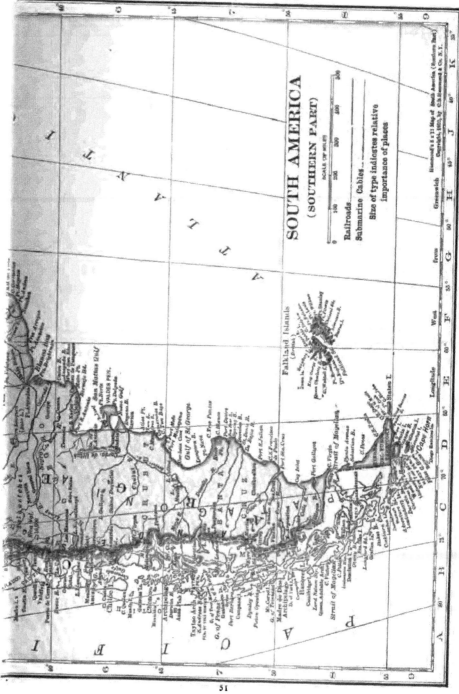

SOUTH AMERICA
(SOUTHERN PART)

SCALE OF MILES

Railroads
Submarine Cables
Size of type indicates relative
importance of places

Hammond's 5.75 Map of South America (Southern Part)
Copyright, 1909, by C.S. Hammond & Co. N.Y.

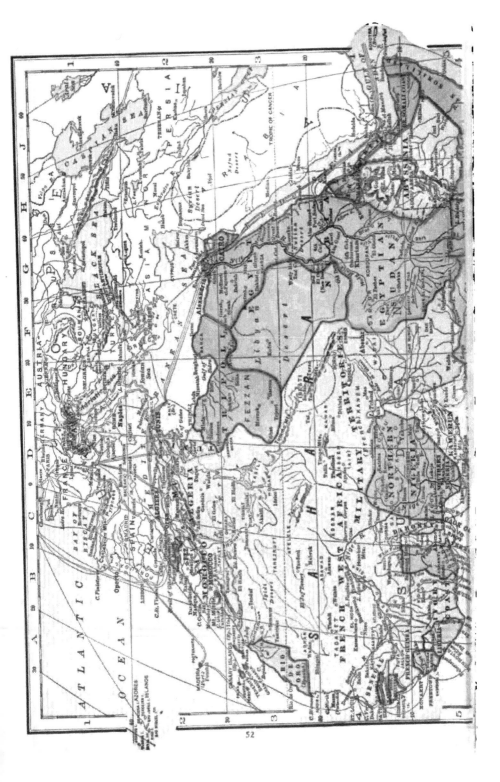

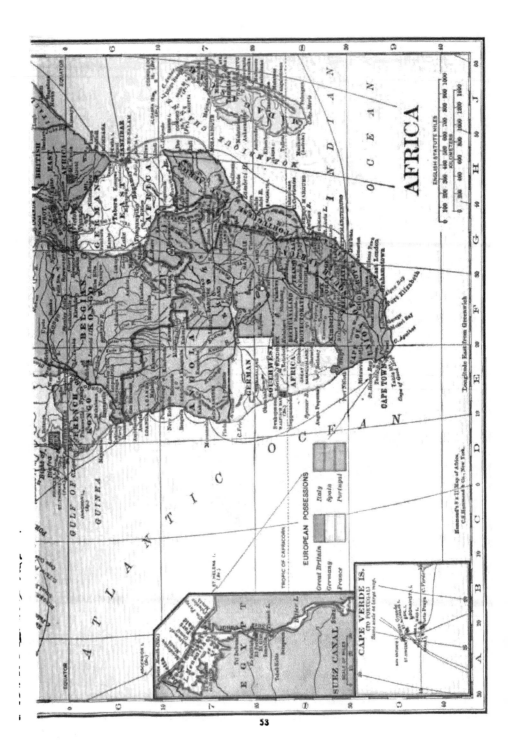

AFRICA

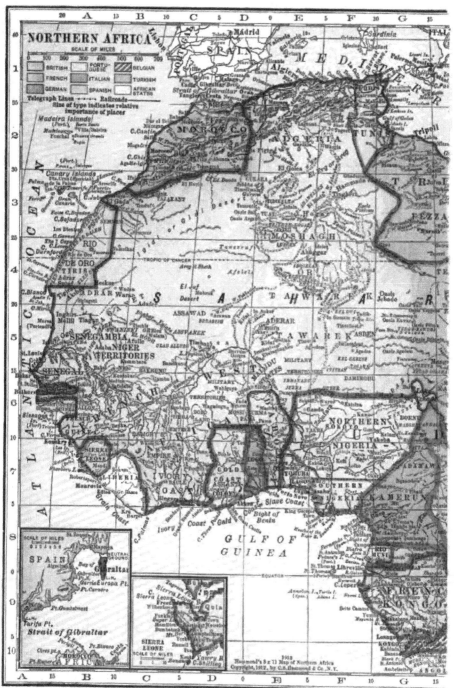

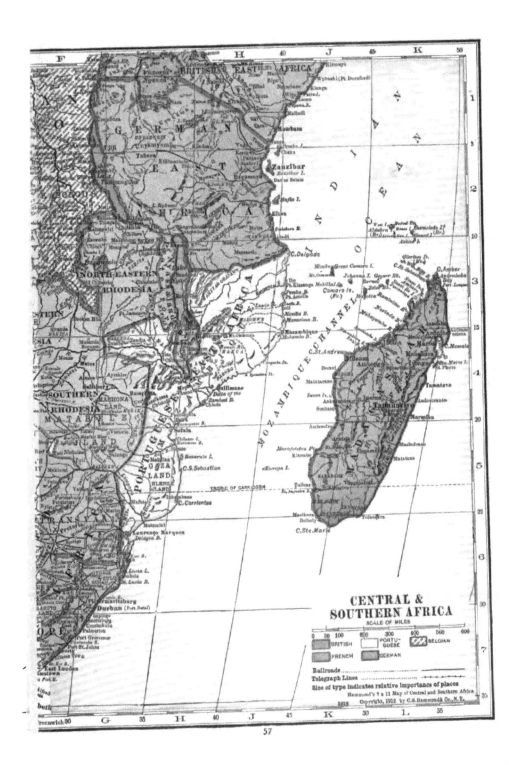

CENTRAL &
SOUTHERN AFRICA

SCALE OF MILES

0 50 100   200   300   400   500   600

BRITISH
PORTU-
GUESE
FRENCH
GERMAN
BELGIAN

Railroads
Telegraph Lines
Size of type indicates relative importance of places

Hammond's 9 x 11 Map of Central and Southern Africa

Copyright, 1912 by C.S.Hammonds Co., N.Y.

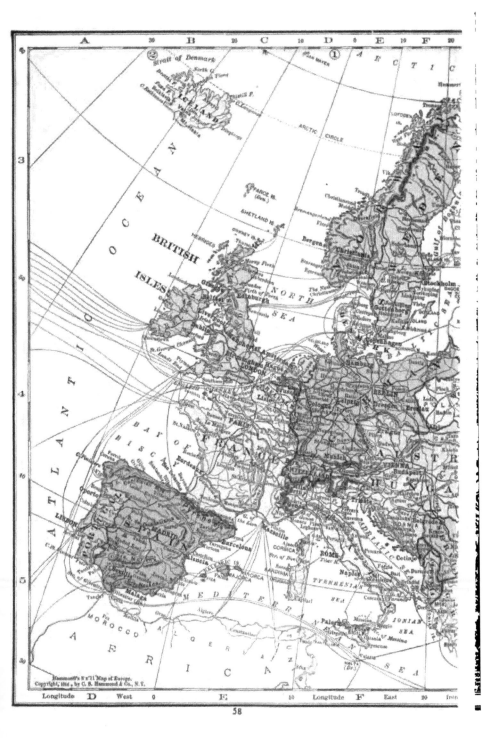

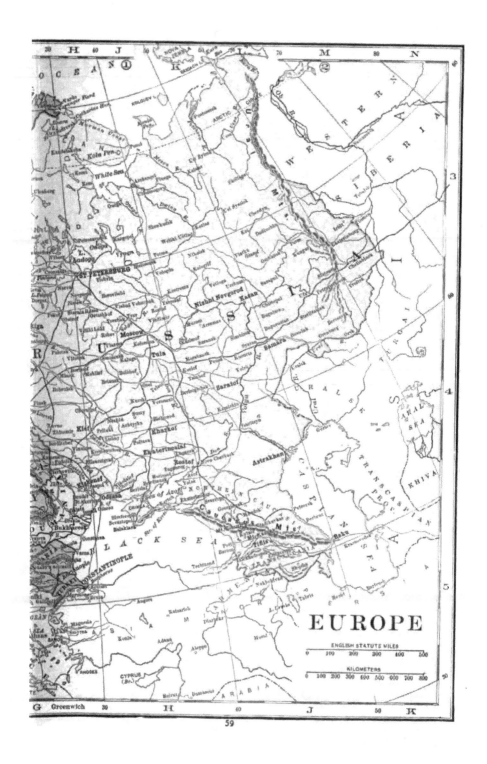

EUROPE

ENGLISH STATUTE MILES

KILOMETERS

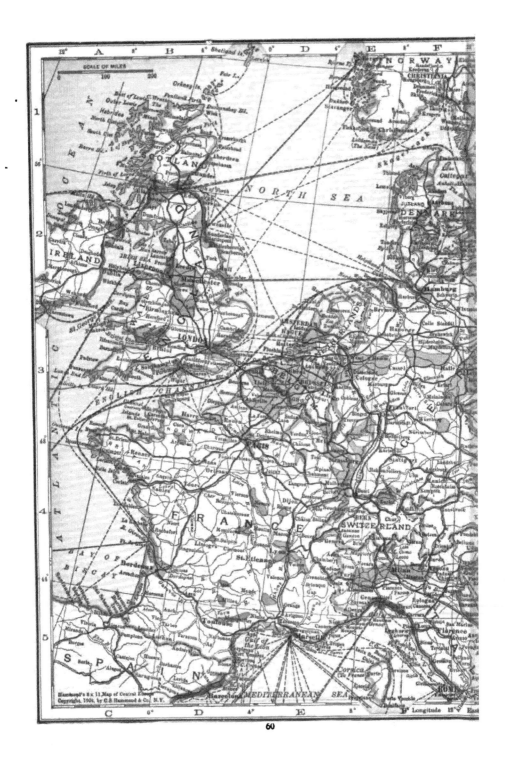

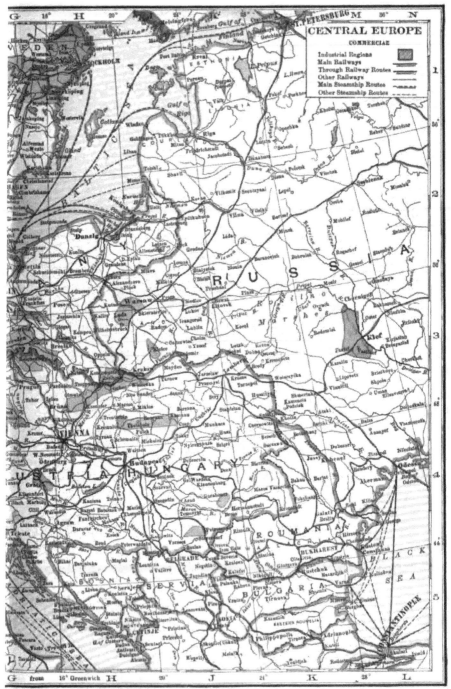

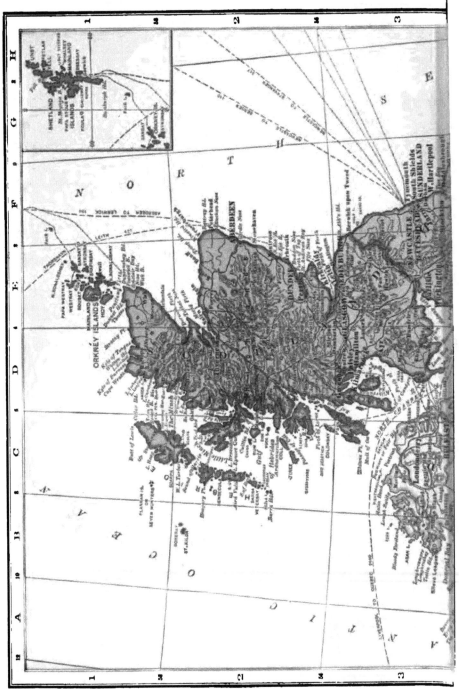

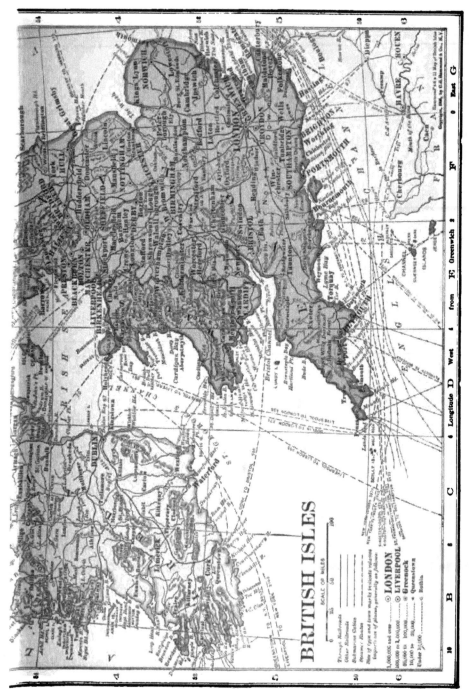

BRITISH ISLES

SCALE OF MILES

LONDON

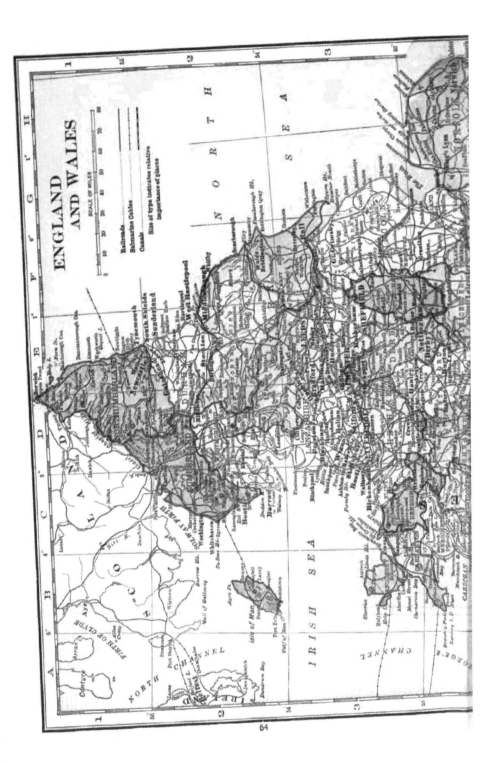

ENGLAND
AND WALES

SCALE OF MILES

Railroads
Submarine Cables
Canals

Size of type indicates relative
importance of places

NORTH SEA

IRISH SEA

CHANNEL

FRITH OF CLYDE

NORTH CHANNEL

ST. GEORGE'S CHANNEL

CARDIGAN

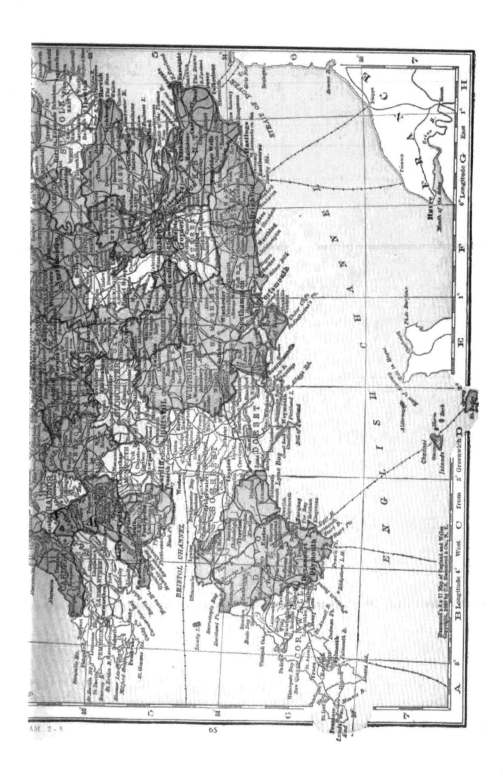

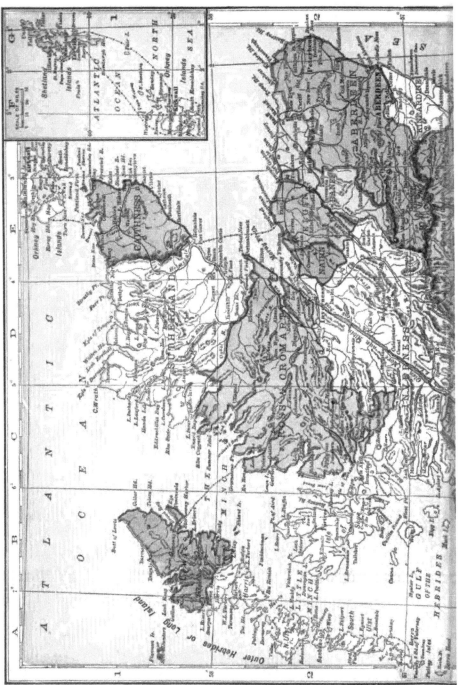

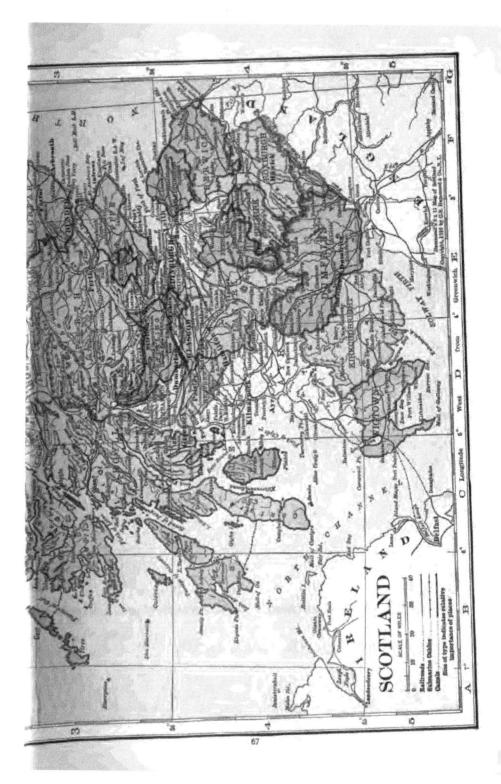

SCOTLAND

SCALE OF MILES

Railroads
Submarine Cables
Canals
Size of type indicates relative
importance of places

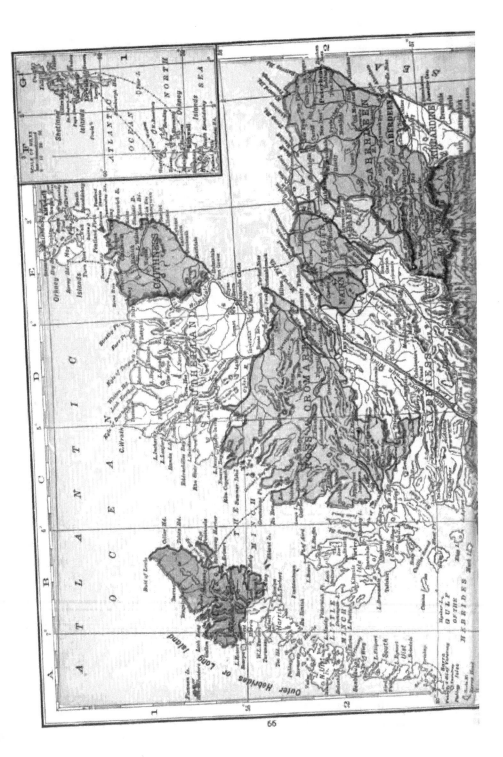

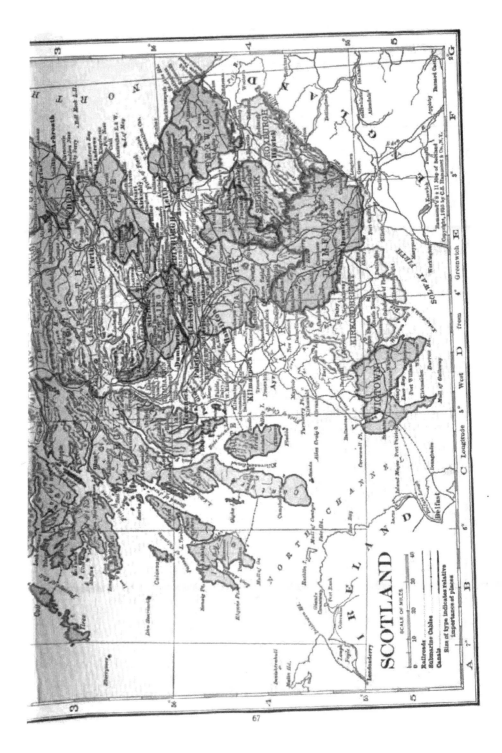

SCOTLAND

SCALE OF MILES

Railroads
Submarine Cables
Canals

Size of type indicates relative
importance of places

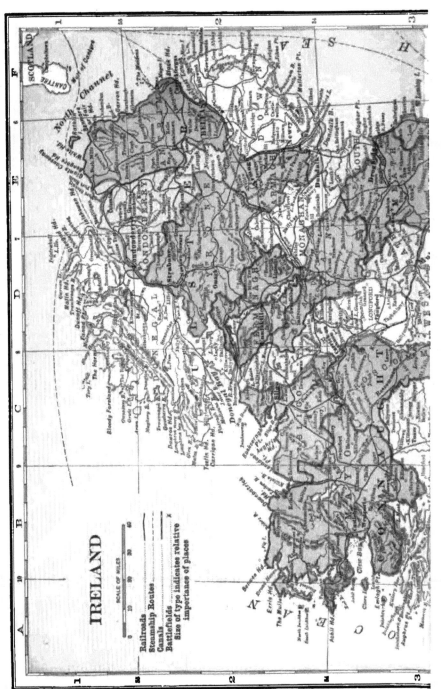

TRELAND

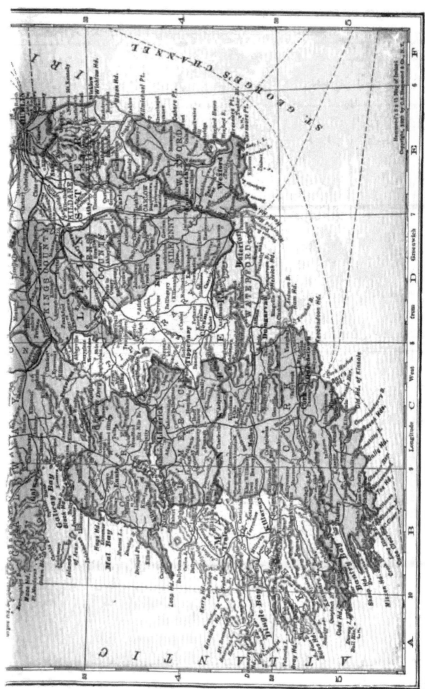

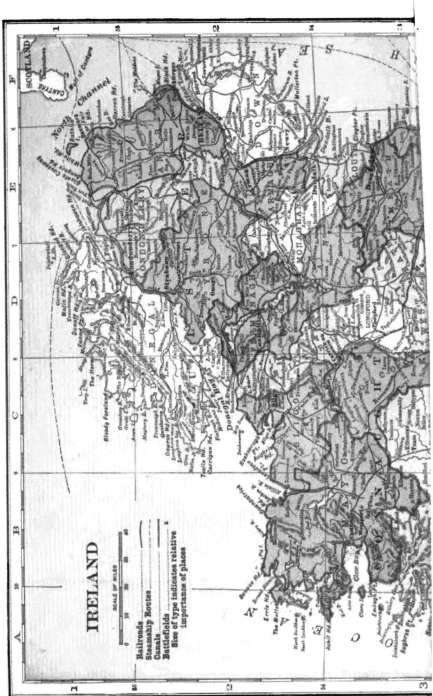

IRELAND

SCALE OF MILES

Railroads
Steamship Routes
Canals
Battlefields
Size of type indicates relative
importance of places

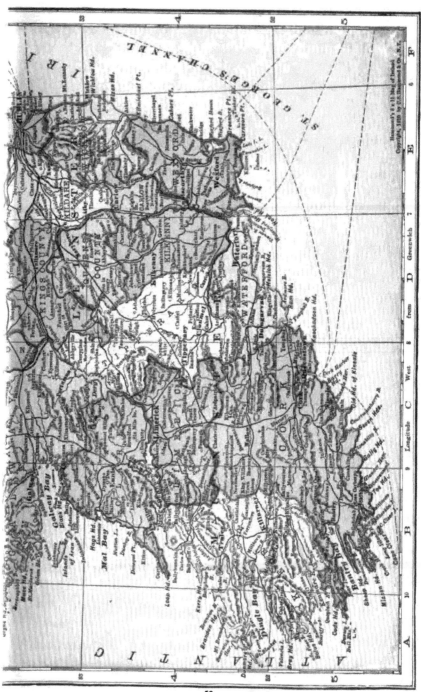

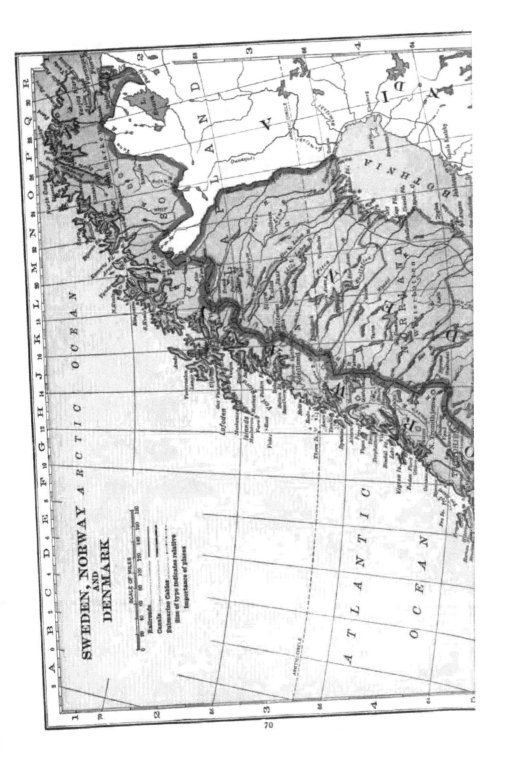

SWEDEN, NORWAY
AND
DENMARK

SCALE OF MILES

Railroads
Canals
Submarine Cables
Size of type indicates relative
importance of places

ARCTIC OCEAN

ATLANTIC

OCEAN

ARCTIC CIRCLE

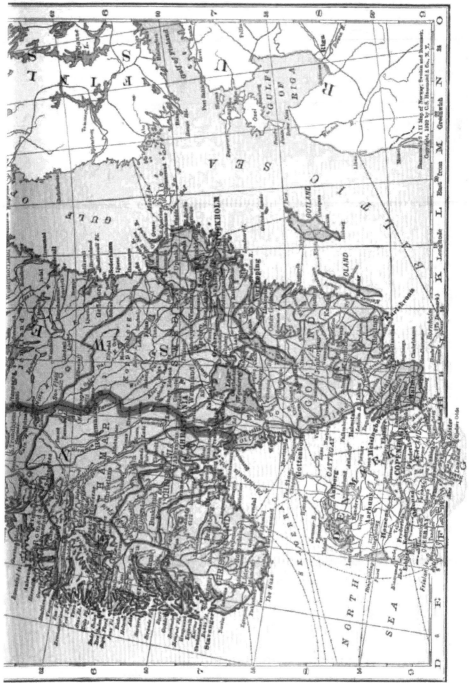

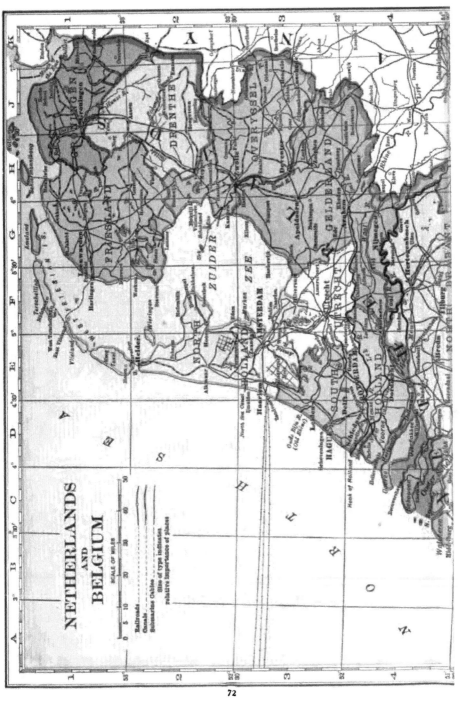

NETHERLANDS
AND
BELGIUM

SCALE OF MILES

0 5 10 20 30 40 50

Railroads
Canals
Submarine Cables

Size of type indicates
relative importance of places

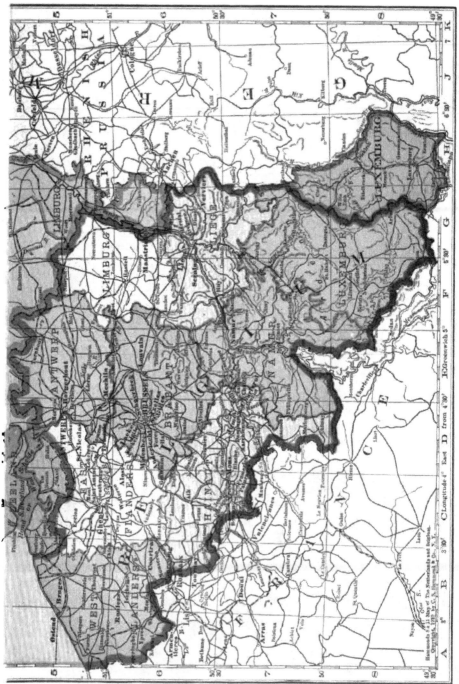

GERMANY

SCALE OF MILES

0   20   40   60   80   100

Railroads
Submarine Cables
Canals
Size of type indicates relative
importance of places

THURINGIAN
STATES

2 Reuss-Greiz.
3 Reuss-Schleiz.
4 Saxe-Altenburg.
5 Saxe-Coburg-
   Gotha.
6 Saxe-Meinin-
   gen.
7 Saxony (Saxe-
   Weimar-Eise-
   nach).
8 Schwarzburg-
   Rudolstadt.
9 Schwarzburg-
   Sondershausen

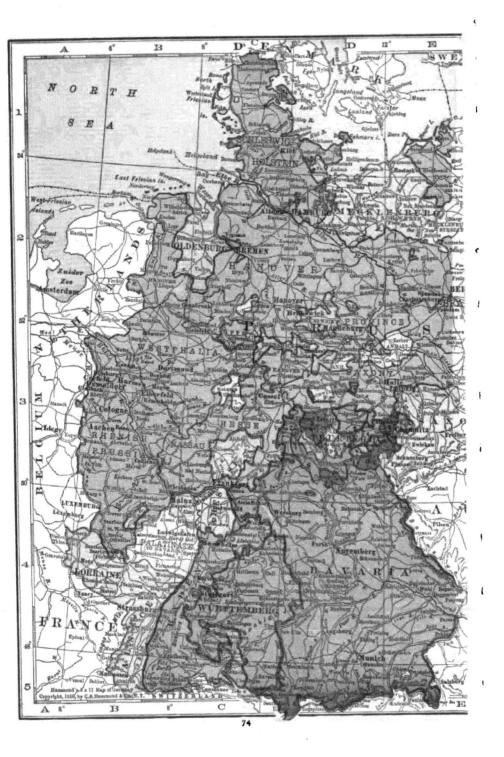

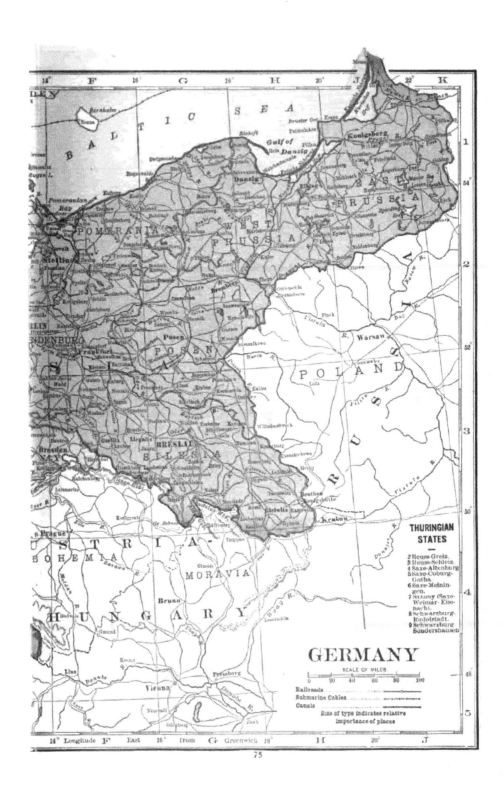

THURINGIAN
STATES

2 Reuss-Greiz.
3 Reuss-Schleiz.
4 Saxe-Altenburg
5 Saxe-Coburg-
Gotha.
6 Saxe-Meinin-
gen.
7 Saxony (Saxe-
Weimar-Eise-
nach).
8 Schwarzburg-
Rudolstadt.
9 Schwarzburg-
Sondershausen

GERMANY

SCALE OF MILES

0   20   40   60   80   100

Railroads
Submarine Cables
Canals

Size of type indicates relative
importance of places

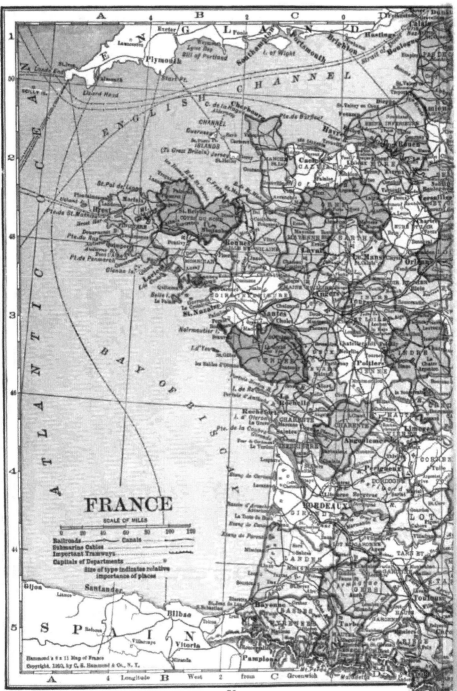

FRANCE

SCALE OF MILES

Railroads ——— Canals ————
Submarine Cables
Important Tramways
Capitals of Departments
Size of type indicates relative
importance of places

Hammond's 8 x 11 Map of France
Copyright, 1910, by C. S. Hammond & Co., N. Y.

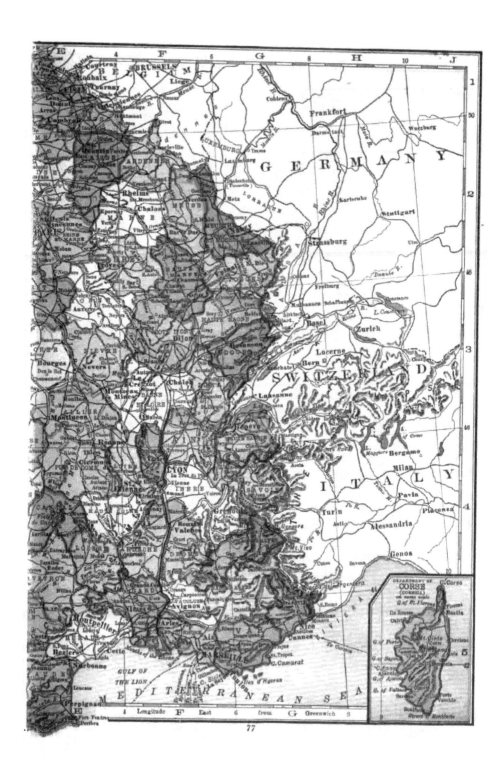

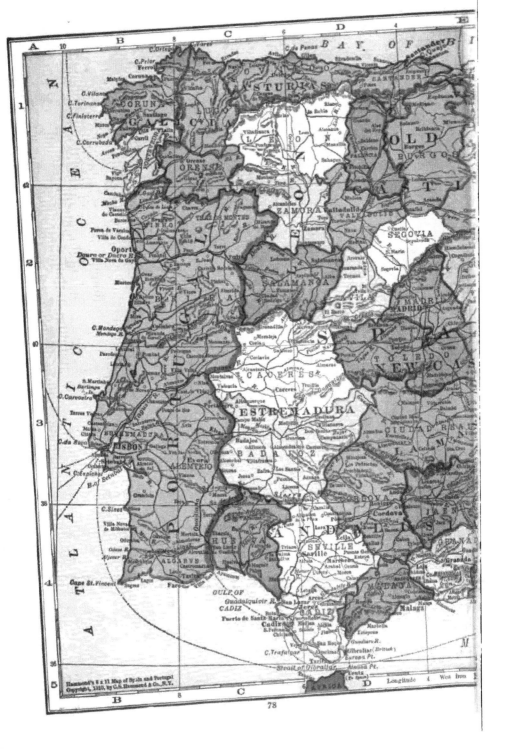

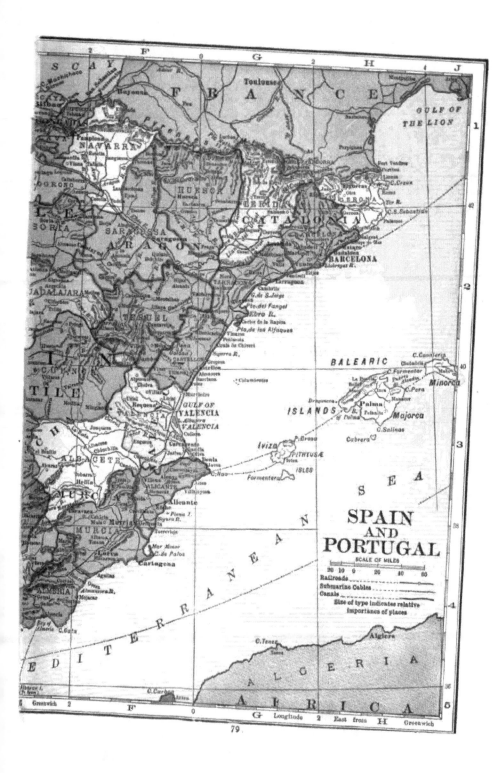

SPAIN
AND
PORTUGAL

SCALE OF MILES

Railroads
Submarine Cables
Canals
Size of type indicates relative
importance of places

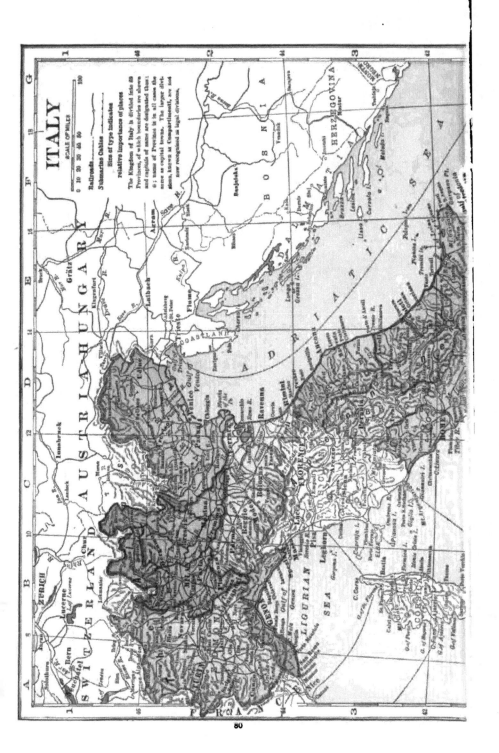

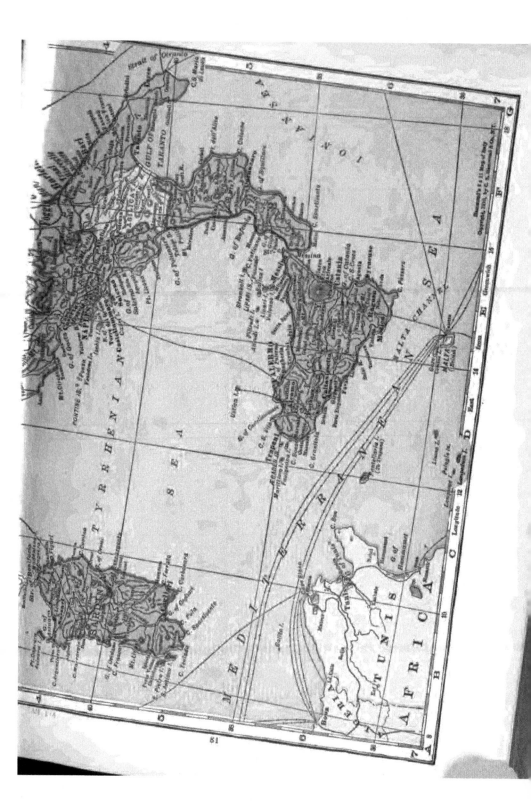

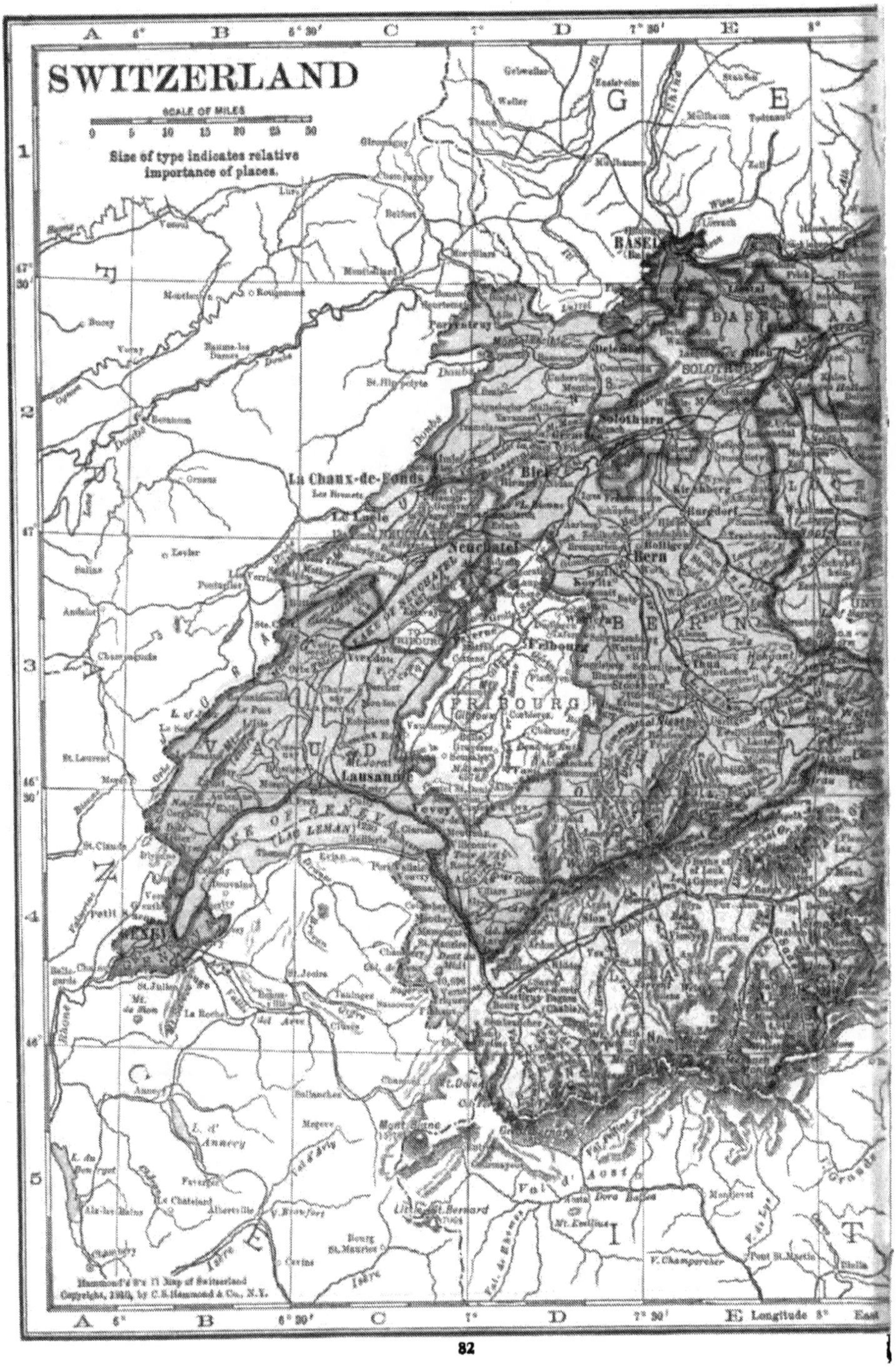

SWITZERLAND

SCALE OF MILES

0   5   10   15   20   25   30

Size of type indicates relative
importance of places.

BASEL

La Chaux-de-Fonds   Biel

Neuchâtel   Bern

Fribourg

FRIBOURG

Lausanne

LAKE OF GENEVA   (LAC LEMAN)

GENEVA

SOLOTHURN

Solothurn

Little St. Bernard

A   6°   B   6° 30'   C   7°   D   7° 30'   E   Longitude 8°   East

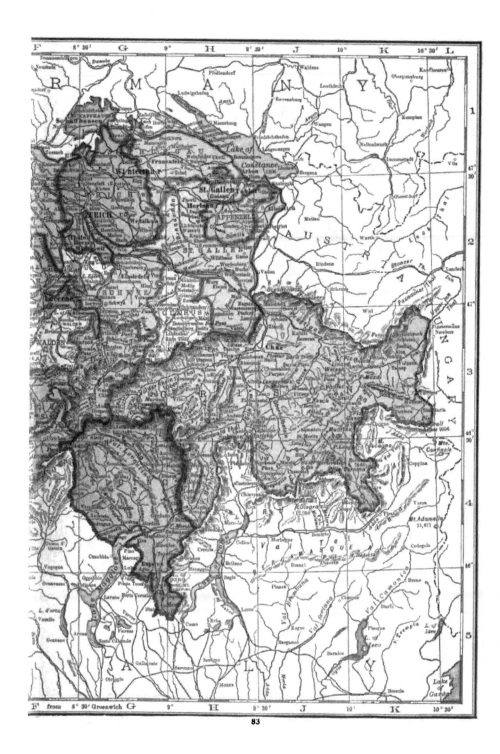

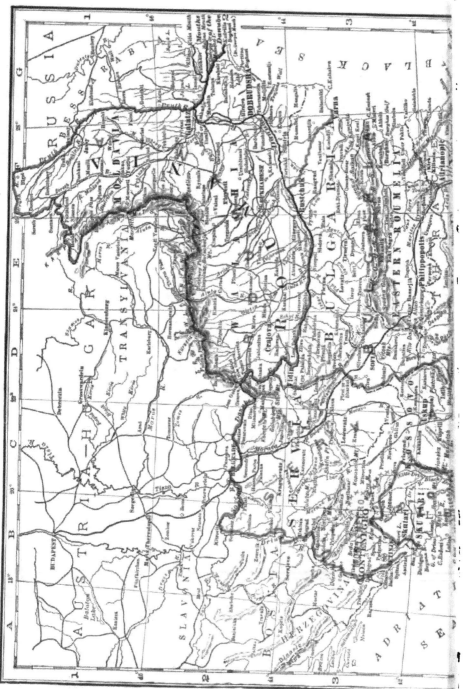

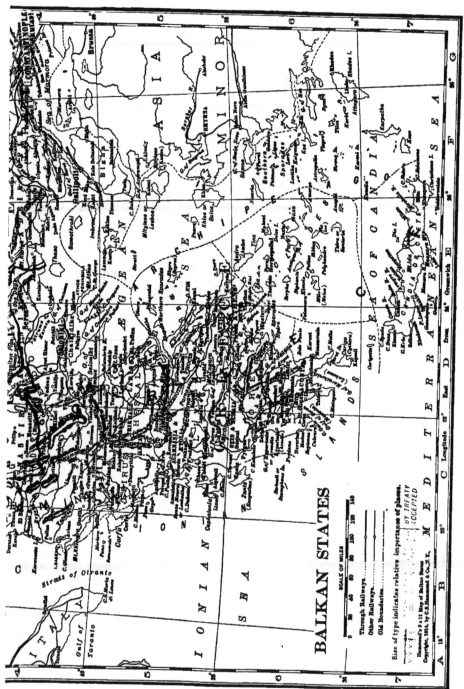

BALKAN STATES

SCALE OF MILES

0 20 40 60 80 100 120 140

Through Railways..........
Other Railways..........
Old Boundaries..........

Size of type indicates relative importance of places.

.... BY TREATY
.... ACCEPTED

Hammond's 11 Map of Balkan States
Copyright, 1914, by C.S.Hammond & Co., N.Y.

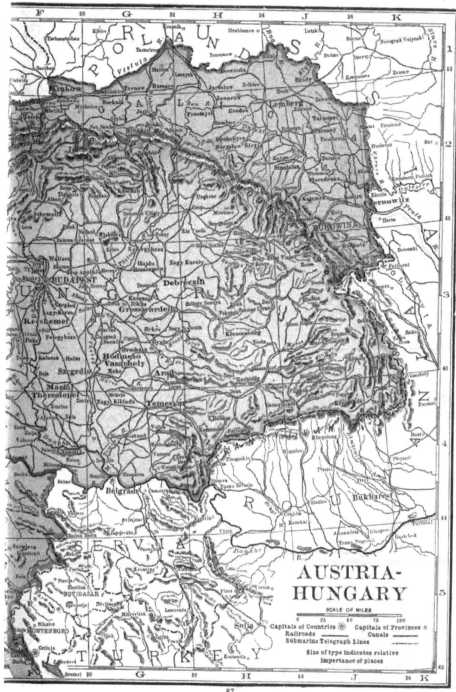

AUSTRIA-
HUNGARY

SCALE OF MILES

0   25   50   75   100

Capitals of Countries ⊛   Capitals of Provinces ⊙
Railroads ———————   Canals ———————
Submarine Telegraph Lines ———————

Size of type indicates relative
importance of places

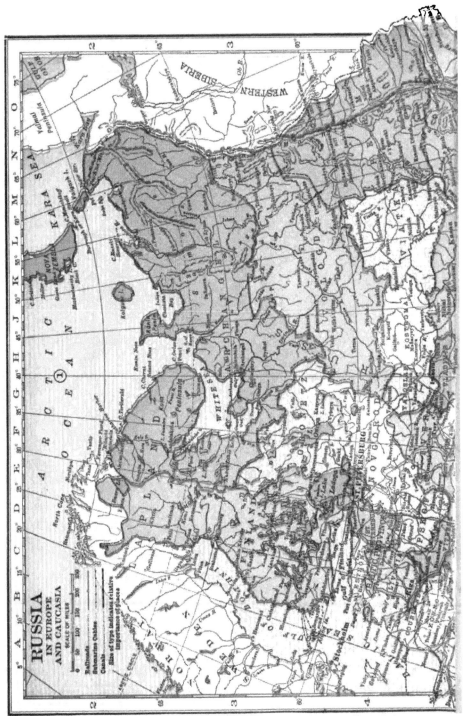

RUSSIA
IN EUROPE
AND CAUCASIA

SCALE OF MILES

Railroads
Submarine Cables
Canals

Size of type indicates relative importance of places

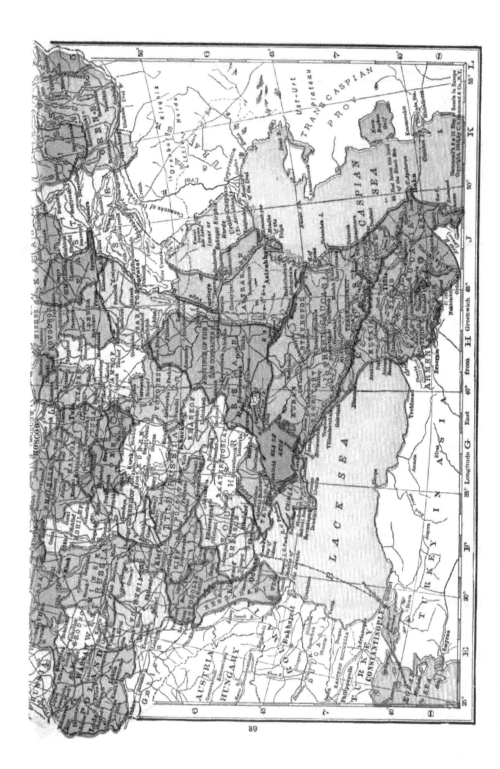

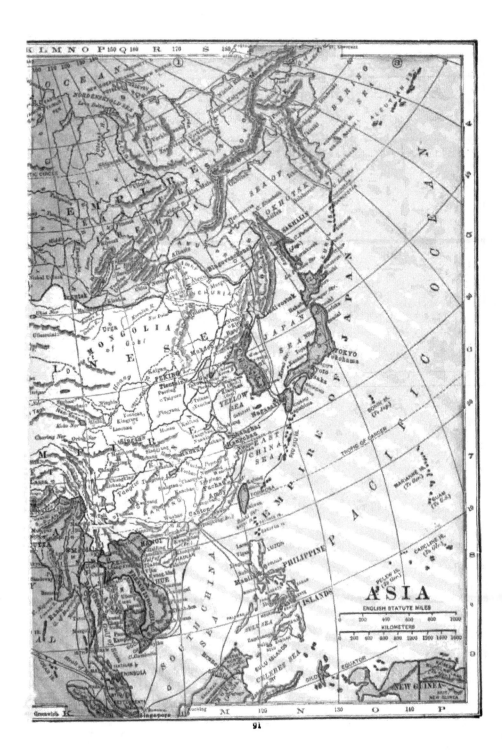

ASIA

ENGLISH STATUTE MILES

KILOMETERS

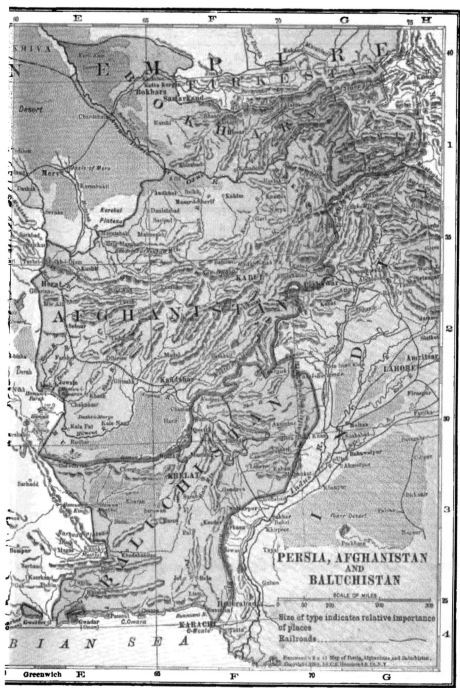

PERSIA, AFGHANISTAN
AND
BALUCHISTAN

SCALE OF MILES

Size of type indicates relative importance
of places
Railroads

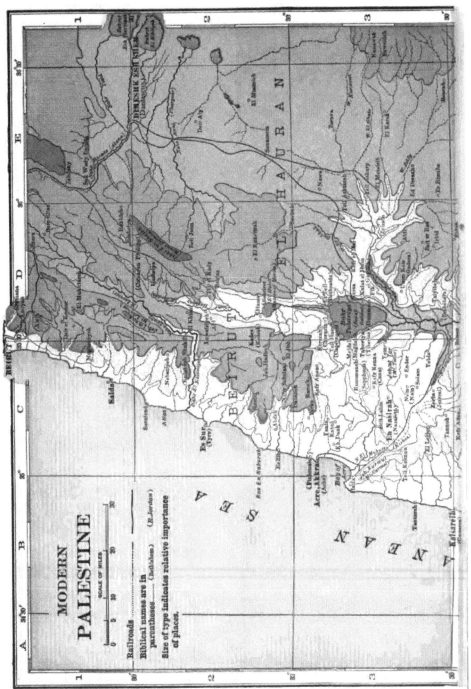

MODERN
PALESTINE

SCALE OF MILES

Railroads
Biblical names are in
parentheses   (B.Jordan)
                      (Bethlehem)
Size of type indicates relative importance
of places.

DIMESHK ESH SHAM
(Damascus)

EL HAURAN

BEIRUT

Acre (Akka)
(Ptolemais)

SEA

JORDAN

94

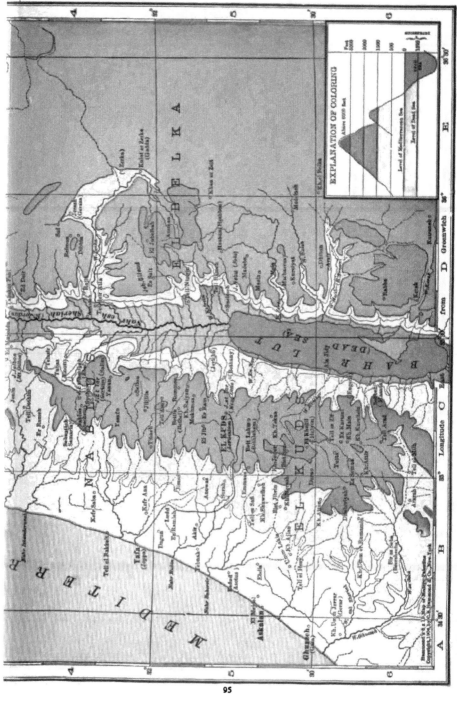

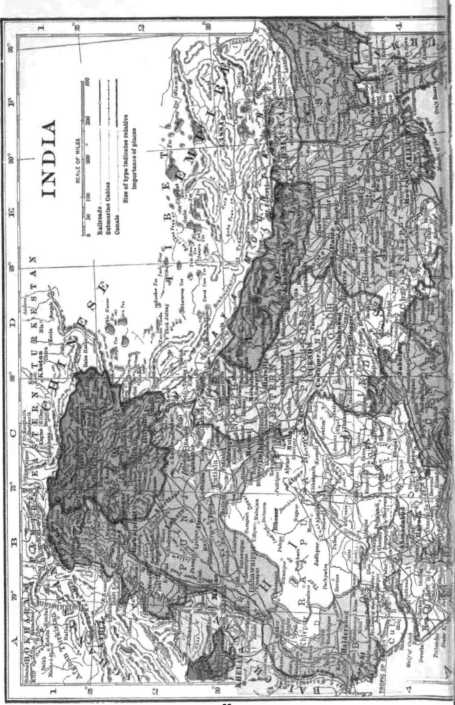

INDIA

SCALE OF MILES

Railroads
Submarine Cables
Canals

Size of type indicates relative
importance of places

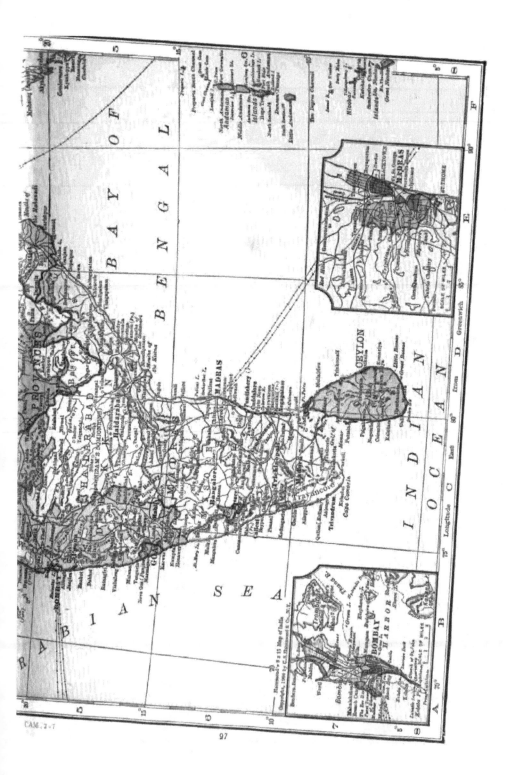

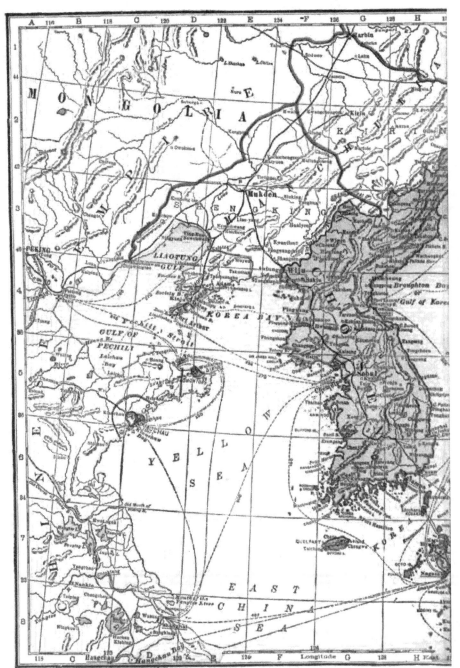

JAPAN

SCALE OF MILES

Capital of Country.....⊛   Treaty Ports.... ⓌⒿⒷ o
Railroads................   Proposed Railroads..........
Fortifications........✛   Navy Yards.... ⚓
Principal Water Routes with Distances in Nautical Miles.........

1912
Copyright 1904, by C.S.Hammond & Co., N.Y.

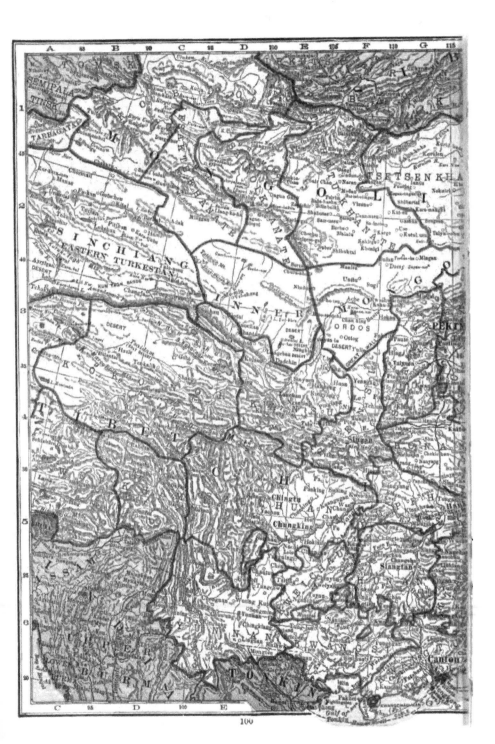

CHINA

SCALE OF MILES

Capital of Country .... ⊛   Treaty Ports .... Amoy
Capital of Province .... ✳   Railroads _____
                        Proposed Railroads _____
                        Submarine Cables _____

Hammond's 8 x 11 Map of China, Japan and Korea
Copyright 1920 by C.S. Hammond & Co. N.Y.

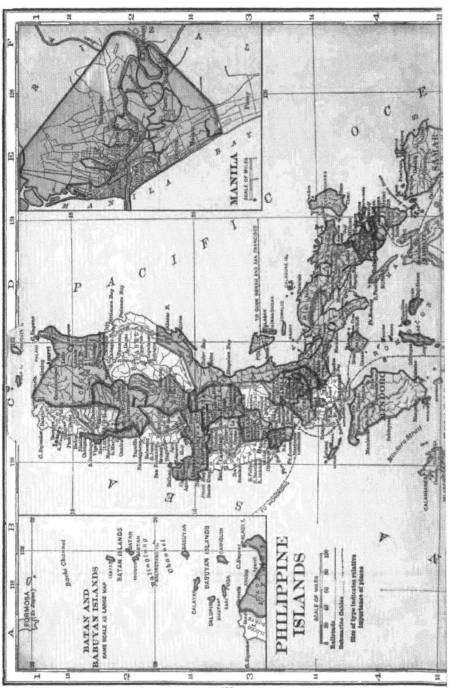

PHILIPPINE ISLANDS

SCALE OF MILES

Railroads
Submarine Cables

Size of type indicates relative
importance of places

BATAN AND
BABUYAN ISLANDS
SAME SCALE AS LARGE MAP

MANILA

SCALE OF MILES

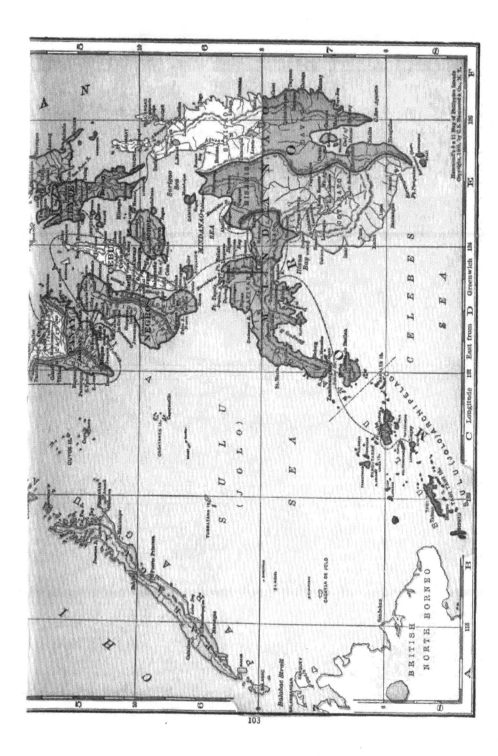

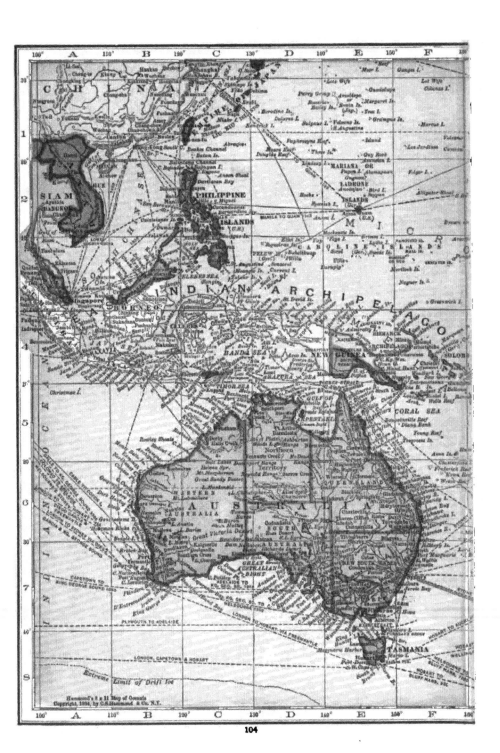

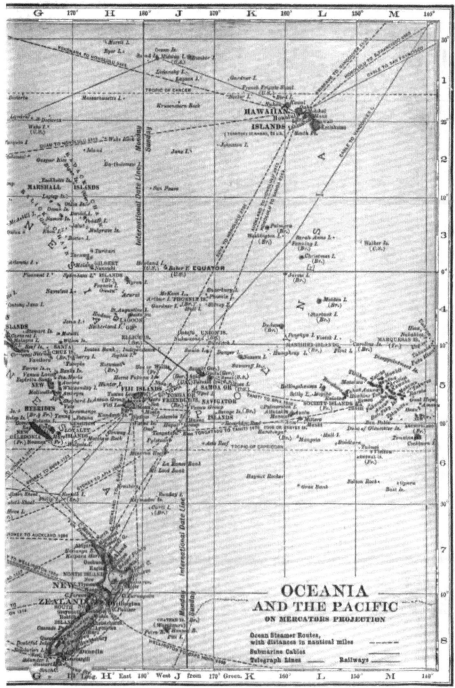

OCEANIA
AND THE PACIFIC
ON MERCATOR'S PROJECTION

Ocean Steamer Routes,
with distances in nautical miles
Submarine Cables
Telegraph Lines   Railways

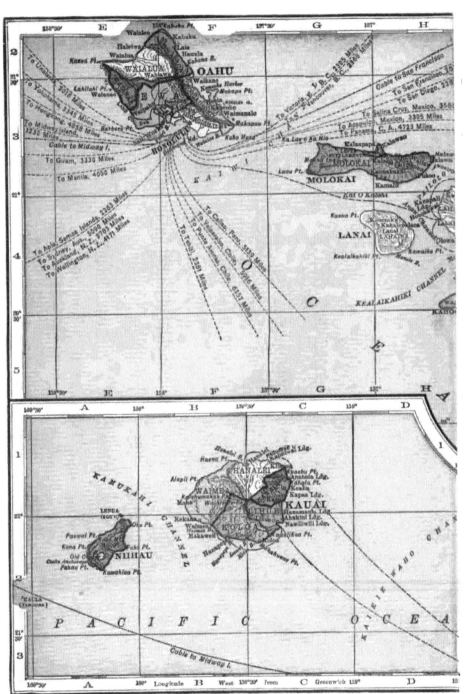

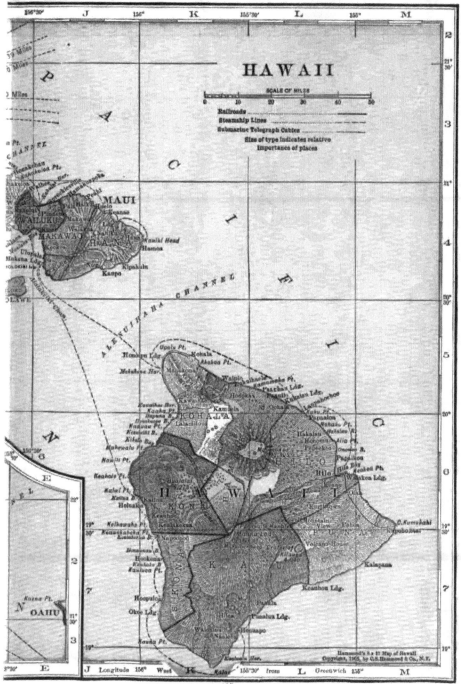

HAWAII

SCALE OF MILES

Railroads
Steamship Lines
Submarine Telegraph Cables

Size of type indicates relative
importance of places

MAUI

WAILUKU

MAKAWAO

HANA

Nauiki Head

Hana

Kipahulu

Kaupo

ALENUIHAHA CHANNEL

Honomu Ldg.

Kohala

Upolu Pt.

Makua Pt.

Mahukona Har.

Mahukona

Waipio

Paauhau Pt.

Paauilo Ldg.

Honokaa

KOHALA

Kawaihae

Kamuela

Ookala

Laupahoehoe Ldg.

Hakalau

Nohu Pt.

Kaupakuea Har.

Honokaa Pt.

Kawaihae Har.

Kamaiki

Puako Pt.

Kiholo

Laimilo

HILO

Honomu

Onomea

Papaikou

Dunan B.

Paauilo

Hilo

Hooked Pt.

Waiakea Ldg.

Mahukona Pt.

Keahole Pt.

Kailua

Holualoa

HAWAII

HONOKOHAU

HILO

C. Kumukahi

Kapoho

Naali Pt.

Kalaoa

Kainaliu

Kelawaha Pt.

Keauhou

Kealakekua

Pahoa

Volcano House

Kaawaloa

Mountain View

Crater of Kilauea

Honaunau

Hookena

Kahaluu

Kealia

Kailua Pt.

KAU

Kalapana

Keauhou Ldg.

Hoopuloa

Okoe Ldg.

Pahala

Honuapo

Punaluu Ldg.

Waiohinu

Naalehu

Nauka Pt.

Kaalualu Har.

Kalae

N

OAHU

Kaena Pt.

Longitude 156° West   155°30'   from   Greenwich   156°

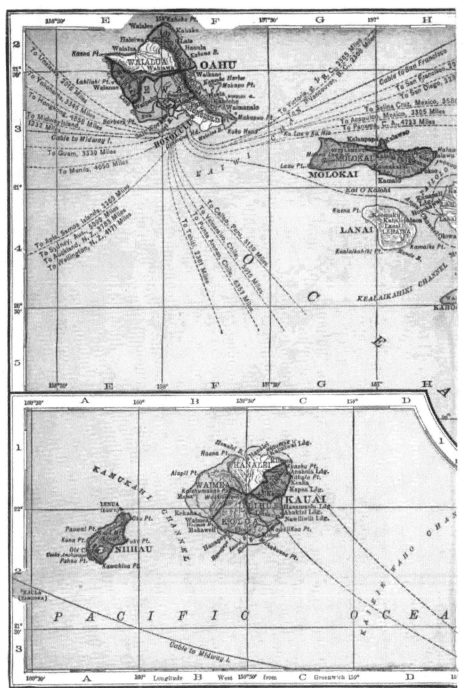

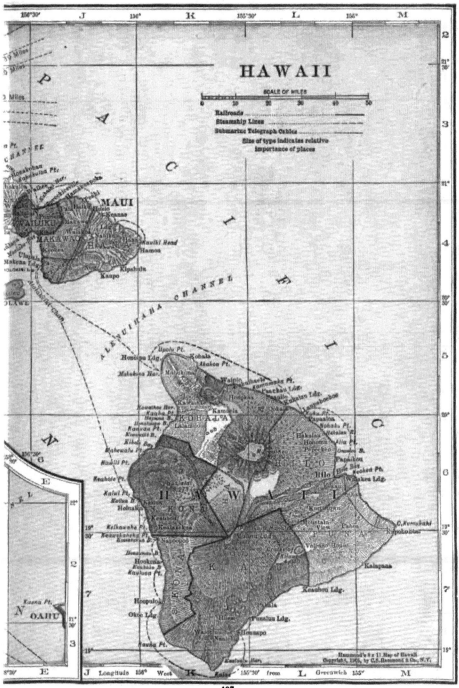

HAWAII

SCALE OF MILES

0   10   20   30   40   50

Railroads
Steamship Lines
Submarine Telegraph Cables

Size of type indicates relative
importance of places

PACIFIC

OCEAN

ALENUIHAHA CHANNEL

MAUI

WAILUKU

MAKAWAO

Upolu Pt.
Kohala
Honoipu Ldg.
Makahoa Har.
Mahukona

Waipio
Kawaihae Har.
Hamoa
Honokaa B.
Kukuihaele
Kailua B.
Kihei B.
Kawela
Bahewala Pt.
Kawaihae Pt.
Honomu

Naalii Pt.

Keahole Pt.

Kaiwi Pt.
Kailua B.
Holualoa

KONA

HAWAII

Kelkawaha Pt.
Keauhou
Kealakekua
Napoopoo
Honaunau Pt.
Hoopuloa
Hookena
Keahuku B.
Kauhuca Pt.

KAU

Hoopuloa

Otoo Ldg.
Hiken
Waiohinu
Nanii Ldg.
Honuapo

Kamaloa Har.
Kaala

Punaluu Ldg.

Kaalaiki

Pahoa Pt.

Kalae

Waipio
Laupahoehoe Pt.
Laupahoehoe Ldg.
Pauchau Ldg.
Ookala Ldg.
Kukaiau Ldg.

HILO

HILO

Papaikou
Hakalau
Honomu
Pepeekeo
Papaikou
Alia Pt.
Hilo

Hooked Pt.
Makea Ldg.

PUNA

Olaa

Kapoho

View
Pahoa
Group of
Craters
Volcano House
Kilauea

C. Kumukahi
Kapoho Pt.

Kalapana

Keauhou Ldg.

OAHU

Kaena Pt.

Honolulu
Pearl Har.

Longitude 156° West   from   Greenwich 155°

107

AUSTRALIA

ENGLISH STATUTE MILES

KILOMETERS

TASMANIA
Same scale as Australia

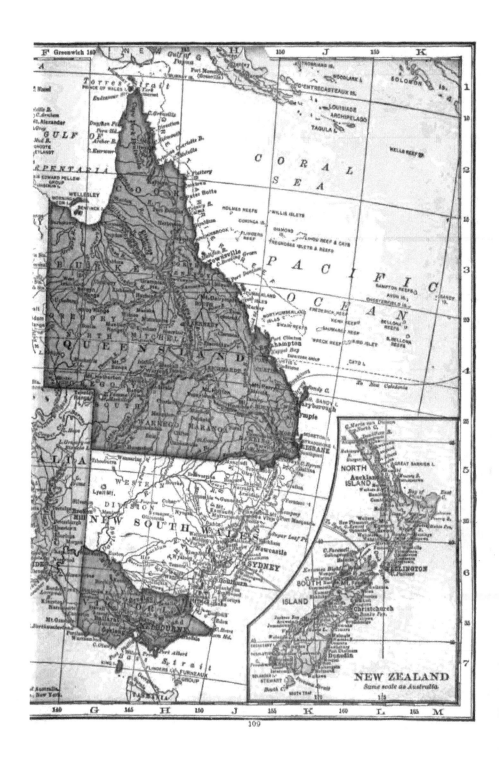

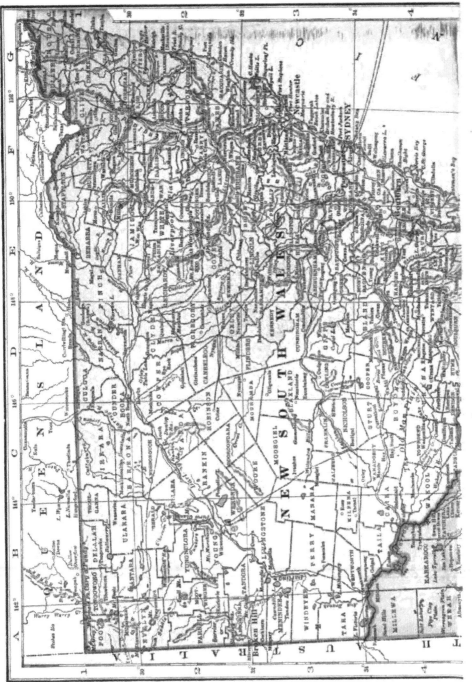

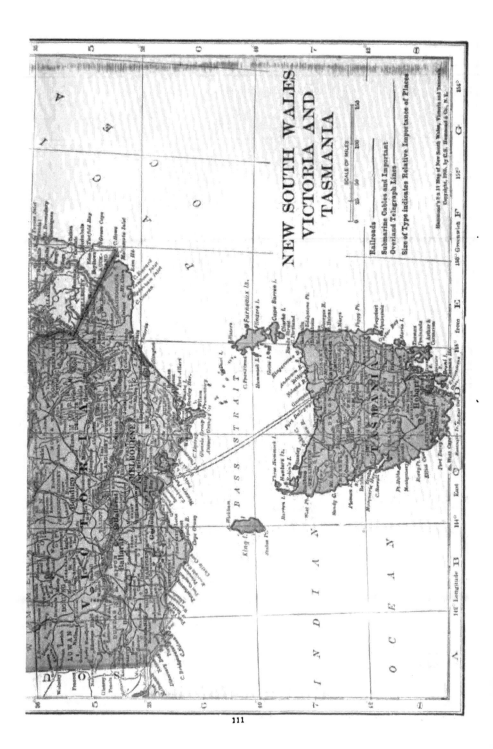

NEW SOUTH WALES
VICTORIA AND
TASMANIA

SCALE OF MILES

Railroads
Submarine Cables and Important
Overland Telegraph Lines
Size of Type indicates Relative Importance of Places

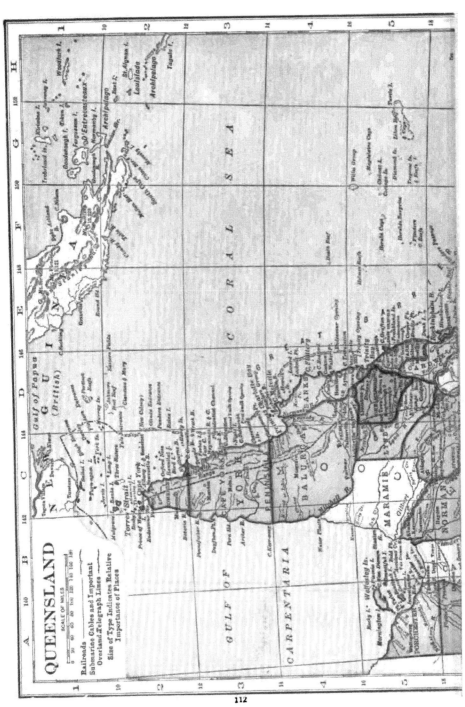

QUEENSLAND

SCALE OF MILES

Railroads
Submarine Cables and Important
Overland Telegraph Lines

Size of Type Indicates Relative
Importance of Places

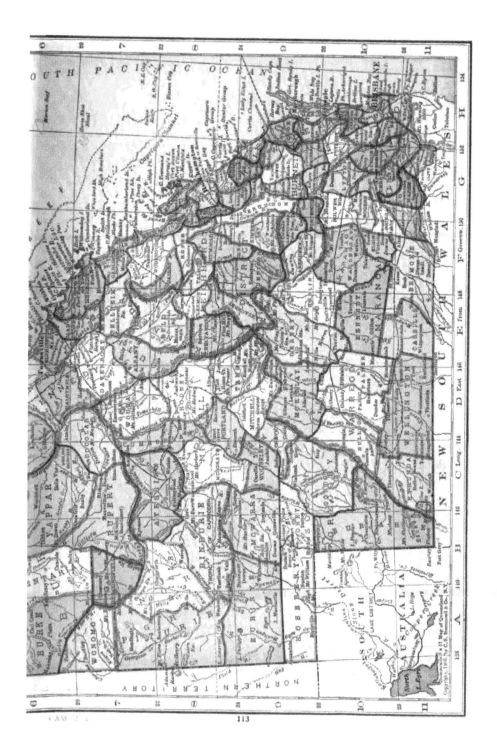

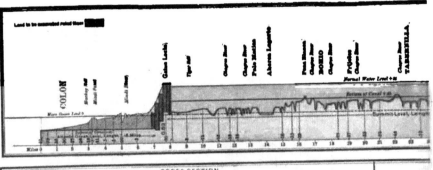

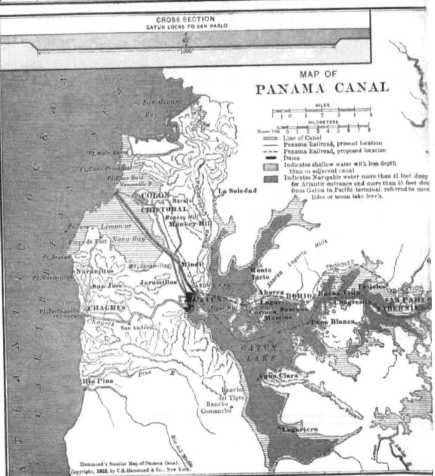

MAP OF
PANAMA CANAL

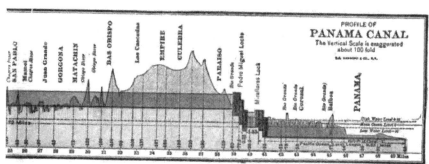

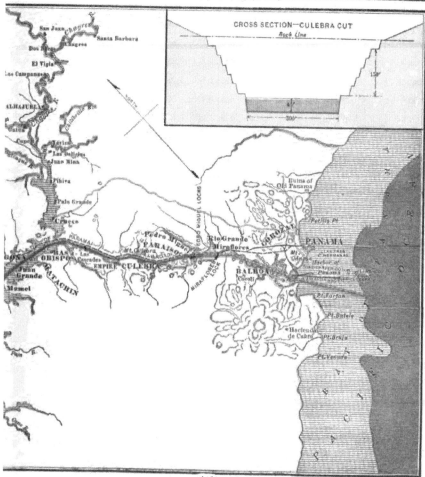

DISTANCES BETWEEN PRINC

The chief ports of the world are named in alphabetical order across the top and down the sides of the tab
names of the selected ports. Thus the distance from Baltimore to Bombay is given as 8,431, and the small "S

	Amsterdam	Antwerp	Barbados	Belfast	Bombay	Boston	Bremen	Buenos Aires	Cape Town	Cherbourg	Genoa	Glasgow	Hamburg
Amsterdam	—	135	3,877	783	6,336	3,164	272	6,370	6,193	303	2,237	818	290
Antwerp	135	—	3,841	697	6,300	3,128	357	6,334	6,157	267	2,201	782	385
Baltimore	3,638	3,602	1,823	3,204	8,431	674	3,851	5,898	6,913	3,357	4,332	3,250	8,866
Barbados	3,877	3,841	—	3,637	8,199	1,880	4,090	4,169	5,284	3,596	4,100	3,719	4,108
Belfast	733	697	3,637	—	6,245	2,730	880	6,271	6,089	459	2,146	113	891
Bermuda	3,248	3,112	1,222	2,958	7,876	707	3,459	5,175	6,181	2,967	3,771	2,885	3,477
Bombay	6,336	6,300	8,199	6,245	—	7,962	6,549	8,349	4,604	6,040	4,473	6,327	6,567
Boston	3,164	3,128	1,880	2,730	7,962	—	3,377	5,804	6,776	2,883	3,863	2,777	3,395
Bremen	272	357	4,090	880	6,549	3,377	—	6,583	6,406	513	2,450	922	171
Buenos Aires	6,370	6,334	4,169	6,271	8,349	5,804	6,583	—	3,778	6,074	6,130	6,350	6,601
Cape Town	6,193	6,157	5,284	6,089	4,604	6,776	6,406	3,778	—	5,897	5,972	6,168	6,424
Charleston	3,846	3,810	1,630	3,561	8,388	845	4,059	5,790	6,828	3,565	4,459	3,498	4,077
Cherbourg	303	267	3,596	459	6,040	2,883	513	6,074	5,897	—	1,941	544	531
Colon	4,818	4,782	1,225	4,543	9,283	2,173	5,031	5,318	6,410	4,537	5,184	4,625	5,049
Constantinople	2,898	3,158	5,057	3,103	3,837	4,820	3,407	7,381	6,137	2,898	1,312	3,185	3,425
Galveston	5,031	4,995	2,240	4,595	9,696	2,118	5,244	6,378	7,470	4,725	5,591	4,641	5,262
Genoa	2,237	2,201	4,100	2,146	4,473	3,863	2,450	6,130	5,972	1,941	—	2,228	2,468
Gibraltar	1,389	1,353	3,252	1,298	4,953	3,015	1,602	5,282	5,124	1,093	854	1,380	1,620
Glasgow	818	782	3,719	113	6,327	2,777	922	6,350	6,168	544	2,228	—	1,049
Hamburg	290	385	4,108	891	6,567	3,395	171	6,601	6,424	531	2,468	1,049	—
Havana	4,335	4,299	1,472	3,899	9,000	1,422	4,548	5,669	6,716	4,029	4,895	3,945	4,566
Havre	267	231	3,662	528	6,115	2,949	480	6,151	5,966	71	2,018	610	498
Hongkong	9,764	9,728	11,627 / 10,989	9,673	4,090	11,390	9,977	10,643	6,898	9,468	7,901	9,755	9,995
Honolulu	13,812 / 9,560	13,776 / 9,533	11,611 / 5,976	13,700 / 9,291	8,375	13,246 / 6,924	14,025 / 9,782	7,709	10,390	13,516 / 9,258	12,386 / 9,953	13,782 / 9,375	14,043 / 9,586
Jamaica	4,314	4,278	1,040	4,039	8,844	1,658	4,527	5,166	6,260	4,033	4,739	4,121	4,545
Lisbon	1,111	1,087	3,085	1,020	5,249	2,797	1,324	5,804	5,148	815	1,150	1,102	1,342

K, via Kiel Canal.　　S, via Suez Canal.　　C, via Cape of Good Hope.

Distances via Panama C:

The distance between any two ports will be found at the intersection of the two columns which contain the hows that the route is via Suez Canal. Distances are for full-powered steam vessels reckoned in nautical miles.

Havana	Havre	Hongkong	Liverpool	London	Melbourne	New Orleans	New York	Plymouth	Rio de Janeiro	San Francisco	Shanghai	Singapore	Southampton	Valparaiso	Yokohama
335	267	9,764	711	201	11,131	4,837	3,346	385	5,280	13,639	10,507	8,324	277	8,869	11,226
299	231	9,728	675	180	11,095	4,801	3,310	349	5,244	13,603	10,471	8,288	238	8,883	11,190
112	3,423	11,859	3,328	3,562	12,703	1,614	410	3,265	4,808	13,167	12,602	10,419	3,383	8,397	13,327
472	3,662	11,627	3,624	3,801	11,047	2,115	1,825	3,549	3,079	11,438	12,370	10,187	3,622	6,668	13,089
899	528	9,673	137	663	11,040	4,401	2,912	360	5,168	13,527	10,535	8,233	488	8,757	11,135
141	3,038	11,627	2,945	3,172	11,972	1,643	699	2,875	4,085	12,444	12,376	9,864	2,956	7,678	13,094
000	6,115	4,090	6,223	6,260	6,535	9,502	8,153	5,997	7,824	9,780	4,633	2,450	6,090	9,875	5,352
422	2,949	11,390	2,854	3,088	12,534	1,924	379	2,791	4,714	13,073	12,133	9,950	2,909	8,303	12,852
548	480	9,977	924	409	11,344	5,050	3,559	598	5,501	13,852	10,720	8,537	479	9,082	11,439
669	6,151	10,643	6,258	6,294	9,099	6,255	5,838	6,029	1,135	7,536	11,386	9,376	6,124	2,766	12,105
716	5,966	6,898	6,076	6,117	5,814	7,347	6,795	5,832	3,267	10,217	7,641	5,631	5,947	5,447	8,360
605	3,631	11,986	3,548	3,770	12,620	1,107	614	3,473	4,700	13,059	12,729	10,546	3,591	8,289	13,448
029	71	9,468	437	227	10,835	4,531	3,065	108	4,984	13,343	10,211	8,028	83	8,573	10,930
987	4,603	12,711	4,530	4,742	12,203	1,380	1,972	4,445	4,228	12,587	13,454	11,271	4,563	7,817	14,173
852	2,975	7,266	3,081	3,118	8,633	6,354	5,011	2,855	5,997	14,352	8,009	5,826	2,948	9,583	8,728
765	4,816	13,124	4,719	4,955	13,263	380	1,893	4,658	5,288	13,649	13,861	11,678	4,776	8,877	14,586
895	2,018	7,901	2,124	2,161	9,268	5,397	4,054	1,898	5,040	13,399	8,644	6,461	1,991	8,626	9,363
047	1,162	8,381	1,276	1,313	9,748	4,549	3,206	1,050	4,192	12,551	9,124	6,941	1,143	7,781	9,843
945	610	9,755	210	745	11,122	4,447	2,959	442	5,158	13,609	10,498	8,315	570	8,839	11,217
566	498	9,995	942	427	11,362	5,068	3,577	616	5,519	13,870	10,738	8,555	497	9,100	11,457
—	4,120	12,428	4,028	4,259	12,512	597	1,197	3,962	4,579	12,974	13,165	10,982	4,080	8,168	13,890
120	—	9,537	503	194	10,904	4,622	3,131	175	5,061	13,412	10,280	8,097	105	8,642	11,999
428	9,537	—	9,651	9,688	5,031	12,924	11,580	9,425	10,118	6,041	853	1,440	9,518	10,536	1,580
111	13,585	4,858	13,690	13,736	4,916	13,697	13,280	13,473	8,587	2,089	4,333	5,925	13,566	5,916	3,445
716	4,099	12,266	4,026	4,238	12,051	1,115	1,457	3,941	4,076	12,471	13,009	10,826	4,059	7,665	13,728
174	884	8,677	998	1,035	10,044	4,676	3,025	770	4,214	12,573	9,420	7,237	865	7,803	10,139

M, via Strait of Magellan. **H, via Cape Horn.** **T, via Torres Strait.**

are shown by red figures

DISTANCES BETWEEN PRINCIPAL

The chief ports of the world are named in alphabetical order across the top and down the sides of the table selected ports. Thus the distance from New York to Melbourne is given as 11,586, and the small "C" show

	Amsterdam	Antwerp	Barbados	Belfast	Bombay	Boston	Bremen	Buenos Aires	Cape Town	Cherbourg	Genoa	Glasgow	Hamburg
Liverpool	711	675	3,624	137	6,223	2,854	924	6,258	6,076	437	2,124	210	94
London	201	180	3,801	663	6,260	3,088	409	6,294	6,117	227	2,161	745	42
Manila	9,667	9,631	11,530 (10,458)	9,576	3,793	11,293 s	9,880 s	10,546 c	6,801	9,871	7,804	9,658	9,89
Melbourne	11,131	11,095	11,047 (9,214)	11,040	6,535	12,534 (10,162)	11,344 s	9,099	5,814	10,835 s	9,268	11,122	11,36
Montreal	3,317	3,281	2,715	2,645	8,141	1,222	3,530	6,421	7,108	3,036	4,042	2,693	3,54
New Orleans	4,837	4,801	2,115	4,401	9,502	1,924	5,050	6,255	7,347	4,531	5,397	4,447	5,06
Newport News	3,514	3,478	1,699	3,080	8,307	550	3,727	5,774	6,789	3,233	4,208	3,126	3,74
New York	3,346	3,310	1,825	2,912	8,153	379	3,559	5,838	6,995	3,065	4,054	2,959	3,57
Panama	11,383 m (4,808)	11,347 m (4,622)	9,182 m (1,265)	11,274 m (4,583)	12,386 m (9,023)	10,817 m (2,213)	11,596 m (5,071)	5,280 m	7,961 (6,450)	11,087 (4,577)	11,143 (5,224)	11,353	11,61
Para	4,319	4,193	1,142	4,056	8,262	2,943	4,442	3,235	4,330	3,933	4,163	4,138	4,46
Philadelphia	3,482	3,446	1,828	3,048	8,284	517	3,694	5,870	6,861	3,201	4,185	8,094	3,71
Plymouth	385	349	3,594	360	5,997	2,791	598	6,029	5,852	108	1,898	442	61
Portland (Me.)	3,086	3,050	1,927	2,652	7,912	98	3,299	5,849	6,787	2,805	3,807	2,698	3,31
Rio de Janeiro	5,280	5,244	3,079	5,168	7,824	4,714	5,501	1,135	3,265	4,984	5,040	5,250	5,51
St. Petersburg	1,070	1,165 k	4,888	1,606	7,353	4,052	951	7,381 k	7,204	1,311	3,248	1,648	86
St. Thomas	3,846	3,810	440	3,586	8,265	1,516	4,059	4,629	5,708	3,570	4,160	3,668	4,07
San Francisco	13,639 m (8,135)	13,603 m (8,099)	11,438 m (4,542)	13,527 m (7,860)	9,780	13,073 m (5,490)	13,852 m (8,348)	7,536	10,217 m (9,727)	13,343 m (7,851)	13,399 m (8,501)	13,609 m (7,942)	13,87
San Juan	3,867	3,831	502	3,600	8,307	1,480	4,080	4,694	5,773	3,583	4,202	3,681	4,09
Savannah	3,916	3,880	1,648	3,443	8,640	928	4,129	5,820	6,860	3,684	4,535	3,489	4,14
Seattle	14,418 m (8,914)	14,382 m (8,878)	12,217 m (5,321)	14,306 m (8,639)	9,515	13,852 m (6,269)	14,631 m (9,127)	8,315	10,996 m (10,366)	14,122 m (8,933)	13,509 m (9,280)	14,388 m (8,721)	14,64
Shanghai	10,507	10,471	12,370	10,535	4,633	12,133	10,720	11,380	7,641 c	10,211	8,644	10,498	10,73
Singapore	8,324	8,288	10,187	8,233	2,450	9,950 s	8,587	9,376	5,631	8,028	6,461	8,315	8,5
Southampton	277	238	3,622	488	6,090	2,909	479	6,124	5,947	83	1,991	570	49
Suez	3,376	3,340	5,239	3,285	2,960	5,002	3,589	8,154	5,259	3,080	1,513	3,867	3,60
Valparaiso	8,869 m (7,473)	8,833 m (7,437)	6,608 m (3,850)	8,757 m (7,198)	9,875	8,303 m (4,525)	9,082 m (7,686)	2,766	5,447 c	8,573 m (7,192)	8,626 (7,539)	8,839	9,10
Yokohama	11,226	11,190	13,089 (8,907)	11,135	5,352	12,852 m (9,915)	11,439	12,105	8,360	10,930	9,363	11,217	11,4

K, via Kiel Canal.   S, via Suez Canal.   C, via Cape Town.

118

Distances via Panama C

PORTS OF THE WORLD—Continued.

The distance between any two ports is found at intersection of the two columns which contain the names of the ... hat the route is via Cape of Good Hope. Distances are for full-powered steam vessels reckoned in nautical miles.

Havana	Havre	Hongkong	Liverpool	London	Melbourne	New Orleans	New York	Plymouth	Rio de Janeiro	San Francisco	Shanghai	Singapore	Southampton	Valparaiso	Yokohama
4,028	503	9,651	—	638	11,018	4,525	3,036	345	5,158	13,517 (7,217)	10,394	8,211	463	8,747 (7,185)	11,113
4,259	194	9,688	638	—	11,055	4,761	3,270	309	5,204	13,568 (8,059)	10,437	8,248	201	8,793 (7,397)	11,150
2,325 (5,420)	9,440	640	9,554	9,591	4,511	12,827 (10,813)	11,584 (11,205)	9,328	10,021	6,238	1,180	1,435	9,421	10,406	1,768
2,512	10,904 (5,976)	5,031	11,018	11,055	—	13,140 (9,369)	12,586 (9,961)	10,792	8,827	6,966	5,234	3,823	10,885	6,280	4,875
2,475	8,102	11,569	2,760	3,241	12,986 (11,132)	2,977	1,451	2,944	5,331	13,690 (6,460)	12,312 (11,855)	10,129	3,062	8,290 (5,798)	13,031 (10,585)
597	4,622	12,924 (5,744)	4,525	4,761	13,140 (9,369)	—	1,699	4,464	5,160	13,524 (4,697)	13,667 (10,092)	11,484	4,582	8,754 (4,035)	14,386 (9,122)
988	3,299	11,735 (11,156)	3,204	3,438	12,579 (9,261)	1,490	286	3,141	4,590	13,048 (5,488)	12,478 (10,984)	10,295	3,259	8,273 (1,127)	13,203 (9,514)
1,227	3,131	11,580 (11,316)	3,053	3,270	11,586 (9,261)	1,699	—	2,973	4,748	13,107 (5,250)	13,324 (10,654)	10,141	3,091	8,337 (4,627)	13,042 (9,714)
0,682 (1,927)	11,156 (4,613)	9,324	11,261 (4,570)	11,807 (4,782)	7,949	11,268 (1,420)	10,851 (2,012)	11,042 (4,485)	6,158 (4,268)	3,277	8,672	13,405	11,137 (4,603)	2,615	7,702
2,577	4,010	11,204	4,043	4,153	10,120	3,136	2,915	8,883	2,145	10,504 (5,421)	11,947 (11,016)	10,250	3,983	5,734 (4,359)	12,666 (10,006)
1,158	3,267	11,712 (11,302)	8,172	8,406	12,652 (9,527)	1,660	234	3,109	4,782	13,141 (5,255)	12,455	10,272	3,227	8,371 (4,593)	13,174 (9,680)
3,962	175	9,425	845	309	10,792	4,464	2,973	—	4,939	13,300 (7,082)	10,168	7,985	129	8,530 (7,100)	10,887
1,470	2,871	11,334	2,776	3,010	12,579 (10,210)	1,972	427	2,713	4,759	13,118 (5,538)	12,077	9,894	2,881	8,348 (4,576)	12,796 (9,903)
4,579	5,061	11,118	5,158	5,204	8,827	5,160	4,748	4,939	—	8,414 (7,245)	10,861	8,851	5,034	3,644	11,580
5,346	1,275	10,755	1,716	1,207	12,142	5,848	4,234	1,396	6,299	14,650 (9,209)	11,518	9,335	1,277	9,880 (8,347)	12,237
1,040	3,639	11,687 (10,384)	3,573	3,770	11,499 (9,609)	1,590	1,435	3,473	3,539	11,898 (4,337)	12,430 (9,722)	10,247	3,591	7,128 (3,675)	13,149 (8,722)
2,974 (1,304)	13,412 (7,920)	6,041	13,517 (7,847)	13,563 (8,059)	6,966	13,524 (4,697)	13,107 (5,289)	13,300 (7,762)	8,414 (7,545)	—	5,491	7,330	13,393 (7,880)	5,140	4,791
975	3,652	11,730 (10,961)	3,588	3,791	11,564 (8,989)	1,525	1,428	3,494	3,604	11,963 (4,217)	12,473 (9,712)	10,289	3,612	7,193 (3,455)	13,192 (8,702)
634	3,701	12,062 (10,927)	3,618	3,840	12,650	1,136	695	3,543	4,730	13,089 (4,580)	12,805 (10,275)	10,622	3,661	8,819 (4,218)	13,524 (9,265)
3,753 (3,483)	14,191 (8,669)	5,779	14,296 (8,626)	14,342 (8,838)	7,326	14,303 (5,476)	13,886 (6,068)	14,079 (8,541)	9,193 (8,324)	799	5,209	7,068	14,092 (8,659)	5,917	4,259
3,165	10,280	853	10,394	10,437	5,234	13,667 (10,092)	12,324 (10,884)	10,168	10,861	5,491	—	2,183	10,267	10,266	1,030
0,982	8,097	1,440	8,211	8,248	3,823	11,484	10,141	7,985	8,851	7,330	2,183	—	8,078	10,899	2,902
4,080	105	9,518	463	201	10,885	4,582	3,091	129	5,034	13,393 (7,880)	10,261	8,078	—	8,623 (7,218)	10,980
6,034	3,157	6,388	3,263	3,300	7,755	6,536	5,193	3,037	6,179	12,278	7,181	4,948	3,130	9,768	7,850
8,168 (3,842)	8,642 (7,258)	10,536	8,747 (7,185)	8,793 (7,397)	6,280	8,754 (4,035)	8,337 (4,627)	8,530 (7,100)	3,644	5,138	10,266	10,899	8,623 (7,218)	—	9,339
3,890 (5,122)	10,999	1,580	11,113	11,150	4,875	14,386 (9,122)	13,043 (9,714)	10,887	11,580	4,791	1,030	2,902	10,980	9,339	—

ARCTIC
REGIONS

120

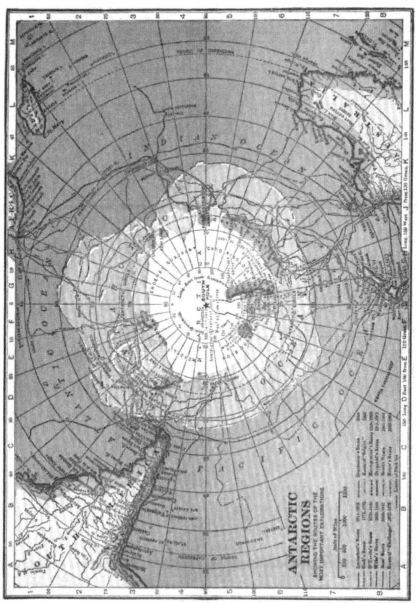

ANTARCTIC
REGIONS

SHOWING THE ROUTES OF THE
MOST IMPORTANT EXPLORATIONS

121

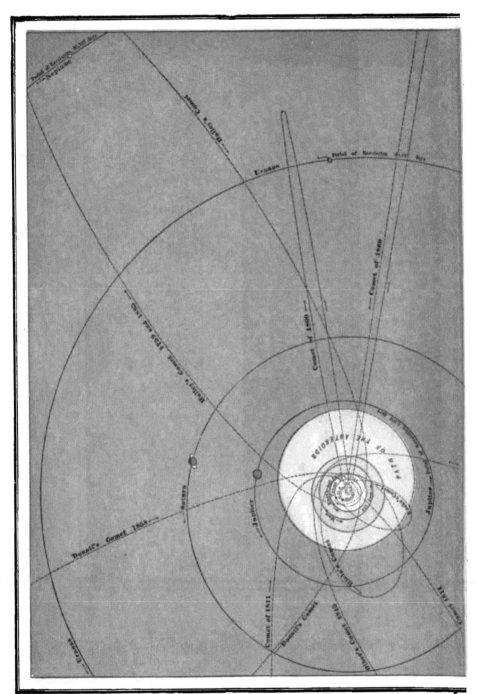

THE SOLAR SYSTEM

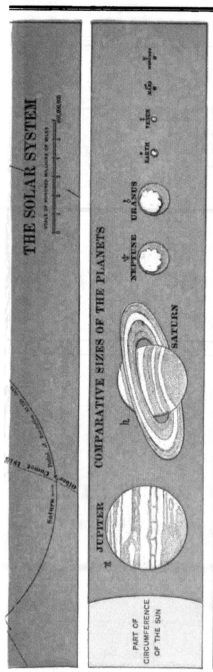

COMPARATIVE SIZES OF THE PLANETS

JUPITER

SATURN

NEPTUNE URANUS

EARTH VENUS

PART OF CIRCUMFERENCE OF THE SUN

APPARENT PATH OF THE SUN

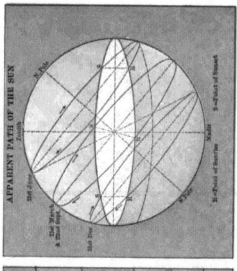

ECLIPSES

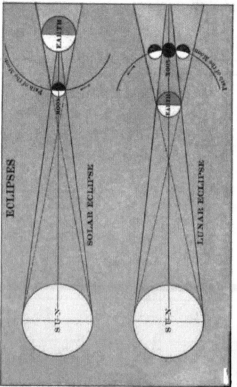

SUN

MOON

EARTH

SOLAR ECLIPSE

SUN

EARTH

MOON

LUNAR ECLIPSE

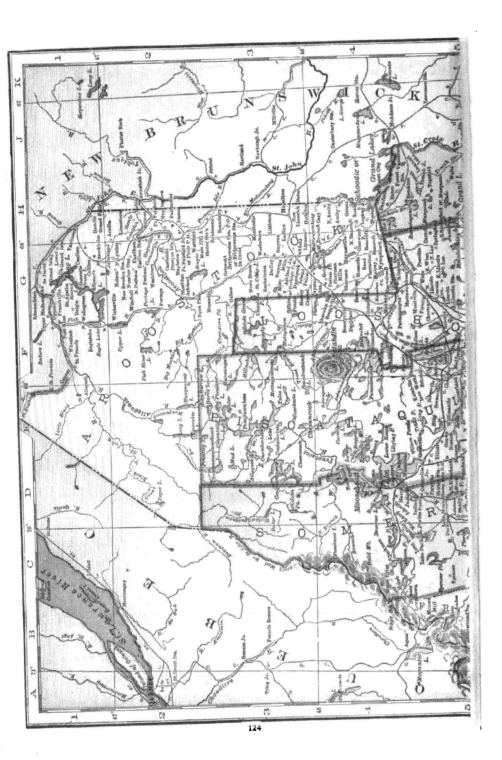

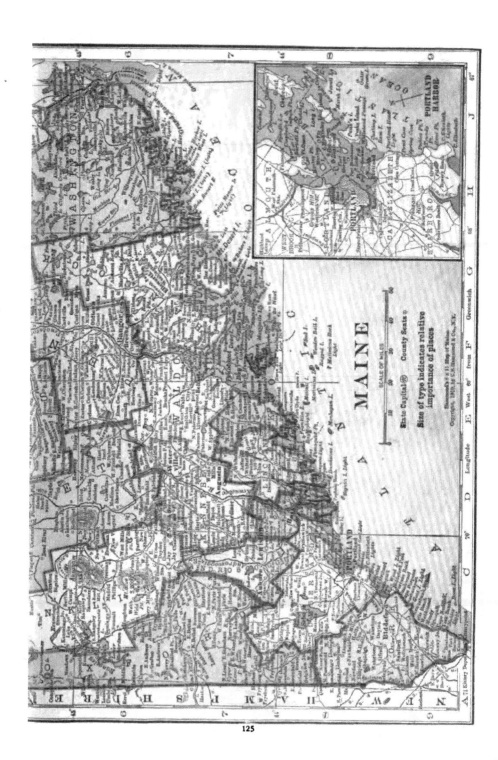

MAINE

State Capital ⊛   County Seats ⊙

Size of type indicates relative
importance of places

SCALE OF MILES

Longitude   West   66°   from   F   Greenwich   G

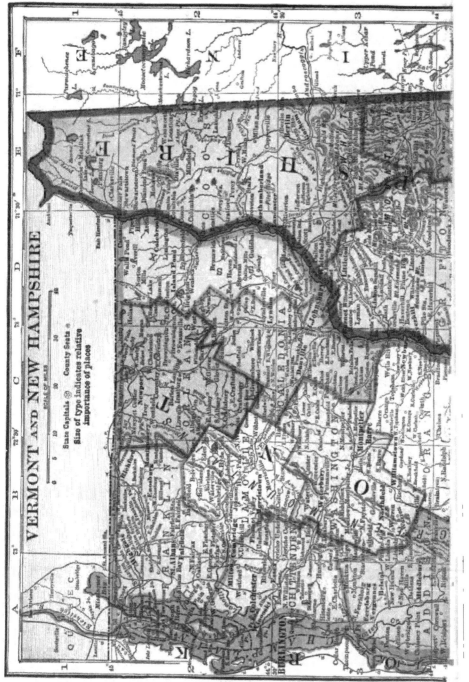

VERMONT AND NEW HAMPSHIRE

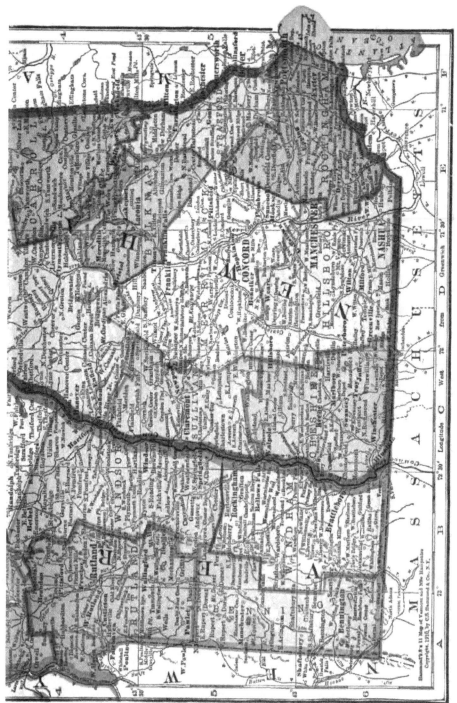

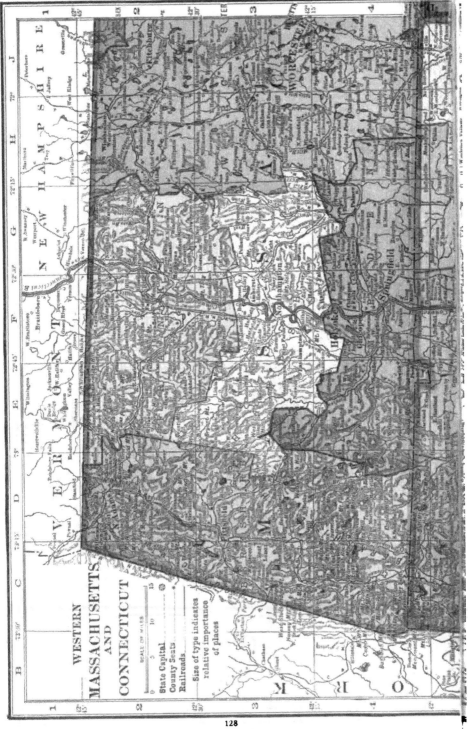

WESTERN
MASSACHUSETTS
AND
CONNECTICUT

SCALE OF MILES

State Capital
County Seats
Railroads

Size of type indicates
relative importance
of places

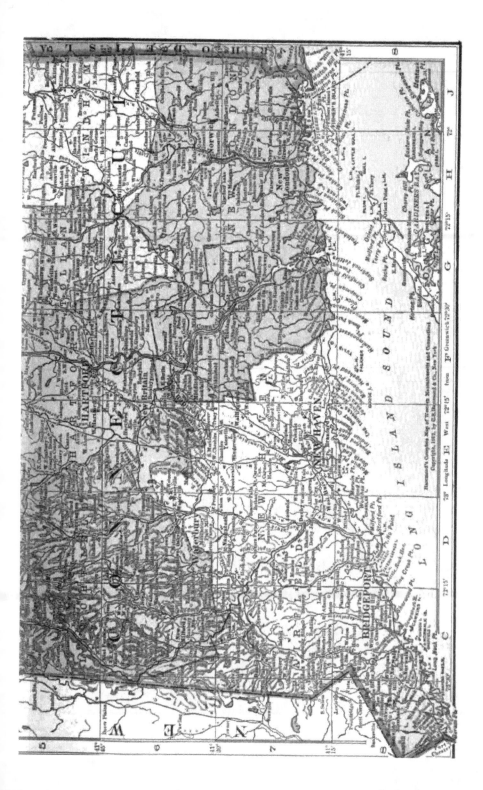

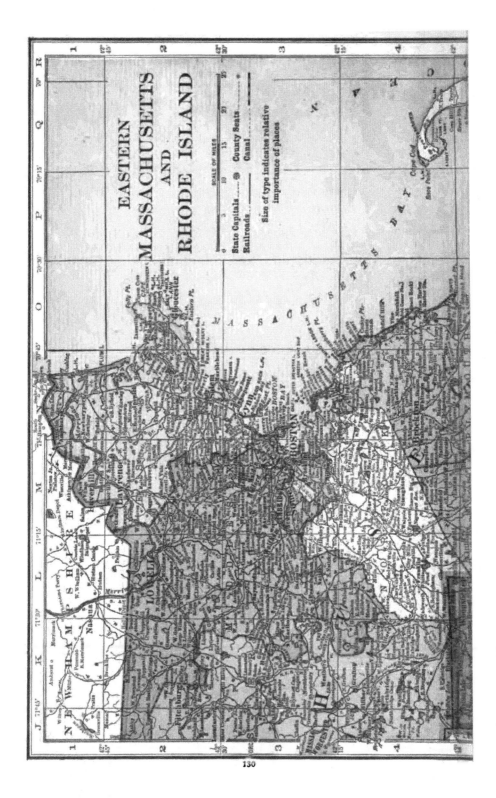

EASTERN
MASSACHUSETTS
AND
RHODE ISLAND

SCALE OF MILES

State Capitals
County Seats
Railroads
Canal

Size of type indicates relative
importance of places

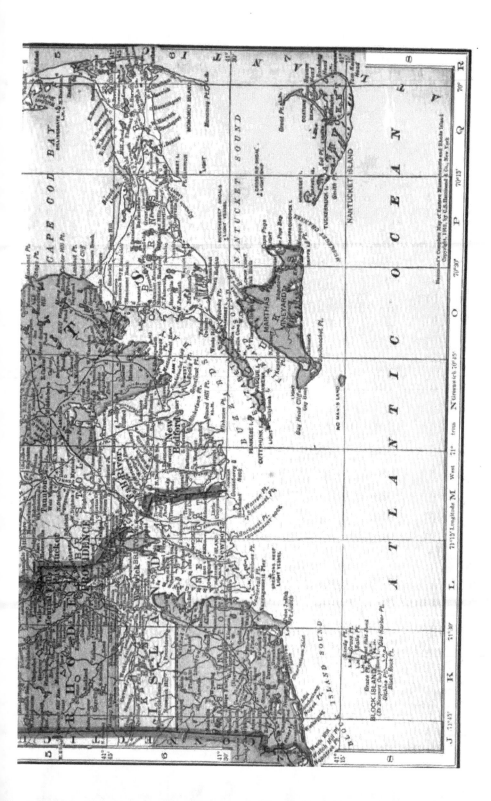

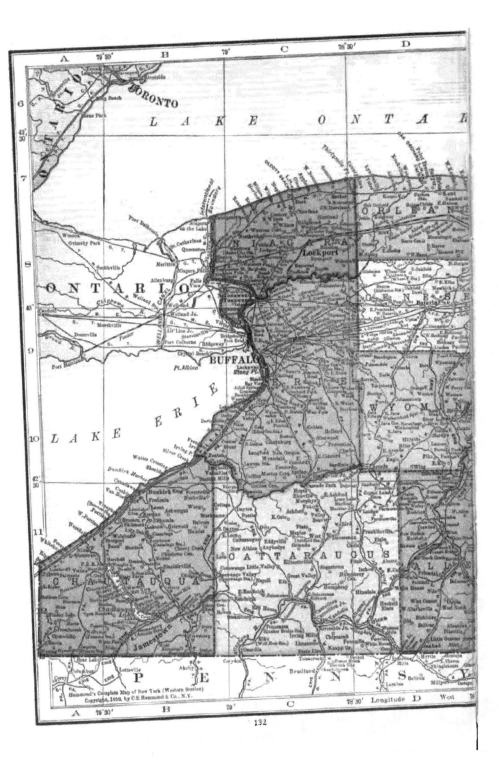

Hammond's Complete Map of New York (Western Section)
Copyright, 1912, by C.S. Hammond & Co., N.Y.

NEW YORK
(Western Section)

SCALE OF MILES

State Capital ◎   County Seats ⊛
Railroads ——   Canals ——
Size of type indicates relative
importance of places.

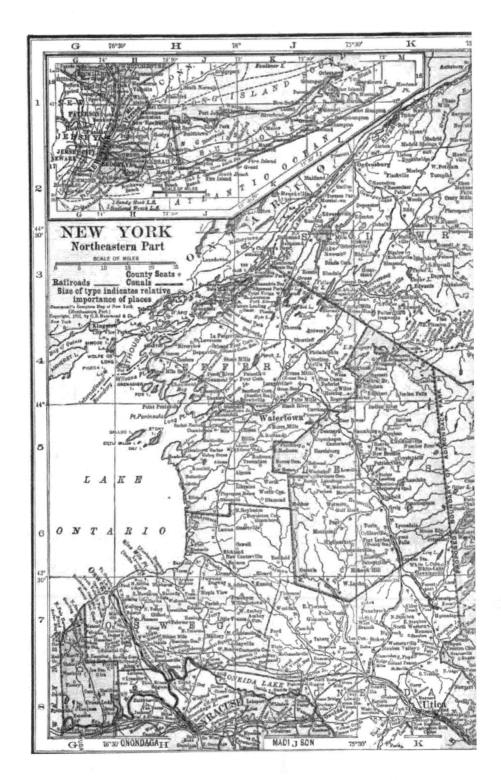

NEW YORK
Northeastern Part

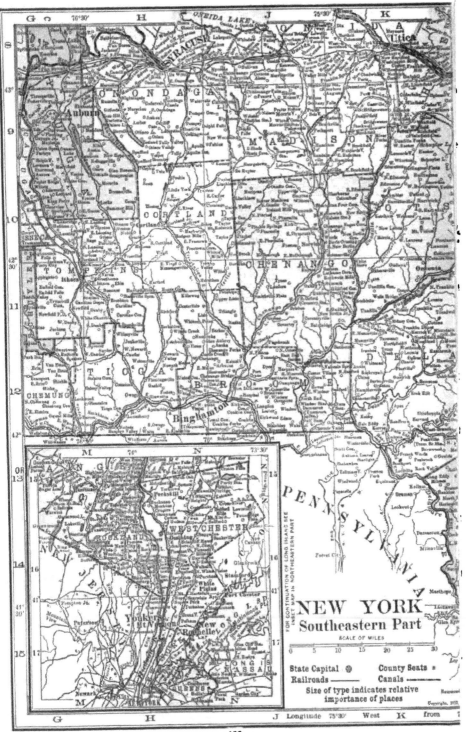

NEW YORK
Southeastern Part
SCALE OF MILES
0   5   10   15   20   25   30

State Capital ⊛   County Seats ∗
Railroads ———   Canals ———
Size of type indicates relative
importance of places

Longitude 75°30′ West from

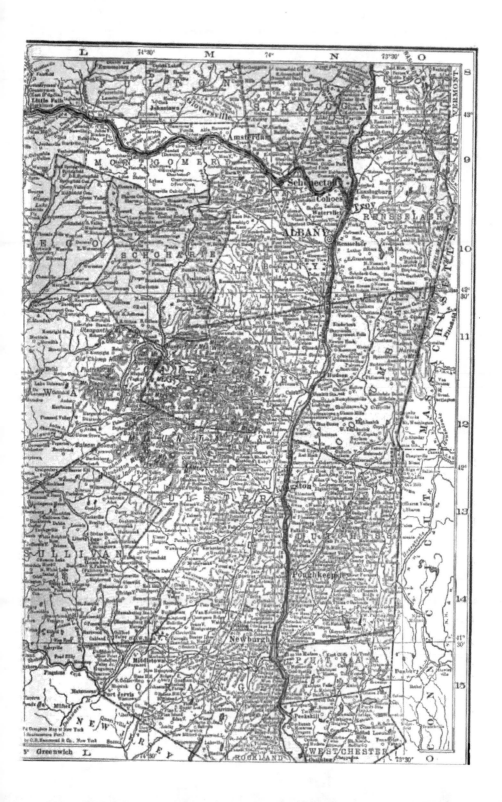

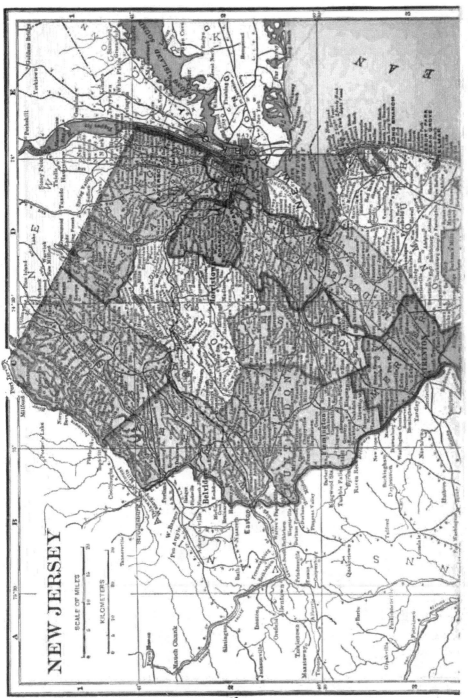

NEW JERSEY

SCALE OF MILES

KILOMETERS

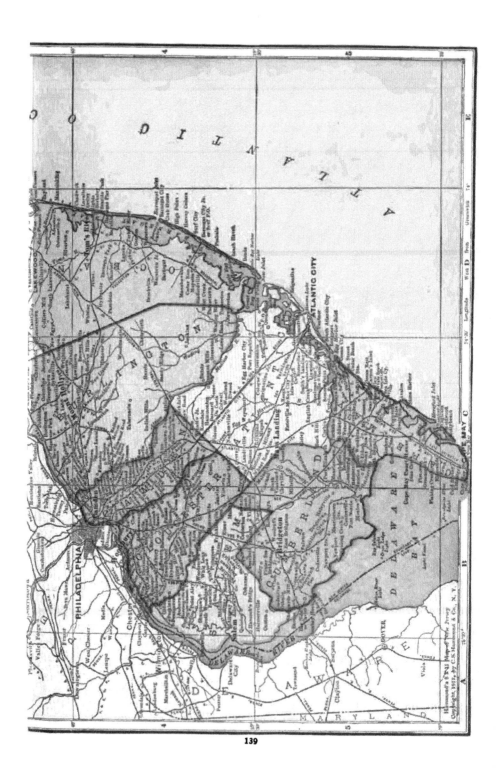

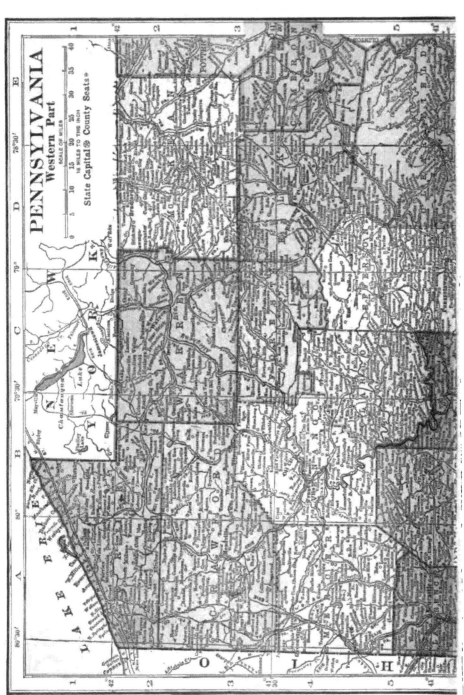

PENNSYLVANIA
Western Part

SCALE OF MILES

0   5   10   15   20   25   30   35   40

16 MILES TO THE INCH

State Capital⊛ County Seats⊙

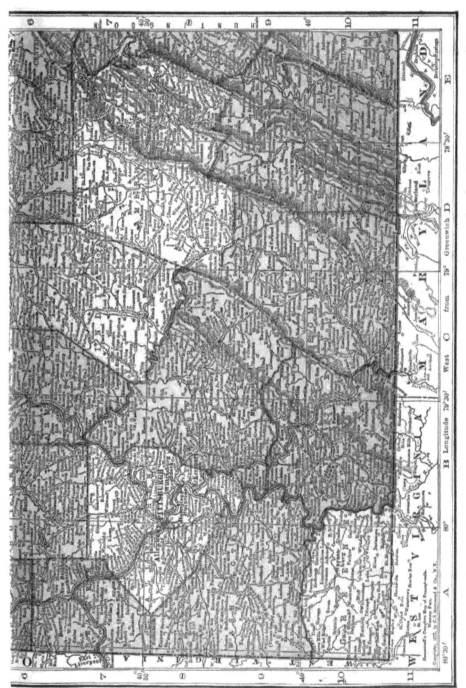

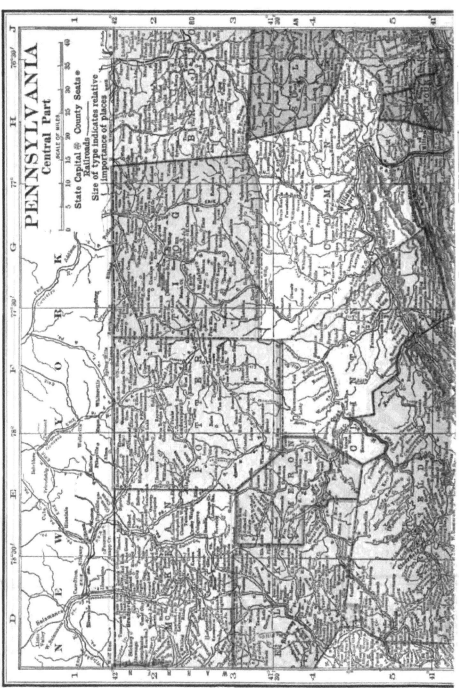

PENNSYLVANIA
Central Part

SCALE OF MILES

State Capital ⊛   County Seats ●
Railroads
Size of type indicates relative
importance of places

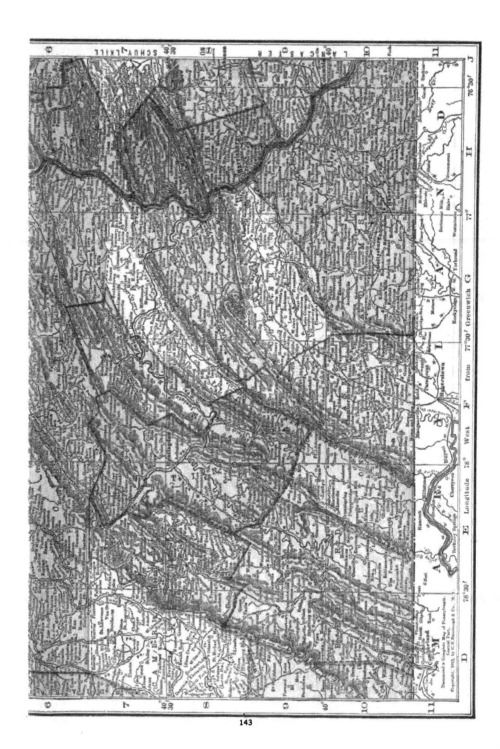

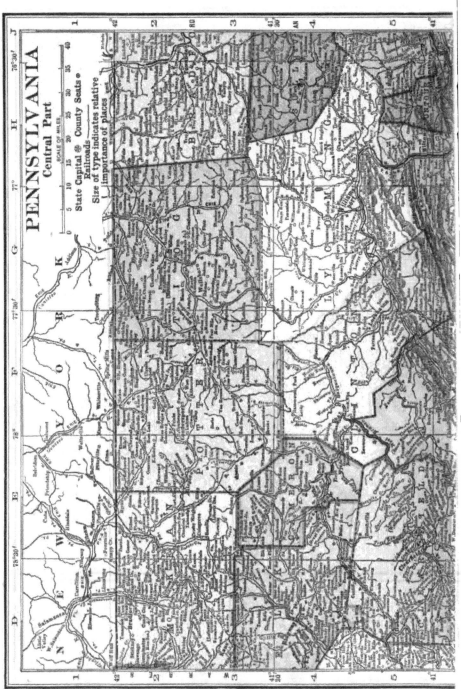

PENNSYLVANIA

Central Part

SCALE OF MILES

State Capital ● County Seats ●
Railroads
Size of type indicates relative
importance of places

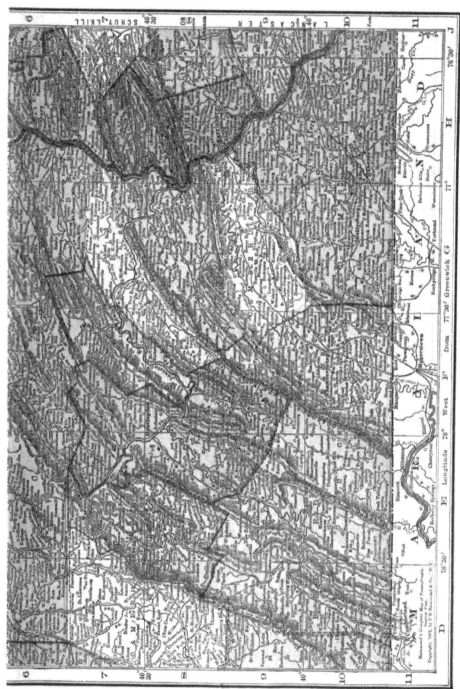

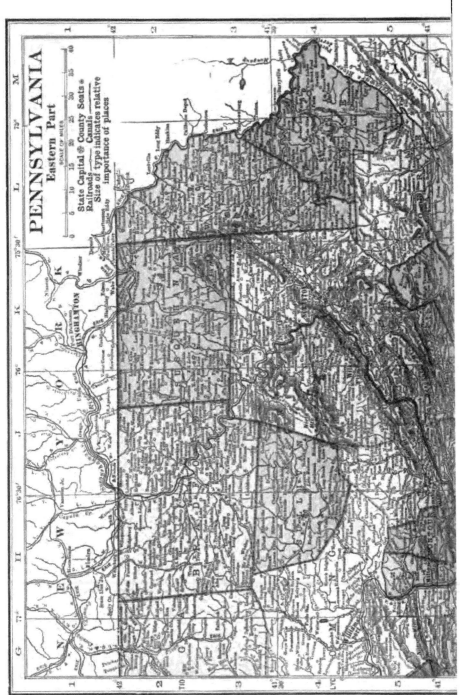

PENNSYLVANIA
Eastern Part

SCALE OF MILES

State Capital ⓢ County Seats ⊛
Railroads ⸻ Canals ⸺
Size of type indicates relative
importance of places

MARYLAND
AND
DELAWARE

SCALE OF MILES

State Capital ⊕   County Seats ⊛

Size of type indicates relative
importance of places

WESTERN PART
OF MARYLAND
Same scale as large map

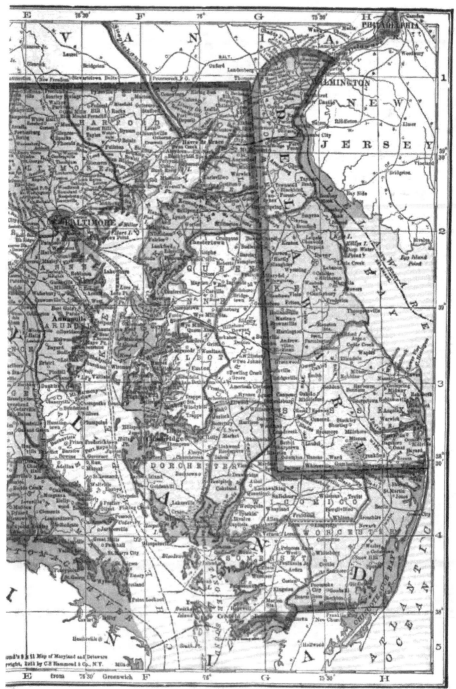

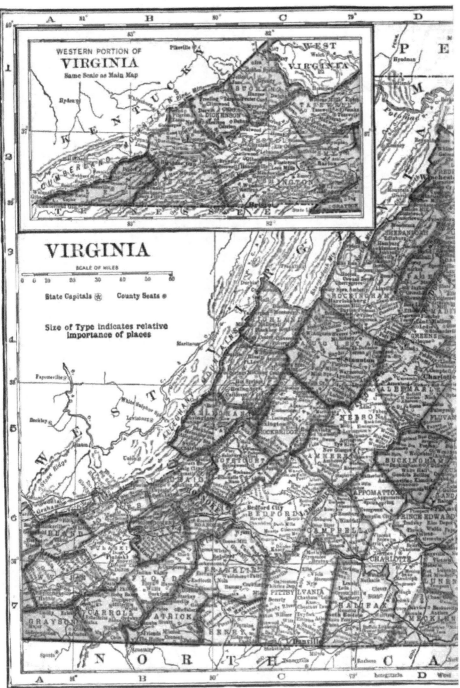

WESTERN PORTION OF
VIRGINIA
Same Scale as Main Map

VIRGINIA

SCALE OF MILES

State Capitals ⊛   County Seats ⊙

Size of Type indicates relative
importance of places

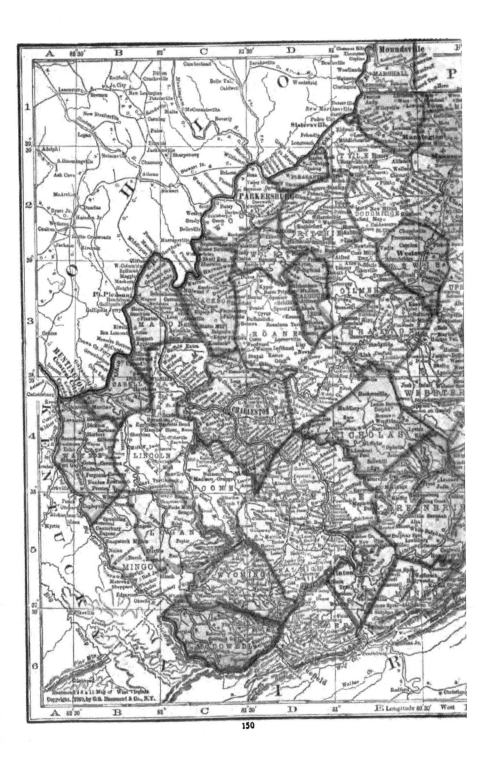

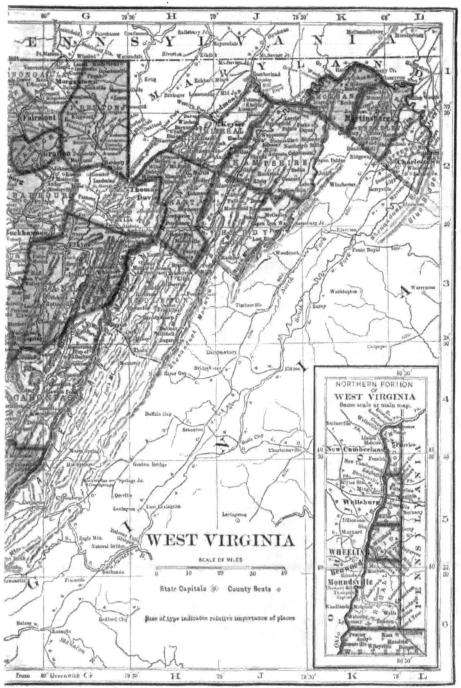

WEST VIRGINIA

SCALE OF MILES

State Capitals ⊛   County Seats ⊕

Size of type indicates relative importance of places

NORTHERN PORTION
OF
WEST VIRGINIA
Same scale as main map.

WESTERN PART
OF
NORTH CAROLINA
Same Scale as Main Map

NORTH CAROLINA

SCALE OF MILES

0   10   20   30   40   50   60   70

State Capital ⊕   County Seats ∗

**Size of type indicates relative
importance of places**

Hammond's 8 x 11 Map of North Carolina
Copyright, 1911, by C.S. Hammond & Co., N.Y.

Long. 78°30' West H from 78° Green. J 77°30' K 77° L 76°30' M 76° N 75°30'

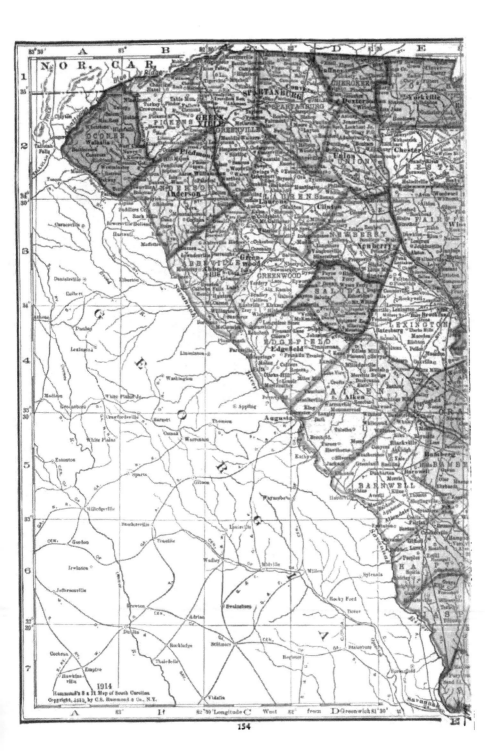

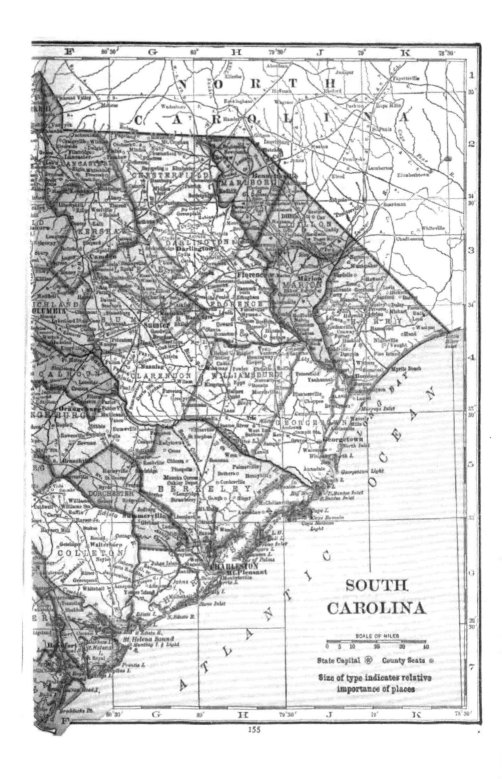

SOUTH
CAROLINA

SCALE OF MILES

State Capital ⊛   County Seats ⊙

Size of type indicates relative
importance of places

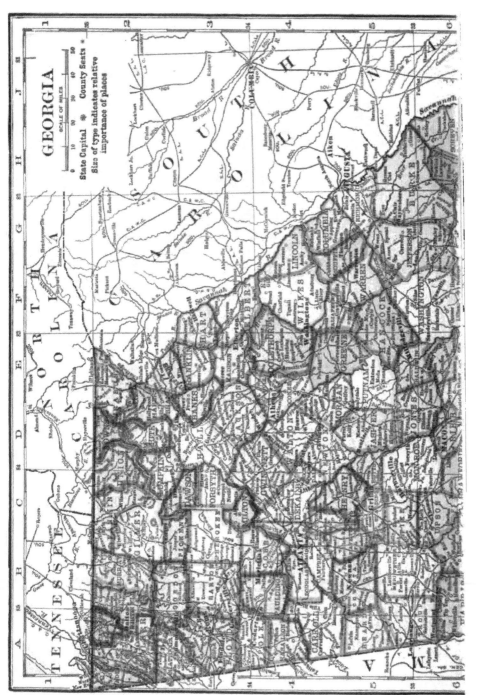

GEORGIA

SCALE OF MILES

State Capital ✱
County Seats ✱
Size of type indicates relative
importance of places

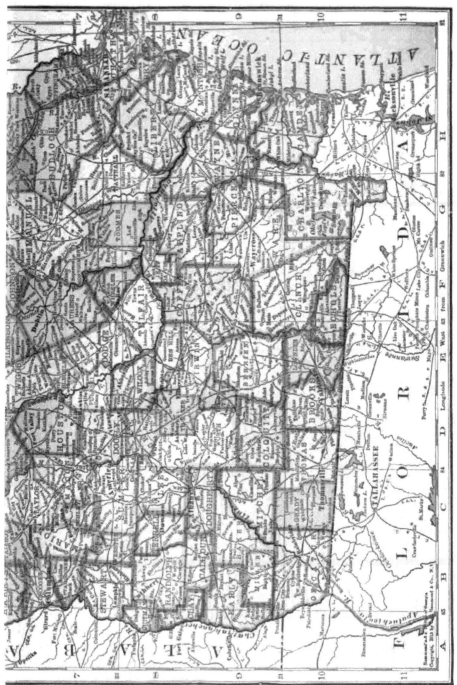

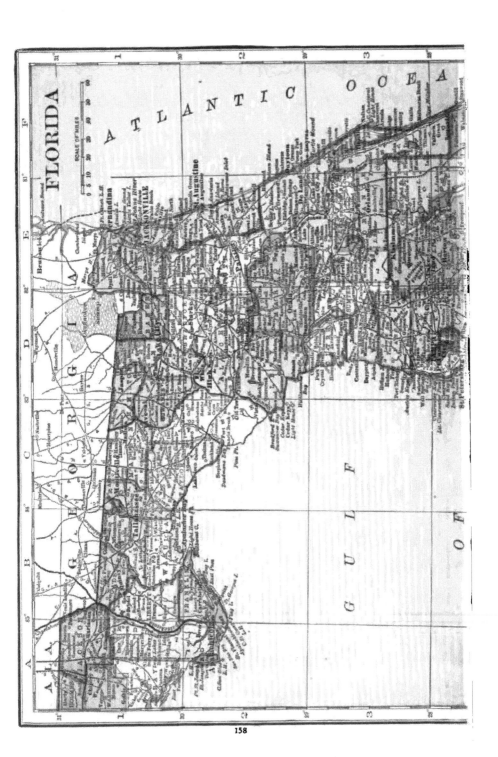

FLORIDA

ATLANTIC OCEAN

GULF OF

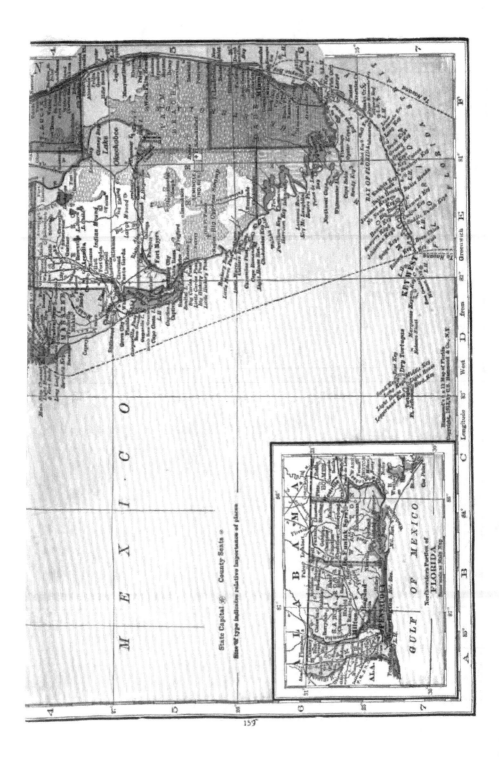

State Capital ✸   County Seats ◦

Size of type indicates relative importance of places

GULF OF MEXICO

Northwestern Portion of
FLORIDA

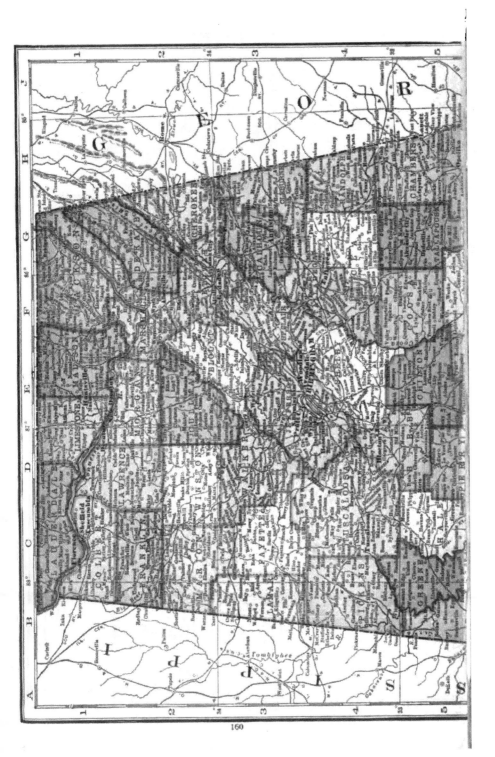

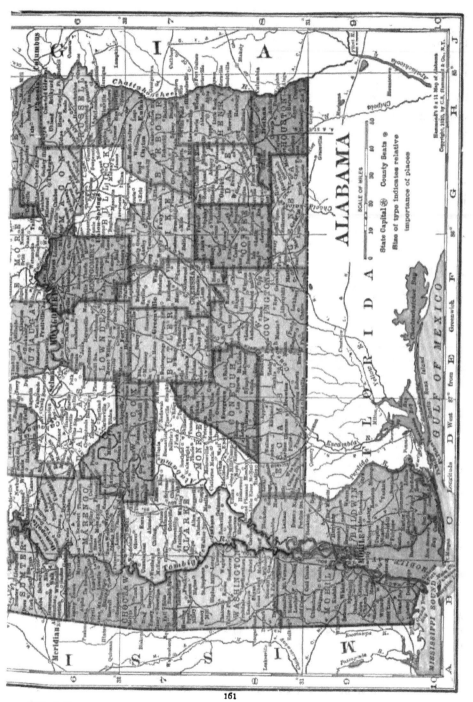

ALABAMA

SCALE OF MILES

State Capital ⊛   County Seats ⊙
Size of type indicates relative
importance of places

CAM . 2 - 11

KENTUCKY AND TENNESSEE

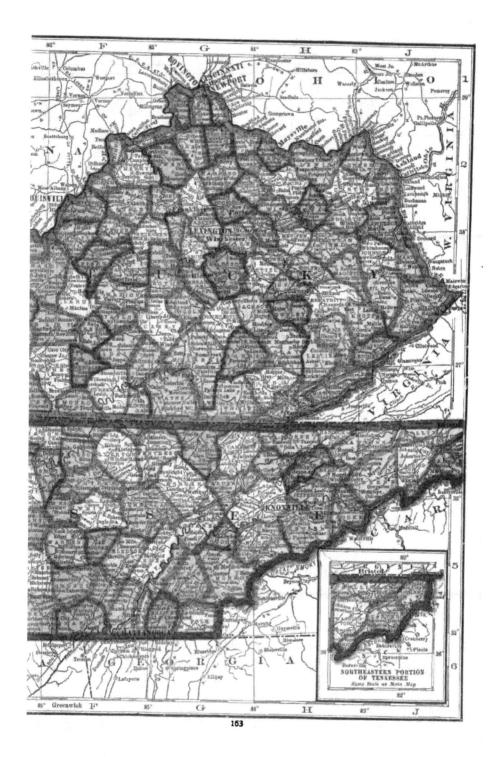

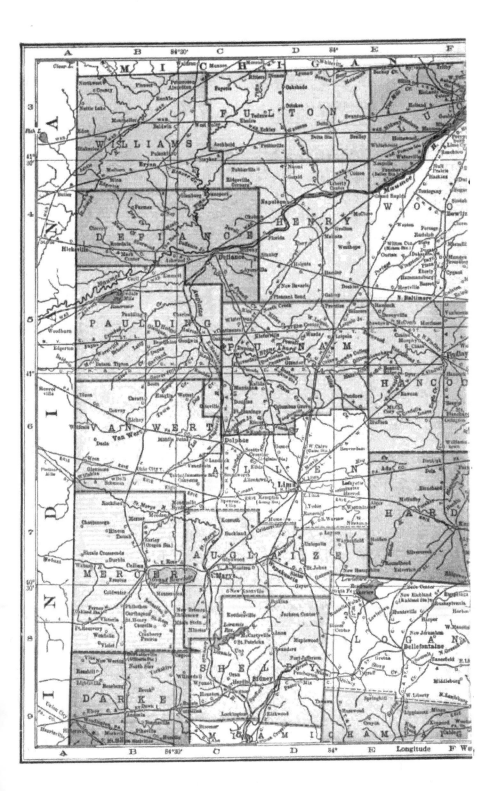

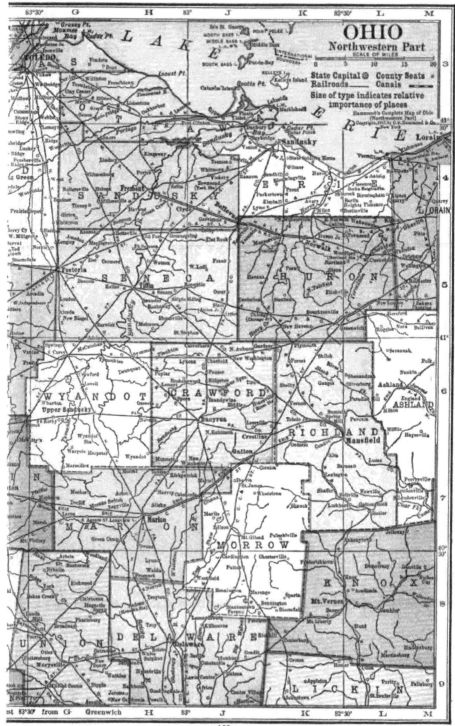

OHIO
Northwestern Part
SCALE OF MILES

State Capital ⊛   County Seats
Railroads ———   Canals ▬▬

Size of type indicates relative
importance of places

Hammond's Complete Map of Ohio
(Northwestern Part)

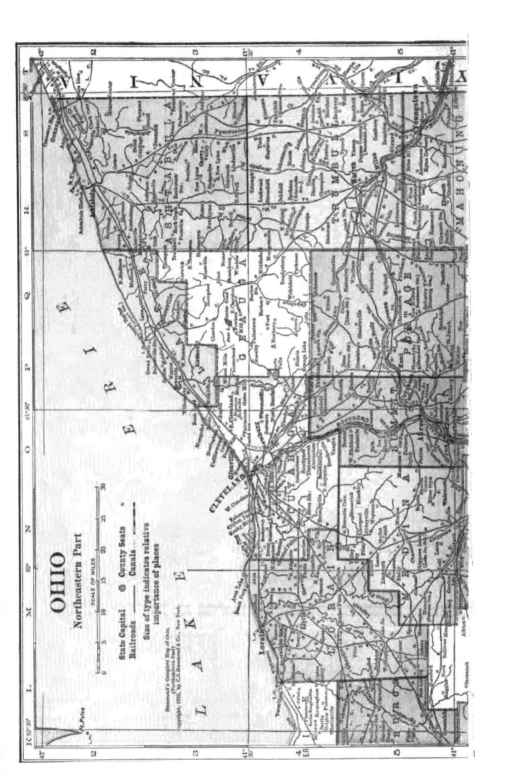

OHIO

Northeastern Part

SCALE OF MILES

State Capital ✱
Railroads ——
County Seats ⊙
Canals ····

Size of type indicates relative
importance of places

Hammond's Complete Map of Ohio.
(For location see Key.)
Copyright, 1911, by C.S. Hammond & Co., New York.

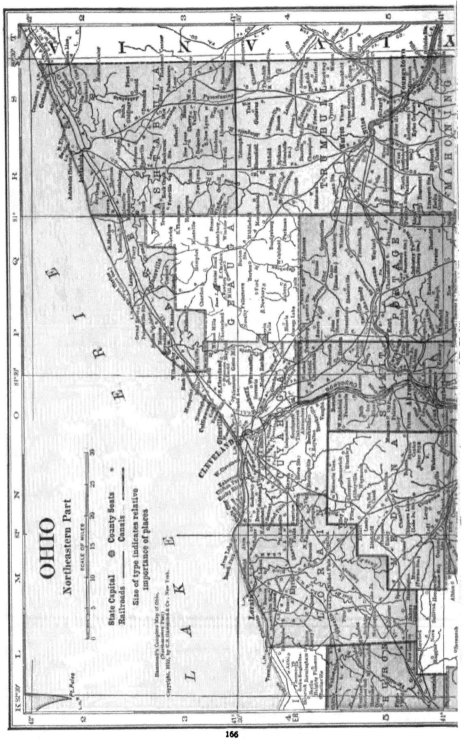

OHIO

Northeastern Part

SCALE OF MILES

State Capital ⊛ County Seats ＊
Railroads ——— Canals ———

Size of type indicates relative
importance of places

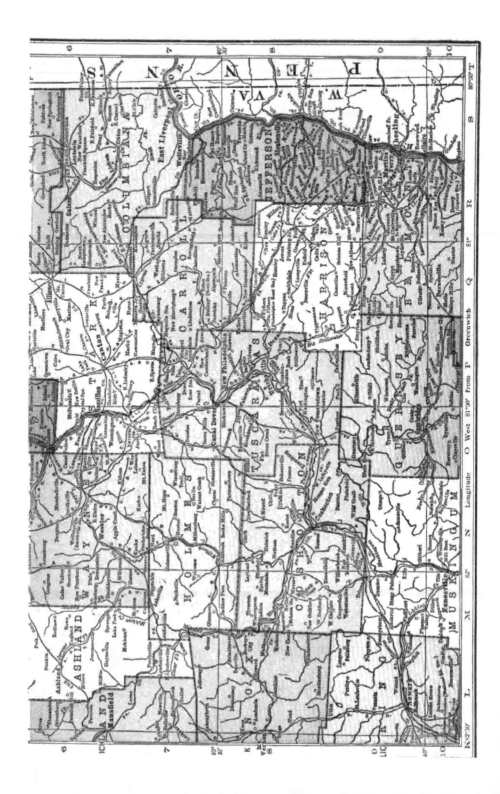

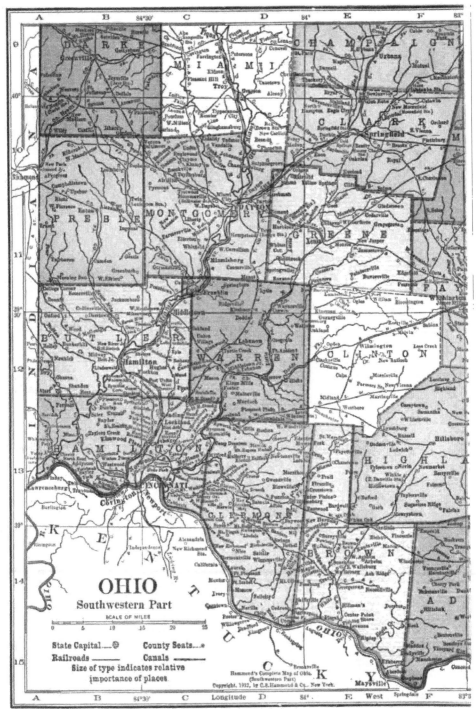

OHIO

Southwestern Part

SCALE OF MILES

0   5   10   15   20   25

State Capital ⊚   County Seats ◦
Railroads   Canals
Size of type indicates relative
importance of places

Hammond's Complete Map of Ohio.
(Southwestern Part)
Copyright, 1912, by C.S. Hammond & Co., New York.

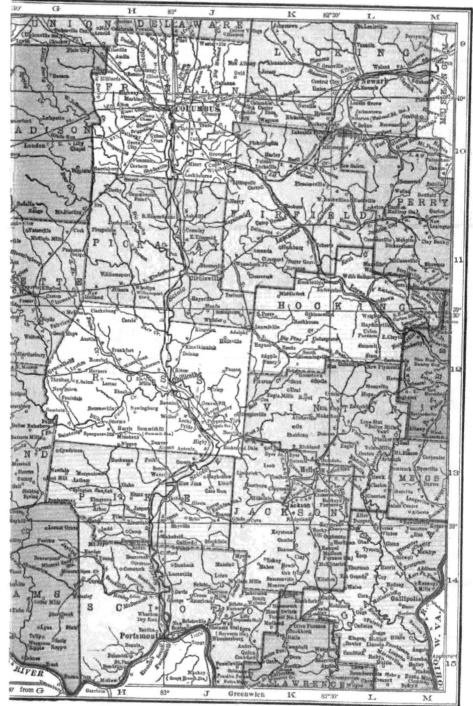

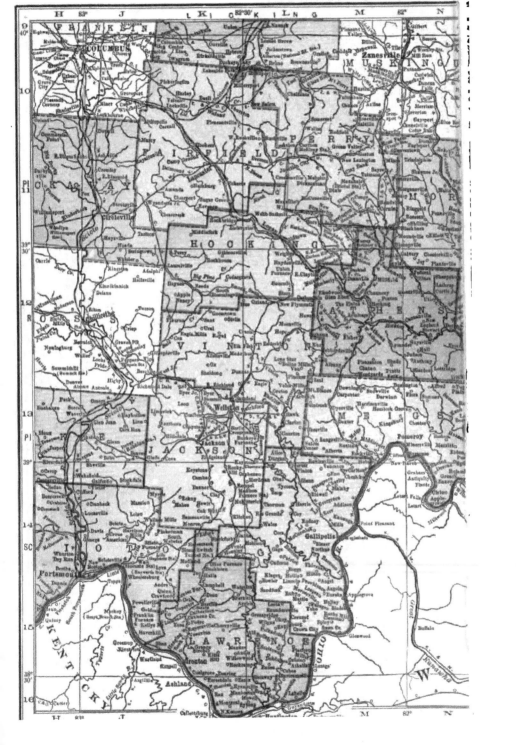

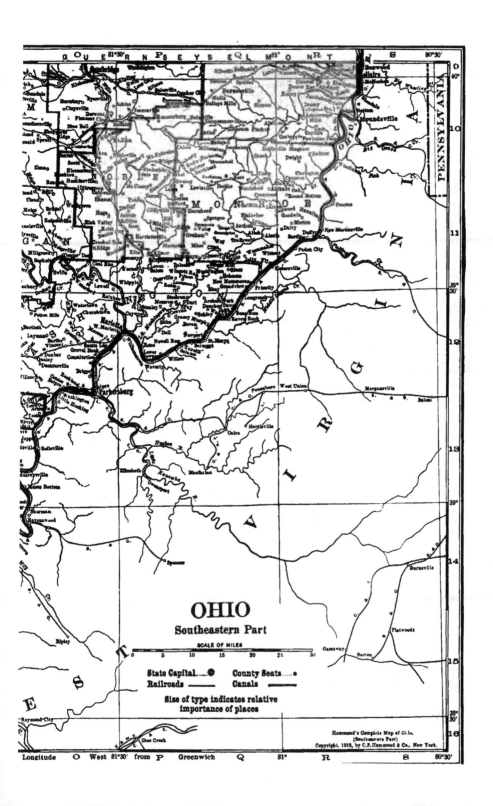

OHIO

Southeastern Part

SCALE OF MILES

State Capital......⊕   County Seats.....•
Railroads ———   Canals ═══

Size of type indicates relative
importance of places

Hammond's Complete Map of Ohio.
(Southeastern Part)
Copyright, 1912, by C.S. Hammond & Co., New York.

Longitude   O   West 81°30′ from   P   Greenwich   Q   81°   R   S   80°30′

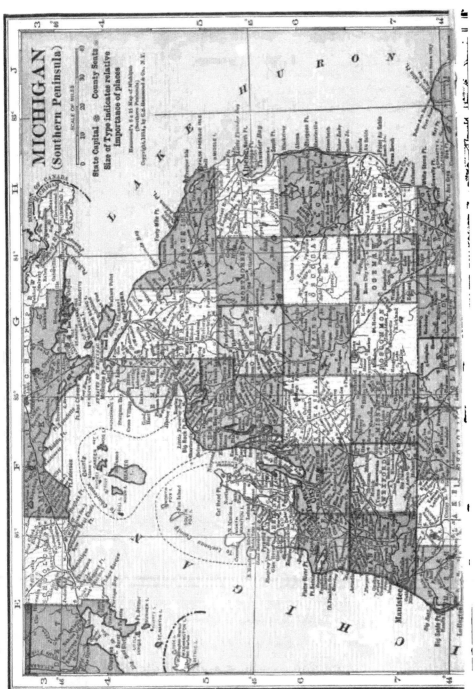

MICHIGAN
(Southern Peninsula)

SCALE OF MILES

State Capital ⊛ County Seats ⊙
Size of Type indicates relative
importance of places

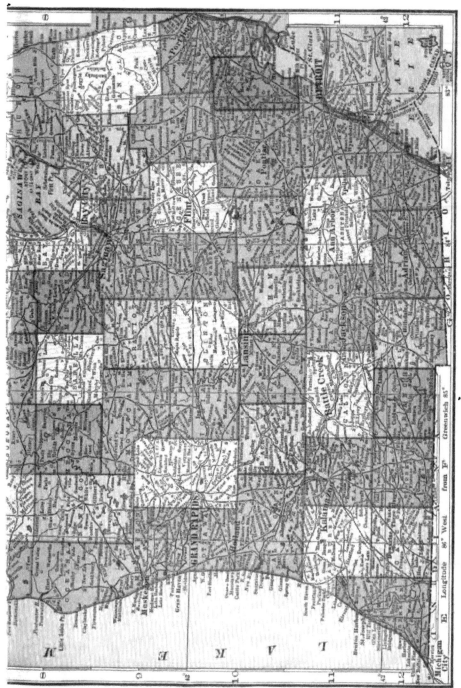

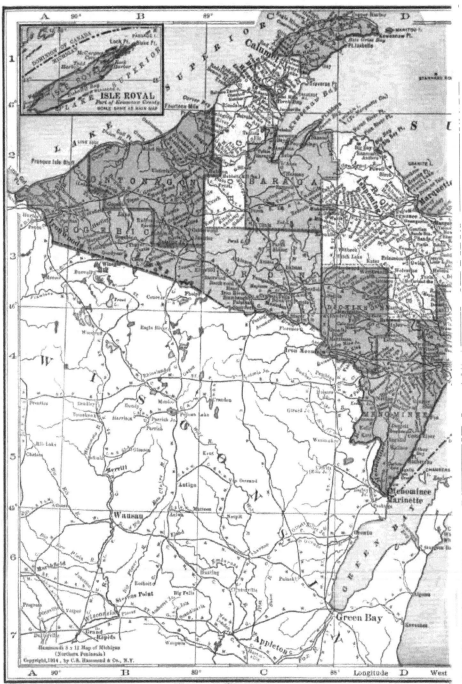

ISLE ROYAL
Part of Keweenaw County
SCALE SAME AS MAIN MAP

Hammond's 8 x 11 Map of Michigan
(Northern Peninsula)
Copyright, 1914, by C.S. Hammond & Co., N.Y.

Longitude West

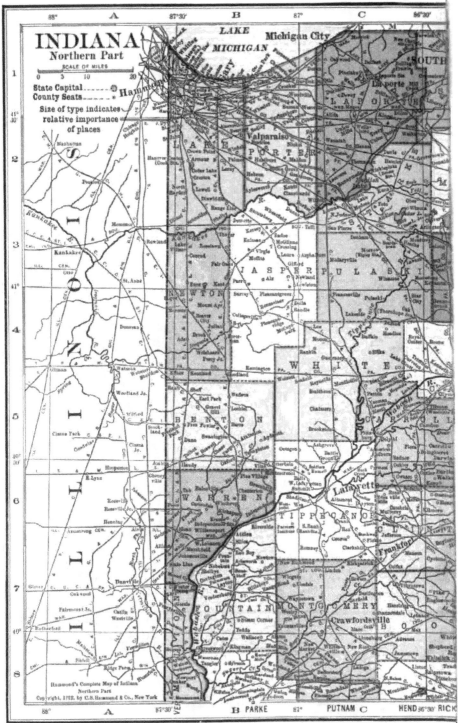

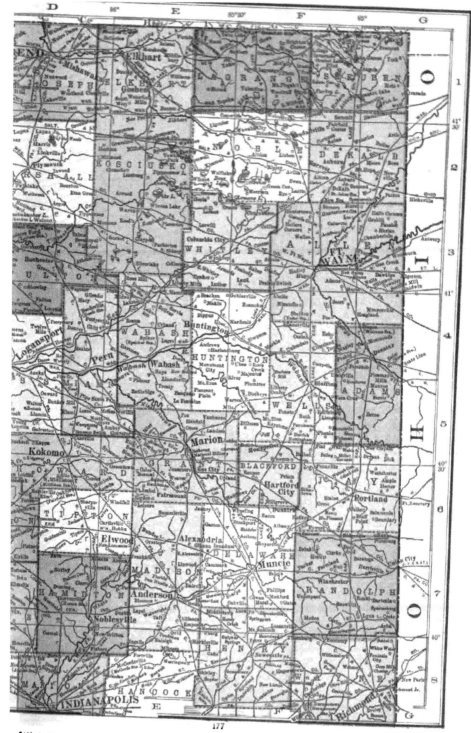

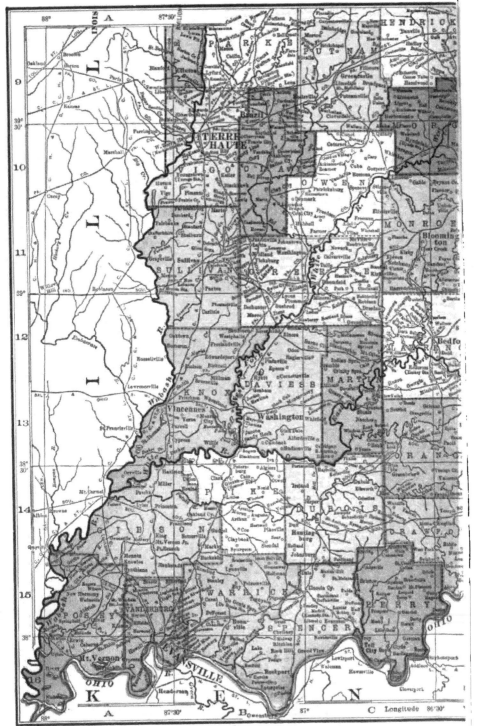

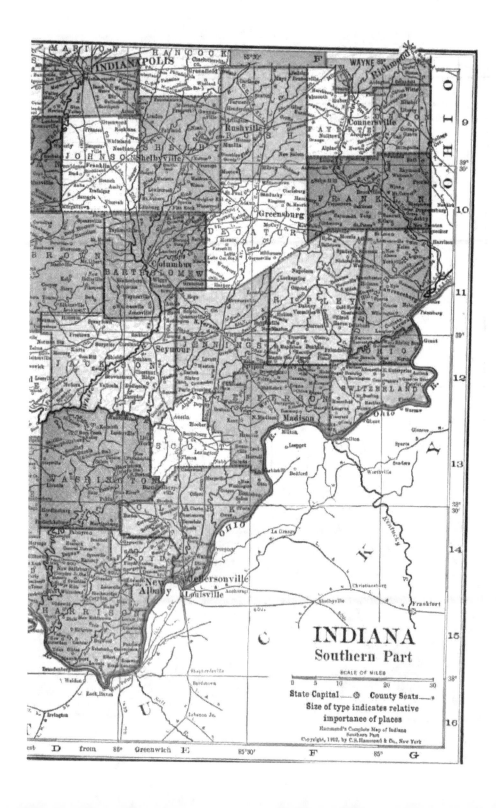

INDIANA
Southern Part

SCALE OF MILES

State Capital .... ⊕   County Seats .... *
Size of type indicates relative
importance of places

Hammond's Complete Map of Indiana
Southern Part
Copyright, 1912, by C.S. Hammond & Co., New York

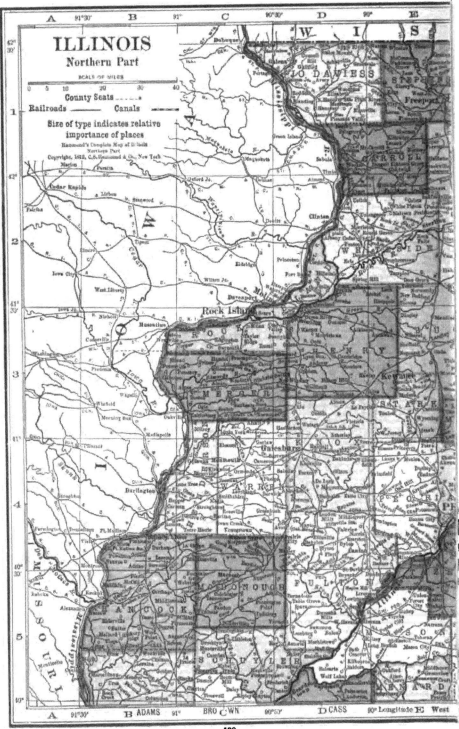

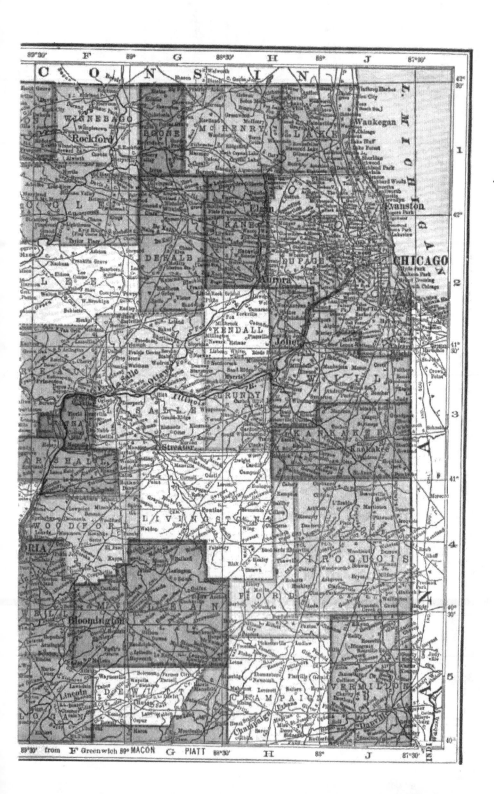

ILLINOIS
Southern Part
SCALE OF MILES

State Capital ⊕   County Seats ✳
Railroads
Size of type indicates relative
importance of places

Hammond's Complete Map of Illinois
Southern Part
Copyright, 1912, C.S. Hammond & Co., New York

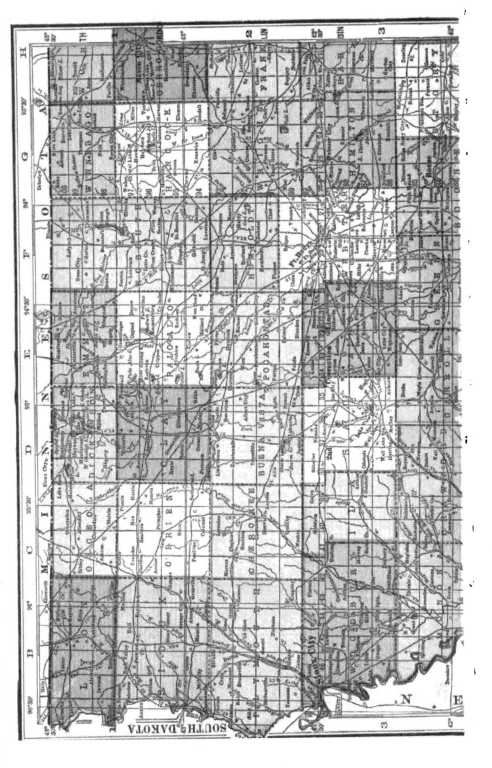

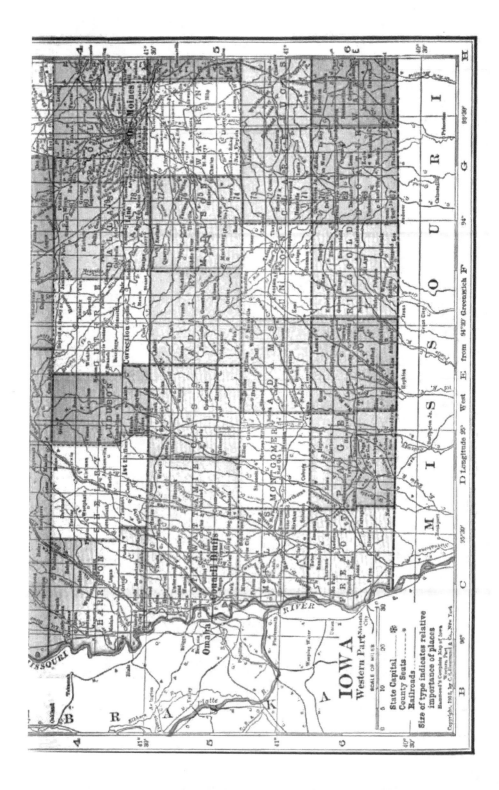

IOWA
Western Part

SCALE OF MILES

State Capital
County Seats
Railroads

Size of type indicates relative
importance of places

Rand's Complete Map of Iowa
Western Part
Copyright, 1911, by C. S. Hammond & Co., New York

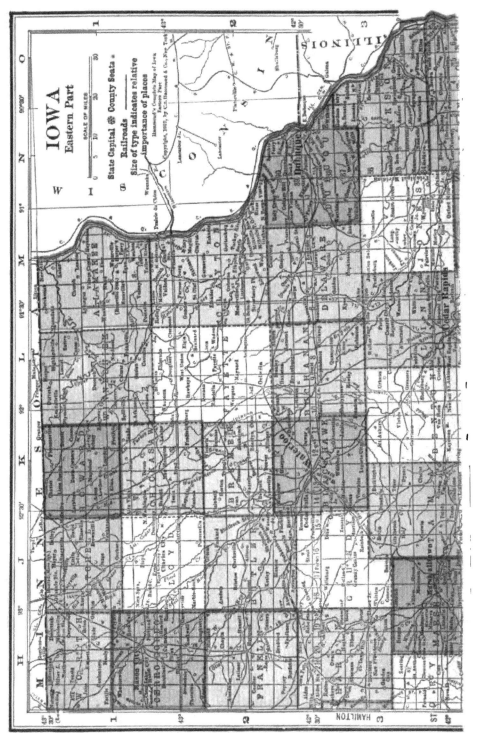

IOWA
Eastern Part

SCALE OF MILES

State Capital ⊛ County Seats ⊙
Railroads
Size of type indicates relative
importance of places

Hammond's County Map of Iowa
Copyright, 1907, by C.S. Hammond & Co., New York

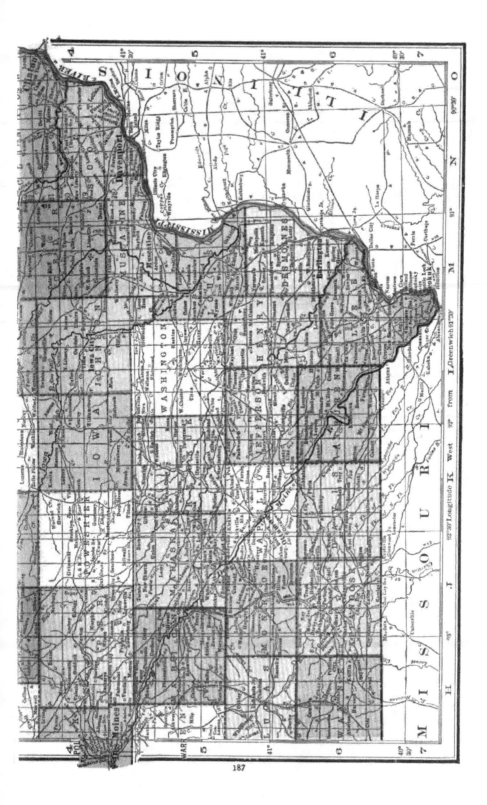

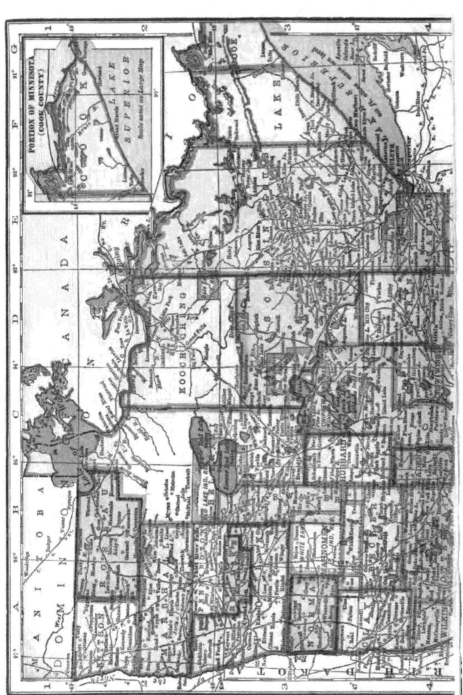

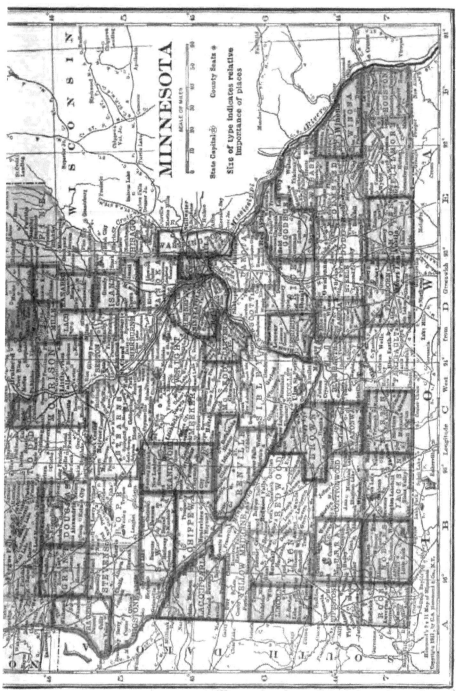

MINNESOTA

SCALE OF MILES

State Capital ⊛   County Seats ⊕

Size of type indicates relative
importance of places

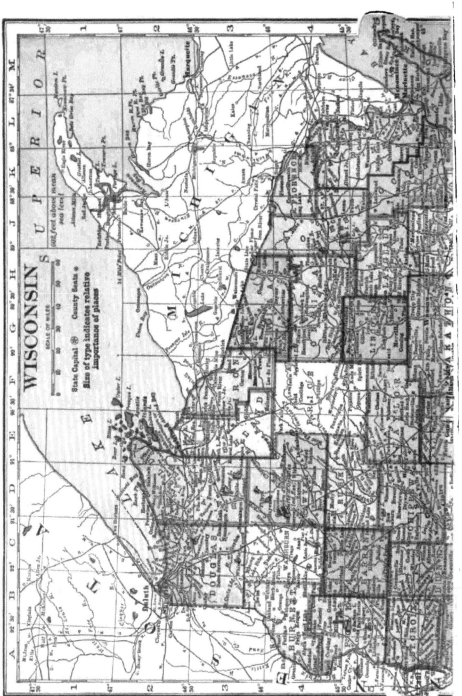

WISCONSIN

SCALE OF MILES

State Capital ⊛   County Seats ⊙
Size of type indicates relative
importance of places

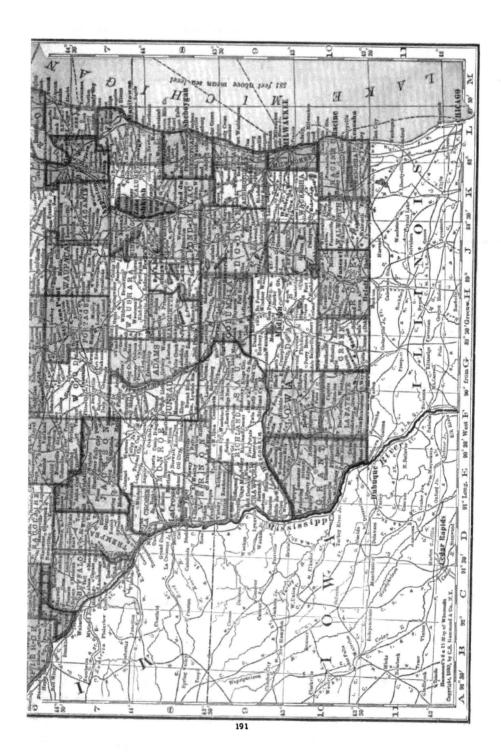

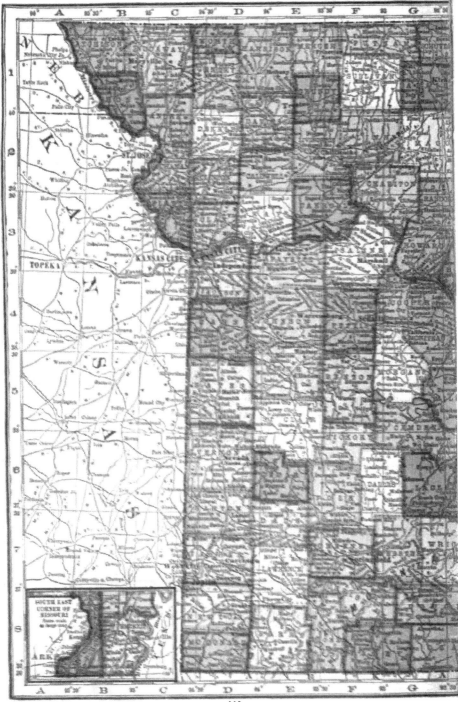

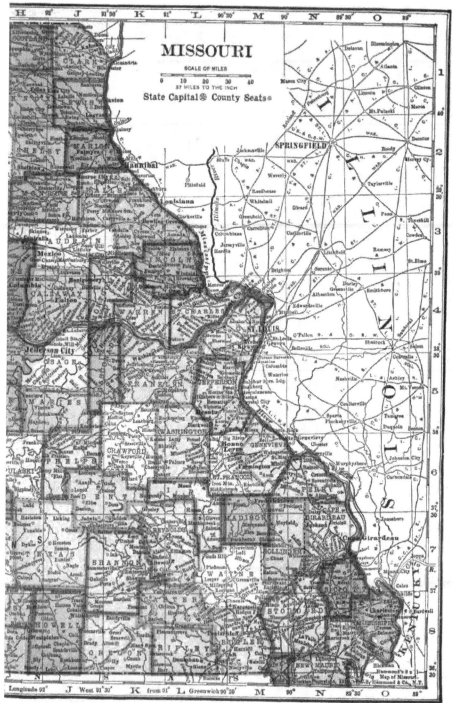

MISSOURI

SCALE OF MILES

0   10   20   30   40
37 MILES TO THE INCH

State Capital ⊛ County Seats ∘

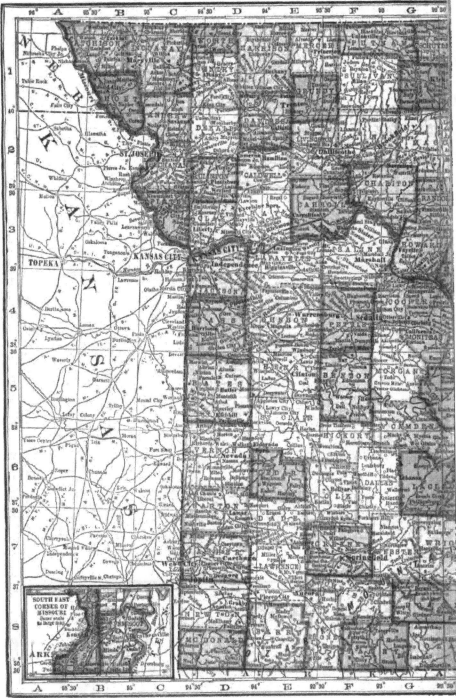

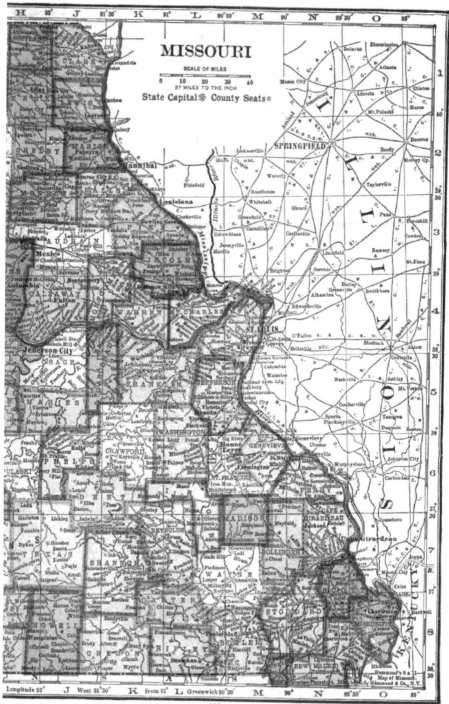

MISSOURI

SCALE OF MILES

State Capital ⊛ County Seats ⊙

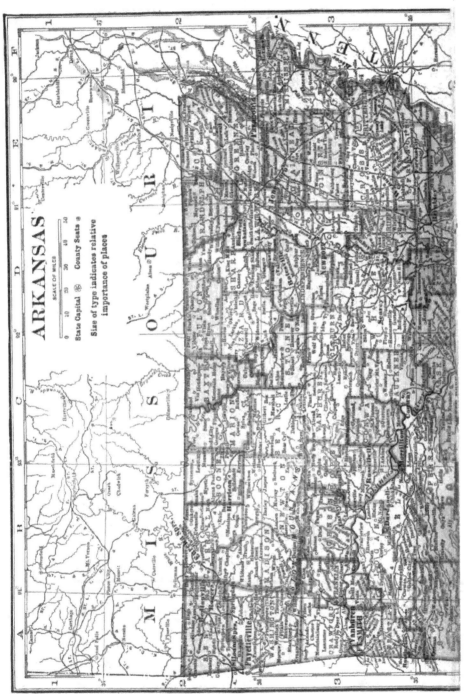

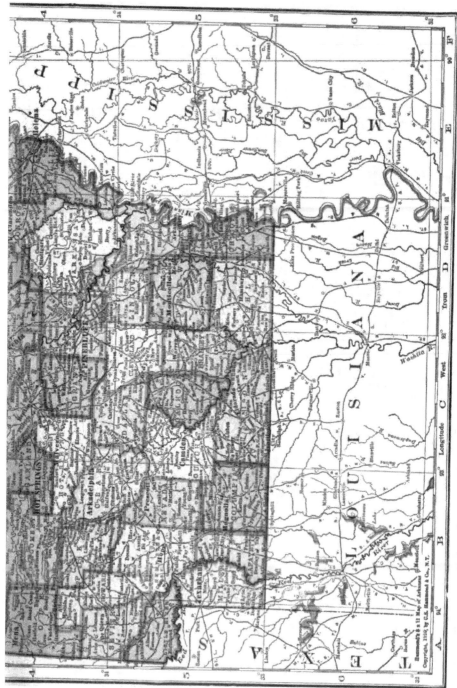

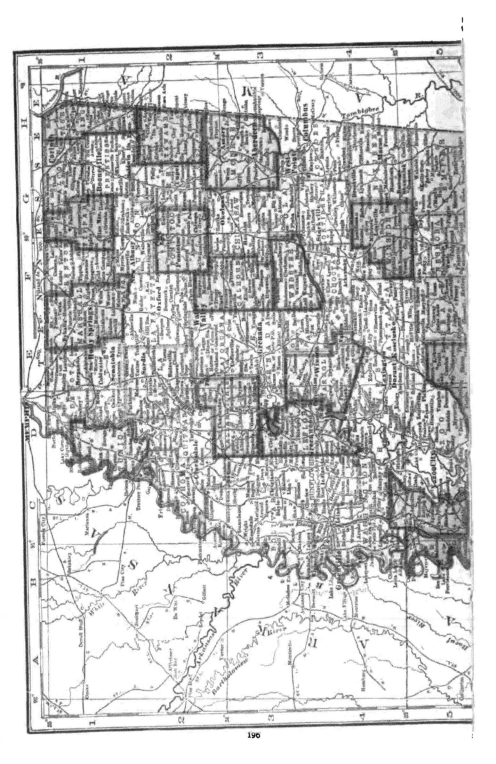

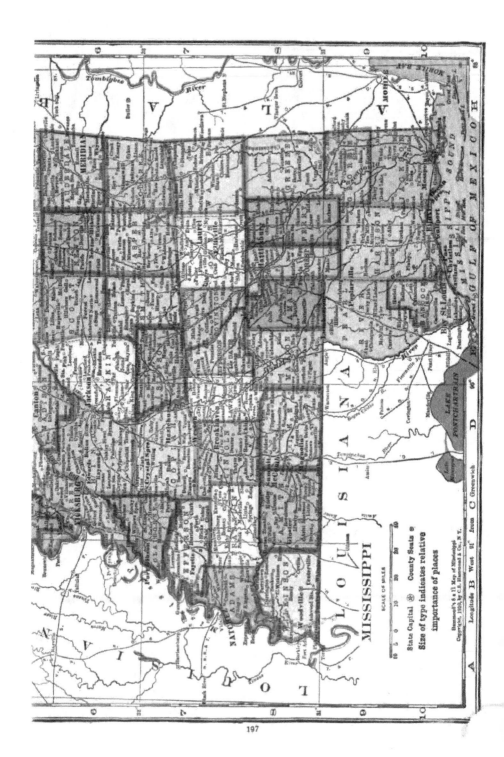

MISSISSIPPI

SCALE OF MILES

State Capital ⊛   County Seats ⊙
Size of type indicates relative
importance of places

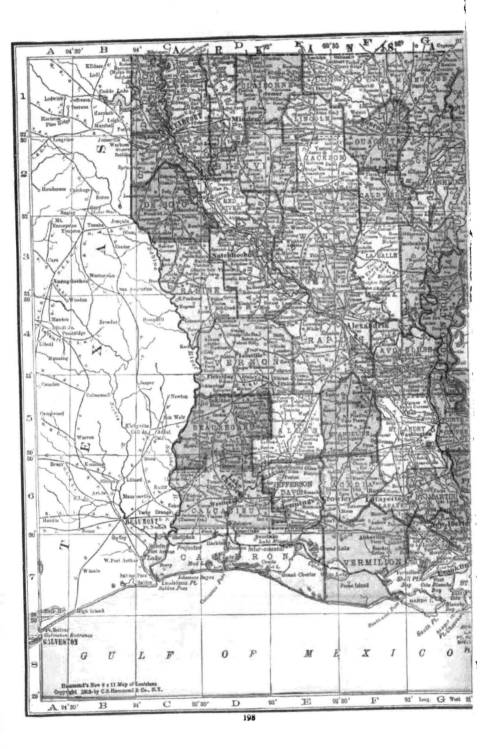

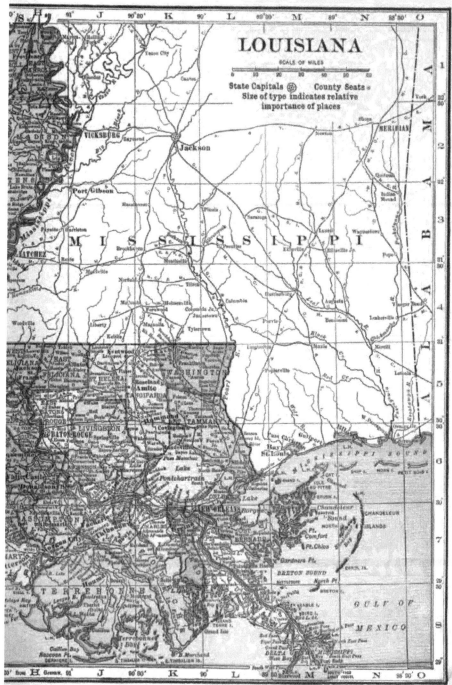

LOUISIANA

SCALE OF MILES

State Capitals ⊛   County Seats ◦
Size of type indicates relative
importance of places

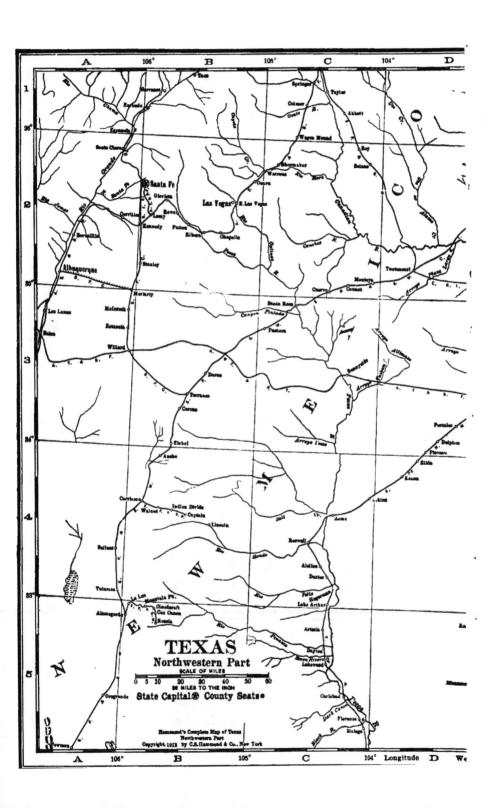

TEXAS
Northwestern Part
SCALE OF MILES

0   5   10   20   30   40   50   60

36 MILES TO THE INCH

State Capital⊛ County Seats•

Hammond's Complete Map of Texas
Northwestern Part
Copyright 1913 by C.S.Hammond & Co., New York

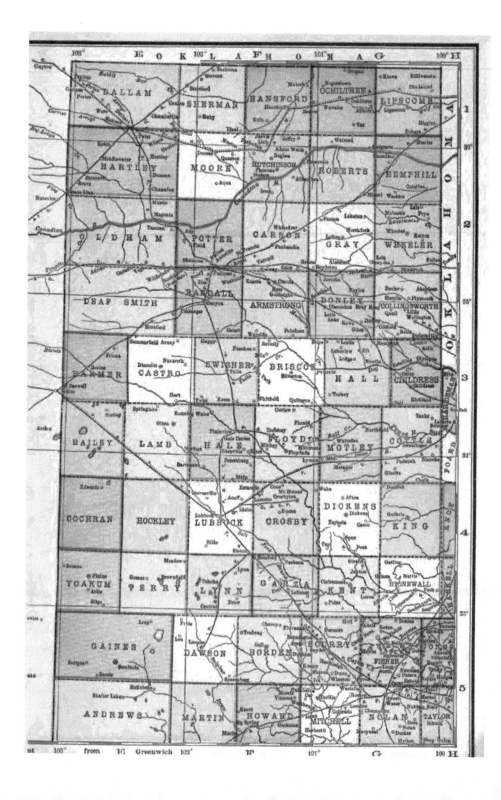

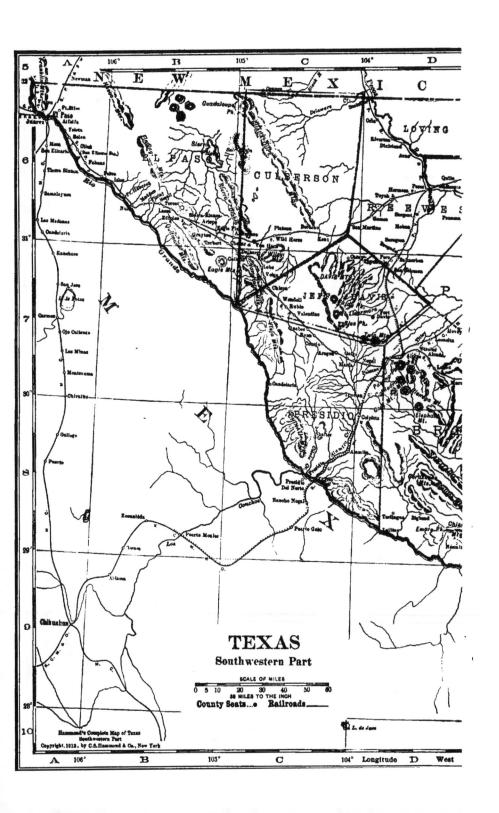

TEXAS

Southwestern Part

SCALE OF MILES

0 5 10   20   30   40   50   60
38 MILES TO THE INCH

County Seats...●   Railroads

A   106°   B   105°   C   104°   Longitude   D   West

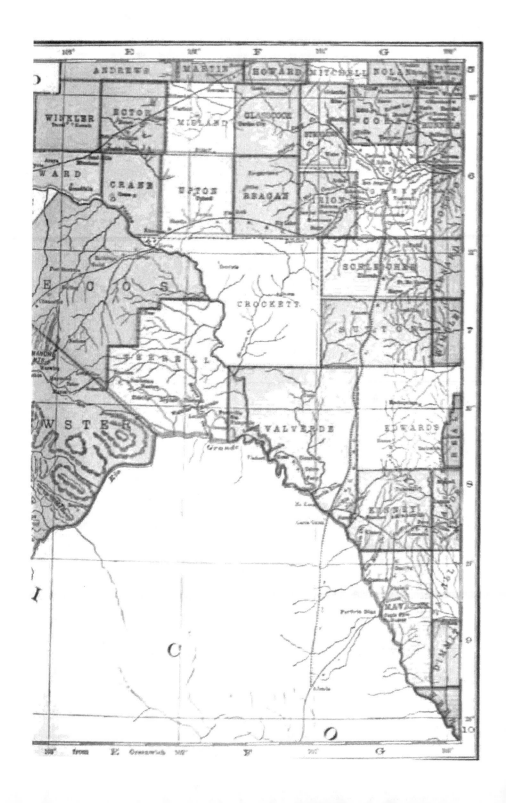

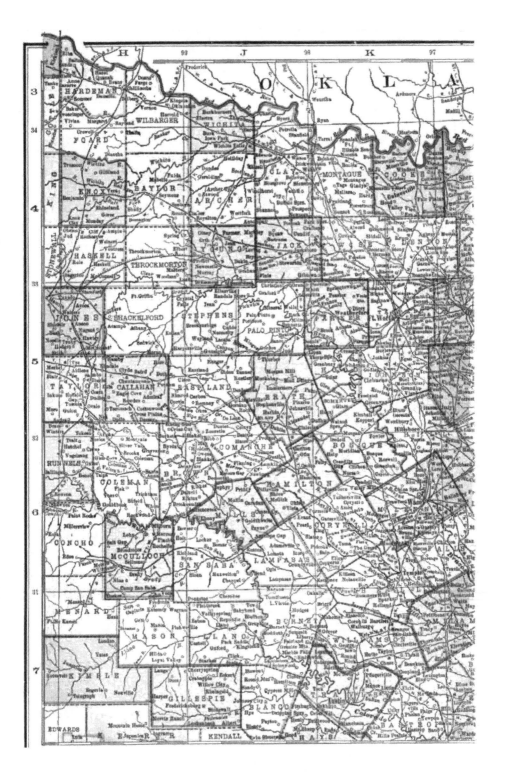

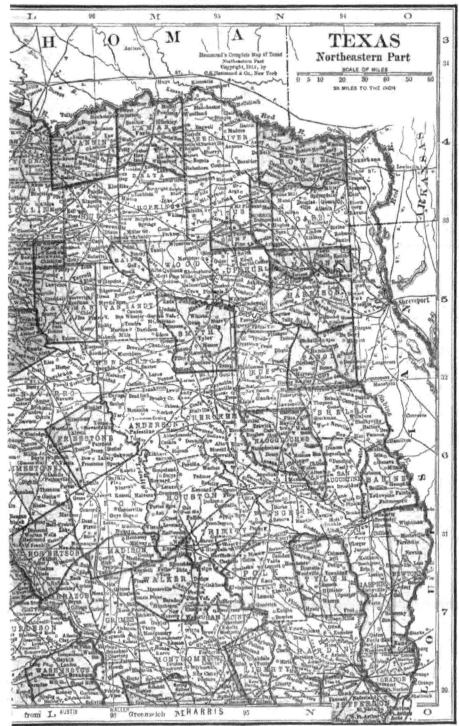

TEXAS
Northeastern Part

SCALE OF MILES
0 5 10 20 30 40 50 60

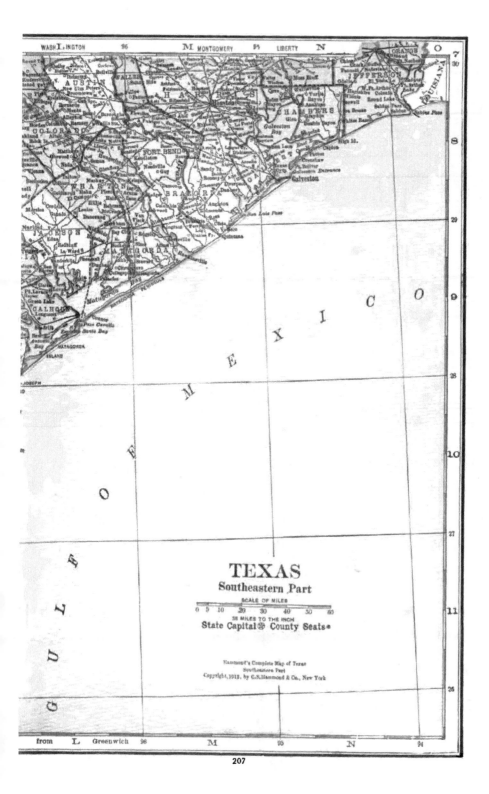

TEXAS

Southeastern Part

SCALE OF MILES

0 5 10   20   30   40   50   60

36 MILES TO THE INCH

State Capital ⊛ County Seats ⊛

Hammond's Complete Map of Texas
Southeastern Part
Copyright, 1913, by C.S.Hammond & Co., New York

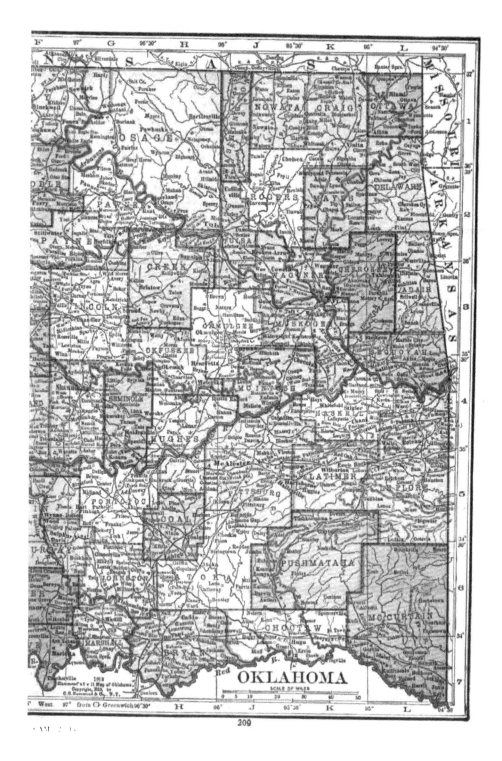

OKLAHOMA

SCALE OF MILES

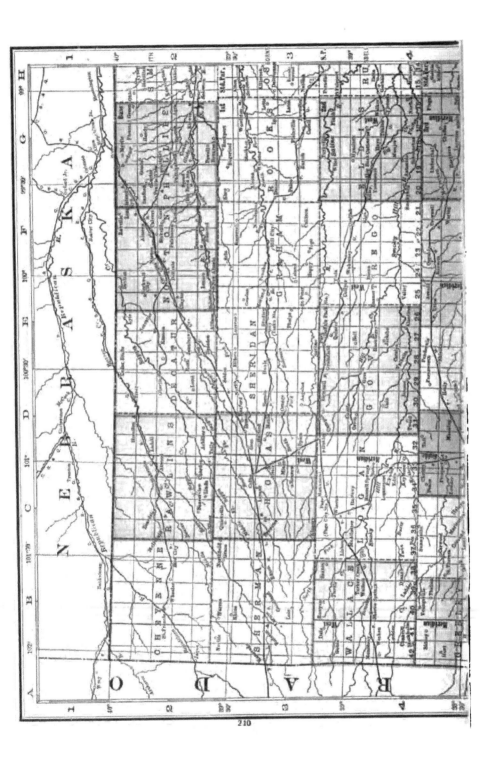

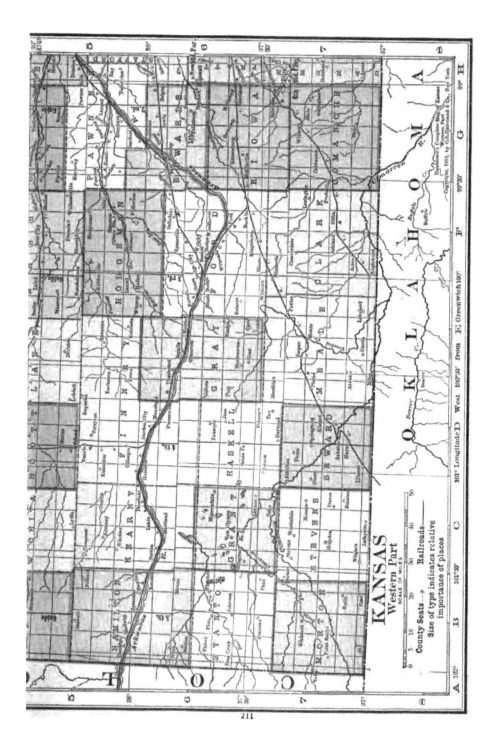

KANSAS
Western Part
SCALE OF MILES

County Seats ···➤   Railroads ————
Size of type indicates relative
importance of places

211

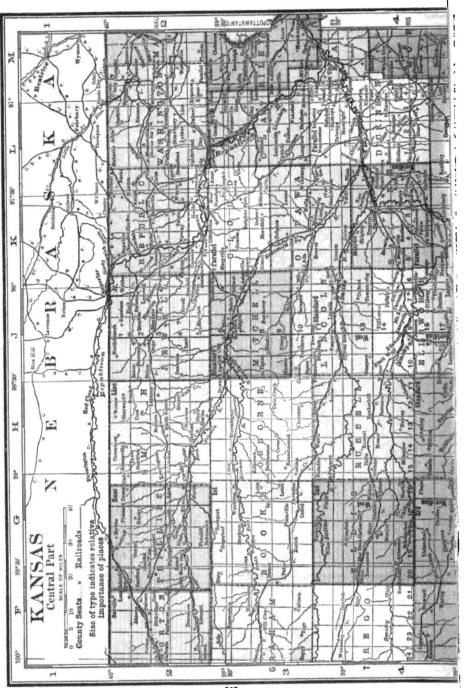

KANSAS
Central Part

SCALE OF MILES

County Seats......○ Railroads ———

Size of type indicates relative
importance of places

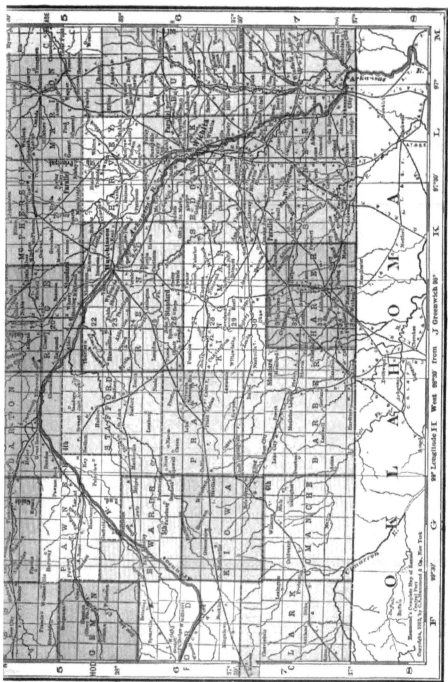

213

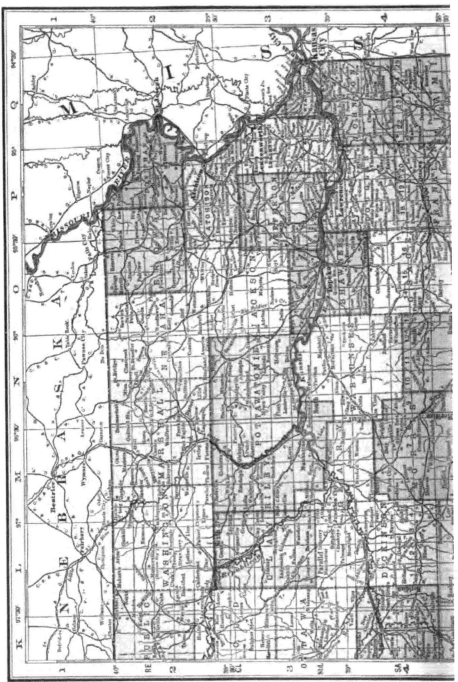

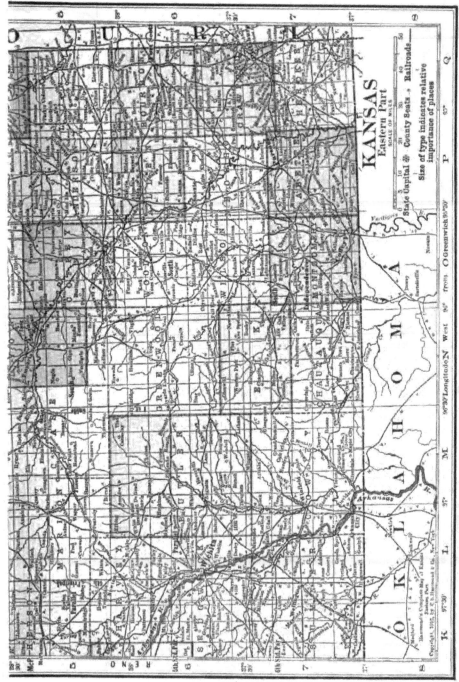

215

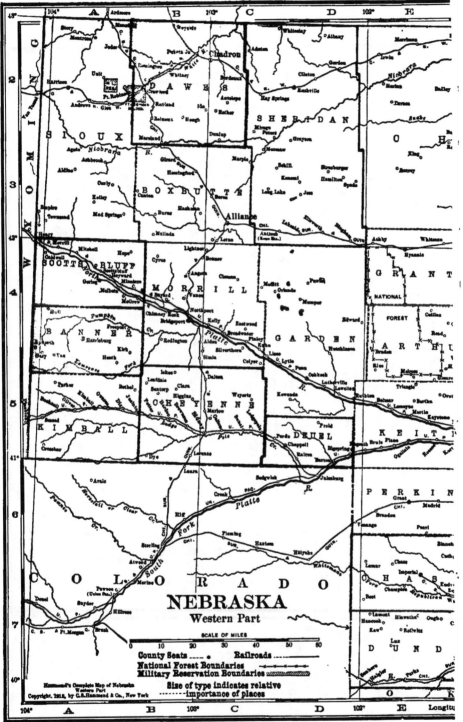

NEBRASKA

Western Part

SCALE OF MILES

0   10   20   30   40   50   60

County Seats ..... •   Railroads ......
National Forest Boundaries
Military Reservation Boundaries
                    Size of type indicates relative
                    .... importance of places

Hammond's Complete Map of Nebraska
Western Part
Copyright, 1912, by C.S. Hammond & Co., New York

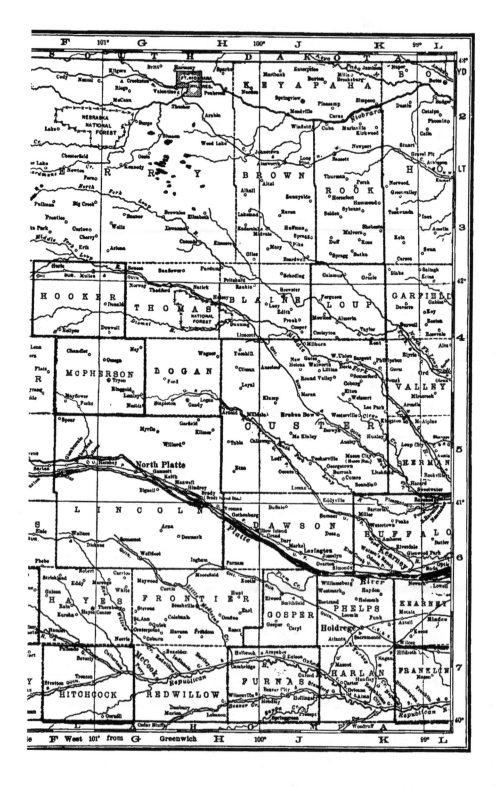

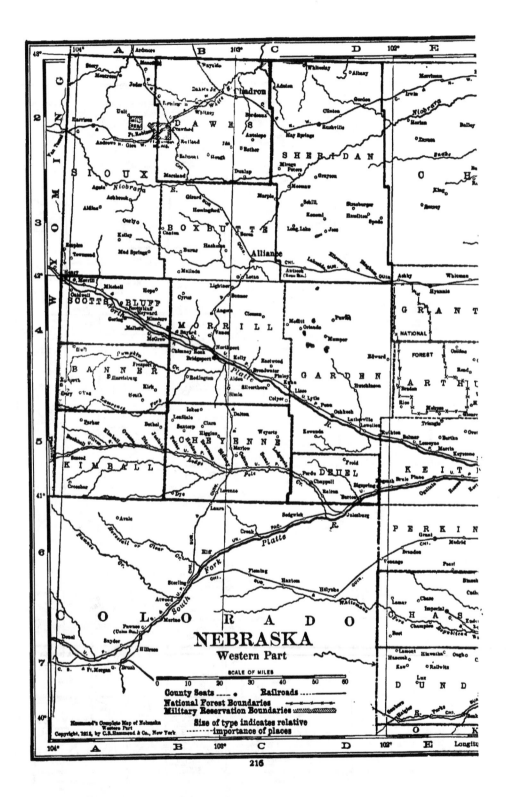

NEBRASKA

Western Part

SCALE OF MILES

0   10   20   30   40   50   60

County Seats _____   Railroads _____
National Forest Boundaries
Military Reservation Boundaries

Size of type indicates relative
importance of places

Hammond's Complete Map of Nebraska
Western Part
Copyright, 1918, by C.S. Hammond & Co., New York

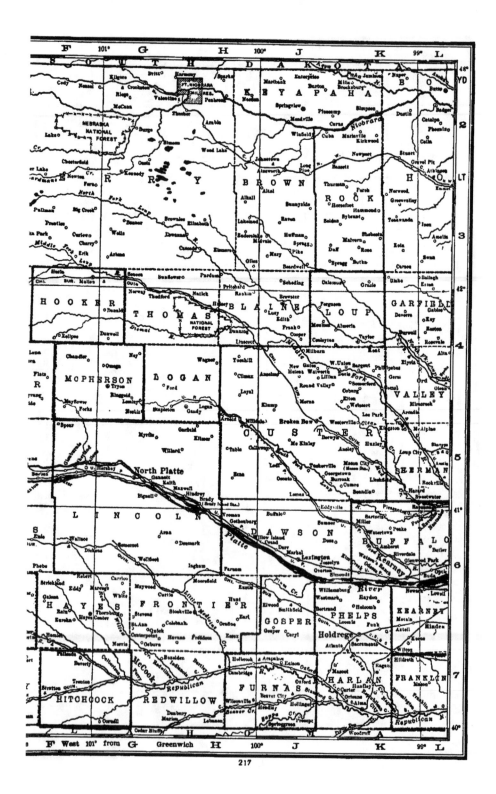

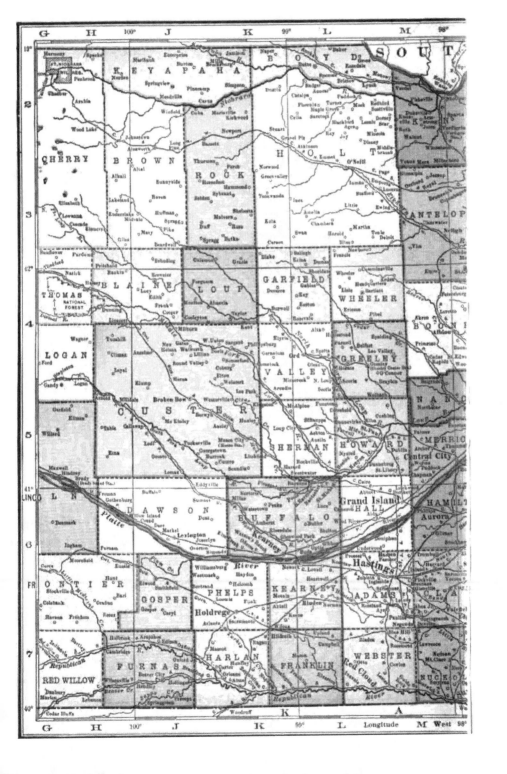

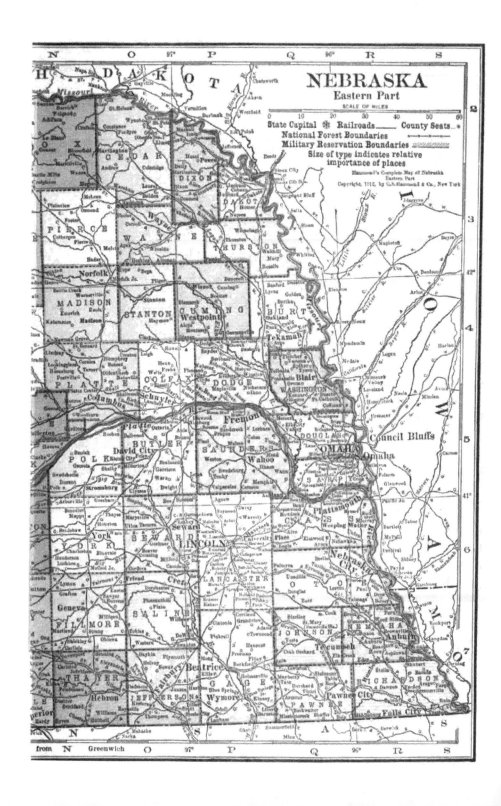

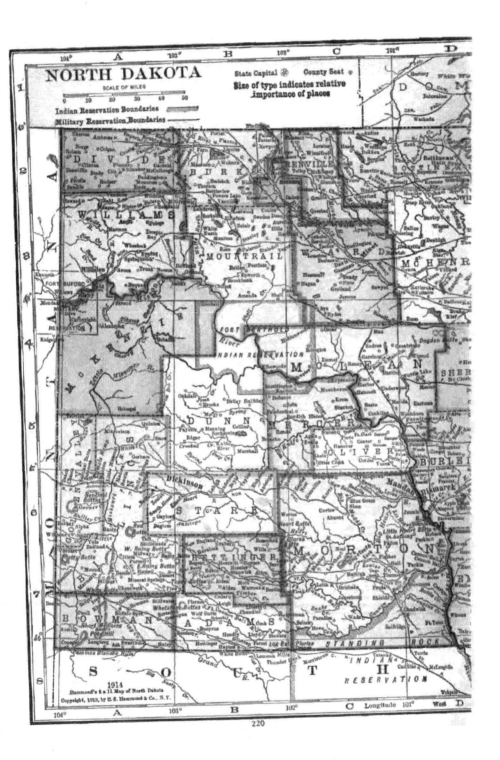

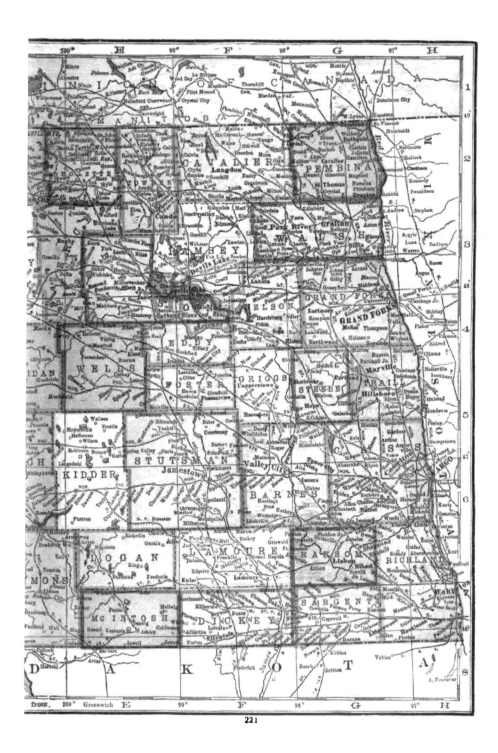

SOUTH DAKOTA

SCALE OF MILES

State Capital   County Seats

Size of type indicates relative
importance of places

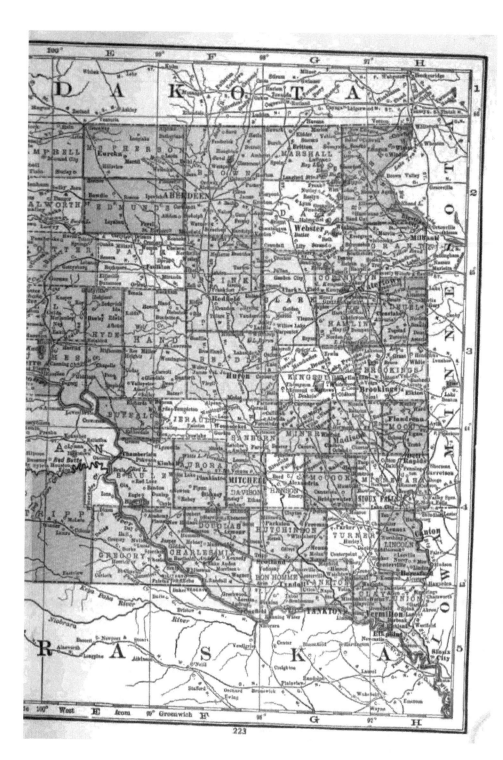

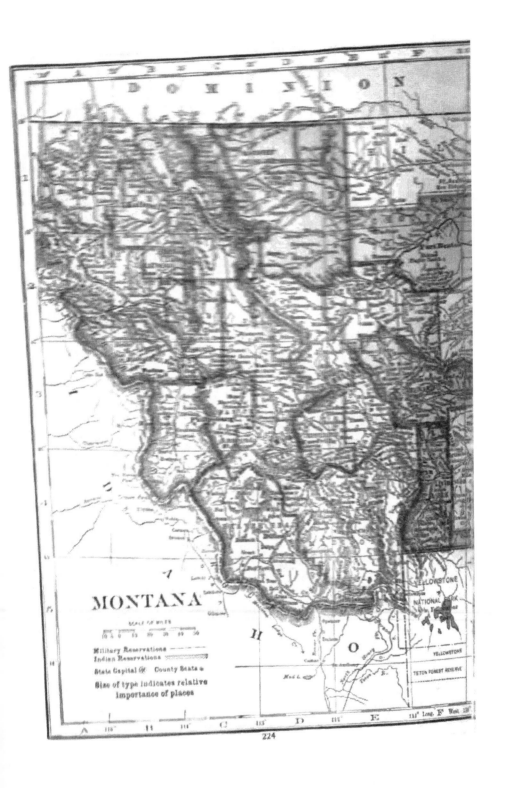

MONTANA

SCALE OF MILES

Military Reservations
Indian Reservations
State Capital ✪   County Seats ●
Size of type indicates relative
importance of places

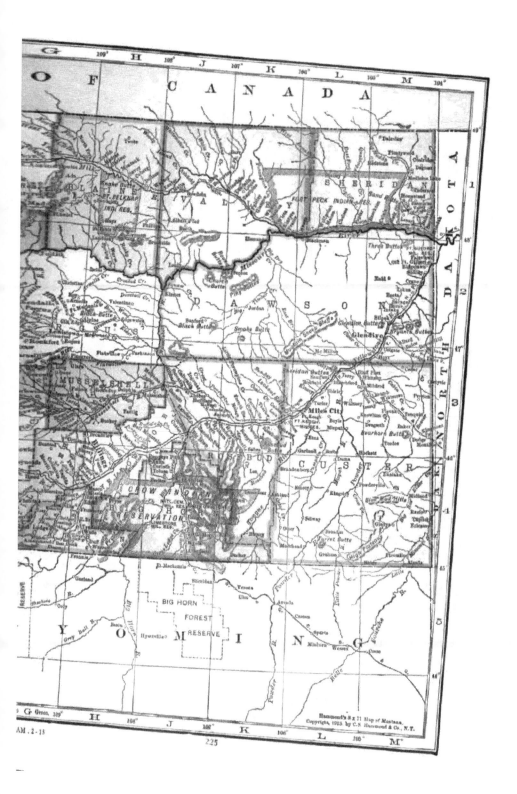

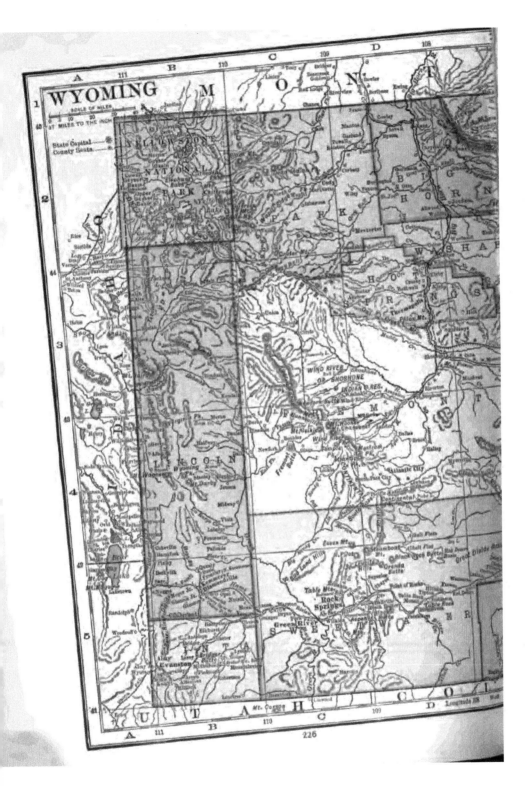

WYOMING

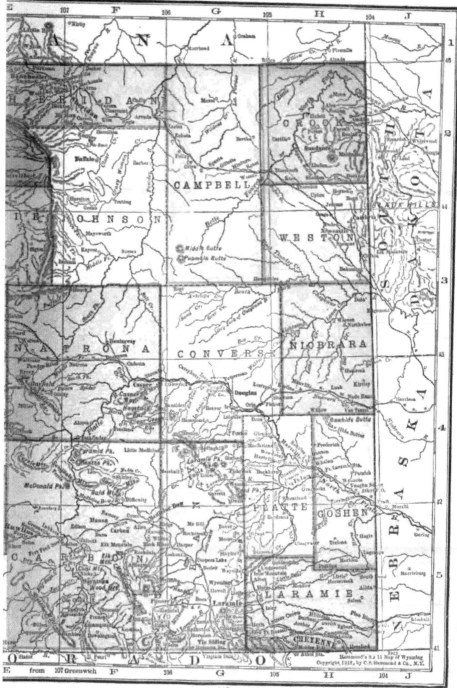

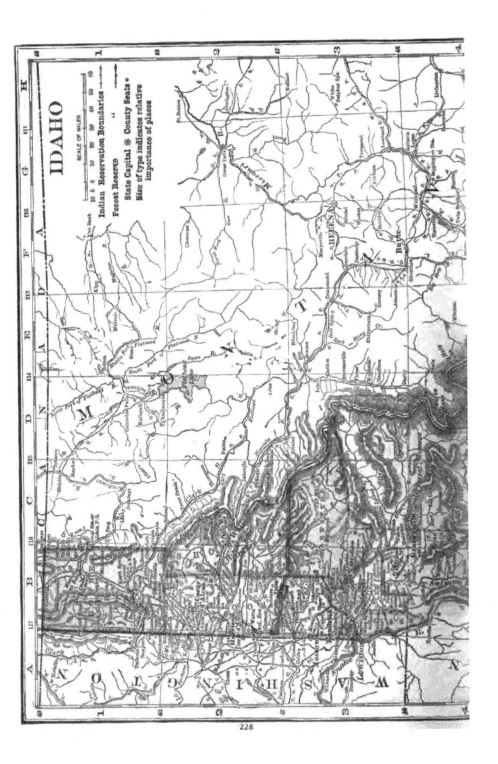

229

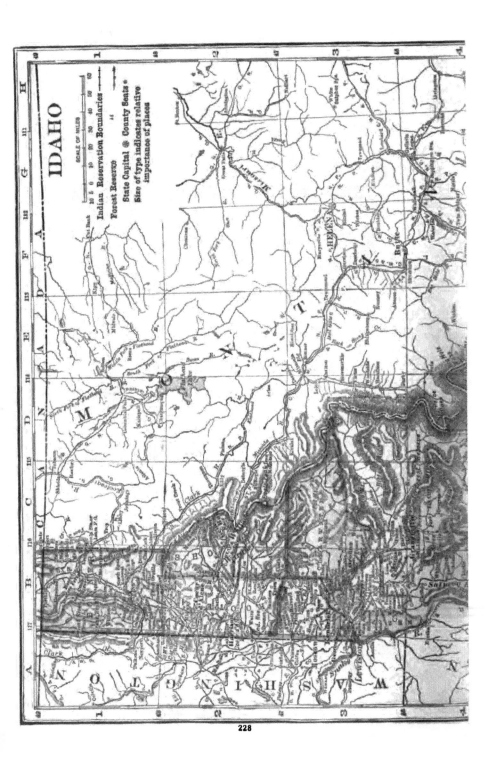

IDAHO

SCALE OF MILES

Indian Reservation Boundaries
Forest Reserve

State Capital ⊛ County Seats ⊙
Size of type indicates relative
importance of places

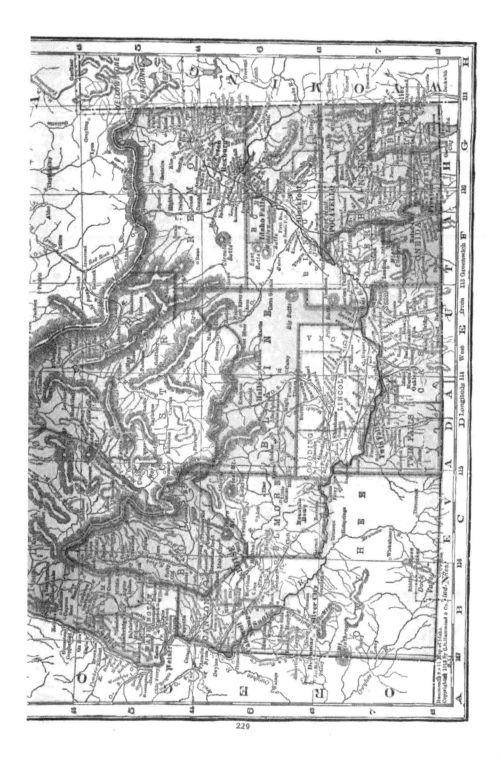

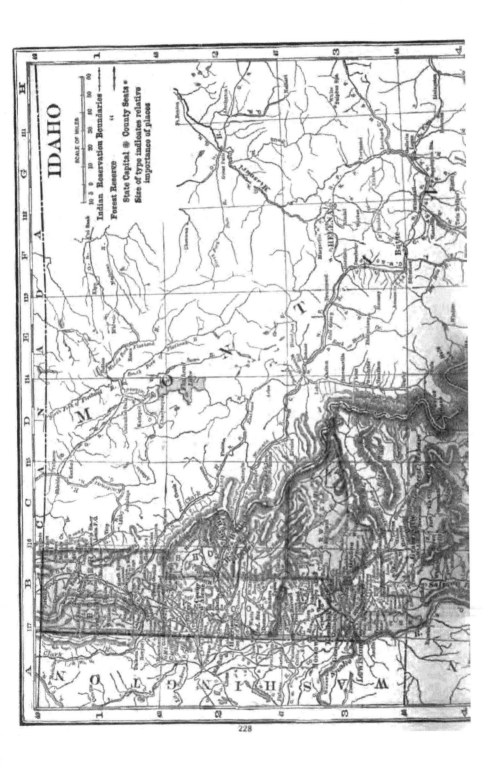

IDAHO

SCALE OF MILES

Indian Reservation Boundaries
Forest Reserve
State Capital ⊛ County Seats ⊙
Size of type indicates relative
importance of places

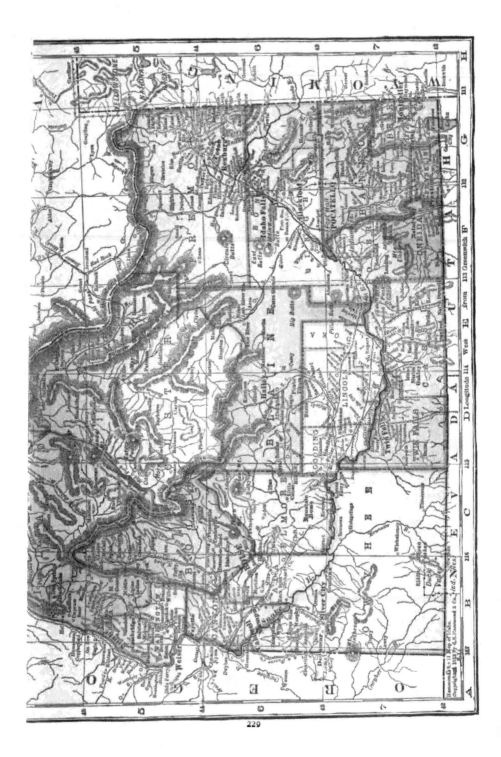

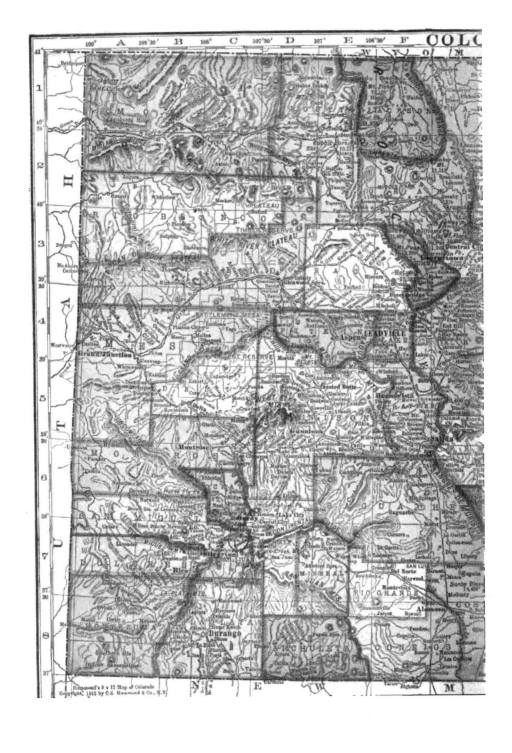

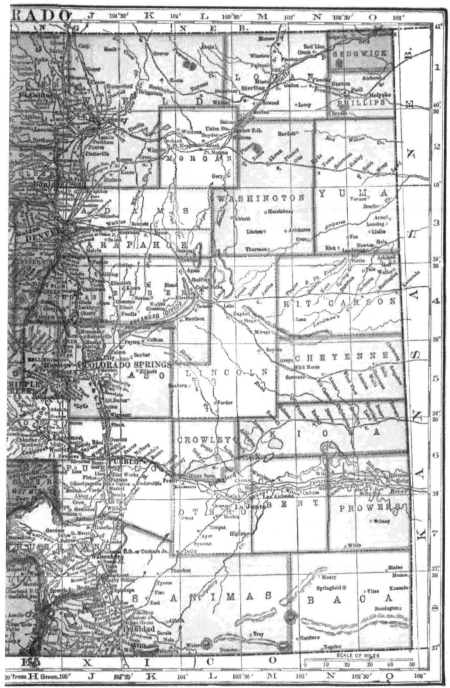

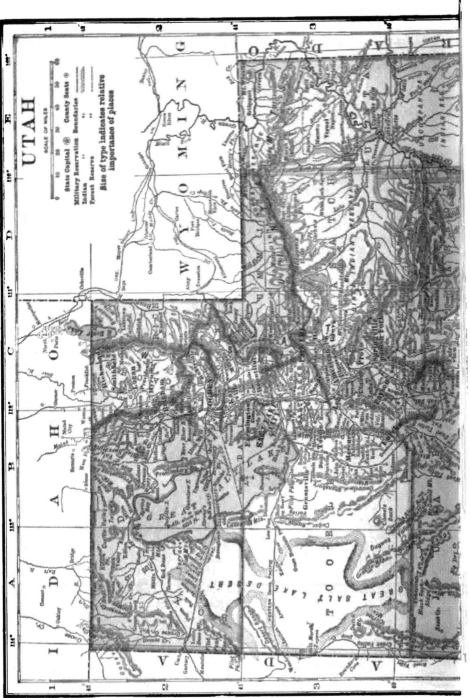

UTAH

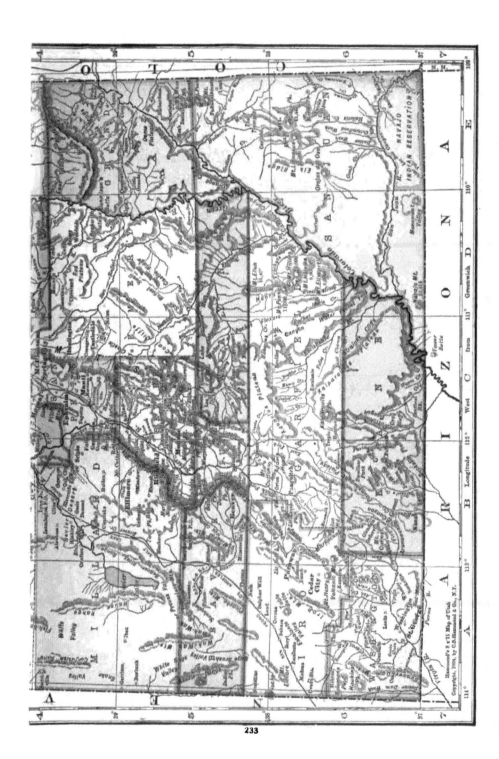

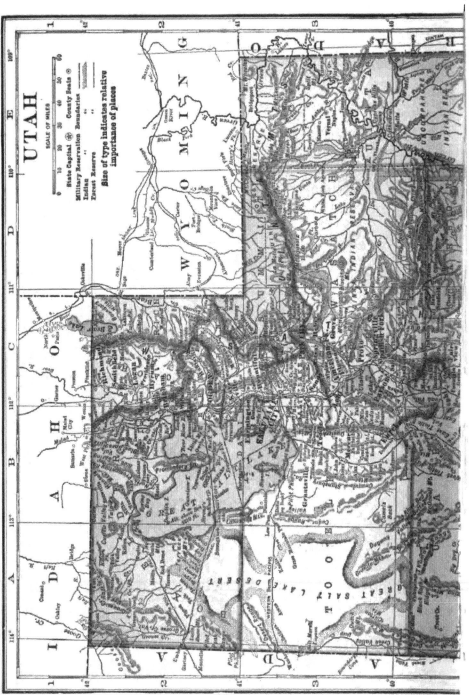

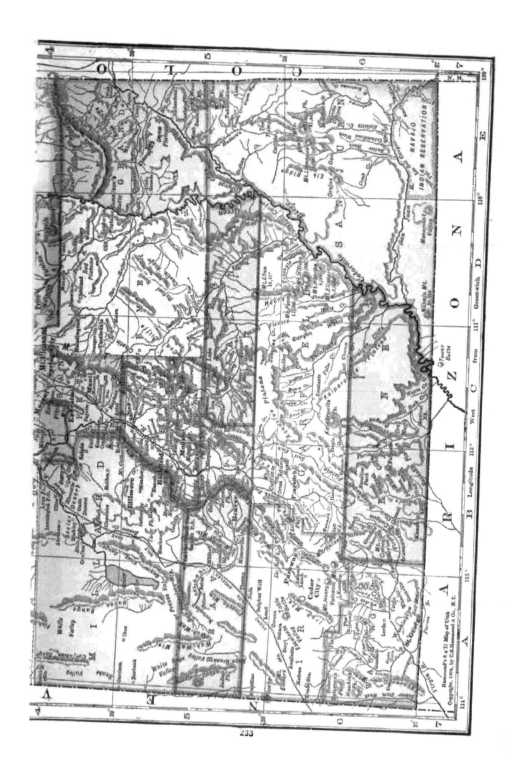

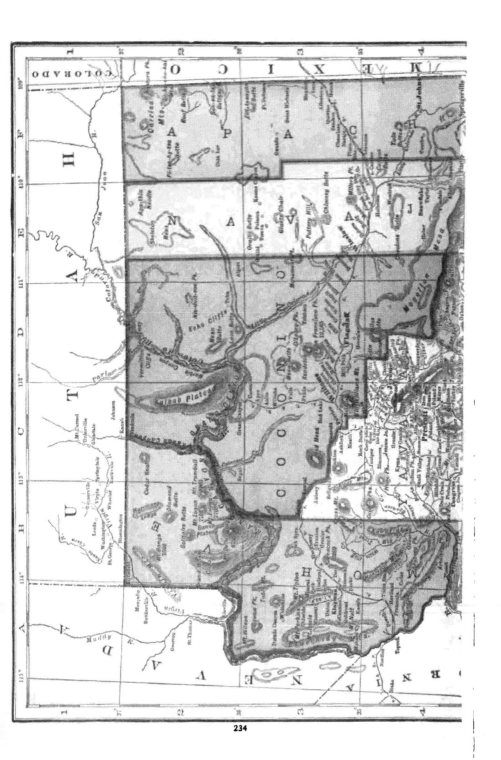

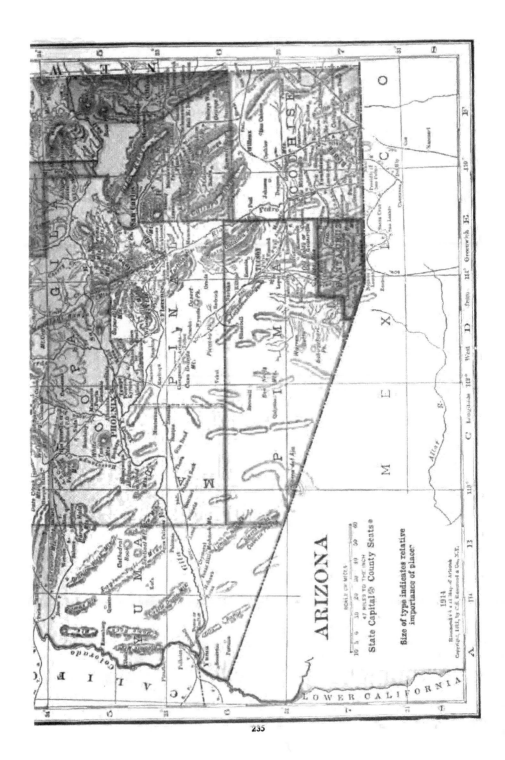

ARIZONA

SCALE OF MILES

State Capital ⊛ County Seats ⊙

Size of type indicates relative
importance of places

1914

Copyright, 1911, by C.S. Hammond & Co., N.Y.

LOWER CALIFORNIA

235

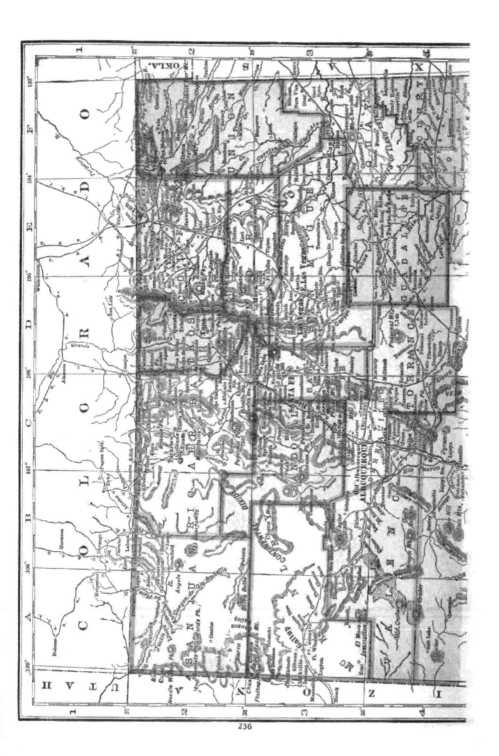

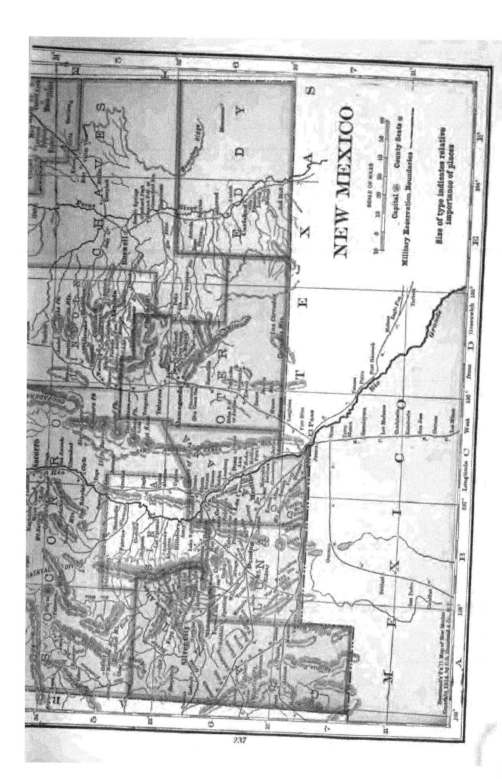

NEW MEXICO

SCALE OF MILES

Capital ⊛   County Seats ◉

Military Reservation Boundaries ———

Size of type indicates relative
importance of places

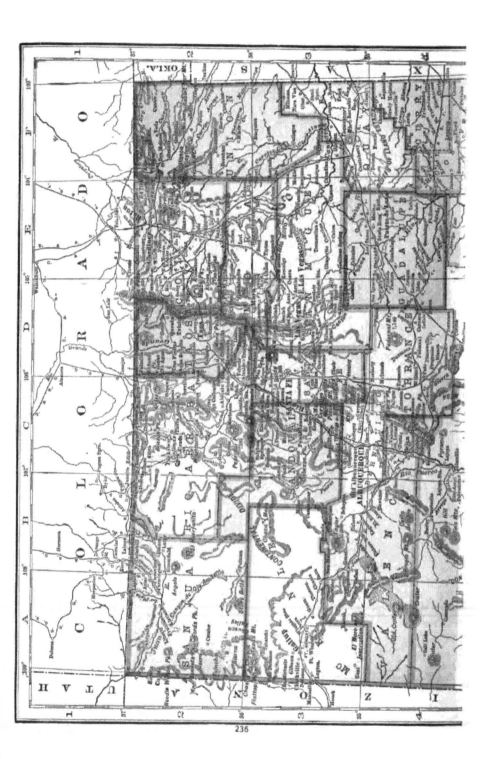

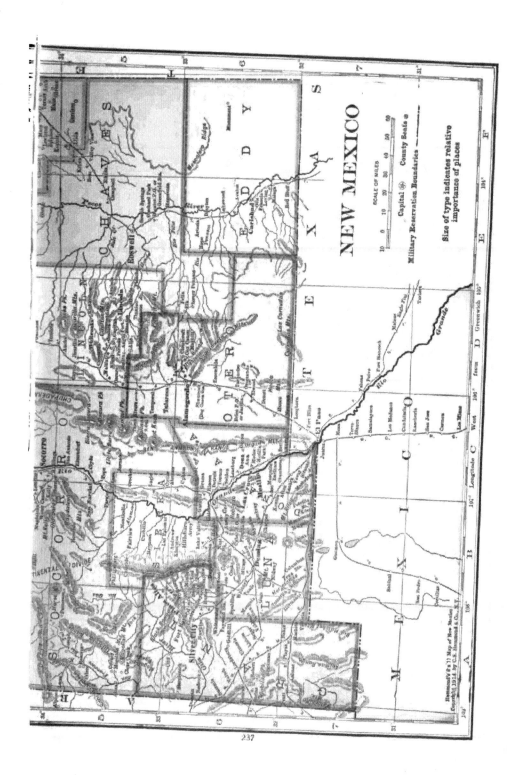

NEW MEXICO

SCALE OF MILES

Capital ⊛   County Seats ▪

Military Reservation Boundaries ----

Size of type indicates relative
importance of places

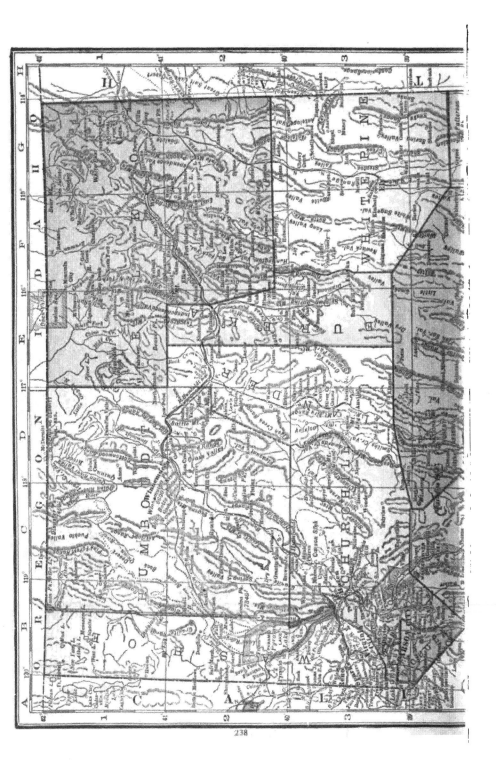

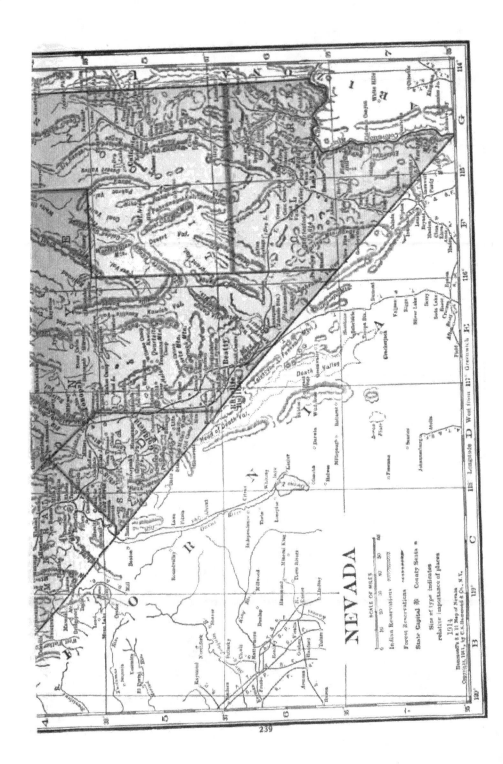

NEVADA

SCALE OF MILES

Indian Reservations
Forest Reservations
State Capital ⊛   County Seats ∘

Size of type indicates
relative importance of places

1914

Reproduced by permission from 1 14 Map of Nevada
Copyright, 1911, by C.S. Hammond & Co., N.Y.

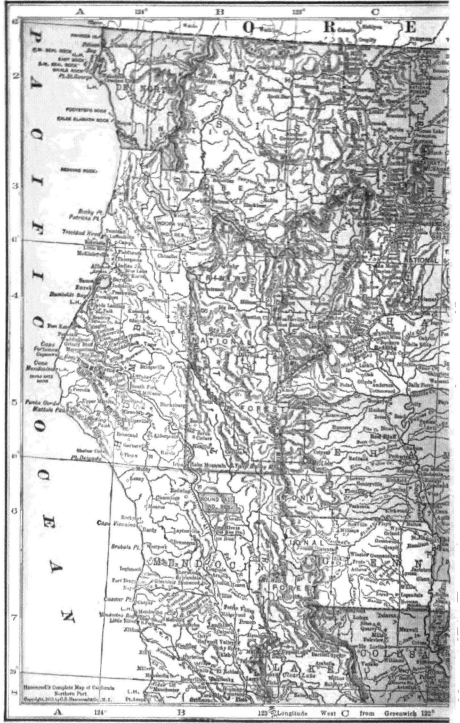

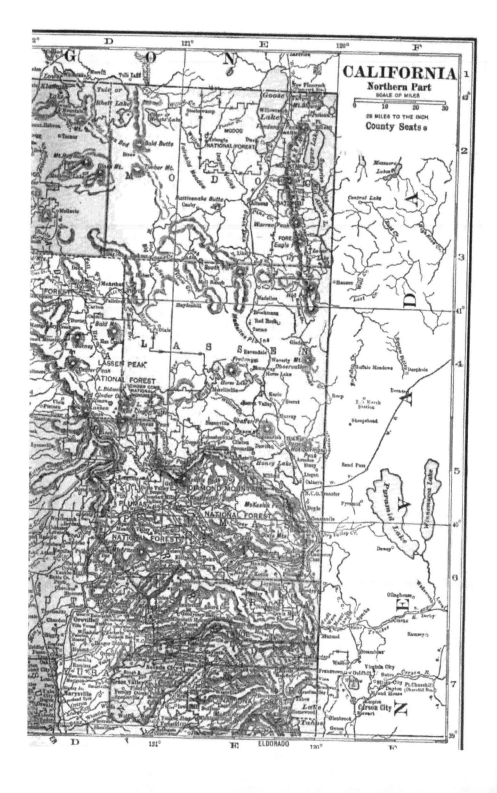

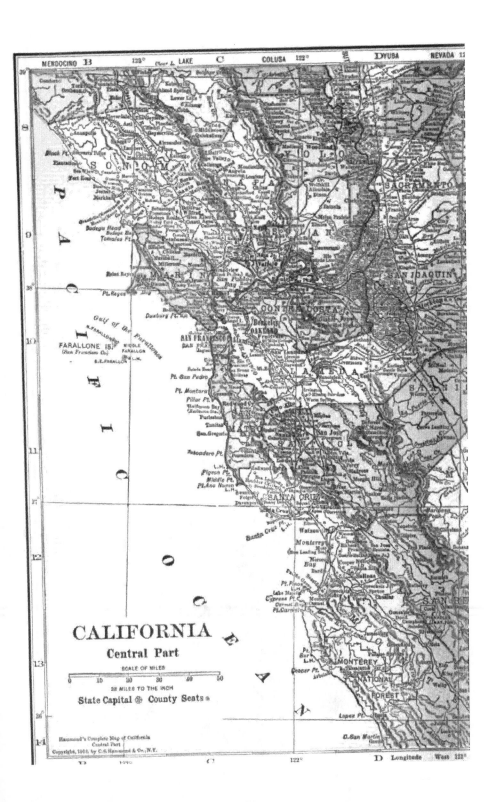

CALIFORNIA

Central Part

SCALE OF MILES

0   10   20   30   40   50

22 MILES TO THE INCH

State Capital ⊕   County Seats *

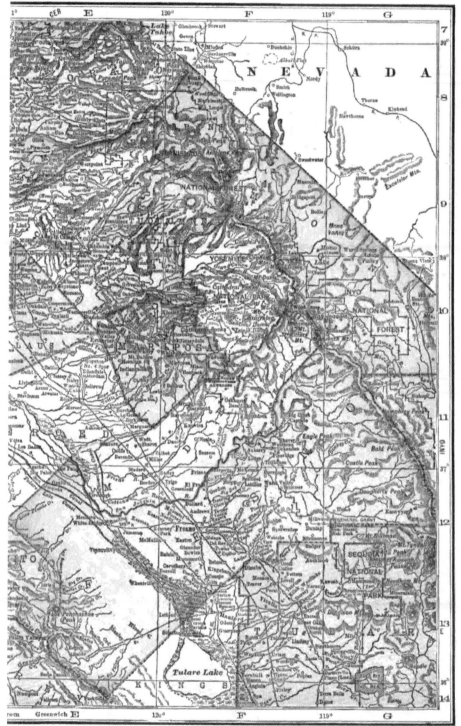

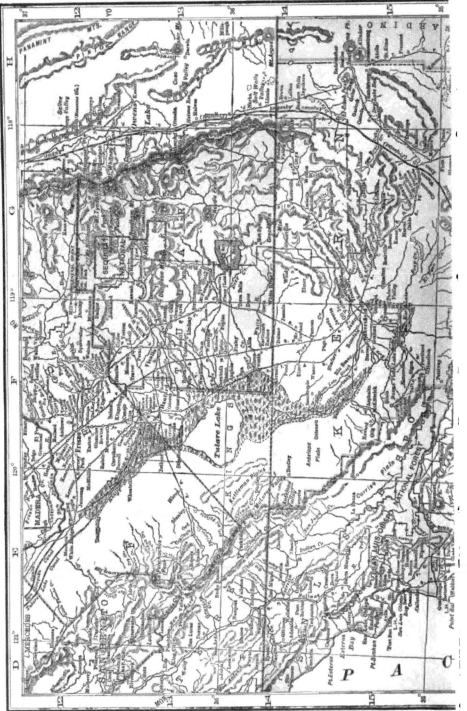

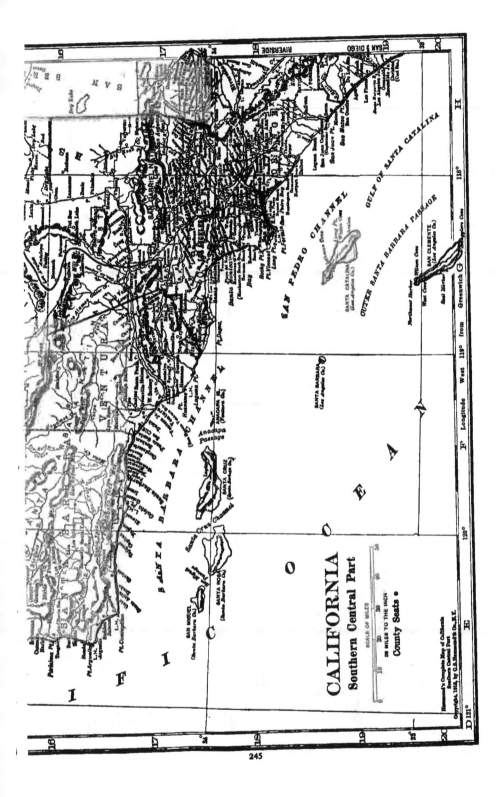

CALIFORNIA
Southern Central Part

SCALE OF MILES

28 MILES TO THE INCH

County Seats •

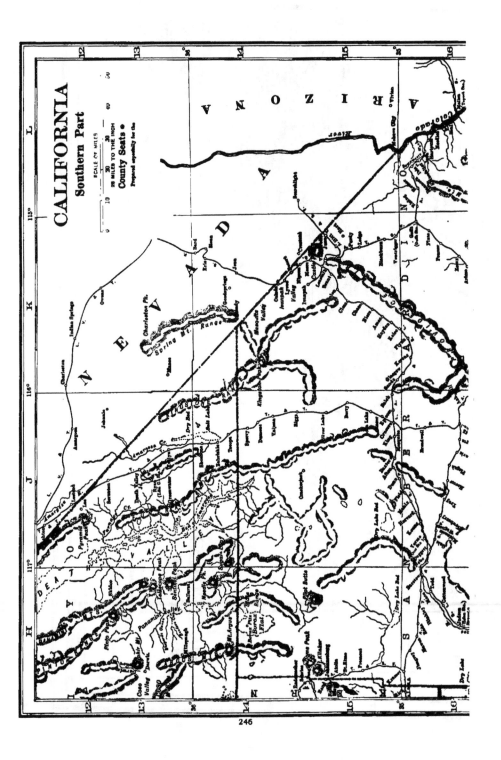

CALIFORNIA
Southern Part

SCALE OF MILES
County Seats ●
Prepared especially for the

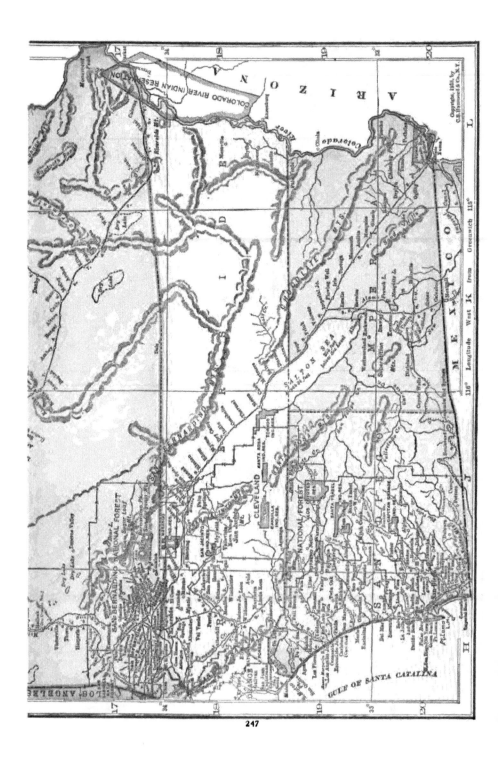

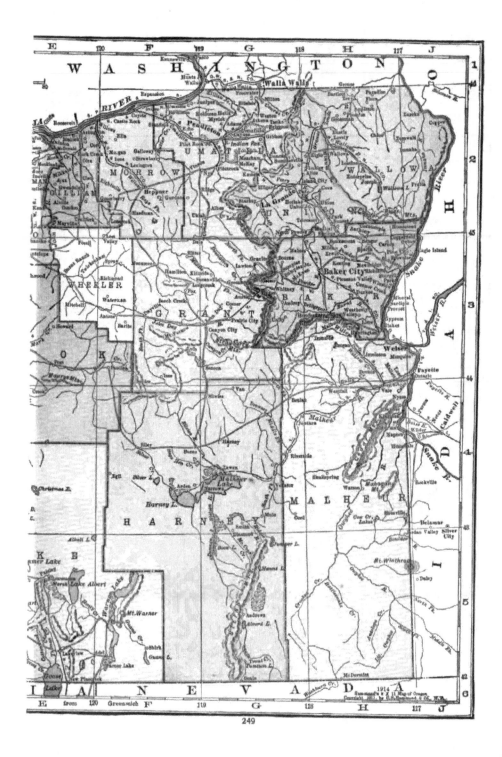

WASHINGTON
Western Part

SCALE OF MILES

Railroads
Railroads Projected
National Parks
National Forests
Indian Reservations

Rand McNally's Complete Map of Washington, Western Part
Copyright, 1914 by Rand McNally & Co., N.Y.

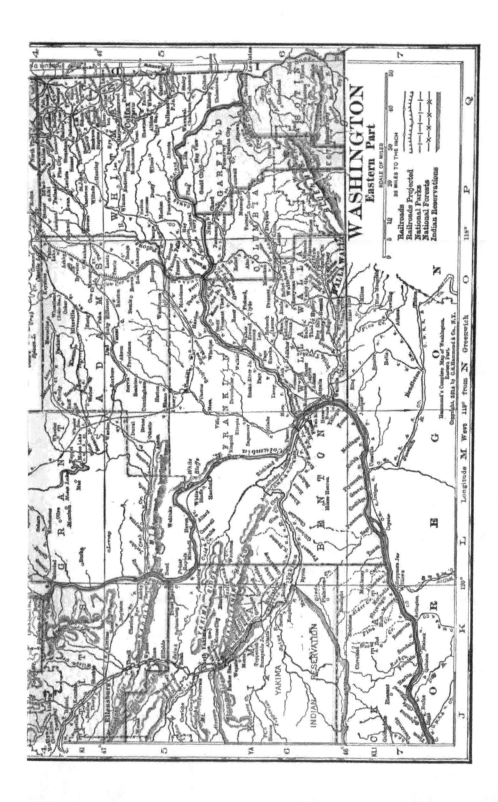

WASHINGTON
Eastern Part

SCALE OF MILES

Railroads
Railroads Projected
National Parks
National Forests
Indian Reservations

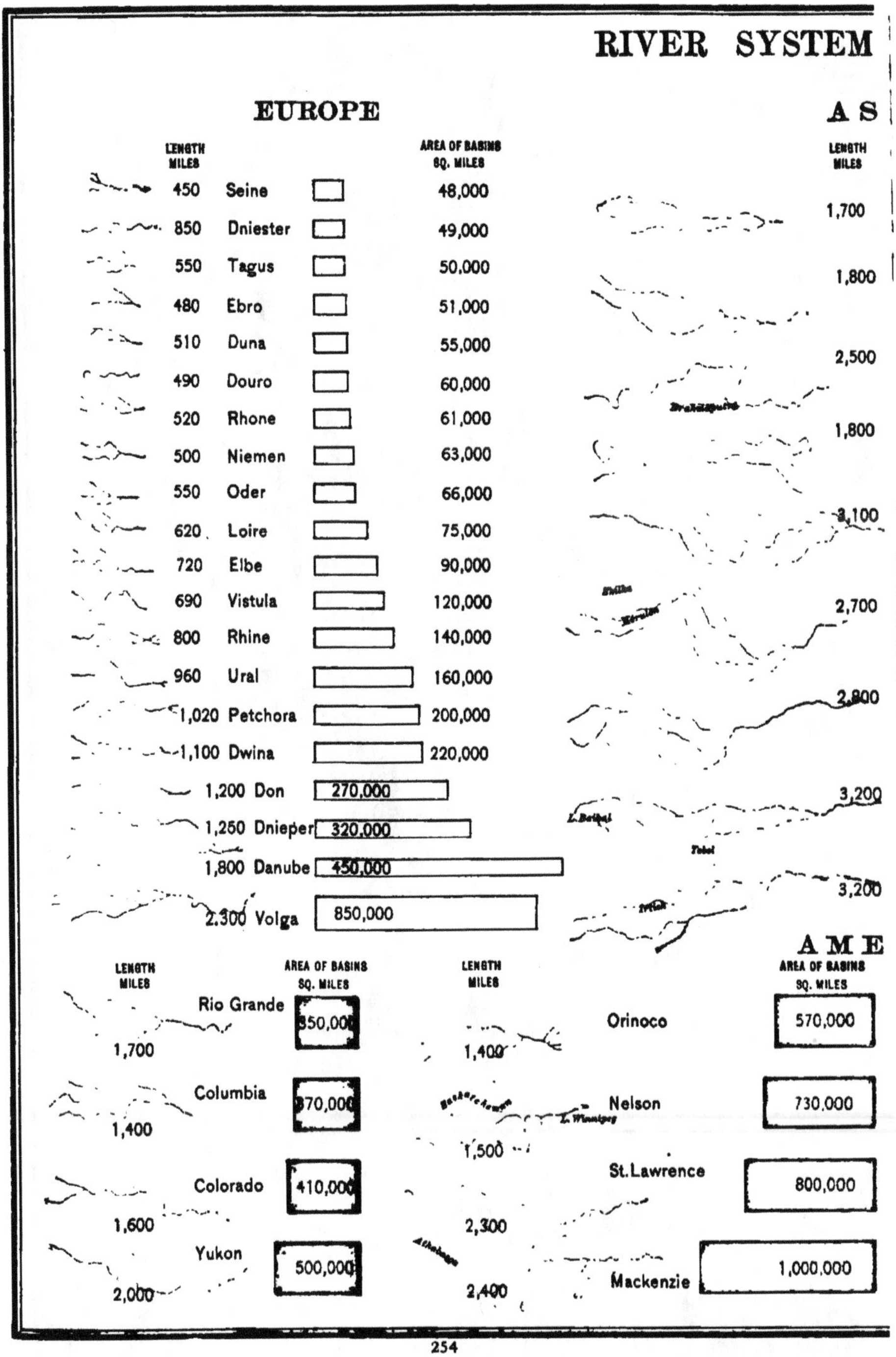

EUROPE

AS

LENGTH MILES		AREA OF BASINS SQ. MILES
450	Seine	48,000
850	Dniester	49,000
550	Tagus	50,000
480	Ebro	51,000
510	Duna	55,000
490	Douro	60,000
520	Rhone	61,000
500	Niemen	63,000
550	Oder	66,000
620	Loire	75,000
720	Elbe	90,000
690	Vistula	120,000
800	Rhine	140,000
960	Ural	160,000
1,020	Petchora	200,000
1,100	Dwina	220,000
1,200	Don	270,000
1,250	Dnieper	320,000
1,800	Danube	450,000
2,300	Volga	850,000

LENGTH MILES

1,700

1,800

2,500

1,800

3,100

2,700

2,800

3,200

3,200

AME

LENGTH MILES		AREA OF BASINS SQ. MILES		LENGTH MILES		AREA OF BASINS SQ. MILES
	Rio Grande	350,000			Orinoco	570,000
1,700				1,400		
	Columbia	370,000			Nelson	730,000
1,400				1,500		
	Colorado	410,000			St. Lawrence	800,000
1,600				2,300		
	Yukon	500,000			Mackenzie	1,000,000
2,000				2,400		

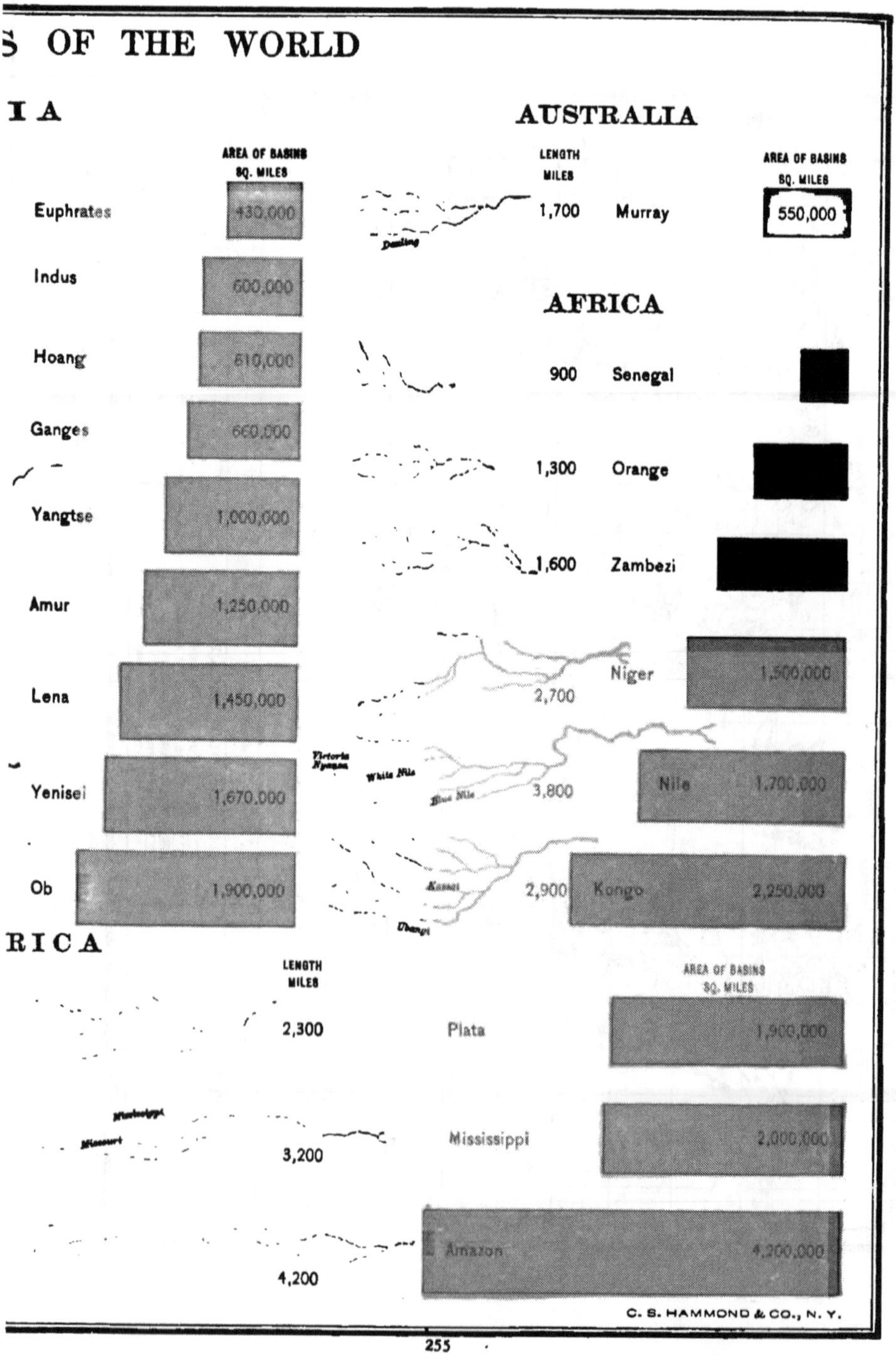

I A

AUSTRALIA

AREA OF BASINS SQ. MILES			LENGTH MILES		AREA OF BASINS SQ. MILES

Euphrates — 430,000

1,700   Murray   550,000

Indus — 600,000

AFRICA

Hoang — 610,000

900   Senegal

Ganges — 660,000

1,300   Orange

Yangtse — 1,000,000

1,600   Zambezi

Amur — 1,250,000

Lena — 1,450,000

Niger   1,500,000

2,700

Yenisei — 1,670,000

Victoria Nyanza
White Nile
Blue Nile   3,800   Nile   1,700,000

Ob — 1,900,000

Kasai   2,900   Kongo   2,250,000

Ubangi

RICA

LENGTH MILES		AREA OF BASINS SQ. MILES

2,300   Plata   1,900,000

Mississippi
Missouri   3,200   Mississippi   2,000,000

Amazon   4,200,000

4,200

C. S. HAMMOND & CO., N. Y.

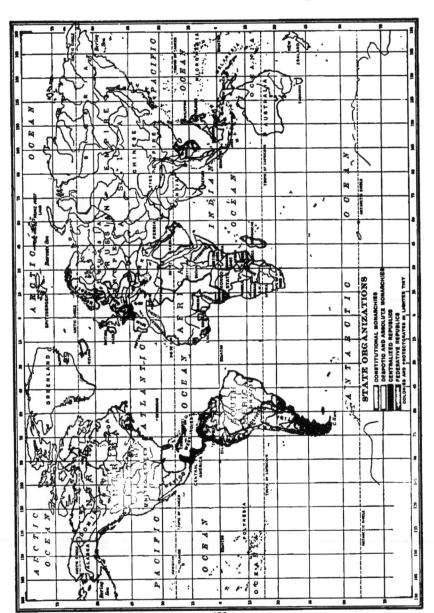

STATE ORGANIZATIONS

▭ CONSTITUTIONAL MONARCHIES
▭ DESPOTIC AND ABSOLUTE MONARCHIES
▬ CENTRALIZED REPUBLICS
▭ FEDERATIVE REPUBLICS
▭ COLONIES AND PROTECTORATES IN LIGHTER TINT

PART II

TEXT, STATISTICS AND INDEX

THE PANAMA CANAL

A HISTORY AND DESCRIPTION

I. THE REPUBLIC OF PANAMA.

HISTORY.—It is probable that the Isthmus of Panama was visited by Alonso de Ojeda in 1499. In 1501 Rodrigo Bastidas skirted the coast from the Gulf of Venezuela to the present Porto Bello. In the following year, 1502, Columbus, on his fourth and last voyage, coasted from Almirante Bay, near the present Costa Rica boundary, to Porto Bello Bay, discovering the "River of Crocodiles," now known as the Chagres River, and founding the short-lived settlement of Nombre de Dios. In 1509 the region which included the present states of Nicaragua, Costa Rica, and Panama, was organized by Spain as the province of Castilla del Oro. In 1513, Vasco Nuñez de Balboa, at that time governor of the province, with a few Indians and Spaniards, cutting his way through the wilderness, crossed the Isthmus and on the 25th of September of that year discovered the Pacific Ocean and named it the South Sea. Balboa was shortly after succeeded in command by Pedro Arias de Avila. In 1510 Martin Fernandez de Encisco, following Ojeda to the New World, had moved the survivors of Ojeda's colony of Nueva Andalucia (near the present Cartagena) to the Tuira River, and founded there the colony of Santa Maria la Antigua del Darien (commonly called Darien). In 1514 Pedro Arias joined this colony to the province of Castilla del Oro and gave to the united colonies the name of Tierra Firma. In 1519 the old city of Panama was founded, the first permanent European settlement on the mainland of the American continent.

The story of the Spanish conquest and settlement of Panama abounds in records of romantic adventure, intrigue, avarice, cruelty, and wastefulness. Lust for gold was joined to the spirit of religious fanaticism. Treasure began flowing into Spanish coffers, but ugly stories of the way in which it was obtained, of persecution, of torture, and of slavery, floated across the sea with the treasure ships. Those stories persisted and the Protestant countries of Europe began to take note. Those were the days of the great English privateersmen—Drake, Parker, Morgan, and many others. Soon the Spanish colonists began to realize that they must suffer for the wrongs they had meted out to others. Protestant Englishmen were not any more merciful than Roman Catholic Spaniards had been. They captured the treasure ships, harried the settlements, and spared not the innocent while punishing the guilty. It remained for Henry Morgan, the able but unscrupulous Welshman, to carry on and eclipse the work of retribution begun by Drake. From 1668 to 1670 he ravaged the settlements from Cartagena to Chagres, and in 1671 he finally captured and destroyed the old city of Panama whose ruins are eagerly sought by the tourists of to-day.

In 1698 a Scotch settlement, under the authority of the Scotch Parliament, was made by William Patterson on the site of the present Porto Escoces in Darien, but in 1700 the Spanish authorities expelled the few colonists remaining there.

Panama was a part of the viceroyalty of New Granada, created in 1719, and it remained a Spanish possession until 1819 when it threw off its Spanish allegiance and became a part of the independent nation of Colombia. In 1831 it joined the Granadine Confederation from which it twice seceded, in 1841 and again in 1857, but in each instance returned to its former allegiance after a brief independence. The completion of the Panama Railway in 1855 led to increased trade and to a more complete knowledge of that part of the Isthmus and its people. Panama had a more or less troubled career from 1885 until November 4, 1903, when it again became an independent state under circumstances which promise well for its permanence and future prosperity. A few weeks after its independence the new Republic made a treaty with the United States of America, which treaty gave this country the right to build the Panama Canal, together with a perpetual leasehold of a strip of territory ten miles

wide across the Isthmus from ocean to ocean. This territory is known as the Canal Zone and, while it divides the domain of Panama into two parts, the arrangement is such that the rights and interests of her people are fully protected. (See History of the Panama Canal).

On January 4, 1904, two months after the declaration of independence, a constitutional assembly was elected. This legislative assembly met on the 15th of January and on the 13th of February of the same year adopted a constitution, and chose a president

The financial condition of Panama is strong. It has no bonded debt upon which to pay interest and it is the only nation of the world collecting interest on its own money instead of paying out interest on loans. It has no army, no navy, and its leading cities are kept clean by the United States.

The Republic of Panama is bound to live and prosper. Peace within her borders is assured by the protection of the United States. The introduction of new racial elements and of modern methods in developing her natural resources can have no other result than an intellectual awakening of her own people and a strengthening of the spirit of national loyalty and patriotism.

GOVERNMENT.—By the constitution, adopted February 13, 1904, the government is a highly centralized republic, with legislative, executive, and judicial departments. All male citizens of twenty-one years of age and over have the right to vote, unless under judicial interdiction or inhabilitated because of crime. The president, who must be at least thirty-five years old, is elected by popular vote for a term of four years and is ineligible to succeed himself. He appoints his cabinet members (secretaries of foreign affairs, government and justice, treasury, interior and public instruction); five supreme court judges; diplomatic representatives; and the governors of the provinces, who are responsible only to him. Three officials, known as designados, elected every two years by the National Assembly are in the line of presidential succession in case of the death or disability of the chief executive. The National Assembly is a single chamber, whose deputies (each at least twenty-five years of age) are elected for terms of four years by popular vote on the basis of one to every 10,000 inhabitants, or fraction over 5,000. The Assembly meets biennially and by a two-thirds vote it may pass any bill over the president's veto. The president has five or ten days, according to the length of the bill, in which to veto legislative acts. At the head of the judiciary is the Supreme Court, the members of which appoint the judges of the

superior court and the circuit courts as well as the justices of the municipal courts.

The Republic is divided into seven provinces, their boundaries representing those of old administrative districts: Panama, with most of the territory east of the Canal Zone and a little (on the Pacific side) west of the Canal Zone; Colon, on both sides of the Canal Zone along the Caribbean; Cocle, west and south of Colon; Los Santos on the Asuero Peninsula, west of the Gulf of Panama; Veraguas, west and northwest of Los Santos, reaching to the Mosquito Gulf on the north; Chiriqui, farthest west, on the Pacific; and Bocas del Toro, farthest west, on

COL. GEORGE W. GOETHALS
Governor of the Canal Zone

the Caribbean. *Towns.* Panama, the capital and largest town, has 37,505 inhabitants; Colon, the chief Atlantic port and the second city in point of numbers, has a population of 17,749. The United States controls the sanitation of these towns but does not interfere in any other way with their administration. David, the capital of Chiriqui province is the most progressive of the provincial towns. Smaller ports are Bocas del Toro and Porto Bello on the Atlantic, and Agua-Dulce and Montijo on the Pacific.

INHABITANTS.—Few of the native inhabitants of Panama are of pure Spanish blood. The great

majority belong to a mixed race comprising Spanish, Indian, and Negro elements. The foreign population is small and includes immigrants from the United States and European countries. Not a few of those who have helped to build the Canal will become citizens of Panama and go in for ranching—agriculture of various kinds—or take advantage of the positions awaiting workers in the mining and forestry industries, now being developed for the first time. The government maintains a system of public schools throughout the provinces, and a university, the Instituto Nacional, has been established in the city of Panama. In addition to the foregoing there are several private educational institutions, and a number of young people, of both sexes, are being educated in the United States and Europe at the expense of the government.

PHYSICAL FEATURES.—*Area, boundaries, and coast-line.* The narrow, sinuous strip of land, running east and west, which joins the continents of North and South America, is occupied by the Republic of Panama and comprises an area of 32,380 square miles. Its extreme length from east to west is 430 miles, and its greatest width from north to south is 118 miles. The points where the isthmus is narrowest are at the Gulf of San Blas (31 m.), the Colon-Panama route (40 m.), and between the gulfs of Darien and San Miguel (46 m.).

The country is bounded on the north by the Caribbean Sea, on the east by Colombia, on the south by the Pacific Ocean, and on the west by Costa Rica. With land frontiers aggregating only 350 m. in length, the little republic boasts a coast line of 1,245 m., of which 478 m. are on the northern or Caribbean side, while 767 m. face the Pacific. Along both the northern and southern shores are many inlets affording good anchorage for vessels engaged in coasting trade.

The main indentations of the northern coast are the Gulf of Darien on the northeast and the Mosquito Gulf on the northwest, while between these, from east to west, there are Caledonia Bay or Harbor; the Gulf of San Blas, 20 m. long and 10 m. wide, with an excellent harbor at Mandinga; Limon Bay, which furnishes the northern approach to the Canal; Chiriqui Bay, with an area of 320 sq. m. and a maximum depth of 120 ft.; and farther west near the Costa Rican boundary is Almirante Bay. Along the north shore there are some 600 islets with a total area of about 150 sq. m. The Pacific side of the isthmus is deeply indented by the Gulf of Panama, more than 100 miles in width between Pt. Garachine on the east and Cape Malo on the west. The chief harbors of this great southern indentation are the Gulf of San Miguel on the east, and the Bay of Parita on the west. Darien

Harbor, formed by the Tuira and Savannah rivers and nearly landlocked, is a part of the Gulf of San Miguel. West of the Asuero Peninsula is Montijo Bay, 20 m. long and 14 m. wide at its mouth. In the Gulf of Panama there are more than 100 islands, most of them very small, with a total area of about 400 sq. miles. The Pearl Islands, named for the pearl-fisheries which are their most valuable resource, are about 60 m. southeast of Panama City. To the west of the Asuero Peninsula are Cebaco and Coiba Islands, the latter, 21 m. long and 4 to 12 m. wide, being the largest island of the Republic.

SURFACE.—Panama is a broken and rugged country, its highlands being either a part of the complex cordillera which runs from Alaska to the Straits of Majellan, or a part of the Antillean system, to which belong the mountains of the West Indies. The former classification is the one generally accepted by those who write of the Panama Canal. Every American child has been taught in school that the cordillera forms the backbone of the American continent. There is something awe-inspiring in the thought of this majestic mountain barrier, with it countless ramifications, offshoots, and lofty summits, stretching through every zone of climate, from Arctic to Antarctic regions, which is so conveniently depressed or weakened in the narrowest part of the continent as to make possible the piercing of the divide and the building of a waterway from ocean to ocean.

The only regular mountain ranges in Panama are in the extreme western part where the Costa Rica divide crosses the boundary and enters its territory. Immediately south of this and parallel to it is the Cordillera of San Blas or the Sierra de Chiriqui, in which range Mt. Chiriqui reaches an altitude of 11,265 feet, while Mt. Blanco, on the Costa Rica frontier, attains an elevation of over 11,700 feet. On the far eastern boundary of the republic is the Serraina del Darien, an Andean range which Panama shares with her Colombian neighbor. Between these highlands of the west and east are many short sierras, groups, and spurs. It should be noted that the ranges lie at right angles to the mountain systems of North and South America, that the summits are "exceedingly irregularly rounded, low-pointed, and densely forested." These characteristics are Antillean, not Andean.

The highest summit, aside from those near the western border and already noted, is Mt. Santiago in the province of Veraguas, which attains an elevation of over 9,000 feet. The Sierra de Panama, in Panama province, is a much-broken range with a maximum height of 1,700 ft. and a minimum, at the Culebra Pass, of about 300 ft., the lowest point in the western continental system, except the interoceanic

water-parting in Nicaragua, which is only 153 ft. above sea level. In considering the construction of the canal at Panama it was the low elevation at Culebra which decided the engineers in favor of the Colon-Panama route instead of the narrower but more elevated San Blas route.

Between the hills and mountains of Panama are richly wooded valleys and occasional stretches of grassy llano. There are a few plains like that of David in Chiriqui province but an irregular surface, the result of heavy rains and marine erosion, is the prevailing characteristic of the country.

DRAINAGE.—There are no lakes in Panama, except the artificial ones in the Canal Zone. A few swampy areas are found along the coasts, but aside from these spots the drainage of the

In the dry season of the year the Chagres was a comparatively placid stream which wound among the hills, first southwest and then north to the Caribbean. It was navigable by small boats for nearly half of its course, about fifty miles. When the rainy season began the Chagres assumed another character. It has been known to rise 35 ft. in 24 hours, overflowing its banks, and destroying all that came in its way. The story of the capture of this river and its utilization in making the Canal forms an interesting part of the history of that achievement. To-day it spreads out into Gatun Lake, from which great reservoir its water flows through the canal channel to both the Atlantic and the Pacific oceans. West of the Canal Zone there are many smaller rivers, rising near the center

A Street View in Panama, showing residence of President of the Republic on right

country is excellent. About 150 streams flow to the Caribbean and more than double that number to the Pacific. East of the Canal Zone are three complicated river systems, of which the largest is the Tuira or Rio Darien, rising near the Caribbean, whose waters reach the Pacific through the Gulf of San Miguel. The Chepo or Bayano is another digitate system with a wide-reaching drainage basin, which enters the Gulf of Panama about 30 m. east of the city of Panama; it is navigable by small boats for a distance of 120 miles.

The Chagres River, now known to all who have followed the building of the Canal, has its origin in the wilderness of southeastern Panama, near the Pacific Ocean. With its numerous tributaries it drains an area of 1,320 square miles.

of the Isthmus and flowing either north or south. Little use is now made of these streams but their banks are heavily wooded, and they are destined to play an important part in the future development of the country.

CLIMATE.—Panama is not far from the equator but it lies between two great oceans whose proximity to all parts of the interior tends to temper the heat. A record of 100° F. is unknown, and the mean temperature varies but little throughout the republic. At Colon, where 68° is a low and 95° a high temperature, the mean is 80°; at Panama the mean is 80.6°. The Caribbean coast, however, is normally a little warmer than the Pacific coast. Moisture loads the air and so great is the humidity in the Zone that it seems much warmer than the thermometer

indicates. In the highland regions in Chiriqui and in some of the other provinces the atmosphere is drier and less enervating. There is a wet and a dry season; in the former, which lasts from April to December, southeast winds prevail and the rainfall equals about 85 per cent. of the total annual precipitation. It comes in the form of short but very heavy rains, which cause the streams to rise suddenly and destructively. In the dry season, from January to March, the northeast wind prevails and the climate becomes more bracing. The rainfall on the north coast varies from 85 to 155 in., with 125 in. as the mean; on the south coast it varies from 47 to 90 in., with 67 in. as the mean.

RESOURCES AND INDUSTRIES.—Two-thirds of the area of Panama is occupied with forests of valuable wood. The Atlantic side of the waterparting and the Darien region are the most densely covered portions of the country. The jungle on the Atlantic slope is so nearly impenetrable that enormous capital will be required to clear the land, a condition which has greatly retarded the development of the lumber industry. On the Pacific side, where the rainfall and humidity are less, the forest is more open and the timber growth smaller. Here are large districts suitable for fruit culture, wheat growing, and cattle raising.

There is no doubt about the richness and fertility of Panama. Part of the province of Chiriqui is splendid farming country, producing wild grasses that are nutritious as well as luxuriant. There are many localities which are suitable for bananas; others that seem to be specially adapted for tobacco culture, and others where coco-nut plantations would thrive. Sugar-cane, coffee, corn, and vegetables and fruits are profitably grown. A soil which gives crops with little work is not an unmixed blessing, for it ministers to the sloth of the natives and prevents the development of the country. Only the most haphazard methods have hitherto been practiced in cultivating the soil, which as a rule is simply scratched with a machete, the seed dropped, and the crop left to care for itself. The present population is sparse, the roads and tracks poor, the settlements scattered, and the majority of the people are lacking in energy and interest.

Wild rubber trees are found throughout the country and the scientific gathering of rubber and the planting of new areas, which have been started by foreign companies, will lead to an important rubber industry.

Cattle raising is the only branch of agriculture which has been conducted with care and intelligence by the natives, and the results have been very gratifying. The industry is bound to grow and there is every probability that in a few years Panama will become a potent factor in the world's cattle markets.

The mineral resources of the country are as yet almost unknown and wholly undeveloped, but they include gold and other metals, and will prove a source of wealth to those who may have capital to develop them.

There is a tradition that the name Panama is an old Indian term meaning "place of abundant fish;" the tradition seems justified, for the waters of the Pacific Ocean at the Isthmus teem with excellent fish, and the oyster and pearl-oyster fisheries have long been known. The pearls from Panama Bay were one of the valued assets of the early Spanish colony.

An English writer, who recently visited the Isthmus, has said that "nothing is manufactured in Panama save revolutions." That may have been true in by-gone days, but there are indications that the Republic has begun to improve and extend the industrial activities of the country in a way which promises benefit to the people. The real prosperity of the country must come from the development of its agricultural resources, but the commercial world has turned its attention to Panama and finds it worth the consideration of those who have money to invest in enterprises calling for large capital.

COMMERCE.—With the exception of fresh meat, fruit, and vegetables, practically everything to be purchased in Panama is imported. The import trade of the Republic has steadily increased since the new government came into existence and, while the exports are still less than one-third the value of the imports, a steady advance is being made towards an equalization of trade. A progressive trade is carried on in bananas and it will not be long before the cacao, sugar, coffee, and rubber produced in the country will add to the number and variety of exports.

The Isthmus is in direct communication with American and European ports by several lines of steamers. The Panama Railroad in the Canal Zone is owned and controlled by the United States. In the province of Bocas del Toro the United Fruit Company owns about 140 miles of narrow gauge track which is used to transport bananas and passengers to the port of Almirante in Chiriqui Bay. This line extends into Costa Rica and is being extended. A national line, 271 miles in length, from Panama City to David, in the province of Chiriqui, has been surveyed and, when built, will hasten the development of the region through which it passes. There are telegraph cables from Panama to North and South American ports, and from Colon to the United States and Europe. The roads throughout the country are being improved and extended.

II. VARIOUS PROJECTS FOR AN ISTH-
MIAN CANAL

EARLY HISTORY.—Columbus sought a western
route from Europe to Asia, to the wonder-lands
portrayed by Marco Polo and other travelers in
the East. He died believing that he had achieved
that result and it was not until long years afterward that the original object of westward ex-
ploration was abandoned. Even when the fact
had been established that a new continent of
enormous extent lay between Europe and Asia,
the search for a passage between the oceans continued, and some of the early maps of America
show an imaginary "Strait of Panama."

When Balboa crossed the Isthmus he
grasped the possibility and significance of an
artificial waterway which should join the two
oceans. With the report of his discovery to
Spain there went also a recommendation that a
canal be immediately dug across the Isthmus.
While the suggestion attracted attention and
even carried conviction to a few minds, Spain
was not yet ready for so daring a venture. But
the vision which Balboa beheld from that lonely
height in the tropical wilderness was not to fade.
It continued to haunt the minds of men for four
hundred years, and then it became a reality.

The discovery and conquest of Peru made a
highway of some sort across the Isthmus im-
perative, and a pathway from ocean to ocean
was cut through the jungle, which was kept open
and in use for nearly 400 years. Over it was soon
passing in ever increasing volume the treasure
brought in vessels from Peru to Panama and
transported thence by mule trains to the Atlantic
port in Limon Bay. There waiting galleons
received it and bore it across the Atlantic to
Spain. The Pacific port became the old city of
Panama, which grew in importance and wealth
until its destruction by Sir Henry Morgan in 1671.

The vision of a canal moved Charles V., in
the early years of his reign, to order a survey
of the Isthmus, but it was not made because the
governor of Darien pronounced the project
impossible. In 1550 the Portuguese navigator
Antonio Galvao published a book to demonstrate
that a canal could be cut at Tehuantepec,
Nicaragua, Panama, or Darien; and the following year F. L. de Gomara, the Spanish
historian, petitioned Philip II. to undertake the
work. But the Spanish government had grown
indifferent to the canal scheme, considering that
a monopoly of communication with their American possessions was of greater importance than
a passage by sea to Cathay. It even discouraged
the improvement of the communications by land,
and decreed that to seek or make known any
better route than the one from Porto Bello to
Panama was an offense punishable by death.

For nearly 150 years following the decree no
serious move was made towards the construction
of a canal. Still the vision persisted, and
Champlain and other travelers who visited the
Isthmus wrote more or less earnestly of the
feasibility and value of such an undertaking.
Spain was fighting to maintain her position in
the New World, and to guard the treasure ships
in their transit across the ocean. Drake and
Parker and Morgan gave her plenty to do in
those days.

In 1698 another advocate arose to plead the
cause of an interoceanic waterway. William
Paterson, a brilliant Scotchman, had spent some
years of his early life in the West Indies, where
his natural ability had made him a man of
influence. He may have been a buccaneer, at
all events he had a wide acquaintance with the
captains of that class, and he became imbued
with a scheme for a canal at Darien. Failing
to interest the English government he went to
some of the continental courts where he met
with no better success. Returning to London
he was drawn into business enterprises and
became the founder of the Bank of England in
1694. A little later Paterson secured from the
Scotch parliament a charter for the "Company
of Scotland trading to Africa and the Indies."
This company, according to Paterson's plan,
should establish a settlement on the Isthmus of
Darien, dig a canal, and "thus hold the key to
the commerce of the world and turn Scotland
from one of the poorest to one of the richest
countries." The enterprise was popular and
funds were freely contributed. The plans were
conceived in a broad and beneficent spirit,
differences of race and religion were to be
ignored, the settlement was to be a harbor of
refuge for the ships of all nations, and world-
wide free trade was to be established. The
settlement was planted in 1698, but its existence
was troubled and it was abandoned before a sod
had been turned toward the digging of a canal.

In 1771 the Spanish government changed its
policy of indifference and ordered a survey of the
Tehuantepec route. When that line was found
to be impracticable attention was turned to
Nicaragua, but before a survey could be made
the political conditions in Europe were such as
to prevent action.

In 1808 the Isthmus was examined by Alexander von Humboldt, who pointed out the lines
which he considered worthy of study. After
the Central American republics acquired their
independence in 1823, interest in the canal
question was greatly increased. In 1825 Nicaragua, having received applications for concessions from citizens of Great Britain, and also
from citizens of the United States, made over-
tures to the United States for aid in constructing

Passing through Gatun Locks

a canal, but they resulted in nothing. In 1830 a concession was granted to a Dutch corporation under the special patronage of the King of the Netherlands to construct a canal through Nicaragua, but the revolution and the separation of Belgium from Holland followed, and the scheme fell through. Subsequently several concessions were granted to citizens of the United States, France, and Belgium, both for the Nicaragua and the Panama routes, but with the exception of those of 1878 and 1887, which are described below, no work of construction was done under any of them.

Knowledge of the topography of the Isthmus remained extremely vague until the discovery of gold in California in 1848 gave a great impetus to travel and rendered improved communications between the Atlantic and Pacific coasts of the United States necessary. A railroad at Panama and a canal at Nicaragua were projected. Instrumental surveys for the former in 1849, and for the latter in 1850, were made by American engineers and, with some small exceptions, were the first accurate surveys made up to that time.

The Panama Railroad, built by Americans under a concession from the republic of New Granada, now known as Colombia, was opened in 1855. For nearly fifteen years, or until the completion of the first transcontinental railroad in the United States, it formed the main line of traffic between the eastern seaboard and California. The building of the railroad had furnished more accurate knowledge of the topography and geology of the Isthmus and gave a new impetus to the consideration of a canal. It was an era of great engineering enterprise and commercial activity. In the United States the transcontinental and other railroads were projected and their construction vigorously pursued. Europe was building the Suez Canal, which would give a short and direct water route to Asia, and the attention of the world was drawn more forcibly than ever before to the Isthmus of Panama.

BEGINNINGS OF AMERICAN INTEREST IN AN ISTHMIAN CANAL.—In 1866 in response to an inquiry from Congress, Admiral Charles H. Davis of the U. S. Navy, reported that "there does not exist in the libraries of the world the means of determining, even approximately, the most practicable route for a ship canal across the American Isthmus." To clear up the difficulty Congress in 1872 appointed an Interoceanic Canal Commission and provided for the sending out of a series of expeditions under officers of the navy, by whom all the routes were examined. These examinations, which were made between 1870 and 1875, proved that the only lines by which a tunnel could be avoided were those of Panama and Nicaragua, and Con-

gress was advised—that the Nicaraguan route possessed greater advantages and offered fewer difficulties than that of Panama. The comparison of the two routes as set forth in the report was as follows: "At Panama the Isthmus is narrower than at any other point except San Blas, its width in a straight line being only 40 miles and the height of the continental divide is only about 300 feet, which is higher than the Nicaragua Summit, but less than half the height on any other route. At Nicaragua the distance is greater, being about 156 miles in a straight line, but more than one-third is covered by Lake Nicaragua, a sheet of fresh water with an area of 3,000 square miles, and a maximum depth of over 200 feet, the surface being about 105 feet above sea level. Lake Nicaragua is connected with the Atlantic by a navigable river, the San Juan, and is separated from the Pacific by the continental divide, which is about 160 feet above sea level. At Nicaragua only a lock canal is feasible, while at Panama a sea-level canal is a physical possibility."

THE FRENCH SCHEME.—While the American Interoceanic Canal Commission was making the investigations relative to the comparative merits of the different isthmian routes, a canal project was started in France. The popular imagination in that country had been stirred to enthusiasm by the part of Count de Lesseps in the making of the Suez Canal. Financiers began the agitation of another enterprise in which France should play the leading role. Their plans were carefully made. In 1875 the subject was discussed by the Congrès des Sciences Géographiques at Paris, and that body recommended the making of surveys with a view to building a canal. Soon after the discussion of that Congress a provisional company of speculators was formed for the purpose of securing a concession from the Republic of Colombia. Acting as agent for this company, Lieutenant L. B. N. Wyse, an officer of the French Navy, secured from the Colombian government in 1878 a concession for the construction of a canal from Colon to Panama. The concession gave the promoters the exclusive privilege of constructing and operating a canal through the territory of the republic, the only restrictive condition being that if the route chosen traversed any portion of the land given to the Panama Railroad Company, the promoters should make a satisfactory settlement with that company before beginning the work of construction. On the part of the company it was agreed that the course of the canal should be determined by an international congress of engineers.

Possessed of this valuable concession the promoting company organized what is generally known as the "Panama Canal Company" and

transferred to it what has come to be known as the Wyse Concession. Count de Lesseps became the chief engineer of the company, a fact which guaranteed the support of all classes of Frenchmen. Invitations were issued for an International Congress to determine the route of the canal.

The Congress convened at Paris in May, 1879, and among the 164 delegates in attendance were representatives from the United States, Great Britain, and Germany; but the French delegates were in the majority. It is said that the proceedings of this congress were pre-arranged; that the opinions of those delegates who were engineers of experience, with knowledge of conditions at the Isthmus, were allowed scant consideration in its deliberations; and that the desires of the unscrupulous promoters, who were back of the whole scheme, dominated the conclusions reached. At the closing session a resolution was adopted (many of the delegates refrained from voting) recommending the construction of a sea-level canal from Limon Bay to the Bay of Panama. The canal was to have a depth of 29½ feet and a bottom width of 72 feet. The cost was estimated by Lesseps at $131,600,000—an estimate far below that given by the International Congress of Paris. In the prosecution of the work itself it appears that miscalculation was followed by mismanagement. Ignorance at the outset, a blind indifference to the lessons taught by their blunders, criminal extravagance and dishonesty, all are a part of the record of the Panama Canal Company, which finally went into the hands of a receiver in February, 1889. The story of the way in which the French people had invested their savings in the enterprise is extremely pathetic and one which it is not pleasant to read. Fully two hundred thousand persons, most of them in moderate circumstances, were involved in the ruin which followed its collapse.

THE NEW PANAMA CANAL COMPANY.—Five years after the appointment of a receiver for the Panama Canal Company, the New Panama Canal Company was formed. It was a private commercial organization, like its predecessor, but it proceeded with more regard for scientific knowledge than had that organization. It made a new study of the entire subject of the canal in both its engineering and commercial aspects.

New and careful surveys were made and at the same time excavation was continued in places where it was certain to be needed, whatever plan might be adopted. The result of this preliminary work was a plan which involved two levels above sea-level, one of them an artificial lake to be created by a dam at Bohio, to be reached from the Atlantic by a flight of two locks, and the other, the summit level, to be reached

by another flight of two locks from the preceding. The summit level was to have its surface at high water, 102 feet above the sea, and to be supplied with water by a feeder leading from an artificial reservoir to be constructed at Alhajuela in the upper Chagres valley; the ascent on the Pacific side to be likewise by four locks. The general plan of the canal was to be about the same as that of the old company.

A NICARAGUAN CANAL PROJECTED.—In 1884 a treaty was made between the United States and Nicaragua, which provided that the United States was to build a canal, without cost to Nicaragua, which after completion was to be owned and managed jointly by the two governments. The treaty was not ratified by the United States Senate and the question was dropped. In 1886 a company of private citizens was formed in New York, known as the "Nicaragua Canal Association." It was to obtain concessions, make surveys, lay out the route, and organize such corporations as should be re uired to construct the canal. Operations upon a moderate scale, mainly of a preliminary character, were continued until 1893. The financial disturbances of that year drove the construction company, known as the "Maritime Canal Company of Nicaragua," into bankruptcy and stopped the work. It has not since been resumed.

In 1895 Congress provided for a board of engineers to inquire into the possibility, permanence, and cost of the canal as projected by the "Maritime Canal Company." The report of this board criticized the plans and estimates of the company, and in 1897 another board was appointed to make additional surveys and prepare new plans and estimates. The report of the second board was not completed when the plans of the new French company attracted the attention of Congress, and led to the creation in 1899 of the Isthmian Canal Commission. This body was to examine all available routes, and to report which was most practicable and feasible for a canal under the "control, management, and ownership of the United States."

THE UNITED STATES DECIDES TO BUILD A CANAL AT PANAMA.—The Isthmian Canal Commission made an investigation of the New Panama Canal Company's project and, learning that the Colombian government would consent to an alienation of the Wyse concession, approached the French company with the object of obtaining an option on its works and property at Panama. After considerable delay and some misunderstandings, the French company offered to sell and transfer its canal rights and property to the United States for $109,141,500. The Commission considering that $40,000,000 was a liberal estimate of its value concluded its report to Congress as follows:

"In view of all the facts, and particularly in view of all the difficulties of obtaining the necessary rights, privileges, and franchises on the 'anama route, and assuming that Nicaragua .nd Costa Rica recognize the value of the canal o themselves, and are prepared to grant concessions on terms which are reasonable and cceptable to the United States, the Commission s of the opinion that 'the mo t practicable and easible route for an isthmian canal, to be under he control, management, and ownership of the Jnited States,' is that known as the Nicaragua oute."

This report brought a new offer from the New 'anama Canal Company to sell out all its assets, ights, and interests in the Isthmus for the sum f $40,000,000. The consequence of this offer vas the issuance of a supplementary report by .he Commission in which it was stated that 'After considering the changed conditions that ow exist, the Commission is of the opinion that the most practicable and feasible route' for an Isthmian Canal to be 'under the control, nanagement, and ownership of the United States' is that known as the Panama route."

Before this supplementary report of the Commission had been filed, the House of Representatives had passed and sent to the Senate the hastily prepared Hepburn Bill, which authorized the President to secure a concession from Nicaragua and to proceed with the construction of a canal by that route. The supplementary report was received while the Senate was considering the Hepburn Bill, and in the debates which followed, it was shown that the Panama route was favored, not only by the most eminent engineers, but, without exception, by ship owners and by the masters and pilots of steamship lines and sailing vessels, whose opinions had been sought by the late Senator Hanna.

Following the debate both branches of Congress passed the Spooner Bill, which became law on June 28, 1902. This authorized the President to acquire all the property of the New Panama Canal Company, including not less than 68,896 shares of the Panama Railroad Company, for a sum not to exceed $40,000,000, and to obtain from Colombia perpetual control of a strip of land 6 miles wide. If the company and Colombia failed to come to terms in a reasonable time, he was by treaty to obtain from Costa Rica and Nicaragua the territory necessary for the Nicaragua canal.

Colombia rejected the terms submitted by the United States. In November, 1903, the Colombian province of Panama declared its independence, and created the Republic of Panama. Within a month after the formation of the new republic, a treaty was negotiated between it and the United States which provided for the construction of the Panama Canal, and gave the United States perpetual control of a strip of territory for that purpose. (See Canal Zone.) The United States was to build a canal at Panama and to make a reality of the vision of the early explorers. The property of the French company was purchased as the Spooner Bill had directed and, in accordance with another of its provisions, the President appointed the Isthmian Canal Commission to take charge of the work of construction.

The American builders found awaiting them the same problems of health, labor, and engineering which had confronted the French companies. When the United States took charge, the workers on the spot numbered less than a thousand men and these were continued in the work of excavation at Culebra. An inventory of the items of the newly acquired property included a hospital and other buildings which were usable, machinery, only part of which was in condition for immediate use, while a considerable part of it was of no practical value whatsoever; and scientific data, surveys, and calculations, which were of inestimable value in the saving of both time and labor. The actual cash value of the various properties and of the work of excavation as computed by the engineers are shown on page 24 of this book.

THE UNITED STATES BEGINS THE WORK OF CONSTRUCTION.—It was not until May 4, 1904, that possession of the Canal Zone was granted to the United States under the Hay-Bunau-Varilla Treaty, and that the preliminary work was put in hand.

For three years the operations were purely of an experimental character, for the fundamental plan of the Canal had not been definitely determined, and the efforts of the staff, warned by the high death-rate among the employees of the old French company, were largely directed to the consideration of those sanitary measures without which it was realized that success could not be attained. As a result of the health campaign thus inaugurated yellow fever has been eradicated, bubonic plauge held at bay, malaria reduced from a virulent to a mild form, and an unhealthy country made into a desirable residential area.

THE CHOICE OF SCHEMES.—The question of whether a sea-level or a high-level canal with locks should be constructed was considered by an international Commission, but, although a majority reported in favor of the sea-level scheme, the minority report, which recommended a canal at an elevation of 85 feet above sea-level, was ultimately adopted. The reasons given for this selection were the reduction in cost which would accompany this scheme, the shorter time in which the canal could be constructed, and the

belief that it would give a better navigable channel than a sea-level canal.

In April, 1907, the work was placed under the control of the Engineer Corps of the U. S. Army. A new Commission was created of which Colonel George W. Goethals was made Chairman and Chief Engineer. At the same time Colonel Goethals was also made the chief executive of the Canal Zone, and President of the Panama Railroad Company.

THE LABOR PROBLEM.—The establishment of sanitary conditions in the zone removed the greatest obstacle to the settlement of the labor problem. The French companies had never had more than 12,000 men working at one any time. To carry out the larger scheme of the United States an army of 30,000 to 40,000 able-bodied men would be necessary. The extra pay and cheap living, which were assured by the government, were powerful inducements to draw the idle and discontented workers of other localities. But the question was not just a case of bringing shiploads of laborers and setting them to work. A colony had to be established and provisions made to feed, clothe, shelter, and govern it. With the physical necessities of life assured, the moral and intellectual well-being of the men was considered. Homesickness is a dread disease which unfits men for doing good work. To combat it many remedies were forthcoming. Men were encouraged to bring their families and comfortable homes were provided for them. The colony was supplied with all the equipment for social life which is to be found in settled communities in the States. The Young Men's Christian Association had branches in each community and took general charge of recreation and club work. Their buildings were also utilized for religious services. The main hospital at Ancon was made a model in its equipment and service. Post offices, banks, schools, and courts were as accessible to the people of the Zone as they are to the people of New Jersey.

The workers in Panama had what we may call a personal grievance court, a luxury not to be found in the States. The presiding judge of this informal court has been called "a beneficent tyrant", but he was regarded by the men as their best friend and counselor. Every Sunday morning Colonel Goethals listened to the petty complaints and grievances of his men and they were satisfied with his judgment in nearly every case. The effect of this outlet for bad feeling can hardly be described. It saved many a good worker from the folly of throwing up his job, while in other cases it inspired an indifferent man with a sense of the importance of his work and the nobility of doing his best.

As a result of the conditions established on the Isthmus the supply of labor was equal to the demand. The workers, generally, were happy and contented, and many have left with genuine regret; while others have taken up permanent residence in Panama.

NATIONALITY ON THE ISTHMUS.—The chiefs of the Canal, the directors, the superintendents, the clerical and medical staffs, the skilled artisans, were all Americans. Of the unskilled laborers, fully half the number employed during the work of construction was of West Indian extraction—Jamaican and Barbadian negroes. The United States, Panama, Spain, and other European countries, contributed the other half. The zone census of 1912 showed that forty nationalities were represented at that time among the workers. The whole world has had a hand in digging the Panama Canal.

CLASSIFICATION OF WORKERS.—The necessity for a distinction between skilled and unskilled labor resulted in the division of employees into the "gold" and "silver" classes. The gold roll included all those drawing over $75. per month, who were paid in gold. The silver roll included all the common and unskilled laborers and these men were paid in the silver money of the Republic of Panama.

ENGINEERING PROBLEMS.—The Chagres River, with a total drainage area of 1,320 square miles, constituted one serious problem, and others were presented by the Gatun Dam and the Culebra Cut. These problems were all interrelated and involved many minor ones. The flood waters of the Chagres, which alone formed an almost insurmountable obstacle to the building of a sea-level canal, have been restrained by the great Dam and made to furnish the water for the lock canal as it is to-day. The making of Culebra Cut involved cutting through the continental divide for nine miles, and the removal of millions of cubic yards of earth and rock.

THE BEGINNING OF THE END.—One by one the problems were met and solved. October 13, 1913, was a great day in the history of the Canal for it witnessed the destruction of Gamboa dike whose removal permitted the water to enter the Cut. From that time on all excavation was done by dredging. The first passage of a self-propelling vessel through the Canal from ocean to ocean was made on January 7, 1914. The vessel thus honored was La Valley, a unit of the old French floating equipment, one of the items acquired by purchase from our brave but unfortunate predecessors on the Isthmus. Thus the Canal is a reality, it is a vision no longer. Soon the ships of all nations will be passing through the channel and the world will grow accustomed to its use and its benefits.

For nine years the world has watched the work progress; thousands have journeyed to the Isthmus to have a glimpse of the Canal in

he making; while those who could not go have
ollowed each step of achievement with no less
f interest. Countless illustrated papers and
nagazines have made the Canal known to all
vho read, and the Cut, the Dam, the Locks, the
.akes, and the Slides, are now familiar terms
1 every community. Americans have a right
o be proud of the Canal and of the men who have
uilt it. The names of Goethals and Gorgas
nd their brave associates are held in highest
onor and can not be forgotten while the Canal
hall serve the world.

III. THE CANAL ZONE

Under the Treaty with Panama the United
States acquired a strip extending for five miles

Of the 436 square miles of Zone territory the
greater part is owned by the United States and the
Panama Railroad Company, the balance being
owned by private individuals and corporations.
The Treaty gives the United States a title in the
nature of an option on privately owned lands, upon
condition that the United States compensate the
owners of the lands in exercising its rights. All
of the privately owned lands in the Canal Zone
are being taken over on behalf of the United
States by virtue of the Executive Order of the
President of December 5, 1912, acting by authority of the Panama Canal Act of August 24,
1912.

The United States has jurisdiction and ownership over the area to be covered by Gatun Lake,
which extends beyond the line of the Canal Zone.

Mandingo Stockade for Canal Zone Convicts, Engaged in Road Building

on either side of the Canal, and all other lands and
and under water needed for Canal purposes. The
Canal Zone contains about 436 square miles,
about 95 of which are under the waters of the
Canal and Gatun and Miraflores Lakes. It begins at a point 3 marine miles from mean low
water mark in each ocean, and extends for 5
miles on each side of the center line of the route
of the Canal. It includes the group of islands in
the Bay of Panama named Perico, Naos, Culebra, and Flamenco. The cities of Panama and
Colon, are excluded from the Zone, but the
United States has the right to enforce sanitary
ordinances in those cities, and to maintain public
order in them in case the Republic of Panama
should not be able, in the judgment of the United
States, to do so.

Gatun Lake will have a summit level of 87 feet,
and the Government of the United States has
taken an additional strip around the lake littoral
up to the 100-foot level, the additional strip being needed for police and sanitary purposes. The
area of this territory is approximately $95\frac{1}{2}$ square
miles.

The population of the Canal Zone, official census, is 62,810; of Panama City, 37,505; of Colon,
17,749.

IV. DESCRIPTION OF THE PANAMA CANAL

The Panama Canal does not cross the Isthmus from east to west, but, as is shown on
the accompanying map, its general direction is

from northwest to southeast, the Pacific entrance ne..r Panama being about 22½ miles east of the Atlantic entrance near Colon. The isthmus at the Canal Zone is about 40 miles wide and the entire length of the Canal from deep water in the Atlantic to deep water in the Pacific is about 50 miles. Of this length about 15 miles are at sea-level, seven miles at the Colon end and eight at the Panama end. The remainder consists of two elevated reaches, the longer of which—that between Gatun and Pedro Miguel locks—is about 32 miles in length and 85 feet above sea-level, while the shorter—between Pedro Miguel and Miraflores—is about two miles long and about 55 feet above sea-level. From Gatun for a distance of 24 miles the Canal follows roughly the channel of the Chagres River,

been made for the largest ocean-going ships.

GATUN DAM.—The Gatun Dam, which forms Gatun Lake by impounding the waters of the Chagres and its tributaries, is nearly 1½ miles long, measured on its crest, nearly ½ mile wide at its base, about 400 feet wide at the water surface, about 100 feet wide at the top, and its crest is 105 feet above mean sea-level, or 20 feet above the normal level of the Lake. It is in reality a low ridge uniting the high hills on either side of the lower end of the Chagres Valley so as to convert the valley into a huge reservoir. Of the total length of the Dam only 500 feet, or one-fifteenth, is exposed to the maximum water head of 85 to 87 feet. The interior of the Dam is formed of a natural mixture of sand and clay, dredged by hydraulic process from pits

Spillway of Gatun Dam

and the greater part of this portion has been converted into Gatun lake.

CHANNEL DIMENSIONS.—The depth of water on the sills of the locks ranges from 41 feet to 45 feet. The widths of the bottom of the Canal channel in its several sections are as follows: The bottom width from the Caribbean Sea to the Gatun locks, a distance of 6-7 miles, is 500 feet. From these locks to a point 23½ miles from the sea the width is 1,000 feet. From this point it narrows to 800 feet, 700 feet, and 500 feet, and for the eight miles through the Culebra Cut the bottom width is reduced to 300 feet. From Pedro Miguel to Miraflores it widens to 500 feet, and that width is maintained to deep water in the Bay of Panama. The minimum depth of water is 41 feet. Thus provision has

above and below the Dam, and placed between two large masses of rock and miscellaneous material obtained from steam shovel excavation at various points along the Canal. The top and upstream slope are thoroughly riprapped. The entire Dam contains about 21,000,000 cubic yards of material.

THE SPILLWAY, GATUN DAM.—The Spillway is a concrete lined channel 1,200 feet long and 285 feet wide, cut through a hill of rock nearly in the center of the Dam, the bottom being 10 feet above sea-level at the upstream end and sloping to sea-level at the toe. Across the upstream or lake opening of this channel a concrete dam has been built in the form of an arc of a circle making its length 808 feet although it closes a channel with a width of only 285 feet.

The crest of the dam is 69 feet above sea-level, or 16 feet below the normal level of the lake which is 85 feet above sea-level. On the top of this dam there are 13 concrete piers with their tops 115.5 feet above sea-level, and between these there are mounted regulating gates of the Stony type. Each gate is of steel sheathing on a framework of girders and moves up and down on roller trains in niches in the piers. They are equipped with sealing devices to make them water-tight. Machines for moving the gates are designed to raise or lower them in approximately ten minutes. The highest level to which it is intended to allow the lake to rise is 87 feet above sea-level, and it is probable that this level will be maintained continuously during wet seasons. With the lake at that elevation, the

at Gatun, was that of the fiscal year, 1912, which was about 132 billion cubic feet. Previous to that year the smallest run-off of record was 146 billion cubic feet. In 1910 the run-off was 360 billion cubic feet, or a sufficient quantity to fill the lake one and a half times. The low record of 1912 is of interest as showing the effect which a similar dry season, occurring after the opening of the Canal, would have upon its capacity for navigation. Assuming that the Gatun Lake was at elevation plus 87 at the beginning of the dry season on December 1st, and that the hydroelectric plant at the Gatun Spillway was in continuous operation, and that 48 lockages a day were being made, the elevation of the lake would be reduced to its lowest point, plus 79.5, on May 7th, at the close of the dry season,

Gatun Lake, the Dying Jungle

regulation gates will permit of a discharge of water greater than the maximum known discharge of the Chagres River during a flood.

GATUN LAKE AND ITS WATER SUPPLY.—This artificial lake, created by building the Gatun Dam, impounds the waters of a basin comprising 1,320 square miles. Its elevation varies with the seasons from 80 to 87 feet above sea-level. When the surface of the water is at 85 feet above sea-level, the lake has an area of about 164 square miles, and contains about 183 billion cubic feet of water. During eight or nine months of the year, the lake is kept constantly full by the prevailing rains, and consequently it is necessary to store a surplus for only three or four months of the dry season. The smallest run-off of water in the basin during the past 22 years, as measured

after which it would continuously rise. With the water at plus 79 in Gatun Lake there would be 39 feet of water in Culebra Cut, which would be ample for navigation. The water surface of the lake will be maintained during the rainy season at 87 feet above sea-level, making the minimum channel depth in the Canal 47 feet. As navigation can be carried on with about 39 feet of water, there will be stored for the dry season surplus over 7 feet of water. Making due allowance for evaporation, seepage, leakage at the gates, and power consumption, this would be ample for 41 passages daily through the locks, using them at full length, or about 58 lockages a day when partial length is used, as would be usually the case, and when cross filling from one lock to the other through the central wall is

employed. This would be a larger number of lockages than would be possible in a single day.

The water level of Gatun Lake, extends through the Culebra Cut, and is maintained at the south end by an earth dam connecting the locks at Pedro Miguel with the high ground to th westward, about 1,400 feet long, with its crest at an elevation of 105 feet above mean tide. A concrete core wall, containing about 700 cubic yards, connects the locks with the hills to the eastward; this core wall rests directly on the rock surface and is designed to prevent percolation through the earth, the surface of which is above the Lake level.

Side Wall of Locks Compared with Six Story Building

THE CULEBRA CUT.—The points of deepest excavation in the Canal are in Culebra Cut, between Gold and Contractor's Hills. The widest part of the Cut is opposite the town of Culebra, where owing to the action of slides on both banks, the top width is about half a mile. This is one of the most important and impressive sections of the work, for it carries the Canal through the watershed of the Isthmus. The material was not difficult to excavate, but as it disintegrated into mud when exposed to the atmosphere and rain, the trouble was not the removal of millions of tons of material but the numerous "slides." These may be described as a glacier-like movement of soft material on layers of slippery clay.

The channel through the Cut is a gently winding stream, shut in by the hills on either side. In a few years the rapid growth of tropical vegetation will cover up the scars made by excavation and the man-made canyon will appear to the tourist as one of Nature's own creations.

As before stated, the water in the Cut is at the same level as Gatun Lake, 85 feet above sea-level, and it comes from the same source, mainly the Chagres River.

MIRAFLORES LAKE AND SPILLWAY.—A small lake between the locks at Pedro Miguel and Miraflores is formed by dams connecting the walls of Miraflores locks with the high ground on either side. The dam to the westward is of earth, about 2,700 feet long, having its crest about 15 feet above the water in Miraflores Lake. The east dam is of concrete, containing about 75,000 cubic yards; is about 500 feet long, and forms a spillway for Miraflores Lake, with crest gates similar to those at the Spillway of the Gatun Dam.

· LOCKS.—There are 6 double locks in the Canal; three pairs in flight at Gatun, with a combined lift of 85 feet; one pair at Pedro Miguel, with a lift of 30½ feet, and two pairs at Miraflores, with a combined lift of 54½ feet at mean tide. The usable dimensions of all are the same—a length of 1,000 feet, and width of 110 feet. Each lock is a chamber, with walls and floor of concrete, and mitering gates at each end.

The side walls are 45 to 50 feet wide at the surface of the floor; are perpendicular on the face, and narrow from a point 24½ feet above the floor until they are 8 feet wide at the top. The middle wall is 60 feet wide, approximately 81 feet high, and each face s vertical. At a point 42½ feet above the surface of the floor, and 15 feet above the top of the middle culvert, this wall divides into two parts, leaving a space down the center much like the letter "U," which is 19 feet wide at the bottom and 44 feet wide at the top. In this center space is a tunnel divided into three stories, or galleries. The lowest gallery is for drainage; the middle, for the wires that carry the electric current to operate the gate and valve machinery installed in the center wall, and the upper is a passageway for the operators.

The lock gates are steel structures seven feet thick, 65 feet long, and from 47 to 82 feet high. They weigh from 390 to 730 tons each. Ninety-two leaves are required for the entire Canal, the total weighing 60,000 tons. The leaves are shells of structural steel covered with a sheathing of steel riveted to the girder framework. Each leaf is divided horizontally into two separate compartments. The lower compartment is watertight, for the purpose of making the leaf so buoyant that it will practically float in the water and thus largely relieve the stress upon the bearings by which it is hinged to the wall. This watertight compartment is subdivided vertically into three sections, each independently watertight, so that if the shell should be broken in any way or begin to leak, only one section would probably be affected. An air shaft 26 inches in diameter runs from the bottom compartment up to the top of the gate, and this also is watertight where it passes through the upper half of the leaf.

Intermediate gates are used in all except one pair of the locks, in order to save water and time, if desired, in locking small vessels through, the gates being so placed as to divide the locks into chambers 600 and 400 feet long, respectively. Ninety-five per cent of the vessels navigating the high seas are less than 600 feet long.

The locks are filled and emptied through a system of culverts. One culvert, 254 sq. ft. in area of cross section, about the area of the Hudson River tunnels of the Pennsylvania railroad, extends the entire length of each of the middle and side walls. From each of the large culverts there are several smaller culverts, 33 to 44 sq. ft. in area, which extend under the floor of the lock and communicate with the lock chamber through holes in the floor. The large culverts are con-

the disturbance in the chamber when it is being filled or emptied.

The average time in filling and emptying a lock is about fifteen minutes, without opening the valves so suddenly as to create disturbing currents in the locks or approaches. The depth of water over the miter sills of the locks is 40 feet in salt water and 41½ feet in fresh water.

TIME REQUIRED FOR PASSAGE THROUGH THE CANAL.—The time required to pass a vessel through all the locks is about 3 hours; one hour and a half in the three locks at Gatun, and about the same time in the three locks on the Pacific side. The time of passage of a vessel through the entire Canal is from 10 to 12 hours, according to the size of the ship, and the rate of speed at which it can travel.

Balboa Terminal

trolled at points near the miter gates by large valves and each of the small culverts extending from the middle wall feeds in both directions through laterals, thus permitting the passage of water from one twin lock to another, effecting a saving of water.

To fill a lock the valves at the upper end are opened and the lower valves closed. The water flows from the upper pool through the large culverts into the small lateral culverts and thence through the holes in the floor into the lock chamber. To empty a lock the valves at the upper end are closed and those at the lower end are opened and the water flows into the lower lock or pool in a similar manner. This system distributes the water as evenly as possible over the entire horizontal area of the lock and reduces

TOWING LOCOMOTIVES.—Ships are not allowed to pass through the locks under their own power, but are towed through by electric locomotives operating on tracks on the lock walls. The system of towing provides for the passing through the locks of a ship at the rate of 2 miles an hour. The number of locomotives varies with the size of the vessel. The usual number required is 4: 2 ahead, 1 on each wall, imparting motion to the vessel, and 2 astern, 1 on each wall, to aid in keeping the vessel in a central position and to bring it to rest when entirely within the lock chamber. They are equipped with a slip drum, towing windlass, and hawser which permit the towing line to be taken in or paid out without actual motion of the locomotive on the track. The locomotives run on a level, except when in

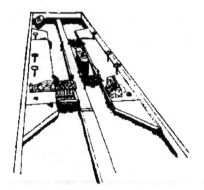

Model of Pedro Miguel Lock

passing from one lock to another they climb heavy grades. There are two systems of tracks, one for towing and the other for the return of the locomotives when not towing. The towing tracks have center racks or cogs throughout, and the locomotives always operate on this rack when towing. At the incline between locks the return tracks also have rack rails, but elsewhere the locomotives run by friction. The only cross-overs between the towing and return tracks are at each end of the locks, and there are no switches in the rack rail.

PROTECTIVE DEVICES.—Several protective devices have been adopted to safeguard the gates in the locks.

First.—Fender chains, 24 in number, each weighing 24,098 pounds, are placed on the upstream side of the guard gates, intermediate and safety gates of the upper locks, and in front of the guard gates at the lower end of each flight of locks. They prevent the lock gates from being rammed by a ship that may approach the gates under its own steam or by escaping from the towing locomotives. In operation, the chain is stretched across the lock chamber from the top of the opposing walls, and when it is desired to allow a ship to pass, the chain is lowered into a groove made for the purpose in the lock floor. It is raised again after the ship passes. The raising and lowering are accomplished from both sides by mechanism mounted in chambers or pits in the lock walls. This mechanism consists of a hydraulically operated system of cylinders, so that 1 foot of movement by the cylinder will accomplish 4 feet by the chain. If a ship exerting a pressure of more than 750 pounds to the square inch should run into the fender, the chain will be paid out gradually by an automatic release until the vessel comes to a stop. Thus, a

10,000-ton ship, running at 4 knots an hour, after striking the fender can be brought to a stop within 73 feet, which is less than the distance which separates the chain from the gate.

Second.—Double gates are provided at the entrances to all the locks and at the lower end of the upper lock in each flight, the guard gate of each pair protecting the lower gate from ramming by a ship which might possibly get away from the towing locomotives and break through the fender chain.

Third.—A dam of the movable type called an emergency dam is placed in the head bay above the upper locks of each flight for the purpose of checking the flow of water through the locks in case of damage, or in case it should be necessary to make repairs, or to do any work in the locks which would necessitate the shutting off of all water from the lake levels. Each dam is constructed on a steel truss bridge of the cantilever type, pivoted on the side wall of the lock approach, and when not in use resting on the side wall parallel to the channel. When the dam is used the bridge is swung across the channel with its end resting on the center wall of the lock. A series of wicket girders hinged to this bridge are then lowered into the channel with their ends resting in iron pockets embedded in the lock floor. After the girders have been lowered into place, they afford runways for gates which can be let down one at a time closing the spaces between the wicket girders. These gates form a horizontal tier spanning the width of the Canal and damming the water to a height of 10 feet. Another series of panels are then lowered, and so on until the dam, constructed from the bottom upward, completely closes the channel. When the dam has checked the main flow, the remainder, due to the clearance between the vertical sides of the gates, may be checked by driving steel pipes between the sides of the adjacent panels. These dams are operated in three movements, and the machinery for operating is, therefore, in three classes, gate-moving, raising and lowering the wicket girders, and hoisting the gates on the girders, all driven by electric motors.

ELECTRICAL CONTROL OF LOCK MECHANISM.—Not the least interesting feature of the Panama Canal from the engineering point of view is the arrangements adopted for working the machinery connected with the locks at Gatun, Pedro Miguel, and Miraflores. The electrical energy employed for this purpose is generated at the hydroelectric station at Gatun Dam, with an emergency Curtis steam turbine plant at Miraflores, and is transmitted across the Isthmus by a 44,000-volt line connecting Christobal and Balboa with those two stations. The extent and intricacy of the installations may be judged from the fact that at Gatun Locks there are 218 electric motors,

at Pedro Miguel 122, and at Miraflores 160, a total of 500. They range in capacity from ¼ to 70 h.p., and their aggregate h.p. is 12,020. At Gatun there are 40 motors of 25 h.p. each for moving the leaves of the gates, at Pedro Miguel 24, and at Miraflores 28. When the gates are closed they are held in position by a device called a mitre forcing machine, the operation of which requires 46 motors of 7 h.p. at the three locks.

The fender chains described in a preceding paragraph are raised and lowered by a method similar to that employed in hydraulic lifts, and two motors are used in each case, one driving the main pump that supplies water under pressure, and the other working the valve which controls the direction of movement of the chain. The pump motors are of 70 h.p. each and the valve motors of ½ h.p., there being in all 48 of each sort at the three locks.

For regulating the flow of water in the culverts by which the locks are emptied and filled a large number of valves are employed, with an elaborate system of duplication to prevent interruption of working. These valves are known as rising stem gate valves, cylindrical valves, guard valves, and auxiliary culvert valves, and for their operation at the three locks are required 266 motors, of 7, 25, and 40 h.p. and with an aggregate capacity of 6,014 h.p.

The length of the various locks and the wide separation of the many operating machines, necessitated the employment of a method of control which would obviate the many opportunities for mistakes certain to occur if a large force of operators was along the locks. The ordinary mechanical devices for central control were likewise out of the question, because of the great distances involved. It was therefore decided to adopt centralized electrical control, so that all of the operations are now effected from control boards located in buildings on the center walls between the lock channels, and at points which command the best possible view of the locks. The operator, however, does not depend on an actual view of conditions in the locks, such as the position of gates, guard chains, and other apparatus, for the control boards are equipped with indicators which reproduce the action of the various lock mechanisms in miniature.

The control boards are of the flat-top benchboard type, 32 inches high by 54 inches wide, and 64 feet long at Gatun; 36 feet at Pedro Miguel, and 52 feet at Miraflores. The water in the locks is represented by blue marble slabs, and the side and center walls by cast iron plates. The indicator for each of the lock gates consists of a pair of aluminum leaves, shaped to correspond to the plan view of the top of the gates, which travel horizontally just above the top of the

board. The position of the chain fenders is shown by small aluminum chains which follow the movements of the chains in the locks, the actual position of the chains from the bottom of the lock being indicated by the angle of semaphore arms attached to them. The position of the rising stem valves is indicated on an arrangement which resembles a small elevator, the index being constituted by a small aluminum cage which moves up and down. The level of the water in the various locks is given on vertical scales and an accuracy of within five-eighths of an inch of the actual level was required by the specifications, but the indicators attained an accuracy even greater than this.

There are some machines which are not represented by working models on the control board. For these, position indicators are not necessary, as all the operator requires to know is whether they are in open or closed position, and they are operated merely by control switches and their open or closed position indicated by various colored lights.

MECHANICAL INTERLOCKING.—It was very necessary to arrange the operating controls so that the operator would maneuver the switch handles in a predetermined sequence and also to prevent the man in charge of one channel interfering with the mechanism under the jurisdiction of the man in charge of the other channel. To this end an elaborate interlocking system was provided. Vertical shafts, worked by connecting rods from the control switches, extend downward past the electrical parts to the interlocking mechanism which is disposed in vertical racks at such a distance below each edge of the control boards that it can be inspected and oiled from a floor which is about 7 feet below the floor on which the switchboard operator stands. It is essentially a bell-crank mechanism, connecting the shaft of the control switch through a movable horizontal bar to a vertical operating shaft which can or cannot move according to the relative positions of the interlocking bars and dogs. Its action depends mainly on the engagement of bevel dogs carried on horizontal and vertical bars, the movement of a horizontal bar tending to lift a vertical bar by the bevels on the dogs. A horizontal bar cannot be moved without raising a vertical bar. Thus if at any time a dog on a horizontal bar rests against the upper end of a dog on a vertical bar, no movement of the horizontal bar where the dog engages with the vertical bar can take place, and the control handle connected to that particular horizontal bar is locked. Over 2¼ miles of interlocking rod are employed in the control boards at the three locks.

Some examples may be given to illustrate the functions of the interlocking mechanism. To

In Gatun Locks

ensure that a chain fender shall always be in the taut or up position to protect a closed gate, interlocks prevent the chain from being lowered until the adjacent gates have been opened, and the gates from being opened until the chain is in the raised position. Again the switch controlling the mitre forcing machine cannot be moved unless the gates are closed, and while the mitre forcing machine is open the rising stem valves of the side wall next above or below a gate must be closed—that is, the valves above or below a gate must remain closed until the gate itself is closed, thus preventing the operator from causing a current of water round the gates while they are open or are being moved for opening or closing. Again there are elaborate interlockings to safeguard the manipulation of the

the ideas of a number of skilled engineers, so that every contingency has been foreseen and everything in the manner of safeguards installed to prevent mistakes and accidents. The greatest credit for these models of engineering skill is due to Mr. Edward Schildhauer, electrical and mechanical engineer of the Isthmian Canal Commission, who was in turn ably assisted by Engineers C. B. Larzelere and W. R. McCann.

PASSING A SHIP THROUGH THE LOCKS.—A ship to be raised to the lake level comes to a full stop in the forebay of the lower lock, prepared to be towed through one of the duplicate locks by electric towing locomotives. The water in the lower lock chamber is equalized with the sea-level channel, after which the miter gates are opened, the fender chain lowered and the vessel

Mindi Excavation, Looking North from East Bank of Canal

various valves. Thus to prevent any possibility of the locks being flooded when the valves are in a certain position, the rising stem valves of the side wall are interlocked with those of the middle wall a lock length away. Such diagonal interlocking when the cylindrical valves are open is wanted to prevent the flow of water from, say, the upper lock by way of a side wall culvert to the middle lock, and thence by way of the middle wall culvert to the lower lock, with the result of flooding the lower lock walls. In some cases an interlock can be released by means of a Yale key, as, for example, when a large vessel is passing through and it is necessary for the intermediate gates to be thrown open to obtain a 1,000 feet level.

These electrical control boards incorporate

passed into the first chamber, where the water is at sea-level. Then the miter gates are closed. The rising stem gate valves at the outlet of the main culverts are closed, while those above are opened, allowing water to flow from an upper level into the lower chamber, which, when filled, raises the vessel 28½ feet, to the second level. This operation is repeated in the middle and upper locks until the ship has been raised to the full height of 85 feet above the level of the sea. At Gatun, in the passing of a large ship through the locks, it is necessary to lower 4 fender chains, operate 6 pairs of miter gates and force them to miter, open and close 8 pairs of rising stem gate valves for the main supply culverts, and 30 cylindrical valves. In all, no less than 98 motors are set in motion twice during each lockage of a

single ship, and this number may be increased to 143, dependent upon the previous conditions of the gates, valves, and other devices.

LIGHTING THE CANAL.—The general scheme of lighting and buoying the Canal includes the use of range lights to establish direction on the longer tangents and of side lights spaced about 1 mile apart to mark each side of the channel. The range lights are omitted in Culebra Cut, where their use is hardly practicable, and on four of the shorter tangents on the remainder of the Canal. In the Cut there are three beacons at each angle, and between these are intermediate beacons in pairs on each side of the Canal. By keeping his ship pointed midway between these beacons, the pilot is able to adhere closely to the center of the Canal. At each tangent it is necessary to have two ranges of two lights each to prolong the sailing line in order that the pilot may hold his course up to the point of turning. These range lights are situated on land. There are three types constructed, all of reinforced concrete. The more elaborate structures are used on the Gatun locks and dam and in the Atlantic and Pacific Divisions, where they are closer to the sailing lines of the vessels, while simpler structures are placed in the Gatun Lake, where they are under less close observation. A light and fog signal on the west breakwater in Limon Bay is also included, and a light on the east breakwater. The illuminants are gas and electricity, the latter being used whenever the light is sufficiently accessible. For the floating buoys, and for the towers and beacons which are in inaccessible places, the system using compressed acetylene dissolved in acetone has been adopted. The buoys are composed of a cylindrical floating body or tank, surmounted by a steel frame which supports the lens at a height of about 15 feet above the water level. The buoys are moored in position along the edge of the dredged channel by a heavy chain and a concrete sinker, and should remain lighted for six to twelve months without being recharged. The candlepower of the range lights varies ac-

cording to the length of the range, from about 2,500 to 15,000 candlepower. The most powerful lights are those marking the sea channels at the Atlantic and Pacific entrances, they being visible from about 12 to 18 nautical miles. The beacons and gas-buoy lights have about 850 candlepower. White lights are used throughout, and, in order to eliminate the possibility of confusing the lights with one another and with the lights on shore, all range lights, beacons, and buoys have individual characteristics formed by flashes and combinations of flashes of light and dark intervals.

BREAKWATERS.—Breakwaters have been constructed at the Atlantic and Pacific entrances of the Canal. That in Limon Bay or Colon Harbor, extends into the bay from Toro Point at an angle of 42° and 53° northward from a base line drawn from Toro Point to Colon light, and is 10,500 feet in length, or 11,700 feet, including the shore connection, with a width at the bottom of fifteen feet and a height above mean sea-level of ten feet. The width at the bottom depends largely on the depth of water. It contains approximately 2,840,000 cubic yards of rock, the core being formed of rock quarried on the mainland near Toro Point, armored with hard rock from Porto Bello. Work began on the breakwater in August, 1910, and on December 1, 1912, the trestle and fill were completed to full length, 11,500 feet. The estimated cost is $5,500,000. A second breakwater is being built in Limon Bay, known as the east breakwater. It will be without land connection and about one mile long. The purpose of the breakwaters is to convert Limon Bay into a safe anchorage, to protect shipping in the harbor of Colon, and vessels making the north entrance to the Canal, from the violent northers that are likely to prevail from October to January, and to reduce to a minimum the amount of silt that may be washed into the dredged channel.

The breakwater at the Pacific entrance extends from Balboa to Naos Island, a distance of about

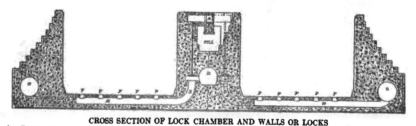

CROSS SECTION OF LOCK CHAMBER AND WALLS OR LOCKS

A.—Passageway for operators.   B.—Gallery for electric wires.   C.—Drainage gallery.   D.—Culvert in center wall.
E.—These culverts run under the lock floor and alternate with those from sidewalls.   F.—Wells opening from lateral culverts
into lock chamber.   G.—Culvert in sidewalls.   H.—Lateral culverts.

17,000 feet, or a little more than three miles. It lies from 900 to 2,700 feet east of and for the greater part of the distance nearly parallel to the axis of the Canal prism; varies from 20 to 40 feet in height above mean sea-level, and is from 50 to 3,000 feet wide at the top. It contains about 18,000,000 cubic yards of earth and rock, all of which was brought from Culebra Cut. It was constructed for a twofold purpose; first, to divert cross-currents that would carry soft material from the shallow harbor of Panama into the Canal channel; second, to furnish rail connection between the islands and the mainland. Work was begun on it in May, 1908, and on November 6, 1912, the last piles were driven connecting Naos Island with the mainland.

V. SAVING OF TIME EFFECTED BY THE PANAMA CANAL

From report by Emory R. Johnson, Special Commissioner on Panama Traffic and Tolls.

"The time and fuel costs which vessels can save by using the Panama Canal instead of some other route are the measure of the Canal's service to commerce. The saving in time of voyage resulting from the reduction which the canal may make in the length of an ocean route will depend upon the speed of the vessel and upon the number of hours required to make the passage through the canal. In the tables found on pages 31 and 32 the number of days that may be saved by using the Panama Canal is calculated for vessels of five different speeds—for steamers of 9, 10, and 12 knots, which are the speeds at which freight vessels are operated, and for ships of 14 and 16 knots, the speed of most passenger vessels. A half day is deducted in each instance to allow for the detention of the vessel in making the transit through the canal. When the comparison is between Panama and Suez, it is assumed that the passage through the Suez Canal will delay a vessel a half day. It will be understood that the length of each route through the Panama or Suez Canal (as from New York via Panama to San Francisco or from New York via Suez to Hongkong) includes the length of the canal—41 nautical miles for the Panama Canal and 87 miles for the Suez—and that the half-day taken from the "number of days saved," as stated in the tables, is the deduction made to allow for the longer time required to make the distance between the canal terminals than would be required to steam the same number of miles at sea.

Southern China and the Philippines—Hongkong and Manila—are near the center of the section whose commerce with New York and other north Atlantic ports of the United States may use the Panama or Suez Canal with equal advantage, as far as distance and time of voyage are concerned. Relative tolls and coal prices via the alternative routes, and the traffic possibilities at intermediate ports, rather than the days to be saved by taking one route rather than the other will determine whether the Panama or the Suez Canal will be used by vessels bound to or from that part of the Orient. Of the commerce of the western side of the Pacific with Europe, only the trade of New Zealand can save time by using the Panama instead of the Suez or some other alternative route. The point on the west coast of South America equally distant from a port of Europe via the Panama Canal and the Straits of Majellan is somewhat south of 40 degrees south latitude. Coronel, the most southerly port mentioned in Table B., is about 37 degrees south of the equator." (See pages 31 and 32.)

VI. ESTIMATED COST OF THE PANAMA CANAL

The cost estimated by the present Commission for completing the Canal is $325,201,000, which includes $20,053,000 for sanitation, and $7,382,-000 for civil administration.

These figures do not include the money paid to the New French Canal Company and to the Republic of Panama for property and franchises. Hence, it is estimated that the total cost of the Canal to the United States will approximate $375,000,000.

VII. CANAL STATISTICS

Length from deep water to deep water (miles)	50
Length from shore line to shore line (miles)..	40
Bottom width of channel, maximum (feet)..	1,000
Bottom width of channel, minimum, 9 miles in Culebra Cut (feet).................	300
Locks, in pairs........................	12
Locks, usable length (feet)...............	1,000
Locks, usable width (feet)...............	110
Gatun Lake, area (square miles)..........	164
Gatun Lake, channel depth (feet)........	85 to 45
Culebra Cut, channel depth (feet)........	45
Excavation, Canal Proper, estimated total (cubic yards).....................	209,668,000
Excavation, Permanent Structures, including terminals, estimated (cubic yards).....	22,685,000
Excavation, grand total, estimated (cubic yards)............................	232,353,000
Excavation, Due to Slides and Breaks, estimated (cubic yards) about............	25,000,000
Excavation accomplished January 1, 1914, (cubic yards).....................	215,458,272
Excavation, Remaining, Canal Proper, January 1, 1914, (cubic yards).......	16,894,728
Excavation by the French (cubic yards)....	78,146,960
Excavation by French, useful to present Canal, (cubic yards).................	29,908,000
Excavation by French, estimated value to Canal................................	$25,389,240
Value of all French property.............	$42,799,826
Concrete, total estimated for Canal (cubic yards)............................	5,208,800
Time of transit through completed Canal (hours)............................	10 to 12
Time of passage through locks (hours)......	3

Culebra Cut, with water at 80-foot level

Relocated Panama Railroad, total cost......	$8,866,392.02
Relocated Panama Railroad, length (miles)..	47.11
Canal Zone, area (square miles)...........	436
Canal and Panama Railroad force actually at work in September, 1913 (about)........	37,000
Canal and Panama Railroad force, Americans, in September, 1913 (about).............	5,000
Cost of Canal, estimated total.............	$375,000,000
Work begun by Americans.................	May 4, 1904
Date of official opening..................	Jan. 1, 1915

VIII. CANAL EXCAVATION TO JANUARY 1, 1914

By French Companies.....................	78,146,960
French excavation useful to present Canal....	29,908,000
By Americans—	
Dry excavation............. 128,747,980	
Dredges.................... 86,710,292	
Total..................	215,458,272
May 4 to December 31, 1904..	243,472
January 1 to December 31, 1905	1,799,227
January 1 to December 31, 1906	4,948,497
January 1 to December 31, 1907	15,765,290
January 1 to December 31, 1908	37,116,735
January 1 to December 31, 1909	35,096,166
January 1 to December 31, 1910	31,437,677
January 1 to December 31, 1911	31,603,899
January 1 to December 31, 1912	30,269,349
January 1, 1913 to January 1, 1914.....................	27,177,960

TOTALS BY SECTIONS AND AMOUNTS TO BE EXCAVATED.

SECTIONS	Amount excavated	Remaining to be excavated
Atlantic— Dry excavation.....	8,854,351 } 48,878,262	649 } 1,288,080 } 4,288,738
Wet.......	40,023,911	
Central— Culebra Cut	98,734,571 } 111,511,643	3,277,429 } 92,072 } 6,369,501
All other points.....	12,777,072	
Pacific— Dry excavation.....	10,372,087 } 55,068,367	415,913 } 3,004,720 } 6,420,633
Wet.......	14,696,280	
Grand total..	215,458,272	17,078,872

IX. VALUE OF THE $40,000,000 FRENCH PURCHASE

A careful official estimate has been made by the Canal Commission of the value to the Commission at the present time of the franchises, equipment, material, work done, and property of various kinds for which the United States paid the French Canal Company $40,000,000. It places the total value at $42,799,826, divided as follows:

Excavation, useful to the Canal, 29,908,000 cubic yards.........	$25,389,240
Panama Railroad Stock...........	9,644,320
Plant and material, used and sold for scrap....................	2,112,063
Buildings, used..................	2,054,203
Surveys, plans, maps and records...	2,000,000
Land...........................	1,000,000
Clearings, roads, etc..............	100,000
Ship channel in Panama Bay, four years' use....................	500,000
Total.....................	$42,799,826

X. PROBABLE EFFECTS OF THE PANAMA CANAL

The Panama Canal has been added to the wonders of the world. A continent has been cut in two and two great oceans have been linked for the purposes of trade. The construction of the Canal has aroused a wider interest than any other undertaking since History began a record of human achievements. This is a time when hope runs high, and, flushed with the successful completion of what is pronounced "the greatest engineering feat of the ages," there is danger of losing sight of the work which yet remains to be done if the world is to realize the full "substance of things now hoped for." The Canal is destined to work changes slowly. Generations will come and go before its full effects will become visible. How much of wisdom and forbearance will be necessary to settle justly the intricate changes in transportation rates, the delicate questions of diplomacy which will arise, and countless other matters upon whose proper adjustment the successful operation of the Canal will, in a great measure, depend! What deepening of harbors, improving of inland waterways, building of railroads and ships must be accomplished in order to secure the benefits for which mankind is looking!

The attention of the great world of business has for years been focused upon the new opportunities which are now offered, and preparations to take advantage of them have been steadily going forward. With the completion of the waterway the general public also turns from the mighty and spectacular feats of engineering, which, during the period of construction, have absorbed its attention, and asks "What is the Canal going to do for the world at large and the United States in particular?" Experts in trade and economic problems, thoughtful men in every branch of commercial activity have been weighing those questions and have tried to settle what the economic effects of the new thoroughfare will be, and the result is disagreement at many points. Absolute agreement of opinion upon a matter involving so many and such complex relations was not to be expected. There have been prophets of both good and evil; but the pessimists form an insignificant minority. The extreme optimist will hardly see all of his predictions verified; the pessimists will have some happy disappointments. Wise men

have known all along, that engineering problems were not the only puzzles involved in this great work and that only time and experience can settle those that attend upon its operation and economic results. The conviction that the completion of the Canal marks the dawn of a new era which will greatly add to the prosperity, happiness, and peace of the world is generally entertained. This healthy optimism will enable the workers of the world, under their great captains of industry, to bring to pass the beneficent results so earnestly desired by all.

POLITICAL SIGNIFICANCE OF THE CANAL.—The primary object of an interoceanic waterway was the shortening of routes and the consequent betterment and extension of the carrying trade of the world. We must, therefore, regard it not as a private enterprise, but as a national and international highway, whose establishment must benefit all the nations using it. To the United States belongs the glory of achieving the great work. In considering the probable effects—political and economic—of the new thoroughfare, there is strong evidence that its greatest benefits will accrue to the nation which furnished the courage, genius, and wealth necessary to its construction. Other nations will gain much. The United States will see that this new highway is kept open to the world under terms which could not offend honorable business men in a private enterprise. The Canal runs through territory owned by the United States, but at a distance of nearly 2,000 miles from the chief ports and the commercial center of the home domain. Therefore the Federal Government, acting upon sound business principles, has taken steps which will enable it to make good its promise to safeguard the interests of the commercial world. It has fortified the approaches and line of the Canal in order that the Canal may be preserved from danger of attack and destruction. Any other course would be inimical, in the long run, to the interests of all nations. The political and strategical aspects of the Canal have been recognized from the beginning. The voyage of the Oregon, during the war with Spain in 1898, was an object lesson that brought home to the nation a realization of the weakness of our defenses in time of war. It undoubtedly crystallized the sentiment, which had been gathering for years, in favor of a canal to be built and operated by the United States. The Panama Canal doubles the power of our navy without increasing the number of our fighting ships; it strengthens the defenses of our long stretches of seacoast by making possible the rapid transfer of forces from one coast to the other; it means greater safety at smaller expenditure of time and money than would otherwise be possible. The tolls of the Canal may not be sufficient for many years to return interest upon the money investment which it represents; but the nation will have compensation for the outlay in the increased security of her domain and the great saving achieved in the movements of her warships.

SOME OF THE ECONOMIC RESULTS WHICH ARE ANTICIPATED.—In addition to the political and strategical benefits gained, the United States expects two great economic results to follow the completion of the Canal: the development of her territory and the extension of her commerce. The changes will come very gradually, but each year will mark an advance towards the attainment of those results. Our ambition is honorable, creditable alike to the United States and to those nations of Europe which are represented in our citizenship.

SETTLEMENT AND DEVELOPMENT OF THE NATIONAL DOMAIN.—European settlement along the coast of the New World met with no great natural barriers when the time came for a westward extension. Following the establishment of independence, a tide of immigration set in which gained in force as the years advanced. As the Republic extended her territory westward there were always hardy and venturesome people from the Eastern States ready to occupy the new lands opened by that extension. Thus the Middle West, the Mississippi Valley, and the Northwest were settled, and all prospered. The wealth of farm and forest, of mine and factory, was theirs. Then appeared the golden vision of the Far West, and the rush to California began. That was in the late 40's. A few years more and the settlement of the great plains region and the inter-mountain region was well under way. The history of it all is wonderful, entrancing. The United States, to-day, is great in territory, in population, and in wealth. What is still better, she has in reserve vast areas which are undeveloped, and room for a population many times greater than that of the present day. Rapid and marvelous as has been the settlement of the Western States, the fact remains that the wealth of their resources has hardly been touched. It is, therefore, in the western third of her continental domain, that the United States is expected to show the earliest and greatest development as a result of the new facilities offered by the Panama Canal. In what way will this new waterway promote the development of the West? It will bring laborers to the fields and mines; it will bring nearer old markets and open new ones where an exchange of products for those of other lands may be profitably carried on. Some idea of the range of the coast products may be gained from the fact that Washington ranks as the fourth wheat-producing state in the Union, while Southern California is finding cotton a profitable crop. Timber, fruit, and

petroleum may be mentioned as the principal products of the group as a whole.

We may gain some conception of the extent of the unoccupied areas in this country by a comparison of the western states with some of the older and more densely populated eastern states. California, Oregon, and Washington have a combined area of 318,095 square miles and a total population (census of 1910) of only 4,192,304, or an average of only 14 people to the square mile. According to the same census, the density of population in New Jersey was 337 to the square mile and that of Rhode Island over 400. Given a density of population of only 300 to the square mile the three Pacific Coast States alone would hold a population of about 95,500,-000, or more people than were found in the entire continental domain of the United States in 1910. A comparison with some of the densely peopled countries of Europe is even more impressive. Germany and Italy have a combined area of 319,440 square miles, and a density of population which equals about 310 to the square mile. With the same density of population as that of these two European countries, the Pacific Coast States, with a slightly smaller area, would show a total population of 98,609,450 as against the total of 99,026,400 which those two countries would contain.

To the east of the Pacific Coast States is a group of eight commonwealths known as the Mountain States. These, roughly speaking, cover the space between the Sierra Nevada and Cascade ranges and the Rocky Mountains, with their eastern and western slopes. The area of these eight states is 859,125 square miles, or more than two and one-half times that of the Pacific States. Each of these states has its own advantages. Montana has great mineral wealth and enormous water power. It is one of the best developed, best equipped, and most promising of the group. Wyoming has a plentiful supply of coal and great agricultural possibilities. Colorado is also splendidly endowed with mineral wealth, and possesses a climate renowned for its health-giving properties. Idaho and Utah have made astonishing growth and are rich in natural advantages. Farther south are Arizona and New Mexico, great in mineral and agricultural advantages. Nevada, although slow in growth, is steadily gaining and doubtless has much hidden mineral wealth.

Millions of acres of unappropriated and unoccupied lands are available for settlers in the Pacific Coast and Mountain States. Besides the government-owned lands, are others, now owned by railroad companies, land companies, irrigation companies, and other corporations which are ready to dispose of their holdings on terms that even a poor man can meet. There are, of course, considerable areas in the mountainous sections of all of these states, which will never be fit for occupation; but, after making wide allowance for the waste places, there will still be room in the productive areas, and for many generations to come, for the overflow from foreign countries and the natural increase of population.

After all, the places which we now designate as waste may prove a source of future wealth exceeding that promised by the fertile areas of the present. To realize that such an event is possible, it is but necessary to reflect that the attractions of mountain scenery are to-day the chief source of wealth to Switzerland, and, in a lesser degree, to other countries of Europe. Here are to be found mountains, plains, thermal springs, geysers, forests of giant trees, prehistoric remains, and natural features which will attract pleasure seekers and students through the coming ages.

The demand for manual labor of all kinds is very great in all parts of the United States and is especially pressing on the Pacific Coast. Emigration from Europe to this section has hitherto been held in check by the distance and expense of the journey. Those obstacles are now removed by the shortened all-water voyage, and the people of the West are scarcely more enthusiastic over the prospect of a growth in trade and in manufactures than they are over the possibility of having their necessities relieved by a new supply of white laborers from abroad who will be content with reasonable wages. The call for workers will be answered. Immigration agents have for some time been offering a continuous ocean passage from southern European ports to the Pacific Coast at extremely low rates. The slight excess in cost over that to Atlantic ports is more than offset by the superior attraction of the new country and the countless opportunities there offered. Immigration will stream out from the Pacific ports over the Coast States and to the Mountain States beyond, where await the home-seekers fertile soils, waterpowers now running to waste, virgin forests of valuable timber, and mines whose wealth is still untouched. With the laborers will go the men of capital who will find openings for investments of many kinds—works of irrigation, agriculture, stock-raising, fruit-growing, mining, lumbering, railroad building, and ship building. The great northwestern territory of Alaska, full of promise for the capitalist and the laborer, will share in the benefits of a shorter route to eastern markets and greater facilities for immigration.

Occupation of territory and development of resources in the United States have moved from East to West. The time is coming when a great wave of immigration and industrial growth will

View from Balboa

move from the Pacific Coast eastward, over the mountains to the plains beyond. As long ago as 1852 Mr. W. H. Seward prophesied that the time would come when Europe would sink into unimportance, and that the Pacific Ocean would become the chief area of events in the world's future. That prophesy has not yet been fulfilled; but its ultimate realization is certain. The West is the Land of Promise for generations to come.

The Panama Canal means other canals, which in coming years will occupy our engineers and give employment to thousands of laborers. They will open a vast net work of waterways over the land, linking the lakes and rivers of the interior. This will give to the inland sections of the country an enlarged commerce at lower rates, thus providing for a greater industrial growth than now seems possible. There are great opportunities for development along the Mississippi and its numerous tributaries.

The industries of the states bordering the Gulf of Mexico and the South Atlantic states will benefit greatly by the opening of new markets on the Pacific. Alabama coal will be extensively used by the coaling stations on the Isthmus. The iron industries of the Birmingham district will find new markets for their output in South American and Oriental countries. The same conditions will hold in regard to raw cotton, cotton textiles, and the products of the lumber and wood industries.

The northeastern section of the United States comprises the oldest states in point of settlement. Their natural resources have been known and utilized for many years. The areas which have not been occupied and developed are few in number and of small extent. Agriculture is still and will continue to be the chief occupation of the people, but this is also a region of great coal-mining and oil industries, and it leads all other sections of the Union in the variety and importance of its manufactures. It is the commercial center of the country. Its foreign and domestic trade exceeds that of all other sections combined. The ship-building plants of the New England and Middle Atlantic States are experiencing an activity such as they have never known before. The advantage, which the new short route to the Pacific gives to the manufacturing and commerical interests of these states, in competing for trade with Western Europe in the markets of South America and Oceania, can hardly be overestimated.

EFFECT OF THE CANAL UPON OUR DOMESTIC COMMERCE.—There can be no doubt whatever of the effect which the Canal is to exert upon the domestic commerce of the country. It shortens by nearly 8,000 miles the water route between the ports of the Atlantic and Pacific coasts, and removes the necessity for a transshipment by rail across the isthmuses of Panama and Tehuantepec. The products of the Pacific and Atlantic states can henceforth be exchanged at such saving in time and transportation charges that the general cost of living in both sections must be affected in favor of the consumer.

EFFECT OF THE CANAL UPON OUR FOREIGN COMMERCE.—It is not strange that there should still be some doubt in the minds of Europeans as to the effects of the Panama Canal upon the foreign commerce of the United States. The Merchant Marine of this country has suffered in the past from legislative restrictions. These have so crippled its operations that the commercial nations of Europe have come to regard it as a negligible factor in the competition for international trade. The American government is removing those restrictions and a great change is anticipated.

Reference to the map will show how far-reaching will be the effects of the Canal upon the foreign commerce of the United States. Taking New York as a representative port on the Atlantic seaboard, the Pacific regions (over and above those of the two Americas) which will be brought nearer by the Canal may be demonstrated as follows:—If a line be drawn from Hong-Kong through British North Borneo, and on through Western Australia to Cape Leeuwin, the south-western extremity of the island continent, it will mark the western limit of the area affected. It will be found that the whole of Japan, the majority of the ports of China, the Philippine Islands, all of Australia, except the western coast regions, New Zealand, and most of the islands of Oceania are included. Broadly speaking, this means that China will be about equi-distant by sea from both New York and London, while the important ports of south-eastern Australia, the ports of New Zealand, and those of Japan, will be nearer to New York via Panama than to London by any route, by distances of from 1000 to 3700 miles. It is obvious that the southeastern and Gulf ports of the country are even more advantageously placed for making use of the Canal than is New York. We have before noted the probable effects which the opening of what to them is a new gateway to the East will have upon our Pacific ports. From San Francisco all the ports of Europe and the whole of the African coast, with the exception of a strip on the Indian Ocean extending from the Strait of Bab-el-Mandeb to Delagoa Bay, will become more accessible.

OUR TRADE WITH THE COUNTRIES OF LATIN-AMERICA.—Great reductions have been made in the steaming distances between the ports of our Atlantic seaboard and those of Latin-America bordering on the Pacific; the same is true of our

Pacific seaboard in its relation to the ports of the West Indies and of the countries of Latin-America bordering on the Caribbean Sea and the Atlantic Ocean. The Pacific countries of Latin-America have a seacoast of more than 8,000 miles and occupy an area of about 2,500,000 square miles. They have a population estimated at 37,000,000. Those countries are bound to grow in wealth and population. Their foreign trade more than doubled in the fifteen years previous to the completion of the Canal, reaching a total of approximately $800,000,000. In this trade the United States had a share of but 37 per cent. By far the greater part has gone to Great Britain and Germany. Those countries will now feel the keen competition of the United States. Mr. John Barrett of the Pan-American Union, who is well fitted to judge of the effects of the Canal, prophesies an immense increase in the total trade of those countries and a more than corresponding increase in the share which will fall to the United States. In his book on the Panama Canal he says: "The opening of the Panama Canal will mean that the Pacific Coast of Latin America will want in increasing quantities our iron and steel manufactures; our steam and electric railway materials; our structural iron and steel; our sewing machines, typewriters, and cash registers; our cotton cloth; our wood and lumber; our flour, butter, cheese, and lard; our agricultural implements; boots and shoes; jewelry; furniture and hardware; drugs and medicines; automobiles; coal; illuminating and crude oils; news print paper; binder twine; clothing; books and maps; and numerous other articles demanded by a developing country and population." The ships which bear our products to the south will bring back for our use sugar, coffee, rubber, bananas, coconuts, nitrates, hides and skins, chinchilla, henequen, sisal, wool, ivory-nuts, tin, copper, quinine, tobacco leaf, and beef.

The development and new life that will come to the West Indies, and to the Latin-American countries bordering on the Gulf of Mexico and the Caribbean Sea, as a result of the new thoroughfare, will bring an increase of trade to both the Pacific and Atlantic ports of the United States. The West Indies are now situated on one of the great highways of the world's trade and those islands will grow both in commercial and strategical importance. Their improved situation is bound to attract both population and capital.

OUR COMMERCE WITH THE FAR EAST.—The United States already enjoys a large and growing commerce with the Far East, whose awakening assumes a new and added interest when considered in connection with the completion of the Panama Canal. China has enormous deposits of coal and iron. The time is coming when she will not lag behind Japan which has become one of the trading nations of the world. The products of the iron mines of China, developed by cheap labor, have entered our markets and they will continue to come in ever increasing volume. Chinese labor is stated to be within ten per cent. of an equality with the best white labor, and its cost is much cheaper. With settled political conditions the Chinese will undoubtedly develop manufactures and the time may come when not only Chinese pig-iron, but Chinese structural steel will invade American markets. This will be beneficial if the Chinese products are brought in American vessels and not in those of Japan.

AN ERA OF AMERICAN SHIP-BUILDING.—American coastwise shipping is already a vast fleet, greater in tonnage and efficiency than ever before. The era of American ship-building has but just begun. American preparation for extensive use of the Panama Canal is ambitious and far reaching. American ship-builders were not idle while the men on the Isthmus were digging the big ditch; they were busy in the construction of ocean-going vessels to be employed upon the isthmian route. The growth of both our domestic and our foreign commerce is as sure as the existence of the Canal itself.

EFFECTS OF THE PANAMA CANAL UPON OTHER COUNTRIES.—Much of that which has been said as to the benefits which will come as a result of the shortened water route between the Atlantic and Pacific seaboards of the United States applies also to Canada. Vancouver of all the ports in the British Empire is the one most likely to be favorably affected by the completion of the Panama Canal. That this is realized is shown by the preparations there being made for increased traffic. Farther north is the fine deep-water harbor of Prince Rupert, the terminus of the Grand Trunk Pacific Company's new transcontinental line, which will become of great importance in the near future. Victoria, on Vancouver Island, and New Westminster on the mainland are fully awake to the potentialities of the change in transport conditions.

The province of British Columbia, of which these ports are the outlets, is one of great and varied resources, which up to the present time have been but slightly developed, although a growing commercial intercourse has been established with China, Japan, Australia, and New Zealand. The water route from Liverpool to Vancouver is shortened 5,921 miles by the Canal. For a vessel steaming 300 knots per day this means a saving of twenty days on each voyage.

The Panama Canal route offers a new outlet for the grain exports of the Canadian Northwest, which, hitherto, have been mainly carried

by rail to eastern seaports. An element in
favor of the Canal route is the ice-free condition
of the harbor at Vancouver throughout the
year, while the ports of the St. Lawrence, though
in a much lower latitude, are closed to commerce
for several months in the year.

The Canal will be a boon to the countries of
Western Europe, whose products destined for the
Pacific States and the Canadian Northwest will
no longer bear the burden of a long rail haul from
Atlantic ports, but will be distributed by rail-
roads from the Pacific ports.

The significance of the new waterway as
affecting European interests in their trade with
the countries of Latin-America which border
the Pacific may have been exaggerated. There
is a point far to the south on the Chilean coast,
which is equidistant by the Panama route and
that by the Straits of Majellan. To reach ports
at no great distance from this line of demarca-
tion, the saving in time and expense would be
so small as to fail to justify the use of the Canal,
with its attendant tolls. From Valparaiso
northwards there is a steady gain in favor of
the Canal route. Thus for a vessel steaming
300 knots per day there will be a saving in time,
which increases from about seven days at Valparaiso to twenty days at Acapulco, Mexico.
Mileage run is an important factor, but the gain
which will be made in distance will to some
extent be offset by the tolls, and by the keener
competition to which European manufacturers
and ship owners will now be subjected.

Excluding the countries of the two Americas
which have Pacific seaboards, there is no region
of commercial importance brought nearer to the
trading nations of Western Europe by the
Panama Canal, with the exception of New Zealand (to which the distance is shortened by only
900 miles), the Hawaiian and Fijiian islands,
and the scattered archipelagoes of Polynesia and
Micronesia. Hawaiian trade is almost wholly
in the hands of the United States and is likely
to remain so. In New Zealand, in Australasia,
in China and Japan, the traders of Europe have
now to meet the competition of those of the
United States.

It is possible that the Panama Canal may
become an alternative route from Europe to the
Far East, even where the distance is greater
than by the Suez route. This will be the case
if the trade to be gained from the ports of
the West Indies, and from the ports of call
on the Pacific seaboard of the Americas, will
make the longer voyage profitable. Another

element in favor of the new route is that it
is available for larger vessels, for whereas the
maximum draught at present allowed through
the Suez Canal is only 28 feet, the Panama
Canal has a minimum depth of over 41 feet.

XI. THE PANAMA RAILROAD

The new or relocated line of the Panama
railroad is 47.11 miles long, or 739 feet longer
than the old line. From Colon to Mindi, 4.17
miles, and from Corozal to Panama, 2.83 miles,
the old line is used, but the remaining 40 miles
is new road. From Mindi to Gatun, the railroad
runs, in general, parallel to the Canal, and as-
cends from a few feet above tidewater elevation
to 95 feet above. At Gatun, the road leaves the
vicinity of the Canal and turns east along Gatun
Ridge to a point about 4½ miles from the center
line of the Canal, where it turns southward
again and crosses the low Gatun Valley to Monte
Lirio, from which point it skirts the east shore of
Gatun Lake to the beginning of the Culebra Cut
at Bas Obispo. In the Gatun Valley section there
are several immense embankments necessary to
place the line above the lake level. Likewise,
near the north end of Culebra Cut, where the
line was located as so to furnish waste dumps
for spoil from the Canal, there are several very
heavy embankments. Originally it was intended
to carry the railroad through Culebra Cut on a
40-foot berm along the east side, 10 feet above
the water level, but the numerous slides made
this plan impracticable and a line was constructed, on a high level around the Cut, known
locally as the Gold Hill Line. Leaving the berm
of the Canal at Bas Obispo, the Gold Hill Line
cuts through a ridge of solid rock, and gradually
works into the foothills, reaching a distance from
the center line of the Canal of two miles opposite
Culebra; thence it runs down the Pedro Miguel
Valley to Paraiso, where it is only 800 feet from
the center line of the Canal. From the south
end of Culebra Cut at Paraiso, the railroad runs
practically parallel with the Canal to Panama.
Where the railroad crosses the Gatun River,
near Monte Lirio, a steel girder bridge has been
erected, the center span of which is in the form
of a lift, to permit access to the upper arm of
Gatun Lake. The Chagres River at Gamboa is
crossed on a steel-girder bridge, one-quarter mile
long, with one 200-foot through truss channel
span. Numerous other rivers and small streams
are crossed on reinforced concrete culverts.
Near Miraflores, a tunnel 736 feet long has been
built through a hill.

TABLE A.—Number of days saved, for vessels of different speeds, by the Panama Canal route between the Atlantic-Gulf ports of the United States and Pacific ports, American and foreign.

To—	New York, for vessels of—					Charleston, for vessels of—					Port Tampa, for vessels of—					New Orleans, for vessels of—					Galveston, for vessels of—				
	9 knots	10 knots	12 knots	14 knots	16 knots	9 knots	10 knots	12 knots	14 knots	16 knots	9 knots	10 knots	12 knots	14 knots	16 knots	9 knots	10 knots	12 knots	14 knots	16 knots	9 knots	10 knots	12 knots	14 knots	16 knots
Sitka	35.9	32.3	26.8	22.9	20.0	37.6	33.8	28.1	24.0	21.0	40.0	35.9	29.9	25.5	22.9	40.2	36.4	30.2	25.9	22.2	40.8	36.8	30.8	26.7	22.7
Port Townsend	35.9	32.3	26.8	22.9	20.0	37.6	33.8	28.1	24.0	21.0	40.0	35.9	29.9	25.5	22.9	40.2	36.4	30.2	25.9	22.2	40.8	36.8	30.8	26.7	22.7
Portland, Oregon	35.9	32.3	26.8	22.9	20.0	37.6	33.8	28.1	24.0	21.0	40.0	35.9	29.9	25.5	22.9	40.2	36.4	30.2	25.9	22.2	40.8	36.8	30.8	26.7	22.7
San Francisco	35.9	32.3	26.8	22.9	20.0	37.6	33.8	28.1	24.0	21.0	40.0	35.9	29.9	25.5	22.9	40.2	36.4	30.2	25.9	22.2	40.8	36.8	30.8	26.7	22.7
San Diego	36.0	32.3	26.8	22.9	20.0	37.6	33.8	28.1	24.0	21.0	40.0	35.9	29.9	25.5	22.9	40.2	36.4	30.2	25.9	22.2	40.8	36.8	30.8	26.7	22.8
Acapulco	36.9	33.2	27.5	23.5	20.5	38.5	34.6	28.8	24.6	21.5	40.9	36.8	30.6	26.1	22.8	41.1	37.3	31.0	26.5	23.0	41.8	37.6	31.2	26.9	23.3
San Jose de Guatemala	38.0	34.2	28.4	24.3	21.2	39.5	35.4	29.4	25.2	22.0	41.9	37.3	31.0	26.5	23.0	42.1	37.6	31.1	26.5	23.0	42.8	38.1	31.7	27.1	24.0
Honolulu	30.1	27.0	22.4	19.1	16.7	31.7	28.5	23.6	20.2	17.6	34.1	30.7	25.2	21.7	19.0	34.3	30.4	25.3	21.9	19.0	35.0	31.0	26.0	22.2	19.5
Guayaquil	33.7	30.3	25.2	21.5	18.7	35.4	31.8	26.4	22.6	19.7	37.8	34.0	28.2	24.1	21.1	38.4	34.4	28.5	24.3	21.2	39.5	35.0	29.0	24.9	21.6
Callao	28.4	25.5	21.2	18.1	15.7	30.1	27.0	22.4	19.2	16.7	32.5	29.2	24.2	20.7	18.1	33.0	29.4	24.4	20.9	18.1	34.0	30.0	24.9	21.3	18.5
Iquique	23.3	20.9	17.3	14.8	12.9	25.0	22.4	18.6	15.9	13.8	27.4	24.5	20.4	17.4	15.2	27.9	25.0	20.7	17.7	15.4	28.9	25.6	21.3	18.1	15.6
Valparaiso	16.8	15.1	12.5	10.6	9.2	18.5	16.6	13.8	11.7	10.2	20.9	18.7	15.5	13.2	11.5	21.6	19.2	16.0	13.6	11.9	22.5	20.0	16.6	14.1	12.0
Coronel	14.7	13.2	10.9	9.3	8.1	16.4	14.7	12.2	10.4	9.0	18.8	16.9	14.0	11.9	10.4	19.4	17.4	14.4	12.3	10.9	20.4	18.1	15.0	12.8	10.9
Yokohama	16.9	15.2	12.6	10.7	9.3	18.5	16.6	13.7	11.7	10.1	20.9	18.8	15.6	13.3	11.5	23.3	19.3	16.0	14.0	14.0	26.3	23.2	19.6	16.5	14.5
Shanghai	8.1	7.3	6.0	5.1		9.7	8.8	7.5	6.5	6.5	12.1	10.8	9.4	8.1	8.1	12.3	11.5	10.8	9.2	9.2	13.0	11.6	11.1	9.9	9.6
Hongkong						3.1	2.8	2.2	1.9	1.5	5.7	5.1	5.8	7.0	4.8	4.2	8.4	7.6	6.2	5.2	4.5	8.7	7.8	6.4	5.4	4.7
Manila						3.4	3.0	2.4	2.0	1.7	5.9	7.2	5.9	7.2	5.0	4.3	8.6	7.7	6.4	5.4	4.7	9.0	8.0	6.6	5.6	4.8
Adelaide	7.5	6.7	5.6	4.6	4.0	10.4	9.3	7.7	6.5	5.6	12.5	10.4	8.8	7.7	6.5	5.6	8.0	14.3	13.3	11.0	9.2	8.0	14.9	13.3	11.0	9.4	8.1
Melbourne	12.3	11.0	9.1	7.7	6.7	15.1	13.5	11.2	9.5	8.3	18.8	16.9	14.0	11.9	10.4	18.3	17.8	16.9	14.0	11.9	10.3	17.6	14.6	12.4	10.8
Sydney	17.7	15.8	13.1	11.2	9.7	20.5	18.4	15.3	13.0	11.3	24.6	22.1	18.4	15.7	13.7	25.0	22.2	18.4	15.7	13.7	25.0	22.1	18.4	15.9	13.8
Wellington	11.0	9.9	8.1	6.9	6.0	12.7	11.4	9.4	8.0	6.9	15.1	13.5	11.2	9.5	8.3	15.6	14.0	11.6	9.9	9.5	16.6	14.3	11.8	10.5	8.7

TABLE B.—Number of days saved, for vessels of different speeds, by the Panama Canal route between European ports and ports of Pacific America and of New Zealand.

To—	Liverpool, for vessels of—					Hamburg, for vessels of—					Antwerp, for vessels of—					Bordeaux, for vessels of—					Gibraltar, for vessels of—				
	9 knots	10 knots	12 knots	14 knots	16 knots	9 knots	10 knots	12 knots	14 knots	16 knots	9 knots	10 knots	12 knots	14 knots	16 knots	9 knots	10 knots	12 knots	14 knots	16 knots	9 knots	10 knots	12 knots	14 knots	16 knots
Sitka	25.7	23.1	19.1	16.3	14.2	25.1	22.5	18.7	15.7	13.9	25.9	23.1	18.5	15.7	13.9	25.4	22.4	18.1	15.5	13.5	22.4	20.1	16.7	14.2	12.3
Port Townsend	25.7	23.1	19.1	16.3	14.2	25.1	22.5	18.7	15.7	13.9	25.9	23.1	18.5	15.7	13.9	25.4	22.4	18.1	15.5	13.5	22.4	20.1	16.7	14.2	12.3
Portland, Oregon	25.7	23.1	19.1	16.3	14.2	25.1	22.5	18.7	15.7	13.9	25.9	23.1	18.5	15.7	13.9	25.4	22.4	18.1	15.5	13.5	22.4	20.1	16.7	14.2	12.3
San Francisco	25.7	23.1	19.1	16.3	14.2	25.1	22.5	18.7	15.7	13.9	25.9	23.1	18.5	15.7	13.9	25.4	22.4	18.1	15.5	13.5	22.4	20.1	16.7	14.2	12.3
San Diego	25.7	23.1	19.1	16.4	14.3	25.1	22.5	18.7	15.7	13.9	25.9	23.1	18.5	15.7	13.9	25.4	22.4	18.1	15.5	13.5	22.4	20.1	16.7	14.2	12.4
Acapulco	26.7	23.9	19.9	17.0	14.4	26.0	23.4	19.4	16.6	14.6	26.0	23.4	19.4	16.6	14.5	26.2	22.8	18.9	16.1	14.1	23.5	21.0	17.0	14.4	12.9
San Jose de Guatemala	27.8	25.0	20.8	17.7	15.4	27.2	24.4	20.3	17.2	15.0	27.2	24.4	20.3	17.3	15.1	25.8	23.8	19.8	16.8	14.7	24.7	22.0	18.3	15.3	13.0
Honolulu	19.8	17.8	14.8	12.6	10.9	19.2	17.2	14.3	12.2	10.6	19.2	17.2	14.3	12.2	10.6	18.5	16.6	13.7	11.7	10.2	16.5	14.8	12.3	10.4	9.1
Guayaquil	23.5	21.1	17.5	14.9	13.0	22.9	20.6	17.1	14.6	12.7	22.9	20.6	17.1	14.6	12.7	22.2	19.9	16.5	14.1	12.2	20.2	18.2	15.0	12.8	11.1
Callao	18.2	16.3	13.5	11.5	10.0	17.6	15.8	13.1	11.1	9.7	17.6	15.8	13.1	11.1	9.7	16.9	15.2	12.5	10.6	9.2	14.9	13.3	11.0	9.4	8.1
Iquique	13.1	11.7	9.7	8.2	7.1	12.4	11.1	9.2	7.8	6.8	12.4	11.1	9.2	7.8	6.8	11.7	10.5	8.7	7.3	6.1	9.7	8.7	7.2	6.1	—
Valparaiso	6.6	5.9	4.8	4.1	3.5	6.0	5.3	4.3	3.5	3.1	6.0	5.3	4.3	3.5	3.1	5.3	4.7	3.8	3.2	2.7	3.8	3.3	2.9	3.3	1.6
Coronel	4.5	4.0	3.3	2.7	2.3	3.9	3.4	2.8	2.3	1.9	3.9	3.4	2.8	2.3	1.9	3.1	2.8	2.3	1.9	1.5	1.8	1.5	1.2	1.0	0.6
Wellington	6.7	6.0	4.9	4.2	3.5	6.0	5.3	4.4	3.7	3.2	6.0	5.3	4.4	3.7	3.2	5.3	4.7	3.8	3.2	2.7	3.8	3.2	2.7	1.7	1.5

PRINCIPAL COUNTRIES OF THE WORLD

COUNTRY	FORM OF GOVERNMENT	CAPITAL	PRESENT HEAD	ACCEDED	SQUARE MILES	POPULA-TION
Abyssinia	Absolute Monarchy	Addis Abeba	Lij Yasu	May 15, '11	432,432	8,000,000
Afghanistan	Absolute Monarchy	Kabul	Habibullah Khan	Oct. 3, '01	250,000	5,900,000
Albania	Limited Monarchy				5,000	2,000,000
Andorra	Republic				175	5,231
Argentina	Republic	Buenos Aires	Roque Saenz Pena	Oct. 12, '10	1,135,119	7,500,000
Australia	Commonwealth	Yass Canberra	Lord Denman	July, '11	3,063,041	4,805,005
Austria-Hungary	Limited Monarchy	Vienna, Budapest	Francis Joseph I	Dec. 2, '48	261,107	51,000,810
Belgium	Limited Monarchy	Brussels	Albert I	Dec. 23, '09	11,373	7,423,784
Belgian Kongo		Boma			909,654	15,500,000
Bhutan	Limited Monarchy	Punakha	Ugyen Wangchuk	Aug. '07	20,000	250,000
Bolivia	Republic	Sucre ; La Paz	Eliodoro Villazon	Aug. 6, '09	515,156	2,049,083
Brasil	Republic	Rio de Janeiro	Hermes da Fonseca	Nov 15, '10	3,270,000	21,500,000
British Empire	Limited Monarchy	London	George V	May 6, '10	13,123,712	434,686,650
Bulgaria	Limited Monarchy	Sofia	Ferdinand	Aug. 11, '87	38,080	5,000,000
Canada	Confederation	Ottawa	Duke of Connaught	Oct 19, '11	3,729,665	7,204,843
Chile	Republic	Santiago	Ramon Barros Luco	Dec. 21, '10	292,580	3,548,394
China	Republic	Peking	Yuan Shi-Kai	Feb 14, '12	4,277,170	439,214,000
Colombia	Republic	Bogota	Carlos E. Restrepo	Aug 15, '10	438,436	5,051,450
Costa Rica	Republic	San Jose	Ricardo Jimenez	May 8, '10	23,000	388,266
Cuba	Republic	Havana	Mario Gracia Menocal	May 20, '13	45,883	2,130,112
Denmark	Limited Monarchy	Copenhagen	Christian X	May 16, '12	15,592	2,757,076
Denmark, Colonies of					87,174	143,143
Dominican Republic	Republic	Santo Domingo	Adolfo Nouel	April 14, '13	19,332	708,000
Ecuador	Republic	Quito	Leonidas Plaza	Sept. 1, '12	116,000	1,500,000
Egypt a	Limited Monarchy	Cairo	Abbas Hilmi	Jan. 8, '92	400,000	11,287,350
France	Republic	Paris	Raymond Poincare	Feb. 17, '13	207,054	39,601,500
France, Colonies of					4,165,815	54,540,700
Germany	Limited Monarchy	Berlin	William II	June 15, '88	208,780	64,925,993
Germany, Colonies of					1,027,820	14,546,000
Great Britain and Ireland	Limited Monarchy	London	George V	May 6, '10	121,391	45,211,838
England		London			58,826	36,070,492
Ireland		Dublin			32,360	4,381,951
Scotland		Edinburgh			30,405	4,759,445
Greece	Limited Monarchy	Athens	Constantine XII	March 18, '13	25,064	4,000,000
Guatemala	Republic	New Guatemala	Manuel E. Cabrera	March 15, '11	48,290	1,992,000
Haiti	Republic	Port au Prince	Michel Oreste	May 4, '13	10,204	2,029,700
Honduras	Republic	Tegucigalpa	Manuel Bonilla	Nov. 3, '11	46,400	553,446
India	Limited Monarchy	Delhi	Charles Hardinge	Nov. 20, '10	1,773,088	316,019,546
Italy	Limited Monarchy	Rome	Victor Emanuele III	July 29, '00	110,660	36,588,683
Italy, Colonies of					601,020	1,767,000
Japan	Limited Monarchy	Tokyo	Yoshihito Harunomiya	July 30, '12	147,655	50,296,279
Japan, Dependencies of					95,696	17,090,000
Liberia	Republic	Monrovia	Daniel Howard	Jan. 1, '13	40,000	2,130,000
Liechtenstein	Monarchy	Vaduz	John II	Nov. 12, '58	68	9,854
Luxemburg	Limited Monarchy	Luxemburg	Marie Adelaide	Feb. 26, '12	998	259,899
Mexico	Republic	Mexico	Victoriano Huerta	Feb. 18, '13	767,274	15,063,207
Monaco	Limited Monarchy	Monaco	Albert Honore Charles	Sept. 10, '89	8	19,121
Montenegro	Limited Monarchy	Centinje	Nicholas I	Aug. 14, '60	3,630	500,000
Nepal	Absolute Monarchy	Khatmandu	Jang	Dec. 11, '11	54,000	5,000,000
Netherlands	Limited Monarchy	The Hague	Wilhelmina, Queen	Nov. 23, '90	18,843	5,980,000
Netherlands, Colonies of					533,473	38,000,000
New Zealand	Commonwealth	Wellington	Lord Liverpool	Aug. 5, '12	104,751	1,041,046
Nicaragua	Republic	Managua	Adolfo Diaz	Dec. 31, '12	49,200	500,000
Norway	Limited Monarchy	Christiania	Haakon VII	Nov. 18, '05	124,130	2,392,698
Oman	Absolute Monarchy	Maskat	Seyyid Feysil	June 4, '88	82,000	500,000
Panama	Republic	Panama	Belisario Porras	Oct. 1, '12	31,970	427,000
Paraguay	Republic	Asuncion	Eduardo Schaerer	Aug. 15, '12	171,304	752,000
Persia	Limited Monarchy	Teheran	Abu'l Kassim Khan	Mar. 6, '11	628,000	9,000,000
Peru	Republic	Lima	Guillermo Billinghurst	Sept. 24, '12	695,733	4,975,000
Portugal	Republic	Lisbon	Manuel de Arriaga	Aug. 24, '11	35,490	5,975,000
Portugal, Colonies of					903,306	9,675,000
Roumania	Limited Monarchy	Bukharest	Carol I	March 26, '81	50,720	6,998,000
Russia	Absolute Monarchy	St. Petersburg	Nicholas II	Nov. 1, '94	8,647,657	168,000,000
Russia, Dependencies of					107,000	1,050,000
Salvador	Republic	San Salvador	Carlos Melendes	March '13	7,225	1,153,004
San Marino	Republic	San Marino			38	11,450
Servia	Limited Monarchy	Belgrade	Peter I	June 15, '03	18,650	4,000,000
Siam	Absolute Monarchy	Bangkok	Chowfa Maha Vajiravudh	Oct. 23, '10	195,000	6,320,000
Spain	Limited Monarchy	Madrid	Alphonso XIII	May 17, '86	194,783	19,588,588
Spain, Colonies of					84,394	276,000
Sweden	Limited Monarchy	Stockholm	Gustaf V	Dec. 8, '07	172,876	5,521,943
Switzerland	Republic	Bern	Edouard Muller	Jan. 1, '13	15,976	3,765,123
Turkey & tributary states b	Limited Monarchy	Constantinople	Mohammed V	April 27, '09	682,960	17,425,000
United South Africa	Confederation	Pretoria, Cape T'n	Viscount Gladstone	May 30, '10	473,184	5,973,394
United States, Continental	Republic	Washington	Woodrow Wilson	March 4, '13	3,026,789	91,972,266
U. S., Dependencies of					716,517	9,143,322
Uruguay	Republic	Montevideo	Jose Battle y Ordones	March 1, '11	72,210	1,140,799
Venezuela	Republic	Caracas	Juan Vicente Gomez	June, '10	393,976	2,713,703

a Under suzerainty of Turkey, but actual administration controlled by Great Britain.   b Exclusive of Egypt and European Turkey.

THE WORLD

SHAPE AND SIZE OF THE EARTH : The earth is very nearly an oblate spheroid, whose shorter axis coincides with its axis of rotation passing through the two poles.

According to Colonel Clarke's calculations, its major axis measures 41,852,124 feet, or 7,926.6 statute miles ; its minor axis 41,710,342 feet, or 7,900 5 statute miles ; its circumference along the Equator measures 24,902 statute miles, or 21,600 geographical miles ; its total area 196,940,400 statute square miles ; and its volume 259,880 million cubic miles.

Of this total area, 8.4 p. c. (16,464,700 sq. m.) are within the Arctic and Antarctic regions ; 51.6 p c. (102,344 554 sq. m.) within the two temperate regions ; 40 p. c. (78,231,046 sq. m.) within the Tropics.

THE LAND : The land covers 54,907,420 sq. m., on the assumption that 250,000 sq. m of land remain to be discovered within the Arctic Regions, and that the supposed Antarctic continent, or "Antarctic," has an extent of 2,500,000 sq. m. within the Antarctic Circle.

THE OCEANS : The oceans, including the inland seas connected with them, cover 142,132,980 sq. m., or 72 p. c. of the total surface of the earth. There are 2.59 sq. m. of ocean to every sq. m. of land. This area is distributed over the different areas as follows : Arctic Ocean, including Hudson Bay, 5,725,000 sq. m.; Atlantic Ocean, 34,301,600 sq. m.; Indian Ocean, 28,615,600 sq. m. ; Pacific Ocean, 67,699,630 sq. m. ; Antarctic Ocean, 5,731,350 sq. m.

HEIGHTS AND DEPTHS : The mean height of the land has been estimated at 2,440 feet ; the mean depth of the sea 11,470 feet (Karstens). The highest mountain (Mt. Everest) rises to a height of 29,000 feet ; the greatest depth of the ocean as yet discovered (in the Pacific, between Guam and Midway) is 31,614 feet. If the whole of the solid crust of the earth were to be levelled, so as to form a spheroid, it would still be covered by an ocean of a uniform depth of 8,000 feet (Prof. Penck).

TABLE B.—Number of days saved, for vessels of different speeds, by the Panama Canal route between European ports and ports of Pacific America and of New Zealand.

To—	From—																								
	Liverpool, for vessels of—					Hamburg, for vessels of—					Antwerp, for vessels of—					Bordeaux, for vessels of—					Gibraltar, for vessels of—				
	9 knots.	10 knots.	12 knots.	14 knots.	16 knots.	9 knots.	10 knots.	12 knots.	14 knots.	16 knots.	9 knots.	10 knots.	12 knots.	14 knots.	16 knots.	9 knots.	10 knots.	12 knots.	14 knots.	16 knots.	9 knots.	10 knots.	12 knots.	14 knots.	16 knots.
Sitka	25.7	23.1	19.1	16.3	14.2	25.1	22.5	18.7	15.5	13.9	25.1	22.5	18.7	15.9	13.9	24.4	21.9	18.1	15.5	13.5	22.4	20.1	16.7	14.2	12.3
Port Townsend	25.7	23.1	19.1	16.3	14.2	25.1	22.5	18.7	15.5	13.9	25.1	22.5	18.7	15.9	13.9	24.4	21.9	18.1	15.5	13.5	22.4	20.1	16.7	14.2	12.3
Portland, Oregon	25.7	23.1	19.1	16.3	14.2	25.1	22.5	18.7	15.5	13.9	25.1	22.5	18.7	15.9	13.9	24.4	21.9	18.1	15.5	13.5	22.4	20.1	16.7	14.2	12.3
San Francisco	25.7	23.1	19.1	16.3	14.2	25.1	22.5	18.7	15.5	13.9	25.1	22.5	18.7	15.9	13.9	24.4	21.9	18.1	15.5	13.5	22.4	20.1	16.7	14.2	12.3
San Diego	25.7	23.1	19.2	16.4	14.3	25.1	22.5	18.7	15.5	13.9	25.1	22.5	18.7	15.9	13.9	24.5	21.9	18.2	15.5	13.5	22.4	20.1	16.7	14.2	12.4
Acapulco	26.7	23.9	19.9	17.0	14.8	26.0	23.4	19.4	16.6	14.4	26.0	23.4	19.6	16.6	14.4	25.3	22.8	18.9	16.1	14.1	23.1	21.0	17.4	14.8	12.9
San Jose de Guatemala	27.8	25.0	20.8	17.7	15.4	27.2	24.0	20.3	17.3	15.1	27.3	24.4	20.3	17.3	15.1	26.6	23.8	19.8	16.9	14.8	24.5	22.0	18.3	15.6	13.6
Honolulu	19.8	17.8	14.8	12.6	10.9	19.2	17.7	14.3	12.2	10.6	19.2	17.4	14.3	12.2	10.6	18.5	16.6	13.7	11.7	10.2	16.2	14.5	12.0	10.4	9.1
Guayaquil	23.5	21.1	17.5	14.9	13.0	22.9	20.6	17.1	14.6	12.7	22.9	20.6	17.1	14.6	12.7	22.2	20.9	16.5	14.1	12.2	20.2	18.2	15.0	12.8	11.1
Callao	18.2	16.3	13.5	11.5	10.0	17.6	15.8	13.1	11.1	9.7	17.6	15.8	13.1	11.5	9.7	16.8	15.1	12.5	10.6	9.2	14.9	13.3	11.0	9.4	8.1
Iquique	13.1	11.7	9.7	8.2	7.1	12.4	11.1	9.2	7.8	6.8	12.4	11.1	9.2	7.8	6.8	11.7	10.5	8.7	7.3	6.3	9.7	8.7	7.2	6.1	5.2
Valparaiso	6.6	5.9	4.8	4.1	3.5	6.0	5.3	4.3	3.5	3.1	6.0	5.3	4.3	3.5	3.1	5.3	4.7	3.8	3.2	2.7	3.3	2.9	2.3	1.9	1.6
Coronel	4.5	4.0	3.3	2.7	2.3	3.9	3.4	2.8	2.3	1.9	3.9	3.4	2.8	2.3	1.9	3.1	2.8	2.2	1.8	1.5	1.2	1.0	0.8	0.6	0.5
Wellington	6.7	6.0	4.9	4.2	3.5	6.0	5.3	4.4	3.7	3.2	6.0	5.3	4.4	3.7	3.2	5.3	4.7	3.8	3.2	2.7	1.7	1.5	1.2	0.6	0.5

PRINCIPAL COUNTRIES OF THE WORLD

COUNTRY	FORM OF GOVERNMENT	CAPITAL	PRESENT HEAD	ACCEDED	SQUARE MILES	POPULATION
Abyssinia	Absolute Monarchy	Addis Abeba	Lij Yasu	May 13, '11	432,439	8,000,000
Afghanistan	Absolute Monarchy	Kabul	Habibullah Khan	Oct. 3, '01	250,000	5,900,000
Albania	Limited Monarchy				8,000	2,000,000
Andorra	Republic				175	5,231
Argentina	Republic	Buenos Aires	Roque Saenz Pena	Oct. 12, '10	1,135,127	7,000,000
Australia	Commonwealth	Yass Canberra	Lord Denman	July 24,	2,963,041	4,455,005
Austria-Hungary	Limited Monarchy	Vienna, Budapest	Francis Joseph I	Dec. 2, '48	261,102	51,390,810
Belgium	Limited Monarchy	Brussels	Albert I	Dec. 23, '09	11,373	7,423,784
Belgian Kongo		Boma			900,654	15,000,000
Bhutan	Limited Monarchy	Punakha	Ugyen Wangchuk	Aug., '07	20,000	250,000
Bolivia	Republic	Sucre, La Paz	Eliodoro Villazon	Aug. 5, '09	515,156	2,268,063
Brazil	Republic	Rio de Janeiro	Hermes da Fonseca	Nov. 15, '10	3,275,060	21,500,000
British Empire	Limited Monarchy	London	George V	May 6, '10	13,193,712	434,686,600
Bulgaria	Limited Monarchy	Sofia	Ferdinand	Aug. 14, '87	33,647	5,000,000
Canada	Confederation	Ottawa	Duke of Connaught	Oct. 13, '11	3,729,665	7,206,643
Chile	Republic	Santiago	Ramon Barros Luco	Dec. 23, '10	307,566	3,543,204
China	Republic	Peking	Yuan Shi-Kai	Feb. 15, '12	4,277,170	439,214,000
Colombia	Republic	Bogota	Carlos E. Restrepo	July 15, '10	438,436	5,031,420
Costa Rica	Republic	San Jose	Ricardo Jimenez	May 8, '10	23,000	388,266
Cuba	Republic	Havana	Mario Garcia Menocal	May 20, '13	45,883	2,150,112
Denmark	Limited Monarchy	Copenhagen	Christian X	May 14, '12	15,582	2,757,076
Denmark, Colonies of					86,174	143,143
Dominican Republic	Republic	Santo Domingo	Adolfo Nouel	April 14, '13	19,325	708,000
Ecuador	Republic	Quito	Leonidas Plaza	Sept. 1, '12	116,000	1,500,000
Egypt	Limited Monarchy	Cairo	Abbas Hilmi	Jan. 8, '92	400,000	11,287,350
France	Republic	Paris	Raymond Poincare	Feb. 17, '13	207,054	39,601,509
France, Colonies of					4,333,154	54,540,709
Germany	Limited Monarchy	Berlin	William II	June 15, '88	208,780	64,925,993
Germany, Colonies of					1,027,820	14,546,000
Great Britain and Ireland	Limited Monarchy	London	George V	May 6, '10	121,391	45,811,558
England		London			50,526	36,070,492
Ireland		Dublin			32,361	4,381,951
Scotland		Edinburgh			30,405	4,759,445
Greece	Limited Monarchy	Athens	Constantine XII	March 18, '13	25,064	4,000,000
Guatemala	Republic	New Guatemala	Manuel E. Cabrera	March 15, '11	48,290	1,992,000
Haiti	Republic	Port au Prince	Michel Oreste	May 4, '13	10,204	2,029,700
Honduras	Republic	Tegucigalpa	Manuel Bonilla	Nov. 1, '11	46,400	553,446
India	Limited Monarchy	Delhi	Charles Hardinge	Nov. 20, '10	1,773,088	315,619,944
Italy	Limited Monarchy	Rome	Victor Emanuele III	July 29, '00	110,659	34,686,693
Italy, Colonies of					601,000	1,367,000
Japan	Limited Monarchy	Tokyo	Yoshihito Haranomiya	July 30, '12	147,655	60,266,279
Japan, Dependencies of					95,826	17,690,000
Liberia	Republic	Monrovia	Daniel Howard	Jan. 1, '12	40,000	2,120,000
Liechtenstein	Monarchy	Vaduz	John II	Nov. 12, '58	65	9,894
Luxemburg	Limited Monarchy	Luxemburg	Marie Adelaide	Feb. 25, '12	998	259,899
Mexico	Republic	Mexico	Victoriano Huerta	Feb. 18, '13	767,274	15,063,207
Monaco	Limited Monarchy	Monaco	Albert Honore Charles	Sept. 10, '89	8	19,121
Montenegro	Limited Monarchy	Cettinje	Nicholas I	Aug. 14, '60	3,630	500,000
Morocco	Absolute Monarchy	Khartoum	Mulai	Dec. 17, '11	56,000	5,000,000
Netherlands	Limited Monarchy	The Hague	Wilhelmina, Queen	Nov. 23, '90	11,543	5,898,000
Netherlands, Colonies of					820,473	38,000,000
New Zealand	Commonwealth	Wellington	Lord Liverpool	Aug. 8, '12	104,751	1,051,068
Nicaragua	Republic	Managua	Adolfo Diaz	Dec. 31, '10	49,200	600,000
Norway	Limited Monarchy	Christiania	Haakon VII	Nov. 18, '05	124,129	2,392,698
Oman	Absolute Monarchy	Maskat	Seyyid Feysil	June 4, '88	82,000	800,000
Panama	Republic	Panama	Belisario Porras	Oct. 1, '12	31,570	427,000
Paraguay	Republic	Asuncion	Eduardo Schaerer	Aug.18, '12	171,804	782,000
Persia	Limited Monarchy	Teheran	Abu'l Kassim Khan	Mar. 4, '11	628,000	9,000,000
Peru	Republic	Lima	Guillermo Billinghurst	Sept. 24, '12	695,730	4,500,000
Portugal	Republic	Lisbon	Manuel de Arriaga	Aug. 24, '11	35,490	5,975,000
Portugal, Colonies of					803,302	9,675,000
Roumania	Limited Monarchy	Bukharest	Carol I	March 26, '81	50,720	6,900,000
Russia	Absolute Monarchy	St. Petersburg	Nicholas II	Nov. 1, '94	8,647,657	168,000,000
Russia, Dependencies of					107,000	1,030,000
Salvador	Republic	San Salvador	Carlos Melendez		7,225	1,125,004
San Marino	Republic	San Marino		March '13	38	11,130
Servia	Limited Monarchy	Belgrade	Peter I	June 15, '03	18,650	4,000,000
Siam	Absolute Monarchy	Bangkok	Chowfa Maha Vajiravudh	Oct. 22, '10	195,000	6,230,000
Spain	Limited Monarchy	Madrid	Alphonso XIII	May 17, '86	194,783	19,588,688
Spain, Colonies of					83,294	576,000
Sweden	Limited Monarchy	Stockholm	Gustaf V	Dec. 8, '07	172,876	5,521,943
Switzerland	Republic	Bern	Edward Muller	Jan. 1, '13	15,976	3,765,122
Turkey & tributary states	Limited Monarchy	Constantinople	Mohammed V	April 8, '09	682,969	17,425,000
Union of South Africa		Pretoria, Cape T'n	Viscount Gladstone	May 30, '10	473,185	5,973,394
United States, Continental	Confederation	Washington	Woodrow Wilson	March 4, '13	3,026,789	91,972,266
U. S., Dependencies of					716,317	9,163,223
Uruguay	Republic	Montevideo	Jose Batlle y Ordones	March 1, '11	72,210	1,140,799
Venezuela	Republic	Caracas	Juan Vicente Gomez	June, '10	393,976	2,713,703

a Under suzerainty of Turkey, but actual administration controlled by Great Britain.   b Exclusive of Egypt and European Turkey.

THE WORLD

SHAPE AND SIZE OF THE EARTH: The earth is very nearly an oblate spheroid, whose shorter axis coincides with its axis of rotation passing through the two poles.

According to Colonel Clarke's calculations, its major axis measures 41,858,124 feet, or 7,926.5 statute miles; its minor axis 41,710,342 feet, or 7,899.5 statute miles; its circumference along the Equator measures 24,902 statute miles, or 21,600 geographical miles; its total area 196,940,400 statute square miles; and its volume 259,880 million cubic miles.

Of this total area, 8.4 p. c. (16,464,708 sq. m.) are within the Arctic and Antarctic regions; 51.6 p.c. (102,244 554 sq. m.) within the two temperate regions; 40 p. c. (78,231,045 sq. m.) within the Tropics.

THE LAND: The land covers 54,807,420 sq. m., on the assumption that 250,000 sq. m. of land remain to be discovered within the Arctic Regions, and that the supposed Antarctic continent, or "Antarctic," has an extent of 2,500,000 sq. m. within the Antarctic Circle.

THE OCEANS: The oceans, including the inland seas connected with them, cover 142,132,980 sq. m., or 72 p. c. of the total surface of the earth. There are 2.59 sq. m. of ocean to every sq. m. of land. This area is distributed over the different areas as follows: Arctic Ocean, including Hudson Bay, 5,755,000 sq. m.; Atlantic Ocean, 34,301,400 sq. m.; Indian Ocean, 28,615,500 sq. m.; Pacific Ocean, 67,699,630 sq. m.; Antarctic Ocean, 5,731,350 sq. m.

HEIGHTS AND DEPTHS: The mean height of the land has been estimated at 2,440 feet; the mean depth of the sea 11,470 feet (Karstens). The highest mountain (Mt. Everest) rises to a height of 29,000 feet; the greatest depth of the ocean as yet discovered (in the Pacific, between Guam and Midway) is 31,614 feet. If the whole of the solid crust of the earth were to be levelled, so as to form a spheroid, it would still be covered by an ocean of a uniform depth of 8,000 feet (Prof. Penck).

PRINCIPAL CITIES OF THE WORLD

The following list contains the principal cities of the world. In it will be found approximately all places of more than thirty thousand inhabitants and many of those of from twenty to thirty thousand. The population figures are from the latest census or are estimates from the most reliable sources. Capitals of countries are indicated by an asterisk (*).

City	Pop.
Aachen, Germany	156,143
Aalborg, Denmark	33,449
Aarhuus, Denmark	61,755
Abbeokuta, Yoruba	200,000
Aberdare, Wales	50,844
Aberdeen, Scotland	163,084
Abersychan, England	24,661
Abertillery, England	35,425
Abo, Russia	48,089
Accrington, England	45,031
Acireale, Italy	35,418
Acton, England	57,523
Adana, Turk. in Asia	45,000
Addis Abeba, Abys.*	35,000
Adelaide, Australia	192,294
Aden, Arabia	46,185
Adrianople, Turkey	81,000
Agra, India	185,449
Agram, Hungary	61,002
Aguascalientes, Mex.	44,800
Ahmedabad, India	215,835
Aidin, Turkey in Asia	38,000
Aintales, Turkey in Asia	43,000
Airdrie, Scotland	24,388
Aix. France	29,418
Ajmere, India	86,222
Akerman, Russia	32,470
Akita, Japan	36,294
Akron, Ohio	69,067
Alais, France	27,435
Albacete, Spain	24,667
Albany, N. Y.*	100,253
Alcamo, Italy	51,809
Alcoy, Spain	33,729
Aldershot, England	35,175
Aleppo, Turk. in Asia	127,150
Alessandria, Italy	75,687
Alexandria, Egypt	332,246
Alexandropol, Russia	30,616
Alexandrov - Grushevsky, Russia in Asia	45,277
Algiers, Algeria*	154,049
Alicante, Spain	51,163
Aligarh, India	64,825
Alkmaar, Netherlands	21,084
Allahabad, India	171,697
Allenstein, Germany	33,077
Allentown, Pa	51,913
Almeria, Spain	45,198
Alost, Belgium	34,309
Altamura, Italy	22,729
Altenburg, Germany	39,977
Altona, Germany	172,533
Altoona, Pa	52,127
Atwar, India	56,771
Amasia, Turkey in Asia	30,000
Ambala, India	80,131
Amberg, Germany	25,222
Amersfoort, Neth.	23,620
Amiens, France	93,207
Amoy, China	114,000
Amritsar, India	152,786
Amsterdam, Neth.	566,131
Amsterdam, N. Y.	31,267
Ancona, Italy	63,145
Anderlecht, Belgium	64,974
Andijan, Rus. in Asia	74,316
Andria, Italy	49,869
Angers, France	83,786
Angora, Turkey in Asia	28,000
Angoulême, France	38,211
Antequera, Spain	31,609
Antigua, Guatemala	14,000
Antofagasta, Chile	32,496
Antwerp, Belgium	320,640
Aomori, Japan	47,206
Apeldoorn, Neth	35,626
Apolda, Germany	22,592
Aquila, Italy	21,929
Arad, Hungary	63,166
Arbroath, Scotland	20,648
Archangel, Russia	35,000
Arequipa, Peru	40,000
Arezzo, Italy	47,498
Argao, P. I.	35,448
Argenta, Italy	20,544
Arles, France	31,030
Armentieres, France	29,418
Arnhem, Neth	64,019
Arras, France	24,921
Aschaffenburg, Ger.	29,831
Aschersleben, Ger.	28,964
Ascoli, Italy	30,631
Ashington, England	24,583
Ashkabad, Russia in Asia	43,729
Ashton-in-Makerfield, England	21,340
Ashton-under-Lyne, England	45,179
Asnieres, France	42,583
Asti, Italy	38,043
Aston Manor, Eng.	75,042
Astrakhan, Russia	149,600
Asuncion, Paraguay	84,000
Athens, Greece*	167,479
Atlanta, Ga.*	154,839
Atlantic City, N. J.	46,150
Aubervilliers, France	37,558
Auburn, N. Y.	34,668
Auckland, N. Zealand	102,676
Augsburg, Germany	102,293
Augusta, Ga.	41,040
Aurora, Ill.	29,807
Aussig, Austria	39,301
Austin, Tex*	29,860
Aux Cayes, Haiti	25,000
Auxerre, France	20,931
Avellino, Italy	23,873
Aversa, Italy	23,477
Avignon, France	49,304
Ayere, France	23,141
Ayr, Scotland	32,985
Ayuthia, Siam	50,000
Azof, Russia	25,124
Barranquilla, Col.	43,849
Barrow-in-Furness, England	63,773
Barry, Wales	33,767
Basel, Switzerland	131,914
Basra, Turkey in Asia	50,000
Bastic, France	27,338
Batangas, P. I.	33,131
Batavia, Java*	138,551
Bath, England	50,729
Batley, England	36,395
Battle Creek, Mich.	25,267
Batum, Russia	30,080
Bauan, Philippine I.	39,094
Bautzen, Germany	32,760
Baybay, P. I.	22,990
Bay City, Mich.	45,166
Bayonne, France	27,601
Bayonne, N. J.	55,545
Beauvene, France	20,395
Beckenham, England	31,693
Beddwelty, England	22,551
Bedford, England	39,183
Bedlingtonshire, Eng.	25,253
Beirut, Turk. in Asia.	118,800
Beken, Hungary	42,599
Belfast, Ireland	388,492
Belfort, France	39,371
Belgrade, Servia*	89,876
Bellary, India	58,247
Bello Horizonte, Brazil	20,000
Belluno, Italy	20,471
Benares, India	203,804
Bendery, Russia	45,716
Bendigo, Australia	42,000
Benevento, Italy	24,314
Benoni, U. S. Africa.	32,412
Berbera, Br. Somaliland	30,000
Berchem, Belgium	31,382
Berdiansk, Russia	29,168
Berditchef, Russia	74,980
Bergamo, Italy	55,489
Bergen, Norway	75,888
Berkeley, Cal.	40,434
Berlat, Roumania	24,484
Berlin, Canada	15,196
Berlin, Germany*	2,071,257
Bern, Switzerland*	85,264
Bernburg, Germany	33,703
Besancon, France	57,978
Beuthen, Germany	67,718
Beziers, France	51,042
Bhagalpur, India	74,349
Bhaunagar, India	56,442
Bhopal, India	56,204
Bialystok, Russia	93,695
Biebrich, Germany	21,199
Biel, Switzerland	23,583
Bielefeld, Germany	78,380
Bikaner, India	53,078
Bilbao, Spain	92,514
Bilston, England	25,681
Bin Diak, Anam	74,400
Binghamton, N.Y.	48,443
Birkenhead, England	130,832
Birmingham, Ala	132,685
Birmingham, England	840,372
Bisceglie, Italy	30,885
Bitlis, Turkey in Asia	38,800
Bitonto, Italy	30,617
Blackburn, England	133,064
Blackpool, England	58,376
Blagoveschensk, Russia in Asia	57,340
Blaydon, England	31,148
Bloemfontein, U. S. Africa	33,890
Blois, France	20,434
Bloomington, Ill.	25,768
Blyth, England	25,490
Bobruisk, Russia	35,177
Bocholt, Germany	26,404
Bochum, Germany	136,931
Bogota, Colombia*	150,000
Bokhara, Rus.in Asia	75,000
Boksburg, U. S. Africa	43,626
Bologna, Italy	172,639
Bolton, England	180,885
Bombay, India	979,445
Bona, Algeria	42,934
Bonn, Germany	87,978
Bootle, England	69,881
Boras, Sweden	21,541
Bordeaux, France	261,678
Borgerhout, Belgium	50,306
Boston, Mass.*	670,585
Botuchany, Roum	34,000
Boulogne-sur-Mer, France	53,128
Boulogne-sur-Seine, France	57,027
Bourg, France	29,248
Bourges, France	45,735
Bournemouth, Eng.	78,677
Bradford, England	288,505
Braga, Portugal	24,202
Braila, Roumania	58,392
Brandenburg, Ger.	53,595
Brandon, Canada	13,839
Brantford, Canada	23,132
Breda, Netherlands	27,389
Bremen, Germany	246,827
Bremerhaven, Ger.	24,140
Brescia, Italy	83,523
Breslau, Germany	512,105
Brest, France	90,540
Brest-Litovsk, Russia	53,244
Bridgeport, Conn.	102,054
Bridgetown, Barbados	35,000
Brieg, Germany	29,035
Brighouse, England	20,845
Brighton, England	131,250
Brindisi, Italy	25,317
Brisbane, Australia	143,077
Bristol, England	357,059
Broach, India	42,000
Brockton, Mass.	56,878
Broken Hill, Austral.	36,600
Bromberg, Germany	57,696
Bromley, England	33,649
Bronte, Italy	20,366
Brookline, Mass.	27,792
Bruges, Belgium	54,015
Brunn, Austria	175,737
Brunswick, Ger.	143,534
Brusa, Turk. in Asia	76,303
Brussels, (with suburbs) Belgium*	717,455
Bruye, France	20,636
Bucaramanga, Col.	30,000
Budapest, Hungary*	880,371
Budweis, Austria	45,137
Buenos Aires, Arg.*	1,333,532
Buer, Germany	61,537
Buffalo, N.Y.	423,715
Bukharest, Roum.*	294,572
Burg, Germany	24,074
Burgos, Spain	31,489
Burlington, Vt.	20,468
Burnley, England	106,337
Burslem, England	38,766
Burton-upon-Trent, England	48,275

City	Pop.
Bury, England	58,649
Butte, Mont	39,165
Byelaya Tserkof, Rus.	54,270
Cadiz, Spain	67,174
Caen, France	46,934
Caerphilly, Wales	32,850
Cagliari, Italy	61,013
Cairo, Egypt*	654,476
Calai, Colombia	30,740
Calais, France	72,322
Calcutta, India	1,222,313
Calgary, Canada	43,704
Calicut, India	78,417
Calino, Peru	31,128
Caltagirone, Italy	44,879
Caltanissetta, Italy	41,320
Calumet, Mich	30,000
Camaguey, Cuba	66,460
Cambrai, France	27,832
Cambridge, England	40,028
Cambridge, Mass	104,839
Cambuslang, Scotland	24,870
Camden, N. J	94,538
Camiling, P. I	25,243
Campeche, Mexico	16,864
Campinas, Brazil	35,000
Campos, Brazil	40,000
Candia, Crete	22,250
Canea, Crete	24,537
Cannes, France	30,420
Cannock, England	28,588
Canterbury, Eng	24,628
Canton, China	1,250,000
Canton, Ohio	50,217
Cape Coast Castle, Gold Coast, Africa	28,948
Cape Haitien, Haiti	29,000
Cape Town, United South Africa*	135,000
Caracas, Venezuela*	72,429
Carcar, P. I	31,895
Carcassonne, France	30,689
Cardenas, Cuba	28,576
Cardiff, Wales	182,280
Carlisle, England	46,432
Carrara, Italy	42,097
Cartagena, Colombia	27,000
Cartagena, Spain	99,871
Casale, Italy	31,793
Caserta, Italy	33,455
Cassel, Germany	153,078
Castellammare, Italy	32,841
Castellon, Spain	30,583
Castleford, England	23,101
Castres, France	27,308
Castrogiovanni, Italy	25,826
Catania, Italy	211,699
Catanzaro, Italy	34,340
Cawnpur, India	178,557
Cayenne, French Guiana*	13,527
Ceara, Brazil	33,000
Cebu, Philippine I	31,079
Cedar Rapids, Iowa	32,811
Celaya, Mexico	25,565
Celle, Germany	23,263
Central Falls, R. I	22,754
Cerignola, Italy	34,195
Cesena, Italy	42,240
Cette, France	33,049
Chadderton, England	28,305
Chalon Sur Marne, Fr.	27,808
Chalon-sur-Saone, France	31,367
Chambery, France	23,027
Chandarnagar, India	26,000
Changchau, China	500,000
Changsha, China	230,000
Chantenay, France	21,671
Chapra, India	45,400
Charleroi, Belgium	28,083
Charleston, S. C.*	58,833
Charleville, France	20,702
Charlotte, N. C.	34,014
Charlottenburg, Germany	305,978
Charlottetown, Can.	11,203
Charters Towers, Australia	20,976
Chartres, France	23,219
Chatham, England	42,250
Chatteaux, France	25,517
Chattanooga, Tenn.	44,604
Chaux de Fonds, Switzerland	37,636
Chefu, China	95,000
Chelsea, Mass	32,452
Cheltenham, Eng	48,944
Chelyabinsk, Russia	61,594
Chemnitz, Germany	287,340
Chengte, China	100,000
Chepping Wycombe, England	20,390
Cherbourg, France	43,731
Chernigof, Russia	32,848
Cherson, Russia	67,237
Chester, England	39,018
Chester, Pa	38,537
Chesterfield, England	37,429
Chicago, Ill	2,185,283
Chicopee, Mass	25,401
Chiengmai, Siam	60,000
Chieti, Italy	25,628
Chihuahua, Mexico	39,061
Chillan, Chile	36,382
Chinandega, Nicr	20,000
Chingleput, China	1,000,000
Chinkiang, China	184,000
Chioggia, Italy	30,563
Chiswick, England	38,705
Chita, Russia in Asia	39,117
Cholet, France	20,427
Cholon, Cochin China	138,000
Chorley, England	30,317
Christchurch, New Zealand	80,193
Christiania, Norway*	242,850
Chungking, China	600,000
Cienfuegos, Cuba	70,416
Cincinnati, Ohio	363,591
Citta di Castello, Italy	26,439
Ciudad Guzman, Mex.	17,596
Ciudad Porfirio Diaz, Mexico	16,000
Cleethorpe with Thrunscoe, England	21,419
Clermont, France	65,386
Cleveland, Ohio	560,663
Clichy, France	46,676
Clinton, Iowa	25,577
Clydebank, Scotland	37,547
Coatbridge, Scotland	43,287
Coban, Guatemala	30,770
Coblenz, Germany	56,487
Coburg, Germany	23,794
Cocanada, India	48,096
Cochabamba, Bolivia	24,512
Coimbatore, India	53,080
Colchester, England	43,463
Colima, Mexico	25,148
Colne, England	25,693
Cologne, Germany	516,527
Colombes, France	29,143
Colombia, S. C.	26,319
Colombo, Ceylon*	158,228
Colon, Panama	15,000
Colorado Springs, Colo.	29,078
Combaconum, India	59,673
Comiso, Italy	21,873
Como, Italy	44,146
Concepcion, Chile	55,330
Concepcion, Paraguay	25,000
Concord, N. H.*	21,497
Conjeevaram, India	46,164
Constantine, Algeria	58,435
Constantinople, Turkey*	1,125,000
Copenhagen, (with suburbs) Denmark*	559,398
Copenick, Germany	30,879
Copparo, Italy	39,267
Corato, Italy	41,573
Cordoba, Argentina	70,380
Cordova, Spain	65,160
Corfu, Greece	28,254
Cork, Ireland	76,632
Corrientes, Argentina	23,904
Coruna, Spain	45,650
Coseley, England	22,841
Cosenza, Italy	24,186
Cothen, Germany	23,617
Council Bluffs, Iowa	29,292
Courbevoie, France	36,138
Courtrai, Belgium	35,547
Coventry, England	106,377
Covington, Ky	53,270
Crajova, Roumania	45,438
Cranston, R. I	21,107
Crefeld, Germany	129,406
Cremona, Italy	40,436
Crewe, England	44,970
Crimmitschau, Ger.	28,804
Cristobal, Canal Zone, Panama	6,000
Cronstadt, Russia	66,624
Croydon, England	169,559
Csaba, Hungary	37,000
Cuddalore, India	52,216
Cuenca, Ecuador	40,000
Cuneo, Italy	27,065
Curitiba, Brazil	12,000
Cuttack, India	51,364
Cuyaba, Brazil	15,000
Cuzco, Peru	30,000
Czegled, Hungary	33,942
Czenstochowa, Rus.	69,525
Czernowitz, Austria	87,000
Dacca, India	108,551
Dagupan, P. I	20,357
Dai Moku, Formosa	87,745
Dalaguete, P. I	21,354
Dallas, Texas	92,104
Dalny, Manchuria	41,333
Daman, India	41,671
Damanhur, Egypt	38,752
Damascus, Turkey in Asia	300,000
Damietta, Egypt	31,515
Danbury, Conn	20,234
Danville, Ill	27,871
Danzig, Germany	170,337
Darbhangah, India	66,244
Darlington, England	55,633
Darmstadt, Germany	87,085
Dartford, England	23,609
Darwen, England	40,334
Davenport, Iowa	43,028
Dayton, Ohio	116,577
Debreezin, Hungary	92,729
Decatur, Ill	31,140
Delft, Netherlands	34,191
Delhi, India*	232,837
Delmenhorst, Ger.	22,480
Denaire, France	24,564
Denver, Colo.*	213,381
Derby, England	123,433
Des Moines, Iowa.*	86,368
Dessau, Germany	56,606
Desterro, Brazil	27,000
Detroit, Mich	465,766
Deventer, Neth	27,787
Devonport, England	81,694
Dewsbury, England	53,358
Diamantina, Brazil	15,000
Diarbekr, Turkey in Asia	38,000
Dieppe, France	20,434
Dijon, France	76,847
Doncaster, England	30,520
Dordrecht, Neth	46,355
Dorpat, Russia	42,421
Dortmund, Germany	214,226
Douai, France	36,314
Douglas, England	21,101
Dover, England	43,647
Drammen, Norway	24,937
Dresden, Germany	546,882
Dublin, Ireland	309,272
Dubuque, Iowa	38,494
Dudley, England	51,092
Duisburg, Germany	229,483
Duluth, Minn	78,466
Dumaujug, P. I	22,203
Dumbarton, Scotland	21,980
Dunaburg, Russia	110,354
Dundee, Scotland	165,006
Dunedin, N. Zealand	64,237
Dunfermline, Scot	28,103
Dunkirk, France	38,891
Durango, Mexico	34,085
Durban, U. S. Africa	72,813
Duren, Germany	32,511
Dusseldorf, Germany	358,728
Ealing, England	61,235
Eastbourne, England	52,544
East Ham, England	133,504
East London, U. S. Africa	21,000
Easton, Pa	28,523
East Orange, N. J	34,371
East St. Louis, Ill	58,547
Ebbw Vale, England	30,559
Eberswalde, Germany	26,075
Eccles, England	41,946
Ecija, Spain	24,395
Edinburgh, Scotland*	320,315
Edmonton, Canada	24,900
Edmonton, England	64,820
Eger, Austria	23,675
Eisenach, Germany	38,353
Eisleben, Germany	24,629
Ekaterinburg, Rus.	56,448
Ekaterinodar, Russia	92,254
Ekaterinoslaf, Russia	195,870
Elberfeld, Germany	170,195
Elbing, Germany	58,636
Elche, Spain	27,380
Elgin, Ill	25,976
Elizabeth, N. J	73,409
Elizavetgrad, Russia	68,710
Elizavetpol, Russia	46,334
Elmira, N. Y	37,176
El Paso, Texas	39,279
Emden, Germany	24,038
Emmen, Netherlands	27,665
Empoli, Italy	20,404
Enfield, England	56,344
Enschede, Neth	34,201
Epernay, France	21,637
Epinel, France	29,058
Erdington, England	32,337
Erfurt, Germany	111,463
Erie, Pa	66,525
Erith, England	27,755
Erivan, Russia	29,033
Erlanger, Germany	24,874
Erzerum, Turk.in Asia	80,000
Eschweiler, Germany	24,718
Eskilstuna, Sweden	28,371
Essen, Germany	294,653
Esslingen, Germany	32,364
Etterbeek, Belgium	32,515
Evansville, Ind	69,647
Everett, Mass	33,484
Exeter, England	48,660
Fabriano, Italy	21,096
Faenza, Italy	40,370
Faizabad, India	54,655
Falkirk, Scotland	33,569
Fall River, Mass	119,295
Farnworth, England	28,142

Farukhabad, India.. 67,318
Felegyhaza, Hungary 33,408
Felling, England..... 25,020
Fermo, Italy........ 20,703
Ferrara, Italy....... 95,196
Ferrol, Spain....... 25,281
Fez, Morocco*..... 140,000
Fianarantsoa, Mada-
  gascar.......... 27,000
Finchley, England... 39,425
Firozpur, India..... 49,302
Fitchburg, Mass.... 37,826
Fiume, Hungary..... 49,806
Flensburg, Germany. 60,922
Flint, Mich......... 38,550
Florence, Italy...... 232,860
Foggia, Italy........ 76,534
Fokthany, Roumania. 25,000
Foligno, Italy........ 26,111
Folkestone, England. 33,495
Forest, Belgium..... 24,398
Forli, Italy......... 45,879
Forst, Germany..... 33,875
Fort-de-France, Mar-
  tinique.......... 27,013
Fort Wayne, Ind.... 63,933
Fort William, Canada 16,499
Fort Worth, Tex.... 73,512
Fougeres, France.... 23,537
Frankfort-on-Main,
  Germany....... 414,576
Frankfort-on-Oder,
  Germany....... 68,277
Frederiksberg, Den.. 97,237
Freetown, Sierra
  Leone*.......... 37,280
Freiberg, Germany.. 36,237
Freiburg, Germany.. 83,328
Freidenau, Germany. 34,806
Fremantle, Australia. 19,346
Fuchau, China...... 624,000
Fukui, Japan....... 50,155
Fukuoka, Japan..... 82,106
Fukushima, Japan... 33,493
Fulda, Germany.... 22,487
Funchal, Madeira... 20,844
Funfkirchen, Hun.... 49,822
Furstenwalde, Ger... 22,626
Furth, Germany.... 66,535

Gainsborough, Eng.. 20,589
Galatz, Roumania... 66,000
Galle, Ceylon....... 48,500
Gallipoli, Turkey.... 30,000
Galveston, Tex..... 36,981
Gardaia, Algeria.... 39,000
Gateshead, England . 116,928
Gaya, India........ 49,921
Gaza, Turkey in Asia. 40,000
Geelong, Australia... 28,800
Geestemunde, Ger... 25,102
Gefle, Sweden...... 35,203
Gelligaer, Wales.... 35,521
Gelsenkirchen, Ger.. 169,513
Geneva, Switzerland. 125,520
Genoa, Italy....... 272,077
Georgetown, British
  Guiana*......... 54,891
Gera, Germany..... 49,283
Germiston, U. S. Af-
  rica............. 54,327
Ghazipur, India..... 39,429
Ghent, Belgium..... 163,149
Giarre, Italy....... 26,000
Gibraltar, Gibraltar.. 23,915
Giessen, Germany... 31,056
Gifu, Japan........ 41,488
Gijon, Spain....... 52,226
Gillingham, England. 52,252
Gilly, Belgium...... 24,306
Gioja das Colle, Italy. 21,721
Girgenti, Italy...... 26,814
Giace Bay, Canada.. 16,562
Gladbach, Germany . 66,414

Glasgow, Scotland... 784,455
Glauchau, Germany.. 25,156
Gleiwitz, Germany... 66,981
Glogau, Ger........ 24,524
Glossop, England.... 21,688
Gloucester, England. 50,029
Gloucester, Mass.... 24,398
Gmund, Germany.... 21,203
Gnesen, Germany... 25,339
Gomel, Russia...... 80,900
Gonaives, Haiti..... 18,000
Goole, England..... 20,332
Goppingen, Germany. 22,361
Gorakhpur, India.... 64,148
Gorlitz, Germany.... 85,806
Gosport and Alver-
  stoke, England... 33,301
Gotha, Germany.... 39,581
Gottenborg, Sweden. 162,813
Gottingen, Germany. 37,594
Gouda, Netherlands.. 24,574
Govan, Scotland.... 89,725
Granada, Nicaragua. 17,000
Granada, Spain..... 77,425
Grand Rapids, Mich. 112,571
Grantham, England.. 20,074
Grasse, France...... 20,437
Gratz, Austria...... 151,781
Graudenz, Germany. 40,325
Gravesend, England. 28,117
Green Bay, Wisconsin. 25,236
Greenock, Scotland.. 75,140
Greifswald, Germany. 24,679
Greiz, Germany..... 23,245
Grenoble, France.... 77,438
Grimsby, England... 74,663
Grodno, Russia..... 50,207
Groningen, Neth.... 74,613
Groshevik, Russia... 42,186
Grosswardein, Hun.. 64,169
Grunberg, Germany.. 23,168
Guadalajara, Mexico. 118,799
Guanajuato, Mexico. 35,142
Guayaquil, Ecuador. 75,000
Gubbio, Italy...... 26,320
Guben, Germany.... 38,593
Guelph, Canada..... 15,175
Guildford, England.. 23,823
Guinobatan, P. I.... 20,027
Gwalior, India..... 89,154
Gyor, Hungary..... 43,300

Haarlem, Neth...... 69,410
Hagen, Germany.... 88,605
Hagony, P. I....... 21,304
Hague, Neth*...... 271,280
Haidarabad, India... 500,623
Haidarabad, India... 69,372
Hakodate, Japan.... 87,875
Halberstadt, Ger.... 46,481
Halifax, Canada.... 46,619
Halifax, England.... 101,556
Halle, Germany..... 180,843
Hamadan, Persia.... 40,000
Hamah, Turkey in
  Asia............ 60,000
Hamborn, Germany.. 101,703
Hamburg, Germany.. 932,166
Hameln, Germany... 22,001
Hamilton, Canada... 81,969
Hamilton, Ohio 35,279
Hamilton, Scotland.. 38,644
Hamm, Germany.... 43,658
Hanau, Germany.... 37,472
Handsworth, Eng.... 68,618
Hangchau, China.... 600,000
Hankau, China...... 870,000
Hanley, England.... 67,998
Hanoi, Anam*...... 103,238
Hanover, Germany... 302,375
Hanyang, China..... 400,000
Harar, Abyssinia.... 40,000
Harbin, China...... 30,000
Harburg, Germany... 67,025

Harrisburg, Pa.*.... 64,186
Harrogate, England.. 33,706
Hartford, Conn.*.... 98,915
Hartlepool, England. 20,618
Haspe, Germany.... 23,476
Hastings, England... 61,146
Hathras, India...... 42,578
Havana, Cuba*..... 319,884
Haverhill, Mass..... 44,115
Havre, France...... 136,159
Hazleton, Pa....... 25,452
Hebburn, England... 21,766
Heidelberg, Germany. 56,010
Heilbronn, Germany. 42,709
Helder, Netherlands. 27,157
Helsingborg, Sweden. 32,763
Helsingfors, Russia.. 137,846
Hendon, England.... 38,806
Hengelo, Netherlands. 20,073
Herat, Afghanistan.. 45,000
Hereford, England... 22,568
Herford, Germany... 32,527
Herne, Germany.... 57,147
Herstal, Belgium.... 22,829
Hertogenbosch('s),
  Netherlands..... 34,928
Heston and Isleworth,
  England......... 43,316
Heywood, England.. 26,698
Hildesheim,Germany. 50,239
Hilversum, Neth..... 31,458
Himeji, Japan...... 41,028
Hindley, England.... 24,106
Hiroeaki, Japan..... 37,487
Hiroshima, Japan.... 142,763
Hirschberg, Germany. 20,564
Hobart, Australia.... 27,719
Hoboken, N. J...... 70,324
Hodmezo-Vasarhely,
  Hungary........ 62,445
Hof, Germany...... 41,001
Hohenfalza,Germany. 25,604
Holyoke, Mass...... 57,730
Homs, Turkey in Asia. 70,000
Honolulu, Hawaii*... 52,183
Horde, Germany.... 32,791
Hornsey, England... 84,602
Horsens, Denmark... 23,843
Horst, Germany.... 20,990
Houston, Tex....... 78,800
Hove, England...... 42,173
Howrah, India...... 157,594
Huddersfield, Eng... 107,825
Hue, Anam........ 65,000
Huelva, Spain...... 27,699
Hull, Canada....... 18,222
Hull, England...... 278,024
Huntington, W. Va.. 31,161
Hyde, England..... 33,444

Ibadan, Yoruba..... 200,000
Ichang, China...... 55,000
Iglau, Austria...... 24,423
Iglesias, Italy...... 21,011
Ilford, England..... 78,205
Ilkeston, England... 31,673
Imola, Italy........ 32,210
Ince-in-Makerfield,
  England......... 21,038
Indianapolis, Ind.*.. 233,650
Indore, India....... 86,686
Ingolstadt, Germany. 23,685
Innsbruck, Austria... 27,056
Inowraslow, Ger..... 26,140
Insterburg, Germany. 31,624
Inverness, Scotland.. 22,216
Ipswich, England.... 73,939
Iquique, Chile...... 40,171
Iquitos, Peru....... 20,000
Irapuato, Mexico.... 19,640
Irkutsk, Rus. in Asia. 108,060
Iserlohn, Germany... 31,274

Ismail, Russia....... 33,607
Ispahan, Persia..... 100,000
Ivanovo-Voznesensk,
  Russia.......... 107,706
Ivry-sur-Seine,France 38,307
Ixelles, Belgium..... 80,439
Izhevsk, Russia..... 41,074

Jabalpur, India..... 100,651
Jackson, Mich...... 31,433
Jacksonville, Fla.... 57,699
Jaen, Spain........ 26,894
Jaffa, Turkey in Asia. 45,000
Jaffna, Ceylon..... 33,879
Jaipur, India....... 137,098
Jalandhar, India.... 67,735
Jalapa, Mexico..... 24,816
Jamestown, N. Y.... 31,297
Janina, Turkey..... 30,000
Janiuay, P. I....... 20,738
Jaroslaw, Austria.... 22,641
Jarrow, England.... 33,732
Jassy, Roumania.... 80,000
Jaunpur, India..... 42,000
Jeleto, Russia...... 51,708
Jena, Germany..... 38,487
Jeremie, Haiti...... 35,000
Jerez, Spain....... 62,628
Jersey City, N. J... 267,779
Jerusalem, Turkey in
  Asia............ 70,000
Jesi, Italy......... 23,028
Jhansi, India....... 55,724
Jodhpur, India..... 60,437
Johannesburg, U. S.
  Africa.......... 237,220
Johnstown, Pa...... 55,482
Jokjokarta, Java.... 58,229
Joliet, Ill......... 34,670
Jonkoping, Sweden.. 26,971
Joplin, Mo......... 32,073
Jumet, Belgium..... 26,924

Kabul, Afghanistan* 70,000
Kagoshima, Japan... 63,646
Kaisarieh, Turkey in
  Asia............ 72,000
Kaiserslautern, Ger.. 53,803
Kalamazoo, Mich... 39,437
Kalgan, China...... 200,000
Kaliaz, Russia...... 46,796
Kaluga, Russia..... 53,854
Kamenetz-Podolsk,
  Russia.......... 46,707
Kanagawa, Japan... 110,994
Kandahar, Afg...... 30,000
Kano, No. Nigeria... 100,000
Kansas City, Kan... 82,331
Kansas City, Mo.... 248,381
Kapstadt, U.S.Africa 67,170
Karachi, India..... 151,903
Karikal, India..... 17,350
Karlskrona, Sweden. 27,448
Karlsruhe, Germany. 134,161
Kars, Russia....... 24,318
Kasan, Russia...... 179,201
Kaschau, Hungary... 44,211
Kashan, Persia..... 40,000
Kashgar, China..... 75,000
Kaslof, Russia...... 45,095
Kasvin, Persia...... 50,000
Kattowitz, Germany. 43,173
Kecskemet, Hungary. 66,834
Keighley, England... 43,490
Kempten, Germany.. 20,885
Keneh, Egypt...... 27,478
Kerbela, Turk.in Asia 65,000
Kerman, Persia..... 60,000
Kermanshah, Persia. 40,000
Kertch, Russia..... 49,708
Kettering, England.. 29,976
Khabarovsk, Russia.. 49,623
Kharkof, Russia..... 221,193
Khartum, Egypt.... 20,956

City	Pop.	City	Pop.
Khartum North,Egypt	36,294	Lassa, Tibet	20,000
Khatmandu, Nepal*.	50,000	Launcestown, Austral.	20,838
Khodjent, Russia in		La Union, Spain	30,275
Asia	31,881	Lausanne, Switz.	63,926
Khoi, Persia	35,000	Laval, France	30,252
Khotin, Russia	30,429	Lawrence, Mass	85,892
Kidderminster, Eng.	24,333	Lecce, Italy	36,310
Kief, Russia	468,712	Le Creuzot, France	35,587
Kiel, Germany	211,627	Leeds, England	445,568
Kilmarnock, Scot	34,729	Leeuwarden, Neth.	36,522
Kimberley, U.S.Africa	30,000	Leghorn, Italy	105,322
King's Lynn, Eng.	20,205	Lehe, Germany	37,457
King's Norton, Eng..	81,163	Leicester, England	227,242
Kingston, Canada	18,874	Leiden, Netherlands	58,253
Kingston, Jamaica	46,542	Leigh, England	44,109
Kingston, N.Y.	25,908	Leipzig, Germany	587,635
Kingston-upon-		Leith, Scotland	80,489
Thames, Eng.	37,977	Leitmeritz, Austria	87,128
Kirin, China	250,000	Le Mans, France	69,361
Kirkcaldy, Scotland	39,600	Lemberg, Austria	206,113
Kishenef, Russia	118,807	Lens, France	31,812
Kiukiang, China	36,000	Leon, Mexico	63,263
Kiungchau, China	42,000	Leon, Nicaragua	62,569
Klausenburg, Hun.	60,808	Le Puy, France	21,420
Knoxville, Tenn.	36,346	Levallois-Perret,	
Kobe, Japan	378,197	France	68,703
Kochi, Japan	38,279	Lewiston, Me.	26,247
Kofu, Japan	49,882	Lexington, Ky	35,099
Kokand, Russia in		Leyton, England	124,736
Asia	112,428	Libau, Russia	64,502
Kokura, Japan	31,615	Licata, Italy	22,931
Kolberg, Germany	24,786	Lichtenberg,Germany	81,199
Kolhapur, India	54,373	Lieben, Austria	21,375
Kolomea, Austria	42,676	Liege, Belgium	174,768
Kom, Persia	40,000	Liegnitz, Germany	66,620
Kongmun, China	62,000	Lierre, Belgium	25,933
Koniah, Turk. in Asia	44,000	Lievini, France	22,070
Konigsberg,Germany	245,994	Lima, Ohio	30,508
Konigshutte, Ger.	72,641	Lima, Peru*	140,884
Konstanz, Germany.	27,582	Limerick, Ireland	38,403
Koslin, Germany	23,236	Limoges, France	92,181
Kostroma, Russia	44,893	Linares, Spain	36,419
Kottbus, Germany	48,643	Lincoln, England	57,294
Kovno, Russia	79,909	Lincoln, Neb.*.	43,973
Krakow, Austria	151,886	Linden,Germany	73,379
Krasnoyarsk, Russia		Lingayun, P.I	21,529
in Asia	62,919	Linkoping, Sweden	22,157
Kremenchug, Russia.	72,991	Linz, Austria	67,817
Kreuznach, Germany	23,167	Lipa, Philippine I	37,934
Kronstadt, Hungary.	41,056	Lisbon, Portugal*	356,009
Krugersdorp, U.S.		Lisle, France	217,807
Africa	53,881	Little Rock, Ark.*.	45,941
Kuching, Borneo	25,000	Liverpool, England	746,566
Kuka, No. Nigeria	60,000	Llandovery, Wales	32,077
Kumamoto, Japan	61,233	Lodi, Italy	27,811
Kumassi, Gold Coast	70,000	Lodz, Russia	393,526
Kure, Japan	100,679	Lokeren, Belgium	22,838
Kursk, Russia	81,527	Lomza, Russia	27,343
Kurume, Japan	35,928	London, Canada	46,300
Kutais, Russia	50,396	London, England	4,522,961
Kwala Kangsa, Straits		London, England	
Settlements	77,234	(Greater)*	7,252,963
Kyoto, Japan	442,462	Londonderry, Ireland	40,799
		Longton, England	35,825
		Lorain, Ohio	28,883
La Crosse, Wis	30,417	Lorca, Spain	69,836
Laeken, Belgium	34,726	Lorient, France	49,039
Lagos, Nigeria	57,000	Los Angeles, Cal	319,198
La Guaira, Ven.	14,000	Loughborough, Eng..	22,992
Lahore, India	228,687	Louisville, Ky	223,928
Laibach, Austria	41,727	Loule, Portugal	22,478
La Linea, Spain	31,861	Louvain, Belgium	41,923
La Louviere, Bel	22,037	Lowell, Mass	106,294
Lambezellec, France.	20,045	Lowestoft, England.	33,780
Lancaster, England.	41,414	Luang-Prabang, Fr.	
Lancaster, Pa.	47,227	Indo China	40,000
Lanchau, China	500,000	Lubeck, Germany	98,620
Landsberg, Germany.	39,339	Lublin, Russia	65,870
Landshut, Germany.	24,799	Lucca, Italy	70,037
Lansing, Mich.*.	31,229	Lucerne, Switzerland.	39,152
Laoag, P.I	34,452	Luckenwalde, Ger.	23,476
La Paz, Bolivia*.	78,856	Lucknow, India	259,798
La Plata, Argentina.	100,608	Ludenscheid, Ger.	32,301
La Rochelle, France.	36,371	Ludwigshafen, Ger.	83,295
Lashkar, India	49,952	Lugo, Italy	27,415

City	Pop.	City	Pop.
Lugo, Spain	26,945	Meerane, Germany	26,466
Lund, Sweden	20,139	Meerut, India	116,227
Luneburg, Germany	27,790	Mehallet-el-Kebir,	
Luneville, France	24,266	Egypt	25,473
Luton, England	50,000	Mehallah el Kebir,	
Luxemburg, Luxem-		Egypt	47,955
burg*	21,024	Meiderich, Germany.	33,690
Lynchburg, Va.	29,494	Meissen, Germany	33,875
Lynn, Mass	89,336	Meknez, Morocco	56,000
Lyon, France	523,796	Melbourne, Australia	591,830
		Memel, Germany	21,470
McKeesport, Pa.	42,694	Memphis, Tenn.	131,105
Maastricht, Neth.	37,483	Mendoza, Argentina.	42,496
Macao, China	78,627	Merida, Mexico	61,999
Macarata, Italy	22,941	Meriden, Conn.	27,265
Macclesfield, Eng.	34,804	Merseburg, Germany	21,226
Maceio, Brazil	33,000	Merthyr Tydfil,Wales	80,999
Macon, Ga.	40,665	Meshed, Persia	80,000
Madison, Wisconsin.	25,531	Messina, Italy	126,172
Madras, India	518,660	Metz, Germany	68,445
Madrid, Spain*	571,539	Mexico, Mexico*	470,659
Madura, India	134,130	Miagao, P.I	20,656
Maebashi, Japan	34,495	Middlesbrough, Eng.	104,787
Maesteg, Wales	24,977	Middleton, England	27,983
Magdeburg, Ger.	279,629	Milan, Italy	599,200
Mahrisch, Austria	30,100	Milwaukee, Wis.	373,857
Maidstone, England.	35,477	Minden, Germany	26,454
Maikop, Rus. in Asia.	34,327	Minneapolis, Minn.	301,408
Mainz, Germany	110,634	Minnia, Egypt	27,221
Maisonneuve, Canada	18,684	Minsk, Russia	97,997
Mako, Hungary	34,918	Mirzapur, India	32,446
Malaga, Spain	133,045	Miskolcz, Hungary.	51,459
Malden, Mass	44,404	Mitau, Russia	35,011
Malmo, Sweden	88,138	Mito, Japan	38,435
Malstatt-Burbach,		Mobile, Ala.	51,521
Germany	38,554	Modena, Italy	70,267
Managua, Nicaragua*	34,872	Modica, Italy	48,962
Manaos, Brazil	50,000	Mohilef, Russia	53,313
Manchester, England	714,427	Moji, Japan	55,682
Manchester, N.H.	70,063	Molenbeek, Belgium.	73,247
Mandalay, India	138,299	Molfetta, Italy	40,135
Mangalore, India	44,108	Mombasa,Br.E.Africa	30,000
Manila, P.I.*	219,928	Monastir, Turkey	50,000
Manipur, India	67,093	Monclova, Mexico	15,000
Manissa, Turkey in		Mongyr, India	35,880
Asia	60,000	Monopoli, Italy	22,545
Manizales, Colombia.	27,375	Monreale, Italy	23,778
Mannheim, Germany	193,379	Mons, Belgium	27,252
Manresa, Spain	25,121	Montauban, France.	30,506
Mansfield, England.	36,897	Montceau les Mines,Fr.	26,306
Mansura, Egypt	40,279	Monterey, Mexico	81,006
Mantua, Italy	32,692	Monte St. Angelo,Italy	21,870
Maracaibo, Ven.	50,000	Montevideo, Uru.*.	317,879
Maranhao, Brazil	32,000	Montgomery, Ala.*.	38,136
Marash, Turk. in Asia	50,000	Montignies -sur-Sam-	
Marburg, Germany.	21,860	bre, Belgium	22,275
Marchienne-au-Pont,		Montlucon, France	33,799
Belgium	21,594	Montpellier, France.	80,230
Margate, England	27,086	Montreal, Canada	470,480
Maria Theresiopel,		Montreuil, France	43,217
Hungary	82,122	Monza, Italy	42,599
Mariupol, Russia	31,116	Moosejaw, Canada	13,823
Marsala, Italy	57,567	Moradabad, India	81,168
Marseille, France	550,619	Morebin, France	21,888
Martina Franca, Italy.	25,007	Morelia, Mexico	39,116
Masaya, Nicaragua	20,000	Morioka, Japan	36,012
Maskat, Arabia	25,000	Morley, England	24,285
Massa, Italy	30,895	Morocco, Morocco*.	60,000
Matanzas, Cuba	64,385	Mors, Germany	23,251
Matsumoto, Japan	35,011	Morshansk, Russia	25,913
Matsuyama, Japan	44,166	Moscow, Russia	1,481,200
Matsuye, Japan	36,209	Mosul, Turk. in Asia.	100,000
Maubeuge, France	21,520	Motherwell, Scotland	40,378
Maulman, India	58,446	Mountain Ash, Wales	42,256
Mayaguez,PortoRico	16,563	Mt. Vernon, N.Y.	30,919
Mazatlan, Mexico	17,852	Mouscron, Belgium.	22,548
Mazzare del Valle,Italy	20,131	Muhlhausen, Ger.	35,091
Mecca, Turk. in Asia	80,000	Mukden, China	158,132
Mechlin, Belgium	59,372	Mulhausen, Germany	94,967
Medellin, Colombia.	60,000	Mulheim-on-Rhine,	
Medford, Mass	23,150	Germany.	53,425
Medina, Tur. in Asia	48,000	Mulheim-on-Ruhr,	
Medinet-el-Fayoum,		Germany	112,580
Egypt	37,320	Multan, India	99,243

City	Population
Munich, Germany	595,053
Munster, Germany	90,254
Murcia, Spain	124,985
Muskogee, Okla	25,278
Mustapha, Algeria	38,327
Muttra, India	60,042
Mysore, India	71,306
Nafa, Japan	47,562
Nagano, Japan	39,242
Nagaoka, Japan	35,376
Nagasaki, Japan	169,941
Nagoya, Japan	378,231
Nagpur, India	101,415
Nagy-Koros, Hun	26,535
Nakhichivan, Russia	40,384
Namangan, Rus. in Asia	73,279
Namdinh, Tonkin	30,000
Namur, Belgium	31,939
Nanchang, China	300,000
Nancy, France	120,213
Nankin, China	276,000
Nantes, France	170,535
Naples, Italy	723,208
Nara, Japan	32,732
Narbonne, France	24,670
Nashua, N. H	26,005
Nashville, Tenn.*	110,364
Nassau, Bahama I	12,000
Natal, Brazil	10,000
Naumburg, Germany	26,962
Navarre, Spain	28,759
Nawangar, India	54,373
Negapatam, India	57,190
Neisse, Germany	25,938
Nelson, England	39,485
Neuchatel, Switz	23,583
Neuilly, France	44,616
Neumunster, Ger	34,555
Neunkirchen, Ger	34,532
Neuss, Germany	37,224
Nevers, France	27,673
Newark, N. J	347,469
Newark, Ohio	25,404
New Bedford, Mass	96,652
New Britain, Conn	41,916
Newburgh, N. Y	27,805
Newcastle, Australia	55,630
Newcastle, England	266,671
New Castle, Pa	36,280
Newcastle under Lyme, England	20,204
New Guatemala, Guatemala*	125,000
New Haven, Conn	133,605
New London, Conn	19,695
New Orleans, La	339,075
Newport, England	83,700
Newport, Ky	30,309
Newport, R. I	27,149
New Rochelle, N. Y	28,867
Newton, Mass	39,806
New Weissensee, Ger	31,944
New Westminster, Canada	13,199
New York, N. Y	4,766,883
Nezhin, Russia	50,737
Niagara Falls, N. Y	30,445
Nice, France	142,940
Nictheroy, Brazil	35,000
Niigata, Japan	61,616
Nijmegen, Neth	54,803
Nikolaief, Russia	95,400
Nimes, France	80,437
Ningpo, China	260,000
Niort, France	23,329
Nissa, Servia	24,949
Niuchwang, China	74,000
Nizhni Novgorod, Russia	92,273
Nizhni Tagilsk, Rus	31,449
Nordhausen, Ger	32,564
Norfolk, Va	67,452
Norkoping, Sweden	46,416
Norristown, Pa	27,875
North Adams, Mass	22,019
Northampton, Eng	90,076
Norwich, Conn	20,367
Norwich, England	121,493
Noto, Italy	22,564
Nottingham, Eng	259,942
Novara, Italy	54,589
Novgorod, Russia	20,972
Novocherkask, Rus	53,473
Novorossijsk, Russia in Asia	44,470
Nowawes, Germany	23,734
Nuka, Russia	24,811
Nuneaton, England	37,083
Nuremberg, Ger	332,651
Nyireghyhaza, Hun	38,198
Oakland, Cal	150,174
Oaxaca, Mexico	37,469
Oberhausen, Ger	89,900
Odenburg, Hungary	33,590
Odenkirchen, Ger	20,060
Odense, Denmark	42,237
Odessa, Russia	478,900
Offenbach, Germany	75,593
Ogden, Utah	25,580
Ogmore and Garw, Wales	26,747
Ohligs, Germany	27,839
Okayama, Japan	93,421
Oklahoma, Okla.*	64,205
Oldbury, England	32,240
Oldenburg, Germany	30,113
Oldham, England	147,495
Old Margelana, Russia in Asia	46,432
Ohmutz, Austria	22,106
Omaha, Neb	124,096
Omdurman, Egypt	42,779
Omsk, Russia in Asia	88,900
Ononichi, Japan	30,529
Oporto, Portugal	167,935
Oppeln, Germany	33,907
Oran, Algeria	106,517
Orange, N. J	29,630
Orebro, Sweden	30,098
Orel, Russia	81,940
Orenburg, Russia	94,979
Orihuela, Spain	28,530
Orizaba, Mexico	32,894
Orleans, France	72,096
Oruro, Bolivia	20,670
Osaka, Japan	1,117,151
Osh, Russia in Asia	43,483
Oshkosh, Wis	33,062
Osnabruck, Germany	65,957
Ostend, Belgium	43,190
Ostuni, Italy	22,997
Otaru, Japan	91,281
Otsu, Japan	42,869
Ottawa, Canada*	87,062
Ouro Preto, Brazil	59,249
Oviedo, Spain	52,874
Oxford, England	53,049
Pachuca, Mexico	38,620
Padang, Sumatra	35,158
Paderborn, Germany	2?,441
Padua, Italy	96,135
Paisley, Scotland	84,477
Pakhoi, China	30,000
Palembang, Sumatra	53,788
Palermo, Italy	341,656
Palma, Spain	68,359
Pamplona, Spain	28,886
Panama, Panama*	40,000
Pantin, France	36,359
Paoting, China	150,000
Para, Brazil	200,000
Parahiba, Brazil	32,000
Paramaribo, Dutch Guiana*	35,082
Parana, Argentina	35,857
Paris, France*	2,888,110
Parma, Italy	51,919
Partick, Scotland	66,843
Partinico, Italy	23,729
Pasadena, Cal	30,291
Passaic, N. J	54,773
Passau, Germany	20,885
Paterson, N. J	125,600
Patiala, India	53,545
Patna, India	136,153
Patras, Greece	37,724
Pau, France	37,149
Pavia, Italy	39,319
Pawtucket, R. I	51,622
Paysandu, Uruguay	28,000
Peking, China*	1,077,209
Pelotas, Brazil	41,591
Pembroke, Ireland	29,260
Penang, Straits Settlements	94,086
Penge, England	22,331
Pensa, Russia	79,552
Peoria, Ill	66,950
Perigueux, France	33,548
Perm, Russia	46,219
Pernambuco, Brazil	150,000
Perpignan, France	39,510
Perth, Australia	54,354
Perth, Scotland	35,851
Perth Amboy, N. J	32,121
Perugia, Italy	63,818
Pesaro, Italy	22,143
Peshawar, India	97,935
Peterborough, Can	18,360
Peterborough, Eng	33,578
Petropolis, Brazil	20,000
Pforzheim, Germany	69,084
Philadelphia, Pa	1,549,008
Philippopolis, East Roumelia	47,924
Piacenza, Italy	38,523
Piatra, Roumania	25,000
Pietermaritzburg, U.S. Africa	30,539
Pilsen, Austria	80,343
Pingyang, Chosen, Jap	74,213
Pinsk, Russia	27,938
Piotegorek, Russia	46,753
Piotrkof, Russia	38,114
Piraeus, Greece	73,579
Pirmasens, Germany	38,460
Pisa, Italy	65,215
Pistoja, Italy	62,606
Pittsburgh, Pa	533,905
Pittsfield, Mass	32,121
Plauen, Germany	121,104
Plevna, Bulgaria	23,081
Plock, Russia	30,612
Ployesti, Roumania	50,000
Plymouth, England	112,042
Pnum Penh, Cambodia	50,000
Poitiers, France	41,242
Pola, Austria	58,051
Poltava, Russia	58,933
Ponce, Porto Rico	35,005
Pont-y-Pridd, Wales	43,215
Poole, England	38,886
Poona, India	158,856
Popayan, Colombo	28,448
Port Arthur, China	15,195
Port au Prince, Haiti*	100,000
Port Elizabeth, U. S. Africa	31,000
Portland, Me	58,571
Portland, Ore	207,214
Port Louis, Mauritius	50,000
Port Maggiore, Italy	20,162
Port Alegre, Brazil	100,000
Port of Spain, Trin	60,000
Porto Novo, Dahome	19,000
Porto Praya, Cape Verde Islands	20,000
Port Said, Egypt	49,854
Portsmouth, Eng	231,165
Portsmouth, Va	33,190
Posen, Germany	156,691
Potosi, Bolivia	23,450
Potsdam, Germany	62,243
Poughkeepsie, N. Y	27,936
Prague, Austria	223,741
Prato, Italy	51,453
Prenzlau, Germany	21,386
Pressburg, Hungary	78,223
Preston, England	117,113
Pretoria, U. S. Africa*	48,609
Prisrend, Turkey	40,000
Prossnitz, Austria	24,343
Providence, R. I.*	224,326
Przemysl, Austria	54,562
Pskof, Russia	32,856
Puebla, Mexico	101,214
Pueblo, Colo	44,395
Puerto Cabello, Ven	14,000
Puerto Plata, Santo Domingo	17,500
Puket, Siam	174,000
Puteaux, France	32,223
Quebec, Canada	78,190
Quedlinburg, Ger	27,233
Queretaro, Mexico	35,011
Quezaltenango, Guate	28,940
Quincy, Ill	36,587
Quincy, Mass	32,642
Quito, Ecuador*	80,000
Racine, Wis	38,002
Radcliffe, England	26,085
Radom, Russia	39,981
Ragusa, Italy	31,922
Rampur, India	74,316
Ramsgate, England	29,605
Randers, Denmark	22,970
Rangoon, India	293,316
Rathenow, Germany	24,891
Rathmines, Irel ind	38,190
Ratibor, Germany	38,424
Ratisbon, Germany	52,540
Ravenna, Italy	71,690
Rawalpindi, India	86,483
Rawtenstall, Eng	30,516
Reading, England	75,214
Reading, Pa	96,071
Recklinghausen, Ger	53,681
Reggio di Calabria, Italy	42,876
Reggio nell' Emilia, Italy	70,499
Regina, Canada	30,213
Reichenbach, Ger	29,681
Reichenberg, Austria	36,350
Reigate, England	38,505
Remscheid, Germany	72,159
Renaix, Belgium	21,686
Rennes, France	79,372
Resht, Persia	60,000
Reus, Spain	26,235
Reutlingen, Germany	29,808
Reval, Russia	72,630
Rheims, France	115,178
Rheydt, Germany	44,003
Rhondda, Wales	152,798
Richmond, England	33,223
Richmond, Va.*	127,628
Riga, Russia	318,400
Rimini, Italy	43,201
Rio de Janeiro, Brazil*	1,000,000
Rio Grande do Sul, Brazil	25,000
Rixdorf, Germany	237,289
Roanne, France	36,697
Roanoke, Va	34,874
Rochdale, England	91,437

Rochefort, France	35,019
Rochester, England	31,388
Rochester, N. Y.	218,149
Rockford, Ill.	45,401
Rodosto, Turkey	42,000
Rome, Italy*	538,634
Roodepoort Maraisburg, U. S. Africa	37,458
Rosario, Argentina	176,076
Rossberg, Germany	20,021
Rostock, Germany	65,377
Rostof, Russia	121,300
Rotherham, England	62,507
Rotterdam, Neth.	417,989
Roubaix, France	122,723
Rouen, France	124,987
Roulers, Belgium	25,870
Rowley Regis, Eng.	37,000
Royal Leamington Spa, England	26,717
Rugby, England	21,762
Rustchuk, Bulgaria	35,825
Rutherglen, Scotland	24,411
Ryazan, Russia	46,122
Saarbrucken, Ger.	105,089
Sacramento, Cal.*	44,696
Saga, Japan	36,051
Sagar, India	42,330
Saginaw, Mich.	50,510
Saharanpur, India	66,251
Saigon, Fr. Cochin China	48,783
St. Brieuc, France	24,081
St. Denis, France	71,759
St. Denis, Reunion I.	24,912
St. Die, France	22,136
St. Etienne, France	148,656
St. Gallen, Switz.	37,657
St. Gilles, Belgium	65,637
St. Helens, England	96,506
St. John, Canada	42,511
St. Johns, Newfoundland	32,292
St. Joseph, Mo.	77,403
St. Josse-ten-Noode, Belgium	33,420
St. Louis, Mo.	687,029
St. Louis, Senegal*	21,020
St. Maur des Fosses, Fr.	28,245
St. Nazaire, France	38,267
St. Nicolas, Belgium	34,206
St. Omer, France	20,993
St. Ouen, France	41,904
St. Paul, Minn.*	214,744
St. Petersburg, Rus.*	1,907,708
St. Pierre, Reunion I.	29,181
St. Quentin, France	55,571
St. Thomas, Canada	14,054
Sakai, Japan	61,103
Salamanca, Spain	26,275
Salem, India	59,153
Salem, Mass.	43,697
Salerno, Italy	43,428
Salford, England	231,380
Salisbury, England	21,217
Saloniki, Turkey	174,000
Salta, Argentine	23,284
Saltillo, Mexico	35,063
Salt Lake City, Utah*	92,777
Salzburg, Austria	36,188
Samara, Russia	95,461
Samarang, Java	96,600
Samarkand, Russia in Asia	80,706
San Antonio, Tex.	96,614
San Carlos, P. I.	27,166
San Diego, Cal.	39,578
San Fernando, Spain	28,951
San Francisco, Cal.	416,912
San Jose, Cal.	28,946
San Jose, Costa Rica*	29,660
San Juan, Porto Rico*	48,716
San Luis Potosi, Mex.	82,946
San Miguel, Salvador	24,768
San Nicolas, P. I.	29,035
San Pablo, P. I.	22,612
San Pier d'Arena, Italy	34,882
San Remo, Italy	21,440
San Salvador, Sal.*	59,540
San Sebastian, Spain	47,894
San Severo, Italy	30,010
Santa Ana, Salvador	48,120
Santa Cruz, Bolivia	20,535
Santa Cruz, Canary Islands	51,403
Santa Cruz, P. I.	35,030
Santa Fe, Argentina	48,000
Santander, Spain	65,209
Santiago, Chile*	332,724
Santiago, Spain	24,927
Santiago de Cuba, Cuba	53,614
Santo Domingo, Santo Domingo*	20,000
Santos, Brazil	35,000
Sao Paulo, Brazil	400,000
Sapporo, Japan	70,084
Saragossa, Spain	105,788
Sarajevo, Austria	51,919
Saratof, Russia	197,882
Sasebo, Japan	93,051
Sassari, Italy	43,240
Savannah, Ga.	65,061
Savona, Italy	38,155
Scarborough, Eng.	37,201
Schaerbeek, Belgium	85,218
Schenectady, N. Y.	72,826
Schiedam, Neth.	32,024
Schneidemuhl, Ger.	26,126
Schoneberg, Ger.	172,823
Schweidnitz, Ger.	31,329
Schweinfurt, Ger.	21,882
Schwelm, Germany	20,439
Schwerin, Germany	42,518
Sciacca, Italy	20,090
Scranton, Pa.	129,867
Seattle, Wash.	237,194
Semipalatinsk, Russia in Asia	36,040
Sendai, Japan	97,944
Senigallia, Italy	23,156
Seoul, Chosen, Jap.*	196,643
Seraing, Belgium	43,208
Seres, Turkey	31,000
Sergievsk, Russia	31,413
Serpukhof, Russia	30,571
Setubal, Portugal	22,074
Sevastopol, Russia	71,100
Seville, Spain	155,366
Shahjahanpur, India	71,778
Shanghai, China	900,000
Shasih, China	96,000
Sheboygan, Wis.	26,398
Sheffield, England	454,653
Shenandoah, Pa.	25,774
Sherbrooke, Canada	16,405
Shidzuoka, Japan	53,614
Shimonoseki, Japan	58,254
Shipley, England	27,710
Shiraz, Persia	30,000
Shirbin, Egypt	23,173
Sholapur, India	61,345
Shreveport, La.	28,015
Shrewsbury, Eng.	29,389
Shumla, Bulgaria	22,225
Shusha, Russia	25,656
Sialkot, India	57,956
Siangtan, China	850,000
Silang, P. I.	25,848
Sidi Bel Abbes, Algeria	29,088
Siegen, Germany	27,416
Siena, Italy	41,659
Simbirsk, Russia	51,959
Simferopol, Russia	62,110
Singan, China	875,000
Singapore, Straits Settlements	193,089
Sioux City, Iowa	47,828
Siut, Egypt	39,642
Sivas, Turk. in Asia	65,000
Skutari, Turk in Asia	80,000
Slivno, E. Roumelia	25,111
Smethwick, England	70,681
Smichow, Austria	47,135
Smolensk, Russia	70,851
Smyrna, Turk in Asia	260,000
Sofia, Bulgaria*	102,769
Solingen, Germany	50,536
Somerville, Mass.	77,236
Southall Norwood, England	26,377
Southampton, Eng.	119,039
South Bend, Ind.	53,684
Southend on Sea, England	62,723
Southgate, England	33,613
South Omaha, Neb.	26,259
Southport, England	51,650
South Shields, Eng.	108,649
Spandau, Germany	81,855
Speyer, Ger.	23,045
Spezia, Italy	65,612
Spokane, Wash.	104,402
Springfield, Ill.*	51,678
Springfield, Mass.	88,926
Springfield, Mo.	35,201
Springfield, Ohio	46,921
Srinagar, India	126,344
Stafford, England	23,385
Stalybridge, England	26,514
Stamford, Conn.	25,138
Stanislau, Austria	29,956
Stanley, England	23,400
Stargard, Germany	27,551
Stolberg, Germany	22,009
Stavanger, Norway	36,641
Stavropol, Russia in Asia	54,834
Stendal, Germany	27,263
Stettin, Germany	236,113
Stirling, Scotland	21,200
Stockholm, Sweden*	341,986
Stockport, England	108,693
Stockton-on-Tees, England	52,158
Stoke-upon-Trent, England	234,553
Stolp, Germany	33,762
Stralsund, Germany	33,988
Strassburg, Germany	178,913
Stratford, England	42,196
Stretford, England	43,000
Stryj, Austria	23,400
Stuttgart, Germany	285,589
Suchau, China	500,000
Sucre, Bolivia*	23,416
Suez, Egypt	24,970
Sumy, Russia	28,511
Sunderland, England	151,162
Superior, Wis.	40,384
Surabaya, Java	150,198
Surakarta, Java	118,378
Surat, India	114,863
Sutton, England	21,275
Sutton in Ashfield, England	21,707
Sutton in Coldfield, England	20,132
Swansea, Wales	114,673
Swatow, China	68,000
Swindon, England	50,771
Swinton and Pendlebury, England	30,759
Sydney, Australia	637,102
Sydney, Canada	17,723
Syracuse, Italy	40,587
Syracuse, N. Y.	137,249
Syzran, Russia	43,977
Szegedin, Hungary	118,328
Szekesfejervar, Hun.	36,625
Szentes, Hungary	31,593
Tabriz, Persia	200,000
Tacoma, Wash.	83,743
Taganrog, Russia	60,089
Taihoku, Formosa, Japan	118,000
Tainan, Formosa, Japan	53,794
Takamatsu, Japan	42,574
Takaoka, Japan	33,603
Takasaki, Japan	39,961
Talca, Chile	43,331
Tambof, Russia	52,942
Tammerfors, Russia	44,473
Tampa, Fla.	37,782
Tampico, Mexico	20,000
Tamsui, Formosa, Japan	100,000
Tananarivo, Madag'r*	94,813
Tangier, Morocco	30,056
Tanjore, India	57,870
Tanta, Egypt	54,437
Taranto, Italy	60,733
Tarbes, France	26,055
Tarnopol, Austria	33,871
Tarnow, Austria	31,700
Tarragona, Spain	26,285
Tashkend, Russia	201,191
Taunton, England	22,563
Taunton, Mass.	34,259
Tegucigalpa, Honduras*	34,692
Teheran, Persia*	280,000
Tembledong, P. I.	20,136
Temesvar, Hungary	72,555
Tempelhof, Germany	20,732
Teplitz, Austria	24,560
Teramo, Italy	24,538
Terni, Italy	30,641
Terre Haute, Ind.	58,157
Theodosia, Russia	27,238
Thorn, Germany	46,227
Tientsin, China	900,000
Tiflis, Russia	196,935
Tilburg, Neth.	50,405
Tilsit, Germany	39,013
Tipton, England	31,763
Tiraspol, Russia	29,323
Tirnovo, Bulgaria	25,295
Tiumen, Rus. in Asia	35,000
Tlemcen, Algeria	39,757
Tobaco, P. I.	21,946
Tobolsk, Russia in Asia	21,401
Todmorden, England	25,455
Tokat, Turk. in Asia	30,000
Tokushima, Japan	65,561
Tokyo, Japan*	2,186,079
Toledo, Ohio	168,497
Toledo, Spain	23,393
Toluca, Mexico	31,247
Tomsk, Russia in Asia	107,162
Tondo, P. I.	39,043
Topeka, Kan.*	43,684
Toronto, Canada	376,538
Torquay, England	38,772
Torre del Greco, Italy	35,299
Torreon, Mexico	13,845
Tortosa, Spain	25,368
Totonicapam, Guatemala	28,310
Tottenham, England	137,457
Tottori, Japan	32,682
Toulon, France	104,582
Toulouse, France	149,576
Tourcoing, France	82,644
Tournay, Belgium	37,976
Tours, France	73,398
Toyama, Japan	57,437
Toyohashi, Japan	43,980
Trani, Italy	31,800
Trapani, Italy	59,375

City	Population
Trebizand, Turkey in Asia	51,000
Tredegar, England	23,604
Trenton, N. J.*	96,815
Treves, Germany	48,975
Treviso, Italy	41,027
Trichinopoli, India	122,028
Trient, Austria	24,868
Trieste, Austria	229,475
Trikhala, Greece	17,809
Tripoli, Tripoli*	35,000
Tripoli, Turkey in Asia	30,000
Trivandrum, India	57,882
Trondhjem, Norway	46,252
Troy, N. Y.	76,813
Troyes, France	55,486
Tsaritzyn, Russia	79,759
Tsu, Japan	41,299
Tucuman, Argentina	74,865
Tula, Russia	130,800
Tunbridge Wells, Eng.	35,703
Tunis, Tunis*	227,519
Turin, Italy	427,733
Turnhout, Belgium	23,613
Tver, Russia	59,083
Twickenham, Eng.	29,374
Tynemouth, England	58,622
Uccle, Belgium	27,251
Udaipur, India	45,595
Udine, Italy	47,626
Ufa, Russia	96,295
Ujpest, Hungary	55,197
Ulm, Germany	55,817
Uman, Russia	31,016
Upsala, Sweden	25,620
Uralsk, Russia in Asia	45,054
Urdaneto, P. I.	20,554
Uria, Turkey in Asia	55,000
Urga, China	30,000
Urmia, Persia	35,000
Uskup, Turkey	26,000
Utica, N. Y.	74,419
Utrecht, Neth.	119,006
Utsunomiya, Japan	47,114
Valence, France	26,964
Valencia, Spain	233,348
Valencia, Venezuela	46,000
Valenciennes, France	34,766
Valetta, Malta	61,268
Valladolid, Spain	67,742
Valparaiso, Chile	162,447
Van, Turkey in Asia	30,000
Vancouver, Canada	100,401
Vannes, France	23,061
Varna, Bulgaria	41,317
Velbert, Germany	23,134
Vellore, India	43,537
Venice, Italy	160,727
Vera Cruz, Mexico	29,164
Vercelli, Italy	54,246
Verdun, France	21,706
Verona, Italy	81,913
Versailles, France	60,458
Verviers, Belgium	47,248
Viborg, Russia	33,525
Vicenza, Italy	44,777
Victoria, Brazil	20,000
Victoria, Canada	31,660
Victoria, Hong Kong	217,773
Vienna, Austria*	2,031,468
Viersen, Germany	30,172
Vigevano, Italy	23,909
Villa Rica, Paraguay	30,000
Villeurbanne, France	42,526
Vilna, Russia	184,582
Vincennes, France	38,568
Vina del Mar, Chile	26,262
Vinnitsa, Russia	34,060
Vitebsk, Russia	101,011
Viterbo, Italy	21,292
Vitoria, Spain	32,577
Vittoria, Italy	32,151
Vladikavkas, Russia	70,369
Vladimir, Russia	32,029
Vladivostok, Russia in Asia	90,162
Voghena, Italy	20,661
Volo, Greece	23,563
Vologda, Russia	31,460
Volsk, Russia	27,572
Voronezh, Russia	74,417
Vyernyi, Russia in Asia	31,317
Vyotka, Russia	42,475
Warsaw, Russia	764,054
Warwick, R. I.	26,629
Washington, D. C*	331,069
Waterbury, Conn	73,141
Waterford, Ireland	27,430
Waterloo, Iowa	26,693
Waterloo with Seaforth, England	26,399
Watertown, N. Y.	26,730
Watford, England	40,953
Wattenscheid, Ger.	27,636
Wattrelos, France	27,503
Wednesbury, Eng.	28,108
Weimar, Germany	34,581
Weissenfels, Ger.	33,581
Wellington, New Zealand*	70,729
Wenchau, China	80,000
Werdau, Germany	20,830
Wesel, Germany	24,441
West Bromwich, Eng.	68,345
West Ham, England	289,102
West Hartlepool, England	63,932
West Hoboken, N. J.	35,403
Weston super Mare, England	23,235
Weymouth and Melcombe Regis, England	22,325
Wheeling, W. Va	41,641
Wichita, Kan.	52,450
Widnes, England	31,544
Wiener Neustadt, Austria	32,874
Wiesbaden, Germany	109,002
Wigan, England	89,171
Wiju, Chosen, Japan.	20,000
Wilhelmshaven, Ger.	35,044
Wilkes-Barre, Pa.	67,105
Willesden, England	154,267
Williamsport, Pa	31,860
Wilmersdorf, Ger.	109,716
Wilmington, Del	87,411
Wilmington, N.C.	25,748
Wimbledon, England	58,876
Winchester, Eng.	23,380
Windsor, Canada	17,829
Winnipeg, Canada	136,035
Winterthur, Switz.	25,066
Wishaw, Scotland	25,263
Wismar, Germany	24,376
Withington, England	36,201
Witten, Germany	37,450
Wittenberg, Ger.	22,419
Wittenberge, Ger.	20,600
Woking, England	24,810
Wolstanton United, England	27,341
Wolverhampton, Eng	95,333
Wood Green, Eng.	49,372
Woonsocket, R. I.	38,125
Worcester, England	47,987
Worcester, Mass	145,986
Workington, Eng.	25,099
Worksop, England	20,387
Worms, Germany	46,821
Worthing, England	30,308
Wuchang, China	800,000
Wuchau, China	59,000
Wuhu, China	129,000
Wursburg, Germany	84,387
Yakoba, No. Nigeria	42,234
Yamagata, Japan	42,234
Yarkand, China	120,000
Yarmouth, England	55,808
Yaroslaf, Russia	71,616
Yeisk, Russia	42,513
Yenhou, China	60,000
Yenpingiu, China	200,000
Yead, Persia	50,000
Yochau, China	20,000
Yokkaichi, Japan	30,140
Yokohama, Japan	394,303
Yokosuka, Japan	24,750
Yonezawa, Japan	35,380
Yonkers, N. Y.	79,803
York, England	82,297
York, Pa	44,750
Youngstown, Ohio	79,066
Yuriep, Russia	42,308
Zaandam, Neth	24,379
Zacatecas, Mexico	25,905
Zagazig, Egypt	35,715
Zanesville, Ohio	28,026
Zanzibar, Zanzibar*	35,000
Zara, Austria	32,506
Zeitz, Germany	33,093
Zhitomir, Russia	90,830
Zittau, Germany	37,084
Zombor, Austria	29,600
Zurich, Switzerland	189,088
Zwickau, Germany	73,538
Zwolle, Netherlands	34,055

DESCRIPTIVE GAZETTEER
OF THE
PRINCIPAL CITIES OF THE WORLD

Abbreviations: Cap=Capital; c.h.=County Seat.

Aachen, (Fr. *Aix-la-Chapelle*), Germany (Prussia), is an important industrial and commercial center, with iron and steel works and noted manufactories of woollen, cloth, chemicals, glass, and machinery. Large coal fields are in the neighborhood. It has several hot and cold mineral springs well-known since the time of Charlemagne. Aachen is rich in historical associations. It has a famous ancient cathedral, which contains the tomb of Charlemagne, who greatly favored the town, and who died here in 814. From 813 to 1531 the German emperors were crowned here, and the coronation hall, where the banquet after the coronation took place and which is located in the city hall, has been now restored according to its original form. Two celebrated treaties of peace were concluded here: one in 1668 between France and Spain, by which France secured a portion of Flanders; and one in 1748, which terminated the war of the Austrian Succession. A famous congress of the principal European Powers took place here in 1818, in order to settle the affairs of Europe after the final overthrow of Napoleon.

Aalborg, Denmark (Jutland), is a seaport with important fisheries and an active export in agricultural produce. It is the seat of a bishop.

Aarhuus, Denmark (Jutland), is a seaport with a large transit trade. It is a bishop's see, and contains a fine Gothic cathedral dating from the 13th century.

Abbeokuta, Southern Nigeria, Africa, is a town with an extensive trade, situated 60 miles from Lagos on the Guinea coast, with which it is connected by railroad.

Aberdare, Wales (Glamorgan), has extensive collieries, and iron and tin works.

Aberdeen, Scotland (Aberdeen), is a seaport and the chief city in the North of Scotland. It is a pretty city, built chiefly of granite and hence called "the Granite City." The principal industries are salmon and herring fisheries, fish-curing, engineering, granite-polishing, and the manufacture of paper, leather, combs, woollens, and cottons. It has a university.

Aberdeen, South Dakota (c. h. Brown Co), is an important railroad and trade center in a prosperous agricultural section.

Abo, Russia (Finland), is a seaport and coaling station, and has large exports of timber and iron. It is the seat of an archbishop.

Accrington, England (Lancashire), has extensive cotton mills, calico-printing works, iron-foundries, and manufacture of chemicals. Coal-mines are nearby.

Acireale, Italy (Catania), is a seaside health resort located at the foot of Mt. Etna, and has several mineral springs. The famous cave of Polyphemus and the grotto of Galatea are in the neighborhood. It is the seat of a bishop.

Acton, England (Middlesex), is a suburb of London, situated 8 miles west of St. Paul's Cathedral.

Adams, Massachusetts (Berkshire Co.), contains manufactories of soap, cotton, warp, etc.

Adana, Turkey in Asia (cap. Adana), has a large trade in wool, wheat, and wine, and is connected by railroad with Mersina, a port on the Mediterranean.

Addis Abeba, cap. Abyssinia (Shoa), has an extensive trade. A treaty of peace was signed here in 1896 between Abyssinia and Italy, by which Italy renounced her claim of a protectorate over Abyssinia.

Adelaide, cap. South Australia, Australia, is spaciously laid out on a large plain at the base of Mount Lofty range, and is about 7½ miles from its seaport, Port Adelaide. It has a university, a botanical and zoological garden, and chief among its public buildings are the parliament houses, government offices, city hall, museum, and library. It is the seat of an Anglican and of a Roman Catholic bishop. Adelaide is the great emporium of trade for South Australia, and exports large quantities of wool, wheat, flour, wine, and copper ore.

Aden, Arabia, is a strongly fortified seaport and coaling station belonging to England. It is of great strategic and commercial importance especially since the opening of the Suez Canal. The principal exports are coffee, gums, spices, pearls, ostrich feathers, and hides.

Adrian, Michigan (c. h. Lenawee Co.), is a trade center and manufacturing town, and has a college and a state industrial school for girls.

Adrianople, Turkey (cap. Adrianople), is one of the principal commercial cities of Turkey in Europe, with manufactures of silk and tobacco, and large trade in perfumes and wine. It was the capital of the Turkish Empire from 1366 to 1453, and contains a beautiful mosque and a palace. It was occupied by the Russians during the Russo-Turkish war of 1829, and the treaty of peace, which concluded that war, was signed here. It was again occupied by the Russians during the Russo-Turkish war of 1877-8.

Agra, India (United Provinces), is a fortified town, a great railroad center, and one of the leading commercial places in the northwestern part of India. The leading manufactures are gold lace and inlaid mosaic work, and the principal articles of trade are cotton, wheat, salt, tobacco, and sugar. Agra was the capital of the Mogul Empire from the beginning of the 16th century to 1658, and it has conserved from that period some magnificent buildings. Of these the Taj Mahal is famous as a gem of Indian architecture in a setting of beautiful gardens. It is a mosque-like mausoleum, built in the 17th century by the Emperor Shah Jehan, in commemoration of his favorite queen. It is of white marble with dome and minarets and is ornamented with the exquisite mosaics for which Agra is still famous. The Moti-Masjid, or "Pearl Mosque," closely rivals in beauty the Taj Mahal.

Agram, Hungary (cap. Croatia and Slavonia), is the chief city of Croatia, and an archiepiscopal see. It has manufactures of linen, tobacco, and hides. It has a university, a national museum, and a fine Gothic cathedral, dating partly from the 11th century.

Aguascalientes, Mexico (cap. Aguascalientes), is famous for the hot springs from which it derives its name. It has large smelters, railway shops, and an active trade.

Ahmadabad, India (Bombay), has manufactures of cotton , gold-brocaded silks, pottery, paper, and tin. It was founded in 1412, and was formerly a city of great splendor, but has now fallen into decay. Its architectural remains, however, are even now magnificent, and often exhibit striking combinations of Jain and Saracenic characteristics. The most important buildings are the Great Mosque, adorned with two superbly decorated minarets; and the Ivory Mosque, so called because, although built of marble, it is lined with ivory and inlaid with gems.

Aidin, Turkey in Asia (Aidin), has manufactures of morocco leather and considerable trade in figs, olives, and grapes.

Aix, France (Bouches-du-Rhône), is the seat of an archbishop and has a university, a large library, and an old cathedral. The town has grown up around a hot saline spring known before the Roman invasion. Produces olive-oil and wine, and manufactures hats.

41

Aix-les-Bains, France (Savoie), is a famous health resort, with numerous warm mineral springs.

Ajaccio, France (cap. Corsica), is famous as the birthplace of Napoleon. It is a seaport with a good harbor, and has a large trade in olive-oil and wine as well as important anchovy and pearl fisheries. It is a favorite winter health resort. It is the seat of a bishop.

Ajmere, India (Rajputana), is an ancient city and famous place of pilgrimage, containing the tomb of a Mahommedan saint of the 13th entury. It exports large quantities of cotton.

Akerman, Russia (Bessarabia), is a fortified seaport with a good harbor. It has large fishing establishments, and exports a good wine produced in the neighborhood.

Akita, Japan (cap. Akita), is a seaport with large exports of rice and manufactures of cloth and silk.

Akron, Ohio (c. h. Summit Co.), has manufactures of cereals, woollens, agricultural implements, and automobile tires; has large printing works and a college.

Alameda, California (Alameda Co.), is a suburb of Oakland, with pleasant gardens and homes. There is some shipbuilding here.

Albany, cap. New York (c. h. Albany Co.), is a city expanded from one of the earliest fore'gn settlements in the United States, having been started in 1540 by the French and carried on by the Dutch, who erected Fort Orange. It was captured by the English in 1664, and then named in honor of the Duke of Albany. Situated near the head of navigation of the Hudson River, connected by canals with the Great Lakes and the Lake Champlain system, and being at the junction of several railroad lines, Albany has commercial importance. Its leading industries are printing, bookbinding, brewing, and the manufacture of stoves and other metal articles. Among its prominent buildings are the State Capitol, completed in 1898, which cost nearly $25,000,000; and the Education Building, which contains, besides the offices of the State Department of Education a valuable Library of over 500,000 volumes, a Library School, and a Museum of Natural History, all under State control. Other prominent buildings are the Cathedral of A l Saints, Dudley Observatory, and the Ci y Hall, designed by H. H. Richardson.

Albuquerque, New Mexico (c. h. Bernalillo Co), is the largest city of New Mexico, and an important railroad and trade center. It is the seat of the University of New Mexico. Gold, silver, and copper mines are in the vicinity.

Alcamo, Italy (Trapani), is an ancient city of Sicily, built on the north slope of Mt. Bonifato, and contains many interesting ruins of mediæval bu ldings.

Alcoy, Spain (Alicante), is an important industrial town, with manufactures of cloth, linen, matches, and paper—especially cigarette-paper.

Aldershot, England (Hampshire), has nearby the largest military camp in England.

Aleppo, Turkey in Asia (cap. Aleppo), is an important and historic trade center of Northern Syria, and one of the prettiest cities of the East, with clean, well-paved streets, stately buildings and numerous mosques with their fine cupolas and m narets. Its Italianized name is remin scent of the trade relations with Venice, active to the close of the 15th century, when the discovery of the Cape route to India deprived Aleppo of its position as the principal emporium of trade between Asia and Europe. Its trade is still very extensive, and it has, besides manufactures of silks, cottons. carpets, gold and silver threadstuffs, soap, and leather goods. Its port is Alexandretta, on the Mediterranean about 70 miles to the west.

Alessandria, Italy (cap. Alessandria), is the strongest fortified town in Northern Italy and an important railroad center. It has manufactures of silk, linen, hats, and macaroni, and carries on an active trade. It is the seat of a bishop. Near Alessandria took place in 1800 the famous battle of Marengo.

Alexandria, Egypt, is the principal port and coaling station of the country, and one of the chief commercial places on the Mediterranean, with large exports of cotton, cotton-seed, sugar, dates, tobacco, etc. It is

the seat of a Roman Catholic archbishop. Alexandria was one of the most famous cities of antiquity. It was founded by Alexander the Great in 332 B. C., and soon became the center of the world's commerce. It reached its highest splendor under the Ptolemies, when it was second only to Rome, and was the principal seat of Greek culture, which spread from here to all parts of the world (3rd century B. C.–4th century A. D.). It contained the famous Alexandrian library, the largest of the ancient world. One of the so-called Cleopatra's Needles, two obelisks of the 16th century B. C. which stood at Alexandria, was presented in 1881 by the Khedive of Egypt to the United States, and was placed in Central Park in New York; the other stands on the Thames Embankment in London.

Alexandria, Virginia (Alexandria Co.), has numerous buildings of historical interest including Christ Church, with the pews in which George Washington and Robert E. Lee worshipped; the old Carlyle House, the headquarters of General Braddock in 1755; and the so-called Lord Fairfax House, a fine example of Colonial architecture. Nearby is Mt. Vernon, the home of George Washington.

Alexandropol, Russia (Erivan), is an important fortress and town of Russian Armenia.

Alexandrovsk, Russia (Ekaterinoslaf), has an extensive grain trade and three annual fairs.

Algiers, cap. Algeria (cap. Algiers), is a strongly fortified seaport and coaling station, which since the French occupation has been transformed from an oriental town to a place of even more modern aspect than many European cities. It has large open squares and gardens, and imposing public buildings, among which are the palaces of the government, the Roman Catholic Cathedral, the military academy, the astronomical observatory, etc It has became lately a favorite winter resort for Europeans. Algiers is the chief commercial center of the country and its principal exports are flour, esparto, wine, olive-oil, and fruit.

Alicante, Spain (cap. Alicante), is a seaport with large exports of lead, copper, almonds, licorice, esparto, and wine.

Aligarh, India (United Provinces), is the seat of a Mohammedan college, and contains a historic fort. The native city is called Koil.

Allahabad, India (United Provinces), is a fortified town, an important railroad center, and being situated on the great waterway between Calcutta and the North of India, is also a very active commercial center, marketing the cotton, sugar, indigo, and other products of the district. The present city with its great citadel was founded by Akbar in 1575. Allahabad is a sacred city of the Hindus and a much-frequented place of pilgrimage. It has a university.

Allegheny, Pennsylvania. See Pittsburg.

Allenstein, Germany (Prussia), has important iron-foundries and machine-shops.

Allentown, Pennsylvania (c. h. Lehigh Co.), has manufactures of iron, steel, silk, clothing, etc. It is the seat of Muhlenburg College, and of a woman's college.

Almeria, Spain (cap. Almeria), is a fortified seaport. It exports grapes, almonds, and oranges.

Alost, Belgium (East Flanders), is a fortified town with manufactures of lace, linen, cottons, and leather.

Alpena, Michigan (c. h. Alpena Co.), is a summer resort and milling town on Thunder Bay. It has extensive shingle and flour mills, tanneries, and cement works.

Altenburg, Germany (cap. Saxe-Altenburg), has manufactures of brushes, woollen goods, gloves, and cigars. Near the town is a famous ducal castle, dating from the 14th century and restored in 1865-70.

Altendorf, Germany (Rhenish Prussia), is a suburb of Essen, and contains numerous industrial colonies.

Alton, Illinois (Madison Co.), has several collegiate institutions and produces building-stone, brick, coal, agricultural implements, etc.

Altona, Germany (Prussia), is a free port and coaling station adjoining Hamburg, and has varied industries, including extensive manufactories of tobacco and cigars.

Altoona, Pennsylvania (Blair Co.), is a mining, manufacturing, and farming trade center. Here the locomotive and car works on the Pennsylvania railroad employ about 12,000 men.

Alwar, India (Rajputana), is a trade center, and contains a notable palace of the Maharajah.

Amarillo, Texas (c. h. Potter Co.), is an important trade center in a rich agricultural section of the "panhandle" of Texas. Dry farming is extensively carried on in the vicinity.

Ambala, India (Punjab), is a trade center and military station.

Amiens, France (cap. Somme), is a large manufacturing center, with extensive manufactories of cotton, velvet, plush, tapestry, linen fabrics, silk, beet sugar, and paper. It is the seat of a bishop and its cathedral, built in the 13th century, is a masterpiece of Gothic architecture. Other noteworthy buildings are the city hall (1600-1760), in which the Peace of Amiens was signed in 1802; the large museum in Renaissance style (1864); and the public library (1791).

Amoy, China (Fokien), is a treaty port and coaling station with a large, well-sheltered harbor for the largest vessels. Exports chiefly tea, sugar, and rice.

Amritsar, India (Punjab), is an important trade center and the religious metropolis of the Sikhs, a distinction which, along with its name (literally, "pool of immortality"), it owes to its sacred tank, in the midst of which stands the marble temple of the Sikh faith. There is also a remarkable moated fortress. It manufactures shawls, cotton and silks.

Amsterdam, Netherlands (North Holland), is a seaport and the largest and most important city of the country. It has an extensive harbor for the largest vessels, where the Amstel joins the "Y",—an enlarged arm of Zuider Zee—and is also connected with the North Sea by a large canal. Amsterdam is the headquarters of the shipping interests and of the financial institutions of the Netherlands. It exports butter, margarin, cheese, gin, linen, Westphalian coal, coke, and iron, and is the principal market for the produce of the Dutch colonies, namely, rice, coffee, sugar, spices, tin, petroleum, and timber. The chief industrial establishments are sugar-refineries, engineering-works, mills for polishing diamonds and other precious stones, breweries, distilleries, etc. Amsterdam is a picturesque city, somewhat semicircular in plan, composed chiefly of red brick houses erected on piles driven through loam and loose sand to a firmer sub-stratum. In the older houses goods were frequently stored in attics, and a characteristic feature is the great hooks and tackle projecting from the quaintly-fashioned gable ends. Many of the streets are threaded by canals, which carry much of the heavy traffic, and are crossed by about 300 bridges. A great palace and some famous old churches are of interest to travellers, but the superb art collections, comprising many paintings by renowned Dutch and Flemish masters, and particularly rich in Rembrandts are the pride of the city. Many of these are housed in the magnificent National Museum built 1877-85. It has a university, an academy of sciences, and botanical and zoological gardens, famous for the excellence and rarity of their exhibits, and the fine condition of the animals. The decline of Antwerp and the closing of the Schelde in the 16th and 17th centuries raised Amsterdam to the rank of the first commercial city in Europe at that time. Its present development dates from the opening of its North Sea Canal in 1876.

Amsterdam, New York (Montgomery Co.) is an industrial center, with extensive manufactures of carpets, knit-goods, silks, etc.

Ancona, Italy (cap. Ancona), is a fortified seaport and naval arsenal, and the most important Italian port on the Adriatic after Venice. It exports grain, fruit, hides, silks, asphalt, and sulphur. It is a bishop's see and has a cathedral dating from the 11th century.

Anderlecht, Belgium (Brabant), is a manufacturing suburb of Brussels with extensive textile works.

Anderson, Indiana (c. h. Madison Co.), has extensive manufactures of iron and steel, glass, brick, etc.

Andijan, Russia (Russian Turkestan), is a trade center situated in the midst of a fruit growing region.

Andria, Italy (Bari), is the seat of a bishop, and has a fine cathedral built in the 11th century. Manufactures majolica, and has large trade in almonds.

Angers, France (cap. Maine-et-Loire), has manufactures of silks, woollens, cottons, iron and copper wares, and a large trade in wine and agricultural produce. It is the seat of a bishop, and has a fine cathedral dating from the 13th century.

Angoulême, France (cap. Charente), has manufactures of paper and a large trade in cognac. It is the seat of a bishop, and has a fine cathedral in the Romanesque style dating from the 12th century. It is the birthplace of Marguerite of Navarre (1492), author of the "Heptameron."

Ann Arbor, Michigan (c. h. Washtenau Co.), has important manufactories of agricultural implements. It is the seat of the state University of Michigan, founded in 1837, and one of the most important educational institutions in the United States; it is attended by about 5,000 students, and is richly endowed.

Ansonia, Connecticut (New Haven Co.), has manufactures of clocks, brass and copper goods, woollens, etc.

Antequera, Spain (Malaga), has a large trade in olive-oil and fruit.

Antigua, Guatemala, was the ancient capital of the country. Its inhabitants are largely engaged in making carved cane heads and dolls, representing very accurately national costumes and customs.

Antofagasta, Chile (Antofagasta), is a seaport with a rough landing through surf. It exports saltpeter, borax, silver and copper ores, etc.

Antwerp, Belgium (cap. Antwerp), is a strongly fortified city, and has one of the finest harbors in the world, made by dredging out the river Schelde. Its quays extend for more than 3½ miles, and it is a coaling station and an important port for the interior, being the terminus of many steamship lines. Diamond-cutting is one of its important trades, and its manufactures include silks, velvets, laces, serges, and flannels; these, with grain, chemicals, coal, iron, glass, and various other manufactured articles are exported. There are also extensive shipbuilding yards and sugar-refineries. While possessing handsome modern residences and business sections, Antwerp is famous for its many ancient edifices; such as the city hall and the guild houses, partially enclosing the square upon which faces the famous cathedral, which was begun in 1352 and completed in 1592, and which is a masterpiece of Gothic architecture. The house and shop of Christopher Plantin, one of the earliest and most famous printers, stands now substantially as it did in the owner's time, having been transformed into a museum. Everywhere are ancient wood-carvings, iron-work, and other metal objects, which are of much interest to students of art and antiquities. There is an art school in Antwerp, and a fine art gallery containing many modern paintings, as well as old masters. The works of Rubens, Van Dyck, and other Flemish painters are treasured in the museums and churches. Rubens is buried in the Church of St. Jacques, but three of his most magnificent religious pictures, the Assumption, the Descent from the Cross, and the Elevation of the Cross, are hung in the cathedral. There is a famous zoological garden in Antwerp. Before its occupation by the Spaniards in the 16th century Antwerp was among the greatest commercial cities of the world. It soon sank into utter ruin, but began to prosper again during the French occupation at the beginning of the 19th century.

Aomori, Japan (cap. Aomori), has a fine natural harbor and considerable trade.

Apeldoorn, Netherlands (Gelderland), has papermills. Nearby is the castle of Het Loo, the summer residence of the royal family.

Appleton, Wisconsin (c. h. Outagamie Co.), has flour, paper, wood-pulp, and woollen mills. It is the seat of several collegiate institutions.

Aquila, Italy (cap. Aquila degli Abruzzi), has manufactories of paper and linen, and a large trade in saffron, which is cultivated in the neighborhood. It is the seat of a bishop, and has a fine cathedral dating from the 13th century, which has been lately restored.

Arad, Hungary, is a fortified town and an important railway center. It has large distilleries and flour-mills, and an active trade in agricultural produce and cattle. It is the seat of a Greek Orthodox bishop.

Arequipa, Peru (cap. Arequipa), is the second largest town in Peru, and has a large trade in wool, alpaca, gold and silver ores, and cattle. It is situated near the volcano El Misti, and has suffered severely from earthquakes. It is the seat of a Catholic bishop. Arequipa was founded by Francisco Pizarro in 1540.

Arezzo, Italy (cap. Arezzo), is an ancient Etruscan city, which abounds in architectural remains of the Middle Ages, including a 13th century cathedral. It is the seat of a bishop. The poet Petrarch was born here in 1304.

Argao, Philippine Islands (Cebu), is a seaport with a good harbor.

Arles, France (Bouches-du-Rhône), is one of the oldest towns of southern France, with famous Roman ruins, the most extensive being the great amphitheatre, which is supposed to have surpassed that of Nimes, but which has since been converted into a bullring. Among the more modern buildings of note are the mediæval cathedral and the Church of Notre Dame. It has shipbuilding yards and manufactures silk and hats.

Armentières, France (Nord), has extensive manufactures of linen and cotton.

Arnhem, Netherlands (cap. Gelderland), has interesting churches and public buildings, which contain valuable art collections. It manufactures furniture, glass, and paper.

Aschersleben, Germany (Prussia), has manufactures of machinery, linen, woollen goods, etc.

Ascoli, Italy (cap. Ascoli-Piceno), is an ancient city with many interesting remains including a very ancient cathedral. It is the seat of a bishop. Manufactures majolica, glass and paper.

Ascot, England (Berkshire), is a famous race-course situated 6 miles southwest of Windsor. The races, which are held in June and are attended by the royal family in semi-state, are one of the social events of the year.

Asheville, North Carolina (c. h. Buncombe Co.), is a widely-known health resort, frequented by southerners for its comparative coolness in summer (mean temp. 72°) and by northern folk for its mildness in winter (39°) and spring (53°). Its climate is dry and bracing, and is said to be pre-eminently suitable for checking the early stages of tuberculosis, while sufferers from asthma, hay fever, etc., derive benefit here. The surroundings are notably picturesque.

Ashkabad, Russia in Asia (Turkestan), has a large distributing trade.

Ashland, Wisconsin (c. h. Ashland Co.), is at the head of navigation of Lake Winnebago and Green Bay waterway, and is a shipping port for the hematite ore of the Gogebic Iron Range. It has enormous ore docks, large charcoal blast-furnaces, and lumber mills. It has several collegiate institutions.

Ashtabula, Ohio (Ashtabula Co.), is an important iron ore shipping point, and has large iron-works, tanneries, etc.

Ashton-under-Lyne, England (Lancashire), has extensive manufactures of cotton, and large dyeing establishments.

Asnières, France (Seine), is a residential suburb of Paris, on the Seine.

Asti, Italy (Alessandria), is the seat of a bishop and has a Gothic cathedral (13th century) with numerous fine paintings. Asti is famous for its wine.

Aston Manor, England (Warwick), has extensive machine-shops and manufactures of arms, toys, etc.

Astrakhan, Russia (cap. Astrakhan), is the chief port of the Caspian Sea, and is situated about 40 miles from the mouth of the Volga. It has a large distributing trade in cottons, fish, caviare, rice, wool, petroleum, timber, and fruit. One of the largest fairs in Russia is held here annually. It is the seat of a Greek Orthodox and of an Armenian bishop.

Asuncion, cap. Paraguay, is at the head of navigation of the Parana River, and has an extensive trade in tobacco, hides, and Paraguay tea. It is one of the oldest towns in South America, having been founded on August 15, 1536. It is the seat of a Catholic bishop.

Atchison, Kansas (c. h. Atchison Co.), has a large trade in grain, flour, and live stock, and more than fifty important manufacturing establishments. It is the seat of the State Soldiers' Orphans' Home, an insane asylum, and several collegiate institutions.

Athens, Georgia (c. h. Clarke Co.), has a large trade in cotton, and numerous manufactories of cotton products. It is the seat of the University of Georgia and of the State College of Agriculture, and has a State Normal School and several other educational institutions.

Athens, cap. Greece (cap. Attica), was anciently the capital of the state of Attica, and the chief center of Greek literature, art, and culture. It is now the capital of the modern kingdom of Greece, and the seat of a Greek Orthodox archbishop. It is a well-built city with a fine royal palace, many handsome private residences, a university, and other educational institutions. It has an important trade, and manufactures of morocco leather, and silks. The city, which takes its name from Athena, "goddess of science, arts and arms," who was its own patron divinity, was originally built on the Acropolis,—a conspicuous limestone rock rising 500 ft. above the Attic plain,—but ultimately extended to the plain itself; while the Acropolis became the citadel and subsequently the site of a group of beautiful temples at the time of Pericles (5th century B. C.). The ruins of the Parthenon, the Erechtheum, the temple of Nike Apteros ("Wingless Victory"), and the Propylaea, still remain to testify to the former glory of the Acropolis. Of the other ancient buildings the most notable are the Theseum (also of the Periclean period, and still almost perfect); the fragments of the vast temple of Zeus (begun in 530 B. C. and finished by the Roman Emperor Hadrian in the 2nd century A D.); the theatre of Dionysus, etc. Not far from the Acropolis rose the hill Lycabettus (911 ft.), and the hillocks or ridges of the Pynx and of the Areopagus, or Mars Hill. At a greater distance the plain is bounded by Hymettus (3,368 ft.), Pentelicus (3,641 ft.), and other ranges. According to tradition Athens was founded by the legendary hero Cecrops. The most brilliant period of the history of Athens was when, after the Persian wars (5th century B. C.), it took the lead amongst the Greek states, became powerful by land and sea, was adorned by Pericles with most glorious buildings, and brought Greek literature, philosophy, and art to their highest development. Its decline dates from the disastrous conclusion of the Peloponnesian war (403 B. C.). It was plundered and ruined by Sulla in 87 B. C., and neither under Byzantine nor Turkish rule ever attained any prosperity. Many of the buildings on the Acropolis have been destroyed since the 15th century, especially during the Venetian raids.

Atlanta, cap. Georgia (c. h. Fulton Co.), is a prosperous commercial and industrial city, an important railway center, and the largest wholesale distributing point of the southeastern states. It has extensive manufactories of cotton, fertilizers, patent medicines, car wheels, flour and iron, and is one of the chief distributing points in the South for northern and western manufactures. The great staples of its trade are tobacco and cotton. The chief point of interest in the history of Atlanta, which was founded in 1840, is its siege and capture (Sept. 2nd, 1864) by Gen. Sherman, who, after holding the city for two months, began here his famous "March to the Sea."

Atlantic City, New Jersey (Atlantic Co.), is the most frequented seaside resort of America, the population of which is increased about fivefold in August by visitors from all over the country. It is also frequented in winter and spring, the climate being comparatively mild and sunny and the air exceedingly bracing. The beach is one of the finest in America, and from 50,000 to 100,000 people often bathe here in one day. It has a board walk 40 ft. wide and over 5 miles long, flanked on the landward side by hotels, shops, and places of amusement.

Attleboro, Massachusetts (Bristol Co.), has exten-

sive manufactories of jewelry and electro-plate, also of woollen and knit goods, boots and shoes.

Aubervilliers, France (Seine), is a suburb of Paris with manufactures of leather, perfumery, chemicals, etc.

Auburn, Maine (c. h. Androscoggin Co.), has extensive manufactories of cotton goods, and boots and shoes.

Auburn, New York (c. h. Cayuga Co.), has large manufactories of agricultural implements, tools, carpets, and shoes. The State Prison, located here, contains over 1,200 convicts.

Auckland, New Zealand (cap. Auckland), is an important seaport and coaling station, possessing a spacious harbor. It exports frozen meats, fish, fruit, flax, wool, hides, tallow, Kauri gum, gold, and timber, and has manufactories of glass, shipbuilding yards, and sugar refineries. It is the seat of an Anglican and of a Roman Catholic bishop. It was formerly the capital of New Zealand.

Augsburg, Germany (Bavaria), is an historic city with many interesting buildings. During the 15th and 16th centuries, Augsburg was one of the trade emporiums between Northern Europe, Italy, and the East. Its name is also prominently associated with the leading events of the Reformation in Germany. It manufactures textiles, machinery, metallic wares, and paper, and has an extensive trade in printing and book-binding. Its principal edifices are the Renaissance city hall (1620), with its splendid "Golden Hall"; the Perlach Tower, dating from the 11th century; the former episcopal palace, where, on the 25th of June, 1530, the Protestant princes presented the Augsburg Confession to Charles V.; and the Gothicised Romanesque cathedral (994-1421) with its bronze doors and early glass-paintings.

Augusta, Georgia (c. h. Richmond Co.), is at the head of navigation of the Savannah River and carries on a large trade in cotton. Its cotton mills, run by a system of water-power canals, produce more unbleached cotton than any other city in America. There is a United States Arsenal at Summerville (3 miles).

Augusta, cap. Maine (c. h. Kennebec Co.), is at the head of tidal navigation of the Kennebec River, one-half mile below the huge Kennebec dam, which affords water-power for its cotton factories, pulp-mills, shoe-manufactories, etc. In addition to the State House, the noteworthy buildings are the Maine Insane Aylum, the United States Arsenal, and the Lithgow Library.

Aurora, Illinois (Kane Co.), has large railroad car shops, cotton-mills, smelting-works, foundries, and machine-shops. It is the center of a prosperous farming section.

Aussig, Austria (Bohemia), has extensive coal-mines, chemical works, and shipbuilding yards.

Austin, cap. Texas (c. h. Travis Co.), is a wholesale supply center for an extensive district. The most prominent building is the Capitol, one of the largest structures of the kind in the United States; it is built of granite and cost $3,500,000. There are also the main building of the State University, which accommodates the law and literary departments; State asylums for the blind, the insane, and deaf-mutes; departments for colored patients of these classes; the State Confederate Home, and numerous collegiate institutions.

Aux Cayes, Haiti, is a seaport with exports of coffee, cocoa, and logwood.

Auxerre, France (cap. Yonne), has manufactures of cloth and trade in wine and leather. It contains a fine Gothic cathedral dating from the 15th century.

Avellino, Italy (cap. Avellino), is situated amidst groves of hazel-nuts, which were famous even in antiquity. It is the seat of a bishop.

Avignon, France (cap. Vaucluse), is celebrated in church history as the temporary seat of the popes, who resided here continuously from 1309 to 1378. From 1378 to 1418 it was the seat of the French antipopes. The most conspicuous buildings are the vast palace of the popes, and the cathedral, which date from the 11th century. During the middle ages the surrounding district formed a county, which was in the possession of the popes until 1790, when it was united with France. Avignon is the seat of an archbishop. Garden produce, fruit, wine, and honey of good quality are produced in the neighborhood.

Ayr, Scotland (cap. Ayr), is a seaport with large docks, and manufactures of carpets, woollens, and chemicals.

Ayuthia, Siam, was formerly the capital of the country.

Azof, Russia (Don Cossacks), is a trade center, of which fish-curing is the leading industry.

Badajoz, Spain (cap. Badajoz), is a bishop's see and has an interesting cathedral. It has extensive leather manufactures. It is a strongly fortified city and was captured by Wellington in 1812.

Baden, Germany (cap. Baden), is a famous health resort, much frequented for its mineral springs and baths.

Bagdad, Turkey in Asia (cap. Bagdad), is a walled city which, before the completion of the Suez Canal, commanded a large part of the traffic between Europe, Persia, and India. It has still a large and growing trade, the principal exports being wool, hides, gums, and dates. Founded in the 8th century it became soon afterwards the capital of the famous caliphate of Bagdad.

Bahia, Brazil (cap. Bahia), is a seaport and coaling station, which exports cocoa, coffee, tobacco, hides, rubber, sugar, cotton, and manganese. Among its public buildings are the government palace, the mint, and a cathedral. It is the seat of a Catholic archbishop.

Bahia Blanca, Argentina (Buenos Aires), is a port with a fine harbor, and a railroad center of considerable importance. Exports wood, sheepskins, hair, and wheat.

Baireuth, Germany (Bavaria), is famous as the place of the Wagner musical festivals, which are held in a national theatre built under the supervision of the composer who is buried in the garden of his villa here. Has manufactories of textiles.

Bakersfield, California (c. h. Kern Co.), is a railroad and trade center in a stock-raising and fruit-growing section. It has oil-refineries, machine shops, car shops and fruit-packing establishments.

Baku, Russia (cap. Baku), is a strongly fortified port and the chief Russian naval station on the Caspian Sea. It is the center of the most productive petroleum district in the world, over 1,500 oil-wells being operated in the neighborhood, and the oil is carried by pipes directly to the refineries. Baku was taken by Russia from the Persians in 1806.

Ballarat, Victoria, Australia, is a gold-mining center, where gold was discovered in 1851. When the surface diggings became exhausted after the first great rush, deposits of gold were found at greater depths, and now there are mines as deep as some coal-pits. The "Welcome Nugget", the largest ever found, was discovered in 1858 at Baker Hill. It weighed 2,217 oz., 16 dwt., and was sold for $52,000. Ballarat is the seat of an Anglican and a Roman Catholic bishop.

Baltimore, Maryland (c. h. Baltimore City Co.), is a port and coaling station and the metropolis of Maryland. It has an excellent harbor, with good anchorage for vessels, and no bar. It exports wheat, flour, cotton, tobacco, oil, lumber, and canned goods. Next to New York, it is the largest grain-market on the Atlantic coast. It is the chief seat of the canning industry in the United States, its materials being the famous oysters from the waters of Chesapeake Bay and fruit from its shores. There are extensive manufactures of iron, steel, and copper, and the Bessemer Steel Works at Sparrow's Point (9 miles) have a daily capacity of 2,000 tons. The cotton-duck mills in and near Baltimore produce three-fourths of the sail-duck made in the United States. Among its educational institutions are the famous Johns Hopkins university, and its adjunct, the Johns Hopkins hospital and medical school. A Roman Catholic Cathedral, containing some interesting paintings, and the house of the Archbishop of Baltimore, who is a cardinal and primate of the United States, are located here.

Bamberg, Germany (Bavaria), is an ancient city with many notable buildings, chief among them being

the cathedral built in the 13th century. Has manufactories of cottons and woollens.

Bangalore, India (cap. Mysore), has manufactures of silk, cotton, cloth, carpet, etc., and is noted for its salubrious climate.

Bangkok, cap. Siam, is a port and coaling station on the Menam, which exports rice, paddy, teak, ebony, sapan-wood, spices, gum, ivory, and hides. The internal traffic is chiefly carried on by means of canals, there being only a few passable streets in the city. A large number of the houses float on rafts moored to the banks of the river and its many canals; and the ordinary houses of the city, which are almost wholly of bamboo or other wood, are raised upon piles. The temples are decorated in the most gorgeous style; the Siamese take great pride in lavishing their wealth on them. In the neighborhood are iron-mines and forests of teak-wood.

Bangor, Maine (c. h. Penobscot Co.), is a port of entry and the chief city of eastern Maine, and is at the head of navigation of the Penobscot River. Its chief industry is the sawing and shipping of timber.

Banjermassin, Borneo (cap. Dutch Borneo), is a fortified city with extensive trade.

Barcelona, Spain (cap. Barcelona), is a seaport and coaling station, and the principal industrial and commercial city in Spain. It has extensive cotton-mills and iron manufactures, and exports cottons, woollens, paper, wines, fruits, and almonds. It is the seat of a bishop, and has a splendid Gothic cathedral built 1298-1448. It possesses a university, a naval institute, an academy of arts, and several large libraries. During the last fifty years Barcelona has been the scene of several revolts against the Spanish government. There was serious rioting in 1909, when many churches and convents were attacked.

Bareilly, India (United Provinces), manufactures fine furniture, cutlery, embroidered silk, and perfumery.

Barfrush, Persia (Mazandaran), is one of the principal trade centers in the country.

Bari, Italy (cap. Bari delle Puglie), is a seaport on the Adriatic, which exports wine, olive-oil, lemons, figs, tartar, and soap. The chief buildings are the Lombardic church of San Nicola (1087), the still older cathedral in Byzantine style, and the old Norman castle. It is the seat of an archbishop.

Barili, Philippine Islands (Cebu), has extensive fisheries and produces woven fabrics.

Barletta, Italy (Bari delle Puglie), is a seaport with a harbor which admits small vessels only, although larger ones find good anchorage ground in a roadstead. Trades in wine, olives, and other agricultural products.

Barmen, Germany (Prussia), is the seat of the Rhenish Missionary Society, which has here a large seminary. It has extensive manufactures of textiles, and is the largest ribbon manufacturing town on the continent.

Barnaul, Siberia (Tomsk), is a great mining and smelting center. There are here a mint for copper coins, a mining school, and an observatory.

Barnsley, England (Yorkshire), has manufactures of linen, iron, steel, and glass, and bleaching and dyeing works. There are numerous collieries in the neighborhood.

Baroda, India (Baroda), is a place of considerable trade and has numerous temples.

Barquisimeto, Venezuela (cap. Lara), is a well-built town in the center of a coffee producing section. Among its prominent buildings are the government palace, barracks, market, and cathedral. It is the seat of a Catholic bishop.

Barranquilla, Colombia (Bolivar), is a port at the head of navigation of the Magdalena River, 15 miles from the sea, and has a good harbor, dry dock, and two slipways. Exports gold, silver ore, hides, rubber, coffee, cattle, tobacco, and balsam.

Barrow-in-Furness, England (Lancashire), is a seaport with a good harbor, and has extensive shipbuilding yards, and large iron and steel works. Exports iron ore, pig iron, and steel rails. Barrow owes much of its prosperity to the discovery of the Bessemer process of steel-making and to the fact that the hema-

tite ores of this district are especially adapted to this process. Interesting ruins of Furness Abbey, which was founded in 1127, lie within two miles of the town.

Basel, Switzerland (cap. Basel), is the second largest city of Switzerland and an important railway center. It has large plants for obtaining electric power from the Rhine and has extensive manufactories of metal wares and of textiles, especially silks and ribbons. It has a university. A great church council, the Council of Basel, was held here, lasting from 1431 to 1443.

Basra, Turkey in Asia (Basra), is a port and coaling station on the Shat-el-Arab, 70 miles from its mouth into the Persian Gulf. Its trade is largely a transit one with Bagdad, and its principal exports are dates and barley.

Batangas, Philippine Islands (cap. Batangas), has a spacious and well sheltered harbor. It is an important center of industry and trade, and during February agricultural and industrial fairs are held. Exports coffee.

Batavia, Java (cap. Batavia), s the capital of the Dutch East Indies and a seaport and coaling station, with anchorage for large vessels. It is the chief mart among the islands of the Asiatic archipelago for the products of the eastern seas and the manufactures of the west. It exports coffee, rice, sugar, spices, indigo, hides, teak, tin, and tamarinds.

Batavia, New York (c. h. Genesee Co.), has extensive manufactories of plows and harvesters, boots and shoes, and contains a state institution for the blind.

Bath, England (Somerset), is famous for its medicinal water, four principal springs rendering no less than 184,000 gallons of water per day. They contain carbonic acid, chloride of sodium and magnesium, sulphate of soda, carbonate and sulphate of lime, etc. The baths are both elegant and commodious.

Bath, Maine (c. h. Sagadahoc Co.), has shipbuilding yards and manufactures of brass and iron goods. It has a coastwise trade in ice and lumber and is a port of entry.

Batley, England (York), manufactures heavy woollen cloths, and has large foundries and collieries.

Baton Rouge, cap. Louisiana (c. h. East Baton Rouge Parish), is the seat of the Louisiana State University, and contains a state institution for the deaf, dumb, and blind, a state agricultural and mechanical college, and an agricultural experimental station. Its manufactures include cotton-seed products, lumber, sugar, molasses, and agricultural implements.

Battle Creek, Michigan (Calhoun Co.), is an attractive industrial city and summer resort, with two large sanitariums and a college. Its manufactures include farm implements, cereals, etc.

Batum, Russia (Kutais), is a fortified port and coaling station on the Black Sea, with a sheltered harbor. Exports petroleum, manganese, walnut wood, licorice, wool, etc.

Bauan, Philippine Islands (Batangas), has a brisk coast and river trade.

Bautzen, Germany (Saxony), has manufactories of hosiery, gloves, and machinery. Here Napoleon won a barren victory over the allied Russians and Prussians on May 20-21, 1813.

Bay City, Michigan (c. h. Bay Co.), is noted for its large shipbuilding plants and extensive trade in lumber, and has many lumber mills.

Bayonne, France (Basses-Pyrénées), is a fortified port on the Adour, 3 miles from the Bay of Biscay with a good harbor. It exports wood, rosin, turpentine, cast iron, zinc, and steel. It is the seat of a bishop, and has a fine cathedral dating from the 13th century.

Bayonne, New Jersey (Hudson Co.), is a suburb of New York City. It has extensive petroleum refineries which are connected by pipe lines with the oil-fields of Pennsylvania, and several of the leading cities of the North Atlantic coast. Other industries are manufactures of chemicals, ammonia, and colors; coal is shipped here.

Beaumont, Texas (c. h. Jefferson Co.), is a railroad and trade center in a petroleum producing section. It has rice mills, car works, machine shops, and other industries. It ships large quantities of lumber, and has

an active wholesale trade in petroleum and general merchandise.

Beaver Falls, Pennsylvania (Beaver Co.), is situated in a region where coal and gas abound, and produces iron, steel, wire, glassware, and pottery.

Bedford, England (cap. Bedford), has extensive manufactories of agricultural implements and straw-plaiting. It is the birthplace of John Bunyan (1628).

Beirut, Turkey in Asia (Syria), is a seaport and coaling station on the Mediterranean, with a tideless harbor and no bar. Its trade is largely a distributing one, Beirut being the port for Damascus and other parts of Syria. Exports silk, oil, gums, and fruit. Much raw silk is produced from the silkworms in the immediate vicinity. It is the center of the Oriental book trade in Syria. Has several collegiate institutions, an observatory and many relics of antiquity. It is the seat of a Greek Orthodox and of a Maronite bishop, and is the headquarters of several foreign missions.

Bekes, Hungary, has a trade in cattle, corn, and honey.

Belfast, Ireland (cap Antrim), is a seaport and coaling station with a safe and easily accessible harbor, and the chief commercial and manufacturing city of Ireland. It is the center of the Irish linen trade, and has two large shipbuilding yards which have turned out some of the largest vessels afloat. It also has extensive breweries, distilleries, flour-mills, rope works, etc. The country around is extremely beautiful and the city is, perhaps, more picturesque than any other large commercial and manufacturing town in the British Isles.

Belfort, France (cap. Haut-Rhin), is a strongly fortified town, the center of the system of fortifications on the eastern frontier of France. It has steel and iron works and manufactures of textiles and hats.

Belgrade, cap. Servia, is a fortified city, which was the scene of many historic sieges. There are no industries of importance, but Belgrade is the chief emporium of trade for Servia.

Bellary, India (Madras), is a fortified city and military headquarters, with two forts. It has an active trade in cotton.

Belleville, Ontario, Canada, is a port on the north shore of the Bay of Quinte and has extensive flour mills, canning factories, cement wroks, etc.

Belleville, Illinois (c. h. Saint Clair Co.), manufactures glass, stoves and machinery. It has one of the largest rolling-mills in the west, and there are also coalmines.

Bellingham, Washington (c. h. Whatcom Co.), is an important trade center and port on Bellingham Bay. It has extensive saw mills, canneries, and other industries.

Bello Horizonte, Brazil (cap. Minas Geraes), is a modern city constructed by the government in 1894, and has some of the most beautiful public buildings in Brazil. There are gold mines in the vicinity.

Beloit, Wisconsin (Rock Co.), is the seat of Beloit College, organized in 1847. It has an extensive woodworking machinery plant and manufactories of gasengines, windmills, agricultural implements, and machinery.

Benares, India (United Provinces), is the most sacred place of the Hindus and one of the principal towns of North India. It contains over 1,400 Hindu temples or shrines and 272 Mohammedan mosques, and draws immense revenues from thousands of pilgrims who visit it from all parts of India. Its brassware, gold cloth, and lacquered toys are famous.

Bendery, Russia (Bessarabia), has manufactures of stoneware, paper, and leather.

Bendigo, Victoria, Australia, is the center of one of the largest gold-mining districts of Australia. It is also called, unofficially, Sandhurst.

Benevento, Italy (cap. Benevento), is an archiepiscopal city with some interesting buildings and antiquities. The well-preserved magnificent triumphal arch of Trajan, built in 114 A.D., deserves particular mention. The cathedral (12th and 13th centuries) is a beautiful building in the Lombardo-Saracenic style.

Berbera, British Somaliland, Africa, is a port on

the Gulf of Aden with a good harbor. and the chief town of British Somaliland. It has large exports of gum, rosin, skins, ostrich feathers, etc.

Berdiansk, Russia (Taurida), is the best port on the Sea of Azof, and exports grain, oil-seeds, and wool. In the neighborhood are salt-mines.

Berditchef, Russia (Kief), has a considerable trade in agricultural products.

Bergamo, Italy (cap. Bergamo), has large manufactures of textiles and iron goods and trades in silk, grain, etc. Has an academy of painting and sculpture, a museum, and is the seat of a bishop.

Bergen, Norway (cap. South Bergenhus), is a fortified seaport and coaling station with a commodious harbor. Exports codfish, cod-oil, cod roes, herrings, bones, and minerals. Manufactures earthenware, porcelain, and leather, and has distilleries and shipbuilding yards. Bergen is the native place of the poet Holberg. It is the seat of a bishop.

Berkeley, California (Alameda Co.), is the seat of the University of California, founded in 1868, an institution which has played a very important part in the educational development of the Pacific slope; it is attended by over 3,000 students. Here are also located the state agricultural college, the state institution for the deaf, dumb, and blind, and several preparatory schools.

Berlad, Roumania (Moldavia), is the trade center of a grain-raising district and has many distilleries.

Berlin, cap. Germany (cap. Prussia), is the third largest city of Europe and by far the largest city in Germany. For the beauty and size of its buildings, the regularity of its streets, the importance of its institutions of science and art, and its activity in industry and trade, it is one of the most imposing of European cities. It is doubtless the greatest manufacturing town on the continent, and is especially prominent in its manufactures of porcelain, clothing, and machinery. The city covers an area of about 40 square miles. The natural waterway of the Spree, which divides into several arms and receives the Panke coming from the north, determines the division of the city; the center consists of Old Berlin, Old Kolln and New Kolln, and the Friedrichswerder. In the heart of the city is the old royal palace, with nearly 700 apartments. Near this are the emperor's palace, the imperial residence; the royal library, which contains upwards of 1,400,000 volumes and 30,000 manuscripts; the old and new museums, the national gallery, the arsenal, the royal theatre, the opera-house, the guard-house, and the university. These are all situated between the Spree and the east end of the street "Unter den Linden" (so called from its double avenue of linden-trees). The city is adorned throughout with numerous statues of national heroes, the Great Elector, Frederick the Great, and many others, while in the Thiergarten is the famous Siegesallee, or Avenue of Victory, lined with 32 marble groups of the rulers of Brandenburg and Prussia, erected by order of Emperor William II. The university, established in 1809, possesses four museums and a library of 300,000 volumes. Other public institutions are the Academy of Sciences, the Military Academy, the Academy of Architecture; the Academic High School (of art); the School of Mines; the School of Agriculture; the Artillery, Technical, and Engineering Colleges; the Industrial (1881), Ethnological (1886), and other museums; the Academy of Music; and the Observatory. The Old Museum contains antiquarian specimens, a collection of 90,000 coins, a gallery of ancient sculpture, and a picture gallery with about 1,300 paintings. The New Museum contains six magnificent mural paintings by Kaulbach in the grand staircase, a very valuable collection of casts, the Egyptian museum, and 500,000 engravings. The National Gallery includes about 1,000 works by modern artists. The celebrated Brandenburg Gate leads from the "Unter den Linden" to the Thiergarten. To the southwest of this lies the Zoological Garden. The Botanical Garden (at Schöneberg) contains 25,000 species. Noteworthy also are the Cathedral (1894-1902), the city hall, the royal chateau of Monbijou, the Ruhmeshalle in the arsenal, the Gothic

monument on the Kreusberg, the Column of Peace in the Belle-Alliance-Plats, the Warriors' Monument, the column of Victory, the War Office, the new building for the Reichstag, the Exchange, and the Reichsbank. A great congress of the principal European Powers took place here in 1878, after the Russo-Turkish war of 1877-8, and resulted in the conclusion of the Treaty of Berlin, which settled many important matters connected with the "Eastern Question".

Berlin, Ontario, Canada, is a trade and manufacturing center in a good agricultural district. Its industries include furniture factories, tanneries, beet sugar works, etc.

Bern, cap. Switzerland (cap. Bern), is one of the finest towns of Switzerland, situated on a lofty sandstone promontory formed by the winding Aar, which surrounds it on three sides. It offers a magnificent view of the lofty peaks of the Bernese Oberland. The principal buildings are the Gothic cathedral (1421-1573), the magnificent Federal council hall (1857), the mint, and the university. It has an interesting museum and a valuable public library. Several important international conventions were held here, and Bern is the seat of several international bureaus, such as the International Postal Union, the International Telegraphic Union, the International Union for the Protection of Literary and Artistic Property, etc.

Bernburg, Germany (Anhalt), manufactures machinery, sugar, spirits, porcelain, etc.

Besançon, France (cap. Doubs), is a fortified city, with extensive manufactures of linen, cotton, woollen and silk goods, but its principal industry is watchmaking in which about 15,000 workmen are employed. It has a university, and is the seat of an archbishop.

Beuthen, Germany (Prussia), is an important center of mining and metallurgy, having iron-works, zinc-works, lead-works, coal-mines, and various industrial establishments.

Beverly, Massachusetts (Essex Co.), is the seat of the New England Institute for the Deaf and Dumb. It manufactures boots and shoes, and has considerable shipping and fishing interests. It is a fashionable summer seaside resort.

Béziers, France (Hérault), has manufactures of woollens, silks, hosiery, chemicals, and spirits. Its most conspicuous edifice is the cathedral, a Gothic structure, crowning the height on which the town stands, and possessing a fine semicircular choir surrounded by columns of red marble.

Bhagalpur, India (Bengal), has some interesting Mohammedan shrines; several indigo works are in the vicinity.

Bhaunagar, India (Bombay), has a good harbor and a large trade in cotton.

Bhopal, India (Bhopal), is a fortified town, and contains an arsenal and mint.

Bialystok, Russia (cap. Grodno), manufactures woollen goods, leather, hats, etc. It has a palace which belonged to Count Branicki, and was once known as the Versailles of Poland.

Biel, Switzerland (Bern), has manufactures of watches, leather, and cotton. It is superbly situated at the foot of the Jura, surrounded by ancient walls with watch towers at intervals. It contains, among other institutions, the West Swiss Technical Institute, with its school for railroad employees and a watchmakers' school.

Bielefeld, Germany (Prussia), is an important linen trade center. It has extensive bleaching grounds, and its meerschaum pipes are celebrated.

Biddeford, Maine (York Co.), has extensive manufactures of cotton goods, lumber, boots and shoes, machinery, etc. There are granite quarries near by.

Bikaner, India (Rajputana), has manufactures of blankets, pottery, etc.

Bilbao, Spain (cap. Vizcaya), is a port and coaling station on the Nervion River. Exports iron ore, pig iron, wine, fruit, etc. It has extensive manufactures of sail-cloth and ropes, large shipyards, and iron and steel works.

Billings, Montana (c. h. Yellowstone Co.), is an important trade center in a region where agriculture has been extensively aided by irrigation. Its leading industry is the manufacture of beet sugar. Wool and hides are extensively shipped.

Binghamton, New York (c. h. Broome Co.), has extensive cigar factories and manufactories of scales, chemicals, sheet-metal work, glass, and refined oils. The state asylum for the insane, and the Commercial Travelers' Home are located here.

Birkenhead, England (Cheshire), is a seaport situated opposite Liverpool. It has extensive docks, with complete railway communication for the shipment of goods, facilities for coaling steamers, and magnificent dock warehouses. Extensive ship-building yards are located here, and there are machine and engineering works, wagon factories, flour-mills, etc. It is connected with Liverpool by tunnel under the Mersey and by ferries.

Birmingham, Alabama (c. h. Jefferson Co.), is the largest city of Alabama, and is situated in the heart of the greatest coal, iron, and limestone district of the south. It has extensive blast-furnaces, coke-ovens, coal-mines, stone-quarries, and rolling-mills. Three huge coal-fields, aggregating over 9,000 square miles, with some sixty seams, more than half of them workable, lie near the city, the nearest deposits being only 4 miles from the city. The city is built partly upon the slope of Red Mountain, named from its outcrop of hematite iron ore, which extends many miles in every direction in a vein from 6 to 26 ft. thick, with an indefinite depth.

Birmingham, England (Warwick), is one of the greatest manufacturing cities in the world, and the commercial capital of the Midlands of England. Its manufactories of firearms, jewelry, glass, chemicals, hardware, steam-engines, railway material, electro-plates, electrical supplies are vast in extent. It is also the birthplace and the chief center of the brass trade. It has a university, numerous collegiate institutions, and a fine museum and art gallery. Considers itself the best governed city in the world. It is the seat of a Roman Catholic bishop.

Bisbee, Arizona (Cochise Co.), has extensive copper mining and smelting industries.

Bisceglie, Italy (Bari), is an episcopal see and has a 12th century cathedral.

Bitlis, Turkey in Asia (cap. Bitlis), has extensive manufactures of cotton, firearms, etc.

Bitonto, Italy (Bari), has extensive wine and oil interests. It has a fine cathedral and old walls.

Blackburn, England (Lancashire), is one of the chief cities of cotton manufacture, and has also extensive works for making cotton machinery and steam-engines. Coal abounds in the vicinity.

Blackpool, England (Lancashire), is a coast resort, with many amusement places, among which are a fine winter garden, aquarium, a great steel tower, etc.

Blagoveschensk, Siberia (cap. Amur), was founded as a military post in 1856, and is now an important trading and gold-mining center.

Bloemfontein, cap. Orange Free State, Africa, has several fine buildings, including the Anglican cathedral, government buildings, and several palaces.

Blois, France (Loir-et-Cher), has a botanical garden founded by Henry IV, and several collegiate institutions. The castle of Blois is rich in historical associations. Gloves, porcelain, and cutlery are manufactured here.

Bloomfield, New Jersey (Essex Co.), manufactures organs, woollen goods, rubber, and brass goods.

Bloomington, Illinois (c. h. McLean Co.), is the seat of the Illinois Wesleyan University and has extensive manufactories of stoves, farming implements, patent medicines, etc. The Illinois State Normal University and State Soldiers' Orphans' Home are located at Normal (2 miles).

Bluefields, Nicaragua, is a seaport with one of the best harbors in Central America.

Bobruisk, Russia (Minsk), a fortified town with trade in cattle and agricultural products.

Bochum, Germany (Prussia), has extensive iron-works and carpet manufactories; large coal-mines are in the neighborhood.

Bogota, cap. Colombia, is situated on an elevated

plain 8,630 ft. above the sea, and despite the fact that it is only 4° 44' north of the equator, it has a delightful climate resembling perpetual autumn. It has a university, a military academy, a mint, an observatory, and a botanical garden. There are manufactures of cloth, cordage, porcelain, and glass. It is the seat of a Catholic archbishop.

Boise, cap. Idaho (c. h. Ada Co.), is the largest city of the state and an important mining center. It has considerable trade in wool, hides, and fruit. A U. S. land-office, and a U. S. assay office are located here.

Bokhara, Russia in Asia (cap. Bokhara), is a fortified city, and has long been famous as a seat of Mohammedan learning, possessing numerous colleges and several beautiful mosques. It is now an important commercial center.

Bologna, Italy (cap. Bologna), is a strongly fortified town and an important railway center. It has an active trade and extensive manufactures of silks, velvet, cloth, and artificial flowers. Bologna is noted for macaroni, sausages, liqueurs, and preserved fruits. It is the site of a famous ancient university and has an Academy of Fine Arts, which contains paintings by most of the masters of the Bolognese school. It is the seat of an archbishop and possesses several palaces and churches, which contain numerous works of art. Bologna was the residence of the Roman Emperors during the later period of the Roman Empire.

Bolton, England (Lancashire), contains some of the largest cotton-mills in the world, and has extensive manufactures of iron, paper, and chemicals. Its town hall in the Grecian style and its fine market hall are notable buildings.

Bombay, India (cap. Bombay), is the second port of India—the first being Calcutta—and an important naval and coaling station. It has a commodius harbor and is provided with large docks, wharves, warehouses, and extensive railway facilities. It is the principal seat of the cotton industry in India, while its other industries include dyeing, tanning and metal working. Principal exports are cotton, wheat, opium, coffee, seeds, gums, hides, and carpets. Bombay possesses a university, numerous colleges, a technical institute, and a famous botanical garden. It is the seat of an Anglican bishop.

Bona, Algeria, is a fortified seaport and naval station, which exports iron, copper, lead, zinc, esparto, phosphates, and tannin. There are iron mines in the vicinity.

Bonn, Germany (Prussia), is famous for its university, which possesses a library of more than 275,000 volumes, an observatory, a botanical garden, etc. It has an academy of arts, and several art museums. Its manufactures comprise carpets, machinery, chemicals, etc.

Bootle, England (Lancashire), is a suburb of Liverpool, and has large docks, jute works, etc.

Bordeaux, France (cap. Gironde), is a port and coaling station on the Garonne River, which has admirable facilities for trade, and exports wines, brandies, fruits, sugar, coffee, gums, porcelain, and glass. The Bordeaux wines have had a great reputation for many centuries. There are manufactories of sugar, chocolate, glass, tobacco, as well as flour-mills, and iron and steel works. It is the seat of an archbishop, and the Gothic cathedral of St. André (13th–15th centuries) is remarkable for its beautiful towers, designed and built by English architects during the English occupation. Bordeaux has a university, an academy of arts and sciences, a naval school, an observatory, and an important art gallery.

Borgerhout, Belgium (Antwerp), is a suburb of Antwerp, with tapestry and bleaching works, and tobacco factories.

Boston, cap. Massachusetts (c. h. Suffolk Co.), is the metropolis of New England and one of the most interesting cities in the United States. A seaport, coaling and naval station, it has an extensive harbor with good anchorage for vessels of any size, and a waterfront lined with wharves and docks. It exports live cattle, provisions, cotton goods, leather goods, etc., but is also the center of higher education in many branches, thus attracting a large number of students. The Public Library, one of the most beautiful buildings in the country, puts the means for scholarly attainments within the reach of all. Harvard University is close at hand at Cambridge and its Medical School is in Boston proper. Here are also the Massachusetts Institute of Technology, Boston University, Boston Museum of Fine Arts, and the New England Conservatory of Music. Among its buildings of historical interest are Faneuil Hall ("cradle of American Liberty"), the Old State House, dating from 1748 and restored as far as possible to its original appearance, and the Old South Meeting House, built in 1729. Trinity Church, designed by H. H. Richardson and completed 1877, is regarded as one of the finest buildings in America. The park system of Boston, with a total area of 2,400 acres and the Metropolitan system forming an outer line of parks with an area of 11,000 acres, are remarkable for beauty, accessibility, and actual benefit to the community.

Botuchany, Roumania, is a trade center.

Boulogne-sur-Mer, France (Pas-de-Calais), is a seaport and the chief station of the French herring and cod fisheries. It exports dried fish, agricultural products, wines, watches, leather, and fancy goods. Has an active coastwise trade, and ranks with Calais as one of the most frequented places of passage between France and England. It is a favorite seaside resort.

Boulogne-sur-Seine, France (Seine et Oise), is a suburb of Paris, which adjoins the famous public park called Bois de Boulogne.

Bourges, France (cap. Cher), is an ancient city with many interesting historical monuments, amongst which is the splendid Gothic cathedral (1220–1538), one of the finest specimens of Gothic architecture in France. It is the seat of an archbishop, and has been the meeting-place of numerous church councils. It has manufactures of cloth, leather, as well as flour-mills and breweries, and a large trade in wool and agricultural products.

Bournemouth, England (Hampshire), is a famous seaside resort on the South coast. It is situated in the midst of pine forests and enjoys a mild climate specially suited for persons suffering from lung diseases.

Braddock, Pennsylvania (Allegheny Co.), is a suburb of Pittsburg and the scene of the memorable defeat of General Braddock on July 9, 1755, on his expedition against Fort Duquesne. It was here that Washington, in rallying the defeated British forces, won his first military laurels.

Bradford, England (Yorkshire), has extensive manufactures of worsted yarn, silk, velvet, and cotton, as well as cotton factories. In the neighborhood are quarries and iron-works.

Bradford, Pennsylvania (McKean Co.), is the center of an extensive coal, oil, and natural gas region, and is principally engaged in industries connected therewith.

Braga, Portugal (cap. Braga), is the seat of an archbishop and contains an archiepiscopal palace, a richly ornamented Gothic cathedral of the 13th century, and a college. Braga is supposed to have been founded by the Carthaginians, and there are remains of a Roman temple, amphitheatre, and aqueduct.

Braila, Roumania, is a port and coaling station on the Danube with large exports of grain, tallow, flour, and cattle.

Brandenburg, Germany (Prussia), manufactures cloth, paper, and leather, and has boat-building yards. It has a castle and a cathedral of the 14th century, with a fine crypt.

Brandon, Manitoba, Canada, is a railroad and trade center, with extensive wheat-shipping interests, and manufactories of machinery, pumps, ale, etc. It is the seat of an Indian industrial school and government experimental farm.

Brantford, Ontario, Canada, is a trade and industrial center, with manufactories of brass, iron, tin, etc. A provincial asylum for the blind is located here.

Breda, Netherlands (North Brabant), has manufactures of woollens, carpets, and hosiery. It was formerly a fortified city and played a prominent part in the military events of the country.

Bremen, Germany (cap. Bremen), is a free city, port and coaling station on the Weser, which exports wool, linen, glass, iron, steelware, and wooden toys. Chief

public buildings include a cathedral (1044), the old Gothic council-house, with the famous wine cellar below it, the modern town hall, and the old and the new exchange. It is well supplied with educational institutions; has a library of 120,000 volumes, and an observatory. The industrial establishments include tobacco and cigar factories, sugar-refineries, rice-mills, iron-foundries, rope and sail works, and shipbuilding yards. It is from its commerce, however, that Bremen derives its importance, since next to Hamburg, it is the principal seat of the export and import trade of Germany, and the greater portion of the German trade with the United States passes through Bremen. It is one of the chief ports of emigration on the continent, and after Liverpool, Bremen is to-day the leading European cotton market.

Brescia, Italy (cap. Brescia), is an ancient city with many interesting buildings, among which are two cathedrals, one dating from the 7th century, and another of white marble, built 1604-1825. Many of the churches contain paintings by celebrated artists, and the picture-gallery also has a valuable collection of paintings, mostly by masters of the Brescian school (16th and 17th centuries). The city is the seat of a bishop. It has a celebrated factory of firearms and manufactures of steel and iron wares, besides cottons, woollens, and silks.

Breslau, Germany (cap. Prussian Silesia), is an important industrial, commercial, and railroad center, with extensive manufactures of linen, cottons, woollens, silks, machinery, pottery, and chemicals. It has a university and among its interesting buildings are a fine cathedral from the 13th century, and a magnificent city hall from the 14th century. It is the seat of a bishop.

Brest, France (Finistère), is a strongly fortified naval harbor, and the chief station of the French navy. It contains an extensive arsenal, large naval stores, and a naval academy. It has manufactures of rope, leather and soap, and trade in agricultural products.

Brest-Litovsk, Russia (Grodno), is a fortified town with a large trade in cloth, leather, and soap.

Bridgeport, Connecticut (c. h. Fairfield Co.), is the third city of the state and a port of entry with a safe harbor which admits quite large vessels. It manufactures sewing machines, automobiles, brass castings and finishings, corsets, cartridges, cutlery, rubber goods, firearms, typewriters, musical instruments, etc.

Bridgeton, New Jersey (c. h. Cumberland Co.), is a very old settlement, having been a place of considerable importance before the Revolutionary War. Manufactures glass, gas-pipe, nails, and machinery. Has numerous educational institutions.

Bridgetown, West Indies (cap. Barbados), is a seaport with a spacious and safe open roadstead, but no harbor. It exports sugar, molasses, and mineral oil. It is the seat of an Anglican bishop.

Brieg, Germany (Prussia), possesses an interesting old castle, and has manufactures of ironware, sugar, leather, tobacco, etc.

Brighton, England (Sussex), is the favorite seaside resort on the south coast of England. Has a massive sea-wall and a promenade and drive over three miles in length, one of the finest in Europe.

Brisbane, cap. Queensland, Australia, is a port and coaling station on the Brisbane River, 25 miles from its mouth. Exports gold, wool, tallow, cotton, frozen and preserved meats, and hides. Among the industrial establishments are sugar-refineries, tobacco factories, flour-mills, etc.

Bristol, England (Gloucester), is a port and coaling station on the Avon, with a harbor which can accommodate the largest vessels. It is one of the leading British ports for foreign trade, and exports coal, which is extensively mined in the neighborhood, cottons, tin-plates, machinery, and chemicals. It has large manufactories of glass, potteries, and chocolate, and shipbuilding yards. Bristol is noted for its fine buildings, among which are the cathedral, the beautiful church of St. Mary Redcliff, the university college, and the museum and art gallery. It is the seat of a bishop.

Broach, India (Bombay), is a trade center, with exports of raw cotton, grain, and seeds.

Brockton, Massachusetts (Plymouth Co.), is one of the largest boot and shoe manufacturing places in the United States, and has extensive manufactories of rubber goods, shoe machinery, tools, etc.

Brockville, Ontario, Canada, is a trade center in an agricultural district. It has manufactories of carriages, agricultural implements, etc.

Broken Hill, New South Wales, Australia, is a mining center with some of the most productive silver mines in the world. Asbestos, gold, and copper are also found here.

Bromberg, Germany (Prussia), has large manufactories of machinery, paper, and leather, iron and steel works, flour-mills, and an active trade in agricultural products.

Brookline, Massachusetts (Norfolk Co.), is a residential suburb of Boston, with many handsome homes.

Brooklyn, New York. See New York.

Brownsville, Texas (c. h. Cameron Co.), is a port of entry, and a trade center and railroad terminus on the Rio Grande, 35 miles from its mouth. It is connected with the Mexican town of Matamoros by a bridge across the Rio Grande. Here was the site of Fort Brown, which the Mexicans attacked without success in May, 1846.

Bruges, Belgium (cap. West Flanders), is a port eight miles inland from the North Sea, and connected by canals with Ostend and other cities. It is a mediæval town with many ancient buildings, chief among them being the famous belfry, the cathedral, the city hall, and the hospital of St. John's, which contains fine paintings by Memling. Its museum and picture-gallery have many notable examples of masters of the Flemish school. During the middle-ages and until the close of the 15th century Bruges was the leading commercial center of Northern Europe. Its trade is now reviving again, and it has manufactures of lace, linen, and woollens. It is the seat of a bishop.

Brünn, Austria (cap. Moravia), is one of the leading industrial towns in the country with extensive manufactures of woollens, cottons, linen, machinery, chemicals, sugar, and leather, and a very active trade. It is the seat of a bishop.

Brunswick, Germany (cap. Brunswick), manufactures machinery, sugar, jewelry, and woollens. It has a cathedral dating from the 12th century, and a ducal palace, which contains valuable collections of majolica, gems, and paintings.

Brusa, Turkey in Asia, has important silk manufactures, and an active trade, being connected by railroad with its port, Mudania, on the Sea of Marmora. Nearby are famous mineral springs. It is the seat of a Greek Orthodox and of an Armenian archbishop.

Brussels, cap. Belgium (cap. Brabant), is the most important city of Belgium and is remarkable for the number and richness of its ancient buildings, grouped about the Grande Place, while, from the elegance of its new quarters, it ranks among the finest cities of Europe. Among its principal buildings are the splendid Gothic cathedral dating from the 13th century, and the magnificent law-courts, (1868-1883), one of the finest buildings in Europe. It is one of the greatest centers of Belgian industry, and is celebrated for its lace, considered the finest in the world. It also manufactures fine linens, silk, woollen and cotton fabrics, gold and silver embroidery, bronzes, leather, gloves, paper, jewelry, machinery, mathematical and musical instruments, and has many large printing and lithographing establishments. Only a few of the so-called Brussels carpets are manufactured here. Its manufactures and trade are greatly promoted by its numerous canals and railways. The scientific, literary, and artistic institutions of Brussels comprise a free university, founded in 1834; a school of geography founded in 1830, with an extensive museum; one of the finest astronomical observatories in Europe; the Belgian Royal Academy of Sciences, Letters and Fine Arts; the Royal Academy of Fine Arts; the public library, containing 350,000 volumes and 30,000 valuable manuscripts; the picture-gallery, with the magnificent examples of Flemish art; and several other valuable museums.

Bucaramanga, Colombia (cap. Santander), is the

center of a region producing coffee and having mines of gold, copper, and iron.

Budapest, cap. Hungary, consists of the united towns of Buda, or Ofen, and Pest, or Pesth, the one on the right and the other on the left of the Danube. Among the chief buildings in Buda are the royal castle, several palaces, the arsenal, town hall, government offices, the Church of St. Matthew, dating from the 13th century, and the finest Jewish synagogue in Austria-Hungary. Pest, which is of modern growth, has a fine frontage on the Danube, and includes the new Houses of Parliament opened in 1896, the academy of science, with a library of 180,000 volumes, the custom-house, and other important buildings. It has a university, an academy of arts and sciences, an art museum, an industrial museum, and numerous other educational institutions. It ranks next to Vienna in commerce and industry. Its chief manufactures are flour-mills, machinery, gold, silver, copper, and iron wares, chemicals, textile goods, brewing and distilling, leather, etc., and it has extensive electrical works. It has a large trade in grain, wine, wool, and cattle. The Elisabeth suspension bridge over the Danube River, completed in 1903, with a span of 951 feet, is notable, as are also the underground trolley lines of the city. The city contains several large parks, among which is the Stadtwäldchen, with an area of about 1,000 acres. Several mineral springs are in the town and neighborhood.

Budweis, Austria (Bohemia), has manufactures of stoneware, lead pencils, and beer. It is the seat of a bishop.

Buenos Aires, cap. Argentina (Buenos Aires), is one of the most extensive capitals of the world, covering an area of 34,829 acres, and the leading South American city. It is a port and coaling station on the Plata, and exports live-stock, hides, horns, bones, tallow, wool, wheat, and corn. It has numerous educational and scientific institutions of a high order, among which are the university, national library, museum of natural history, zoölogical garden, and an observatory. On the Plaza de Mayo face the government palace, House of Congress, the exchange, cathedral, and municipal buildings. Its street railway system is remarkably in advance of other cities of its size. It has an extensive park system and is one of the cleanest and healthiest cities in the world. It is the seat of a Catholic archbishop.

Buer, Germany (Westphalia), has extensive coal mining interests.

Buffalo, New York (c. h. Erie Co.), is the second city of the state and one of the most important ports of the great inland lakes, having a safe and capacious harbor and several miles of water front. It has an extensive distributing trade in lumber, iron ore, grain, coal, live-stock, and meat, and is the western terminus of the Erie Canal and a great railroad center. It has extensive manufactures of iron and steel, metal goods, soap, starch, flour, railway cars, and elevators. It has numerous educational institutions, and a public library containing over 300,000 volumes.

Bukharest, cap. Roumania, is, after Constantinople and Budapest, the most populous city of southeastern Europe and is spoken of by Roumanians as the "Paris of the East." It has a university, museum of natural history, and many antiquities, and, among the public buildings the most conspicuous are the royal palace and the new palace of justice. It has magnificent public gardens. Its manufactures are not of importance, but it has an active trade in grain, wool, and cattle. It is the seat of a Greek Orthodox archbishop.

Bullfrog, Nevada (Nye Co.), is a gold-mining camp of recent growth.

Burgos, Spain (cap. Burgos), is a bishop's see. Its most remarkable structure is the cathedral, one of the finest buildings of the kind in Europe, begun in 1221 and not finished for several centuries, which is built of white marble in the form of a Latin cross, both exterior and interior being of great magnificence, adorned with fine carvings and paintings; it contains numerous monuments. Burgos is a great wool market and has some woollen manufactories.

Burlington, Iowa (c. h. Des Moines Co.), has a con-

siderable river trade and is a port of delivery. It manufactures lumber, carriages, wagons, foundry and machine-shop products.

Burlington, Vermont (c. h. Chittenden Co.), is the chief city of Vermont and one of the largest lumber marts in America, the lumber coming chiefly from Canada. It is the seat of the University of Vermont, and of the state agricultural and medical colleges. It is noted for its benevolent and educational institutions.

Burnley, England (Lancashire), manufactures cotton goods, worsted, and machinery; there are collieries and quarries in the vicinity.

Burslem, England (Stafford), has extensive manufactures of china and earthen wares, and coal is mined.

Burton-upon-Trent, England (Stafford), is celebrated for its ale, of which vast quantities are made for home consumption and export. Its malting and brewing establishments employ over 10,000 men and boys.

Bury, England (Lancashire), has large cotton and woollen factories, bleaching and printing works, dye-works, foundries, etc. In the vicinity are extensive coal-mines.

Butler, Pennsylvania (c. h. Butler Co.), is the center of a region having oil and natural gas. It has manufactures of glass, flour, and lumber.

Butte, Montana (c. h. Silverbow Co.), is the largest mining town in the world, having extensive mines of copper, gold, and silver. It is the trade and jobbing center for southern and western Montana.

Cadiz, Spain (cap. Cadiz), is a fortified seaport and naval station and one of the handsomest cities in Spain. It ranks as one of the first ports of the country, and was the center of the Spanish-American trade before the separation of the colonies. It exports sherry, salt, olives, fruit, cork, etc. It has a university.

Caen, France (cap. Calvados), is a port situated on the river Orne, 7 miles from the sea, with which it is connected by a canal. It has manufactures of lace, cottons, and woollens, as well as ship-building yards, and a large trade in agricultural produce, cattle, and horses. Among its churches are St. Etienne and La Sainte Trinité, both founded in 1066 by William the Conqueror and his queen Matilda, and containing their tombs. It has a university.

Cagliari, Italy (cap. Cagliari), is a seaport and the principal town in the island of Sardinia. It has manufactures of woollens, cottons, soap, and a very active trade in salt and wine, which is produced in the neighborhood. Cagliari is an ancient town and contains many interesting Roman remains. It is the seat of an archbishop, and has a cathedral dating from the 14th century. It contains a university, and a museum with valuable collections of antiquities found in the island.

Cairo, cap. Egypt, is the largest town in Africa, situated on the Nile, about seven miles above the upper point of its delta. It consists practically of two towns, the modern European town, with its broad streets, handsome buildings, and fashionable hotels and shops, and the native town. The latter presents a picturesque Oriental appearance with its narrow streets, crowded bazaars and numerous mosques. It has a Mohammedan university, founded 988, which is one of the leading seats of Mohammedan learning in the world, and a library which contains many valuable manuscripts of the Koran. One of its museums—the Bulak Museum—contains most of the antiquities of ancient Egypt, which have been excavated in various parts of the country. Cairo is the seat of a Roman Catholic and of a Greek Orthodox bishop, as well as of a bishop of the Coptic Church. It is now a great winter resort, and has famous mineral springs and baths in the neighborhood. The commerce of Cairo is large, the town being a great emporium of trade for Central Africa.

Cairo, Illinois (c. h. Alexander Co.), is a port of entry and has a large trade in agricultural products and lumber. The United States marine hospital is located here.

Calais, France (Pas-de-Calais), is a fortified port and coaling station situated on the Strait of Dover, about 26 miles from Dover, England. It is one of the principal ports for debarkation of travellers from England. It has manufactures of tulle, lace, hosiery, and gloves, and

exports, beside these articles, wines, spirits, and dairy products, mostly to England. It is an important fishing station.

Calcutta, India (cap. Bengal), is a port and coaling station on the Hugli River, 80 miles from the sea and a terminus of numerous canals and railroad lines. It has ample docks and slips. It is the seat of an immense trade by sea and river, being the natural outlet of the Ganges and the Brahmaputra. Exports jute, cotton, indigo, wheat, rice, opium, spices, sugar, hemp, hides, and india-rubber. It monopolizes the foreign trade of this part of India, and next to Bombay, is the greatest commercial emporium of the country. The manufacture of jute and cotton is extensively carried on. It has numerous educational institutions, including a university, several colleges, supported mainly by missionary efforts, four government colleges, etc. Here are located the government dockyards, with the arsenal and Fort William, the largest fortress in India. Among the other fine buildings are the government houses, the principal residence of the viceroy of India.

Calgary, Alberta, Canada, is the largest city of Alberta and an important railroad and trade center. It is situated in a prosperous agricultural region, producing wheat and other grains, and having large stock growing interests. In the neighborhood are extensive coal beds (lignite and anthracite), lime, brick clay, and building stone. A normal school and two colleges are located here.

Calicut, India (Madras), is a seaport and the first place in India visited by Europeans. Covilhao, the Portuguese adventurer, landed here about 1486, and Vasco da Gama in 1498. In 1792 it came into the possession of the British. The cotton cloth called calico, introduced into Europe by the Portuguese, is still largely manufactured here.

Callao, Peru (Lima), is the chief seaport of the country, with large exports of hides, copper, guano, nitrates, sulphur, and wool.

Caltagirone, Italy (Catania), is a town which manufactures celebrated pottery and statuettes. It is the seat of a bishop.

Caltanissetta, Italy (cap. Caltanissetta), is a fortified town and has manufactures of potteries, and sulphur mines. It is the seat of a bishop and contains a cathedral.

Camaguey, Cuba (cap. Camaguey), is the largest inland city of Cuba and the center of a great grazing district.

Cambridge, England (cap. Cambridge), is the seat of Cambridge University, which, dating from about the 12th century, comprises the following colleges given in the order of their antiquity: St. Peter's, Clare, Pembroke, Caius, Trinity Hall, Corpus Christi, King's, Queen's, St. Catharine's, Jesus, Christ's, St. John's, Magdalene, Trinity, Emmanuel, Sidney Sussex, Downing, Cavendish, Selwyn, Ayerst. The teaching staff number 120, and the students close on 3,000. The chief among the college buildings are King's (1441), with its noble perpendicular chapel; Trinity, with its courts, its hall, and its library by Wren; and St. John's, with its splendid new chapel (1869) by Scott. There are also the library, senate house, Fitzwilliam museum, observatory, etc. Although no part of the university, Girton and Newnham, colleges for women, are also situated here. Cambridge is an ancient place, on the site of a Roman station, and contains a guildhall, and several interesting old churches, such as St. Benedict's, with a tower in the Saxon style of architecture, and the round church of the Holy Sepulchre.

Cambridge, Massachusetts (Middlesex Co.), is a suburb of Boston and the seat of Harvard University, the oldest and most famous of American educational institutions, and of Radcliffe College. As far as historical and literary associations are concerned, Cambridge is one of the most famous cities in the United States. It has extensive printing establishments, and manufactories of glass, furniture, organs, etc.

Camden, New Jersey (c. h. Camden Co.), is situated on the Delaware River opposite Philadelphia and is an important railroad terminus. It is a port of entry, and has extensive manufactures of worsted goods, oilcloth,

boots, shoes, and textiles, as well as large shipbuilding yards.

Campeche, Mexico (cap. Campeche), is a seaport with large exports of logwood, termed Campeche wood, hemp, wax, hides, cocoa, and salt. It has manufactories of cigars and shipbuilding yards. It still preserves the old fortifications erected in colonial times to protect it against the English and Dutch buccaneers.

Campinas, Brasil (Sao Paulo), is a modern commercial city in the midst of coffee plantations.

Candia, Crete, is the largest city of the island, and has an export trade in oil, soap, raisins, wine, almonds, etc. It has an important museum of antiquities.

Canea, Crete, is the capital of the island, situated on the northwest coast, and occupies the site of the ancient Cydonia; but the present town is due to the Venetians, from whom it was wrested by the Turks after a two-years' siege in 1669. The small mediæval, artificial harbor is much silted. Several Venetian monuments survive.

Cannes, France (Alpes-Maritimes), is one of the most fashionable winter resorts of Europe and is famed for its salubrious climate. Among its prominent modern buildings are the new town hall, the English church, and the principal casino. Other interesting structures are the tower of a mediæval castle, and the bridge called Pont-de-Rion. It has some industry in perfumes, soap, and fisheries, and a large trade in flowers.

Cannstadt, Germany (Württemberg), is a growing industrial place, with iron-works and manufactures of machinery, zinc wares, electrical apparatus, motors, and cloth. It has celebrated and much frequented mineral springs.

Cantanzaro, Italy (cap. Cantanzaro), has manufactures of silk and velvet. It is the seat of a bishop.

Canterbury, England (Kent), is famous as the ecclesiastical metropolis of England. The Cathedral, rebuilt by Archbishop Lanfranc on the site of the more ancient monastery church of St. Augustine (burnt down 1067), is a magnificent doubly cruciform edifice, presenting fine examples of Norman and later styles of architecture, the Bell Harry tower being a prominent feature. Connected with the east nave are several chapels: Trinity Chapel formerly contained the shrine of Thomas à Becket, and at the extreme east is the chapel known as Becket's Crown containing the ancient stone chair on which the archbishops are enthroned. There are numerous interesting ancient monuments, including those of Henry IV., and of his queen, Joan of Navarre, the Black Prince, and Cardinal Pole. A handsome monument has recently been erected to Archbishop Benson (d. 1896). St. Augustine's missionary college occupies some of the restored buildings of the ancient monastery. Extensive barracks for cavalry and infantry are located here.

Canton, China (Kwangtung), is a port of the Canton River, 70 miles from its mouth. It was the first Chinese port opened to European trade and is now the second port in China. The special feature of the city is the great number of the population who live in boats on the river. It has manufactures of silks, fireworks, wood and iron wares, and exports tea, silk, matting, tobacco, sugar, cassia, and camphor.

Canton, Ohio (c. h. Stark Co.), has extensive manufactures of watches, agricultural implements, steel cars, safes, and locks. Important steel and bridge works are located here and it is a coal-shipping point. It was the home of President McKinley, to whom an elaborate monument has been erected, enshrining the bodies of himself and his wife.

Cape Coast Castle, Gold Coast Colony, Africa, is a port and naval coaling station and the mart for native trade. Exports gold-dust, palm-oil, pepper, rubber, skins, ivory, etc.

Cape Haitien, Haiti, is a port with a commodious harbor, and exports coffee, cocoa, logwood, honey, etc. It is the seat of a Catholic bishop.

Cape Town, cap. Cape Colony, Africa, is a seaport, coaling-station, and the metropolis of South Africa. It has fine government buildings, a cathedral, the South African college, a museum, a library, botanic gardens, and the finest observatory in the southern hemisphere.

Exports wool, diamonds, ostrich feathers, gold, ivory, hides, and copper ore.

Caracas, cap. Venezuela (cap. Caracas), is the seat of a Roman Catholic archbishop and the chief city of the country. It is connected with the port of La Guayra by a railroad 24 miles long. The chief buildings are the federal palace, the university, the museum, the library, a cathedral, and the residence of the president.

Carbondale, Pennsylvania (Lackawanna Co.), is a busy coal-mining city, with large deposits of anthracite.

Carcar, Philippine Islands (Cebu), is a port with exports of sugar and extensive fisheries.

Carcassonne, France (cap. Aude), is an ancient city and the seat of a bishop. It has manufactures of cloth, linen, paper, and soap.

Cardenas, Cuba (Matanzas), is a seaport with large exports of sugar and molasses.

Cardiff, Wales (Glamorgan), is a seaport and coaling station with large dock accommodations, ranking as the third port in the United Kingdom. It exports large quantities of coal and iron, and its industries include smelting, ship-building, engine-building, iron-founding, brewing, and the manufacturing of paper and chemicals. Here are located the university college of South Wales and other numerous collegiate institutions. The free libraries have a museum and art gallery connected with them.

Carlisle, England (cap. Cumberland). is an ancient city, and has an interesting cathedral dating from the 12th century. It has manufactories of cottons and woollens, iron and steel works, and extensive railway shops.

Carlisle, Pennsylvania (c. h. Cumberland Co.), has manufactures of shoes, silk, and railroad cars. It is the seat of the Dickinson Methodist College, founded in 1783, and of the United States Indian industrial training-school.

Carrara, Italy (Massa e Carrara), is a town which contains famous quarries of fine-grained marble, mostly white, but also black, yellow, and green. About 600 separate quarries are worked, which yield from 150,000 to 180,000 tons annually, and employ between 6,000 and 7,000 workmen. It has an academy of fine arts, and also a school of sculpture.

Cartagena, Colombia (cap. Bolivar), is a port with a fine harbor, and has a fair export trade in cotton, sugar, hides, coffee, etc.

Cartagena, Spain (Murcia), is a fortified port, and the principal Spanish naval arsenal and dockyard. It has an extensive coastwise trade, and exports silver, lead, zinc, and iron ores, esparto, fruits, etc. Nearby are extensive silver-lead mines. It is the seat of a bishop.

Casale, Italy (Alessandria), has a cathedral dating from 8th and 12th centuries, and other interesting churches, among which is San Domenico, begun in 1489.

Caserta, Italy (cap. Caserta), has manufactures of silk. It is the seat of a bishop.

Cassel, Germany (cap. Hesse-Nassau), is one of the handsomest towns of its size in Germany, the architectural and sculptural adornments of former ages being richly supplemented by the works of recent art. The principal square is flanked by the former palace of the electors (1769 and 1821), and a couple of museums, but the most imposing buildings are those of the administration and the law courts (1876–80). Another fine new structure is the picture gallery. The palace in which Jerome, brother of Napoleon, lived when king of Westphalia, is now given up partly to the academy of fine arts, and partly to military offices. The Karl Park (Aue) contains the Orangery palace, with famous marble baths. About 3 miles west is the castle of Wilhelmshöhe, placed on a terrace of the Habichtwald, and surrounded by woods and fine gardens. Here Napoleon III. was confined (1870–71), after the battle of Sedan, and here the Emperor William II. and his family frequently spent part of the summer. The industries include iron-works, engineering shops, and factories for railway carriages, mathematical instruments, tobacco, small metal fittings, paper, and pianofortes, also lithography and gardening.

Castellamare, Italy (Naples), is noted for sea-bathing and sulphur baths. Its industries include ship-building, fishing, macaroni, cotton, soap, needles, etc. It exports macaroni, fruit, cheese, and olive oil.

Castellon, Spain (cap. Castellon de la Plana), is the center of a fruit-growing section, and has manufactories of flax goods of all sorts.

Castres, France (Tarn), has manufactures of cottons, leather, soap, and paper. During the 16th century it was an important Huguenot stronghold, but was destroyed in 1629.

Catania, Italy (cap. Catania), is a port at the foot of Mt. Aetna, with an extensive trade in sulphur, oranges, lemons, and dried fruit. Cotton, silk, bentwood furniture, hats, and soap are manufactured; olive-oil is distilled and sulphur refined. Catania has frequently been visited by earthquakes. It is a bishop's see, and the seat of a university, and a famous academy of natural sciences.

Catanzaro, Italy (cap. Catanzaro), has silk and velvet manufacturing and olive-oil industries. It is a favorite place of residence for wealthy Calabrians.

Cawnpur, India (United Provinces), is a large military cantonment. The town was rendered infamous during the Sepoy mutiny by the treacherous massacre of European women and children which was ordered by Nana Sahib. Has manufactories of boots, leather, and cottons.

Cayenne, cap. French Guiana, is a seaport with exports of coffee, cocoa, pepper, rice, tobacco, indigo, and vanilla. It contains a college and museum.

Ceara, Brazil (cap. Ceara), is a port and coaling station, sometimes called Fortaleza, and is regarded as one of the most beautiful cities of Brazil. It exports cotton, india-rubber, skins, hides, coffee, sugar, etc.

Cebu, Philippine Islands (cap. Cebu), is the oldest town in the Philippines, and the chief commercial city of the island of the same name. It is the seat of a Roman Catholic bishop. It has an extensive trade in sugar, hemp, etc.

Cedar Rapids, Iowa (Linn Co.), is an important railroad center, and has extensive manufactories of flour, machinery, agricultural implements, etc., as well as large pork-packing establishments. It is the seat of a college.

Celaya, Mexico (Guanajuato), has manufactures of cloth, carpets, and sweetmeats. It has several handsome churches, among which is Our Lady of Carmen, a magnificent structure.

Central Falls, Rhode Island (Providence Co.), has cotton, woollen, silk, and haircloth mills, and manufactures leather, paper, boxes, and machinery.

Cerignola, Italy (Foggia), is a bishop's see. Nearby the French, under Duc de Nemours, were defeated by the Spaniards in 1503.

Cesena, Italy (Forli), is famous for its wine and hemp, and the people spin silk and mine sulphur.

Cette, France (Hérault), is a port with a good harbor, and a large trade in wine. It has manufactures of liqueurs, beer, brandy, and wine, and there is a considerable fishing industry. It is a favorite seaside resort, and has several mineral springs.

Chalon-sur-Saône, France (Saône-et-Loire), manufactures hats, gloves, iron goods, tiles, and glass. Its library, museum, bridge of 1415, and church of St. Vincent (1386) are of interest.

Chandarnagar, India (Bengal), belongs to France, and has only a moderate trade.

Changchau, China (Fokien), is a seat of silk manufacture, and has extensive iron-works. It has an active trade in tea and other products of the province.

Changsha, China (Hunan), is a large depot for timber poles and a place of considerable trade. A famous college, Yo-lo, is located here, and the silk industry flourishes.

Chapra, India (Bengal), is a place of much wealth, but its trade has declined.

Charleroi, Belgium (Hainaut), is one of the centers of the coal and iron industries of Belgium. Machinery, cutlery, and glass are extensively manufactured.

Charleston, South Carolina (c. h. Charleston Co.), is the largest city of South Carolina and a port and coaling station, seven miles from the sea, with a fine harbor, which is a naval station for the South Atlantic. Exports cot-

ton, rice, phosphates, and general merchandise. There are extensive beds of excellent phosphates near by. It manufactures cotton, flour, carriages, and machinery. A medical college, and the South Carolina military academy are located here, and among its historic buildings of note is St. Michael's Church, originally built in 1752-61. Ft. Sumter, in which the first shot of the Civil War was fired, occupies a small island in the middle of the entrance to the harbor. An earthquake in 1886 destroyed property and lives in this city.

Charleston, cap. West Virginia (c. h. Kanawha Co.), is the center of an important bituminous coal region. There are in the vicinity deposits of salt, iron, oil, and natural gas, and it has lumber mills, distilleries, packing-houses, and many other industries.

Charlotte, North Carolina (c. h. Mecklenburg Co.), is an important railroad center and cotton manufacturing city, with various industries and numerous educational interests. A monument in front of the court house celebrates the signing of the Mecklenburg Declaration of Independence, May 21st, 1775, which, according to a strong tradition, substantially anticipated Jefferson's.

Charlotte Amalie, Danish West Indies, St. Thomas Island, is a seaport and coaling-station with a good harbor. It has an extensive trade, mostly with the United States and England.

Charlottenburg, Germany (Prussia), is a residential suburb of Berlin. It owes its name and its existence to the royal castle which was built in 1695-1707 for Sophia Charlotte, the wife of Frederick, first king of Prussia. In the park is the mausoleum of Frederick William III. and his queen, Louise, and of the Emperor William I. and his queen, Augusta. It also contains two palaces, the Emperor William memorial church (1891-5), and the Trinity church (1896-8). The Berlin water-works are in Charlottenburg, and there is a very famous technical high school. The chief manufactures are machinery, porcelain, pottery, paper, chemicals, and soap.

Charlottetown, cap. Prince Edward Island, Canada, is a port with an excellent harbor on the Hillsboro River. It has pork-packing, condensed milk, and lumber products industries. It has the parliament buildings, a provincial lunatic asylum, and the College of St. Dunstan.

Chatham, England (Kent), is one of the chief ship-building towns of England, and the naval and military establishments here, or in the immediate vicinity, are very extensive. The royal dockyard now extends for two miles along the river, and is most thoroughly equipped for the building, fitting out, and repairing of war vessels, while the largest ironclads are built here. The military establishments include infantry, royal marine, and naval barracks.

Chatham, Ontario, Canada, is a trade center in a good farming and fruit region. It has flour-mills, machine shops, engine works, canneries, etc.

Chattanooga, Tennessee (c. h. Hamilton Co.), is the center of a district rich in iron, coal, and timber, and has extensive cotton-mills, iron-foundries, blast-furnaces, steel and rail works, tar-works, etc. It is the seat of the Grant University and the Chattanooga College. Near by is the National Military Park, which embraces the Chickamauga battlefield and the National Cemetery. Lookout Mountain, the scene of the dramatic "Battle above the Clouds," gives fine views of the surrounding country. The city is surrounded by picturesque mountains.

Chaux de Fonds, Switzerland (Neuchâtel), is one of the chief centers of watchmaking in Switzerland.

Chefu, China (Shantung), is a treaty port on an open bay at the entrance of the Gulf of Pechili. It exports straw-braid, beans, ground-nuts, bean-cake, silk, and vermicelli. This port became of considerable importance during the Russo-Japanese War.

Chelsea, Massachusetts (Suffolk Co.), is a suburb of Boston, with United States marine hospital, naval hospital, and a soldiers' home. Has art-tile works and other industries.

Cheltenham, England (Gloucester), owes its importance to its mineral springs, accidentally discovered in 1716, and is a favorite residential town for Anglo-Indians. The parish church of St. Mary is an ancient cruciform structure. It is an educational center. The college (1840) is an important public school, and the ladies' college (1854) ranks with the first educational institutions for women in the country. There are also two training colleges for elementary teachers.

Chemnitz, Germany (Saxony Kg.), is the chief manufacturing town of Saxony and one of the principal manufacturing towns of Germany. Its specialties are the manufacture of locomotives, agricultural implements, and tools, but it has also important industries of cotton-spinning, hosiery, gloves, dyeing and bleaching, ribbons, leather goods, stoneware, chemicals, and brewing. The town is surrounded by a close ring of industrial villages.

Chengtu, China (Chili), has manufactures of inlaid wares, and considerable trade.

Cherbourg, France (Manche), is a port, and naval and coaling-station on the English Channel. The naval dock-yard has three great basins hewn out of the solid rock. The exports are chiefly dairy and agricultural products.

Chernigof, Russia (cap. Chernigof), is the commercial center of a rich agricultural region. It is the seat of a Greek Orthodox archbishop.

Cherson, Russia (cap. Cherson), is a fortified town with an active trade.

Chester, England (Cheshire), is one of the most ancient towns of England, an outgrowth of a Roman camp, famous for its frequently-discovered relics of Roman and Norman occupation, and for the examples of mediæval architecture still existing. Its unique "Rows" on the principal streets, are continuous galleries penetrating the fronts of the houses on the first floor above the ground, and roofed by the second stories projecting over them. They are reached by staircases, and are lined with shops. Ancient walls still encircle the town, the walk on the top forming an admirable promenade. The cathedral was originally the monastery church of St. Werburgh. The present edifice, erected in the 16th century and restored in the 19th, contains portions of the ancient abbey. Among the modern buildings is the Grosvenor museum, which contains Roman and other antiquities. Industries include the manufacture of patent shot, paint, lead piping, leather, etc. Cheese fairs are held every month.

Chester, Pennsylvania (Delaware Co.), is a port of delivery, and has ship-building works and manufactures of cotton, woollen, and worsted goods. It is the seat of the Pennsylvania Military College and the Crozier Theological Seminary. Swarthmore College is near by. It was settled by the Swedes in 1643 under the name of Upland, and is the oldest town in the state.

Cheyenne, cap. Wyoming (c. h. Laramie Co.), is one of the chief centers of the cattle industry of the northwest, and an important railway center and shipping point. Fort Russell, a United States military post, lies four miles to the north of Cheyenne, and is connected with it by electric railroad. Elevation 6,075 ft.

Chicago, Illinois (c. h. Cook Co.), is the second city and the largest railway center of the United States. Its position gives it a great share of the commerce by way of the Great Lakes, and its manufactures and trade have grown to enormous magnitude. The manufacture of iron and steel in South Chicago, the agricultural implement works, and the beef and pork-packing establishments are among the largest. It is the greatest grain market and also the largest lumber mart in the world. The city covers an area of nearly 200 square miles. Its park system forms a girdle around the city, from Lincoln Park on the lake shore on the north to Jackson Park on the lake shore to the south; the total area of parks and boulevards is over 3,000 acres. It is the seat of the University of Chicago, Rush Medical College, College of Physicians and Surgeons, and numerous other educational institutions. The city library contains over 350,000 volumes, and there are, in addition, the Newberry library on the north side, the John Crerar library on the south side, confined to branches of science, and numerous others. The Chicago Art Institute, an imposing building of semi-classical style, contains a valuable collection of paintings, sculpture, and other objects of art, and is one of the largest and most comprehensive art schools in America.

Chicopee, Massachusetts (Hampden Co.), has manufactories of cotton, artillery, bronze, rifles, swords, and automobiles, run by power furnished by the Chicopee River.

Chiengmai, Siam, is the chief trade center of northern Siam; it has a large trade in tea.

Chieti, Italy (cap. Chieti), has manufactories of hats, woolens, soap, etc. It contains the ruins of a Roman amphitheatre and of a Roman castle. It is the seat of an archbishop, and has a cathedral dating from the 11th century.

Chihuahua, Mexico (cap. Chihuahua), is the metropolis of the northwestern section of Mexico, and supplies outfits for the majority of mining camps and prospecting expeditions in this section. It manufactures cottons, carpets, blankets, machinery, and has large iron-foundries. It has a magnificent cathedral, and its mint, established a few years ago, is now the third in the republic.

Chillan, Chile (cap. Nuble), is frequented for the sulphur waters in its neighborhood. It was long a mission center of the Jesuits.

Chillicothe, Ohio (c. h. Ross Co.), was the first capital of the state, and is the trade center for an agricultural and coal-mining region.

Chinandega, Nicaragua, has considerable trade in sugar, cotton, bananas, etc. In 1849, it was the capital of the united republics of Honduras, Nicaragua, and Salvador.

Chingtu, China (cap. Ssechuan), is situated in one of the largest fertile plains in China, and is surrounded by mountains rich in various minerals. The walls that surround the city are twelve miles in extent. It has an arsenal with modern equipment, and goods of European manufacture are found in some shops.

Chinkiang, China (Kiang Su), is a treaty port at the mouth of the Yangtze River. Vessels of the largest tonnage can enter at all states of the tide. It exports beans, bean-cake, peas, ground-nuts, silks, hides, dried flowers, etc., and its imports exceed those of any other Chinese port except Shanghai.

Cholon, Cochin China, has a large trade in rice.

Christchurch, New Zealand (cap. Canterbury), is a railway and commercial center. The chief public buildings are a fine cathedral, Canterbury College, with engineering school and observatory, a museum, a school of art, and Canterbury Hall. Its industries include iron-foundries, agricultural implement works, etc. It is the see of the primate of New Zealand.

Christiania, cap. Norway (cap. Christiania), is a seaport and coaling station on Christiania Fjord. The principal public buildings are the Houses of Parliament (the Storthing), where the national archives are also kept; the palace, built 1821-8; the university (with 1,400 students), built in 1841-51, the library of which contains 350,000 volumes; the national museum; the museum of antiquities; the museum of sculpture, including the national gallery; the observatory; Akers Church, built in the 11th century and restored in 1860; and the Cathedral of Our Saviour. Its industries include weaving, spinning, sailmaking, iron-foundries, paper-mills, some shipbuilding, and large manufactures of tobacco. The waterfalls of the Akers provide the chief motive power. Exports timber, wood-pulp, oil-cakes, fish, cod-liver oil, iron nails, etc.

Chungking, China (Ssechuan), is a treaty port on the Yangtze River and a great distributing center for the province. Its chief exports are opium, silks, wax, porcelain, and musk. Though 1,325 miles from the sea, it is accessible by steamer, but with difficulty on account of the rapids in the Ichang gorge. The river, about 900 yards wide, is subject to sudden rises; in 1892 it rose to 97 feet above low water mark. Valuable coal-mines are in the vicinity.

Cienfuegos, Cuba (Santa Clara), is the second seaport in Cuba with a magnificent landlocked harbor, and the center of the sugar trade of the south side of the island.

Cincinnati, Ohio (c. h. Hamilton Co.), is one of the most important manufacturing and commercial centers of the middle west, and has a frontage of 14 miles on the Ohio River. It has an extensive river trade in ore, coal, iron, lumber, salt, etc., and an immense railroad business; not less than sixteen railroads enter the city. The slaughtering and packing of meats has long been and still is one of the leading industries of the city. It holds first position in this country in the sale of pig iron, and over 9,000 industrial establishments are engaged in a multiplicity of manufactures. It is especially prominent in iron-work, men's clothing, boots and shoes, tobacco products, and malt liquors. Here are located the University of Cincinnati, the Ohio Mechanics' Institute, the Lane Theological Seminary, and numerous other educational institutions. The Cincinnati Society of Natural History has a museum of interesting relics, and the public library contains over 300,000 volumes. The art museum and art school have a valuable collection of paintings, sculpture, etc. The zoological garden contains a fine collection of animals and is a favorite resort.

Ciudad Porfirio Diaz, Mexico (Coahuila), is a frontier trade center on the Rio Grande, opposite Eagle Pass, Texas.

Clermont-Ferrand, France (cap. Puy-de-Dôme), is an ancient town and is the seat of a bishop. Its cathedral (1248-1346), built of lava from the surrounding district, was finished in the 19th century. The first crusade was preached by Peter the Hermit in its fine Notre Dame Church in 1096, after a great church council held here in the year preceding. It contains a university. It has manufactures of cottons, chemicals, rubber goods, and food preserves. Near by are famous mineral springs.

Cleveland, Ohio (c. h. Cuyahoga Co.), is the largest city of the state and the second on the Great Lakes, one of the greatest ore markets in the world, and a leading coal port. It is one of the chief markets for grain and petroleum, and one of the greatest in the world for fresh-water fish. It is the second manufacturing place on the lakes and the second in the state; while, as regards the manufacture of iron and steel it is among the foremost in the country; it is also the center of the malleable iron trade and steel shipbuilding yards. It has an endless variety of other manufactures, such as bridge castings, car-wheels, stoves, electrical apparatus, automobiles, etc. In the manufacture of cloth, malt liquors, paints, and chemicals, and in petroleum refining, it also ranks high, and it has, besides, extensive stock-yards and meat-packing establishments. The Western Reserve University and numerous other educational institutions are located here. Cleveland has one of the finest park systems in the country, covering upwards of 1,500 acres.

Clichy, France (Seine), is a suburb of Paris, and has starch, oil, candle, and chemical works.

Clinton, Iowa (c. h. Clinton Co.), is a trade center with a large radius of activity. Its industries include lumber mills, iron bridge works, railroad machineshops, etc.

Clinton, Massachusetts (Worcester Co.), has manufactories of wire cloth, ginghams, and carpets.

Coatbridge, Scotland (Lanark), is the center of the most important iron and coal district in Scotland, and has numerous blast-furnaces, tube works, rolling-mills, boiler- and tin-plate works. Fire-bricks, tiles, etc., are made in the immediate neighborhood. A mining college and technical school are located here.

Cobalt, Ontario, Canada, is a mining center in a rich silver district.

Coban, Guatemala, is the center of the finest coffee district of the republic. Chalk is mined here and made into crayons.

Coblenz, Germany (Prussia), is a fortified city, anciently called Ad Confluentes, from its seat at the confluence of the Rhine and Moselle. It is connected by a pontoon bridge over the Rhine with the fortress of Ehrenbreitstein, which, with its other fortifications, renders Coblenz one of the strongest places in Germany and capable of accommodating 100,000 men. Its industries embrace cigars, machinery, wines, and pianos.

Cocanada, India (Madras), is a seaport, with exports of cotton, rice, sugar, and cigars.

Cochabamba, Bolivia (cap. Cochabamba), has large manufactures of woollens, leather, and pottery, and an extensive trade in wheat. It is an ancient city, founded in 1563, and is the seat of a bishop.

Coffeyville, Kansas (Montgomery Co.), has varied manufacturing interests. There are gas-wells in the vicinity.

Cohoes, New York (Albany Co.), has manufactures of cotton, woollen and worsted knit goods, boots, shoes, etc. The Mohawk River has a fall of over 70 ft. at this point and supplies a large amount of power.

Coimbatore, India (Madras), is the center of a sugar, cotton, tobacco, and rice producing section. It commands the Palghat Gap and the Gazalhatti pass.

Colchester, England (Essex), is an ancient town with relics of a castle built in the reign of William Rufus. Ruins also remain of the 12th century Augustinian priory of St. Botolph, and a gateway represents the 11th century abbey of St. John. The church of St. James dates from before the Norman conquest. Other public buildings are the Albert School of Science and Art, the 16th century Grammar School, and Eastern Counties Asylum. Industrial establishments include iron-foundries, clothing and boot factories, breweries, and flour-mills. The town has considerable shipping trade, and there are extensive oyster-beds on the coast.

Colima, Mexico (cap. Colima), has several cotton mills.

Colmar, Germany (Alsace), has spinning-mills and manufactures of textiles, thread, starch, foodstuffs, machinery, etc. Among the interesting buildings is the mediæval church of St. Martin (13th century), with the famous Madonna in a rose-arbor by Martin Schongauer.

Cologne, Germany (Prussia), is a strongly fortified city, and, because of its history, the magnitude of its commerce, its position as a fortress of the first class, and its archiepiscopal see, is one of the most important towns of Rhenish Prussia. In the heart of the city are many houses of the 15th and 16th centuries and even earlier, but the glory of Cologne is its cathedral, one of the noblest and most impressive structures of Gothic architecture in existence. Its foundations were made in 1248 and the building went on down to the year 1447. Then nothing was done till 1842, when the work was again taken up, and in 1880, the structure was completed according to the original plan. It is a cruciform basilica; the choir is surrounded by eight chapels, in which are the tombs of several archbishops; the west front is adorned by two towers, each 512 ft. high. The reputed bones of the legendary three kings of the East are preserved in a gold shrine in the cathedral treasure-house. There is also a museum of ecclesiastical furniture, antiquities, etc., in a former chapel of the archiepiscopal palace. Some interesting examples of ancient architecture are in the Rathhaus, the Gurzenich, the church of St. Maria in Capitol, the Apostles Church, and the Jesuits' church. The Wallraf-Richarts, or Municipal Museum (1855-61), is especially noteworthy for its collections of pictures, and Roman and mediæval antiquities. The principal industries are printing, and the manufacture of woollens and cottons, sugar, tobacco, chocolate, porcelain, carpets, machinery, bricks, perfumes, etc.

Colombo, Ceylon (cap. Colombo), is a port and coaling station, with an extensive and newly improved harbor. It has mills for the manufacture of oils, fibers, desiccated cocoanut, and conducts a large trade in tea, coffee, cinchona, cocoanut-oil, pearls, tobacco, cinnamon, ivory, satin-wood, and other woods.

Colon, Panama, is a seaport, once called Aspinwall, in honor of one of the founders of the Panama Railroad, which has here its northern terminal. Exports mahogany, cedar, mother-of-pearl, hides, and silver. It has a large transshipment trade.

Colorado Springs, Colorado (c. h. El Paso Co.), is the principal health resort of Colorado, and has become the permanent residence of many who are unable to bear the changeable climate of the Atlantic coast. It is situated on an elevated plateau (6,000 ft.) near the eastern base of Pike's Peak, and is surrounded by beautiful scenery. The climate is serene, mild, and healthy. Here are located Colorado College, the oldest institution of its kind in the state; the state institution for the blind and mute, and numerous sanatoriums.

Columbia, Pennsylvania (Lancaster Co.), has manufactories of stoves, engines, wagons, and silk.

Columbia, cap. South Carolina (c. h. Richland Co.), is noted for its sanatoriums. It is the seat of the University of South Carolina, possesses large car, machine, and iron works, and is one of the handsomest cities of its size in the country. The most important public building is the state-house. Other large edifices are the state penitentiary, and the lunatic asylum.

Columbus, Georgia (c. h. Muscogee Co.), has extensive manufactures of cotton, numerous cotton compressors, ginning works, cottonseed-oil mills, etc. It also has a large iron industry and manufactories of agricultural implements. One of the largest syrup and sugar-refining works of the south is located here.

Columbus, cap. Ohio (c. h. Franklin Co.), is an important railroad and manufacturing center, with extensive steel plants, blast-furnaces, malleable iron-works, manufactories of agricultural implements, shoes, medicines, automobiles, etc. Among the institutions for higher education is the Ohio State University. The Ohio penitentiary and institutes for deaf-mutes and the blind, and the state hospital for the insane are located here.

Combaconum, India (Madras), has a government college, and is an important center of Hindu learning.

Como, Italy (cap. Como), is beautifully situated on the south end of Lake Como. Is famous for its manufactures of silk and velvet, though hosiery, gloves, machinery, and statuettes are also made. The chief architectural features of the city are the marble cathedral (1396-1519), with pictures by Luini, Gaudenzio, Ferrari, and others; the town hall; the mediæval building called Il Broletto (1215); the ancient churches of San Fedele and Sant' Abondio, of the 10th and 11th centuries; and the ruins of the citadel of Baradello.

Concepcion, Chile (cap. Concepcion), is the third city of Chile and an episcopal see. Coal-mines are near by, and it is the chief manufacturing center of the republic.

Concepcion, Paraguay, is the second city in the republic and is a port of entry and delivery. It has considerable trade in Paraguay tea.

Concord, cap. New Hampshire (c. h. Merrimack Co.), has an abundance of water-power, and manufactures carriages, shoes, twine, electrical apparatus, silverware, leather goods, and machine-shop products. Near by are extensive quarries of fine-grained white granite. The noteworthy buildings include the state-house, a fine building of Concord granite, the state prison, the state insane asylum, etc. It has numerous collegiate institutions.

Conjevaram, India (Madras), is one of the seven holy cities of India, and is called the Benares of the South. It has several temples and pagodas.

Constantine, Algeria (cap. Constantine), is a fortified town, and has tanneries, manufactures of woollens and cloth, and an active trade in wheat, which it carries on by rail and caravan.

Constantinople, cap. Turkey (cap. Constantinople), is a port, naval and coaling station, and the chief commercial center of the Levant, famous for its Oriental aspect, and for the beauty of its site, on a peninsula jutting into the Strait of Bosphorus, with the inlet called the Golden Horn on its north. The chief quarters of the city are Pera, Gaalata, Tophane, and Scutari. The old Seraglio occupies a large space at the outer tip of the peninsula. At one time it was entered from the quays, through a great gate—destroyed by fire in the sixties—the "Sublime Porte", from whence the cognomen of the Ottoman government is derived. The old Tcheragan Serai, the chief of the imperial palaces, was finished in 1857 by Abdil-Medjid in the style of the Turkish Renaissance. It is a marble building of great size, and its interior decorations and arrangements are of a luxury and magnificence which are unexcelled in Europe. Its chief facade, about 2,400 ft. long, is mirrored in the Bosphorus. Constantinople is the seat of the Sheik-ul-Islam, or the head of the Mohammedan religion, and of a Greek Orthodox archbishop. Of its numerous and beautiful mosques the principal is the great Byzantine mosque of Santa Sophia, originally a Greek church built by Justinian, famous for its interior decorations of colored marble and

mosaics on a gold ground; it has forests of exquisite columns, and is lighted by many hanging lamps. Constantinople has a university founded in 1900, and numerous other educational institutions. It also possesses several libraries, museums, and scientific, and art institutions. Its great bazaars, housing the varied and picturesque industries and merchandise of the East, are the liveliest part of the city. Ships of the large class find anchorage in the harbor, and there are large graving and dry docks. Exports grain, mohair, opium, silks, carpets, wool, drugs, dye-woods, and hides. The city was founded by Constantine the Great 330 A. D., was the capital of the Eastern Empire after 305; was unsuccessfully besieged by the Arabs, especially in 718; and was taken by the Latins in 1203-04, by Michael Palaeologus in 1261, and by the Turks, May 29, 1453.

Copenhagen, cap, Denmark (Copenhagen), is a fortified seaport, naval and coaling station, and the largest city of Denmark. Its harbor is one of the best seaports on the Baltic, to which fact the prosperity of the city is largely due. Its public institutions include a university (1479), a royal palace, a royal library, a valuable national museum, a picture gallery, a polytechnic school, a cathedral (1811-29), and the imposing Frederick, or Marble, Church. The chief industries are ship-building, brewing, distilling, sugar-refining, and the manufacture of soda, porcelain, machinery, and textile fabrics. Exports dairy products, pork, horses, cattle, sheep, yarn, paper, etc.

Cordoba, Argentina (cap. Cordoba), is one of the most progressive inland cities of the republic, and has an extensive trade in hides and wool. Four miles above the city is the San Roque dam, the largest work of the kind in South America, which supplies power for several industries and for irrigation. It has a university, an astronomical observatory, and is the seat of a Roman Catholic bishop.

Cordova, Spain (cap. Cordova), is an ancient city, famous as the capital of the Caliphate of Cordova (8th-11th century), which embraced the whole of Mohammedan Spain. During that time it was one of the chief commercial centers of the world, and an important seat of learning, containing a public library of 600,000 volumes. It retains but few marks of Moorish remains, except its world famous cathedral, which was formerly a mosque, and is one of the most beautiful examples of Moorish architecture in the world. It has manufactures of leather, cloth, silk, paper, and hats, besides the ancient silver filigree industry for which Cordova is famous. It is the seat of a bishop.

Corinto, Nicaragua, is the chief Pacific port of Nicaragua. Exports coffee, hides, rubber, cedar, and hardwoods.

Cork, Ireland (cap. Cork), is a port, naval and coaling station, and the third city of Ireland. It is situated on the river Lee, 15 miles from its mouth, and besides an upper harbor at the city itself, and quays extending over 4 miles in length, has a lower harbor at Queenstown, 11 miles below. The chief industries are tanning, distilling, brewing, iron-founding, and the manufacture of woollen goods. Exports grain, cattle, dairy produce, and provisions. The principal buildings are the Protestant and the Roman Catholic cathedrals, exchange, custom-house, chamber of commerce, and Queen's College. It is the seat of a Roman Catholic bishop.

Corning, New York (Steuben Co.), has extensive foundries, glass factories, and railroad car works.

Corrientes, Argentina (cap. Corrientes), is a trade center on the Parana River, with large exports of timber and oranges.

Cortland, New York (c. h. Cortland Co.), is a farming and manufacturing trade center, and has extensive manufactories of wire, carriages, steel ware, etc. A state normal school is located here.

Coruna, Spain (cap. Coruna), is a fortified seaport, with a safe and commodious harbor, and was formerly the principal port of departure for England. The Great Armada sailed from here in 1588. It exports cattle, minerals, and dairy produce, and has manufactures of cigars, paper, ropes, as well as large sardine fisheries.

Council Bluffs, Iowa (c. h. Pottawattamie Co.), is located at the foot of the bluffs of the Missouri, nearly opposite Omaha, and is an important railroad center and terminus.

Courbevoie, France (Seine), is an industrial suburb of Paris, with manufactures of chemicals, white lead, and carriages. It has numerous handsome villas, and extensive barracks.

Courtrai, Belgium (West Flanders), is famous for its lace, and linen bleacheries. One of its churches, adorned with a masterpiece of Van Dyck, was the burial-place of the counts of Flanders.

Coventry, England (Warwick), is one of the oldest cities of England, and was formerly a walled town. A Benedictine monastery was founded here in 1044. It is now one of the greatest industrial centers of England, with manufactures of bicycles, ribbons, sewing-machines, and automobiles.

Covington, Kentucky (c. h. Kenton Co.), has extensive distilleries, cotton and woollen mills, rolling-mills, and tobacco factories. It is a residence town for Cincinnati business men, and is the see of a Roman Catholic bishop.

Crajova, Roumania (Wallachia), is the most important town in the western portion of Roumania.

Cranston, Rhode Island (Providence Co.), has manufactures of cotton goods and other industries.

Crefeld, Germany (Prussia), is the chief city of Germany for the manufacture of silks and velvets, and exports these fabrics extensively. It also has extensive railway car shops, engineering-works, and manufactories of sugar, and chemicals.

Cremona, Italy (cap. Cremona), is an ancient town. Its chief ornament is its cathedral (1107-90), with frescoes by masters of the Cremona school of painting, and the tall bell-tower. There are several other interesting churches and some old palaces. It has manufactures of silk, cotton, woollens, pottery, and stringed instruments. Was formerly famous for its violins, the chief makers of the Cremona school being Amati, Guarneri, and Stradivarius.

Crewe, England (Cheshire), has extensive railway works, and is an important railway center.

Cripple Creek, Colorado (c. h. Teller Co.), is a trade center for the Cripple Creek mining district, one of the chief gold-mining sections of the state. Here are located several cyanide mills, smelters, and other mining industries.

Cristobal, Panama (Canal Zone), is an American port of entry and post-office, adjoining the city of Colon, and situated near the northern entrance to the Panama Canal.

Cronstadt, Russia (St. Petersburg), is a port, and a naval and coaling station. Has extensive fortifications protecting St. Petersburg, and naval works, arsenals, barracks, cannon foundries, and shipyards. Here are schools for naval instruction, a maritime hospital, a cathedral, and an interesting summer garden, originally planted by Peter the Great and attached to the small palace in which he lived. Exports lumber, hemp, leather, and cordage.

Croydon, England (Surrey), is a residential suburb of London, and has large clock factories, and some other industries.

Csaba, Hungary, has flour mills and distilleries, and a trade in cattle, corn, hemp, and wine.

Cuddalore, India (Madras), has an extensive trade in grain.

Cuenca, Ecuador (cap. Azuay), is the center of a region fertile in corn, cotton, sugar, and fruit. Interesting Aztec remains are in the neighborhood.

Cumberland, Maryland (c. h. Allegheny Co.), is the shipping point for large quantities of semi-bituminous coal, mined in this section. Its manufactures include cement-works, flour-mills, tanneries, steel-works, iron-foundries, and numerous other industries. Here, moreover, are located large rolling-mills for the manufacture of rails, bars, and other materials of railway supply, as well as railway car and repair shops.

Cuneo, Italy (cap. Cuneo), has manufactures of silk, cotton, and paper. It is the seat of a bishop, has a cathedral, a 12th century church, and many mediaeval houses.

Curitiba, Brazil (cap. Parana), is a modern and at-

tractive city, with a large trade in corn, beef, fruit, tobacco, and Paraguay tea. There are gold-mines in the neighborhood.

Cuttack, India (Bengal), is an important trading center situated at the apex of the Manamadi River, and is noted for its filigree work in gold and silver.

Cuyaba, Brasil (cap. Matto Grosso), is an important trading center in a region formerly famous for its gold and diamond mines.

Cuzco, Peru (cap. Cuzco), was formerly the capital of the Incas, and is the most ancient of the Peruvian cities, having been founded, according to tradition, about 1020 by Manco Capac, the first Inca of Peru. A municipal government was established here by Francisco Pizarro in 1534. It is the residence of a bishop, and the seat of a university (1692), and of a meteorological station erected by Harvard University. Cocoa leaves are exported.

Czegled, Hungary, is an important market town in the center of a section producing large quantities of grain and wine.

Czenstochowa, Russia (Poland), has important industries of cotton, cloth, paper, etc. It contains a convent of St. Paul, visited yearly by about 400,000 pilgrims.

Czernowitz, Austria (cap. Bukowina), is the seat of a Greek Orthodox archbishop, and has a university. It is an important trade center, and has machine-shops, saw-mills, and breweries.

Dacca, India (Bengal), was selected about 1610 as the seat of the Mohammedan government of Bengal. In the 18th century its fine muslin was famous. Gold and silver utensils, and filigree work are made here.

Dallas, Texas (c. h. Dallas Co.), is the leading manufacturing and commercial city of northern Texas. It is situated in a fertile agricultural region, of which the chief products are cotton, wheat, corn, and fruits. It has extensive lumber interests, and its principal manufacturing industries include saddlery, in which Dallas leads all other places in the United States, and cotton-gin machinery. Other prominent manufactures are iron and metal works, woollen mills, cottonseed-oil mills, and cotton compresses. It has a large trade in agricultural implements. Numerous collegiate institutions are located here.

Daman, India (Daman), is a seaport with important fisheries.

Damanhur, Egypt (cap. Behera), occupies the site of the ancient Hermopolis Parva, and has cotton factories.

Damascus, Turkey in Asia (cap. Syria), is one of the holy cities of the Mohammedans, and is regarded by the Arabs as one of their four terrestrial paradises. Its caravan trade is but a shadow of what it was in former days. Is noted for its gold and silver embroidered stuffs, confectionery, cotton and silk fabrics, saddlery, perfumes, carpets, etc. One of the most interesting features of the city is the world-renowned Ommiad mosque, which figures in Arabic literature as one of the wonders of the world; it was built early in the 5th century, but was several times destroyed and rebuilt. Among other places of interest is the citadel built early in the 13th century. The city has in all about 250 mosques and Mohammedan schools.

Damietta, Egypt, has manufactures of cotton and silk fabrics, and pottery.

Danbury, Connecticut (c. h. Fairfield Co.), is the greatest hat-making city in the United States. It has also extensive manufactures of iron, brass, and silver-plated ware.

Danville, Illinois (c. h. Vermilion Co.), has large railroad-car and machine-shops, iron-foundries, carriage and wagon factories, organ and furniture factories, etc. Coal-mining, which is the chief industry, is carried on extensively in the vicinity.

Danville, Virginia (Pittsylvania Co.), is one of the oldest cities of the South, having been incorporated in 1792. Has cotton-mills, flour-mills, foundries, etc. It is the largest loose tobacco market in the world. It has a Baptist woman's college, a Methodist institution for young ladies, and a military institute.

Danzig, Germany (Prussia), is a naval and coaling station on the Vistula, 3 miles from the Baltic. It has ship-building, sugar-refining, and distilling industries, also iron-works, rolling-mills, and machine factories. It exports sugar, grain, timber, spirits, oils, oil-cakes, etc. The most interesting part of the city is the old town, with the town hall (14th and 16th centuries); the Artushof (15th century), formerly the rendezvous of the Danzig patrician families, now the exchange; the Green Gate (of the town), converted into the W. Prussian Provincial Museum; the High Gate (16th century); the church of St. Mary, rebuilt in 1400-1502, one of the largest and handsomest churches along this part of the Baltic coast; and the old armory (1602-5). A technical high school was opened in 1901. Danzig is the birthplace of Fahrenheit and Chodowiecki. From 1308 to 1454 it was subject to, or in alliance with, the knights of the Teutonic Order. Down to 1772 it was Polish; it became part of Prussia in 1793.

Darbhangah, India (Bengal), has a trade in oil-seeds, food-grains, and timber.

Darlington, England (Durham), has extensive local railroad locomotive works, also rolling-mills, wagon-works, iron- and brass-foundries, besides worsted-mills, malt-works, and tanneries. Among the public buildings may be noted the technical institute, the Edward Pease library, and the church of St. Cuthbert, which was erected in the 12th century.

Darmstadt, Germany (Hesse), has a grand-ducal castle dating from the 15th century, which shelters a library of over 350,000 volumes, a picture gallery, and museums. It has also a 16th century town hall. The industries are varied,—none of commanding importance.

Darwen, England (Lancashire), has large paper works. Coal is found in the vicinity.

Davenport, Iowa (c. h. Scott Co.), has many industries and an active river trade. Among its numerous manufactured articles are lumber and planing-mill products, brick and stone, carriages, agricultural implements, woollen goods, glucose and its products, cordage, and cooperage products. It has a Roman Catholic college, and numerous other educational institutions. The surrounding region is important both for its agriculture and its coal mines.

Dayton, Ohio (c. h. Montgomery Co.), is an important railway center, and has widely diversified industries. Among the more important manufactures are cash registers, railroad-cars, agricultural implements, sewing-machines, etc. It has numerous collegiate institutions, and the training college of the United Brethren denomination. A National Soldiers' Home with 5,000 inmates is located near this city.

Debreczin, Hungary, produces flour, cigars, brushes, and machinery, and has famous cattle markets. It is the center of the Hungarian Protestants, and possesses a theological (Calvinist) college, and an agricultural academy.

Decatur, Illinois (c. h. Macon Co.), is situated in the midst of the famous Illinois corn belt, and is the trade center of several counties. Its corn-mills are among the largest in the United States. It has railroad-shops, iron-works, agricultural implement works, casket and coffin factories, etc. The principal departments of the James Milliken University are located here.

Delft, Netherlands (South Holland), was famous in the 17th and 18th centuries for its porcelain, and of late years the industry has been revived. It also manufactures carpets, firearms, etc. The Oude Kerk contains monuments to Admiral Piet Hein and A. van Leeuwenhoek (born here in 1632), also the tomb of Admiral Maarten Tromp. In the Nieuwe Kerk (1384-96), is the grand monument which the United Provinces erected (1616-21) to William the Silent, who was assassinated (1584) in Delft. It was also for generations the burial-place of the princes of the House of Orange.

Delhi, cap. India (Punjab), contains a grand palace, and the Jama Maajid—the largest cathedral mosque in India—monuments of the Emperor Shah Jehan, who, in the 17th century, founded the present city. Beyond the walls can be traced the sites of ancient Delhis. Of these memorials the most striking are the deserted fortress of Taghlakabad, the mausoleum of the Emperor Humayun, and the Kutab Minar. Once the seat of Buddhist religious life and culture, Delhi became in the 12th century the splendid capital of the Mogul empire and

the center of Mohammedan power. In 1857, it was the scene of a frightful massacre of Europeans, during the Sepoy Rebellion, but was subsequently captured after a long siege. On New Year's Day, 1903, Delhi furnished the setting of the pageant of the coronation durbar. The principal industries are ivory-carving, miniature-painting on ivory, and pottery.

Denison, Texas (Grayson Co.), is a cattle-shipping place and trade center. Has manufactories of cotton, iron, machinery, etc.

Denver, cap. Colorado (c. h. Denver Co.), is the metropolis of Colorado and the Rocky Mountain region, and is the principal smelting center of the world. It lies within sight of the mountains, at an altitude of one mile above the sea, and its climate is one of the most delightful and remarkable in the country. The mean annual temperature is 49°, and the annual rainfall about 15 inches. Its jobbing trade extends over a wide area of territory and is very extensive. The University of Denver, a coeducational institution, has departments here and in a suburb, University Park. Here are also located the College of the Sacred Heart, a Baptist woman's college, and numerous other educational institutions. Its manufacturing industries include canning and packing, iron- and steel-works, glass-works, etc. It is the emporium of the rich gold and silver mining-districts of the state, and also the chief center of the coal trade.

Derby, England (Derby), is rich in churches, and has manufacture of silk, cotton, and lace goods, porcelain, patent shot, and paper; also extensive railway shops, brass- and iron-foundries, etc.

Des Moines, cap. Iowa (c. h. Polk Co.), is the largest city in the state and the center of a rich coal-mining region. It has an extensive jobbing trade, and the principal industries, besides coal-mining, include pork-packing, the manufacture of starch, pipe, cotton and machine products, and glass goods. Here are located Des Moines College, Drake University, and a normal college.

Dessau, Germany (cap. Anhalt), has a ducal palace (1748) containing valuable pictures and interesting relics, and several art galleries. Sugar, machinery, cloth, and carpets are the chief industrial products.

Desterro, Brasil (cap. Santa Catharina), lies on an island about two miles from the mainland, and is a port with exports of coffee, sugar, rice, and dairy products.

Detroit, Michigan (c. h Wayne Co.), is the largest city of the state, and has one of the finest harbors on the Great Lakes. It is the terminus of a large number of lake steamer lines and a great railroad center. Its extensive parks and boulevards make it a city remarkable for its beauty. It has a large trade in grain, wool, hides, and copper. The staple manufactures are iron and steel goods, boilers and engines, cars and car-wheels, automobiles, stoves, tobacco, and varnish. It has important lumbering interests and large tanneries. Its educational institutions include a Jesuit college, three medical colleges, schools of pharmacy and dentistry, a college of law, etc.

Deventer, Netherlands (Overyssel), has iron-foundries and carpet factories, and is famous for a kind of gingerbread cake called "Deventer koek."

Devonport, England (Devon), is one of the chief naval arsenals of Great Britain, with a dockyard covering over 70 acres, and other works of importance. Here are located the Royal Naval Engineering College, and barracks.

Diamantina, Brasil (Minas-Geraes), is the chief center of the diamond fields. It has iron-works, tobacco factories, and cotton-mills.

Diarbekr, Turkey in Asia (Kurdistan), is the seat of a Greek bishop, and has numerous mosques. Morocco leather, filigree-work, cotton, and silk are exported.

Dijon, France (Côte-d'Or), is the center of the upper Burgundy wine trade, but mustard, black-currant liqueurs, oil, vinegar, candles, hats, and shoes are also manufactured. It is the seat of a bishop, and possesses a university, a museum, and a library. Among its notable buildings are the cathedral Ste. Benigne (1271); Notre Dame, a fine example of the 13th-14th century

architecture, with a clock from Courtral (1383); St. Michel (1529); and the city hall, formerly the palace of the dukes of Burgundy.

Dordrecht, Netherlands (South Holland), is a mediæval town, with a good picture-gallery, and has oil-mills, sugar-refineries, engineering shops, and ship-building yards.

Dorpat, Russia (Livonia), has a university and is one of the principal seats of learning in Russia, with a library, an observatory, and a botanical garden. Has considerable trade and important fairs.

Dortmund, Germany (Prussia), is one of the greatest industrial centers of Westphalia, and has coal-mines, iron- and steel-works, breweries, etc. It also produces machinery, hardware, zinc, and flour. It possesses the fine old churches of St. Reinold and St Mary, both 13th century; and a town hall of the same date, rebuilt in 1899.

Douai, France (Nord), has sugar-refineries, woollen, and paper-mills, leather factories, and glass-works.

Dover, England (Kent), is one of the Cinque Ports, and a strongly fortified military station. It is the chief port of departure for the continent. The national harbor encloses an area of nearly 700 acres. A large part of the castle buildings date from the Norman period, and the keep from the time of Henry II.

Dover, New Hampshire (c. h. Strafford Co.), has abundant water-power, and its industries include several large woollen-mills and extensive print-works, also the manufacture of boots and shoes, oilcloth, hats and caps, sand-paper and glue. It has also several tanneries, brass- and iron-foundries, and machine-shops. It is the oldest city in the state, having been settled in 1623.

Drammen, Norway (Buskerud), is a port and one of the oldest towns in Norway. It is a great timber emporium, and has iron-foundries, breweries, and knitting-works.

Dresden, Germany (cap. Kg. Saxony), was by reason of its situation and its art treasures called by Herder, the "German Florence." The handsomest part of the city is the portion on the south bank which faces the river, namely the Brühl Terrace and the square at its west end where are clustered together the Albertinum (1884-9), which contains a valuable archæological museum; statues of Semper (1891), Ludwig Richter (1898), and the Elector Maurice of Saxony (1895); the sumptuous Art Academy (1890-4); the court (Roman Catholic) church; the royal palace (first built in 1530-5, and rebuilt 1890-1902), with its "green vaults," which contain numerous gems of artistic work in enamel, crystal, gold, ivory, bronze, silver, wax, etc.; the handsome royal opera-house (1869-78); and the vast pile known as the Zwinger, in which is preserved a choice and very valuable gallery of old masters and other collections. Dresden also possesses the Johanneum museum, the national industrial art museum, a municipal museum, fine monuments of Carl Maria von Weber (who is buried in Dresden), Goethe and Schiller, Luther, King John, Körner, and other celebrities, by such artists as Rietschel, Hähnel, Schilling, and Kircheisen. Also, on the south side, but more distant from the river, are the law-courts, synagogue, Schilling museum, and the large royal garden, embracing within its confines two royal palaces, an exhibition hall, and zoological and botanical gardens. On the opposite bank—i. e. the right bank of the Elbe—there is another frontage of handsome edifices, including the Japanese Palace (1715), which shelters the royal library; the Körner museum, so named after Theodore Körner, who was born in Dresden; and the ministries of justice, war, and finance. Dresden's fame as a center of artistic and intellectual activity is supported by its conservatory of music and by the polytechnical high school. It has considerable industries, including tobacco, machinery, chemicals, paper, piano, and sugar works, etc.

Dublin, cap. Ireland (cap. Dublin), is a port and coaling-station, and the metropolis of Ireland. It is one of the handsomest capitals of Europe. The splendid Phoenix Park has an area of nearly 2,000 acres, and includes a well-known zoological garden. Among the institutions of note are Trinity College, founded in 1591, a

Roman Catholic university, the Royal College of Surgeons, and the Royal University of Ireland. Leinster House, once the town mansion of the Dukes of Leinster, now the home of the Royal Dublin Society, has been enlarged by the erection of a national art gallery and a museum of natural history. New buildings for a science and art museum and a national library were opened in 1890. Among the other public edifices may be mentioned the Bank of Ireland (formerly the Houses of Parliament), the custom-house, and the Four Courts. The Castle (the Lord Lieutenant's official residence) has no pretensions to architectural beauty. The Chapel is interesting and contains some fine carved work of Grinling Gibbons. Dublin is remarkable in possessing two Protestant cathedrals. St. Patrick's, founded in 1190, was restored in 1865, and Christ Church, dating from 1038, but not raised to cathedral rank till 1541, is a smaller but more beautiful edifice. The chief manufacture is porter, of which nearly half a million hogsheads are annually exported, "Guinness" being, of course, the most important. Next in order is whiskey, and then the textile called poplin.

Dubuque, Iowa (c. h. Dubuque Co.), is the principal business center of the lead and zinc region of the Northwest, and has a large trade by river and rail. Its packing interests, lumber-mills, brewing, carriage and wagon factories are important. It has several collegiate and scientific institutions, the Iowa Institute of Science and Art, and various theological seminaries.

Dudley, England (Worcester), has workings of ironstone, coal, and fire-clay; other industries are brassfounding, brick, tile, and cement making. In the vicinity are chalybeate springs.

Duisburg, Germany (Prussia), is the seat of large iron-works and cotton-factories, and has manufactures of machinery, chemicals, and tobacco, as well as shipbuilding yards and coal shipping.

Duluth, Minnesota (c. h. Saint Louis Co.), is the western terminus of the marine traffic of the Great Lakes. It leads all other ports on the Great Lakes in point of vessel tonnage enrolled, and is one of the greatest iron ore shipping ports of the world. It has an extensive trade in grain, its elevator storage capacity ranking high; transhipment trade in lumber and coal; and a large trade in iron and steel manufactures, being an outlet of the important mineral region of northeastern Minnesota. It has extensive stock-yards, slaughtering and cold storage establishments, machine-shops, blast-furnaces, flouring-mills, etc. It has a state normal school.

Dumbarton, Scotland (cap. Dumbarton), has Dumbarton Castle, a fortress at least a thousand years old. Its iron and steel ship-building yards are among the most important on the Clyde. Other industries are engine and boiler-making, rope and sail-making, etc.

Dünaburg, Russia (Vitebsk), is an important strategetical and commercial center of northwestern Russia. Has an important trade in linen, hemp, and wood.

Dundee, Scotland (Forfar), is a port and coaling-station and the third largest city of Scotland. Among its institutions are the University College, affiliated with St. Andrew's University, and the Technical Institute. The Scottish Arctic and Antarctic seal and whale fishing fleet makes Dundee its headquarters. The staple industries are the manufacture of jute, flax, canvas sails, sacks, ropes, and sheeting. Its fruit-preserves are famous, while ship-building, dyeing, brewing, and distilling are also carried on.

Dunedin, New Zealand (Otago), is a port at the head of a lake-like harbor, and has a university with schools of medicine and mines. Here are woollen factories, railway workshops, meat-refrigerating works, and various other industries.

Dunkirk, France (Nord), is a fortified seaport and coaling-station. It is a fortress of the first class and all its approaches can be submerged. Its harbor, recently enlarged and improved, is now a rival of that of Bordeaux. Ship-building is carried on, and it has manufactories of chemical manures, flax, hemp, and jute spinning. Exports sugar, phosphates, manures, rails, and agricultural products.

Dunkirk, New York (Chatauqua Co.), has extensive locomotive works and various other factories. It has a safe and commodious harbor, and close connection with the coal and oil fields of Pennsylvania.

Dunmore, Pennsylvania (Lackawanna Co.), has rich mines of anthracite coal, and manufactures of silk and iron. A state school for the deaf and dumb is located here.

Durango, Mexico (cap. Durango), has manufactures of wool, cotton, sugar, etc. A fine cathedral, and a government assay office are among its public buildings.

Durban, Natal, Africa, is a seaport and naval coaling-station on a land locked bay. Its gardens are especially beautiful, revealing the horticultural wealth of the "Garden Colony" of South Africa. Exports sugar, gold, ivory, coffee, hides, tanning materials, feathers, tea, etc.

Duren, Germany (Prussia), is the seat of various large industries, including the manufacture of cloth, iron, machinery, paper, carpets, etc.

Dusseldorf, Germany (Prussia), has been since 1819 the seat of one of the most influential of the German schools of art, the academy, founded in 1767. It also possesses two picture-galleries, the meeting-hall of the Rhineland provincial estates, and the town hall dating from 1570–3. It has extensive manufactories of machinery and explosives, cotton-spinning, book-printing, calico-printing, dyeing, and locomotive and car shops. It is an important banking center.

Ealing, England (Middlesex), is a suburb of London. The Royal India Asylum is near by.

Eastbourne, England (Sussex), is a fortified town and fashionable watering-place. Handsome promenades, consisting in places of two and three tiers of broad walks, extend along the sea-front for about 3 miles, with a cliff drive above on the west side.

East Ham, England (Essex), is an eastern suburb of London.

East Liverpool, Ohio (Columbiana Co.), has extensive china, earthenware, terra-cotta, and glass works. Its pottery works are the largest in the United States. Natural gas is used for light and fuel.

East London, Cape Colony, Africa, is a seaport which ranks third in importance in the colony. Exports wool, mohair, hides, gold, etc.

Easton, Pennsylvania (c. h. Northampton Co.), is the seat of Lafayette College, founded in 1826, and has manufactures of mining implements, railroad supplies, boots, shoes, hosiery, etc.

East Orange, New Jersey (Essex Co.), is a residential suburb of New York, with important pharmaceutical works.

East Providence, Rhode Island (Providence Co.), is a suburb of Providence. Has chemical, electrical, and wire works, bleacheries, etc.

East Saint Louis, Illinois (Saint Clair Co.), is an important railroad terminus and center. Has extensive breweries, rolling-mills, locomotive works, glass factories, and large coaling and coking industries. Its stock-yards are among the largest in the United States, and it has extensive packing-houses.

Eau Claire, Wisconsin (c. h. Eau Claire Co.), is the outlet of the Chippewa lumber district, and has extensive water-power. It also has manufactures of iron, linen, furniture, machinery, and shoes.

Eccles, England (Lancashire), is a residential suburb of Manchester. It has cotton, thread, silk, fustian, and gingham factories.

Ecija, Spain (Seville), has manufactures of textiles, and an extensive shoe industry. Was an ancient Roman colony, and in 1220 was taken from the Moors by Ferdinand the Saint.

Edinburgh, cap Scotland (cap. Edinburgh), is the second city of Scotland, and stands 2 miles from the Firth of Forth on a series of ridges. It has many buildings famous in history that are important for their architectural merit. Edinburgh has long been known for its educational institutions, which draw many to the city for the advantages they offer. At the head of these is the Edinburgh University, founded in 1582; it has six branches, arts, science, divinity, law, medicine, and music; its library contains over 200,000 volumes. The

most remarkable public building in Edinburgh is the castle, a mediæval royal residence, surmounting an eminence 430 ft. high. The famous royal palace of Holyrood occupies the site of Holyrood Abbey. St. Giles' Church, dating from the 15th century, the Episcopal Cathedral of St. Mary's, which is one of the largest churches built in Great Britain since the Reformation, the Parliament House built in 1633 and now used as the "Outer House" of the Supreme Courts, and the handsome building of Heriot's Hospital, erected in the 17th century and now the seat of a technological school, are of interest. Chief among the industries of Edinburgh are brewing, printing, and type-founding; paper and ink manufacture are conducted on a large scale. Its distilleries and rubber works are also of importance. It derives a certain commercial distinction from being the headquarters of six Scottish banks, as well as of the law and the church.

Edmonton, Alberta, Canada is a trading center of an immense region, which is rich agriculturally, and has valuable deposits of the precious metals. The neighborhood supplies coal, brick clay, timber, marble, asphaltum, graphite, and mica. The town was originally a settlement of the Hudson Bay Company, and one of the most important fur trading depots of the Far West. It is now a rapidly growing railroad, banking and outfitting center. It is the seat of Alberta college with 400 students.

Edmonton, England (Middlesex), is a residential suburb of London. Was for some time the residence of Cowper and Keats; Charles Lamb died here in 1834.

Eger, Austria (Bohemia), has extensive breweries and manufactories of textiles, shoes, and machinery. The town hall, in which Wallenstein was assassinated (Feb. 25, 1634), is partly converted into a museum. The old imperial citadel, built by Frederick I. in 1157-9, is now in ruins.

Eisenach, Germany (Saxony, Grandduchy of), is a famous summer resort, and has manufactories of woollen, pottery, leather, and colors. Its most conspicuous building is the Wartburg, formerly the residence of the ancient landgraves of Thuringia but now occupied by the grand duke of Weimar,—the scene of the legendary contest of the early German poets or Minnesingers, and of Luther's voluntary imprisonment.

Ekaterinburg, Russia (Perm), has besides two cathedrals, an old mint and the central assay for the gold of the Ural mines. The machinery for most of the Ural mining work is made here. The cutting, polishing, and engraving of precious stones is a prominent industry.

Ekaterinodar, Russia (cap. Kuban), has a large trade in corn and flour, a cathedral, and a museum of natural history.

Ekaterinoslaf, Russia (cap. Ekaterinoslaf), is one of the greatest iron-working centers in Russia; tobacco and beer are also manufactured. It has four colleges, including the orthodox seminary.

Elberfeld, Germany (Prussia), is the seat of extensive manufactures of cotton, woollens, and silk. It also has iron- and steel-works, and various other industries.

Elbing, Germany (Prussia), has one of the largest ship-building yards in Germany. It also has iron- and tin-works.

Elche, Spain (Alicante), is a famous ancient city with Roman ruins. Its products are subtropical, including figs, dates, carobs, etc. The date-palm groves are without a rival in Europe; they supply the leaves for use on Palm Sunday to all of Spain. A large hemp industry also exists here.

Elgin, Illinois (Kane Co.), is famous for its watches and dairy products, and has many other industries, including two large publishing houses. The Illinois Northern Hospital for the Insane, and several important educational institutions are located here.

Elizabeth, New Jersey (c. h. Union Co.), is a residential suburb of New York, and near it is located the largest sewing-machine manufactory in the world. It also has large steel-works, a ship-yard, and many other industries. Elizabeth was settled in 1665, and was the capital of New Jersey from 1755 to 1757.

Elizavetgrad, Russia (Cherson), is an important trade center. Lignite is extensively worked.

Elizavetpol, Russia (cap. Elizavetpol), is noted for its grapes and other fruits. It has a citadel and mosque built about 1620. There are numerous remains of antiquity in the vicinity.

Elkhart, Indiana (Elkhart Co.), is a railroad center and shipping point for a large agricultural region. Extensive railroad shops and manufactories of brass, carriages, automobiles, and paper are located here.

Elmira, New York (c. h. Chemung Co.), has railway-car construction and repair shops, steel-plate works, boot and shoe manufactories, glass-works, rolling-mills, fire-engine construction works, tobacco warehouses, bridge works, silk-mills, etc. Stone-quarries are in the vicinity. Elmira Reformatory, which takes the place of a State prison for male offenders, who have not become hardened in crime, is located here.

El Paso, Texas (c. h. El Paso Co.), is a trade center of western Texas, an important gateway to Mexico, and has ore-smelters, iron-foundries, and extensive cattle interests. A school of mines and several collegiate institutions are located here.

Elwood, Indiana (Madison Co.), is an important trade center and manufacturing place in a natural gas belt. Its industries are chiefly lumber, flour, and tin-plate mills, and it also has window plate glass and lamp chimney factories.

Enfield, England (Middlesex), is a residential suburb of London, and has a government manufactory of rifles, swords, and machine guns.

Enschede, Netherlands (Overyssel), is the center of the cotton trade of the Netherlands.

Erfurt, Germany (Prussia), is famous for the growing of flower seeds, flowers and vegetables. Other industries include the manufacture of dress goods, machinery, firearms, lamps, boots, malt, woollens, cottons, and silks. Its chief ornament is the cathedral dating from the 12th to the 15th centuries. In the monastery of St. Augustine, now an asylum, is the cell once occupied by Luther.

Erie, Pennsylvania (c. h. Erie Co.), is a port of entry on Lake Erie, with a safe landlocked harbor and numerous industrial establishments. It has oil-refineries, tanneries, breweries, and extensive manufactories of iron, steel, and lumber. The Pennsylvania Soldiers' and Sailors' Home is located here.

Erivan, Russia (Erivan), has extensive fruit-growing and gardening interests; its peaches are famous. There are several Armenian churches and mosques.

Erzerum, Turkey in Asia (Armenia), was the bulwark of the Armenians in Byzantine times, and was formerly an important trading center. The most important industries now are iron and copper working. It exports cattle, sheep, and furs.

Escanaba, Michigan (c. h. Delta Co.), is an important ore-shipping point, and has an active trade in lumber, and fish. It manufactures furniture and woodenware, and has large railroad machine-shops.

Essen, Germany (Prussia), stands in the middle of the Uhr coal-field region, and is the seat of some of the largest iron- and steel-works (including the cannon-foundries and other workshops of Krupp) in Germany. It has besides large factories for machinery and tobacco. The minster church is one of the oldest in the empire, and presents several architectural features of singular interest. From 1275 to 1803 there was here a Benedictine monastery, the abbess of which ranked as a princess of the empire.

Esslingen, Germany (Württemberg), manufactures machinery, cottons, lithographs, tin wares, woodenwares, etc. The town is famous for its wine and fruit. It possesses several old churches, including the Church of Our Lady (1324-1420), a 15th century town hall, and an old citadel.

Eureka, California (c. h. Humboldt Co.), is a port of entry and trade center on Humboldt Bay. It is the largest coast city of northern California, and has an extensive trade in lumber and lumber products.

Evanston, Illinois (Cook Co.), is a residential suburb of Chicago, and the seat of the Northwestern University founded in 1854. Here are also located the Garrett Biblical Institute, and numerous other educational institutions.

Evansville, Indiana (c. h. Vanderburg Co.), is the

second city in Indiana, and has an important wholesale and shipping trade. Has extensive manufactures of cigars, flour, woollen and cotton goods, furniture, stoves, machinery, saddlery, and farming implements. The Southern Indiana Hospital for the Insane is located here.

Everett, Massachusetts (Middlesex Co.), is a residential suburb of Boston, and has manufactures of steel and structural iron, automobiles, boots and shoes, varnishes, etc.

Everett, Washington (c. h. Snohomish Co.), is a port on Puget Sound with large manufactories of flour, paper, and lumber. It is the center of an extensive mining region producing gold, silver, lead, and copper.

Exeter, England (Devonshire), is one of the chief cities of western England which still retains to a great extent its antique appearance. Considerable portions of the ancient walls remain, but the principal object of interest is the cathedral, begun in 1112 and dedicated in 1351. The other ancient buildings are the bishop's palace (14th century), and the Hall of the College of Vicars, incorporated in 1401. The chief manufactures are paper, iron and brass, tanneries and breweries.

Fairbanks, Alaska, is the largest city of Alaska and an important mining center for a region vast in extent and rich in precious metals.

Faizabad, India (United Provinces), is an important railway center, and occupies the site of an ancient city, Ajodhya.

Falkirk, Scotland (Stirling), is famous for its cattle markets. In and near the town are several iron-works, and coal is mined extensively in the vicinity. Brewing, distilling, tanning, brick and tile making, and the manufacture of chemicals are among the other industries. It was the scene of the defeat of Wallace by Edward I. in 1298, and of the victory of Prince Charles Edward over General Hawley in 1746.

Fall River, Massachusetts (c. h. Bristol Co.), is a port of entry, and one of the chief cotton-manufacturing places in New England. It has large manufactories of cotton, thread, woollens, bobbins, shuttles, rubber, rope, wire, machinery, etc. The river, to which it owes its name, rises a little to the east, and falls about 140 feet within one-half mile, affording admirable water-power for the mills. Granite is quarried in the vicinity.

Fargo, North Dakota (c. h. Cass Co.), is the largest city in the state, and has a considerable jobbing trade. Has large railroad car-shops, flour- and planing-mills, grain elevators, and packing industries. It has an extensive trade in agricultural implements. The state agricultural and mechanical college, and a United States land-office are located here.

Farukhabad, India (United Provinces), has a government gun-carriage factory, and a large trade.

Felegyhaza, Hungary, has fruit, wine, and tobacco industries.

Fernie, British Columbia, Canada, is an important trade and coal-mining center with extensive collieries and coke ovens. It is an outfitting point for hunters in the East Kootenai game reserve. It has large saw-mills, railway-shops, and other industries.

Ferrara, Italy (cap. Ferrara), as the capital of the powerful family of Este, was from the 14th to the 17th century a large and prosperous city. Now its wide deserted streets and decaying palaces give it an air of desolation. Nevertheless it possesses a fine Lombardesque cathedral (12th and 13th centuries), the castle, or old ducal palace, a good picture-gallery, a university (founded in 1391), a famous library, and the houses of the poets Ariosto and Guarini. The last named was born here, as was also Savonarola.

Ferrol, Spain (Coruna), is a fortified seaport and the chief naval arsenal of Spain, with extensive dockyards, and also a naval school. There are manufactures of leather, naval stores, and textiles. Ferrol was twice attacked by the English: unsuccessfully in 1799, and successfully in 1805. It fell before the French in 1809.

Fez, cap. Morocco, is a holy city, the chief commercial center of the country, and one of the capitals of the sultan. Situated on a plateau, the city is surrounded by walls. Old Fez contains mosques, bazaars, and caravanserais, while New Fez is the official district,

containing the palace, and the "Mellah", or Jews' quarter. The most beautiful of the one hundred and thirty mosques is that of Bu Ainan; to the largest, which is also the largest in Africa, the Kairuin, is attached the Kairuin University. The Sanctuary of Mulai Idris, the most holy place in Morocco, contains the tomb of the founder of the city. The chief industries are leather work, and the manufacture of colored tiles, carpets, and saddlery. Raw silk, paper, weapons, drugs, and spices are also produced. In the 13th century Fez was the capital of an independent kingdom; three centuries later it was conquered and annexed by Morocco.

Findlay, Ohio (c. h. Hancock Co.), is in the heart of the oil- and gas-fields of Ohio, and is widely known on account of its yield of natural gas. In the vicinity are rich beds of clay. It has manufactures o. glass, pressed brick, furniture, wooden implements, and nails, as well as oil-refineries, extensive potteries, and rolling-mills. Findlay College, a coeducational institution, is located here.

Firozpur, India (Punjab), contains the chief arsenal in the Punjab.

Fitchburg, Massachusetts (c. h. Worcester Co.), has manufactures of firearms, cotton and woollen goods, pianos, paper, electrical apparatus, tools, etc. Quarries of granite are extensively worked. A state normal school is located here. The Walker Free Library, with its art collection, is notable.

Fiume, Hungary, is a free city, and the chief seaport and coaling-station of Hungary. Paper, torpedoes, tobacco, and chemicals are manufactured. There are also rice-mills and petroleum refineries. The cathedral, the naval academy, and the governor's residence are the principal architectural features.

Flensburg, Germany (Prussia), is a seaport, and has a large trade in agricultural and forest products. It has shipyards, breweries, distilleries, iron-foundries, and various other manufacturing establishments.

Flint, Michigan (c. h. Genesee Co.), has a large number of sawmills, carriage and wagon factories, woollen-mills, bicycle works, etc. A state institution for the deaf and dumb is located here.

Florence, Italy (cap. Florence), ranks as the intellectual capital of Italy, a position which it owes not only to the natural gifts of its inhabitants, but also to the memory of its greatest citizen, Dante, and to the fact of its being the seat of the national Accademia della Crusca, and the vigorous Institute for Advanced Studies, which virtually fulfils the offices of a university. Florence contains a great number of magnificent edifices and squares, its principal architectural feature being the Duomo, a cathedral, one of the largest churches in Italy, erected mainly between 1296 and 1436. Its interior is adorned with several fine pieces of sculpture by Michael Angelo, Ghiberti, Sansovino, Della Robbia, and others, and it possesses two fine bronze doors. The trio of bronze doors closing the little octagonal Baptistery, near the cathedral, are however still more famous. The 13th century campanile is Giotto's masterpiece. Among the other famous churches should be named Santa Croce (1294-1442), the Pantheon of Florence, adorned with frescoes by Giotto, the Gaddis, and others; Santa Maria Novella (1278-1350), with frescoes by Ghirlandajo, and Orcagna; Santa Maria del Carmine, which contains frescoes by Filippino Lippi, Masolino, and others. San Lorenzo, originally founded at the end of the 4th century, shelters the Laurentian Library, which contains many very valuable classic MSS. These, and many other edifices, testify to the prominence of Florence in the world of art, a position which is still further enhanced by the glorious canvases and sculptures of the Academy of Fine Arts; by the treasures of the Uffizi Palace (1560), which holds one of the most valuable collections of paintings and sculptures in the world, as well as the national library with its most printed books; and by the even more valuable collection of the Pitti Gallery (1440). Among the remaining treasures of the city are the castle-like Palace of the Priors (1298-1314), for ages the seat of the city's government; the cloistered hall of Dei Lanzi (1637), with Benvenuto Cellini's Perseus, and other masterpieces of sculpture (it was here that Savonarola was

burned in 1496); the archæological museum, exhibiting notable Etruscan collections; the national museum of the middle ages and the Renaissance, sheltered in the Bargello (13th century), the ancient chief law court of the republic; and a number of interesting palaces, including the Riccardi (1430), Strozzi (1489–1533), Rucellai (1450), and Corsini (with a good picture-gallery). Florence is an archiepiscopal see. The straw-plaiting industry of the vicinity is declining, but the manufacture of majolica flourishes. Otherwise Florence is a place of little industrial importance.

Foggia, Italy (cap. Foggia), is a bishop's see, and has a Norman cathedral; a famous fair takes place in May.

Folkestone, England (Kent), is a seaport and a fashionable resort. The parish church is an ancient cruciform structure, and the Leas is a beautiful promenade along the cliffs. In the 18th century and the early part of the 19th century, it suffered much from encroachments by the sea.

Fond du Lac, Wisconsin (c. h. Fond du Lac Co.), has railroad shops, manufactories of automobiles, gas-engines, wagons, furniture, and chemicals, and wood-working industries. A state insane asylum is located here. Steamboats ascend from Green Bay to this place, via the Fox River and Lake Winnebago.

Forli, Italy (cap. Forli), is a bishop's see, and has several imposing palaces. Its manufactures include silk, hats, and shoes.

Forst, Germany (Prussia), has extensive manufactures of woollen cloth, and tanning industries.

Fort-de-France, West Indies (cap. Martinique), is a fortified seaport, and has a naval arsenal. The governor-general of the French West Indies resides here.

Fort Dodge, Iowa (c. h. Webster Co.), is an important grain market and has quarries of building-stone and gypsum. Coal is also mined in the vicinity. It has railroad repair-shops, and manufactories of woodenware, cereals, etc.

Fort Scott, Kansas (c. h. Bourbon Co.), is one of the most prosperous towns in the southeastern part of Kansas and an important market. The mining and shipping of bituminous coal is the leading industry. It has paper-and flour-mills, brick-works, railroad-shops, etc. The Kansas Normal College is located here.

Fort Smith, Arkansas (c. h. Sebastian Co.), is the second city of the state, and has cotton and cottonseed-oil industries, manufactures of bricks, furniture, etc., as well as a considerable jobbing trade.

Fort Wayne, Indiana (c. h. Allen Co.), is a railroad center in a rich agricultural region. It has large manufactories of car-wheels, engines, boilers, machinery, iron and steel bars, railway-cars, locomotives, electrical machinery, automobiles, furniture, etc. Concordia College, founded in 1850, is maintained here under the auspices of the German Lutheran church, and there are numerous other educational institutions. It takes its name from the fort built on a part of the present site of the city by General Anthony Wayne in 1794.

Fort William, Ontario, Canada, is a port and railroad center on Thunder Bay, Lake Superior, with a large transhipment trade. Its industries include large elevators, flour mills, iron foundries. Nearby are large blast furnaces, and silver and copper mines.

Fort Worth, Texas (c. h. Tarrant Co.), is the second largest city of northern Texas and the center of the cattle trade of the state. It has extensive packing-houses, flour-mills, foundry and machine works, etc. Its stockyards at the suburb, North Fort Worth, are the largest in the United States south of Kansas City. Fort Worth Universtiy, with upwards of 1,000 students, is located here.

Framingham, Massachusetts (Middlesex Co.), has extensive woollen goods interests, and manufactories of boots and shoes, rubber and straw goods, etc. A state normal school is located here.

Frankfort-on-Main, Germany (Prussia), is an important financial and commercial center. Among its most noteworthy edifices and institutions are the house in which Goethe was born, with a Goethe museum attached; the Römer, a group of ancient buildings that serve as town hall, and contain not only the hall in which the German emperors were formerly elected, but that in which they dined in state with the electors, as described in Gothe's "Wahrheit und Dichtung"; the cathedral, founded in 850, but mostly rebuilt at subsequent periods; the Saalhof, which stands on the site of the palace of the Carlovingian emperors; the Academy of Social Sciences, opened in 1901, with a university character; the house in which Schopenhauer lived; the ancestral house of the Rothschilds; the old Main bridge; and the Justitia fountain (1543). The Senckenberg Foundation (1763); the Thurn and Taxis palace (1731) now used by the post-office, but down to 1866 the meeting-place of the German Imperial Diet; two conservatories of music; the commercial high school; and, in the suburb of Sachsenhausen, on the opposite bank of the Main, the Städel Institute (which embraces a picture gallery, art collection, and an art school) all figure prominently.

Frankfort-on-Oder, Germany (Prussia), has manufactures of machinery, iron and steel goods, chemicals, tobacco, etc. The town hall (1607–10), the church of St. Mary (13th century), and the government buildings (1900) are the most notable features. Frankfort was famous for its three fairs, and as a center of the trade with Poland. From 1506 to 1811 it was the seat of a university (now at Breslau).

Fredericton, cap. New Brunswick, Canada, is the seat of the University of New Brunswick, a normal school, an infantry school, and an Anglican cathedral. Its industries include lumber mills, boat factories, etc.

Frederiksberg, Denmark, is a residential suburb of Copenhagen. It has a military college housed in the old palace.

Freeport, Illinois (c. h. Stephenson Co.), is an important trade center in a rich agricultural section, and has railroad-shops, and manufactories of hardware, windmills, pumps, organs, wagons, leather, etc.

Freetown, cap. Sierra Leone, Africa, is a fortified seaport and a coaling-station of the British fleet, the largest town of the colony, and a Protestant missionary station. Furah Bay College is close by. India-rubber, palm-oil, and hides are exported.

Freiberg, Germany (Kg. of Saxony), is the center of the Saxon mining industry, and has a famous mining academy, founded in 1765. The cathedral (1484) contains the burial vaults of the electors of Saxony from 1541 to 1694. Gold, silver, bismuth, zinc, nickel, cobalt, sulphuric acid, and arsenic are produced.

Freiburg, Germany (Baden), is the seat of a university founded in 1460, and of a Roman Catholic archbishop. The cathedral is one of the most remarkable in Germany. The town is backed by the castle hill, formerly surmounted by two castles, which were destroyed by the French in 1744. It was besieged more than once in the 17th century, and passed to Baden in 1806. It has a trade in timber and wine, and produces silks, cottons, buttons, machinery, etc.

Fremantle, Western Australia, Australia, is a seaport and coaling-station. It is the first and last port of call in Australia for all European mail steamers. It has manufactures of soap, tin, leather, and tobacco, and boat-building yards. It exports gold, sandalwood, copper and lead ores, gum, wool, pearl-shell, manna, cattle, etc.

Fresno, California (c. h. Fresno Co.), is an important fruit-growing center, and has the largest raisin trade of any city in the United States. Other important industries are the cultivation and exporting of oranges, olives, and grapes, and wine-making. It also has a large live-stock trade, and there are petroleum interests near by.

Fuchau, China (cap. Fokien), is a treaty port, with a large export trade in tea. Timber and bamboos are also largely exported. Cotton goods and porcelain are manufactured.

Fukui, Japan (cap. Fukui), is noted for its paper and tinned crabs. At the time of the renascence Fukui became an important educational and ecclesiastical center.

Fukuoka, Japan (cap. Fukuoka), is chiefly famous for its silk and interwoven picture fabrics.

Funchal, cap. Madeira Islands, is a seaport, coaling-station, and health resort. Although it has only an open roadstead, it has a large export trade in wine, embroidery, and fruits. It is much visited in winter by convalescents from pulmonary and other diseases.

Fünfkirchen, Hungary, has a fine 12th century cathedral. The name is derived from five Turkish mosques, three of which are now in ruins and two in use as churches. Manufactures leather, cloth, and pottery.

Fürth, Germany (Bavaria), has manufactures of mirrors, gold-leaf, bronzes, toys, pencils, fancy ornaments, picture-books and chromo-lithographs, gold and silver wares, chicory, machinery, etc.

Galatz, Roumania (Moldavia), is a river port, with a large export trade in wheat, corn, lumber, and cattle. It is the seat of the Danube Commission. Here in 1789 the Turks defeated the Russians, and in 1828 were in turn defeated by them. It is the seat of a Greek Orthodox bishop.

Galesburg, Illinois (c. h. Knox Co.), is the seat of Knox College, founded in 1837, where took place the famous Lincoln-Douglass debate in 1858. Lombard University (Universalist) was established here in 1852. The St. Joseph Academy and the Ryder Divinity School are also located here. The city has extensive railroad shops, stock-yards, brickmaking plants, boiler- and engine-works, farm machinery works, and carriage factories.

Galle, Ceylon, is a port and coaling-station. Exports tea, coffee, and plumbago.

Gallipoli, Turkey (Adrianople), is the see of a bishop of the Greek church. Here, in 1294, the Venetians were defeated by the Genoese, but in 1416 won a naval battle here over the Turks.

Galveston, Texas (c. h. Galveston Co.), is a port of entry and the most important shipping point on the Gulf of Mexico west of New Orleans. It has a large export and coastwise trade. Its foreign exports are principally cotton, cottonseed-oil, oil-cake and meal, cattle, wheat, and provisions. It has a large jobbing trade. Among its industries are bagging mills, rope mills, cement and pipe works, cotton-presses, rice-mills, and breweries. On September 8, 1900, Galveston was almost entirely destroyed by a tidal wave, causing the death, direct or indirect, of 6,000 to 8,000 people; the loss of property was also immense, but the city has rallied bravely from the blow, and is rapidly increasing in commercial importance. It claims to be the first port in the United States for cotton and grain shipping. Its sea-wall, made of crushed granite sand and cement, is 3½ miles long, 17 ft. high, 16 ft. wide at its base, and 5 ft. wide at the top, curving from the top to the base. A walk 13 ft. wide and a driveway 38 ft. wide are on the filled-in lands inside this wall. The school of medicine of the State University, a Roman Catholic university, and numerous other educational institutions are located here. Galveston was the first city in the U. S. to inaugurate the commission form of government.

Gardaia, Algeria, has a caravan trade in oil, dates, ostrich feathers, and pottery.

Gardner, Massachusetts (Worcester Co.), is the trade center of an agricultural region, and has a large chair-manufacturing industry, in which almost every known kind of chair is made. It also has manufactures of machinery, toys, etc.

Gary, Indiana (Lake Co.), is an industrial suburb of Chicago, with a good harbor and the largest steel-mills in the world.

Gateshead, England (Durham), manufactures locomotives, anchors, chain-cables, etc. It also has large chemical and glass works. Excellent grindstone is quarried, and coal is mined. Among its notable ancient buildings are the Church of St. Mary, originally erected in the 12th century, but almost entirely rebuilt in 1854, and Trinity Chapel, which is supposed to occupy the site of an ancient monastery.

Gaya, India (Bengal), is annually visited by many pilgrims. Six miles to the south is Buddha Gaya, the site of a temple dating from B. C. 543, and of a bo-tree sprung from that under which Buddha attained Nirvana.

Gefle, Sweden (cap. Gefleborg), is a port on the Gulf of Bothnia, with a large export in bar iron, timber, tar, and cellulose. It has ship-building yards, and manufactories of machinery, cotton, tobacco, and sail-cloth.

Gelsenkirchen, Germany (Westphalia), has large coal-mines and iron-works.

Geneva, New York (Ontario Co.), is the seat of Hobart College (1822) and its co-ordinate institution, the William Smith Hall for Women. The Delancey Divinity School is located here, also a state agricultural experiment station. It has extensive manufactories of cereals, stoves, steam-boilers, canned goods, etc. There are large nurseries and green-houses in the vicinity.

Geneva, Switzerland (cap. Geneva), is an important educational center, and also an industrial city renowned for its clock, watch, music box, scientific instrument, and jewelry manufactures. Diamond-cutting and enameling are also carried on. It affords fine views of Lake Geneva, and of the Alps, and of Mt. Blanc in particular. Its cathedral is Byzantine in character, and is said to have been built in 1124. Other places of interest are Calvin's house; the 16th century city hall; the botanic gardens and university, opposite the interesting Corraterie, formerly one of the ramparts of the city; several museums, including the Musée Rath; and the Ile Jean Jacques Rousseau. The historical fame of the city rests chiefly on the severe rule of John Calvin (1541–64), the reformer. In 1798 the republican army marched in and annexed it to France; but in 1815 the city joined the Swiss confederacy.

Genoa, Italy (cap. Genoa), is a fortified port and coaling-station and an archiepiscopal city. It is the principal commercial port of Italy and one of its busiest industrial centers. The chief branches of activity are ship-building, iron-founding, iron-works, sugar-refining, tanning, cement-making, the manufacture of cotton, macaroni, ornaments for personal wear, preserved fruits, etc. Of the palaces the most famous are the former palace of the doges, now the meeting-place of the senate; and the Doria, presented in 1520 to the great Genoese citizen, Andrea Doria. Foremost amongst the churches stands the cathedral, a grand 12th century pile in the Italian Gothic style. The university, originally built in 1623, reorganized in 1812, has a library of 116,000 volumes. To Columbus, Genoa's most famous son, there is a fine monument (1862) by Lanzio. A great mediæval republic, the rival of Pisa and Venice, Genoa in 1768 ceded Corsica to France.

Georgetown, cap. British Guiana, is a port with good anchorage in the harbor. Timber, gold, sugar, and small quantities of balata (very like gutta-percha), cocoa, coffee, etc., are exported. There are some fine buildings, such as the Roman Catholic cathedral, the Anglican cathedral, and the government buildings. The town also contains good botanical gardens. It is an episcopal see.

Gera, Germany (cap. Reuss, Younger Branch), has large woollen factories, iron-works, manufactories of accordions, concertinas, etc., besides dye-works.

Ghazipur, India (United Provinces) is the headquarters of the opium trade, and manufactures rosewater and attar of roses.

Ghent, Belgium (cap. East Flanders), is a river port with a large transport trade. Exports sugar, chicory, flax, hops, and other agricultural products, and has large manufactures of cotton, linen, and lace. It is noteworthy as a center of floriculture. It is an episcopal see, and among the chief buildings are the splendid cathedral of St. Bavon, of the 13th and 14th centuries, containing the "Adoration of the Lamb," by the brothers Van Eyck; the belfry (1183–1339), 280 ft. high, or 375 with the iron spire of 1855; the new citadel (1822–30); the Hôtel de Ville (1480–1628); the Palais de Justice (1835–43); the university (1816); the Béguinage; and the Academy of Painting.

Gibraltar, Spain, is a British fortress, a naval port, and coaling-station. Being a free port, it is much frequented as a depot and port of call. It exports, in transit from Spain, wine, fruit, and cork. It is situated on a bold rocky promontory at the extreme southern point of Spain, the entire length at the base of the rock

being 2½ miles, and its greatest breadth three-quarters of a mile. The place was celebrated in the times of the Phœnicians and Greeks, and was fortified by Tarik, the Berber leader, who invaded and conquered Spain in 711. It was finally captured by the Spaniards in 1462. After its capture by the British in the war of the Spanish Succession, it was repeatedly attacked by the Spaniards, and underwent a long siege in 1726. From 1779 to 1783 it withstood the greatest siege in its history, under the gallant Eliott (Lord Heathfield), against the Spaniards and French. Gibraltar is the see of an Anglican bishop.

Giessen, Germany (Hesse), is the seat of a university founded in 1607, and is the site of the chemical laboratory rendered famous by Liebig. It has manufactures of tobacco, cigars, beer, etc., and there are lignite mines.

Gifu, Japan (cap. Gifu), has manufactures of silk and paper.

Gijon, Spain (Oviedo), is one of the best Spanish ports on the Bay of Biscay, and in summer is a popular sea-bathing resort. Exports zinc, copper, manganese, quicksilver, and agricultural produce. Gaspar de Jovellanos was a native and founded the Instituto Asturiano, with its fine art collection. Other features of the town are the Campos Eliseos, a bull-ring, a fine 15th century church, and two palaces. It was the capital of the Asturian princes in the 8th century.

Gillingham, England (Kent), has an important torpedo factory, also brick and cement works, and is the center of an extensive fruit district.

Girgenti, Italy (cap. Girgenti), is the successor of the ancient Acragas, or Agrigentum, which was founded in 582 B. C., flourished between 560 and 406 B. C., and again during the first and second Punic wars. From 828 to 1086 it was in the hands of the Saracens. Fine examples of Greek temples remain, while the modern buildings include a cathedral, library and museum.

Glace Bay, Nova Scotia, Canada, is a coal mining center on Glace Bay. It also has fishing industries, machine works, etc.

Gladbach, Germany (Prussia), is the chief center of the Rhenish cotton industry. It has also dye-works and calico-printing establishments, iron-foundries, machinery factories, silk and woollen mills, stationery factories, breweries, etc. It derives its name from an abbey of Benedictine monks, founded here in 972, and dissolved in 1802. The minster has a Gothic choir of the 12th century and an 8th century crypt.

Glasgow, Scotland (Lanark), is the industrial and commercial metropolis of Scotland, and, next to London, is the most populous city of Great Britain. It is a port and coaling-station on the Clyde, and the water area of its harbor and tidal docks is 206 acres. The most important edifice is the cathedral, an early English structure, containing a small part of the previous cathedral consecrated in 1197; the present building was commenced in 1238 and is now in almost the same state as when at the Reformation, though stripped of its ornaments, it was saved by the citizens from the fury of the mob. Other notable structures are the municipal buildings, in the style of the Italian Renaissance, the Royal Infirmary, the Western and Victoria infirmaries, the city art galleries (Spanish Renaissance), a people's palace, and the University of Glasgow, founded in 1451. The university is attended by over 2,000 students annually, and includes the Hunterian Museum, with its famous anatomical collection. The Glasgow and West of Scotland Technical College, founded in 1886, has a large attendance. Among other noteworthy institutions are St. Mungo's College and Anderson's College, the Mitchell Library, containing about 150,000 volumes, the Corporation Public Libraries and the Glasgow Ar Gallery and Museum. Glasgow commenced very early to manufacture woollens, and acquired a fame for its plaids. In the 18th century were added muslin-weaving, distilling, ink-making, of which the secret was stolen from Holland, and the manufacture of tobacco, shalloons, cotton, leather, and furniture. The first cotton-mill was erected in 1792; and the greater part of the textile factories in the city are concerned with this fiber. There are also numerous calico-printing and bleaching works. Linen

comes next in importance. The iron trade, introduced in 1732, is now the dominant one in Glasgow. The city contains blast-furnaces, iron- and steel-works and forges, and engineers' and boiler-makers' shops. The manufactures of chemicals, leather, clothing, whiskey, furniture, and timber are of the first importance. Shipbuilding is a leading industry at Clyde ports, the annual output aggregating half a million tons. Glasgow is the seat of an Episcopalian bishop and of a Roman Catholic archbishopric. It was in this city that Watt conceived the idea of the steam-engine; on the Forth and Clyde Canal, Symington ran the first steamboat in 1801; and ten years later Henry Bell built the steamship "Comet" which plied between Glasgow and Helensburgh.

Glauchau, Germany (Kg. Saxony), is one of the busiest manufacturing centers of the kingdom, its specialties being woollen dress stuffs, though it has also dye-works, calico-printing works, iron-foundries, breweries, flour- and sawmills.

Gleiwitz, Germany (Prussia), is the seat of iron industries; produces also bricks and glass.

Glens Falls, New York (Warren Co.), is a manufacturing and trade center on the Hudson River, where there is a succession of falls, with a descent of about 60 ft. The water-power is utilized by a number of manufactories, making flour, lumber, lime, cement, paper, etc. There are quarries of stone, limestone, and black marble. The island below the falls is the scene of some well-known incidents in Cooper's "Last of the Mohicans."

Gloucester, England (Gloucester), is a port on the Severn River, connected with the Bristol Channel by a canal. It contains a famous cathedral, built around a Norman core, with a stately tower, vast eastern window, and the canopied tomb of the murdered Edward II. Other public buildings are St. Mary le Crypt, the bishop's palace, shire hall and guildhall; the first prison built on Howard's plan; the public library and technical schools

Gloucester, Massachusetts (Essex Co.), is a port of entry with an extensive fishing trade. It is one of the most important fishing ports and fish-markets in the world; cod, haddock, halibut, herring, and mackerel are the principal catches. It also has large manufactories of machinery, oil, fish-glue, shoes, twine, etc. It is a great resort of artists, attracted partly by the picturesqueness of the town itself and partly by the fine scenery of Cape Ann.

Gloversville, New York (Fulton Co.), is the greatest glove-manufacturing center in the United States and one of the largest in the world. It also has manufactories of gauntlets, mittens, leather goods, etc.

Goldfield, Nevada (Esmeralda Co.), is the center of a comparatively new and very productive gold-mining section. It has a large distributing trade, especially in lumber and mining machinery.

Gomel, Russia (Mohilef), has sugar-refineries and some ship-building.

Gonaives, Haiti, is a port on the Gulf of Gonaives, with exports of cotton, coffee, and mahogany.

Gorakhpur, India (United Provinces), contains a fine 17th century mosque, and carries on a river trade in grain and timber.

Gorlitz, Germany (Prussia), manufactures cloth, mixed woollen goods, machinery, glass, bricks, and other commodities. The town possesses some fine Gothic churches (especially St. Peter and St. Paul, built in 1423-97, one of the noblest churches in Eastern Germany), and a 14th century town hall. Boehme, the mystic, lived and died in Gorlitz.

Gotha, Germany (Saxe-Coburg-Gotha), is one of the busiest industrial towns in Thuringia, and produces sausages, toys, shoes, rubber tubes, sugar, etc. Here, too, is the geographical institute of Justus Perthes. The famous Almanach de Gotha, and Petermann's Mitteilungen are published in the town. The dominant feature is the ducal castle of Friedenstein (1643), crowning a hill south of the old town. Here also are the former ducal palace and the castle of Friedrichsthal.

Gottenborg, Sweden (cap. Gottenborg och Bohus), is a port and the second largest city of Sweden, remark-

able for its beautiful parks and pleasure-grounds. It is the seat of a bishop and has a cathedral built in 1815; a museum, founded in 1833, is housed in the old East India Company's buildings. There is also a famous free technical school. The city has ship-building and machinery factories, sugar-refineries, breweries, tanneries, weaving and spinning factories, etc. Owing to its excellent harbor, which is almost always free from ice, it has a large trade. The principal exports are iron, timber, wood-pulp, corn, butter, and fish. Gottenborg was founded by Gustavus Adolphus in 1619.

Göttingen, Germany (Prussia), is the seat of a famous university founded in 1737. Göttingen manufactures sausages and other eatables, mathematical and scientific instruments, cloth, chemicals, sugar, and beer. In 1626 the town was taken by Tilly. Göttingen is famous for its Academy of Sciences, founded by Haller in 1751.

Govan, Scotland (Lanark), has ship-building yards which are among the largest on the Clyde. It has also engineering-, railway locomotive-, and iron-works.

Granada, Nicaragua (cap. Granada), is the seat of a considerable trade in dyewoods, indigo, hides, cacao, and has manufactories of gold wire chains.

Granada, Spain (cap. Granada), is an ancient Moorish city, the seat of the last Moslem kings, and was conquered by Ferdinand and Isabella in 1492. It possesses the famous palace of the Alhambra, and other picturesque structures of Moorish character. It is the seat of an archbishop, and has a university and a richly decorated Renaissance cathedral (1529), containing the tombs of Ferdinand and Isabella, and of Philip I., and his consort Juana. In the church of San Geronimo is the cenotaph of Gonsalvo de Cordova. Liqueurs, textiles, paper, and soap are manufactured.

Grand Forks, North Dakota (c. h. Grand Forks Co.), is an important railroad center and the second city of the state. It has large lumber-mills, flour-mills, steamboiler works, and a large trade in live-stock, grain, and agricultural produce. It is the seat of the State University of North Dakota and other educational institutions.

Grand Junction, Colorado (c. h. Mesa Co.), is the largest city of western Colorado, and an important railroad and trade center in the famous fruit producing valley of the Grand River. It has brass- and ironfoundries, planing-mills, etc.

Grand Rapids, Michigan (c. h. Kent Co.), is a port of entry, the second city of the state, and the third city in the United States in the manufacture of furniture. The Grand River has here a fall of 18 feet, supplying excellent water-power. In addition to furniture, there are extensive manufactures of other lumber products, flour, plaster, carriages, wagons, bicycles, brass goods, automobiles, and agricultural implements. The gypsum quarries here have the largest output in the world. There are numerous educational institutions and a state soldiers' home.

Gratz, Austria (cap. Styria), is an episcopal see and the favorite place of residence for retired Austrian officials. It is the seat of a university founded in 1586, which has some 1,500 students, and of a polytechnic school. Its manufactures are machinery, steel and iron wares, soap, chemicals, pottery, confectionery, beer, etc. In addition there are large railway-car works. The central feature is the Castle Hill, which was fortified until 1809. There is also a 15th century cathedral.

Graudenz, Germany (Prussia), manufactures machinery, cigars, tapestry, flour, etc. About one mile north is the fortress, constructed by Frederick the Great in 1772-6, famous for its defence by the Prussians against the French in 1807, but dismantled in 1874.

Great Falls, Montana (c. h. Cascade Co.), is the second city of the state, near an important mining region, and has large smelting works. Gold, silver, copper, lead, bituminous coal, iron, and sandstone are found in the vicinity. It has manufactories of flour, furniture, and mining and agricultural implements. It derives its name and importance from the falls formed here by the Missouri, with a total descent of 500 ft.

Great Grimsby, England (Lincoln), is a seaport and coaling-station and the largest fishing-port of England. Its industries comprise ship-building, tanning, brewing, and rope-making, and it exports cottons, woollens, leather, fish, etc.

Greeley, Colorado (c. h. Weld Co.), is a trade center in a rich agricultural region largely under irrigation and is famous for its potatoes and alfalfa. It has a large trade in produce and agricultural implements, and manufactories of machinery, beet sugar, etc.

Green Bay, Wisconsin (c. h. Brown Co.), is a port of delivery, with an extensive lake traffic. It has a large export trade in grain and lumber, and extensive manufactures of paper, sulphite, lumber products, etc. It is the oldest town in the state and was first visited in 1634 by Jean Nicollet, who had been sent by Champlain, the governor of New France, to find the rumored short route to China.

Greenock, Scotland (Renfrew), is a port on the Clyde. Ship-building is the chief industry, and there are sugar and glucose refineries, iron-foundries, wool-spinning, aluminum-rolling, and paper-mills. The harbor works date from 1707.

Greensboro, North Carolina (c. h. Guilford Co.), is the center of a tobacco, fruit, and grain region. It also has steel works, cotton-mills, and other industries. Has a large trade in tobacco, coal, and iron. It is the seat of the state agricultural and mechanical college, and other educational institutions.

Greenville, Mississippi (c. h. Washington Co.), is a trade center in a region producing a large amount of cotton. It has cotton-compresses, cottonseed-oil mills, manufactories of lumber products, tile and brick works, etc.

Greenville, South Carolina (c. h. Greenville Co.), is the seat of Furman University (Baptist), organised in 1851, the Chicora Female College (Presbyterian), a military institute, and numerous other educational institutions. It has cotton-mills, iron-works, and flour-mills.

Grenoble, France (cap. Isère), is an episcopal see and a great tourist center. The manufacture of kid gloves is the chief industry. Among other manufactures are leather-dressing, cement, paper flowers, silk and linen, furniture, buttons and fasteners, hosiery, and straw hats. Walnuts are largely exported. Grenoble is the seat of a university.

Grodno, Russia (cap. Grodno), contains an ancient palace of the Polish kings (12th and 13th centuries); a more modern palace built by Augustus III., now converted into a military hospital; the Greek house of St. Basil; the former Jesuit College; and the Carmelite monastery. The academy of medical science was founded by King Stanislaus Augustus. It has manufactories of tobacco, soap, candles, vinegar, bricks, machinery and carriages, with breweries, distilleries, sawmills, iron-foundries, etc.

Groningen, Netherlands (cap. Groningen), is the largest town in northern Netherlands and the seat of a university, founded in 1614. It possesses many 17th century quaint old houses and is famous for its corn and oil-seed markets. It was taken from the Spaniards by the Dutch, after a stubborn defence, in 1594.

Grosswardein, Hungary, possesses two cathedrals—one belonging to the Roman, the other to the Greek Catholics; also a Roman Catholic episcopal palace, a law academy, and portions of a former citadel. Milling, distilling, brickmaking, etc., are carried on, and good wine is made. The city was destroyed by the Mongols in 1241, and from 1663 to 1692 was in the power of the Turks.

Guadalajara, Mexico (cap. Jalisco), has a magnificent cathedral containing a famous "Assumption" by Murillo, and a theatre, the Degollado, one of the largest on the continent. Here are also a university, an art academy, and a mint. Coffee is grown; paper, Panama hats, and leather are manufactured, and artistic terra-cotta ware is made. Elevation 5,185 ft.

Guanajuato, Mexico (cap. Guanajuato), is the center of one of the richest mining districts in Mexico, and is noted for its pottery. It produces silver, gold, mercury, tin, lead, and copper. Elevation 6,750 ft.

Guayaquil, Ecuador (cap. Guayas), is a port and coaling-station, and, next to Quito, the most important town in Ecuador. It is an episcopal see and contains a fine cathedral built of wood, government buildings, and a university. It has a good harbor and exports cocoa, coffee, cinchona, gold, silver, and hides. It was founded in 1535 and after Valparaiso is the most populous port on the west coast of South America. Yellow fever is very prevalent.

Guben, Germany (Prussia), manufactures woollens, hats, leather, and machinery.

Guelph, Ontario, Canada, is a trade center in one of the richest agricultural sections of Canada. It has a piano and organ factory, knitting mills, flour mills, axle works, etc. Guelph is the seat of the Ontario Agricultural College and Macdonald Institute, with over 1,200 students. An annual fat stock show is held here.

Guinobatan, Philippine Islands (Albay), is the center of a region producing a large amount of hemp.

Guthrie, cap. Oklahoma (c. h. Logan Co.), has a considerable distributing trade. It has cotton-gins, and manufactures of cottonseed-oil, lumber, flour, etc.

Gwalior, India (cap. Gwalior), contains a large palace of the Maharajah Scindia, and a huge fortress.

Haarlem, Netherlands (cap. North Holland), is an ancient city with many interesting buildings, among which are the town hall, a former palace of the Counts of Holland, which contains a small collection of paintings by Dutch artists, including eight large canvases by Frans Hals. Of its churches the finest is the Groote Kerk, a Late Gothic basilica, built in the 15th century, one of the largest churches in Holland, noted for its tower 260 feet high, and its large organ. Haarlem was in the 17th century the center of a famous school of painting, to which belonged such artists as the Wouwermans, the Ostades, Frans Hals, Ruisdael, and the Everdingen, most of them natives. It is also the birthplace of Laurens Coster, to whom the Dutch ascribe the invention of printing. One of the oldest and most extensive industries is the cultivation of flowers, Haarlem exporting bulbs to many European countries. It has, besides, manufactures of machinery, cottons, and woollens, and a large trade in dairy produce.

Hackensack, New Jersey (c. h. Bergen Co.), is a residential suburb of New York and has brick, silk, and other manufacturing industries. Was settled in the later part of the 17th century and, during the Revolution, was occupied in turn by the British and the American armies.

Hagen, Germany (Prussia), is a growing industrial center, with ironworks, cotton and cloth mills, tanneries, breweries, distilleries, and manufactures of cigars and paper. It has a school of engineering and several technical colleges.

Hagerstown, Maryland (c. h. Washington Co.), is a trade center of western Maryland, and has extensive manufactures of knit goods, machinery, steam-engines, and lumber. Was a center of military operations in the Civil War.

Hague, The, cap. Netherlands (cap. South Holland), is one of the prettiest towns of Holland, and among its most important edifices are the royal palace, and the Binnenhof, a large irregular building, founded in 1249 and containing the hall of assembly of the States-General, and various government offices where the International Peace Conference met in 1899. Other buildings are the provincial government house, a large roomy edifice; the town hall; the ministry of justice; the municipal museum, containing pictures and antiquities; the royal library; a cannon foundry, one of the largest and most conspicuous structures in the town; the colonial and war office; the national monument, etc. The royal collection of pictures, in the Mauritshuis, is confined chiefly to Dutch masters and is famous for the magnificent exhibition of Rembrandt's masterpieces. Mr. Andrew Carnegie has undertaken to build at The Hague a Palace of Peace. Not far from the city and connected with it by trolley car is Scheveningen, the most famous seaside resort of the Netherlands.

Haidarabad, India (cap. Haidarabad), is one of the most important strongholds of Mohammedanism in India, and has several fine mosques.

Hakodate, Japan (Hokkaido), is a treaty port and coaling-station, with exports of edible seaweed, sulphur, dried fish, rice, salt, straw ropes, and bags.

Halberstadt, Germany (Prussia), is an ancient city, and among its many interesting buildings are the cathedral (13th and 14th centuries); the churches of Our Lady (1135-46) and St. Martin (1350); the 14th century town hall; and the former episcopal residence. It has manufactures of sugar and leather, and railway repairing works.

Halifax, England (York), has large manufactures of cottons, woollens and carpets. Its chief buildings are the parish church of St. John (12th century), the town hall, the public library and museum, the Akroyd museum and art gallery, the mechanics' institution, and the piece-, or cloth-hall (1779).

Halifax, cap. Nova Scotia, Canada, is a fortified port, naval and coaling-station, and an important railroad terminus. Its harbor is one of the finest in the world, being six miles long and one mile wide. It has a large export trade in coal, timber, cattle, furs, oils, and agricultural produce, and extensive manufactures of iron and steel, sugar, woollens, paper, and leather, as well as shipbuilding yards and large fishing interests. Halifax is the seat of an Anglican bishop and a Roman Catholic archbishop, and has a Roman Catholic and an Anglican cathedral. St. Paul's church is the oldest church building in North America. Among the higher educational institutions are the non-sectarian Dalhousie university and college, the Roman Catholic college of St. Mary, and the Presbyterian theological college. The city has become a favorite summer resort, owing to the beauty of the surroundings, its bracing climate, and its sanitary conditions.

Halle, Germany (Saxony), is an ancient city with many medieval buildings, and contains a famous university, founded in 1694, with which the University of Wittenberg was incorporated in 1817, and which is attended by some 1,750 students. Here also are the asylums and schools founded (1698) by Francke. In the market-place are the red tower (276 ft.), and a monument to Handel, a native of the town. The handsomest church is that of St. Maurice (12th to 16th century), but Halle also possesses a cathedral (16th century). The most notable of the secular buildings are the partly ruined citadel of St. Maurice (1484-1503), formerly a residence of the archbishops of Magdeburg; the ruined castle of Giebichenstein; and the university (1834). Halle has engineering works, sugar factories, printing-works, malt-kilns, and manufactories of confectionery, oils, chicory, and salt.

Hamadan, Persia, has manufactures of leather, carpets, silks, and copper work, and considerable trade. It is supposed to stand on the site of the ancient Ecbatana, and among its tombs the Jews still show the reputed burial-places of Esther and Mordecai.

Hamah, Turkey in Asia (Syria), has manufactures of yarn and coarse woollens, and a general domestic and caravan trade. It is a very ancient city, the Hamath of the Bible, and was one of the chief towns of the Hittites.

Hamburg, Germany (cap. Hamburg), is a coaling-station and the most important seaport on the continent of Europe, especially for the embarkation of emigrants. It is the terminus of many steamship lines going to all parts of the world, and the chief articles of trade are coffee, grain, iron, fancy goods, butter, hides, and sugar. It is the seat of extensive and varied industries, mostly connected with the needs of a great seaport, besides controlling the trade in animals for zoological gardens. The more noteworthy among the public buildings and institutions are the churches of St. Michael (1750-62), St. Peter (1842-9), and St. Nicholas (1846-63); the town hall (1886-97), marine office, the picture gallery, the museum of arts and crafts (1878), the natural history museum (1891), the commercial and municipal libraries, the hygienic institute, a fine hospital at Effendorf, a new observatory, and a famous zoological garden. One of the most prominent features of the town is a lake-like

able for its beautiful parks and pleasure-grounds. It is the seat of a bishop and has a cathedral built in 1815; a museum, founded in 1833, is housed in the old East India Company's buildings. There is also a famous free technical school. The city has ship-building and machinery factories, sugar-refineries, breweries, tanneries, weaving and spinning factories, etc. Owing to its excellent harbor, which is almost always free from ice, it has a large trade. The principal exports are iron, timber, wood-pulp, corn, butter, and fish. Gottenborg was founded by Gustavus Adolphus in 1619.

Göttingen, Germany (Prussia), is the seat of a famous university founded in 1737. Göttingen manufactures sausages and other eatables, mathematical and scientific instruments, cloth, chemicals, sugar, and beer. In 1626 the town was taken by Tilly. Göttingen is famous for its Academy of Sciences, founded by Haller in 1751.

Govan, Scotland (Lanark), has ship-building yards which are among the largest on the Clyde. It has also engineering-, railway locomotive-, and iron-works.

Granada, Nicaragua (cap. Granada), is the seat of a considerable trade in dyewoods, indigo, hides, cacao, and has manufactories of iron wire chains.

Granada, Spain (cap. Granada), is an ancient Moorish city, the seat of the last Moslem kings, and was conquered by Ferdinand and Isabella in 1492. It possesses the famous palace of the Alhambra, and other picturesque structures of Moorish character. It is the seat of an archbishop, and has a university and a richly decorated Renaissance cathedral (1529), containing the tombs of Ferdinand and Isabella, and of Philip I., and his consort Juana. In the church of San Geronimo is the cenotaph of Gonsalvo de Cordova. Liqueurs, textiles, paper, and soap are manufactured.

Grand Forks, North Dakota (c. h. Grand Forks Co.), is an important railroad center and the second city of the state. It has large lumber-mills, flour-mills, steam-boiler works, and a large trade in live-stock, grain, and agricultural produce. It is the seat of the State University of North Dakota and other educational institutions.

Grand Junction, Colorado (c. h. Mesa Co.), is the largest city of western Colorado, and an important railroad and trade center in the famous fruit producing valley of the Grand River. It has brass- and iron-foundries, planing-mills, etc.

Grand Rapids, Michigan (c. h. Kent Co.), is a port of entry, the second city of the state, and the third city in the United States in the manufacture of furniture. The Grand River has here a fall of 18 feet, supplying excellent water-power. In addition to furniture, there are extensive manufactures of other lumber products, flour, plaster, carriages, wagons, bicycles, brass goods, automobiles, and agricultural implements. The gypsum quarries here have the largest output in the world. There are numerous educational institutions and a state soldiers' home.

Gratz, Austria (cap. Styria), is an episcopal see and the favorite place of residence for retired Austrian officials. It is the seat of a university founded in 1586, which has some 1,500 students, and of a polytechnic school. Its manufactures are machinery, steel and iron wares, soap, chemicals, pottery, confectionery, beer, etc. In addition there are large railway-car works. The central feature is the Castle Hill, which was fortified until 1809. There is also a 15th century cathedral.

Graudenz, Germany (Prussia), manufactures machinery, cigars, tapestry, flour, etc. About one mile north is the fortress, constructed by Frederick the Great in 1772-6, famous for its defence by the Prussians against the French in 1807, but dismantled in 1874.

Great Falls, Montana(c. b. Cascade Co.), is the second city of the state, near an important mining region, and has large smelting works. Gold, silver, copper, lead, bituminous coal, iron, and sandstone are found in the vicinity. It has manufactories of flour, furniture, and mining and agricultural implements. It derives its name and importance from the falls formed here by the Missouri, with a total descent of 500 ft.

Great Grimsby, England (Lincoln), is a seaport

and coaling-station and the largest fishing-port of England. Its industries comprise ship-building, tanning, brewing, and rope-making, and it exports cottons, woollens, leather, fish, etc.

Greeley, Colorado (c. h. Weld Co.), is a trade center in a rich agricultural region largely under irrigation and is famous for its potatoes and alfalfa. It has a large trade in produce and agricultural implements, and manufactories of machinery, beet sugar, etc.

Green Bay, Wisconsin (c. h. Brown Co.), is a port of delivery, with an extensive lake traffic. It has a large export trade in grain and lumber, and extensive manufactures of paper, sulphite, lumber products, etc. It is the oldest town in the state and was first visited in 1634 by Jean Nicollet, who had been sent by Champlain, the governor of New France, to find the rumored short route to China.

Greenock, Scotland (Renfrew), is a port on the Clyde. Ship-building is the chief industry, and there are sugar and glucose refineries, iron-foundries, wool-spinning, aluminum-rolling, and paper-mills. The harbor works date from 1707.

Greensboro, North Carolina (c. h. Guilford Co.), is the center of a tobacco, fruit, and grain region. It also has steel works, cotton-mills, and other industries. Has a large trade in tobacco, coal, and iron. It is the seat of the state agricultural and mechanical college, and other educational institutions.

Greenville, Mississippi (c. h. Washington Co.), is a trade center in a region producing a large amount of cotton. It has cotton-compresses, cottonseed-oil mills, manufactories of lumber products, tile and brick works, etc.

Greenville, South Carolina (c. h. Greenville Co.), is the seat of Furman University (Baptist), organized in 1851, the Chicora Female College (Presbyterian), a military institute, and numerous other educational institutions. It has cotton-mills, iron-works, and flour-mills.

Grenoble, France (cap. Isère), is an episcopal see and a great tourist center. The manufacture of kid gloves is the chief industry. Among other manufactures are leather-dressing, cement, paper flowers, silk and linen, furniture, buttons and fasteners, hosiery, and straw hats. Walnuts are largely exported. Grenoble is the seat of a university.

Grodno, Russia (cap. Grodno), contains an ancient palace of the Polish kings (12th and 13th centuries); a more modern palace built by Augustus III., now converted into a military hospital; the Greek house of St. Basil; the former Jesuit College; and the Carmelite monastery. The academy of medical science was founded by King Stanislaus Augustus. It has manufactories of tobacco, soap, candles, vinegar, bricks, machinery, and carriages, with breweries, distilleries, sawmills, iron-foundries, etc.

Groningen, Netherlands (cap. Groningen), is the largest town in northern Netherlands and the seat of a university, founded in 1614. It possesses many 17th century quaint old houses and is famous for its corn and oil-seed markets. It was taken from the Spaniards by the Dutch, after a stubborn defence, in 1594.

Grosswardein, Hungary, possesses two cathedrals—one belonging to the Roman, the other to the Greek Catholics; also a Roman Catholic episcopal palace, a law academy, and portions of a former citadel. Milling, distilling, brickmaking, etc., are carried on, and good wine is made. The city was destroyed by the Mongols in 1241, and from 1663 to 1692 was in the power of the Turks.

Guadalajara, Mexico (cap. Jalisco), has a magnificent cathedral containing a famous "Assumption" by Murillo, and a theatre, the Degollado, one of the largest on the continent. Here are also a university, an art academy, and a mint. Coffee is grown; paper, Panama hats, and leather are manufactured, and artistic terracotta ware is made. Elevation 5,185 ft.

Guanajuato, Mexico (cap. Guanajuato), is the center of one of the richest mining districts in Mexico, and is noted for its pottery. It produces silver, gold, mercury, tin, lead, and copper. Elevation 6,756 ft.

Guayaquil, Ecuador (cap. Guayas), is a port and coaling-station, and, next to Quito, the most important town in Ecuador. It is an episcopal see and contains a fine cathedral built of wood, government buildings, and a university. It has a good harbor and exports cocoa, coffee, cinchona, gold, silver, and hides. It was founded in 1535 and after Valparaiso is the most populous port on the west coast of South America. Yellow fever is very prevalent.

Guben, Germany (Prussia), manufactures woollens, hats, leather, and machinery.

Guelph, Ontario, Canada, is a trade center in one of the richest agricultural sections of Canada. It has a piano and organ factory, knitting mills, flour mills, axle works, etc. Guelph is the seat of the Ontario Agricultural College and Macdonald Institute, with over 1,200 students. An annual fat stock show is held here.

Guinobatan, Philippine Islands (Albay), is the center of a region producing a large amount of hemp.

Guthrie, cap. Oklahoma (c. h. Logan Co.), has a considerable distributing trade. It has cotton-gins, and manufactures of cottonseed-oil, lumber, flour, etc.

Gwalior, India (cap. Gwalior), contains a large palace of the Maharajah Scindia, and a huge fortress.

Haarlem, Netherlands (cap. North Holland), is an ancient city with many interesting buildings, among which are the town hall, a former palace of the Counts of Holland, which contains a small collection of paintings by Dutch artists, including eight large canvases by Frans Hals. Of its churches the finest is the Groote Kerk, a Late Gothic basilica, built in the 15th century, one of the largest churches in Holland, noted for its tower 260 feet high, and its large organ. Haarlem was in the 17th century the center of a famous school of painting, to which belonged such artists as the Wouwermans, the Ostades, Frans Hals, Ruisdael, and the Everdingen, most of them natives. It is also the birthplace of Laurens Coster, to whom the Dutch ascribe the invention of printing. One of the oldest and most extensive industries is the cultivation of flowers, Haarlem exporting bulbs to many European countries. It has, besides, manufactures of machinery, cottons, and woollens, and a large trade in dairy produce.

Hackensack, New Jersey (c. h. Bergen Co.), is a residential suburb of New York and has brick, silk, and other manufacturing industries. Was settled in the later part of the 17th century and, during the Revolution, was occupied in turn by the British and the American armies.

Hagen, Germany (Prussia), is a growing industrial center, with ironworks, cotton and cloth mills, tanneries, breweries, distilleries, and manufactures of cigars and paper. It has a school of engineering and several technical colleges.

Hagerstown, Maryland (c. h. Washington Co.), is a trade center of western Maryland, and has extensive manufactures of knit goods, machinery, steam-engines, and lumber. Was a center of military operations in the Civil War.

Hague, The, cap. Netherlands (cap. South Holland), is one of the prettiest towns of Holland, and among its most important edifices are the royal palace, and the Binnenhof, a large irregular building, founded in 1249 and containing the hall of assembly of the States-General, and various government offices where the International Peace Conference met in 1899. Other buildings are the provincial government house, a large roomy edifice; the town hall; the ministry of justice; the municipal museum, containing pictures and antiquities; the royal library; a cannon foundry, one of the largest and most conspicuous structures in the town; the colonial and war office; the national monument, etc. The royal collection of pictures, in the Mauritshuis, is confined chiefly to Dutch masters and is famous for the magnificent exhibition of Rembrandt's masterpieces. Mr. Andrew Carnegie has undertaken to build at The Hague a Palace of Peace. Not far from the city and connected with it by trolley car is Scheveningen, the most famous seaside resort of the Netherlands.

Haidarabad, India (cap. Haidarabad), is one of the most important strongholds of Mohammedanism in India, and has several fine mosques.

Hakodate, Japan (Hokkaido), is a treaty port and coaling-station, with exports of edible seaweed, sulphur, dried fish, rice, salt, straw ropes, and bags.

Halberstadt, Germany (Prussia), is an ancient city, and among its many interesting buildings are the cathedral (13th and 14th centuries); the churches of Our Lady (1135-46) and St. Martin (1350); the 14th century town hall; and the former episcopal residence. It has manufactures of sugar and leather, and railway repairing works.

Halifax, England (York), has large manufactures of cottons, woollens and carpets. Its chief buildings are the parish church of St. John (12th century), the town hall, the public library and museum, the Akroyd museum and art gallery, the mechanics' institution, and the piece-, or cloth-hall (1779).

Halifax, cap. Nova Scotia, Canada, is a fortified port, naval and coaling-station, and an important railroad terminus. Its harbor is one of the finest in the world, being six miles long and one mile wide. It has a large export trade in coal, timber, cattle, furs, oils, and agricultural produce, and extensive manufactures of iron and steel, sugar, woollens, paper, and leather, as well as shipbuilding yards and large fishing interests. Halifax is the seat of an Anglican bishop and a Roman Catholic archbishop, and has a Roman Catholic and an Anglican cathedral. St. Paul's church is the oldest church building in North America. Among the higher educational institutions are the non-sectarian Dalhousie university and college, the Roman Catholic college of St. Mary, and the Presbyterian theological college. The city has become a favorite summer resort, owing to the beauty of the surroundings, its bracing climate, and its sanitary conditions.

Halle, Germany (Saxony), is an ancient city with many medieval buildings, and contains a famous university, founded in 1694, with which the University of Wittenberg was incorporated in 1817, and which is attended by some 1,750 students. Here also are the asylums and schools founded (1698) by Francke. In the market-place are the red tower (276 ft.), and a monument to Handel, a native of the town. The handsomest church is that of St. Maurice (12th to 16th century), but Halle also possesses a cathedral (16th century). The most notable of the secular buildings are the partly ruined citadel of St. Maurice (1484-1503), formerly a residence of the archbishops of Magdeburg; the ruined castle of Giebichenstein; and the university (1834). Halle has engineering works, sugar factories, printing-works, maltkilns, and manufactories of confectionery, oils, chicory, and salt.

Hamadan, Persia, has manufactures of leather, carpets, silks, and copper work, and considerable trade. It is supposed to stand on the site of the ancient Ecbatana, and among its tombs the Jews still show the reputed burial-places of Esther and Mordecai.

Hamah, Turkey in Asia (Syria), has manufactures of yarn and coarse woollens, and a general domestic and caravan trade. It is a very ancient city, the Hamath of the Bible, and was one of the chief towns of the Hittites.

Hamburg, Germany (cap. Hamburg), is a coaling-station and the most important seaport on the continent of Europe, especially for the embarkation of emigrants. It is the terminus of many steamship lines going to all parts of the world, and the chief articles of trade are coffee, grain, iron, fancy goods, butter, hides, and sugar. It is the seat of extensive and varied industries, mostly connected with the needs of a great seaport, besides controlling the trade in animals for zoological gardens. The more noteworthy among the public buildings and institutions are the churches of St. Michael (1750-62), St. Peter (1842-9), and St. Nicholas (1846-63); the town hall (1886-97), marine office, the picture gallery, the museum of arts and crafts (1878), the natural history museum (1891), the commercial and municipal libraries, the hygienic institute, a fine hospital at Effendorf, a new observatory, and a famous zoological garden. One of the most prominent features of the town is a lake-like

able for its beautiful parks and pleasure-grounds. It is the seat of a bishop and has a cathedral built in 1815; a museum, founded in 1833, is housed in the old East India Company's buildings. There is also a famous free technical school. The city has ship-building and machinery factories, sugar-refineries, breweries, tanneries, weaving and spinning factories, etc. Owing to its excellent harbor, which is almost always free from ice, it has a large trade. The principal exports are iron, timber, wood-pulp, corn, butter, and fish. Gottenborg was founded by Gustavus Adolphus in 1619.

Göttingen, Germany (Prussia), is the seat of a famous university founded in 1737. Göttingen manufactures sausages and other eatables, mathematical and scientific instruments, cloth, chemicals, sugar, and beer. In 1626 the town was taken by Tilly. Göttingen is famous for its Academy of Sciences, founded by Haller in 1751.

Govan, Scotland (Lanark), has ship-building yards which are among the largest on the Clyde. It has also engineering-, railway locomotive-, and iron-works.

Granada, Nicaragua (cap. Granada), is the seat of a considerable trade in dyewoods, indigo, hides, cacao, and has manufactories of gold wire chains.

Granada, Spain (cap. Granada), is an ancient Moorish city, the seat of the last Moslem kings, and was conquered by Ferdinand and Isabella in 1492. It possesses the famous palace of the Alhambra, and other picturesque structures of Moorish character. It is the seat of an archbishop, and has a university and a richly decorated Renaissance cathedral (1529), containing the tombs of Ferdinand and Isabella, and of Philip I., and his consort Juana. In the church of San Geronimo is the cenotaph of Gonsalvo de Cordova. Liqueurs, textiles, paper, and soap are manufactured.

Grand Forks, North Dakota (c. h. Grand Forks Co.), is an important railroad center and the second city of the state. It has large lumber-mills, flour-mills, steam-boiler works, and a large trade in live-stock, grain, and agricultural produce. It is the seat of the State University of North Dakota and other educational institutions.

Grand Junction, Colorado (c. h. Mesa Co.), is the largest city of western Colorado, and an important railroad and trade center in the famous fruit producing valley of the Grand River. It has brass- and iron-foundries, planing-mills, etc.

Grand Rapids, Michigan (c. h. Kent Co.), is a port of entry, the second city of the state, and the third city in the United States in the manufacture of furniture. The Grand River here has a fall of 18 feet, supplying excellent water-power. In addition to furniture, there are extensive manufactures of other lumber products, flour, plaster, carriages, wagons, bicycles, brass goods, automobiles, and agricultural implements. The gypsum quarries here have the largest output in the world. There are numerous educational institutions and a state soldiers' home.

Gratz, Austria (cap. Styria), is an episcopal see and the favorite place of residence for retired Austrian officials. It is the seat of a university founded in 1586, which has some 1,500 students, and of a polytechnic school. Its manufactures are machinery, steel and iron wares, soap, chemicals, pottery, confectionery, beer, etc. In addition there are large railway-car works. The central feature is the Castle Hill, which was fortified until 1809. There is also a 15th century cathedral.

Graudenz, Germany (Prussia), manufactures machinery, cigars, tapestry, flour, etc. About one mile north is the fortress, constructed by Frederick the Great in 1772-6, famous for its defence by the Prussians against the French in 1807, but dismantled in 1874.

Great Falls, Montana (c. h. Cascade Co.), is the second city of the state, near an important mining region, and has large smelting works. Gold, silver, copper, lead, bituminous coal, iron, and sandstone are found in the vicinity. It has manufactories of flour, furniture, and mining and agricultural implements. It derives its name and importance from the falls formed here by the Missouri, with a total descent of 500 ft.

Great Grimsby, England (Lincoln), is a seaport

and coaling-station and the largest fishing-port of England. Its industries comprise ship-building, tanning, brewing, and rope-making, and it exports cottons, woollens, leather, fish, etc.

Greeley, Colorado (c. h. Weld Co.), is a trade center in a rich agricultural region largely under irrigation and is famous for its potatoes and alfalfa. It has a large trade in produce and agricultural implements, and manufactories of machinery, beet sugar, etc.

Green Bay, Wisconsin (c. h. Brown Co.), is a port of delivery, with an extensive lake traffic. It has a large export trade in grain and lumber, and extensive manufactures of paper, sulphite, lumber products, etc. It is the oldest town in the state and was first visited in 1634 by Jean Nicollet, who had been sent by Champlain, the governor of New France, to find the rumored short route to China.

Greenock, Scotland (Renfrew), is a port on the Clyde. Ship-building is the chief industry, and there are sugar and glucose refineries, iron-foundries, wool-spinning, aluminum-rolling, and paper-mills. The harbor works date from 1707.

Greensboro, North Carolina (c. h. Guilford Co.), is the center of a tobacco, fruit, and grain region. It also has steel works, cotton-mills, and other industries. Has a large trade in tobacco, coal, and iron. It is the seat of the state agricultural and mechanical college, and other educational institutions.

Greenville, Mississippi (c. h. Washington Co.), is a trade center in a region producing a large amount of cotton. It has cotton-compresses, cottonseed-oil mills, manufactories of lumber products, tile and brick works, etc.

Greenville, South Carolina (c. h. Greenville Co.), is the seat of Furman University (Baptist), organized in 1851, the Chicora Female College (Presbyterian), a military institute, and numerous other educational institutions. It has cotton-mills, iron-works, and flour-mills.

Grenoble, France (cap. Isère), is an episcopal see and a great tourist center. The manufacture of kid gloves is the chief industry. Among other manufactures are leather-dressing, cement, paper flowers, silk and linen, furniture, buttons and fasteners, hosiery, and straw hats. Walnuts are largely exported. Grenoble is the seat of a university.

Grodno, Russia (cap. Grodno), contains an ancient palace of the Polish kings (12th and 13th centuries); a more modern palace built by Augustus III., now converted into a military hospital; the Greek house of St. Basil; the former Jesuit College; and the Carmelite monastery. The academy of medical science was founded by King Stanislaus Augustus. It has manufactories of tobacco, soap, candles, vinegar, bricks, machinery and carriages, with breweries, distilleries, sawmills, iron-foundries, etc.

Groningen, Netherlands (cap. Groningen), is the largest town in northern Netherlands and the seat of a university, founded in 1614. It possesses many 17th century quaint old houses and is famous for its corn and oil-seed markets. It was taken from the Spaniards by the Dutch, after a stubborn defence, in 1594.

Grosswardein, Hungary, possesses two cathedrals—one belonging to the Roman, the other to the Greek Catholics; also a Roman Catholic episcopal palace, a law academy, and portions of a former citadel. Milling, distilling, brickmaking, etc., are carried on, and good wine is made. The city was destroyed by the Mongols in 1241, and from 1663 to 1692 was in the power of the Turks.

Guadalajara, Mexico (cap. Jalisco), has a magnificent cathedral containing a famous "Assumption" by Murillo, and a theatre, the Degollado, one of the largest on the continent. Here are also a university, an art academy, and a mint. Coffee is grown; paper, Panama hats, and leather are manufactured, and artistic terracotta ware is made. Elevation 5,185 ft.

Guanajuato, Mexico (cap. Guanajuato), is the center of one of the richest mining districts in Mexico, and is noted for its pottery. It produces silver, gold, mercury, tin, lead, and copper. Elevation 6,759 ft.

Guayaquil, Ecuador (cap. Guayas), is a port and coaling-station, and, next to Quito, the most important town in Ecuador. It is an episcopal see and contains a fine cathedral built of wood, government buildings, and a university. It has a good harbor and exports cocoa, coffee, cinchona, gold, silver, and hides. It was founded in 1535 and after Valparaiso is the most populous port on the west coast of South America. Yellow fever is very prevalent.

Guben, Germany (Prussia), manufactures woollens, hats, leather, and machinery.

Guelph, Ontario, Canada, is a trade center in one of the richest agricultural sections of Canada. It has a piano and organ factory, knitting mills, flour mills, axle works, etc. Guelph is the seat of the Ontario Agricultural College and Macdonald Institute, with over 1,200 students. An annual fat stock show is held here.

Guinobatan, Philippine Islands (Albay), is the center of a region producing a large amount of hemp.

Guthrie, cap. Oklahoma (c. h. Logan Co.), has a considerable distributing trade. It has cotton-gins, and manufactures of cottonseed-oil, lumber, flour, etc.

Gwalior, India (cap. Gwalior), contains a large palace of the Maharajah Scindia, and a huge fortress.

Haarlem, Netherlands (cap. North Holland), is an ancient city with many interesting buildings, among which are the town hall, a former palace of the Counts of Holland, which contains a small collection of paintings by Dutch artists, including eight large canvases by Frans Hals. Of its churches the finest is the Groote Kerk, a Late Gothic basilica, built in the 15th century, one of the largest churches in Holland, noted for its tower 260 feet high, and its large organ. Haarlem was in the 17th century the center of a famous school of painting, to which belonged such artists as the Wouwermans, the Ostades, Frans Hals, Ruisdael, and the Everdingen, most of them natives. It is also the birthplace of Laurens Coster, to whom the Dutch ascribe the invention of printing. One of the oldest and most extensive industries is the cultivation of flowers, Haarlem exporting bulbs to many European countries. It has, besides, manufactures of machinery, cottons, and woollens, and a large trade in dairy produce.

Hackensack, New Jersey (c. h. Bergen Co.), is a residential suburb of New York and has brick, silk, and other manufacturing industries. Was settled in the later part of the 17th century and, during the Revolution, was occupied in turn by the British and the American armies.

Hagen, Germany (Prussia), is a growing industrial center, with ironworks, cotton and cloth mills, tanneries, breweries, distilleries, and manufactures of cigars and paper. It has a school of engineering and several technical colleges.

Hagerstown, Maryland (c. h. Washington Co.), is a trade center of western Maryland, and has extensive manufactures of knit goods, machinery, steam-engines, and lumber. Was a center of military operations in the Civil War.

Hague, The, cap. Netherlands (cap. South Holland), is one of the prettiest towns of Holland, and among its most important edifices are the royal palace, and the Binnenhof, a large irregular building, founded in 1249 and containing the hall of assembly of the States-General, and various government offices where the International Peace Conference met in 1899. Other buildings are the provincial government house, a large roomy edifice; the town hall; the ministry of justice; the municipal museum, containing pictures and antiquities; the royal library; a cannon foundry, one of the largest and most conspicuous structures in the town; the colonial and war office; the national monument, etc. The royal collection of pictures, in the Mauritshuis, is confined chiefly to Dutch masters and is famous for the magnificent exhibition of Rembrandt's masterpieces. Mr. Andrew Carnegie has undertaken to build at The Hague a Palace of Peace. Not far from the city and connected with it by trolley car is Scheveningen, the most famous seaside resort of the Netherlands.

Haidarabad, India (cap. Haidarabad), is one of the most important strongholds of Mohammedanism in India, and has several fine mosques.

Hakodate, Japan (Hokkaido), is a treaty port and coaling-station, with exports of edible seaweed, sulphur, dried fish, rice, salt, straw ropes, and bags.

Halberstadt, Germany (Prussia), is an ancient city, and among its many interesting buildings are the cathedral (13th and 14th centuries); the churches of Our Lady (1135-46) and St. Martin (1350); the 14th century town hall; and the former episcopal residence. It has manufactures of sugar and leather, and railway repairing works.

Halifax, England (York), has large manufactures of cottons, woollens and carpets. Its chief buildings are the parish church of St. John (12th century), the town hall, the public library and museum, the Akroyd museum and art gallery, the mechanics' institution, and the piece-, or cloth-hall (1779).

Halifax, cap. Nova Scotia, Canada, is a fortified port, naval and coaling-station, and an important railroad terminus. Its harbor is one of the finest in the world, being six miles long and one mile wide. It has a large export trade in coal, timber, cattle, furs, oils, and agricultural produce, and extensive manufactures of iron and steel, sugar, woollens, paper, and leather, as well as shipbuilding yards and large fishing interests. Halifax is the seat of an Anglican bishop and a Roman Catholic archbishop, and has a Roman Catholic and an Anglican cathedral. St. Paul's church is the oldest church building in North America. Among the higher educational institutions are the non-sectarian Dalhousie university and college, the Roman Catholic college of St. Mary, and the Presbyterian theological college. The city has become a favorite summer resort, owing to the beauty of the surroundings, its bracing climate, and its sanitary conditions.

Halle, Germany (Saxony), is an ancient city with many mediæval buildings, and contains a famous university, founded in 1694, with which the University of Wittenberg was incorporated in 1817, and which is attended by some 1,750 students. Here also are the asylums and schools founded (1698) by Francke. In the market-place are the red tower (276 ft.), and a monument to Handel, a native of the town. The handsomest church is that of St. Maurice (12th to 16th century), but Halle also possesses a cathedral (16th century). The most notable of the secular buildings are the partly ruined citadel of St. Maurice (1484-1503), formerly a residence of the archbishops of Magdeburg; the ruined castle of Giebichenstein; and the university (1834). Halle has engineering works, sugar factories, printing-works, maltkilns, and manufactories of confectionery, oils, chicory, and salt.

Hamadan, Persia, has manufactures of leather, carpets, silks, and copper work, and considerable trade. It is supposed to stand on the site of the ancient Ecbatana, and among its tombs the Jews still show the reputed burial-places of Esther and Mordecai.

Hamah, Turkey in Asia (Syria), has manufactures of yarn and coarse woollens, and a general domestic and caravan trade. It is a very ancient city, the Hamath of the Bible, and was one of the chief towns of the Hittites.

Hamburg, Germany (cap. Hamburg), is a coaling-station and the most important seaport on the continent of Europe, especially for the embarkation of emigrants. It is the terminus of many steamship lines going to all parts of the world, and the chief articles of trade are coffee, grain, iron, fancy goods, butter, hides, and sugar. It is the seat of extensive and varied industries, mostly connected with the needs of a great seaport, besides controlling the trade in animals for zoological gardens. The more noteworthy among the public buildings and institutions are the churches of St. Michael (1750-62), St. Peter (1842-9), and St. Nicholas (1846-63); the town hall (1886-97), marine office, the picture gallery, the museum of arts and crafts (1878), the natural history museum (1891), the commercial and municipal libraries, the hygienic institute, a fine hospital at Effendorf, a new observatory, and a famous zoological garden. One of the most prominent features of the town is a lake-like

expansion of the Alster, a little stream which intersects the city, and which is flanked by handsome residences and parks, constitutes one of its prettiest quarters.

Hamilton, Ohio (c. h. Butler Co.), has extensive manufactures of paper, tools, machinery, agricultural implements, iron and woollen goods, etc. It has excellent water-power.

Hamilton, Ontario, Canada, is a port of entry at the head of navigation on Lake Ontario, and an important railroad center. It is in the middle of a fine grain and fruit producing region, and has considerable manufacturing and shipping interests. It has extensive iron and tool works, cotton and tobacco factories, etc. It is the seat of an Anglican and of a Roman Catholic bishop. The chief educational institution is the Wesleyan Female College.

Hamilton, Scotland (Lanark), is the center of a very rich coal and iron field, and has extensive iron-works, iron and brass factories, and cotton mills, etc. Hamilton Palace, the seat of the dukes of Hamilton, stands near the town.

Hamm, Germany (Prussia), has manufactures of iron and steel, machinery, furniture, and leather.

Hammond, Indiana (Lake Co.), is an industrial suburb of Chicago. It has large slaughtering and meat-packing establishments, printing, chemical, steel spring, carriage, and glue works.

Hanau, Germany (Prussia), is famous for its jewelry work and diamond cutting. It also has engineering works, iron-foundries, textiles and gunpowder works. In 1813 this place was the scene of a series of engagements between the Bavarian general Wrede and Napoleon.

Handsworth, England (Stafford), is a suburb of Birmingham, with large industrial works.

Hangchau, China (cap. Chehkiang), is one of the most beautiful cities of China. It is a great center of the silk trade, and fans and tinfoil are its special industries. It was opened to foreign trade in 1896. From 1127 to 1278 A. D., it was the capital of the Sung dynasty, and is identified with the Kinsay of Marco Polo.

Hankau, China (Hupeh), is a treaty port on the Yangtse River, 700 miles from its mouth. It is an important center of Chinese trade, and the central market of the tea districts of the Yangtse. It also exports hides, vegetable tallow, rhea fiber, hemp, bristles, and Chinese medicines. It is connected by railroad with Peking.

Hanley, England (Stafford), is a modern town known as "The Metropolis of the Potteries," and is famous for its manufactures of earthenware and fine porcelain. It has also large iron and steel works, and coal and iron mines in the neighborhood.

Hannibal, Missouri (Marion Co.), is an important trade center, with excellent railroad and river facilities. The chief manufactures are foundry and machine-shop products, flour, lumber, cigars, lime, cement, car-wheels, and furniture.

Hanoi, Anam (cap. Tonkin), is a port on the river Songkoi, or "Red River", and a great commercial center. It has manufactures of silks, embroideries, furniture, and jewelry, and an extensive trade in rice.

Hanover, Germany (Prussia), is an important railway, commercial, and manufacturing center. It has large factories for india-rubber and gutta-percha goods, for iron-founding and making machinery and hardware, and for producing linens, chemicals, tobacco, books, and furniture. Among the objects of interest are the Waterloo column, war monument, Kestner museum, palace, Markt-kirche, museum, picture gallery, town hall, and theater. Near the city is the royal palace of Herrenhausen with beautiful gardens. Hanover was an ancient Hanseatic town and a former ducal and royal capital.

Harar, Abyssinia, is the most important place in the eastern portion of Abyssinia. The country around produces coffee, tobacco, and bananas.

Harbin, China (Manchuria), is an important junction on the trans-Siberian railroad, and is in the center of a rich agricultural and grazing district. Has large mineral fields not yet developed, while its industries include flour-mills, brick works, distilleries, and meat-packing establishments.

Harburg, Germany (Prussia), is a port on the Elbe, with a large distributing trade and manufactures of rubber goods, chemicals, linseed and cocoanut oil, glass, and machinery.

Harrisburg, cap. Pennsylvania (c. h. Dauphin Co.), is an important railroad, agricultural, and industrial center. The iron, steel, lumber, and railroad interests here are of great importance. There are also extensive manufactories of machinery, malt liquors, boilers, cars, leather, lumber, cotton goods, silk goods, etc. The new State Capitol was erected at a cost of over $13,000,000, and has a dome adorned with paintings by Edwin Abbey. The State Library, founded in 1790, contains over 100,000 volumes. Here are also located the state arsenal, the state lunatic asylum, and numerous educational and charitable institutions. John Harris, an adventurous English trader, built the first house here in 1726, and the settlement was incorporated as a borough in 1791.

Harrison, New Jersey (Hudson Co.), is a suburb of Newark, and has the largest hydraulic pump works in the world, as well as extensive manufactures of wire cloth, marine engines, etc.

Hartford, cap. Connecticut (c. h. Hartford Co.), is a port of entry and at the head of navigation of the Connecticut River, 50 miles from Long Island Sound. It is the second city of the state, is noted for its wealthy and powerful insurance companies, and has extensive manufactures of automobiles, steam-engines, small arms, bicycles, rubber goods, etc. The State Capitol is the most conspicuous building and has fine structural embellishments. The Wadsworth Atheneum contains a library of 150,000 volumes and the collections of the State Historical Society. Hartford is the seat of Trinity College (Episcopal), established in 1823, a theological seminary (Congregational), the American Deaf and Dumb Asylum, the Connecticut Retreat for the Insane, and other educational and charitable institutions. It was settled in 1635, and was the scene of the attempt of Andros to secure the colonial charter which was hidden in the "Charter Oak," in 1687. It was sole capital 1665-1701, and capital jointly with New Haven 1701-1873.

Hastings, England (Sussex), is a favorite seaside resort, with extensive pleasure-grounds and parks. The churches of All Saints and St. Clements date respectively from the 11th and 13th centuries, and on West Hill are the ruins of the castle built by William the Conqueror; its dungeons were discovered in 1894.

Hathras, India (United Provinces), is a trade center with exports of sugar, grain, and oil seeds.

Havana, cap. Cuba (cap. Havana), is a port, coaling station, the chief commercial city of the West Indies, and a great tobacco and sugar center. It has a spacious harbor with good anchorage for the largest vessels. Exports tobacco, cigars, sugar, molasses, beeswax, honey, etc. Its manufactures are mainly tobacco products; its cigar factories, of which there are over 100 of the first rank, are the largest in the world. The Tacon theatre is very large and beautiful. The principal church is the Merced, one of the alleged burial-places of Columbus, although his alleged remains were in 1898 removed to Spain. The principal educational institutions are the university and the Jesuit College de Belen. The city was founded in 1519, and was taken several times by buccaneers in the 16th century and by the English in 1762, but restored to Spain in 1763. The blowing up on February 15, 1898, of the United States battleship Maine, anchored in Havana harbor, led to the Spanish-American War, in which the city and harbor were blockaded.

Haverhill, Massachusetts (Essex Co.), is at the head of navigation of the Merrimac River. It is noted for its shoe manufactories, and manufactures also boot and shoe machinery and supplies, slippers, paper, woollens, leather, etc. It was the birthplace of Whittier.

Havre, France (Seine-Inférieure), is a fortified seaport and coaling station. The chief exports are wines, textiles, and "articles de Paris." The principal industries are shipbuilding, copper and nickel founding, distilling, leather, soap, and candle manufacture.

Hazleton, Pennsylvania (Luzerne Co.), is situated

in the anthracite coal region and its industrial interests are largely connected with the mining and shipping of coal. Manufactures foundry and machine-shop products, carriages, lumber, caskets, and has knitting and silk mills. A state hospital for miners is located here.

Heidelberg, Germany (Baden), is noted for its picturesque situation, its university, and its castle. The university, founded in 1386, is the oldest in the present German empire. From 1556 it came under the control of the leaders of the Reformation, and was reorganized by the elector Charles Frederick of Baden in 1803. The library was plundered and sent to Rome in 1623, and partially returned in 1816; it now consists of over 320,000 volumes. The castle, whose ruins are the most imposing in Germany, is a famous monument founded at the end of the 13th century by the count palatine Rudolf I., and enlarged and strengthened by succeeding electors. During the 16th century it received the architectural development which, despite disaster, makes it still one of the richest productions of the German Renaissance. In 1689 and 1693 it was ruined by the generals of Louis XIV., but was subsequently restored, only to be destroyed by fire from a lightning stroke in 1764. Heidelberg was the capital of the Palatinate from the 13th century to 1720. It was sacked by Tilly in 1622, by the French in 1689, and was nearly destroyed by the latter in 1693. It passed to Baden in 1803.

Heilbronn, Germany (Wurttemberg), has manufactures of silversmiths' ware, iron and steel goods, paper, sugar, salt, chicory, and chemicals. Wine and fruit are produced, and sandstone is quarried in the neighborhood.

Helder, Netherlands (North Holland), is a seaport strongly fortified both by sea and land. It contains several naval establishments, including an arsenal and a naval cadet school. Just outside the harbor the English fleet was defeated by De Ruyter and Tromp in 1673.

Helena, cap. Montana (c. h. Lewis and Clark Co.), is situated in the heart of one of the richest mining districts in the country and claims to be one of the wealthiest cities of its size in the world. It is said that gold amounting to $40,000,000 has been taken from the Last Chance Gulch, which runs through the city, and all around the city are valuable gold and silver bearing veins of quartz, besides deposits of copper, iron and galena. The State Capitol, costing $400,000, and the Montana Wesleyan University (M. E.) are among the prominent buildings. Broadwater Natatorium, fed by a hot spring, the temperature of which at its source is about 160°, is one of the places of interest. Elevation, 4,200 feet.

Helsingborg, Sweden (Malmohus), is a port and coaling station, and one of the oldest towns in Scandinavia. Exports timber, butter, wood pulp, paper, and cattle.

Helsingfors, Russia (cap. Finland), is a strongly fortified seaport, naval and coaling station, and has considerable trade in timber and agricultural produce, as well as manufactures of linen, carpets, iron goods, sugar, and tobacco. It has a university (transferred from Abo in 1827), an astronomical observatory, and numerous other educational institutions, and is the seat of a Greek Orthodox archbishop. It is a favorite seaside resort.

Henderson, Kentucky (c. h. Henderson Co.), is one of the oldest settlements on the Ohio River, but it was not incorporated until 1797. It is situated in a fertile agricultural region, rich in timber and coal, and has a large river trade. The chief manufactures are cotton and woollen goods, tobacco, hominy, flour and agricultural implements.

Herat, Afghanistan, is a fortified city, which at one time was one of the most flourishing places in Asia. Has manufactures of silks, carpets, sheepskin caps and cloaks. It is supposed to have been founded by Alexander the Great, and is regarded as the "gateway" to Afghanistan and India.

Herne, Germany (Prussia), has coal mining and coke industries, and manufactures of machinery.

Hertogenbosch, Netherlands (cap. North Brabant), is an ancient city, and contains the Gothic cathedral of St. John (1312-1498), one of the finest mediæval churches in the Netherlands. It is the seat of a bishop. It has manufactures of linen, ribbons, woollens, and cutlery.

Hildesheim, Germany (Prussia), is an old town, and is renowned for its specimens of mediæval and German Renaissance buildings. Its cathedral, built in the 11th century, is an early Romanesque monument with a late-pointed south aisle and north transept. St. Michael's, formerly the Benedictine abbey church, built early in the 11th century, and somewhat modified in the 12th and 13th centuries, is one of the noblest Romanesque monuments in Germany; the church of St. Godehard, built early in the 12th century, is also a splendid example of Romanesque architecture. Among the secular buildings, the old town hall (15th century) deserves mention. Its industries include iron-foundries, textile-mills, bell-foundries, and sugar-refineries. In the 15th and 16th centuries, it was famous for its goldsmiths' work.

Himeji, Japan (Hyogo), has manufactures of cotton and stamped leather goods.

Hirosaki, Japan (Aomori), manufactures lacquered ware, and there are manganese mines near by.

Hiroshima, Japan (cap. Hiroshima), has an extensive trade in lacquered, bronze, and other artistic wares. Opposite the city is a sacred island dedicated to the Goddess Benten, one of the "three chief sights" of Japan.

Hobart, cap. Tasmania, Australia, is a port and coaling station on the Derwent River, 12 miles from its mouth. The harbor is of easy access, well sheltered, and of great depth of water. The city is much frequented in summer by visitors from New South Wales and Victoria, who are attracted hither by its agreeable climate. The finest public buildings are the government house, the parliament houses, and cathedrals. Hobart is the seat of an Anglican bishop, and of a Roman Catholic archbishop. There are breweries, jam-factories, flour-mills, iron-foundries, tanneries, etc.

Hoboken, New Jersey (Hudson Co.), is a suburb of New York and a terminus of several trans-atlantic steamship lines. It is the seat of Stevens Institute of Technology and of several lesser educational institutions. It has extensive coal, lumber, and brick yards, and manufactures of iron products, leather, silk, lead pencils, caskets, as well as ship-building yards.

Hódmező-Vásárhely, Hungary, has several well-frequented fairs and manufactures of oil, tobacco, and wine.

Hof, Germany (Bavaria), has manufactures of woollens, cottons, linen, dyeing and stamping, chemicals, machinery, leather, hardware, and beer.

Holyoke, Massachusetts (Hampden Co.), is noted for its manufactures, especially of paper, being one of the chief paper-manufacturing cities of the world. It also manufactures thread, cotton and woollen goods, silk, automobiles, bicycles, machinery, and school supplies. This city possesses the greatest water-power in New England. The Connecticut River has a fall here of 60 ft. and is bridled by a huge dam 1,000 ft. across (30,000 horse-power). It has a College of Music and many other educational institutions. It was settled in the last half of the 17th century by people from Ireland and for some time it was called "Ireland Parish." It has a large foreign population.

Homestead, Pennsylvania (Allegheny Co.), has extensive manufactures of steel products, glass, machinery, etc. Its steel plants employ over 6,000 men. It was the scene of a serious strike, which lasted from July to November 1892, and was attended by such rioting as to necessitate the presence of State troops.

Hongkong, China, is an island off the coast of Kwangtung, forming, with the peninsula of Kowloon on the mainland, a British crown colony and naval station. Possessing a magnificent harbor, 10 square miles in extent, and, enjoying the privilege of a free port, it is one of the principal ports of the world. It is also the principal European-Chinese financial center for southeastern Asia. It has sugar-refineries, glass, match, and other factories. Its trade is chiefly transit, with large exports

of silk, tea, rice, camphor, etc. On Victory Peak (altitude 1809 ft.) a large town of fine residences, including a magnificent sanitarium for the European garrison, is rapidly extending.

Honolulu, cap. Hawaii (c. h. Honolulu Co.), is a port and coaling station, the principal city of the Hawaiian Islands, and the commercial metropolis of Polynesia. It has a safe harbor formed by a natural breakwater of coral reefs, pierced by a broad opening. It has a large export trade in sugar, rice, canned pineapple, coffee, hides, tallow, wool, oranges, and molasses, and extensive rice-mills, sugar-mills, shipbuilding and ice-making plants. It is a port of call for several great trans-pacific steamship lines, and has regular communication with San Francisco, Seattle, Los Angeles, Salina Cruz, Panama, and the principal Asiatic and Australian ports. The climate is mild and equitable, the extreme range of temperature being 52° to 88°, and the average, 77°, with an annual rainfall of 40 to 60 inches. The city has nearly 200 acres of public parks, a "tent" city resort on the beach, and tropical gardens, and is much frequented by tourists and health-seekers from the continental portion of the United States. The public edifices include the former royal palace, now the executive building, the judiciary and other government buildings, and an interesting museum containing many curious relics of early Hawaiian history. It is the seat of a Roman Catholic and of an Anglican bishop.

Hornell, New York (Steuben Co.), is a railroad center in a fertile agricultural region, and has manufactures of furniture, leather, wire fencing, gloves, and agricultural implements.

Hornsey, England (Middlesex), is a suburb of London.

Houston, Texas (c. h. Harris Co.), is one of three largest cities of the state, and is the highest inland point of Texas permanently accessible by water from the Gulf of Mexico, being situated on the Buffalo Bayou, 50 miles from the Galveston entrance. It has a large export trade in cotton, sugar, timber, and cottonseed-oil, and is claimed to be the largest spot-cash market in the world, barring Liverpool, besides having a large distributing trade. It is the seat of the William M. Rice Institute, a coeducational industrial or polytechnic school, and of many other educational institutions. The city was named after Gen. Sam Houston and was settled shortly after the battle of San Jacinto, which was fought April 21, 1836, within a few miles of its site. It was made the seat of government for the Republic of Texas, and so remained until 1840.

Hove, England (Sussex), is a seaside resort adjacent to Brighton.

Howrah, India (Bengal), is practically a suburb of Calcutta, situated on the opposite bank of the Hugli river. It has jute and oil mills, iron-foundries, a government engineering college, and botanical gardens.

Hubli, India (Bombay), is a center of the cotton industry.

Huddersfield, England (York), is one of the chief centers of the woollen and cloth manufacture in England. It is situated in a rich coal region, and has also large iron-foundries and manufactories of machinery. It has a technical college, an art gallery, and a spacious cloth-hall, dating from the early part of the reign of George III.

Hudson, New York (c. h. Columbia Co.), is a trade center of an agricultural section, and has manufactures of ale, machinery, foundry products, knit-goods, car-wheels, and creamery products. The State House of Refuge for Women and the State Volunteer Firemen's Home are located here. The town was settled in 1783.

Hué, cap. Anam, is a fortified city on the Hué river, 9 miles from the sea.

Hull, England (York), is a port and coaling station, with extensive docks and important coastwise and foreign trade, as well as large fishing interests. The manufacture of oil and oil-cake is a special industry; others are flour-milling, shipbuilding, engineering works, tanneries, chemical works and breweries. The city exports coal, coke, oil-cake, textiles, machinery, etc. Trinity Church is one of the greatest of English parish churches, in the Decorated and Perpendicular styles, exhibiting highly interesting tracery. Hull is the seat of an Anglican bishop. It was the birthplace of William Wilberforce.

Hull, Quebec, Canada, is an important center of lumber industries and has large pulp and paper mills. It was swept by fire in 1900, but has been rebuilt.

Huntington, West Virginia (c. h. Cabell Co.), is an important industrial center, and has car-manufacturing shops, lumber and planing mills, glass-works, meat-packing establishments, etc. It has a state normal school, a state asylum for incurables, etc.

Hutchinson, Kansas (c. h. Reno Co.), has one of the largest salt interests in the west. It is an important meat-packing and shipping center and has manufactures of lumber, machinery, boilers, etc. A state reformatory is located here.

Hyde, England (Chester), has manufactures of cottons, felt hats, engineering works and foundries.

Hyde Park, Massachusetts (Norfolk Co.), has manufactures of rubber goods, paper, morocco, and machinery.

Ibadan, Southern Nigeria Colony, Africa, is an important trading center about 90 miles from the port of Lagos, with which it is connected by railroad.

Ichang, China (Hupeh), is a treaty port on the Yangtse, 965 miles from the sea and 10 miles below the Ichang gorges. It is a transhipment port for cargoes to and from Szechuan, and steamers of light draught run to and from Hankau. It has an extensive transit trade and also a large export trade in fruit grown in the neighborhood.

Iglau, Austria (Moravia), is a very ancient town and, after Brünn, the largest of the province. It has extensive manufactures of woollens and cottons, besides glass, pottery, flour, and cigars.

Ilford, England (Essex), is a residential suburb of London, and has paper mills and photographic dry-plate works. A large insane asylum is situated here.

Independence, Kansas (c. h. Montgomery Co.), is a trade center and distributing point for a large agricultural section. It has cotton-mills, paper-mills, window-glass factories, and other industrial works. Natural gas and oil wells are numerous near the city.

Indianapolis, cap. Indiana (c. h. Marion Co.), is the largest city in the State, a great railroad center, and the seat of extensive manufactures. Among its most important industries are slaughtering and meat packing, iron work of all kinds, flouring and grist mill products, carriages and wagons, furniture, malt liquors, clothing, lumber, and lumber mill products. Indianapolis is noted for its beautiful residential quarters, and amongst its important buildings are the State Capitol, which cost $2,000,000; the Propylaeum, devoted to literary and social purposes, and owned and controlled by a stock company of women, the deaf and dumb asylum, the blind asylum, the court-house, the post-office, and the city hall. The Soldiers' and Sailors' Monument, designed by Bruno Schmitz of Berlin, is the central and most notable decorative feature of the city. It is 285 ft. high, including the bronze statue, and its base is embellished with groups of symbolic statuary and reliefs in stone and bronze. Among the leading educational institutions are the University of Indianapolis, an organisation formed in 1896 to unite several of the older institutions of the city; The John Herron Art Institute, which contains a school of art, and a collection of modern paintings; and several other collegiate institutions. Indianapolis was laid out in 1821 and chartered as a city in 1847.

Indore, India (Central India), is the capital of a native state of the same name, and has manufactures of cotton and a large trade in grain.

Innsbruck, Austria (cap. Tyrol), is beautifully situated at the foot of the Alps in the Inn valley, with a background of lofty snow-crowned summits. It has various establishments connected with the textile industry and a factory for staining glass and making mosaic. It has a university, founded in 1672, attended by upwards of 1,000 students. A colossal marble sarcophagus, erected between 1509 and 1593, in the Franciscan church to the memory of Emperor Maximilian I, is notable. The chief public buildings are the imperial

castle (1766-70), the Ambras castle, and the Tyrolese museum (1842).

Inowraslaw, Germany (Prussia), has saline springs and beds of rock salt.

Insterburg, Germany (Prussia), has an agricultural experimental station, linen and machinery factories, tanneries, and iron-foundries.

Iowa City, Iowa (c. h. Johnson Co.), is a trade center in a rich agricultural region, and is the seat of the State University, organized in 1860, having about 2,000 students, and departments of arts, science, law, medicine, dentistry, and pharmacy.

Ipswich, England (Suffolk), has extensive manufactories of agricultural implements, engineering works, chemical works, tanneries, and other industries. It has a grammar-school, founded by Elizabeth, and still contains several picturesque old buildings, such as Sparrowe's House, a fine example of 16th century architecture. It is the birthplace of Cardinal Wolsey.

Iquique, Chile (Tarapaca), is the most important northern seaport of the country, with a large export trade in nitrate of soda, borax, copper and silver ores. The mines lie on the coast cordillera, at an elevation of about 3,300 ft.

Iquitos, Peru (Loreto), is the third port of Peru in foreign commerce and the center of the rubber trade. It is on the Maranon River, and all products go via the Amazon to the Atlantic. It was founded in 1858 as a strategic outpost. The possession of this port gives Peru the unique advantage of having an outlet to both the east and west coasts of South America.

Irkutsk, Siberia (cap. Irkutsk), is a fortified town, an important station on the Trans-Siberian R.R., and has considerable export business in tea, silks, porcelain, and other Chinese products, as well as furs and ivory.

Ironton, Ohio (c. h. Lawrence Co.), is situated in a section of the country notable for its pottery clay, iron ore and bituminous coal. Its chief industrial establishments are foundries, rolling-mills, blast-furnaces, machine-shops, nail works, furniture factories, and planing-mills.

Ironwood, Michigan (Gogebic Co.), is situated in a region rich in iron ore and timber, known as the "Gogebic iron region." It is the trade center for the greater part of the mining and lumbering business of the county.

Iserlohn, Germany (Prussia), has numerous iron and steel works, also manufactures of chemicals, cutlery, and other metal goods, and calamine mines.

Ishpeming, Michigan (Marquette Co.), is an important mining center, and produces a large amount of iron ore. Gold, marble, and an excellent building-stone are found in the vicinity. The manufactures are chiefly machinery used in mining.

Ismail, Russia (Bessarabia), is a river port with a large trade in grain and fruit.

Ispahan, Persia, has manufactories of calico, armory, tiles, and it exports opium, tobacco, carpets, and rice. It is the residence of an Armenian bishop, and possesses a great mosque, built by Shah Abbas in the 17th century, when Ispahan was the capital and an important city of 600,000 inhabitants. It was sacked by the Afghans in 1722.

Ithaca, New York (c. h. Tompkins Co.), is the seat of Cornell University (coeducational), one of the leading colleges of America, organized in 1865, and endowed with funds amounting to several millions of dollars. It occupies several fine stone edifices situated on Cornell Heights and has an attendance of about 3,000 students. The city has manufactures of flour, glass, typewriters, drop forgings, calendar clocks, firearms, salt, agricultural implements, etc. The Taughannoc Fall, nine miles from the city, has a vertical plunge of 215 ft. and is said to be the highest waterfall east of the Rocky Mountains.

Ivanovo-Voznesensk, Russia (Vladimir), is an important center of the cotton and textile industry of the country. It is often called the Russian Manchester.

Ivry-sur-Seine, France (Seine), has manufactures of earthenware, glass, and chemicals; also steel-works and breweries.

Ixelles, Belgium (Brabant), is a manufacturing suburb of Brussels.

Jabalpur, India (Central Provinces), is an important trading center, and has manufactures of cottons and carpets.

Jackson, Michigan (c. h. Jackson Co.), is the center of a rich agricultural region, and an important industrial city, with manufactures of flour, paper, machine-shop products, corsets, carriages and wagons, agricultural implements, and sewer pipe. A state prison is located here.

Jackson, cap. Mississippi (c. h. Hinds Co.), is a trade center in a cotton growing section. It has extensive lumber interests, manufactories of lumber products, cottonseed-oil mills, etc. The Capitol is an imposing building.

Jackson, Tennessee (c. h. Madison Co.), is an important cotton market, and has manufactures of engines, boilers, cotton goods, lumber, cottonseed oil, agricultural implements, etc. Here are located the Southwestern Baptist University, the Memphis Conference Female Institute, and Lane University.

Jacksonville, Florida (c. h. Duval Co.), is the largest city in the state, one of the chief southern railroad centers, and a port of delivery on the St. John's River, 24 miles from its mouth. It has extensive manufactories of lumber products, carriage, saddlery, cigars, and breweries, as well as shipyards and engineering works, and is a busy shipping point for fruit, cotton, phosphates, naval stores, lumber, vegetables, etc. It is much frequented by visitors from the north on account of its dry and equitable winter climate, the mean winter temperature being 55°F.

Jacksonville, Illinois (c. h. Morgan Co.), is the seat of Illinois College, the oldest college in the state. It also has a Woman's College, a conservatory of music, a state asylum for the insane, a school for the blind, and an institution for the deaf and dumb. Among its industrial establishments are woollen mills, car shops, foundries, paper-mills, flour-mills, etc.

Jaffna, Ceylon, is the chief city of the island of Jaffna, with extensive fishing interests.

Jaipur, India (Rajputana), is the capital of a native state of the same name, and the most important commercial center of Rajputana. Muslins, cloths, and jewelry are the chief manufactures.

Jalandhar, India (Punjab), is a very ancient city, and is now the site of a British cantonment.

Jalapa, Mexico (cap. Vera Cruz), is famous for the production of jalap, which grows wild in the district.

Jamestown, New York (Chautauqua Co.), is an important trade center and summer resort on a navigable outlet of Chautauqua Lake. It has manufactures of furniture, woollen goods, metallic goods, boots and shoes, agricultural implements, and brooms; and also saw-mills, canning factories, brick-yards, and knitting works.

Janesville, Wisconsin (c. h. Rock Co.), is the trade center of an agricultural region noted for the amount and quality of tobacco raised. The chief manufactures are agricultural implements, foundry products, wagons and carriages, furniture, cotton and woollen goods, boots and shoes, fountain pens, etc. Here are located the state school for the blind, and numerous charitable and educational institutions.

Janina, Turkey (cap. Janina), has manufactures of gold ware and silk goods. From 1788 to 1818 it was the stronghold of Ali Pasha, the tyrant of Epirus. It is the seat of a Greek Orthodox archbishop.

Jaroslaw, Austria (Galicia), has manufactures of cloth, pottery, brandy, confectionery, and an active trade.

Jarrow, England (Durham), has iron ship-building, engine works, iron-foundries, rolling-mills, large paper-mills, chemical works, and other industries. The church of St. Paul, which contains a "Crucifixion" by Vandyck, formerly belonged to the Benedictine monastery, founded in the 7th century, and is famous as the scene of the labors of the Venerable Bede.

Jassy, Roumania (Moldavia), is the see of a Greek Orthodox metropolitan and of a Roman Catholic bishop,

and possesses a university. The industries are of no importance, but the town has a lively trade in grain, wine, and cattle. Down to 1859 it was the capital of Moldavia, and here in 1792 peace was concluded between Turkey and Russia. Alexander Ypsilanti began here the struggle for Greek independence in 1821.

Jaunpur, India (United Provinces), is famous for its manufacture of perfumes.

Jeffersonville, Indiana (c. h. Clark Co.), is on the Ohio River opposite Louisville, Ky., with which it is connected by railroad bridges. It contains the state penitentiary, a United States quartermaster's supply depot, and numerous industries, among which are car works, iron-foundries, steamboat yards, chain works, etc. It has excellent water-power furnished by the Ohio Falls.

Jelets, Russia (Orel), has tanning, iron-founding, and lace-making industries.

Jérémie, Haiti, is noted as the birthplace of Alexandre Dumas the elder. It exports cocoa, coffee, and logwood.

Jerez, Spain (Cadiz), is celebrated for the production and export of sherry wine.

Jersey City, New Jersey (c. h. Hudson Co.), is the second largest city of the state and the most important suburb of New York, now connected with the latter by tunnels under the Hudson River. It occupies about 5 miles of the Hudson River frontage opposite New York, and is an important railroad and transatlantic steamship terminus. It has large grain elevators, slaughtering and meat-packing industries, varnish works, and iron and steel works. Among the other manufactures are locomotives, boilers, heating apparatus, planing-mill products, lead pencils, sugar refineries, soap and perfumes, paints, and roofing materials. It was formerly called Paulus Hook, and was incorporated as the City of Jersey in 1820, and as Jersey City in 1838.

Jerusalem, Turkey in Asia (Palestine), is the ancient capital of Palestine, still regarded by the Jews as their sacred city, and as a holy city by both Christians and Mohammedans. It is the scene of the most important events described in the Bible. The city stands on a rocky plateau at an elevation of 2,500 ft. above the sea. The city proper is surrounded by a wall of hewn stone 2½ miles in circumference and was probably built by the Sultan Solyman the Magnificent. The noble Mosque of Omar, in the center of the temple area, supposed to occupy the site of Solomon's temple; the Church of the Holy Sepulchre, on the reputed site of our Lord's tomb; and the Jews' Wailing Place, are among the more interesting features. Jerusalem is the seat of Roman Catholic, Greek Orthodox, and Anglican bishops (the last since 1842), and of an Armenian patriarch. The manufactures are mostly articles of mother-of-pearl and olive wood for pilgrims and tourists. Since 1892 a narrow-gauge railway, 54 miles long, has connected Jerusalem with its port, Jaffa, or Joppa.

Jhansi, India (Gwalior), is an important railroad center. It has a fort on a rocky eminence commanding the city and the surrounding country.

Jodhpur, India (Rajputana), is the capital of the native state of the same name, and has manufactures of ivory and hardware.

Johannesburg, Transvaal, Africa, is the largest city of the country, and the center of the rich gold-mining district, Witwatersrand, South Africa. It contains many handsome government buildings, a chamber of mines, several fine hotels, clubs, etc. The great Uitlander agitation which culminated in the Transvaal war 1899-1902, centred in Johannesburg, which was occupied by Lord Roberts in 1900. Since the close of the war, important public improvements to cost $30,-000,000 have been started. These include many miles of well-paved streets, an elaborate electric car system, new waterworks, sanitation system, a factory section, and handsome suburbs. Mementos of the former Boer rule are an imposing, but dismantled, fortress which dominates the town, and the monument near Krugersdorp, which commemorates the declaration of Boer independence in 1880.

Johnstown, New York (c. h. Fulton Co.), is a manufacturing and trade center which produces a large quantity of gloves and mittens, knit goods of different kinds, leather, gelatine, grist-mill and machine-shop products, and lumber.

Johnstown, Pennsylvania (Cambria Co.), is situated in a coal and iron region, which, combined with the water-power furnished by the Conemaugh River, has made it an important manufacturing center. It has several large steel-works, one of which employs over 10,000 men, iron-works, iron-plate mills, street-car rail factory, cement-works, furniture factories, potteries, wire-works, etc. It was founded in 1791 by a German pioneer named Joseph Jahns. The disastrous Johnstown Flood, which occurred in May 31, 1889, caused the loss of 2,235 or more lives, and of property estimated at about $10,000,000.

Jokjokarta, Java (cap. Jokjokarta), is the seat of a native sultan, a vassal of the Dutch. Its principal feature is a vast walled enclosure, the citadel of the native prince.

Joliet, Illinois (c. h. Will Co.), is situated in a rich agricultural region, and there are large limestone quarries in the vicinity. It has extensive manufactures of steel, wire, tin-plate, agricultural implements, foundry and furniture products, boots and shoes, horseshoes and stoves. The state penitentiary is located here.

Jonkoping, Sweden (cap. Jönköping), is a port on Lake Wetter with a good shipping trade, and has manufactures of matches, carpets, paper, wood-pulp, tobacco, arms, and machinery. Here, in 1809, peace was concluded between Sweden and Denmark.

Joplin, Missouri (Jasper Co.), is the center of the zinc and lead fields in southwestern Missouri. It has smelting-works, foundries, machine-shops, paint-works, white lead works, flour mills, and lumber mills.

Jumet, Belgium (Hainaut), is a mining town with extensive glass works and gold mines.

Juneau, cap. Alaska, has a considerable trade in furs and blankets, and a woollen mill. On Douglas Island, nearly opposite Juneau, is the famous Treadwell Gold Mine, which has one of the largest quartz-crushing mills in the world.

Kabul, cap. Afghanistan, has an arsenal and a mint, and trades in carpets, shawls, silk, and cotton goods. Much fruit is grown in the vicinity. The Bala Hissar a former residence of the Ameer, dominates the city. Kabul was in 1879 the scene of the murder of the British envoy, Sir Louis Cavagnari. It was from Kabul that Lord Roberts set out, in August, 1880, on his memorable march to Kandahar.

Kagoshima, Japan (cap. Kagoshima), is a seaport with manufactures of Satsuma ware, arms, cottons, and cigarettes. It was bombarded by the British on Aug. 15, 1863, and was the head of the Satsuma rebellion in 1877.

Kaisarieh, Turkey in Asia (Angora), is the seat of a Greek Orthodox and a Roman Catholic bishop, and of an Armenian archbishop. It has extensive bazaars, and there are large orchards and vineyards in the vicinity.

Kaiserslautern, Germany (Bavaria), has manufactures of cottons, woollens, furniture, sewing-machines, tobacco, iron and steel, beer, and bricks, also railway works and sawmills. The French army suffered three defeats by the Prussians near here, in 1793-4.

Kalamazoo, Michigan (c. h. Kalamazoo Co.), is the center of a rich agricultural region, the chief products of which are celery, fruits, and grain. It has manufactures of paper, automobiles, patent medicines, clothing, playing cards, wagons and carriages, machinery, caskets, agricultural implements, and flour. A Baptist college, a woman's seminary, a Roman Catholic academy, and an asylum for the insane are located here.

Kalgan, China (Chili), is a fortified town, with considerable trade in tea.

Kaluga, Russia (cap. Kaluga), has manufactures of soap, agricultural machinery, bricks, and pottery; also tanneries, breweries, and iron-foundries, with leather, fur, and confectionery industries. It is the seat of a Greek Orthodox bishop.

Kamenets-Podolsk, Russia (cap. Podolia), is the

seat of a Greek Orthodox archbishop. It has manufactories of tobacco, wadding, and mineral water.

Kanagawa, Japan (Kanagawa), is a seaport, whose trade is declining, since it was superseded in 1858 as a treaty port by Yokohama.

Kandahar, Afghanistan (cap. Kandahar), is a fortified town, and the center of a fertile fruit-growing region. It has also large products of silk and felt. Traditionally founded by Alexander the Great, it was occupied by the British in 1839, and was successfully defended by General Nott in 1842. It was again entered by the British in 1879, and the following year was besieged by Ayub Khan, being relieved by Lord Roberts in August 1880, after a magnificent march from Kabul.

Kankakee, Illinois (c. h. Kankakee Co.), is an important manufacturing and trade center, situated in an agricultural region. The Kankakee River furnishes excellent water-power, and the manufactures include agricultural implements, pianos, furniture, sewing-machines, knit goods, starch, and wagons. A state hospital for the insane is located here.

Kano, Northern Nigeria, Africa, has an extensive trade in kola-nuts, salt, iron, ivory, ostrich feathers, and cattle, and manufactures of silk, cotton cloths, and leather.

Kansas City, Kansas (c. h. Wyandotte Co.), is the largest city in the state, and is closely allied with the commercial interests of the city of the same name in Missouri. Its stock-yards and packing-houses are among the largest in the country. It has railroad car shops, grain elevators, iron and steel works, flour-mills, soap and candle factories and a large trade in groceries, lumber, etc. A university, a state institution for the blind, and a college of medicine and surgery are located here.

Kansas City, Missouri (c. h. Jackson Co.), is the second city of Missouri, and an important industrial, commercial, and railway center. Over thirty different railway lines enter the city and it has an extensive wholesale trade. It has a large number of grain elevators, and is claimed to be the greatest winter wheat market in the world. Its trade in corn and oats is also important, and it has an extensive distributing trade in cereals. It has a large number of packing-houses and manufactures of agricultural implements, railroad supplies, iron, furniture, paints, linseed-oil, car-wheels, flour, etc. In its vast slaughtering and packing industries, it is closely associated with the adjoining Kansas City, in Kansas. It has a school of law, a medical college, and other important educational interests. Among the public buildings is Convention Hall, which will seat 25,000 persons. It has an extensive system of parks and boulevards.

Karachi, India (Bombay), is a port and coaling station with an excellent harbor and good railway accommodations. It has an active inland trade with Kashmir, Afghanistan, Turkestan, and Tibet. The exports include hides, tallow, oil, cotton, wheat, and tea, and the principal manufactures are carpets and silver wares.

Karikal, India (Madras; French prov. of Pondicherry), is a seaport with a large trade in rice and saffron.

Karlskrona, Sweden (cap. Blekinge), is a fortified seaport, and the chief station of the Swedish fleet. It has ship-building yards, an arsenal, a naval school, and manufactures of tobacco, hats, cloth, and matches.

Karlsruhe, Germany (cap. Baden), is an industrial center, producing railway engines and carriages, machinery, firearms and explosives, cigars, furniture, silver wares, leather, cement ware, beer, carpets, and perfumery. It possesses an interesting ducal castle (1750-82), a polytechnic school, an academy of art, and a picture gallery with canvases by many native artists.

Kasan, Russia (cap. Kasan), is a great river port, and its principal industries are tanneries, breweries, and distilleries, leather and cloth works, iron and copper foundries, tallow, soap, candle, and sugar manufactories, and naphtha refineries.

Kaschau, Hungary, is the seat of a Roman Catholic bishop, and has a remarkably fine cathedral (14th to 15th century). The museum for Upper Hungary, and mineral springs are other principal features. It has

flour-mills, and manufactures of tobacco, machinery, furniture, and textiles. One of the chief national strongholds, it was captured by the Austrians in 1848.

Kashan, Persia, manufactures silks, satins, brocades, copper ware, glazed tiles, and carpets. An earthquake in 1895 caused great destruction.

Kashgar, Turkestan, stands at the meeting-place of several ancient and important routes, and thus has considerable strategical, commercial, and social importance. It has manufactures of cotton goods, carpets, and articles of gold and jasper.

Kaslof, Russia (Tambof), has an active trade and manufactures of machinery, leather, and tallow.

Kasvin, Persia, manufactures cotton and ironware, and exports large quantities of raisins to Russia. Its breeds of camels and horses are celebrated.

Kattowitz, Germany (Prussia), has iron and other metal works, saw-mills, brick works, etc.

Kearny, New Jersey (Hudson Co.), is a suburb of Newark, with extensive linoleum works, celluloid works, and thread mills. It has a state soldiers' home and an industrial school for boys. It was named in honor of Gen. Philip Kearny, who once lived there.

Kecskemét, Hungary, has a large trade in corn, fruit, wine, tobacco, and cattle.

Keighley, England (York), has manufactures of machinery and tools, and iron and steel foundries.

Kenosha, Wisconsin (c. h. Kenosha Co.), is a port on Lake Michigan and a trade center for a rich agricultural region. It has manufactures of leather, furniture, wagons, automobiles, iron and brass goods, and flour.

Keokuk, Iowa (Lee Co.), is at a point on the Mississippi River where it falls about 21 ft. in 11 miles, furnishing great water-power, and a ship canal has been built around the rapids at a cost to the Federal government of about $8,000,000. The chief manufactures are stoves, tin cans, cereals, boots and shoes, clothing, gunpowder, lumber, canned fruits, etc. It has a medical college founded in 1849, a dental college, and many benevolent institutions.

Kerbela, Turkey in Asia, is noted for the tomb of Hussein, the son of Ali, which is a place of pilgrimage for Shiite Mohammedans, who also carry their dead there for burial. The mosque, which contains the tomb of Hussein, has its domes and minarets plated with gold. Dates and cereals are exported, and sacred bricks, and shrouds stamped with verses from the Koran are made.

Kerman, Persia, is a fortified town and the largest of southeastern Persia. It has manufactures of shawls and carpets.

Kermanshah, Persia, is an important caravan center on the road between Teheran and Bagdad. The chief exports are silk yarns, opium, raw hides, gum, carpets, and spices.

Kertch, Russia (Taurida), is a fortified seaport and watering-place. It is a very ancient city, and many valuable remains of antiquity have been discovered here.

Key West, Florida (c. h. Monroe Co.), is a port of entry and naval station on one of the coral islands called the Florida Keys. It is the farthest south of any city in the continental portion of the United States, and has an excellent harbor, at the main entrance of which is located Ft. Taylor. Its chief industries are the manufacture of cigars, sponge gathering, and deep-sea fishing. It has a large trade in fish, fruit, vegetables, turtles, salt, and tobacco. Government buildings, barracks, machine-shops, and a marine hospital are located here. It has a Methodist seminary and a Roman Catholic academy. Its pure air and mild climate have made it a popular health and winter resort.

Kharkof, Russia (cap. Kharkof), is an important industrial and commercial center of Southern Russia, with a large trade in wool and horses. The industrial establishments include woollen manufactures, dyeing and cleaning works, distilleries and breweries, iron and copper foundries, the making of machinery and agricultural implements, tobacco, candle, gingerbread, and sweetmeat manufactures. It has a university, founded in 1804, also a technological institute and a seminary for priests. Kharkof was founded in 1650 by Cossacks.

Khatmandu, India (cap. Nepal), has many Buddhist temples, but the maharaja's palace is the chief building.

Khotin, Russia (Bessarabia), is a trade center.

Kief, Russia (cap. Kief), is the chief city of Little Russia and a fortified town. The most important of its industries are sugar and tobacco, brick and machine works, distilleries, breweries, tanneries, flour-mills, and manufactures of macaroni, gingerbread, and chemicals. Among its antiquities are the "Golden Gate", a relic of the medieval fortifications; the church of St. Sophia, built in 1037, with the tomb of its founder, Prince Yaroslav; and the churches of St. Basil and of the Nativity. Still older are the tomb of Askold, the Russo-Varangian leader of the 9th century, and the old wooden church of St. Nicholas the wonder-worker; but the most famous building is the Pecherskoi monastery, said to have been founded by St. Anthony in the 9th century, with catacombs, in which many venerated Russians are buried. Every year from 200,000 to 350,000 pilgrims visit this greatest center of Russian devotion. It also contains the church of St. Vladimir, ornamented with remarkable fresco pictures by leading Russian artists. Kief has a university, founded by Nicholas I. in 1834; and also an academy of theology (1631), a cadet school, and a technical school. The city was the capital of the grand princes of Kief from the 10th century; was sacked by the Mongols in 1240; passed later to Lithuania and Poland; and was annexed to Russia in the 17th century.

Kiel, Germany (Prussia), is a strongly fortified seaport and coaling station, with a splendid harbor, situated on an inlet of the Baltic, and at the terminus of a ship-canal between the North Sea and the Baltic. It is the chief naval station of Germany, where the Government builds its war-ships, and has extensive machine-shops, a naval academy, naval hospital, and schools for officers and engineers. The Thaulow Museum, the former castle of the Dukes of Holstein-Gottorp, the provincial museum of antiquities, the university, and the ancient church of St. Nicholas, are the principal edifices. Its industries include ship-building, flour, oil, and sawmills, engineering works, and breweries.

Kilmarnock, Scotland (Ayr), manufactures tweeds, carpets, and shoes, and has engineering shops and foundries.

Kimberley, Cape Colony, Africa, is the greatest diamond-mining center in the world, its mines yielding about 98 per cent. of the total output. The chief mines are the Kimberley Mine, De Beers Mine, Bultfontein, and Du Toit's Pan. On October 15, 1899, Kimberley was besieged by the Boers, and was not relieved until Feb. 16, 1900.

King's Norton, England (Worcester), is a suburb of Birmingham, and has manufactures of paper, metal, and screw nails. The model village of Bournville, with great cocoa and chocolate works, is situated in this parish.

Kingston, cap. Jamaica, is a port and British coaling station, and the principal commercial city of the colony. Its harbor admits the largest vessels. Four miles southwest is the naval station of Port Royal, the headquarters of the British naval forces in the West Indies, near which is the site of old Port Royal, once the most flourishing English city of the New World. In 1693 it was destroyed by an earthquake. On Jan. 14, 1907, Kingston was partially destroyed in the same way.

Kingston, New York (c. h. Ulster Co.), is an important river port and is noted for the manufacture of cement. It has an extensive trade in farm produce, lime, lumber, grain, and bluestone. The Senate House, built in 1676, and the first home of the New York State Legislature, contains an interesting collection of Dutch and Revolutionary relics. Kingston was burned by the British in 1777.

Kingston, Ontario, Canada, is a port of entry at the outlet of Lake Ontario and an important trade center. It has manufactories of machinery, locomotives, cars, cotton, woodenware, etc. Ship-building is extensively carried on. It is the seat of a university.

Kingston-upon-Thames, England (Surrey), has a parish church dating from the 14th century. The old royal chapel, in which several of the Saxon kings were crowned, fell in 1730, but the coronation stone is preserved opposite the Court-House. Its benevolent institutions include the Royal Cambridge Asylum for Soldiers' Widows.

Kirin, China (Manchuria), is at the head of navigation of the Sungari River. It has a mint, and junks and boats are built here.

Kirkcaldy, Scotland (Fife), is a seaport with extensive manufactories of linoleum and linen; bleaching, engineering, iron-founding, pottery-making and brewing are other industries.

Kishenef, Russia (cap. Bessarabia), has manufactures of leather, soap and candles, and woollen stuffs. Jews form an exceptionally large element of the population, and were the victims of massacres in 1904, and again in 1905.

Kiukiang, China (Kiangsi), is a treaty port on the Yangtze River, with exports of tea and porcelain. A large sanatorium is situated in the neighboring mountains.

Kiungchau, China, constitutes with its port, Hoihow, a treaty port on the island of Hainau. Its chief exports are sugar, sesamum, grass-cloth, pigs, and poultry.

Klausenburg, Hungary, is a town founded in 1272 by Saxon colonists, and was for a long time the capital of Transylvania. It is the seat of two bishops, one of the Unitarian, the other of the Reformed Church. The chief features of the place are its old churches, its citadel (1715), and university.

Knoxville, Tennessee (c. h. Knox Co.), is the chief city of east Tennessee, and the center of a coal-mining region and of an extensive marble district. It has a large trade in country produce and manufactures of woollen and cotton goods, furniture, bar iron, boilers, foundry and machine-shop products, flour, etc. It is the seat of the University of Tennessee, the East Tennessee Female Institute, the Tennessee Normal College, the Knoxville College for colored students, and numerous other educational institutions. Knoxville was founded in 1791 and was the first capital of Tennessee. In 1863 the city and the adjoining Fort Sanders were unsuccessfully besieged by the Confederates.

Kobe, Japan (Hyogo), is a port and coaling station, with a harbor that is both deep and capacious. It also possesses an imperial ship-building yard, and ranks first amongst Japanese ports, both in number of ships and in volume of trade.

Kochi, Japan (cap. Kochi), is noted for its coral, and is the center of the Japanese paper-making industry.

Kofu, Japan (cap. Yamanashi), manufactures silk and wine. A kind of sweetmeat, consisting of grapes coated with sugar, is made in large quantities, and rock crystals are cut and polished.

Kokand, Russia (Russian Turkestan), is the seat of an extensive trade, and is developing into a modern city.

Kokomo, Indiana (c. h. Howard Co.), is a manufacturing city and a trade center, with extensive manufactories of glass, potteries, stove works, rubber works, automobile factories, and pulp and paper mills.

Kokura, Japan (Fukuoka), is a maritime town.

Kolhapur, India (Bombay), is the capital of a native state of the same name, and contains the remains of several Buddhist shrines dating from the 3rd century B. C.

Kolomea, Austria (Galicia), has petroleum, pottery, and candle industries.

Kom, Persia, contains the tomb of Fatima, sister of Imam Riza, and is much visited by pilgrims. Next to Meshed it is considered the most sacred place in Persia. Cotton is largely cultivated in the neighborhood.

Koniah, Turkey in Asia, has manufactures of carpets, woollen goods, and leather, and an active trade. It is the seat of a Greek Orthodox archbishop. From the 11th to the 14th century it was the capital of the Seljuk (Turkish) sultans.

Konigsberg, Germany (Prussia), is a fortified port, the second capital and place of residence of the kings of Prussia, and the most important town in the northeastern part of the monarchy. The royal palace was formerly a castle of the Teutonic knights, and houses the provincial supreme court, the archives, and the

museum of antiquities. In the palace chapel Frederick I. crowned himself first king of Prussia in 1701. Königsberg has a Gothic cathedral, begun in 1333, and a university, founded in 1544, in which the famous philosopher Kant was a professor. It has manufactures of iron and steel, machinery, chemicals, tobacco, cloth, linen, leather, and paper.

Königshütte, Germany (Silesia), is situated in a rich coal region, and has great iron and steel works.

Kostroma, Russia (cap. Kostroma), is an important port on the Volga, and its chief industries are flax spinning, tanning, brickmaking, distilling, and the manufacture of cloth, tobacco, and cement. The most famous building is the Ipatskoi monastery (1330), and its cathedral is among the most picturesque in Russia.

Kottbus, Germany (Prussia), has manufactures of cloth, woollens, linens, carpets, hats, and jute, as well as iron-founding, tanning, brewing, and distilling.

Kovno, Russia (cap. Kovno), is a fortified town, and an important commercial center. It has manufactures of iron, nails, lace, pottery, leather, flour and tobacco.

Krakow, Austria (Galicia), is a fortified city, and a large trade center at the head of navigation of the Vistula. It is the seat of an archbishop, and among the churches are the glorious Gothic "castle", or cathedral church, built on the site of an 11th century edifice in 1320-59. It contains monuments of Polish kings, patriots, and poets, and is adorned by some of Thorwaldsen's sculptures and other fine works of art. Scarcely less interesting is the Gothic church of St. Mary, built in 1226, restored in 1889-93, and adorned with a splendid altar-piece (1477-89) by Veit Stoss and ornate tombs of Polish magnates. The most imposing of the secular buildings is the castle, or citadel, first built in the 14th century, used since 1846 as barracks, military hospital, etc. There are also the university, founded in 1364, frequented by about 1,550 students; the bazaar known as the Cloth-House; the Museum Czartoryski, and the municipal theatre (1891-3). The educational, artistic, and other institutions of the city include the national Polish art museum (1883), the picture gallery, the university library with 320,000 volumes and 7,500 MSS., valuable for Polish literature; the academy of sciences (1873), the botanic garden, observatory, industrial and technical museum, archæological museum, and academy of art. Krakow is the center of a large trade in agricultural produce, salt, timber, cloth, and linen, and manufactures machinery, chemicals, beer, sausages, oils, and tobacco. Krakow was the capital of Poland from 1320 to about 1609, and the place of coronation of her kings till the 18th century. It came to Austria in the last partition of Poland in 1795. By the Congress of Vienna (1815) it was made the capital of the Republic of Krakow, but after the insurrection of 1846 it was again annexed to Austria.

Krasnoyarsk, Siberia (cap. Yeniseisk), has an extensive trade. It is the seat of a Greek Orthodox bishop.

Kremenchug, Russia (Poltava), is one of the principal commercial centers of Little Russia, and one of the chief river ports on the Dnieper. It has iron-foundries, tanneries, sugar-refineries, sawmills, and manufactures of agricultural implements, tobacco, carriages, and hats. Its liqueurs and preserved fruits are also famous.

Kronstadt, Hungary, is a fortified town, with an extensive trade, and manufactures of cloth, leather, cement, and candles, also petroleum refineries. It has a beautiful Protestant cathedral in the Gothic style, which dates from 1385, and several other mediæval buildings. In the 16th century it became the center of Protestantism in Transylvania.

Kuching, Borneo (cap. Sarawak), is a port, which exports rice, timber, edible birds'-nests, bees wax, pepper, canes, and camphor.

Kuka, Northern Nigeria, Africa, was formerly an important slave mart.

Kumamoto, Japan (cap. Kumamoto), is a fortified town, with an extensive trade. Outside the town is a much-frequented Buddhist temple.

Kumassi, Gold Coast Colony, Africa, has an active trade with the interior, and is situated about 180 miles from the port of Sekondi on the Gulf of Guinea, with which it is connected by railroad.

Kure, Japan (Hiroshima), is an important naval port.

Kursk, Russia (cap. Kursk), is an important trading center, and has carriage works, tobacco, soap, and wax candle manufactures, distilleries, breweries, tanneries, iron-foundries, and flour-mills.

Kurume, Japan (Fukuoka), is a trade center in a silk producing section.

Kustendji, Roumania (Dobrudsha), is the principal port of the country on the Black Sea, and has a large export trade in wheat, corn, flour, timber, wool, and petroleum.

Kutais, Russia (cap. Kutais), is a modern town built on the supposed site of Kutatision of the Argonauts. Several ancient remains have been found here.

Kyoto, Japan (cap. Kyoto), was until 1868 the capital of Japan, and is noted for its magnificent temples, monuments, and Buddhist monasteries. The Midako's, or Imperial, Palace covers an area of 26 acres. The most important industries are connected with the making of porcelain, faience, embroidery, brocades, and bronzes, while the weaving and dyeing of silks is also largely carried on. Kyoto has a university founded in 1875.

La Crosse, Wisconsin (c. h. La Crosse Co.), is an important manufacturing and trade center of western Wisconsin, and has very large lumber mills and manufactures of agricultural implements, lumber products, knit goods, flour, etc. A United States marine hospital and an asylum for the chronic insane are located here.

Laeken, Belgium (Brabant), is a suburb of Brussels, and the usual summer residence of the Belgian royal family. It is noted for its manufactures of carpets.

La Fayette, Indiana (c. h. Tippecanoe Co.), is the seat of Purdue University, and of the state college of agriculture and technology, established in 1862. With the university, which has a fine library, is connected an agricultural experiment station. The city has manufactories of boots and shoes, carpets, railway cars, woollen goods, flour, and agricultural implements. It stands on the site of the old French fort built in 1720, called Post Oniatanon. It was surrendered to the British in 1760 and, in the same year, was captured by the Indians.

Lagos, Africa (cap. Lagos), is a port and naval station that is accessible to vessels with draught of about 10 ft.; surf boats are used in loading and discharging. Exports palm-oil, nuts, cotton, ivory, gum, copal, etc.

La Guaira, Venezuela, is the chief port of Venezuela, and is one of the most picturesque ports of the world. Exports coffee, cocoa, hides, cotton, and rubber. In 1903 the port was bombarded by the British and German fleets to enforce the settlement of claims against the government of Venezuela.

Lahore, India (Punjab), is an important railway and trade center, noted for its carpets, silks, and woollen goods. It contains numerous educational institutions and notable buildings. It is the seat of an Anglican bishop. Lahore is a very ancient city, and its chief period of glory was under Akbar (1556-1605), when it became the capital of the Mogul empire. It was annexed by Britain in 1849.

Laibach, Austria (cap. Carniola), has manufactures of iron, cotton, tobacco, and pottery. The town was severely injured by an earthquake in 1895. It is the seat of a Roman Catholic bishop.

Lake Charles, Louisiana (c. h. Calcasieu Parish), is a trade center in a rice growing region, and has large rice and lumber mills, shipyards, etc.

Lancaster, England (Lancashire), is an old city, and contains several ancient buildings, such as the church of St. Mary and the castle, which was erected during the Norman period on the site of a Roman station and was begun by John of Gaunt; it serves now as county court-house and jail. Lancaster has manufactures of furniture, linoleum, and railway carriages.

Lancaster, Pennsylvania (c. h. Lancaster Co.), is a manufacturing and trade center of a densely populated section, and an important tobacco market. It has large manufactures of cigars, cotton goods, iron and steel goods, shoes, combs, etc. It is the seat of Franklin

and Marshall College. The city was founded in 1718 by Mennonites and was called Hickory Town until 1730. In 1777 Congress sat here for a few days and, from 1799 to 1812, it was the capital of the state. It became a borough in 1742 and a city in March, 1818.

Lanchau, China (Kansu), manufactures cloth and camel's-hair goods, and has a trade in silk, fur, metal and wooden articles, grain, and tea.

Landsberg, Germany (Prussia), has manufactures of machinery and furniture, sawmilling and brickmaking.

Lansing, cap. Michigan (Ingham Co.), is a farming trade center, and has extensive manufactories of automobiles, agricultural implements, flour, stoves, machinery, beet sugar, canned goods, knit goods, etc. In addition to the State Capitol, erected at a cost of $1,500,000, it has a state hospital, a state library containing 105,000 volumes, a state school for the blind, a state industrial school, a state agricultural college, etc.

Laoag, Philippine Islands (cap. Ilocos Norte), exports rice, corn, tobacco, and sugar, and has manufactures of cotton.

La Paz, Bolivia (cap. La Paz), is the largest town of Bolivia, and has an extensive trade in agricultural and mining products. It has a cathedral, a university, a military school, a seminary for priests, and a museum. The city was founded in 1548, after the struggle between the Almagros and Pizarros, and named Nuestra Senora de la Paz; but at the declaration of independence the name was changed to La Paz de Ayacucho, in memory of the decisive victory over the Spaniards. Elevation 12,307 feet.

La Plata, Argentina (cap. Buenos Aires), is a port with large exports of cattle and agricultural products. It has a university, observatory, and museum.

Laredo, Texas (c. h. Webb Co.), is the trade center for a large section of the southern part of Texas, and has extensive stock-raising interests. It is in the Rio Grande coal belt and there are valuable iron ore deposits in the vicinity. It has extensive concentrating and sampling works, brickyards, furniture factories, stockyards, grain elevators, large coal yards, etc. It is the seat of Laredo Seminary, founded in 1822.

La Rochelle, France (cap. Charente-Inférieure), is a fortified port and coaling station, which has long been connected with the Newfoundland fishing trade, and has also industries of ship-building, cotton yarns, glass, sugar-refining, and distilling. The finest building is the Hôtel de Ville (1486-1607), in the Renaissance style, but there are also a cathedral and an episcopal palace. In 1628 the town, a Huguenot stronghold, offered stubborn resistance to the royalist army, commanded by Richelieu in person.

Lasalle, Illinois (Lasalle Co.), is a trade center in a rich bituminous coal region, and is engaged in coal-mining, zinc-smelting, and the manufacture of sulphuric acid, hydraulic cement, sewer pipes, bottles, clocks, and brick. It is the seat of Saint Bede college.

Lassa, China (cap. Tibet) is celebrated as the residence of the Dalai Lama, or Grand Lama, the Pope of the Buddhist religion, and as a place of pilgrimage. It is remarkable for the number of its convents, but the chief buildings are the grand temple and the palace of the Grand Lama.

Las Vegas, New Mexico (c. h. San Miguel Co.), is an important wool market. Near by are the Las Vegas Hot Springs, a health resort with curative waters, mudbaths, etc.

Lausanne, Switzerland (cap. Vaud), has a cathedral church (dating from the 13th century), which is perhaps the finest medieval building in Switzerland. It has a university, a very interesting museum, and numerous other excellent educational institutions; and on that account is a favorite place of residence of the English. Has manufactures of machinery, tobacco, and chocolate, and large vineyards are in the neighborhood.

Laval, France (cap. Mayenne), has manufactures of "tickings", cotton goods, paper, leather, and machinery, and there are marble quarries in the neighborhood. It contains a Gothic cathedral dating from the 12th century, and an ancient ducal castle.

Lawrence, Kansas (c. h. Douglas Co.), is an impor-

tant trade and manufacturing center, with excellent water-power. It is the seat of the State University, Haskell Institute, and a United States industrial school for Indians.

Lawrence, Massachusetts (c. h. Essex Co.), is an important cotton and woollen milling center on the Merrimac River, which here has a gradual descent of 28 ft. in a distance of one-half mile, affording excellent water-power. It is estimated that the river here develops 12,000 horse-power, and that it turns more spindles than any other stream in the United States.

Leadville, Colorado (c. h. Lake Co.), is one of the most celebrated mining centers in the world. It has sampling, refining, and reduction works, smelting furnaces, etc. It also has iron-foundries, manufactures of machinery, and a government fish-hatchery. Leadville was founded in 1859 under the name of "California Gulch", and was for several years one of the richest gold-washing camps in Colorado. In 1876 the great carbonate beds of silver were discovered, and the population rose for a time to 30,000. The annual yield of silver in the Leadville mines amounts to $15,000,000, and its gold-mining has again become profitable. The total yield of its mines has been over $350,000,000. Elevation, 10,200 feet.

Leavenworth, Kansas (c. h. Leavenworth Co.), is an important railway and trade center in a farming and coal-mining region. An inexhaustible coal deposit, underlying the city at a depth of 700 ft., gives employment to over 1,000 miners and yields 60,000 bushels of coal daily. Its manufacturing industries are extensive, and include flour-mills, woollen mills, iron-foundries, manufactures of mill machinery, mine machinery, steam-engines, lumber, wagons, shoes, bicycles, patent medicines, dye-works, etc. Fort Leavenworth, one of the oldest and most important military depots on the Missouri River, and a national soldiers' home are near by. Leavenworth is also the seat of the Kansas State Orphan Asylum, a state normal school, and numerous other educational institutions.

Lebanon, Pennsylvania (c. h. Lebanon Co.), is a trade and manufacturing center in a valley noted for the fertility of its soil, but the largest part of its wealth comes from the quarries and mines in the vicinity. The Cornwall iron mines, about 5 miles distant from the city, the limestone, brownstone, brick-clay, and iron ore found in the vicinity contribute to its industrial wealth. It has many furnaces and foundries, rolling-mills, steel plants, machine-shops, large knit and boot works, chain works, etc. It was settled in 1700 by German immigrants.

Lecce, Italy (cap. Lecce), has a large trade in Lecce oil, tobacco, cotton, wool, soap, and leather, and in fruits and grain grown in the district.

Le Creusot, France (Saône-et-Loire), has one of the largest iron- and steel-works in the world. Its coal-fields are extensive, and locomotives, heavy guns, and armor-plates are extensively manufactured.

Leeds, England (York), is the chief seat of the English woollen manufacture, and an important railway center. The leading manufactures are woollen, flax, iron, machinery, clothing, caps, leather, and boots. Among its principal buildings are the town hall, Yorkshire College, which is now a university, the art gallery and museum, and the market hall. Near the city is Kirkstall Abbey, one of the best preserved monastic houses in England.

Leeuwarden, Netherlands (cap. Friesland), contains a Frisian museum, and a royal palace (1587-1747). It has manufactures of linen, musical instruments, vehicles, and glass, and large cattle and fruit markets.

Leghorn, Italy (cap. Leghorn), is a fortified seaport and coaling station, with large exports of hemp, hides, marble, olive-oil, coral, candied fruit, wine, soap, boracic acid, and hats; and imports of coal, fish, tobacco, wheat, and rawhides. Ship-building, glass-making, copper and brass-founding, are the principal industries. It is the seat of a bishop, and has a beautiful 17th century cathedral, a large naval academy, and famous lazarettos.

Leicester, England (cap. Leicester), manufactures worsted hosiery, boots and shoes, elastic webbing, and

agricultural implements. The town occupies the site of the Roman Ratæ and fragments of Roman pavement have been unearthed. Also, a part of the Jewry wall, a Roman barrier in the ancient Jewish quarter, still exists. Leicester was one of the "five burghs" of the Danes. A castle erected here early in the 12th century, on the site of an earlier fortress, became afterwards a residence of John of Gaunt. Two gateways and the great hall of the castle still remain, also ruins of an Augustinian abbey.

Leiden, Netherlands (South Holland), is the seat of a famous university, founded in 1575. In the 14th century the town was famous for its cloth and baize, which are still made here, as well as woollens, cottons, leather, soap, and salt. In 1573-4, it heroically resisted the Spaniards, being relieved by the Prince of Orange, who made an opening in one of the dikes, thus partially submerging the camps of the enemy, and gaining access to the town by boat. It is the birthplace of Rembrandt and of some of the Elsevirs, the printers.

Leigh, England (Lancashire), has extensive coal-mining, cotton and silk manufactures, brewing, agricultural implements, and foundries.

Leipzig, Germany (Saxony Kg.), is an important industrial, commercial and banking center, and is also the seat of the Reichsgericht, the highest court of justice within the German empire. Paper-making, printing, and bookbinding are among the principal industries, the city being one of the leading publishing centers of Germany. Auerbach's Keller, immortalized in Goethe's Faust, consisting of wine vaults dating from 1530, still serves as a restaurant. The churches of St. Nicholas (1017) and St. Thomas (1222) are the oldest. The famous university, founded in 1409, is attended by 3,600 students. Leipzig suffered severely during the Thirty Years', Seven Years', and the Napoleonic wars. The most celebrated of the battles fought here was that between the French, under Napoleon, and the allied armies of Austrians, Russians, Prussians, and Swedes (Oct. 16-19, 1813), the result of which effectually shattered Napoleon's power.

Leith, Scotland (Edinburgh), is a port and coaling-station on the Firth of Forth, 2 miles north of Edinburgh.

Le Mans, France (cap. Sarthe), manufactures iron products, machinery, clocks and watches, linen goods, and chemicals, especially sulphuric acid. The cathedral contains the tomb of Berengaria, queen of Richard Cœur de Lion. The town was the birthplace of Henry II. of England.

Lemberg, Austria (cap. Galicia), is the seat of an archbishop of the Roman Catholic, the Greek Catholic, and the Armenian churches. The chief buildings are the cathedrals; provincial house of assembly; university, founded in 1784; Ossolinski National Institute; and a polytechnic (1873-7). Machinery, beer, leather, matches, and candles are manufactured.

Lens, France (Pas-de-Calais), is situated in an important coal-field, and has iron and steel works, also sugar and soap factories.

Leominster, Massachusetts (Worcester Co.), is a busy industrial center, with extensive manufactures of combs, pianos, paper, cement, jewelry, toys, buttons, cabinet work, etc. It was settled in 1725.

León, Mexico (Guanajuato), has copper and silver mines in the neighborhood, and manufactures leather.

León, Nicaragua, is the largest town of Nicaragua and the seat of a university. It was at one time the capital of Nicaragua. Its public edifices are considered the finest in Central America, and it has manufactories of dressed leather and cutlery.

Lethbridge, Alberta, Canada, is an important railroad and coal mining center with large collieries nearby. Live stock, wool, flax and sugar beets are produced in the neighborhood.

Lewiston, Maine (Androscoggin Co.), is a trade and manufacturing center, with extensive water-power furnished by the Androscoggin River, which here has a fall of about 50 ft., and the power of which is utilized by means of a distributing dam and canal. The chief industries are cotton and woollen goods, foundry products, engines, boilers, lumber, boots and shoes, etc. It

also has dye-works and bleacheries. It is the seat of Bates College, which includes the Cobb Divinity School.

Lexington, Kentucky (c. h. Fayette Co.), is a trade center in the famous "blue-grass" region and is noted for the stock-farms in its vicinity. The chief manufactures are Bourbon whiskey, harnesses, saddlery, flour, canned goods, lumber, and carriages. It is the seat of Kentucky University, a state agricultural and mechanical college, a state reform school, and numerous other educational institutions. The state asylum for the insane is located here. Lexington was settled in 1775, and in 1792, when Kentucky became independent of Virginia, was made the capital of Kentucky.

Leyton, England (Essex), is a suburb of London and has Roman antiquities.

Libau, Russia (Courland), is a port with an artificial harbor and an important railway terminus. Its chief exports are grain, flax, hemp, linseed, petroleum, fish, meat, wool, leather, and skins. There are manufactures of rope, matches, agricultural implements, furniture, amber, and soap It has naval dockyards, and a large meat-freezing establishment.

Lichtenberg, Germany (Prussia), is a suburb of Berlin.

Lieben, Austria (Bohemia), is a suburb of Prague and has manufactures of chemicals, machinery, sugar, and beer.

Liège, Belgium (cap. Liège), is an important industrial town, with manufactures of machinery, tools, bicycles, railway material, firearms, cannon, linen, woollens, leather, zinc goods, sugar, beer, and spirits. The law courts built in 1505-40, were long the residence of the prince-bishops. The city is famous for its university, attended by over 1,000 students. Liège fell before Marlborough in 1702. It is defended by a ring of modern forts. It is the seat of a bishop and has several ancient churches, some of them dating from the 10th century.

Liegnitz, Germany (Prussia), has manufactures of cloth, machinery, pianos, shoes, tobacco, oil, woollens, and pottery. The town, which dates from the 11th century, was the residence of the dukes of Lower Silesia after 1163. Near the town, Silesians, Poles, and Teutonic Knights fought a fierce engagement with the Mongols in 1241.

Lima, Ohio (c. h. Allen Co.), is an important trade center in the natural gas and petroleum belt of western Ohio, the oil fields of which extend into six counties. It has manufactures of locomotives and railway cars, machine-shops, petroleum refineries, etc. It is the seat of Lima College.

Lima, cap. Peru, is the commercial center of the country. Its manufactures include pottery, iron, copper, and furniture. More than half of the inhabitants are Indians, half-breeds, negroes, and Chinese. The cathedral is the most imposing building. Overthrown by the earthquake of 1746, which destroyed the greater part of the city, it has only recently been completely restored. The University of San Marcos was founded in 1551 and claims to be the oldest in America. There are also numerous handsome public buildings and the city preserves the air of the old vice-regal days better than any other city of South America. It has many pleasant suburbs and seaside resorts. Lima, a corruption of Rimac, was founded in 1535 by Pizarro.

Limerick, Ireland (cap. Limerick), has, among its principal buildings, the cathedral of St. Mary, a Gothic edifice founded in the 12th century; St. Mainchin's; and the castle built by King John, a fine example of Norman architecture. The "treaty stone" is preserved on a pedestal beside Thomond Bridge. Bacon-curing, flour-milling, and the manufacture of army and police clothing are the principal industries.

Limoges, France (cap. Haute-Vienne), was in the middle-ages celebrated for its enamel work and is now the principal seat of the porcelain manufacture, employing over 6,000 hands. Other manufactures are cloths and druggets, nails, knives, gloves, and paper. The cathedral of St. Etienne dates from 1273. There are also remains of a Roman fountain and amphitheatre.

Limon, Costa Rica, is the chief port of the country

on the Caribbean Sea and the eastern terminus of the Costa Rica Railway. It exports coffee, tropical fruits, rubber, and dyewood.

Linares, Spain (Jaén), is the center of a great silver-lead mining district, which produces over 80,000 tons of silver and copper ore annually. Manufactures sheet lead, pipes, dynamite, and rope.

Lincoln, England (cap. Lincoln), has extensive manu-factures of agricultural implements and holds annually important horse and cattle fairs. Lincoln Cathedral is a noble edifice, doubly cruciform, with central tower (271 ft.) and two western towers. The castle was erected by William the Conqueror; its principal remains are the gateway and two towers. Other buildings of special interest are the Stone Bow, a 15th century town gate; the High Bridge, with houses thereon; the site of John of Gaunt's Palace; St. Mary's Conduit (16th century); and the Jews' House (12th century). Lincoln was an important Roman station and colony.

Lincoln, cap. Nebraska (c. h. Lancaster Co.), is an important railroad and trade center. It is the seat of the State University and Agricultural College, with 2,500 students; of the Wesleyan University, with 650 students, situated in University Place, a suburb of Lincoln; of Union College (Seventh Day Adventists); and of Cotner University. It contains a penitentiary and an insane asylum. It has extensive manufactures of horse-collars, harness, oils and paints, also railway-car and repair shops, and carries on a large trade in dairy and farm products.

Linden, Germany (Hanover), has manufactures of machinery, velvets, woollen goods, carpets, chemicals, sugar, rubber, and artificial manures.

Linz, Austria (cap. Upper Austria), is a river port, and has boat building yards, cotton works, breweries, tobacco works, and carpet mills. The principal build-ings are the Museum Francisco Carolinum, the house of the provincial diet, and a 17th century cathedral. It is the seat of a bishop.

Lipa, Philippine Islands (Batangas), is the trade center of a region producing sugar, cacao, tobacco, and indigo.

Lisbon, cap. Portugal (cap. Estremadura), is a port and coaling station, with a large trade, especially with Great Britain and Brazil. The celebrated aqueduct of the Aguas Livres, finished in 1749, crosses the valley of Alcantara on a bridge of 35 pointed arches, the largest 204 ft. high, with a span of 95 ft. The cathedral was originally a fine Romanesque building, but has been disfigured by earthquakes and modernization. The royal palace of Ajuda is a large building in a commanding situation above the Tagus, with a library considered the finest in Portugal. Lisbon was an ancient Roman city; was captured by the Saracens about 716; was taken from them by Alfonso I. in 1147; was made the capital in 1422; was in its most flourishing state about 1500; was occupied by the Spaniards 1580-1640; was nearly destroyed by an earthquake Nov. 1, 1755 (with a loss of about 40,000 lives); and was held by the French 1807-08.

Lisle, France (cap. Nord), is a first-class fortress, and is one of the chief industrial towns of France, spe-cially noted for its textile factories, in which linen, cottons, velvets, ribbons, and woollen goods are pro-duced. There are also sugar, soap, and tobacco fac-tories, dye-works, chemical works, and distilleries, while large bleach-fields are found in the outskirts. Among its buildings may be noted the citadel (designed by Vauban) in the northwest, the church of Notre Dame de la Treille, the town hall, the bourse, and the museum, which contains exceptionally rich art collections. Lisle was taken by the Duke of Marlborough (1708). In 1792 it successfully withstood a terrible bombardment by the Austrians.

Little Falls, New York (Herkimer Co.), has exten-sive manufactures of knit goods, paper, carriages, bicy-cles, leather, knitting machinery, foundry and creamery products, etc. The Mohawk River furnishes abundant water-power, since it falls here about 44 ft. in less than a mile. It was settled about 1770, and in 1782 the settlement was destroyed by Indians and tories. No

successful efforts were made to rebuild it until 1790, when a colony of Germans took possession.

Little Rock, cap. Arkansas (c. h. Pulaski Co.), is the largest city of the state, and has extensive manufac-tories of cotton, cottonseed-oil, foundry and machine-shop products, railway shops, etc. In addition to the State Capitol, it has the Arkansas Military Academy, the Philander Smith School for colored youths, a state school for the blind, a lunatic asylum, and the state penitentiary. The University of Arkansas has branch schools of law and medicine located here.

Liverpool, England (Lancashire), is a port and coaling station on the Mersey, 3 miles from the Irish Sea. It is the second city of England, one of the most impor-tant ports of the world, the terminus of many trans-atlantic steamship lines, and has extensive ship-build-ing, manufactures of rope, sugar, iron, chemicals, etc. It has a very extensive export trade in cotton, iron and steel manufactures, woollens, and machinery, and is the chief raw cotton importing port of Great Britain. Its harbor has 36 miles of quayage, and a total dock area of 1,614 acres. It is the seat of an Anglican and of a Roman Catholic bishop. At the head of the numerous educational institutions stands the University, founded in 1903, which contains besides all the usual depart-ments of a complete university, a school of tropical medicine. Among the city's architectural features the first place is given to St. George's Hall. This modern basilica is of the Corinthian order of architecture, and possesses a dignity, refinement, and style which makes it one of the finest buildings of the classical Renaissance. The town hall is also in the classic style, as are most of the public buildings of the city. The municipal offices form imposing buildings in the Palladian style. Among other important municipal buildings are the library and museum, erected by Sir William Brown; the Picton Reading Room, an extension of the library, erected by the city; the Walker Art Gallery, erected by Sir Andrew B. Walker; the technical schools; and the Exchange Buildings, in the style of the Flemish Renaissance. Liverpool received a charter from King John in 1207; was incorporated in 1229; and was taken by Prince Rupert in 1644. The commencement of its prosperity dates from the last half of the 17th century. It was largely engaged in the African slave-trade and in smug-gling. It developed greatly in the 18th and still more in the 19th century. Liverpool was the birthplace of W. E. Gladstone, Mrs. Hemans, and Mrs. Oliphant.

Lockport, New York (c. h. Niagara Co.), is an im-portant trade and manufacturing center, with exten-sive water-power obtained from the Erie Canal, which has here 10 large locks making a descent of 66 ft. There are large limestone and sandstone quarries in the vicinity.

Lodz, Russia (Poland), is one of the largest industrial towns in Russia, and its population has grown faster than any other town in the empire. The chief industry is cotton; after this come silk, wool, linen, cloth, flour, beer, spirits, and iron. There are also extensive dye-works, flour mills, and agricultural implement manu-factures. It has several good technical schools. The population is mainly composed of Poles, Germans, and Jews, each amounting to nearly a third of the whole.

Logansport, Indiana (c. h. Cass Co.), is an impor-tant trade center in a fertile agricultural region. It has extensive manufactories of automobiles, car-trucks, motors, baskets, brooms, flour, lumber products, wagons, carriages, lime and cement. The Northern Indiana Hospital for the Insane is located here.

London, cap. England, is the largest city in the world, the seat of government of the British Empire, a port and coaling station on the Thames about 50 miles from the North Sea. Foremost among its many structures and institutions are St. Paul's Cathedral, designed by Sir Christopher Wren, erected in 1675-1710, and measur-ing 500 ft. in length and 364 ft. in height to the top of the cross on its vast dome; Westminster Abbey, the Walhalla of the English nation, dating in its present form (in part) from the later half of the 13th century, and measuring 513 ft. in length, one of the most famous churches of the world; Buckingham Palace, the resi-dence of the King of England; St. James's Palace, the

residence of the Prince of Wales; the Houses of Parliament, a vast Gothic pile dating from 1840, which extends along the left bank of the Thames for a distance of 940 ft., and whose square Victoria Tower is 340 ft. high; the Mansion House, the official residence of the Lord Mayor; the Guildhall, or Council Hall of the City; the New Law Courts, a magnificent Gothic edifice with a frontage of 500 ft., opened in 1882; Lambeth Palace, the seat of the Archbishop of Canterbury; the British Museum, with a vast collection of Egyptian, Assyrian, Greek, and Roman antiquities, the famous Elgin Marbles, British prehistoric remains, an unrivalled collection of original drawings, engravings, and etchings, and a library of upward of 2,000,000 volumes; the Natural History Museum, containing the natural history collections properly belonging to the British Museum, and occupying a building that is the largest in the world devoted to collections of its class; the South Kensington Museum (or Victoria and Albert Museum), containing magnificent collections of ornamental or applied art, oriental and other collections, the National Art Library, etc.; the National Gallery, one of the greatest galleries of paintings in the world; the Bank of England, founded in 1694 and occupying a building which covers about 4 acres,—a huge but dreary edifice, in whose vaults are usually housed twenty million pounds sterling in gold and silver; and the Tower, dating from the time of William the Conqueror, historically the most interesting structure of all England,—at first a royal palace and stronghold, and afterwards a gloomy dungeon, or state prison, of London, and now containing the crown jewels and a large collection of old armor. The most famous monument of the city is the Albert Memorial, erected to the memory of Albert, Prince Consort, and adorned with reliefs in marble of 178 figures. Among its institutions of learning may be mentioned the University of London, founded as an examining body in 1836, and reorganized in 1900-01 into a body which, in addition to its examining functions, has control of the higher education in the metropolis, the principal institutions, including most of the medical schools, being subject to it; University College, opened in 1828; King's College, opened in 1831; Royal College of Science, founded as the Royal School of Mines in association with the Geological Survey of Great Britain; Central Technical College; Gresham College; St. Bartholomew's Hospital (founded in 1123) and College; London Hospital Medical College; Guy's Hospital Medical School; Royal College of Surgeons (with the famous Hunterian Museum); the schools of art of the Royal Academy and South Kensington Museum; the Inns of Court; the Imperial Institute of the United Kingdom, the Colonies and India, with a huge building erected in 1887-93 at South Kensington; British Academy (founded in 1901); British Association for the Advancement of Science; Institute of Architects; Institute of Civil and Mechanical Engineers; Royal Geographical Society; Royal Asiatic Society; the Geological, Zoological, Botanical, Linnean, Astronomical, Archæological, Chemical, and Physical societies; Society of British Artists; Society of Painters in Water Colors; and Society of Arts. Other institutions are denominational colleges for theology, sometimes combined with general education; the Royal Naval College at Greenwich; the Royal Military Academy at Woolwich; the Royal College of Science; Royal Academy of Music; Royal College of Music; Trinity College, chiefly for music; several colleges for women, etc. Among the grammar and secondary schools are: St. Paul's School, founded in 1509, which provides a free education for 153 boys, with scholarships to Oxford and Cambridge; the Merchant Tailors' or Charterhouse School; Westminster School, founded by Queen Elizabeth in 1560; University College School; King's College School; City of London School; Mercer's School; and schools of the several city companies. Christ's Hospital School, founded by Edward VI. and commonly known as the Blue Coat School, from the uniform of its scholars, has been removed to Sussex. In Regent's Park there is a most famous zoological garden, and Hyde Park and Kensington Gardens, are of international fame. The Kew Botanic Gardens, with immense greenhouses, mu-

seums and the like, and Hampton Court are among the many park spaces on the Thames. The manufacturing industries of London are on a vast scale. It contains the largest breweries, distilleries, and sugar-refineries in the kingdom; was long the principal seat of silk-weaving; has extensive manufactures in metal, including machinery of all kinds, plate, jewelry, watches, and brass work, and an enormous production of books and prints. Millinery and the making of clothes, also of boots and shoes, are extensive branches of industry. Besides these, there are cabinet-making, coopering, coach-building, rope-making, mast-making, and much ship-building. London appears to have been resettled by the Romans about 43 A. D., and Londinium (called also Augusta) was the capital of Britannia in the last part of the Roman period. After the departure of the Romans (about 412) and in the early Saxon period, its history is obscure, though there were bishops of London from the 7th century. It was plundered by the Danes and rebuilt by Alfred and Athelstane. It received a charter from William I. and many privileges from Henry I. By the 14th century its commerce had greatly developed. The insurrection of Wat Tyler occurred in 1381. London sided with the Yorkists in the Wars of the Roses, and with the Parliamentarians in the Civil War. It was scourged by the plague in 1665, and was almost entirely destroyed by the great fire of 1666. A financial panic occurred in 1720, and the "No Popery" riots in 1780. London is the seat of an Anglican bishop and of a Roman Catholic archbishop.

London, Ontario, Canada, is the center of a rich agricultural district and has iron-foundries, chemical works, and extensive petroleum refineries. The sulphur springs in the neighborhood are a favorite resort. It is the seat of the Western University, of Huron and Hellmuth colleges, and of other collegiate institutions.

Londonderry, Ireland (cap. Londonderry), is the seat of a bishop, and has a cathedral, built in the 17th century. Shirt-making is the principal industry, and there are distilleries, foundries, tanneries, and ship-building yards. In 1688-9 took place the memorable siege by the forces of James II.

Long Beach, California (Los Angeles Co.), is a residential suburb of Los Angeles and a resort on San Pedro Bay.

Long Branch, New Jersey (Monmouth Co.), is one of the older summer resorts of the Atlantic Coast visited by many thousands of people during the hot summer months. The avenue along the bluff is a favorite walk, and on the beach there are excellent bathing facilities. It was settled about 1607.

Longton, England (Stafford), manufactures china and earthenware, and there are valuable coal and iron mines in the vicinity.

Lorain, Ohio (Lorain Co.), is a trade center in an agricultural and natural gas region, and a shipping port for central Ohio coal, iron ore, and lumber. Its chief industries are ship-building, coal-shipping, steel manufacturing, and fishing.

Lorca, Spain (Murcia), was the scene of much strife between Moors and Christians in the 12th and 13th centuries. Wine is produced, and there are some lead mines in the neighborhood.

Lorient, France (Morbihan), is a seaport and fortified naval arsenal. The dockyard is used for the construction and equipment of men-of-war, and is one of the finest in France. The exports are wheat, wines, sardines, etc. The town was founded in 1670 by the French East India Company, and on the dissolution of the company in 1770 it became a naval station.

Los Angeles, California (c. h. Los Angeles Co.), is the largest city of Southern California, the second city of the state, and one of the greatest winter resorts in the world. Its phenomenal growth is largely due to the influx of tourists, health-seekers, and others from the eastern states. It has a semi-tropical climate, remarkable for its squability and dryness, with little frost, no snow, and moderate winter rains. It is an important railroad and banking center, and has a large wholesale and distributing trade with southern California, Arizona, New Mexico, and old Mexico. The manufac-

turing industries have been greatly stimulated by the supply of petroleum fuel, also by the bringing of electricity from the mountain streams, a distance of over 100 miles. It has a large transhipment trade in oranges and other citrus fruits. It was founded by the Spaniards in 1781 and passed into American possession in 1846. It was, however, of no great importance till after 1880, when it underwent an almost unprecedentedly rapid increase in wealth and population. Its population rose from 11,183 in 1880 to 50,395 in 1890, to 102,479 in 1900, and to 319,198 in 1910.

Louisville, Kentucky (c. h. Jefferson Co.), is the largest city of the state, and one of the most important manufacturing and jobbing centers of the south. It has extensive manufactures of whiskey, cement, wagons, agricultural implements, leather, etc. It has large pork-packing establishments, and is the largest market in the world for Bourbon whiskey and leaf tobacco. It is also an important educational center, being the seat of the University of Louisville (800 students) with its law and medical departments, a manual training school, a normal school, several medical colleges, theological seminaries, and other educational institutions. It is the seat of the State Blind Asylum, containing the American Printing House for the Blind. Louisville was founded in 1778 and named in honor of Louis XVI. of France. In March, 1890, this city was visited by a terrific tornado, which destroyed property valued at over $3,000,000 and killed 76 persons.

Louvain, Belgium (Brabant), is famous for its university, founded in 1426, now attended by some 1,600 students. Its town hall (one of the finest on the Continent) and the church of St. Gertrude, both of the 15th century, are the chief features of the town. It manufactures beer, lace, starch, and tobacco.

Lowell, Massachusetts (c. h. Middlesex Co.), is noted for its great number of manufactories and its annual output of manufactured articles. Its power is obtained from the falls of the Merrimac, which here has a descent of 32 ft., and from the Concord River, but considerable steam-power is also used. It has a canal system, completed in 1825, through which water-power is furnished to manufactories in Lowell and then returned to the Merrimac to turn the wheels of mills in cities further east. Its woollen and cotton factories, hosiery and knitting mills, carpet and felt factories, bleacheries, dye-works, machine-shops, patent medicine works, and furniture factories are among the leading industries. It has an old ladies' home.

Lubeck, Germany (Lübeck), is a free city, one of the three remaining Hansa towns, a port on the Trave River, 10 miles from its mouth. Of its numerous churches the Marienkirche, founded in 1170, contains valuable works of art, and its dome, enlarged during the 13th century, possesses an altar painting by Hans Memling. The town hall (1250) is built of black glazed bricks in the style of the Renaissance period. The principal shipping trade is with Denmark, Sweden, Russia, and Finland, chiefly in machinery, chemicals, preserved food, linen goods, and cigars. Lübeck was made a free imperial city in 1226, and became the leader of the Hanseatic league in 1241. In 1866 it joined the North German Confederation, and in 1870 became, with its adjacent territory, one of the states of the new empire.

Lublin, Russia (Poland), has considerable trade, especially in corn, wine, and linen cloth. Distilleries, breweries, tanneries, brick works, soap, tobacco, candle manufactures, and flour-mills are the main industries. It has a 13th century cathedral and a palace of the Polish king John Sobieski.

Lucca, cap. Italy (Lucca), is an ancient city and has several interesting buildings. Its cathedral of San Martino (11th century) is rich in paintings and sculptures, and many of its churches, which are fine examples of mediæval architecture, are built of Carrara marble. The ducal palace possesses a fine picture gallery. It is the seat of an archbishop. The town has extensive silk-mills; jute, velvets, tobacco, and cottons are also manufactured.

Lucerne, Switzerland (cap. Lucerne), is picturesquely situated on the shores of the Lake of Lucerne, and is now the chief center of foreign tourists in summer. To the east rises the celebrated view of the Rigi, and to the southwest that of Pilatus. The main features of interest in the town are mediæval towers and walls; the 5 bridges, including the covered wooden bridge, with its paintings representing scenes from the lives of patron saints and a "Dance of Death"; the Quai National and the Schweizerhof Quai, the latter with a fine avenue of chestnut trees; the Hofkirche, erected in 1506; the town hall, containing antiquarian and art collections; and the "Lion of Lucerne," a rock monument chiseled by Thorwaldsen, commemorating the heroic defence of Louis XVI.'s Swiss Guards during the attack on the Tuileries (Aug. 10, 1792). The glacier garden exhibits remarkable pot-holes.

Lucknow, India (United Provinces), is an important commercial and railroad center, with manufactures of muslins and shawls, gold and silver embroidery, glass and pottery ware. Its chief architectural features are the fort, the Imambara, or mausoleum of Asaf-ud-Daula, and the Jama Masjid. It is an important educational center, having, besides the Canning and Martinière Colleges, numerous missionary schools. Lucknow was defended (at first under Sir Henry Lawrence) against the Indian mutineers, July-Sept., 1857; was relieved by Havelock, Sept. 26; again relieved by Campbell, Nov. 16; and finally captured by Campbell, March, 1858.

Ludwigshafen, Germany (Bavaria), has very extensive manufactures of aniline dyes and soda, and considerable trade in timber, iron, and coal.

Luneburg, Germany (Hanover), has several historic churches and public buildings dating from the 14th and 15th centuries. Its manufactures include chemicals, ironware, and carpets.

Luton, England (Bedford), has brass and iron works, cocoa works, and manufactures of felt hats and straw plait.

Luxemburg, cap. Luxemburg, has manufactures of gloves, pottery, vinegar, and machinery and large tanning industries.

Lynchburg, Virginia (Campbell Co.), is an important trade and manufacturing center, and has a large trade in tobacco, both raw and manufactured. It has manufactures of iron and brass products, cotton goods, agricultural implements, shoes, dyes, hardware, lumber, and flour. There are granite quarries near the town and large coal-fields and iron ore in Campbell and the neighboring counties. It is the seat of a woman's college and other educational institutions.

Lynn, Massachusetts (Essex Co.), is noted for its shoe factories, which employ over 25,000 people, and for its large electrical works. It also has manufactures of cut leather, shoe machinery, electrical supplies, lamps, morocco, patent medicines, etc. It was settled in 1629 and incorporated in 1630.

Lyon, France (cap. Rhône), is the third city of the country, and is a strongly fortified town, a leading industrial and commercial center, and an important railway and waterway junction. The silkworms reared in the Rhone valley and the proximity of coal and iron, at St. Etienne, have made Lyon the first town in France in silk manufacture, an industry first established here in 1450 by Italian refugees. Other manufactures are those of cotton, hardware, dyes, sulphuric acid, chemical manure, starch, candles, soap, paper-hangings, and machinery, and much business is done in chestnuts, cheese, and wine. Lyon is adorned with numerous interesting and beautiful buildings, among which may be mentioned the cathedral of St. Jean (12th to 15th century); the archiepiscopal palace (15th century); the church of St. Martin d'Ainay, which is built on the site of a Roman temple, and contains the remains of a Roman votive altar; the City Hall (1646); and several extensive and admirably arranged picture-galleries and museums, in addition to a state university ranking next to that of Paris in the number of its students (2,500), and a public library, which boasts some of the earliest specimens of printing. Lyon is the seat of a bishop. It was founded by Greeks in 560 B. C.; was greatly developed under the Roman emperor Augustus; was the capital of the first Burgundian kingdom (478),

afterward passing to the Franks; was plundered by the Saracens in the 8th century; and was united to France at the beginning of the 11th century. Two important councils were held here in 1245 and in 1247.

McAlester, Oklahoma (c. h. Pittsburg Co.), is a trade center in a rich agricultural region. Coal-mining is the leading industry.

McKeesport, Pennsylvania (Allegheny Co.), is a trade center, originally noted for its fields of bituminous coal and natural gas. It has extensive manufactories of steel and iron, and tube works which employ 12,000 men; also railroad shops, glass-works, locomotive works, etc. It has a large trade in coal and lumber.

Maastricht, Netherlands (cap. Limburg), is an ancient city, and contains the church of St. Servatius, founded in the 6th century. South of the town are the tufa quarries of Petersberg, worked since Roman times; they have yielded many fossils, including saurians. Glass, pottery, and carpets are manufactured. Maastricht was, till 1878, one of the strongest fortresses in Europe, and suffered many sieges.

Macao, China, is a port belonging to Portugal, situated on an island at the mouth of the Canton River. Exports tobacco, preserves, and essential oils, and has a transit trade in tea, silks, etc. The old Collegiate church of the Jesuits is the most striking architectural feature. The town is notorious for its gambling-houses.

Macclesfield, England (Cheshire), is the chief silk-manufacturing center in England, producing brocades, plain and fancy silks and satins, ribbons, gimps, and fringes. Silk-throwing is an important branch of the industry.

Maceió, Brazil (cap. Alagoas), is a seaport with extensive exports in cotton, sugar, cotton-seed, and hides; and manufactures of cottons and sugar.

Macon, Georgia (c. h. Bibb Co.), is an important trade and manufacturing center on the Ocmulgee River, which has a fall of about 90 ft. seven miles above the city, furnishing immense power. It is a railroad center and lies in the midst of a fast developing cotton district. It is one of the most important inland cotton markets in the United States, and is at the edge of the Georgia peach belt. It has extensive cotton manufactories, and yarn, duck, cordage, twine, hosiery, and knit underwear works. It also has extensive manufactures of cotton-seed-oil, cottonseed meal and cake for live-stock, agricultural implements, brick, and fertilizers. Here also are railroad shops, lumber-mills, etc. Near by is a large vein of kaolin.

Madison, Wisconsin (c. h. Dane Co.), is situated in a rich agricultural region, and is the seat of the University of Wisconsin (4,000 students). Its observatory is one of the best in America, and its departments of history, economics, geology, agriculture, and engineering are well equipped and have a national reputation. In addition to the State Capitol, containing the State Law Library of 40,000 volumes, another building of note is that of the Wisconsin Historical Society, the most important institution of the kind west of the Alleghanies, possessing a reference library of nearly 250,000 volumes and an historical and ethnological museum. A soldiers' orphan home is located here, and a state fish-hatchery is near the city. The chief manufactures are boots and shoes, agricultural implements, tools, flour, electrical machinery, wagons, and carriages.

Madras, India (cap. Madras), is a port and coaling station, where loading and discharging is done largely by lighter; but nevertheless it is one of the foremost seaports of India. It exports cotton, sugar, indigo, rice, hides, cocoanut-oil, oil-seeds, pepper, etc. It has a number of handsome edifices, such as the Government House, St. George's cathedral, the Scottish church of St. Andrew, the university buildings, the new law courts, Patcheappah's Hall, Memorial Hall, and the Chepauk Palace. There is a large arsenal. Besides the university, an examining institution, Madras possesses Presidency College, schools of law, medicine, agriculture, engineering, and art, Victoria Technical Institute, a teachers' college, the Madras Christian College, a botanical garden, and a valuable museum. Madras is the seat of an Anglican and of a Roman Catholic bishop. It

manufactures cottons, known as "Madrases", and has dye-works, tanneries, potteries, glass-works, etc. In 1504 the Portuguese founded here the town of St. Thomé. In 1639 the rajah of Chandragiri granted to the East India Company a site close to St. Thomé, where the town of Madras was founded.

Madrid, cap. Spain (cap. Madrid), has a university, a normal school, veterinary school, schools of commerce, engineering, architecture, music, and the fine arts. The National Library contains 600,000 volumes and 30,000 manuscripts; while the National Museum of Painting and Sculpture has a fine collection of the works of Raphael, Titian, and Rubens, in addition to the most famous productions of Velasques. The Church of San Francisco, finished in 1784, is a great rotunda, with a dome 163 ft. high, an apse, and three domed chapels radially arranged on each side. The interior is remarkable for its spaciousness, and for its profuse decorations with sculpture and painting by modern masters. The royal palace, begun in 1737, is imposing from its great size and its fine situation on a lofty terrace. In the royal armory is a unique collection of splendid mediæval and Renaissance armor, arms, banners, and trappings, a large proportion of which was actually used by some of the most famous personages in Spanish history. Industrially it is not very active, tobacco being the chief manufacture. Madrid was a Moorish outpost; was taken from the Moors in 1083; became the favorite residence of Charles V.; and was made the capital by Philip II. in 1560.

Madura, India (Madras), was for centuries the religious and political capital of South India, and contains some of the finest extant examples of Hindu architecture, including the granite temple of Minarchi, or the Fish Mother. Rebuilt in the 2nd or 3rd century, it was nearly destroyed during the Mohammedan conquest of the 14th century. Its present splendor is due to Tirumulla Nayak (1623-59). The town has coffee and cotton-mills and cigar factories. Brass ware and dyed cotton cloth are also made. Madura is the center of the American missionary effort in South India.

Magdeburg, Germany (Prussia), is one of the leading commercial centers of Germany and a strongly fortified city. Its industries comprise ship-building, and the construction of engines, machinery, armor-plate, and ordnance. The principal articles of commerce are sugar, chicory, and tobacco. It has a 13th century cathedral and a municipal museum of industries.

Mahanoy City, Pennsylvania (Schuylkill Co.), is a trade center in the anthracite coal region, and in the vicinity are also found fire-clay and an excellent building stone. The chief manufactures are pottery, foundry products, flour, hosiery, and lumber.

Mährisch-Ostrau, Austria (Moravia), has iron foundries, rolling-mills, chemical works, oil refineries, etc. There are coal mines in the vicinity.

Maidstone, England (cap. Kent), has among its principal buildings the mediæval church of All Saints, the College of All Saints, and the curious 16th century manor-house, which contains a valuable museum. Its industries include paper-mills, breweries, malt-kilns, and agricultural implement works. Stone is quarried, and hops and fruit are grown. The town was taken by Fairfax in 1648.

Maikop, Russia (Kuban Territory), is a fortified town and trade center.

Mainz, Germany (Hesse), is a fortified town and one of the most important commercial centers on the Rhine, and carries on a brisk shipping trade with Holland and Belgium. Its principal exports are leather goods, furniture, and wine, while printing, lithographic and otherwise, is a flourishing industry. Gutenberg, the inventor of movable type for printing, was born here. The picturesque cathedral dates back to 978. The old castle of the electors, built in 1627-78, contains rich collections of Roman and Germanic antiquities, and the Gutenberg museum was established in 1901.

Makó, Hungary, has oil-mills and a fine episcopal palace.

Malaga, Spain (cap. Malaga), is a port on the Mediterranean, with export trade in fruits and wine, and

cane-sugar and distilling industries. The cathedral is a vast structure, mainly Gothic. The climate in winter is beautifully mild, and favorable for invalids. Of Phoenician origin, Malaga was for centuries a Moorish city and one of the principal ports of the kingdom of Granada. Ferdinand the Catholic captured it, after a long siege, in 1487.

Malden, Massachusetts (Middlesex Co.), is a suburb of Boston, with extensive manufactures of rubber boots and shoes, boot-trees, boot and shoe lasts, wire cord, leather, sand and emery paper, cotton goods, hosiery, knit goods, furniture, and soap.

Malmo, Sweden (cap. Malmöhus), is a fortified port and coaling station, with one of the largest artificial harbors in Scandinavia. Malmö has iron-works, woollen mills, dockyards, machinery, sugar, steam-mill factories, and breweries.

Malstatt-Burbach, Germany (Prussia), has coal-mines and iron-works.

Managua, cap. Nicaragua, has a national palace, and an industrial, commercial, and scientific museum. It exports coffee.

Manaos, Brasil (cap. Amazonas), is a port on the Rio Negro, a few miles above its junction with the Amazon. It has modern docking facilities and storage houses for the immense trade of the region in rubber, Brazil-nuts, cocoa, and copaiva oil. It has a cathedral, a magnificent theater, and all modern improvements.

Manchester, Connecticut (Hartford Co.), has extensive silk-mills, cotton and woollen mills, needle works, paper-mills, and electrical supply works. A large quantity of tinware is made here.

Manchester, England (Lancashire), is the greatest industrial town of England, and the center of its cotton industry, chiefly engaged in the spinning and weaving of cotton, and in the bleaching, printing, and making up of "Manchester goods." There are also a vast number of engineering works. The shipping houses of Manchester export cottons, silk and woollen goods, steam, gas, and electrical machinery, chemicals, India-rubber, iron, steel, and copper goods. It is now a port accessible for the largest sea-going vessels by means of a ship-canal, 35½ miles long, which connects it with the river Mersey, near its estuary. It is the seat of a university and of an Anglican bishop. It was the first town in England to adopt free libraries, and has numerous educational institutions, an art gallery containing a noteworthy collection, one of the finest town halls (1877) in England, etc. It was known as a manufacturing place as early as the 14th century; developed rapidly during the last half of the 18th century; was a leading center of the reform agitation in the early part of the 19th century (the scene of the "Peterloo massacre" in 1819); and became the center of the anti-corn-law and free trade movements under the lead of Cobden and Bright.

Manchester, New Hampshire (c. h. Hillsboro Co.), is an important trade and manufacturing center, the Amoskeag Falls above the city providing extensive water-power, which is made available by canals. It has extensive cotton-mills, and manufactures foundry and machine-shop products, hosiery, paper, boots and shoes, woodenware, needles, knit goods, woollen goods, carriages and furniture. A state industrial school is located here.

Mandalay, India (Upper Burma), is composed of the original town, a square encompassed by a wall, and an open city. Its gates are surmounted by curious wooden towers, and in the center of the town is the picturesque palace of the former kings of Burma. It has a famous pagoda containing an image of Buddha that attracts thousands of pilgrims. Silk-weaving is the chief industry.

Mangalore, India (Madras), is a seaport with exports of coffee, pepper, etc. It is the seat of a Roman Catholic bishop and the headquarters of the Basel Lutheran mission in India.

Manila, cap. Philippine Islands (Rizal), is a port, coaling station, the commercial metropolis of the Philippine Islands, and the seat of the United States military, naval, and civil administration in the Far East. It is an archiepiscopal see and possesses a cathedral and a university (1,200 students). Its industries are extensive, especially in tobacco, cigars, cheroots, cord, rope, thread, and ice. There are also sugar-refineries, breweries, and manufactures of machinery. The climate of Manila is enervatingly tropical and generally uniform in temperature, which, as a mean for the year, is about 80°. The principal public buildings are palaces of the governor and the archbishop, the town house, churches of the different religious orders, monasteries, convents, arsenal, barracks, supreme court, prison, civil and military hospitals, and custom-house. Manila was founded by the Spaniards in 1571; was taken by the English in 1762; and has often been devastated by earthquakes. The bay was the scene of the first naval battle of the Spanish-American war of 1898, in which Dewey destroyed a Spanish fleet (May 1). The city surrendered to the United States' forces on August 13th of the same year.

Manipur, India (Assam), is the capital of a native state of the same name, and has a trade in opium, indigo, silk, rice, and tea.

Manissa, Turkey in Asia, is a busy trading place and has many mosques. Among the notable buildings are the palace of Kara Osman Oglu and the government house.

Manistee, Michigan (c. h. Manistee Co.), has extensive manufactories of timber products, leather, etc. There are large salt deposits, and the vacuum evaporating salt plant produces annually about 6,300,000 bushels. It has extensive shipping interests, and its transportation facilities make it an excellent distributing point for the interior of the northern part of the upper peninsula of Michigan.

Manitowoc, Wisconsin (c. h. Manitowoc Co.), is an important lake shipping port, with manufactures of leather, dairy products, flour, foundry and machine-shop products, agricultural implements, furniture, cigars, beer, etc.

Mankato, Minnesota (c. h. Blue Earth Co.), is an important trade and manufacturing center, with manufactures of knit goods, lime, cement, dairy products, flour, lumber, foundry and machine-shop products, and candy. Valuable stone quarries are in the vicinity. A state normal school is located here. Mankato was the scene of several battles during the Sioux Indian war, 1862-3.

Mannheim, Germany (Baden), is one of the principal trading centers of South Germany. Its industrial establishments include iron-foundries, machine-shops, saw-mills, and chemical, woollen, carpet, and glass-works. A large palace, built in 1720-9, formerly the residence of the Elector Palatine, faces the Rhine to the southwest of the town.

Manresa, Spain (Barcelona), has manufactures of textiles. The convent of Santo Domingo, where Loyola lived for a time, is located here and is the objective point of pilgrimage.

Mansfield, Ohio (c. h. Richland Co.), has extensive manufactories of boilers, agricultural machinery, pumps, soil pipe, steel harrows, brass goods, electrical supplies, stoves, wagons, etc. A state reformatory is located here.

Mantua, Italy (cap. Mantua), is a fortified city, and has many interesting mediæval palaces and churches, containing art treasures by Mantegna and Giulio Romano, both natives. It has an academy of sciences and arts, a theological seminary, an interesting museum, a botanical garden, and an observatory. It manufactures leather and textiles. Mantua is the birthplace of Vergil, the Roman poet.

Maracaibo, Venezuela, is the principal port of the country and has a large transit trade in merchandise destined for Colombia. It exports coffee, cocoa, box-wood, etc., and has some ship-building. It has a university and many fine buildings. Near by are asphalt deposits.

Maranhao, Brasil (cap. Maranhao), is a port and coaling station, situated on an island first settled by the Dutch. It exports cotton, sugar, rice, rubber, manioc hides, ginger, gum, etc. It is the seat of a bishop and has a cathedral.

Margate, England (Kent), is a favorite seaside resort, with a sea-front of nearly three miles, and is noted for its bracing air. It contains "The Grotto," a curious artificial cave adorned with shell mosaics, discovered in 1837.

Maria Theresiopel, Hungary, is an important trade center, whose industries include weaving of linen, tanning, soap-making, ship-building, etc.

Marietta, Ohio (c. h. Washington Co.), is a trade center in a coal, iron, petroleum, and gas region, with fertile agricultural land in the valleys. It has extensive chair factories, glass-works, tool shops, brick works, boat yards, foundries, and other manufactories. It is the seat of Marietta College, and numerous other educational institutions. The site of Marietta was once part of a remarkable group of the mound-builders' prehistoric works, which consisted of two sections, one containing about 40 acres and the other about 20 acres. The remains of mounds, truncated pyramids, walls, and other ancient works still exist, although the city covers a large part of the original enclosures. Fragments of the walls, about 5 or 6 ft. high by 20 or 30 ft. base, may still be seen.

Marinette, Wisconsin (c. h. Marinette Co.), is the seat of an extensive lumber industry. The large lumber mills are the chief manufacturing establishments in the city, and there are also pail factories, paper and pulp mills, box and broom factories, engine and iron-works, threshing-machine factories, furniture factories, and cabinet shops. The city carries on a large commerce with all the important lake ports. In the vicinity are assembly grounds, where various religious and educational conventions are held each summer.

Marion, Indiana (c. h. Grant Co.), is a trade and manufacturing center in a fertile agricultural region, and a natural gas belt. The chief manufactures are flour, lumber, pulp, paper, brick, foundry products, window-glass, bottles, glass jars, furniture, linseed-oil, and rolling-mill products. It has a normal college, and three miles south of the city is a national soldiers' home.

Marion, Ohio (c. h. Marion Co.), is a trade and manufacturing center in a fertile agricultural section. It has extensive steam shovel works and manufactures of agricultural implements, carriages, buggies, mattresses, engines, etc.

Mariupol, Russia (Ekaterinoslaf), is a port on the Sea of Asof, with large exports of cereals, linseed, hides and tallow.

Marlboro, Massachusetts (Middlesex Co.), has manufactures of boots and shoes, shoe-making machinery, automobiles, automobile tires, bicycles, carriages, wagons, lamps, electrical machines, woodenware, and machine-shop products. During King Philip's war (1676), the Indians destroyed nearly the whole town.

Marquette, Michigan (c. h. Marquette Co.), is a port on Lake Superior having a fine harbor with the best of facilities for loading steamers with the iron ore and other minerals which are shipped from here in large quantities. Near the city are large quarries of brownstone. The chief industrial establishments are planing-mills, blast-furnaces, and stone-quarries. A state normal school, the Upper Peninsula State Prison, and a house of correction are located here.

Marsala, Italy (Trapani), is a fortified seaport, noted for its wines. It has a cathedral and occupies the site of the ancient Lilybæum, the chief fortress of the Carthaginians in Sicily.

Marseille, France (cap. Bouches-du-Rhône), is the leading port in the Mediterranean, a coaling station, and the second city of the country. The manufactures yield sugar, soda, metals, the products of lead-smelting, macaroni, and leather; but the city's specialties are oil-refining and soap-making. The fishing industry, especially for tunny, is flourishing. Marseille exports wine, fruits, cork, anchovies, silks, cotton, wool, etc., but the leading articles of trade are coal, oil-seeds, grain, soap, and petroleum. Of the older structures, the most interesting is the church of St. Victor, with its 11th century crypt. Originally perhaps settled by Phœnicians, Marseille owes its historical foundation to a Greek colony of Phocæans (600 B. C.). Having espoused the cause of Pompey, it was besieged and captured by Cæsar (B. C.

49), and under Roman domination was famous alike for its commerce and its culture. Semi-independent through the middle ages, Marseille was deprived (1660) of many of its privileges as a free port by Louis XIV. In 1720 it was fearfully ravaged by the plague.

Marshalltown, Iowa (c. h. Marshall Co.), is a trade center in an agricultural and stock-raising region. It has flour-mills, grain elevators, glucose factories, meat-packing plants, furniture factories, carriage works, foundry and machine-shops, etc. The Iowa State Soldiers' Home is located here.

Masaya, Nicaragua, Central America, is a trade center, which has tobacco interests.

Maskat, Arabia, is a seaport and the capital of the independent state of Oman, which is under British political influence. It exports dates, hides, horses, asses, pearls, and drugs. It was taken by the Portuguese in 1508 and remained under their rule till the 17th century.

Massa, Italy (cap. Massa e Carrara), has a fine ducal palace, and is a center of the famous Carrara marble industry. Its manufactures include silk, paper, olive-oil, cotton, and tobacco.

Massillon, Ohio (Stark Co.), is a railroad and trade center in a section noted for its large bituminous coal-fields and for its excellent farm lands. There are quarries of white sandstone in the vicinity. Among its industrial establishments are foundries, rolling-mills, machine-shops, bridge works, steel-works, glass-works, potteries, flour-mills, and creameries. It has a large trade in coal, sandstone, grain, and live stock. The State Hospital and Asylum for the Insane is located here.

Matanzas, Cuba (cap. Matanzas), is a fortified port with a capacious harbor, and the third city of Cuba. It exports sugar, and has rum-distilling, sugar-refining, and manufactures of guava jelly.

Matsumoto, Japan (Nagano), manufactures silk, baskets, and bamboo boxes.

Matsuyama, Japan (Ehime), has a large trade in fruits, vegetable wax, etc.

Matsuye, Japan (cap. Shimane), has extensive manufactures of paper.

Maulmain, India (Lower Burma), is a seaport and military cantonment. It exports timber, rice, cotton, horns, ivory, wax, lead, and copper, and has ship-building yards.

Mayaguez, Porto Rico (cap. Mayaguez), is the third city of Porto Rico, and has exports of sugar, coffee oranges, and pineapples.

Mayebashi, Japan (cap. Gumma), has an important silk trade.

Mazatlán, Mexico (Sinaloa), is the chief port of Mexico on the Pacific Coast, and a coaling station. It exports gold, silver, dyewood, hides, cotton, coffee, and sugar.

Meadville, Pennsylvania (c. h. Crawford Co.), is a trade center in a fertile agricultural region, and in the vicinity of extensive oil fields. It has large railroad car shops, iron-works, silk-mills, chocolate works, engine works, etc. It is the seat of Allegheny College, established in 1815, a theological school, a college of music, and other educational institutions. It was settled in 1788, became a borough in 1823, and a city in 1866.

Mecca, Turkey in Asia (Arabia), is the holy city of the Mohammedan world, being the birthplace of Mohammed and the goal of the annual pilgrimage to his shrine. Every Moslem is bound to undertake once in his life a pilgrimage to Mecca, and in the rites performed on this occasion are included the circuit around the Kaaba and the kissing of the black stone. Mecca was the residence of the first five caliphs. It is now governed by a sheriff, who is chosen by the people from the descendants of the prophet, but holds his authority from the Turkish sultan. It has always been a center of trade, but its industries are now confined to pottery and articles for sale to pilgrims. Mecca was sacked by the Carpathians in 930 and passed to the Turks in 1517.

Mechlin, Belgium (Antwerp), is the see of the cardinal-primate of Belgium, and is noted for its vast Gothic cathedral, dating partly from the 13th century and containing Van Dyck's Crucifixion and Rubens' Last Supper. Other works of Rubens are in the churches

of Notre Dame and St. John. Among other interesting buildings are the cloth hall, the law courts, and the archbishop's palace. Mechlin was formerly important for lace, shawls, and linen, but now its chief manufactures are woollen goods and "Gobelin" tapestry, with cabinet-making and carpentry.

Medellin, Colombia, is the second city of Colombia and its most important commercial and mining center. It has cotton factories. It is a bishop's see, and has a theological seminary, a school of mines, and a school of art and technology.

Medford, Massachusetts (Middlesex Co.), is a residential suburb of Boston, with manufactures of machinery, cotton goods, dyes, rum, chemicals, carriages, brick, and novelties. It is the seat of Tufts College (1120 students). The Craddock House, built in 1634, is one of the oldest, if not the oldest building, in the United States; it retains its original form.

Medina, Turkey in Asia (Arabia), is a walled city and is second only to Mecca in sanctity, being the scene of Mohammed's work after the Hejira or flight from Mecca Friday, July 16, 622 A. D. It contains the tomb of Mohammed and attracts many pilgrims.

Medinet-el-Fayoum, Egypt, manufactures woollen stuffs. It is the site of ancient Arsinoe.

Meerut, India (United Provinces), has a large military cantonment. It was the scene of the outbreak of the mutiny in 1857.

Mehallet-el-Kebir, Egypt, is a trade center and has manufactures of cotton.

Meiderich, Germany (Prussia), has iron and steel works.

Meissen, Germany (Saxony), is noted for the manufacture of "Dresden china." Its cathedral dates from the 13th century and is one of the finest Gothic structures in Germany.

Mekines, Morocco, contains the summer residence of the sultan and the Mulai Ismael mosque, a royal burial-place much visited by pilgrims. Leather and earthenware are manufactured.

Melbourne, cap. Victoria, Australia, is a port and coaling station on Port Phillip Bay, and the largest city of Victoria. It is the seat of an Anglican bishop and of a Roman Catholic archbishop. It contains magnificent buildings, such as the Government House, the two cathedrals, the government offices, Houses of Parliament, law courts, and the university. Exports wool, gold, wheat, flour, frozen mutton, wine, etc.

Melrose, Massachusetts (Middlesex Co.), is a residential suburb of Boston, and has manufactures of rubber boots and shoes. Middlesex Fells, 1,800 acres, a part of the metropolitan system of parks, is near by.

Memphis, Tennessee (c. h. Shelby Co.), is the first city of Tennessee, and the most important on the Mississippi River between St. Louis and New Orleans. It is an important railroad center and is of great importance as a distributing point for cotton, lumber, groceries, shoes, hardware, and other commodities. It has the only bridge spanning the Mississippi River south of St. Louis. It has large cotton compresses, cottonseed oil-mills, flour-mills, pulp- and paper-mills, fiber plants, car works, pump works, machinery works, etc. It has a medical college, a college of the Christian Brothers, a normal institute for colored pupils, and other educational institutions. A national cemetery, containing over 20,000 graves, is 5 miles distant from the city.

Mendoza, Argentina (cap. Mendoza), is a trade center in a rich mining and agricultural region. Vineyards abound in the neighborhood, and wine is extensively manufactured. It is a fine city with many handsome buildings, and has an agricultural college. It was founded in 1559, and destroyed by an earthquake in 1861, after which it was rebuilt a mile nearer the mountains.

Menominee, Michigan (c. h. Menominee Co.), is a trade center in an extensive lumber region, and is one of the largest shipping ports on the Great Lakes. Its chief industries are manufactures of lumber, paper, shoes, machinery, beet sugar, electrical machinery, etc.

Mentone, France (Alpes-Maritimes), is a favorite winter resort on the Riviera, with a fine climate and surrounded by orange and lemon groves.

Merida, Mexico (cap. Yucatan), is a bishop's see, and has a cathedral dating from 1598, a university, etc. Exports sisal hemp, hides, sugar, indigo, logwood, etc.

Meriden, Connecticut (New Haven Co.), is noted for its large manufactures of cutlery, silver and plated ware, steel pens, hardware, screws, glassware, cut glass, malleable iron, brass castings, curtain fixtures, self-playing attachments for pianos and organs, and agate ware. The Connecticut School for Boys, and the Curtis Home for Orphan Children are located here.

Meridian, Mississippi (c. h. Lauderdale Co.), is a trade center in a cotton-producing district. The chief manufactories are cotton-mills, cotton seed oil-mills, lumber mills, railroad shops, cotton-gins, and cotton compresses. It is the seat of the East Mississippi Female College, opened in 1869, and other educational institutions.

Merthyr Tydfil, Wales (Glamorgan), is the great center of the iron and steel industry of South Wales, and has extensive coal mines.

Meshed, Persia, is a famous place for pilgrimages, the attraction being the tomb of Imam Riza, son of Ali, founder of the Shiites, located in a magnificent and richly adorned mosque. Its trade, once considerable, has decreased since the completion of the Russian railway from the Caspian Sea to Samarkand. Fine silks, carpets, shawls, and sword blades are manufactured.

Messina, Italy (cap. Messina), manufactures silks, muslin, linen, and coral ornaments. The principal exports are wine, oranges, lemons, almonds, licorice, and pistachios. It is the seat of an archbishop, and has a cathedral, commenced in 1098, and the church of Sta. Annunsiata, dating from the Norman period. Its university contains a valuable library collected by the Jesuits, and has nearly 700 students. In 829 Messina was taken by the Saracens, who, in turn, were expelled by the Normans in 1072. The town suffers periodically from earthquakes, and was nearly destroyed in December, 1908.

Metz, Germany (Alsace-Lorraine), is a fortified town, with a 14th century cathedral, a museum, and several learned societies. It has trade in leather goods, preserved fruit, and wine. Known to the Romans as Divodurum, it was called Mettis in the 5th century, when it became the capital of the kingdom of Austrasia. It was frequently besieged, notably, and in vain, by the Emperor Charles V. in 1552. Metz remained in the possession of the French, to whom it was secured by the Peace of Westphalia in 1648, until 1870, when it capitulated to the Germans.

Mexico, cap. Mexico (Federal District), is the most imposing city of Latin America, and the metropolis of Mexico. Its famous cathedral, occupying the site of the chief Aztec temple (Teocalli), begun in 1573 and dedicated in 1667, is the second largest in the world. The National Museum contains valuable and interesting historical collections. The government offices, houses of Congress, Museum of Fine Arts, mint, school of mines, industrial museum, school of medicine, and conservatory of music are among the many notable buildings and institutions. The Palace of Chapultepec occupies the site of Montezuma's Palace, and dates from 1783-5, with later additions; it is now occupied by President Diaz and by the National Military School. The commerce of the city is mainly in transit. Its manufactures include cigars and cigarettes, gold and silver work, pottery, feather-work, saddlery, paper, religious pictures, hats, and beer. The present city was founded in 1522 on the site of the Aztec capital, Tenoctitlan, then situated in the lake Texcoco. The water of the lake afterwards retired, but the valley was subject to serious inundations until a few years ago. Numerous attempts have been made at various epochs to drain the valley of Mexico, but none of these proved successful until the completion in 1898 of the great Drainage Canal, constructed at a cost of $10,000,000. It is 30 miles long and crosses the mountains by a tunnel 6 miles in length. Its width at the top varies from 45 ft. to 168 ft., and it is crossed by numerous bridges of stone and iron. Elevation 7,850 ft.

Miagao, Philippine Islands (Iloilo), has manufactures of hemp.

Michigan City, Indiana (Laporte Co.), is a port at

the southern end of Lake Michigan, and has manufactures of hosiery, knit underwear, chairs, lumber, railroad cars, and furniture. It has a large trade in iron ore, salt, lumber, and farm products. A state prison is located here.

Middlesborough, England (York), is a port with a large harbor, in the Cleveland iron district. Has large iron and steel-works, blast-furnaces, foundries, rolling-mills, tube works, and wire mills. Rock-salt was discovered in 1882, and large quantities are prepared, chiefly from brine.

Middletown, New York (Orange Co.), is a trade center in a fertile agricultural and dairying section, which has large railroad shops, and manufactures of shirts, leather, condensed milk, etc. A state homœopathic hospital for the insane is located here.

Milan, Italy (cap. Milan), is in size the second city of Italy, and is the chief commercial and financial city of the country. It is the center of the Italian book-trade, and the printing of music is important; other industries are silk reeling and weaving, and the manufacture of velvet, machinery, chemicals, and metal work. The center of the city is occupied by the famous cathedral, built of white marble, commenced in 1386 and completed under Napoleon I., who was crowned King of Italy in 1805 with the iron crown of Lombardy. The town is exceptionally rich in monuments and in public gardens. There are numerous museums, picture galleries, and educational institutions. Among its theaters the famous Teatro della Scala is of world-wide renown. The city dates from the 4th century B. C., and was named Mediolanum. It was taken by Scipio (222 B. C.), and under Roman rule became a seat of learning. In 303 A. D. the Emperor Maximilian I. chose it as his residence, and made it the capital of Northern Italy, which it remained for about a century. After numerous wars with the German emperors and much internal strife, Milan became a duchy in 1395 under Gian Galeazzo Visconti. Having been held by Spain (1545) and Austria (1714), it was occupied by Napoleon Bonaparte in 1796, who made it the capital of the kingdom of Italy in 1805. Ten years later it reverted to Austria. It was the scene of great disturbance during the troubles of 1848-9 and 1853. After the peace of Villafranca (1859), it was ceded to Piedmont.

Milford, Massachusetts (Worcester Co.), is a trade center in an agricultural region and has considerable manufacturing interests. Its chief manufactures are foundry and machine-shop products, silk, boots and shoes, thread, straw goods, etc.

Millville, New Jersey (Cumberland Co.), is a manufacturing city, whose chief industrial works are foundries, glass factories, dyeworks, bleacheries, cotton-mills, machine-shops, etc.

Milwaukee, Wisconsin (c. h. Milwaukee Co.), is the largest city of Wisconsin, one of the chief manufacturing and commercial cities of the northwest, and one of the most important lake ports, with an excellent harbor, formed by the erection of huge breakwaters. It has an extensive commerce in grain, flour, coal, lumber, hides, and the products of its great manufacturing industries. The leading industry is the making of iron, steel, and heavy machinery. The flour-mills are very large, with a daily capacity of over 9,000 barrels, and the grain elevators have a capacity of about 6,000,000 barrels. Milwaukee lager beer is known all over the United States, and is produced annually to the amount of over 3,500,000 barrels. Pork-packing is extensively carried on, and the staple manufactures include leather, clothing, and tobacco. It is the seat of a state normal school, and the Milwaukee-Downer College for young women. The Layton Art Gallery has some interesting pictures and statues, and the magnificent public library contains over 200,000 volumes. Milwaukee was settled about 1818, became a village in 1835, and received the city charter in 1846.

Minneapolis, Minnesota (c. h. Hennepin Co.), is the largest city of Minnesota, and the chief flour-making place in the world. It is situated on the Mississippi River at the point where the river descends over the Falls of St. Anthony, about 50 ft., yielding over 50,000 horse-power. Its flour-mills have a daily capacity of

about 85,000 barrels, and produce about 16,000,000 barrels annually. Its lumber mills cut 600,000,000 ft. of timber annually. It also has extensive manufactures of iron and steel goods, machinery, street-cars, woollens, boots and shoes, clothing, furniture, linseed-oil, paper, etc. It has an extensive trade in agricultural implements, and a large jobbing trade in general merchandise. It is the seat of the University of Minnesota, with an annual enrollment of about 4,000 students, and many other educational institutions. Minneapolis owes its prosperity and rapid growth to the extensive agricultural district tributary to it. It was settled in 1856 and became a city in 1867.

Minsk, Russia (cap. Minsk), is the seat of a Greek Orthodox and of a Roman Catholic bishop, and has cathedrals of Orthodox and Roman Catholic communions. Soap, tobacco, matches, woollen cloth, hats, and leather are manufactured, and there are breweries, distilleries, saw-mills, iron-foundries, and tile-works.

Mirzapur, India (United Provinces), has manufactures of carpets and brassware. Its ghats and temples are specially noteworthy.

Miskolcz, Hungary, has a Gothic church dating from the 13th century. It manufactures flour, pottery, and porcelain.

Missoula, Montana (c. h. Missoula Co.), is the largest city of western Montana, and the seat of the state university. It has iron-works, tile- and brick-works, lumber mills, etc.

Mitau, Russia (cap. Courland), has tanneries and cloth-works, and manufactures tin articles, linen, and white lead. From 1561 to 1795 it was the capital of the duchy of Courland. It was the favorite residence of Marshal Biron, ruling courtier of Empress Anne, who built there a fine palace.

Mito, Japan (cap. Ibaraki), manufactures cloth, paper, and cigarettes, and exports fish.

Mobile, Alabama (c. h. Mobile Co.), is a port and coaling station on Mobile Bay, 33 miles from the Gulf of Mexico. It is one of the leading cotton markets and shipping points in the country, and the natural center of the Alabama-Tombigbee cotton region. It exports cotton, cottonseed-oil, fruits, coal, lumber, live stock, meat, tar, turpentine, and rosin. It has extensive manufactures of lumber and grist-mill products, foundry and machine-shop works, ship and boat building, etc. A United States marine hospital is located here and there are numerous educational institutions. Mobile was founded in 1702 by the Sieur de Bienville. It was the capital of Louisiana down to 1723. In 1763 it passed, with part of Louisiana, to Great Britain; in 1780 it was handed over to Spain, and in 1803 it became part of the United States by the Louisiana Purchase, although the government did not take possession until 1813. It was incorporated as a city in 1819. In 1864 the harbor was attacked and closed by Admiral Farragut, but the city itself did not surrender to the Federal troops until April 12, 1865.

Modena, Italy (cap. Modena), is the seat of an archbishop and is especially rich in churches, palaces and public buildings. The cathedral, with a marble tower, was commenced in 1099 by the countess Matilda of Tuscany; while the palace, built by Francis II. in the 17th century, contains a valuable library, an important picture gallery, and a rare collection of coins. Modena has a university, a military college, an academy of arts and sciences, an observatory, and a botanical garden. It has manufactures of woollen and hempen cloths, hats, and leather. It was from the end of the 13th century until 1860 the capital of the duchy of Modena, ruled by the Este family.

Modica, Italy (Syracuse), trades in grain, wine, oil, and cotton. There are remarkable prehistoric caves in its vicinity.

Mohilef, Russia (cap. Mohilef), possesses a fine town hall (1679), a Roman Catholic cathedral (1692), and an Orthodox cathedral, founded in 1780 by Catherine II. The Roman Catholic archbishop is primate of all Roman Catholics in Russia. Leather, tobacco, and pottery are manufactured.

Molenbeek, Belgium (Brabant), has machine-works

dye-works, tanneries, and manufactures of textiles, carpets, soap, and margarine.

Molfetta, Italy (Bari delle Puglie), is a port with a commodious harbor. The principal exports are wine, almonds, olive-oil, and niter.

Moline, Illinois (Rock Island Co.), is an important manufacturing city, with good water-power secured by damming a part of the Mississippi River between the shore and an island at this point. There are coal-fields near by, which also contribute to the industrial development of the city. The chief manufactures are wagons, carriages, agricultural implements, steel, steam-engines, pumps, flour, and foundry and machine-shop products.

Mombasa, cap. British East Africa, Africa, is a port and coaling station, with a harbor well sheltered by a coralline bar of depth sufficient for largest vessels. Exports gum, copra, cattle, ivory, grain, etc. Mombasa dates from the end of the 15th century, when the Portuguese acquired it.

Monastir, Turkey (cap. Monastir), has tanneries and manufactures of silver filigree, woollen stockings, and carpets.

Monclova, Mexico (Coahuila), is a trade center, and has large railroad shops. It was founded in 1685.

Moncton, New Brunswick, Canada, is a railway and trade center with car shops, machine shops, wire fence works, etc.

Monghyr, India (Bengal), has manufactures of cotton, cloth, and shoes.

Monroe, Louisiana (c. h. Ouachita Parish), is the largest city of northeastern Louisiana, and has extensive shingle-mills, and other lumber interests.

Mons, Belgium (cap. Hainaut), is the center of a very rich coal-mining region, and its industries include engineering works, cotton-spinning, and sugar-refining. It contains the church of Sainte Waudru (1450-1589), one of the finest Gothic edifices in Belgium, and a 15th century town hall.

Montauban, France (cap. Tarn-et-Garonne), has manufactures of textiles, and a large trade in wine, grain, and agricultural produce. It was one of the strongholds of the Huguenots, and even now half of its inhabitants are Protestants. It possesses a Protestant theological college, the only one of the kind in France.

Montclair, New Jersey (Essex Co.), is a residential suburb of New York and is noted for its healthy climate. It has a state normal school, a military academy, and electrical construction works.

Monte Carlo, Monaco, is one of the chief resorts on the Riviera, and is famous for its casino with its gaming tables of roulette and trente-et-quarante. It is charmingly situated, and has an exquisite climate, beautiful gardens, and a well-known opera house.

Monterey, Mexico (cap. Nuevo Leon), is a winter resort with mineral springs, and the most important manufacturing city of Northern Mexico, with large smelting-works, iron-foundries, cotton factories, and silver mining. It has a cathedral. Monterey was taken from the Mexicans by the United States army under General Taylor in 1846.

Montevideo, cap. Uruguay, is the chief port of Uruguay and has the best harbor on the Plata River, which at this point is 50 miles wide. The most imposing buildings are the senate-house, the cathedral, the university (over 400 students), the national library and museum, the great Solis Theatre, the school of art, etc. The principal exports are hides, wool, and preserved beef. The port is noted for salubrity and cleanliness, and has all modern improvements. Until 1814 Montevideo was in the possession of the Spaniards. In 1828 it became the capital of the republic.

Montgomery, cap. Alabama (c. h. Montgomery Co.), is at the head of navigation of the Alabama River, 410 miles from the Gulf of Mexico, and is one of the chief cotton marts and distributing points of the south. There are coal and iron fields near by, and the vast forests of yellow pine on the south add to its great natural advantages for manufacturing. Its industries include car shop and foundry works, factories for boilers and other iron goods, cotton factories, cordage factories, ginning and compress plants, cottonseed-oil and cake works,

wood-working and lumber plants, roofing material plants, etc. There are located here a state normal school and other educational institutions. In the State House the Confederate Government was inaugurated by Jefferson Davis on Feb. 18th, 1861, whence Montgomery has been called the "Cradle of the Confederacy". It was also the scene of Yancey's celebrated secession speech on Jan. 11th, 1861.

Montlucon, France (Allier), has blast-furnaces and iron-works, and engineering shops. Mirrors are a specialty.

Montpellier, France (cap. Hérault), is chiefly important as a wine center and for silk culture. There are also manufactures of blankets, soap, and candles. Montpellier possesses not only the oldest botanical garden in Europe, but one of its most famous universities, constituted (in 1289) from the already existing schools of medicine, law, and arts. Moorish physicians founded the school of medicine and science, with which are associated the names of De Villeneuve, Rabelais, Rondelet the anatomist, Bauhin, Magnol, Tournefort, and De Jussieu. At the university, too, Petrarch was a student and Casaubon a professor. The town has a cathedral.

Montreal, Quebec, Canada, is a port at the head of navigation of the St. Lawrence River, nearly 1,000 miles from the ocean, and is the largest city of Canada. It is the chief commercial and financial center of Canada, and has extensive manufactures of woollen and cotton, boot and shoe factories, breweries, iron-foundries, railway works, clothing, sugar, and tobacco factories. More than half of its population are French. The city, which is predominatingly Roman Catholic, contains many fine churches, notably the cathedral of Notre Dame, St. James' Cathedral, St. Patrick's, and the Jesuit Church, noted for its frescoes. The Anglican cathedral, Christ Church, is said to be the finest specimen of Decorated Gothic in America. Mention must also be made of the fine group of buildings in which McGill University is housed. When Jacques Cartier sailed up the St. Lawrence in 1535, he found here the Indian town of Hochelaga. In 1760 Montreal surrendered to the English under Lord Amherst. During the war of independence, it was occupied by the Americans (1775-6). The riot of 1847 resulted in the destruction of the parliamentary buildings by fire, and the removal of the seat of government to Quebec.

Montreuil, France (Seine), is a suburb of Paris, and contains a beautiful 13th century church. It is celebrated for its culture of peaches, and has manufactures of colors, glue, varnish, and soap.

Monza, Italy (Milan), is an ancient city, and has a cathedral, founded in 595, where the famous iron crown of Lombardy is preserved. The town hall dates from 1293, and the royal palace from 1777. It has manufactures of coarse cottons, hats, leather, and silk; also dye-works. In 1900 King Humbert of Italy was assassinated here.

Moosejaw, Saskatchewan, Canada, is in the midst of a prosperous agricultural district, and has large flour-mills and stock-yards.

Moradabad, India (United Provinces), has manufactures of cotton, and is noted for engraved metal ware. It has ruins of Rustram Khan's fort, dating from the foundation of the town in 1625.

Morelia, Mexico (cap. Michoacan), has manufactures of cottons, pottery, and sugar. It possesses a cathedral and two colleges.

Morioka, Japan (cap. Iwate), is noted for silks and kettles.

Morocco, cap. Morocco, is an inland walled town, of oriental aspect, which contains the imperial residence of the Sultan. It was formerly important as the place of manufacture of morocco leather.

Morristown, New Jersey (c. h. Morris Co.), is a residential suburb of New York, and is noted for its beautiful homes, country-seats, etc. It was settled in 1709-10 and was called West Hanover. In 1740 the name was changed to Morristown in honor of William Morris, then colonial governor of New Jersey. It was incorporated in 1865. This place figured prominently in the Revolution; in 1777, from January to May, and from December, 1779 to June, 1780, Morristown was Washington's head-

quarters. An historical museum, in the house occupied by Washington, is now owned by the Washington Association, who preserve here many mementoes of Revolutionary and pre-Revolutionary times.

Morshansk, Russia (Tambof), carries on a large trade, and has manufactures of soap, tallow, spirits; also breweries, saw-mills, and fisheries.

Moscow, Russia (cap. Moscow), is the ancient and still, in a sense, joint capital of Russia, the second largest city of the empire, and the chief railway and manufacturing center. The Kreml or Kremlin, Moscow's acropolis, is entirely surrounded by a wall (40 ft. high), enclosing a space of more than 80 acres. Inside is the cathedral of the Assumption, built in 1474-9, in Lombardo-Byzantine style, and here the Russian emperors are crowned. The cathedral of the Annunciation, founded at the beginning of the 14th century, contains 15th century paintings. The cathedral of St. Michael contains life-sized frescoes of early Russian sovereigns round the walls. The isolated bell-tower of Ivan Velikii has a head of gilded copper, some 260 ft. above the pavement. One of the bells weighs almost 64 tons. The Czar Kolokol, or King of Bells, stands on a pedestal at the foot. It is the largest bell in the world, and weighs some 200 tons. Cast in 1735, it was broken in 1737 without having once sounded. The old imperial palace, the former residence of the Czars, was built in 1487; the Great Palace, an immense pile built in 1806 by Alexander I., contains a collection of ancient robes and armor; the Little Palace, built by Nicholas I., contains valuable pictures and a great collection of books relating to Moscow. Among other wonders of the Kremlin are the Treasury of the Patriarchs, the Chudov Convent, the House of the Holy Synod, and the Saviour Gate. The cathedral of St. Basil (built in 1554), is one of the strangest erections of Christian art. Moscow is the seat of the metropolitan, the chief prelate of the Russian Orthodox Church. The university, founded in 1755, has a library of 300,000 volumes. The city has woollen, silk, and metal manufactures, cotton-mills, tanneries, and candle factories. The principality of Moscow was united with that of Vladimir, and Moscow became the capital of the grand principality of Moscow and the seat of the metropolitan in the first part of the 14th century. It was taken and burned by Lithuanians and Tatars in the 14th century, was nearly destroyed by fire in 1547, and burned by the Khan in 1571. The capital was removed to St. Petersburg by Peter the Great in 1711. Moscow was burned by its inhabitants during its occupation by Napoleon in Sept., 1812.

Mosul, Turkey in Asia, was very prosperous between the 10th and 13th centuries, and was for long afterwards a great trading center, noted particularly for its manufactures of muslin, to which it gave its name. Its importance has greatly declined since the opening of the Suez Canal. Near Mosul are the mounds which mark the site of ancient Nineveh, the capital of Assyria.

Motherwell, Scotland (Lanark), is situated in a rich coal-mining district, and has iron- and steel-works. A growing industry is iron bridge-building.

Mountain Ash, Wales (Glamorgan), has coal-mines and iron-works.

Mount Carmel, Pennsylvania (Northumberland Co.), is a trade center in the midst of valuable coal-fields and nearby are a number of anthracite mines. It has manufactures of mining implements, miners' lamps, hats, caps, men's clothing, flour, cigars, etc.

Mount Clemens, Michigan (Macomb Co.), is a celebrated health resort with several mineral springs.

Mount Vernon, New York (Westchester Co.), is a residential suburb of New York, and has several industrial establishments and a considerable trade. It was founded in 1852 and chartered in 1892.

Mukden, China (cap. Manchuria), is an important banking and commercial center, especially for furs. It has an arsenal and a mint, and outside its walls four conspicuous Lamaist monasteries. During the Boxer rising in 1900 the city was much injured by fire. Inside the east gate is the famous "Fox" temple, much frequented by those who seek to have their diseases miraculously cured. After a fiercely contested battle of four-

teen days' duration, the Russians were here defeated by the Japanese, who entered the city on March 10, 1905.

Mulhausen, Germany (Alsace-Lorraine), is a most important manufacturing center, principally for cotton goods and chemicals. Machinery and railway materials are also produced.

Mulheim-on-Rhine, Germany (Prussia), has chemical and engineering works, and manufactures silks, satins, and plush.

Mulheim-on-Ruhr, Germany (Prussia), has ironworks, engineering shops, and glass-works. A large coal trade is carried on.

Multan, India (Punjab), is a railroad center, with extensive trade and important bazaars and banking business. It manufactures textiles, including brocades.

Muncie, Indiana (c. h. Delaware Co.), is a trade and manufacturing center in the natural gas belt of Indiana. In the vicinity are coal-fields and glass sand, and the White River furnishes good water-power. The chief industrial establishments are iron- and steel-works, glass-works, machine-shops, canneries, pulp and paper-mills, manufactories of silver and plated goods, flour, etc. It has numerous educational institutions.

Munich, Germany (cap. Bavaria), contains a royal residence, a magnificent pile, commenced in the beginning of the 17th century, under Maximilian I. The cathedral dates from 1368. Among other buildings of great architectural beauty are the Glyptothek, the new Pinakothek (both picture and sculpture galleries), the National Theatre, the new Courts of Justice (1897), and the Imperial Bank (1901). In addition to the university (4,500 students), and the famous Academy of Arts, there are numerous educational institutions. The royal library is one of the most valuable in Germany, while the art collections are unrivalled. Munich is the home of countless literary, scientific, geographical, and other societies. The principal industry is brewing, but leather goods, gloves, machinery, artificial flowers, embroideries, gold and silver articles, and scientific instruments form other branches of commerce. The town was founded by Henry the Lion, Duke of Saxony, in the middle of the 12th century. It was taken by the Swedes under Gustavus Adolphus in 1632. King Louis I. (1825-48), a great patron of the arts, did much towards embellishing the town. In his reign the university, originally established at Ingolstadt in 1472, and removed to Landshut in 1800, was transferred to Munich. King Maximilian II. (1848-64) followed in his father's footsteps, but devoted more care to the development of the sciences. He founded the Maximilianeum and the National Museum, and gathered round him many of the principal scientists of the day.

Munster, Germany (Prussia), has numerous mediæval buildings, including the 14th century Gothic church of St. Lambert. In the town hall, built in 1335, was signed the Peace of Westphalia on October 24, 1648, which terminated the Thirty Years' war. The university was founded in 1771. The principal manufactures are linen and cotton goods, while Westphalian hams and pumpernickel, a coarse, black rye-bread, are exported. Distilling, brewing, wood-carving, and glass-painting are thriving industries.

Murcia, Spain (cap. Murcia), is the seat of a bishop and has a cathedral in the Gothic Romanesque style begun in the 14th century, with a striking tower completed in 1766. Cloth, gunpowder, and flannel are manufactured.

Muscatine, Iowa (c. h. Muscatine Co.), is a trade center in a fertile agricultural and manufacturing region. Its chief manufactures are farm and machine-shop products, oatmeal, flour, wagons, carriages, pottery, rolling-mill products, lead-works, and lumber.

Muskegon, Michigan (c. h. Muskegon Co.), is a port on Lake Michigan and an important trade and manufacturing center. It has extensive manufactures of furniture, curtain rollers, refrigerators, flour, beer, knit goods, pianos, iron products, chemical engines, tin-plate products, electric cranes, leather, and cutlery.

Muskogee, Oklahoma (c. h. Muskogee Co.), is in size the second city of Oklahoma, and an important trade center in a rich agricultural region. It has an extensive

trade in grain, lumber, and general merchandise. Its industries are chiefly machine and boiler works, planing-mills, iron-works, etc. It is the seat of an Indian university.

Muttra, India (United Provinces), was once the center of the Buddhist faith. It is the reputed birthplace of Krishna and his brother Balarama, and is consequently a great pilgrim resort.

Mysore, India (Madras), is the capital of a native state of the same name, and has manufactures of carpets, silks, and jewelry.

Nafa, Japan (cap. Okinawa), is a port with exports of sugar, cotton, and silks.

Nagano, Japan (cap. Nagano), has a Buddhist temple, founded 670 A. D., which is one of the most famous in Japan.

Nagasaki, Japan (Nagasaki), is one of the five treaty ports opened in 1859, and a coaling station with important coal-mines near by. It has large ship-building yards and engine works, and a large European and Chinese trade, exporting coal, rice, sugar, flour, etc. The manufactured goods include enamelled pottery and lacquer, and tortoise-shell articles. Nagasaki is noted for its temples and its festivals, the chief among the former being the Shinto "Bronze Horse Temple." It has also a college modelled on European lines.

Nagoya, Japan (cap. Aichi), is the emporium for the pottery of Seto. It is also noted for cloisonné enamels. Cotton and silk fabrics are manufactured. The castle, built in 1610, is one of the wonders of Japan.

Nagpur, India (cap. Central Provinces), was an important town in the days of the Mahratta empire. Its chief industry is cotton-weaving.

Nagy-Koros, Hungary, is a market-town in a wine-growing district. It is noted for melons.

Nakhitchivan, Russia (Don Cossacks), has manufactures of textiles and a trade in gems and jewels. Tradition affirms that it is the place where Noah settled after the flood.

Namangan, Russia (Russian Turkestan), is a market for sheep, cotton, and fruit, and has deposits of petroleum and coal.

Namdinh, Tonkin, French Indo-China, is a fortified town with a trade in silk and cottons.

Namur, Belgium (cap. Namur), manufactures cutlery, pottery, porcelain, and glass, and produces coal, iron, and limestone. The cathedral (18th century) and the church of St. Loup (17th century) are handsome edifices.

Nanaimo, British Columbia, Canada, has a fine harbor and is a trade center in a fruit growing and farming section. It has extensive fishing and fish curing interests and ships considerable coal.

Nanchang, China (cap. Kiangsi), is the center of the porcelain trade of the country.

Nancy, France (cap. Meurthe-et-Moselle), was formerly the capital of the duchy of Lorraine, and it possesses a university, a large public library, and a famous school of forestry. It is an important railway center, and its chief manufactures are embroidery, cambric, muslin, jaconets, cotton and woollens, and artificial flowers. Outside the town Charles the Bold of Burgundy was killed in the battle against the Swiss (1477).

Nankin, China (cap. Kiangsu), is a port on the Yangtse River, 130 miles from its mouth. Its silks and satins are famous in China, its nankeens in Europe. It contained the remarkable "porcelain tower", and the tomb of the first Ming emperor (Hung Wu) is near the city. An arsenal stands outside the city.

Nantes, France (cap. Loire-Inférieure), exports wine, preserved provisions, and hardware, and has manufactures of sugar, leather, soap, preserved meats, fruits, and sardines, as well as cotton-mills, iron-foundries, and ship-building yards. In its 14th century castle Henri IV. signed the Edict of Nantes (1598). The somewhat unsightly cathedral (1434-1852), contains an exquisite Renaissance monument by Colomb to the last Duke and Duchess of Brittany.

Nanticoke, Pennsylvania (Luzerne Co.), is a trade center in an anthracite coal region. Has extensive water-power, which is utilized in manufacturing mining and agricultural implements, hosiery, knit goods, lumber, flour, etc. The industry contributing most to the wealth of the town is coal-mining.

Naples, Italy (cap. Naples), is a port, coaling station and the largest city of Italy, famous for its beautiful situation. It has extensive ship-building, is an important port of embarkation of emigrants, and has a large export trade in wine, silk, hemp, flax, fruit, etc. It is the seat of an archbishop, and has a cathedral, commenced in 1272 and completed in 1316. The university (5,550 students) was founded by the Emperor Frederick II. in 1224. Among other public buildings are the castle of St. Elmo (1334); the new castle, built by Charles I. in 1283; the Borbonico (now National Museum), founded in 1586; the royal palace, dating from 1600; and the San Carlo theatre, opened in 1737. Naples was a Greek colony from Cumæ; flourished under Roman rule; suffered in the barbarian invasions; was taken by Belisarius in 536, by Totila in 543, and by the Normans in 1130; and became the capital of the kingdom of Naples and of the Two Sicilies in 1139.

Nara, Japan (cap. Nara), was from 709 the first fixed capital of Japan. It has beautiful temples, a Shinto dating from 767, and a Buddhist from 752; also a huge bell, weighing 37 tons (cast in 732), and a gigantic image of Buddha. Fans and toys are manufactured.

Narbonne, France (Aude), is an ancient city, and has a cathedral dating from the 13th century and a town hall, also from the 13th century, which is now a museum. It makes brandy and wine, and is famous for its heather honey.

Nashua, New Hampshire (c. h. Hillsboro Co.), is an important manufacturing city. Its water-power is obtained from the Nashville River by means of a canal 3 miles long, 60 ft. wide and 8 ft. deep. The chief manufactures are cotton goods, paper, shoes, iron and steel products, edged-tools, saddlery, refrigerators, registers, sash, doors, and blinds, ice-cream freezers, etc. It has a United States fish-hatchery. The first settlement was made in 1655.

Nashville, cap. Tennessee (c. h. Davidson Co.), is the second city of the state, and the most important educational and publishing center in the South. It is the seat of Vanderbilt University, attended by upwards of 1,000 students; the University of Nashville, attended by about 1,500 students; and the Peabody teachers' college, 1,000 students. Fisk University, 600 students, the Roger Williams University, 250 students, and Walden University, 1,000 students, are the leading seats of learning for colored persons. There are numerous other educational institutions. The most prominent public building is the State Capitol. The state library contains 40,000 volumes, and the Tennessee Historical Society has a large and valuable library, with many rare MSS., portraits, etc. The city has manufactures of hardware, large flour-mills, copper mills, tobacco works, etc; also a large and extensive wholesale trade in dry goods and other merchandise; and is a leading lumber-market. Several large sectarian publishing houses are located here. Nashville was settled in 1780 and became the permanent state capital in 1843. The Federal army occupied the city in 1862 and around it, in 1864, was fought one of the greatest battles of the Civil War, in which Gen. Hood, at the head of the Confederate army, was defeated by Gen. Thomas.

Nassau, cap. Bahama Islands, is a port of entry on New Providence Island, and is a favorite winter resort for English and Americans. It has a trade in sponges, cotton, fruit, and salt.

Natal, Brasil (cap. Rio Grande do Norte), has a poor harbor, defended by an enormous fortress on a natural reef. It exports sugar and cotton.

Natchez, Mississippi (c. h. Adams Co.), is an important river port and has a large trade in cotton, rice, and sugar. It is the seat of Stanton College and other educational institutions. A national cemetery, con-

taining over 5,000 graves, is nearby. The city was founded by Bienville, who built here Fort Rosalie in 1716, but the place was destroyed and many inhabitants murdered in 1729 by the Natchez Indians. The fort came into possession of the English in 1763, when the name was changed to Fort Parmure. In 1779 the Spaniards took possession, and in 1798 the United States became undisputed owner of the land east of the Mississippi, which included Natchez and much of the adjacent territory. From 1798 to 1820 Natchez was the capital of Mississippi.

Naugatuck, Connecticut (New Haven Co.), has extensive manufactures of rubber goods, knit goods, paper boxes, etc. It was incorporated as a town in 1844 and as a borough in 1893.

Nawanagar, India (Bombay), is the capital of a native state of the same name, and a port on the Gulf of Cutch. It manufactures gold and silk embroidery and perfumed oils.

Negapatam, India (Madras), was one of the earliest settlements of the Portuguese on the east coast. It was taken by the Dutch in 1660, and by the British in 1781. Oil is extracted.

Neisse, Germany (Prussia), has manufactures of furniture, lace, wire netting, and machinery.

Nelson, British Columbia, Canada, is a trade center in a mining and fruit growing section, and has considerable wholesale trade. The neighborhood supplies timber, marble, iron, silver, gold, lead, and other minerals.

Nelson, England (Lancashire), manufactures cotton and other textiles.

Neuchatel, Switzerland (cap. Neuchâtel), is the chief entrepôt of the canton, and has large manufactories of watches and jewelry. It is dominated by its castle and collegiate church, and is noted for its educational institutions.

Neuilly, France (Seine), is a suburb of Paris. It has machine-shops and manufactures of embossed leather and carpets. Here stood, till it was burnt in the revolution of 1848, the beautiful Château de Neuilly, built by Louis XV. and the favorite residence of Louis-Philippe.

Neumunster, Germany (Prussia), has tanneries, paper and cotton mills, railway repairing shops, and breweries.

Neunkirchen, Germany (Prussia), has coal-mines and iron-works.

Neuss, Germany (Prussia), has manufactories of screws, nails, paper, soap, and starch.

Nevers, France (cap. Nièvre), manufactures chains, cables, and agricultural implements; but the most famous product is majolica pottery, introduced from Italy in 1565. It is the seat of a bishop, and the chief buildings are the cathedral (11th to 15th century), with an apse at either end, and the Law-Courts (1475), the residence of the former dukes of Nevers.

New Albany, Indiana (c. h. Floyd Co.), is an important railroad center and manufacturing city on the Ohio River, nearly opposite Louisville. It has large glass-works, pork-packing establishments, tanneries, furniture factories, rolling-mills, lumber mills, and foundries. It is the seat of De Pauw College for women and other educational institutions. It was incorporated in 1839.

Newark, New Jersey (c. h. Essex Co.), is the largest city of New Jersey, and the third insurance center in the United States. It has extensive manufactories of jewelry, patent enameled leather, celluloid, hats, automobile bodies, varnish, cutlery, malt liquors, clothing, etc. Essex County Court-House has mural decorations by Blashfield, Walker, Cox, Maynard, Pyle, Turner, Millet, and Low. Newark was settled in 1666 and chartered as a city in 1836.

Newark, Ohio (c. h. Licking Co.), has manufactures of glassware, chemicals, instruments, carriages, locomotives, cars, flour and lumber products. In the vicinity are valuable deposits of sandstone, coal, and natural gas. Near by are two large mounds which belong to the works of the "Mound Builders." It was settled in 1802.

New Bedford, Massachusetts (c. h. Bristol Co.), was formerly an important whaling port, and is now a busy center in the manufacture of cotton. It has upwards of fifty cotton-mills, employing many thousand operators, large cordage works, large shoe factories, woollen mills, glass-works, paint works, and other industries. It ranks first among the cities of the United States in the production of fine cotton yarn and second in the number of spindles. It has a state textile school. It was settled in 1652, incorporated in 1787, and chartered as a city in 1847. On September 5, 1778, it was attacked by a British fleet, captured and almost destroyed.

New Britain, Connecticut (Hartford Co.), has extensive manufactures of stamped ware, hardware, knit goods, and cutlery, and is the seat of a state normal school and other educational institutions. It was settled in 1687 and granted a city charter in 1871.

New Brunswick, New Jersey (c. h. Middlesex Co.), is the seat of Rutgers College, a state agricultural and mechanical college, and other educational institutions. It has extensive manufactures of rubber, wall-paper, cigars, etc. It was settled in 1681 and chartered as a city in 1784. It was the scene of much of the trouble during the Revolutionary War, and was occupied by the British during the winter of 1776-7.

Newburgh, New York (Orange Co.), has extensive manufactures of cotton and woollen goods, silk, powder, paper, carpet, flour, lumber, and cigars. It has shipyards and extensive coal, brick and lumber-yards. Large quantities of coal from Pennsylvania are brought here to be loaded on coasting vessels and barges. The Hasbrouck Mansion, which was Washington's headquarters from 1782-3 and dates in part from 1750, contains interesting relics; it was here that Washington was offered the title of king by the officers of the army. In the ground surrounding this building is the so-called Tower of Victory with a statue of Washington. The city was incorporated in 1865.

Newburyport, Massachusetts (c. h. Essex Co.), has boot and shoe factories, cotton factories, hat shops, etc. It was settled in 1635 and chartered as a city in 1851.

Newcastle, England (Northumberland), is a port and coaling station near the mouth of the Tyne. Has large ship-building yards, engineering and steel-works, locomotive works, glass-works, and other extensive industries. It is the largest coal market in the world. It is the seat of a bishop, and the Church of St. Nicholas, dating from the 14th century, is now the cathedral. The Norman castle, built in 1080 and rebuilt by Henry II., was long a noted stronghold. Newcastle was a Roman and a Saxon town; was taken by the Scots in 1640 and 1644; and long held an important place in border warfare.

Newcastle, New South Wales, Australia, is a port with an extensive export trade in coal, wool, and frozen meats.

New Castle, Pennsylvania (c. h. Lawrence Co.), is a trade center in a productive agricultural region, and in the vicinity are extensive coal-fields and deposits of iron, sandstone, limestone, and fire-clay. The chief manufactures are glass, iron, flour, paper, agricultural implements, stoves, nails, tin-plate, and lumber. It was settled in 1712 and received its city charter in 1869.

New Guatemala, cap. Guatemala, Central America, manufactures muslins, cotton yarns, silver articles, and embroidery, and is the center of the trade of the entire republic. It has most of the government offices, a museum, a cathedral, and a university.

New Haven, Connecticut (c. h. New Haven Co.), is the largest city of Connecticut, and the seat of Yale University, attended by over 3,000 students and having, besides the academic department, schools of science, theology, medicine, law, forestry, music, and fine arts. A state normal school, a manual training school, and other educational institutions are located here. It has extensive manufactures of clocks, watches, hardware, firearms, ammunition, carriages, engines, automobiles, rubber goods, etc. New Haven was settled in 1638 and incorporated as a city in 1784.

New London, Connecticut (c. h. New London Co.), is a port on the Thames River, three miles from its entrance into Long Island Sound. It has large manufactories of silk, woollen goods, furniture, foundry and machine-shop products, and shipyards. A small government naval station is near by. It was settled in 1646 and incorporated in 1784. On September 6, 1781, a British force under Benedict Arnold attacked the town, destroyed the wharves and many of the buildings, and killed a number of the people.

New Orleans, Louisiana (c. h. Orleans Parish), is the largest city of the Southern States, the second largest cotton market in the world, and ranks next to New York City in value of exports. It is a port and coaling station on the Mississippi River, 107 miles from its mouth. It is the outlet of the greatest agricultural valley in the world and exports large quantities of cotton, sugar, molasses, rice, pork, corn, wool, timber, hides, and tobacco. It manufactures extensively cottonseed oil, machinery, timber products, flour, rice, tobacco, and sugar. It is the seat of Tulane University, an important and well-equipped institution with nearly 2,000 students, and of other educational institutions. It has an archiepiscopal palace dating from 1737. The Cabildo, now occupied by the state Supreme Court, was erected in 1795 and was formerly the government building; here the cession of Louisiana from France to the United States took place in 1803. A United States mint and a marine hospital are located here. A national cemetery nearby contains 12,000 graves of Union soldiers of the Civil War. New Orleans was settled in 1718, and in 1726, it was made the capital of the French colony of Louisiana, which then included nearly all of the Mississippi Valley. It was incorporated as a city in 1804. When, in 1812, Louisiana was admitted to the Union as a state, this city was its capital and remained such until 1852. In 1865-80 it was again the state capital. On January 8, 1815, occurred the battle in which Gen. Andrew Jackson defeated the British army under Gen. Pakenham. Soon after the secession of Louisiana from the Union and the establishment of the Confederate government, New Orleans was blockaded by the Federal fleet under Farragut and was captured in April, 1862.

Newport, England (Monmouth), has manufactures of steam-engines, railway material, agricultural implements, glass, pottery, chemical manures, shipbuilding, and brewing. Coal, iron, and steel are exported. It is an ancient city, and many Roman remains have been discovered here.

Newport, Kentucky (c. h. Campbell Co.), is a residential suburb of Cincinnati, and has manufactures of cigar boxes, iron, carriages, watch-cases, and furniture. Fort Thomas, a government military post, is nearby.

Newport, Rhode Island (c. h. Newport Co.), is a port and famous summer resort on Narragansett Bay. It is a favorite watering-place of people of great wealth who have here many costly summer residences. Overlooking and guarding the harbor are Fort Greble and Fort Adam, and a United States torpedo station, war college, and naval hospital are near by. It has a large trade in fish. Newport was settled in 1639 by William Coddington and other dissenters from the Puritan church of Massachusetts, and a century later had about 5,000 inhabitants. In 1770 Newport was surpassed by Boston only in the extent of its trade, which was considerably greater than that of New York. It suffered greatly during the Revolution, however, and never recovered its commercial importance, so that in 1870 its population was no larger than in 1770. During part of the Revolutionary struggle Newport was occupied by the French allies of the Americans, who were so favorably impressed with Rhode Island, that they sought to have it ceded to France.

Newport News, Virginia (Warwick Co.), is a port on Hampton Roads and an important railroad terminal. It has an extensive export trade in grain, a large shipbuilding plant, and other industries.

New Rochelle, New York (Westchester Co.), is a residential suburb of New York, settled in 1687 by Huguenots and named after La Rochelle, France.

Newton, Massachusetts (Middlesex Co.), is a suburb of Boston, and has manufactories of boots and shoes, silkmills, rubber works, printing works, worsted mills, etc. It has a theological seminary and other institutions of learning. It was settled in 1631 and was incorporated in 1688 under the name of New Cambridge, which name it retained until 1692. The city charter was granted in 1873.

New Westminster, British Columbia, Canada, is a fresh water port on the Fraser river, 13 miles from the Strait of Georgia and is a center of lumbering, farming and fishing industries. Its industries include large sawmills, many fish canneries, iron works, railway car building works, etc.

New York, New York (c. h. New York Co.), is the largest city of the state and of the United States, the second largest city of the world, a port and coaling station, and the chief portal of the foreign commerce of the United States. It is a great manufacturing as well as trading center, with extensive manufactures of sugar, tobacco, clothing, brass and copper, chemicals, patent medicines, iron and steel, boats and sails, glass and glassware, india-rubber and leather products, instruments for scientific purposes, gold and silver wares, malt liquors and distilled spirits, products of cotton, wool, and wood. It has large printing, lithographing, engraving, and map-making works, in which industries New York holds the first place in America. It is noted for its magnificent public and commercial buildings, especially for those of extraordinary height called "sky-scrapers." Among its prominent buildings are the Custom-House, with its splendid sculptured groups emblematic of the continents, a large quadrangular granite building in the French Renaissance style, erected in 1902-7, and occupying the site of Fort Amsterdam; the Stock Exchange, the Produce Exchange, the United States Sub-Treasury, the Singer Building, the Metropolitan Life Building, the Hudson Terminal Building, the City Hall, the Hall of Records, and the Appellate Court-House. Trinity Church, a handsome Gothic edifice of brownstone, dates from 1839 and occupies the site of a church of 1696. Grace Church, with its rectory, chantry, and church-house, forms, perhaps, the most attractive ecclesiastical group on the continent. St. Patrick's Cathedral, a white marble building in the decorated Gothic style, is doubtless the most important church edifice in the United States. It was designed by James Renwick and erected in 1850-79 at a cost of $3,500,000. The Cathedral of St. John the Divine, designed by Heins and La Farge, the corner-stone of which was laid in 1892, is in course of construction and is intended to be the largest and most important cathedral of the Episcopal church. The Metropolitan opera-house is a large and handsome building. New York is the seat of Columbia University, attended by upwards of 5,000 students and ranking with the foremost universities of America; it was founded in 1754 as King's College; in 1890, the institution was reorganized on a broad university basis and now consists of Columbia College proper for men, Barnard College for women, and schools of law, medicine, mines, engineering, chemistry, architecture, music and design, education (teachers' college), pharmacy, political science, philosophy, and pure science. Its library contains 427,000 volumes. Among the other educational institutions are the Normal College, the Union Theological Seminary, the College of the City of New York, attended by 4,000 students, and New York University, also attended by 4,000 students. The Metropolitan Museum of Art contains, perhaps, the foremost collection of paintings, antiquities, and statuary in the western hemisphere. The American Museum of Natural History is the most important institution of its kind in America, and its buildings, planned to cover an entire city block of 18 acres, house one of the largest and most interesting collections of the kind in the world. The New York Public Library contains over 1,000,000 volumes and over 400,000 pamphlets. The first settlement of Manhattan Island

was made by the Dutch in 1624 and the little town was christened Amsterdam, which, by 1650, had about 1,000 inhabitants. In 1664 the town was seized during a time of peace by the English, and, though retaken by the Dutch in 1673, it passed permanently into English possession by a treaty in the following year. The name of the town was then changed to New York in honor of the Duke of York. In 1765 delegates from nine of the thirteeen colonies met in New York to protest against the Stamp Act and to assert the doctrine of no taxation without representation, and the first actual bloodshed of the Revolution took place here in 1770. The town was occupied by Washington in 1776 but, after the battles of Long Island and Harlem Heights, the Americans retired, and New York became the British headquarters for seven years. The British troops evacuated the city on November 25, 1783. From 1785 to 1790 New York was the seat of the Federal government, and it was the state capital down to 1797. In 1807 the first steamboat was put on the Hudson River and in 1825 a great impulse was given to the city's commerce by the opening of the Erie Canal. In July, 1863, the opposition in New York to the draft act culminated in a riot, which lasted three days, cost over 1,000 lives, and destroyed property worth about $1,500,000. In March, 1888, New York was visited by a terrible blizzard. In 1890 a Commission was appointed by the state legislature to inquire into the expediency of the consolidation of New York with Brooklyn and other contiguous towns and cities, and in 1896 the act to make this consolidation became a law. In 1897 the charter of Greater New York was finally passed. In 1909 New York was connected with New Jersey by underground tunnels.

Nezhin, Russia (Chernigof), is a trade center in a tobacco growing district.

Niagara Falls, New York (Niagara Co.), is famous on account of the Falls, one of the great scenic wonders of the world, and is visited annually by thousands of tourists. The power of the falls is utilized for extensive manufactories of paper, flour, cereals, electrochemical products, machinery, lumber products, etc.

Niagara Falls, Ontario, Canada, is a popular summer resort and manufacturing center. Power is supplied by natural gas and electric generating plants. Its industries include cereal foods, carborundum, wire chain, etc.

Nice, France (cap. Alpes-Maritimes), is one of the largest winter health resorts of the Riviera, picturesquely situated at the foot of the Maritime Alps. It exports wine and liqueurs, olive-oil, glass, pottery, fruits, and soap, and manufactures art pottery and olive-wood inlay work. It is the birthplace of Garibaldi.

Niigata, Japan (cap. Niigata), manufactures coarse lacquer ware, and there are coal and petroleum deposits in the vicinity.

Nijmegen, Netherlands (Gelderland), produces tobacco, metal wares, and beer. The principal church (1272) contains a fine monument to Catherine of Bourbon. Here treaties of peace were signed between Spain, France, Austria, and the Netherlands (1678-9).

Nikolaief, Russia (Cherson), is a fortified port and naval station on the Bug River, 20 miles from its mouth. It has an Orthodox cathedral, Admiralty buildings, an observatory (1821), and a school of navigation. The dock-yards and naval arsenal are important, and there are industries of rope, tallow, tobacco, and candies. It is, with Sevastopol, the principal naval station of Russia on the Black Sea.

Nimes, France (cap. Gard), manufactures shawls, carpets, silks, wine, and brandy. It is an episcopal see, and possesses a modern cathedral. Nîmes is noted for its Roman antiquities, among which are the amphitheater, in excellent preservation; the Maison Carrée one of the finest Corinthian temples of the Roman period, now a museum; and an aqueduct. The town was the birthplace of the historian Guizot (1787) and of the novelist Daudet (1840).

Ningpo, China (Chehkiang), is a treaty port with large exports of rice, cotton, varnish, oils, sepia, and bamboo. It is noted for its wood-carving and em-

broideries, and is the meeting-place of large fishing fleets.

Nissa, Servia, is a fortified city and the seat of a bishop. It was the birthplace of Constantine the Great, and was long a bone of contention between Hungarians and Turks.

Niuchwang, China (Shengking), is a treaty port in South Manchuria, on the Liao River, 25 miles from its mouth. Exports beans and bean-cake, castor-oil, tobacco, silk, etc.

Nishni Novgorod, Russia (cap. Nishni Novgorod), is a port on the Volga and the seat of one of the greatest European fairs, held annually, from July to September, to which some 400,000 visitors and traders resort. In the Kremlin are the cathedrals of the Transfiguration (founded in 1227) and of the archangel Michael (founded in 1222), the governor's palace, the tower of Prince Dmitri (1374, rebuilt in 1869), the tower of the Czar Ivan III. (1500), the arsenal, a military cadet school, and the law-courts. The principal industrial establishments are flour-mills, metal foundries, machine-works, and distilleries. Nishni Novgorod was founded in 1212, and became the capital of a separate principality in 1350. It was sacked by the Tartars in 1377, and united to Moscow in 1392.

Nishni Tagilsk, Russia (Perm), is a mining and iron and steel manufacturing city. Iron, copper, gold, and platinum are found in the vicinity.

Nome, Alaska, is the center of an important mining district on Seward Peninsula, and, next to Fairbanks, is the largest place in Alaska. In 1898 gold was discovered here in the beach sands of the coast and shortly afterwards rich placer deposits in the interior were found.

Nordhausen, Germany (Prussia), has distilleries, and manufactures of tobacco, cigars, chicory, sugar, leather, wall-paper, and machinery.

Norfolk, Virginia (Norfolk Co.), is the second largest city of Virginia, also a port and coaling station, situated on an arm of the Chesapeake Bay. It is an important railroad and steamship terminus, and has an extensive trade in lumber, coal, grain, cotton peanuts, oysters, fruit, and vegetables. The manufacturing industries include cotton knitting mills, cotton compress mills, fertilizer factories, shipyards, tobacco and sugar factories, iron-foundries, machine-shops, steel-works, lumber mills, silk-mills, etc. It was settled in 1608. The Norfolk navy yard, located at Portsmouth, on the opposite side of the Elisabeth River, is one of the largest naval stations in the United States. It was bombarded and nearly destroyed by the British on Jan. 1, 1776. Norfolk was incorporated as a borough in 1736 and chartered as a city in 1845.

Norristown, Pennsylvania (c. h. Montgomery Co.), is a manufacturing center in a rich agricultural section. It has manufactories of knitting-machines, hosiery, glass, iron, wire, agricultural implements, furniture, etc. Valley Forge and other places of historical interest are in the vicinity. It was settled in about 1688 and incorporated as a borough in 1812.

Norrkoping, Sweden (Oster Gotland), is a port with a good harbor, situated at the head of Braviken Bay. It manufactures cottons, paper, tapestries, and sugar, and has ship-building yards.

North Adams, Massachusetts (Berkshire Co.), has extensive manufactures of cotton and woollen goods, boots and shoes, cigars, etc. It is the seat of a state normal school, and was settled in 1765.

Northampton, England (cap. Northampton), is one of the oldest towns in the country, and has several ancient interesting churches, such as St. Sepulchre's (12th century), one of the few round churches still remaining in England, and St. Peter's, a fine example of Norman architecture. It is the center of the boot and shoe manufacture of England.

Northampton, Massachusetts (c. h. Hampshire Co.), is a trade center in an active agricultural section, and has extensive manufacturing and commercial interests. A state asylum for the insane, Smith College, and a classical school for girls are located here. The "Home Culture Club", founded by George

W. Cable and endowed by Andrew Carnegie, gives instruction in household arts and amusements. The city was settled in 1654.

North Tonawanda, New York (Niagara Co.), is a manufacturing suburb of Buffalo. It has large lumber yards and manufactories of steam-pumps, merry-go-rounds, pig-iron, and lumber products.

North Yakima, Washington (c. h. Yakima Co.), is the largest city of central Washington and the trade center of the rich Yakima Valley, an agricultural region largely under irrigation. It has a large trade in fruits, grains and agricultural produce. Its industries are chiefly canneries, flour-mills, and wood-working establishments.

Norwich, Connecticut (c. h. New London Co.), is at the head of navigation of the Thames River, about 14 miles from Long Island Sound. Its exceptional water-power has contributed to extensive development of manufacturing industries. The chief manufactures are firearms, leather, silk fabrics, cotton and woollen goods, machinery, stoves, iron products, and furniture. A state armory and a state insane hospital are located here.

Norwich, England (cap. Norfolk), has manufactures of boots and shoes, ready-made clothing, and agricultural implements, iron-foundries, tanneries, breweries, and flour-mills, as well as mustard, starch, and vinegar works. The cathedral begun in 1096 is said to preserve its Norman plan with less alteration than any other English cathedral. The largest church, St. Peter Mancroft, is a handsome 15th century edifice; St. Andrew's was built in 1506; St. Michael Coslany is an example of fine flint and stone work; St. Andrew's Hall, a handsome Gothic structure, was formerly the nave of the Blackfriars monastery church. Norwich was a British and Roman town; was burned by Sweyn; became the seat of the bishopric of East Anglia in 1094; received a colony of Flemish weavers in the 14th century and became an important center for cloth manufactures. It was one of the leading towns in England in the 17th century.

Nottingham, England (cap. Nottingham), is the center of the English lace and hosiery manufacture, and has also manufactories of silk, leather, and machinery. It has a castle, built by William the Conqueror, which contains now a museum and an art gallery, a University College, and a very large market-place. It is the seat of a Roman Catholic bishop. Nottingham was one of the Five Boroughs of the Danes, and was reconquered by Edward the Elder. Here Mortimer and Queen Isabella were captured in 1330. Charles I. raised his standard here, in 1642, at the beginning of the civil war.

Novara, Italy (cap. Novara), has manufactures of silks, cottons, and linen. It has several fine churches, notably a Romanesque cathedral, dating from the 14th century and rebuilt in 1870. The Italian army was totally defeated here by the Austrians under Radetsky in 1849.

Novgorod, Russia (cap. Novgorod), is one of the oldest, and in mediæval times was one of the largest cities of Russia. It was one of the leading commercial centers of Europe, and was a member of the Hanseatic League. It was the capital of an independent state, but was brought under the dominion of Moscow about 1478, and was sacked by Ivan the Terrible in 1570. Its commercial importance has been entirely destroyed by the foundation of St. Petersburg, and by the building of railroads. It is the seat of a Greek Orthodox archbishop. The Cathedral of St. Sophia, within the walls of the highly picturesque Kremlin, or citadel, was built in the middle of the 11th century by workmen from Constantinople; and, despite several restorations, it retains in great measure its Byzantine character.

Novocherkask, Russia (cap. Don Cossacks), has extensive vineyards, fisheries, and flour-mills. An Orthodox cathedral has been building since 1893; the library and museum illustrate the history of the Don Cossacks. There are coal-mines in the vicinity.

Nuka, Russia (Elisavetpol), has silk raising and spinning interests.

Nuremberg, Germany (Bavaria), is one of the richest towns on the continent in mediæval buildings and works of art, and still retains its ancient walls and moat. The churches are full of priceless paintings, statuary, and carvings. Of its many famous collections the Germanic museum is the most valuable, and a remarkable library, dating from 1445, is preserved in the old Dominican monastery. The castle, dating from 1050, was enlarged by Frederick Barbarossa, and has served as residence for many German emperors. The principal industries are the manufacture of toys, optical and other scientific instruments, cycles, automobiles, bronzes, and the brewing of beer. Nuremberg was made a free city in 1219, and retained its independence till 1803, when Napoleon I. bestowed it upon the king of Bavaria.

Nyireghyhaza, Hungary, has manufactures of soda, saltpeter, and matches, and there are large vineyards in the neighborhood.

Oakland, California (c. h. Alameda Co.), is situated on the east shore of San Francisco Bay, and is connected with San Francisco by ferries. Its harbor accommodates the largest ocean vessels and the city has 15 miles of water frontage. High-tide jetties, forming the harbor entrance, are built of rubble stone. Its excellent facilities for transportation are greatly increasing the amount of its shipments each year, and it is the commercial center of a large agricultural region. The principal industrial establishments are iron-works, ship-building works, canneries, flour-mills, planing-mills, and lumber-yards. It has numerous educational institutions, including a Baptist college, a Roman Catholic academy, etc.

Oaxaca, Mexico (cap. Oaxaca), has sugar-mills, breweries, and fiber factories. It has handsome churches and other public buildings. Near the city stood Uaxyaca, the ancient capital of the Zapotecs.

Oberhausen, Germany (Prussia), is the center of a mining district, and has foundries, chemical works, and porcelain and glass factories.

Odenburg, Hungary, manufactures sugar, preserved fruit, agricultural implements, etc.

Odense, Denmark (cap. Fünen), exports butter, cheese, hides, bacon, corn, and molasses. It is the seat of a bishop and contains the churches of St. Knud (11th century) and Our Lady (12th century). It was the birthplace of Hans Andersen.

Odessa, Russia (Cherson), is a coaling station, the chief seaport and commercial center of southern Russia, and one of the largest cities of the realm. It is especially noted for its export of grain; but has also large exports of sugar, flour, wool, hides, flax, tallow, caviare, fish, and cattle. It has a university with a good library and a museum; a cathedral; a museum of antiquities, which contains many interesting Greek and Scythian remains, and various educational and scientific institutions. It was founded in 1794 and was bombarded by the English and French forces in 1854.

Offenbach, Germany (Hesse), is a great industrial center, and exports leather goods, carriages, chemicals, soap, carpets, varnish, etc.

Ogden, Utah (c. h. Weber Co.), is an important railroad center in a fertile agricultural region and in the vicinity of productive mines. The chief industrial establishments are canneries, beet sugar factories, and woollen mills. A state industrial school, a state institution for the deaf, dumb, and blind, a Mormon academy, and a Catholic academy are located here. It was founded in 1848, and, in 1850, under the direction of Brigham Young, was laid out as a city.

Ogdensburg, New York (St. Lawrence Co.), is a port of entry on the St. Lawrence River and is connected by ferry with Prescott, Ontario. It is at the foot of deep water navigation of the Great Lakes and has excellent transportation facilities. Shipping and wholesale dealing in lumber, grain, and coal are the most prominent industries. It also has manufactories of silk, flour, gloves, lumber, and lumber products. A state armory and a state hospital for the insane are located here.

Oil City, Pennsylvania (Venango Co.), is a trade

center in the midst of the most productive oil fields of Pennsylvania, and many of its industries are connected with the marketing of petroleum. On June 5, 1892, a terrible fire swept over the city, caused by burning oil coming down Oil Creek from Titusville, and more than 100 persons were killed, while property exceeding $1,000,000 in value was destroyed.

Okayama, Japan (cap. Okayama), contains an important mission and is noted for its castle and superb gardens.

Oklahoma, Oklahoma (c. h. Oklahoma Co.), is the largest city of Oklahoma and an important trade center in a rich agricultural region. The chief manufacturing establishments are cotton-mills, flour-mills, iron-works, woodworking shops, etc. It has an extensive trade in grain, cotton, and live stock, and is the seat of the Epworth University, a college for girls, and the Oklahoma Military Institute. It was settled on April 22, 1889, when, by proclamation, the country was declared open for settlement.

Oldenburg, Germany (cap. Oldenburg), exports leather goods, soap, machinery, and musical instruments.

Oldham, England (Lancashire), is the chief center of the cotton-spinning industry, over 12,000,000 spindles being employed, more than one-fourth of the total for Lancashire and neighboring district. The fabrics produced include velvets and velveteens, fustians, sheetings, nankeens, and sateens. There are also large engineering works. In the district are important collieries.

Olean, New York (Cattaraugus Co.), is a trade center in a fertile agricultural region and near the Pennsylvania oil fields. It has a large distributing trade in petroleum, which is brought by pipe lines from the oil fields south of the city. The chief manufacturing establishments are lumber mills, tanneries, oil refineries, glass factories, wagon and carriage factories, brick-yards, flour-mills, and furniture factories. A state armory is located here. It was settled in 1804.

Olmutz, Austria (Moravia), manufactures malt, beer, sugar, starch, and alcohol. It is an archiepiscopal see and has interesting churches and a university founded in 1581. Lafayette was imprisoned here in 1794, and Ferdinand I. here resigned the crown to his nephew in 1848.

Olympia, cap. Washington (c. h. Thurston Co.), is a port of entry at the head of navigation of Puget Sound. It has fishery interests and a large trade in fruit, agricultural produce, etc. The State Capitol is an imposing structure built of native sandstone.

Omaha, Nebraska (c. h. Douglas Co.), is the largest city of the state and one of the most important railroad centers of the middle west. Among the important industries are extensive breweries and distilleries, smelting and refining works. It has an extensive trade in wheat and corn and is an important cattle and hog market. Its wholesale and jobbing trade in merchandise extends over a large area. The city owes its commercial importance to its position as one of the chief gateways to the west, and has grown rapidly since its foundation in 1854. Omaha is the headquarters of the military department of the Missouri, and Fort Omaha, just north of the city, is the chief signal-service, balloon, and "wireless" experimental station of the United States army.

Omdurman, Egypt (Sudan), was the Dervish capital and here, on Sept. 2, 1898, the Anglo-Egyptian troops, under Lord Kitchener, defeated the Dervishes, completely destroying the Mahdi's power.

Omsk, Russia (cap. Akmolinsk), is a fortified town and contains a cathedral and governor-general's palace. Its industries include brick and pottery works, breweries and distilleries, tobacco, oil, and soap manufactories.

Onomichi, Japan (Hiroshima), is a port on the Inland Sea and has manufactures of fancy mats. It has two interesting old temples.

Oporto, Portugal (cap. Oporto), is the second city of the country and the center of the port wine trade. It is also a great manufacturing center, the chief industries being the spinning and weaving of cotton,

woollen, and silk, distilling, sugar-refining, and tanning. There is an ancient Gothic cathedral, modernised in the 18th century, and an interesting municipal museum with many notable paintings.

Oppeln, Germany (Prussia), exports leather goods, cement, wood, and cattle.

Oran, Algeria, is a fortified seaport and coaling station, with exports of esparto-grass, iron ore, grain, wool, hides, and wine, most of the trade being with the Spanish coast. Originally built by the Moors, Oran was captured by the Spaniards in 1509 and held by them, with a break of twenty-four years, till 1792. It came into the possession of France in 1830.

Orange, New Jersey (Essex Co.), is a residential suburb of New York, which has extensive hat manufactories and is the site of the Edison laboratory. It was settled in 1666.

Orebro, Sweden (cap. Orebro), has a trade in mining products. At Orebro in 1529, Lutheranism was established as the state religion.

Orel, Russia (cap. Orel), has manufactures of candles, ropes, oil, and flour; and also distilleries, breweries, brick works, and tobacco factories. Here Turgenev was born.

Orenburg, Russia (cap. Orenburg), has an extensive trade with Central Asia, and exports frozen meat, tallow, skins, wool, butter, and cheese. Between June and November caravans arrive from all parts of Russian Central Asia and West Siberia. It contains military schools, an Orthodox seminary, and an arsenal.

Orihuela, Spain (Alicante), has silk and other textile industries

Orizaba, Mexico (Vera Cruz), is the trading center of a rich sugar region, and has manufactures of flour, tobacco, and textiles.

Orleans, France (cap. Loiret), is an important railroad and commercial center, and has also small industries of cottons, linen, and leather. It contains a late-Gothic cathedral built in 1601; a city hall, a museum (15th century); and the houses of Joan of Arc and of Agnes Sorel. It was besieged in 1428-9 by the English, but was delivered by Joan of Arc, who is therefore called the Maid of Orleans.

Oruro, Bolivia, has mines of silver and tin. Altitude 12,117 feet.

Osaka, Japan (cap. Osaka), has cotton spinning, glass and iron-works, and ship-building. Foreign trade has been allowed since 1868. Its temples and castle (1583) are famous,and it has an arsenal and a mint.

Osh, Russia in Asia (Ferghana), has on the west Takht-i-Suleiman, or the "Throne of Solomon", a prominent rock, famous in Oriental legend, and an object of pilgrimage to Moslems of Central Asia.

Oshkosh, Wisconsin (c. h. Winnebago Co.), is a trade center in a region having extensive lumbering interests. The chief manufactures are lumber and lumber products, grass twine, and agricultural implements. It is the seat of a state hospital for the insane and a state normal school. It was settled in 1836.

Oskaloosa, Iowa (c. h. Mahaska Co.), is a trade center in a fertile agricultural region in which stock-raising receives much attention. The chief industries are flour- and grist-milling, and the manufacture of vitrified brick, wagons, iron and brass goods, and woollen goods. It has a large meat-packing plant. In the vicinity are large coal-fields and deposits of limestone and fire-clay. Penn College, opened in 1873 by the Society of Friends, and the Oskaloosa College, under the auspices of the Disciples of Christ, are located here.

Osnabruck, Germany (Prussia), has a fine cathedral and a 14th-century Gothic church It manufactures cigars, chemicals, nails, machinery, and musical instruments.

Ostend, Belgium (West Flanders), is a port, coaling station and a famous seaside resort. Public buildings include the handsome casino, royal chalet, town hall, and the fine hotels facing the sea. The fisheries are important, and there are large oyster parks. It exports rabbits to England.

Oswego, New York (c. h. Oswego Co.), is a port of entry and has an extensive lake trade. It has a large starch factory, machine-shops for oil-well supplies, boiler and engine works, knitting mills, car spring works, match factories, etc. A state arsenal and a state normal and training school are located here. It was founded in 1724 as a trading post and military station, and was chartered as a city in 1848.

Otaru, Japan (Hokkaido), is a seaport and a center for herring-fishing.

Otsu, Japan (cap. Shiga), was anciently a capital of the mikados and has a citadel, palaces, and wide streets. It manufactures especially the abacus (or calculating-frame). On May 11, 1891, it was the scene of an attempt on the life of Nicholas II. of Russia (then Czarewitch).

Ottawa, Illinois (c. h. Lasalle Co.), is a trade center in a productive agricultural region, in which are extensive deposits of fire-clay, glass sand, and bituminous coal. It has extensive glass-factories, potteries, agricultural implement shops, carriage factories, organ and piano factories, etc. A Lutheran college and a Roman Catholic college are located here.

Ottawa, Ontario, cap. Canada, is not only a most important town socially and politically, but is the center of a great lumber trade, due to its position on the Ottawa River, which brings down logs from the interior. This river forms the great Chaudière Falls at Ottawa, and, with its branches, the rapidly flowing Gatineau and Rideau rivers, furnishes motive power for many lumber- and flour-mills, and for factories in which agricultural implements, machinery, ironware, bricks, etc., are made. The very handsome Parliamentary buildings, including the circular library, stand in spacious grounds on a bluff overlooking the river, and crowned by the fine Victoria Tower (180 ft.). Rideau Hall, the governor-general's residence, two cathedrals, the national art gallery, and the museum of the Geological Survey are noteworthy buildings. There are various educational institutions, including the Ottawa University and a normal school. The city was founded under the name of Bytown in 1829, but was not incorporated until 1854, when its name was changed to Ottawa. In 1858 it was selected by Queen Victoria as the capital of the two Canadas.

Ottumwa, Iowa (c. h. Wapello Co.), is a trade center in an agricultural region, in which there are extensive coal-fields. It has good water-power, which, combined with the abundance of coal, has developed extensive manufacturing industries. The chief industrial establishments are mining and agricultural implement works, foundries, iron-works, pork-packing plant, flour-mills, etc. It was settled in 1849.

Ouro Preto, Brazil (Minas Geraes), has mines of iron, manganese, asbestos, and gold near by. A school of mines is located here.

Oviedo, Spain (cap. Oviedo), was the ancient capital of Asturias, and has an ancient cathedral, where some of the kings of Asturias are buried. It has a university, a national factory of arms, and metallurgical industries.

Owensboro, Kentucky (c. h. Daviess Co.), is a trade center in a farming and stock-raising region, and there are coal and iron fields near by. It has an extensive river trade. The chief manufactures are tobacco products, cellulose, whisky, brandy, brick and tile, and iron products.

Owen Sound, Ontario, Canada, is a port at the head of Owen Sound, an inlet of Georgian Bay. It has cement works, furniture factories, tanneries, etc.

Oxford, England (cap. Oxford), is chiefly noted as the seat of Oxford University, which dates from the 12th century and comprises 21 colleges. Among Oxford's numerous buildings are All Souls College (founded 1437); the Ashmolean Museum (1682); Balliol College (c. 1268); the Bodleian Library (1602; 500,000 books, 30,000 MSS.); Brasenose College (1509); Christ Church College (1525-46; its chapel, the "cathedral", dating from 1120 and onwards); the Clarendon Building (1712-30, till 1830 the University Press); Corpus Christi College (1516); the Divinity Schools (1445-80); the Examination Schools

(1882); Exeter College (1314); Hertford College (1874); the Indian Institute (1884); Jesus College (1571; still partly Welsh); Keble College (1870); Lincoln College (1429); Magdalen Conege (1458); Manchester College (1893); Mansfield College (1886); the Martyrs' Memorial (1841); St. Mary's Church (1300-1488), with a spire 180 ft. high; Merton College (1264); the New Museum (1856-60); New College (1379); Oriel College (1326); Pembroke College (1624); Queen's College (1340); the domed Radcliffe Library (1749; since 1861 a reading-room for the Bodleian); the Radcliffe Observatory (1795); St. John's College (1555); the Sheldonian Theatre (1669; in which "Commemoration" is held); the Taylor Institution (1843); Trinity College (1554); the Union Society (1823; new building 1859); University College (1249); the University Press (1830); Wadham College (1613); and Worcester College (1714). To these may be added Somerville Hall (1879), Lady Margaret Hall, and St. Hugh's Hall, all for women. The authentic annals of Oxford begin in 912, when it was annexed by Edward the Elder, king of the West Saxons. It was a place of strategical importance and one of the political centers in the middle ages; it was a meeting-place of the witenagemot. Harold Harefoot was proclaimed king there in 1036, and died there in 1039. The population in the time of Edward the Confessor is estimated at 3,000; in 1086 it was only 1,700. The castle was besieged by Stephen in 1141-42, Matilda escaping then over the frozen river. The city was the Royalist headquarters in the Civil War, and was taken by Parliamentarians under Fairfax in 1646. Oxford is the seat of an Anglican bishop.

Pachuca, Mexico (cap. Hidalgo), has rich silver mines and a meteorological observatory.

Padang, Sumatra, is a seaport with large export trade in coffee, spices, rubber, tobacco, hides, and copra.

Padua, Italy (cap. Padua), is the seat of a bishop, and has a cathedral (16th century), with a collection of valuable miniatures. The Palazzo della Ragione was built as law-courts (1172-1219); and the university, especially famous in the 13th century, was founded in 1222. Padua, the ancient Patavium, was the birthplace of Livy, and an important Roman commercial center.

Paducah, Kentucky (c. h. McCracken Co.), has extensive manufactures of pig iron, lumber products, stoneware, pottery, saddles, tobacco products, knit goods, etc. It has a large ore milling plant and in the vicinity are deposits of fire-clay, glass sand, zinc, etc. It was settled in 1821.

Paisley, Scotland (cap. Renfrew), has manufactures of cotton thread, carpets, textiles, chemicals, corn-flour, starch, and preserves, engineering and electric works, and ship-building. The scholastic buildings include the John Neilson Educational Institution (1852) and the technical school (1899). There was here a Cluniac priory, raised to an abbey (1219), burnt by the English (1307), and rebuilt between 1445-1525. The present edifice formed the nave of the sanctuary, which was restored in 1902.

Pakhoi, China (Kwangtung), is a treaty port, opened to foreign trade in 1876, and has large exports of indigo.

Pa'embang, Sumatra (cap. Palembang), is a port on the Palembang River, 60 miles from the sea. It is a health-recruiting station for the troops of the Dutch East Indies. It contains a fine mosque (1740), and the ancient palace and tombs of the former native dynasty. Cotton, sago, pepper, honey, rattans, dyewoods, and gutta-percha are exported.

Palermo, Italy (cap Palermo), is a seaport, and the principal town of Sicily. It is the seat of an archbishop, and has a magnificent cathedral in the Gothic style, begun in 1180, which contains the tombs of Roger the Norman and Emperor Frederick II. It has a uni ersity, founded in the 14th century; a famous observatory, located in the royal palace; and large barracks, which occupy the former Palazzo Sclafani. The principal exports are olive-oil, wines, sulphur, and fruits. Palermo is a very ancient town, and was founded by the Phoenicians. It was occupied by the Romans in 254 B. C., and became one of their principal naval stations.

Palestine, Texas (c. h. Anderson Co.), is a trade center in a rich agricultural region, of which cotton is the leading product. It has cotton gins, lumber mills, machine shops, etc.

Palma, Spain (cap. Balearic Islands), is a fortified seaport, and has manufactures of flour, soap, starch, alcohol, liqueurs, glass, and leather, as well as shipbuilding yards and exports of fruit, wine, and olive-oil. It is the seat of a bishop, and has a fine cathedral in the Gothic style. It has a seminary for priests, a normal school, a school for painting, and a fine picture gallery.

Pamplona, Colombia, is the seat of a Roman Catholic bishop. It has breweries and match factories.

Pamplona, Spain (cap. Navarra), is a strongly fortified town, and has a cathedral dating from the middle ages. It has manufactures of paper, leather, etc.

Panama, cap. Panama, is a port and coaling station, and exports pearls, mother-of-pearl, rubber, and mahogany. It has a large transit trade. The original Panama, founded in 1518, lay 4½ miles to the northeast, and was destroyed in 1671 by the buccaneer Morgan. The present town dates from 1673. Its most noteworthy sights are the cathedral, the bishop's palace, and a few famous ruins.

Pantin, France (Seine), is an industrial suburb of Paris, and has glass-works, sugar-refineries, and railway shops.

Paoting, China (cap. Chili), is a trade center, which figured prominently in the Boxer insurrection of 1900, and was captured by the French.

Pará, Brazil (cap. Pará), also called Belem, is a port on the estuary of the Para River, 85 miles from the Atlantic Ocean, and forms the principal entrance for ships into the Amazon River. It is, therefore, a great trading center, and has large exports of rubber, cocoa, sugar, coffee, cinchona, and hides. Its exports of rubber exceed that of any other port in the world. Pará is the seat of a bishop, and has a cathedral dating from 1720. It has a museum, botanical gardens, a fine theater, and several educational establishments. Pará was founded in 1615, but its actual prosperity dates from the last quarter of the 19th century.

Paramaribo, cap. Dutch Guiana, is a port and the center of trade of the colony. Coffee, cocoa, sugar, and rum are exported.

Parana, Argentina (cap. Entre Rios), was the capital of Argentina from 1852 to 1861, and is now the residence of a bishop.

Paris, cap. France (cap. Seine), is the third largest city of the world, and as much of a center for commercial and manufacturing occupations as it is for art, literature, science, fashion, and kindred interests. The fancy goods industry takes the first place, including gold and silver-work, jewelry, dressmaking, high-class cabinet-making, and all manufactures known as French goods, and employs 500,000 work-people. Paris is strongly fortified, and extends along both northern and southern banks of the Seine, in which lies the Ile de la Cité, the earliest original site of the city. On this small island, the heart of Paris, are some ancient buildings; one group includes the Palais de Justice, the Conciergerie, a prison which won evil fame during the Revolution, and the exquisite Sainte Chapelle. At the eastern end of the island stands the noble cathedral of Notre Dame. The two banks of the Seine are connected by numerous bridges, some of them old and of handsome design. On the left, or southern, bank are the Quartier St. Germain, the home of the French aristocracy, and the Diplomatic Corps; and the Quartier Latin, famous for its picturesque student life. Here are located schools of art, medicine, law, science, literature, and the like, the Institute of France and the University of Paris, including the Sorbonne, for the faculties of literature and science, with about 12,000 students. Here, also, is Mont Ste. Geneviève, crowned by the Panthéon, set apart as a mausoleum for Frenchmen considered worthy of fame, the interior of which is decorated by well-known artists depicting episodes in France's history. The gardens of the Luxembourg containing the Luxembourg Palace, where the Senate convenes, and the noted gallery of modern painting and sculpture; the Observatory; the Chamber of Deputies; the Hôtel des Invalides, containing a museum of artillery, and Napoleon I.'s tomb; the great Salpêtrière hospital; the factory of the Gobelins; the Odéon theater; and the Jardin des Plantes are among the more celebrated institutions on the south bank of the Seine. The Hôtel Cluny, an old mansion containing collections of mediæval objects, lace, carriages, etc., is built on the site of the palace of the Frankish kings, and shows ruins of Roman baths. On the right bank of the Seine lies the magnificent open space known as the Place de la Concorde, flanked by the broad tree-bordered Champs Elysées, near which is the Elysée Palace where the President of the Republic resides, and by the gardens of the Tuileries, a royal palace destroyed during the Commune. These gardens extend to the Louvre, one of the most beautiful palaces in the world, which was at first a fortress and was rebuilt by Francis I. in 1541. It contains priceless treasures of art, both paintings and sculpture. The social life of Paris and much of its business concentrate about the grand boulevards, broad tree-shaded avenues which girdle the older part of the city. Two triumphal arches, erected by Louis XIV., the Porte St. Martin, and the Porte St. Denis, divide the stream of traffic on the busiest of these boulevards, and in their vicinity are the Madeleine, a church fashioned like a Greek temple, and the magnificent Opéra, celebrated alike for its music and its architecture, and for its museum of dramatic history. The Palais Royal, built by Cardinal Richelieu, the scene of many notable gatherings of historical personages and containing the Théâtre Francais; the Hôtel de Ville, a reproduction of the one burned in 1871, splendidly decorated; the national library, of about 2,500,000 books, and considered to be the richest library of the world as regards rare books and old and modern bindings; the Bourse; the quaint little church of St. Germain l'Auxerrois, dating from the 13th century, from whose belfry sounded the signal for the massacre of the Huguenots of France on St. Bartholomew's Day; and the curious tower of St. Jacques, a fragment of an ancient church are all clustered on the right bank. There, also, are the Halles Centrales, the great markets of Paris. There are numerous columns, monuments, and triumphal arches, the largest of which is the Arc de Triomphe de l'Etoile commenced by Napoleon I., commemorating various events in the turbulent history of France. The Palace of the Trocadéro, erected for the Universal Exhibition of 1878, has a hall capable of holding 5,000 spectators; and across the Seine the Eiffel tower, erected for the exhibition of 1889, reaches a height of 984 feet. There are many churches of interest in the city, both for age or architecture or for decoration, but the most prominent is the church of the Sacré Cœur, which stands alone on the top of Montmartre, and commands the whole capital. It is modern, and as yet unfinished, and forms a vast basilica in the Roman style, with a Byzantine dome 272 ft. high. The Bois de Boulogne is a great park in the city, affording a fashionable driving place and promenade and there are other splendid parks in and about Paris, including the grounds about the palaces of St. Cloud, Chantilly, and the more distant Fontainebleau and Versailles. Even the cemeteries of Paris are renowned, especially that of Père Lachaise, crowded with mortuary chapels and sculptured memorials. Other educational institutions of Paris are the Collège de France; Musée d'Histoire Naturelle; Conservatoire des Arts et Métiers (applied sciences); Ecole Normale Supérieure; Ecole des Ponts et Chaussées (for civil engineers); Ecole des Beaux-Arts, founded in 1648, one of the most famous schools of painting, sculpture, engraving, and architecture in the world; Ecole Polytechnique; and the Conservatory of Music and Declamation. Paris was the ancient capital of a small Gallic tribe, the Parisii; was the capital of Constantius Chlorus (292-306); was made the capital of the Frankish kingdom by Clovis in 508; was ruled by counts under the Carlovingians; became again the capital under the Capetians (10th century); was largely developed under Philip Augustus and St. Louis; suffered from civil strife under Charles VI.; was entered

by Henry V. of England in 1420, but expelled the English in 1436; was the scene of the massacre of St. Bartholomew in 1572; and was the scene of many of the leading events in the first revolution and in those of 1830 and 1848, as well as during the Commune in 1871. International expositions were held here in 1855, 1867, 1878, 1889, and 1900.

Parkersburg, West Virginia (c. h. Wood Co.), has oil-refineries, oil-well supply works, flour and lumber mills, and an extensive trade in manufactured goods, farm products, and coal. In the vicinity are valuable clay and coal deposits, gas and petroleum wells and several medicinal springs noted for their quality. It was settled in 1773.

Parma, Italy (cap. Parma), is the seat of a bishop, and has a cathedral dating from the 11th century, which contains Correggio's "Assumption." It possesses a university founded in 1599, and among other notable buildings are the 12th century baptistery; a former ducal palace; a library; a museum of antiquities; and a picture-gallery, with famous paintings by Correggio, Giulio Romano, and Parmigiano. Parma has a seminary for priests, an agricultural college, a school of fine arts, and an academy of music. It has manufactures of silk, woollens, linen, hats, and tobacco, and large printing establishments. The town, known to the ancient Romans as Gallia Cispadana, was colonized in 183 B. C. The Roman road Via Æmilia crosses it, dividing it into two equal parts. In 1346 it passed into the possession of the Visconti family, and until 1512 its fate was linked with that of the duchy of Milan.

Parsons, Kansas (Labette Co.), is a trade center in a productive agricultural region. It has extensive grain elevators, creameries, agricultural implement works, and railroad shops. A state hospital for the insane is located here.

Partick, Scotland (Lanark), has ship-building yards, brass and other foundries, cotton factories, and flour mills.

Pasadena, California (Los Angeles Co.), is a health resort and residential city, famous for its beautiful homes and gardens. Its annual floral parade attracts thousands of visitors.

Passaic, New Jersey (Passaic Co.), is a manufacturing suburb of New York. It has good water-power and the chief industrial establishments are large rubber works, silk-mills, woollen mills, chemical works, etc. It was settled in 1692, and is near the scene of several of the engagements of the Revolutionary War. Washington crossed the Passaic River here in 1776, when he was retreating through New Jersey.

Paterson, New Jersey (c. h. Passaic Co.), is the third city of the state, and one of the most important manufacturing centers in the country, with abundant water-power afforded by the Passaic River, which here makes a descent of about 70 ft. over the Passaic Falls. It is the principal silk manufacturing center of the western hemisphere, this industry alone employing 16,000 wage-earners. It also has extensive works for dyeing and finishing textiles. Steel and iron manufactures rank next in importance. There are large breweries and many smaller industries. Paterson was founded in 1791; on February 8, 1902, the heart of the business section of the town was destroyed by fire, causing a loss of over $6,000,-000. Less than a month later, the greatest flood ever known in the Passaic Valley caused great loss of property, and, in October, 1903, another great flood occurred, with heavy loss. Notwithstanding these calamities, the city is prosperous and progressive.

Patiala, India (Punjab), is the capital of a native state of the same name, and has marble quarries and copper mines.

Patna, India (Bengal), has a government college and a Roman Catholic cathedral. Patna rice is famous, and other articles of export are cotton, oil-seeds, and salt. Opium is manufactured.

Patras, Greece (cap. Achaia), is a fortified seaport on the north coast of Morea. The principal buildings are the citadel, the cathedral, and the castle. Its principal exports are currants, olives, olive-oil, and lemons.

Pau, France (cap. Basses-Pyrénées), is a favorite winter resort in the foothills of the Pyrenees. It possesses golf links, laid out in 1854, which, except Blackheath, are the oldest links in the world outside Scotland. Pau was the ancient capital of French Navarre and Béarn. Its history centers round the castle, rebuilt in 1363 by Gaston Phoebus, Comte de Foix. Pau was the birthplace of General Bernadotte, who became King of Sweden.

Pavia, Italy (cap. Pavia), is the seat of a bishop, and has a cathedral dating back to 1488. The university (1,500 students), said to have been founded by Charlemagne in 774, was restored by Galeazzo Visconti, Count of Pavia, in 1361. The municipal museum, a valuable collection of paintings and engravings, is housed in the Palazzo Malaspina. Charlemagne took the town, which was known to the ancient Romans as Ticinum, in 774. Church councils were held here in 1061, 1160, and 1423. Here also was fought in 1525 the great battle which resulted in the defeat of the French and the capture of their king, Francis I., by the troops of the Emperor Charles V. The town was annexed by Austria in 1714, and was joined to the kingdom of Italy in 1859.

Pawtucket, Rhode Island (Providence Co.), is an important manufacturing city with extensive water-power furnished by the Blackstone River, which, at this place, has a fall of 50 ft. It has extensive cotton and woollen mills, silk factories, foundries, machine-shops, electrical supply works, wire-works, yarn mills, hosiery mills, knit goods factories, textile dyeing and finishing establishments, lumber products and factories, boot and shoe factories, etc. It has a state armory and a state emergency hospital. It was settled in 1654, and cotton manufacturing was first introduced into the United States at Pawtucket.

Paysandu, Uruguay, is the second port of the country, and has a large export trade in preserved meat.

Peabody, Massachusetts (Essex Co.), has extensive manufactories of leather, shoes, electrical supplies, thermometers, and glue. It contains the Peabody Institute, founded in 1852 by George Peabody.

Peekskill, New York (Westchester Co.), has knit goods factories, brick works, foundries, machine-shops, etc. A Roman Catholic home for children has over 1,000 inmates. The state military camp ground is just north of the village. It was settled by the Dutch in 1664.

Peking, cap. China (Chili), is the largest city of China and consists of an inner and outer—that is, a Manchu and a Chinese—city. The former is 14½ miles in circuit, and has 3 gates on the south and 2 on the northwest and east sides. The wall is about 45 ft. high and 47 ft. thick. Within the Manchu city lie the imperial palace and parks, the palaces of the nobles, the public offices and barracks, the examination hall, the Confucian and other large temples, and many private residences. In the Chinese city, which projects on the south of the Manchu city, and is enclosed by a wall 22 ft. high with 7 gates, are shops, theatres, restaurants, club-houses, the temples of heaven and agriculture; but large spaces are unoccupied, and other ground is under cultivation or used for desiccating manure. The summer palaces of Yuan-ming-yuan, Wan-shou-shan, and I-ho-yuan lie from 5 to 6 miles north of the city; and in the plain surrounding the city are many picturesque graveyards belonging to noble families. The tombs of the last dynasty lie about 20 miles north of the city, and those of the present dynasty from 2 to 3 days' journey east and west of Peking. Peking has, except for short intervals, been an imperial residence for more than 900 years, and was an important place long before that date. It probably owes its choice as a capital in large measure to its favorable position from a geomantic point of view. As the country around it is too poor to support its population, supplies have to be imported from the south. This need led to the construction of the Grand Canal, and of a shorter canal across the promontory of Shan-tung. Peking became one of the capitals of the Khitan Tatars at the end of the 10th century; was rebuilt by Kublai Khan; and has been sole capital since the beginning of the 15th

century. It surrendered to the Anglo-French force in 1860, and was taken by the allied forces sent in 1900 for the relief of the foreign legations.

Pelotas, Brazil (Rio Grande do Sul), has a trade in jerked meat, and has flour-mills and glass factories.

Pensa, Russia (cap. Pensa), has breweries and tanneries, iron and bell foundries, brick works, sawmills, and flour-mills, and manufactures soap, candles, wax, cloth, paper, and celebrated camel's-hair stuffs. It is the seat of a Greek Orthodox bishop and possesses a fine cathedral. It has a botanical garden, a school of agriculture, and several technical colleges.

Pensacola, Florida (c. h. Escambia Co.), is a port of entry on Pensacola Bay, 6 miles from the Gulf of Mexico, and an important railroad terminus. It has large domestic trade in lumber, fish, naval stores, phosphate, cotton, and coal. A United States navy yard and a state armory are located here. Pensacola was settled in 1696.

Peoria, Illinois (c. h. Peoria Co.), is the commercial and industrial center of a large extent of country. The chief manufacturing establishments are meat-packing houses, distilleries, strawboard factories, breweries, flour and lumber mills, glucose works, wagon and carriage factories, wire fence works, and agricultural implement works. Peoria ranks first among the cities of the Union in the products of distilleries; it is also noted for the quality and annual output of fine wine. It has a large trade in grain, live stock, and manufactures. A polytechnic institute and a Roman Catholic academy are located here. Near by are large bituminous coalfields.

Périgueux, France (cap. Dordogne), contains the remarkable cathedral of St. Front, in the Byzantine style (984-1047). Périgueux was the Vesunna of the Romans, and has a large amphitheatre and the circular tower of Vesone, 89 ft. high. It trades in buildingstone, figs, and truffles; has industries of chinaware, iron, and woollens; and is famous for its pâtés de foie gras and truffled partridges.

Perm, Russia (cap. Perm), is a river port, and has tanneries, brick fields, distilleries, machinery and chemical works, sawmills, copper foundries, soap, candle, rope, and pottery manufactures, and a cannon foundry. It is the seat of a Greek Orthodox bishop, and has a fine cathedral.

Pernambuco, Brazil (cap. Pernambuco), also called Recife, is a port and coaling station, with exports of cotton, sugar, carnauba wax, hides, and skins. The public buildings include the palaces of the governor and bishop, and the Pernambuco Archæological and Geographical Institute.

Perpignan, France (cap. Pyrénées-Orientales), is a fortress of great strength, commanding the passage from Spain across the east Pyrenees. Perpignan did not become French till 1642, and is still half-Spanish, half-Moorish in appearance, while its people resemble those of Catalonia. It has a 14th century cathedral, and from 1349 to the revolution had a university. Trades in red wine, brandy, cork, silk, and wool.

Perth, Scotland (cap. Perth), is, on account of the charming scenery of its surroundings, called the "Fair City." It is the chief center of the Scottish dyeing industry, and has manufactures of linen, carpets, glass and ink. It has also large cattle markets. The salmon fisheries of the Tay are very important. Perth has been prominent in Scottish history. The kings of Scotland frequently resided here. James I. was murdered here in 1436. It was taken by Bruce in 1311, by Montrose in 1644, by Cromwell in 1651, and by the Jacobites in 1715 and 1745.

Perth, cap. West Australia, Australia, has numerous fine buildings, including a town hall, Anglican and Roman Catholic cathedrals, council chambers, banks, museum, library and art gallery, mechanics' institute, and hospital. It is the seat of a Roman Catholic and of an Anglican bishop.

Perth Amboy, New Jersey (Middlesex Co.), is a port of entry with an extensive shipping trade. The chief manufacturing establishments are smelting and refining plants, ship-building yards, brick, terra-cotta, iron and steel works, chemical works, and railroad shops. In the vicinity are valuable deposits of fire-clay. The first settlement was made in 1683.

Perugia, Italy (cap. Perugia), is an ancient town, and contains many interesting mediæval buildings. It is the seat of a bishop, and has an unfinished Gothic cathedral, begun in the 15th century, which contains a library with a valuable collection of manuscripts. Other interesting buildings are the Collegio del Cambio, with the celebrated frescoes of the Chief Virtues by Perugino; and the Palazzo del Municipio, which contains a picture gallery with fine paintings by Perugino and Pinturrichio, a valuable collection of Umbrian remains, and a library with rare mediæval manuscripts. It has a university founded in 1308, which contains a valuable collection of Etruscan, Roman, and mediæval antiquities; a seminary for priests; an academy of arts; and an agricultural college. It has manufactures of silk, and a trade in wine and olive-oil. Perugia was an important Etruscan town, and was taken by the Romans in 309 B. C. It was incorporated with the Papal states in 1512, and annexed to Italy in 1860.

Pesaro, Italy (cap. Pesaro e Urbino), has manufactures of majolica ware, silk and woollen goods, sealingwax, and cream of tartar. Figs, for which the neighborhood is famous, wine, oil, silk, soap, wax, iron, and lead are exported. It contains a ducal palace, built in 1455 by Francesco Sforza, and an academy of music. It is the seat of a bishop.

Peshawar, India (cap. Northwest Frontier Province), is an important strategic point near the Khyber Pass, on the route from India to Kabul.

Peterborough, England (Northampton), is a town grown up around a Benedictine monastery, founded in 655, destroyed by the Danes in 870, rebuilt in 970, burnt in 1116, and refounded in 1117. The greater part of the building was completed previous to 1200. The magnificent west front of 3 arches, the distinguishing feature of the cathedral, was erected between 1200 and 1238, and the eastern chapel in the 15th century. During the restoration of 1887-8 the foundations of the second Saxon church (burnt in 1116) were discovered. Two queens were buried here—Catherine of Aragon, and Mary Queen of Scots, the latter only till 1612. Portions of the cloisters and of other monastic buildings still remain, and within the precincts are the museum and Young Men's Institute. The town has trade in agricultural produce, manufactures bricks, and has large railway locomotive works.

Peterborough, Ontario, Canada, is a trade center noted for its canoes. It has manufactures of sawmill machinery, packing houses, etc.

Petersburg, Virginia (Dinwiddie Co.), is a trade center in an agricultural region of which tobacco is one of the chief products. Owing to its good water-power and the abundance of raw material near at hand, it has large manufacturing interests. The chief manufacturing establishments are tobacco factories, cotton-mills, knitting mills, silk factory, machine-shops, and lumber mills. A state hospital for the colored insane, a state normal and collegiate institute for colored students, and several other educational institutions are located here. It was founded in 1733. It was the scene of several engagements during the Revolutionary War, and twice it was occupied by the British troops under Gen. Philips. During the Civil War it was the scene of many of the engagements of the famous Virginia campaign.

Petropolis, Brazil (cap. Rio di Janeiro), is a popular watering-place, and has cotton and other factories.

Pforzheim, Germany (Baden), has manufactures of gold and silver ornaments, chemicals, paper, and machinery.

Philadelphia, Pennsylvania (c. h. Philadelphia Co.), is the third city of the United States in extent and population. Its commerce by sea and land is very large, and, as a manufacturing center, it ranks next to New York and Chicago. It has on its 2 rivers over 30 miles of frontage, docks and wharfage. The chief manufactures are machinery, locomotives, ironware, ships, cotton and woollen goods, silk, leather, cigars, drugs, and chemicals. It is the seat of the University of Pennsyl-

vania with about 3,000 students, which maintains in its college department, besides its usual arts course, courses in architecture, mechanical engineering, music, etc. Its medical school has about 500 students and its law school about 400 students. Girard College, founded in 1831 by Stephen Girard, for the education of poor white male orphans, is one of the richest and most notable philanthropic institutions in the United States, It now accommodates over 1,500 boys, and the value of Mr. Girard's bequest of $5,260,000 has been increased to about $35,000,000. Hahnemann College and Hospital is one of the chief homœopathic institutions of the kind in the country. The Drexel Institute, founded by A. J. Drexel and opened in 1892, is an institution for "the extension and improvement of industrial education as a means of opening better and wider avenues of employment to young men and women." The Free Library of Philadelphia contains 280,000 volumes and has 19 branches in different parts of the city. Ridgway Library contains 120,000 volumes, including many rarities. The Pennsylvania Academy of the Fine Arts, founded in 1805, has an extensive collection of over 500 paintings, sculptures, casts, and engravings, and maintains an important art school. The Academy of Natural Sciences, founded in 1812, contains large collections of natural history and a library of 50,000 volumes. Independence Hall is, in some respects, the most interesting building in the United States; here the Continental Congress met during the American Revolution (1775-81) and here, on July 4, 1776, the Declaration of Independence was adopted. A United States Navy Yard, with a dry dock measuring 750 ft. by 134 ft., is located on League Island. The Philadelphia Mint, established here in 1792, is the parent mint of the United States. Philadelphia was founded in 1682 by a Quaker colony under William Penn, who purchased the site from its Indian owners. The city received its charter from Penn in 1701, when it had about 4,500 inhabitants. From that time to about the 19th century, it rivaled Boston as the leading city of the country, and was the scene of the most important official acts in the Revolution. The first Continental Congress assembled here in 1774; the Declaration of Independence was signed here on July 4, 1776. The Constitution of the United States was drawn up and promulgated here in 1777. The first President of the United States resided here, and here Congress assembled until 1797. From September, 1777, to June, 1778, the city was in the possession of the British. During the 19th century, its history was one of quiet and rapid growth. In 1876 Philadelphia was the scene of the Centennial Exhibition, held in honor of the 100th anniversary of the Declaration of Independence.

Philippopolis, Bulgaria (Eastern Roumelia), manufactures silk, cloth, woollens, leather, and tobacco, and prepares and exports large quantities of attar of roses. It is the seat of a Greek Orthodox archbishop. In Roman times it was the capital of Thracia.

Phillipsburg, New Jersey (Warren Co.), has large railroad shops, foundries, and machine-shops, horseshoe works, stove works, silk-mills, etc. It was settled in 1749.

Phœnix, cap. Arizona (c. h. Maricopa Co.), is the trade center of the fertile Salt River Valley, and one of the most important commercial cities of Arizona. It has creameries, flour-mills, olive-oil mills, etc. The remains of several prehistoric towns have been found in the vicinity.

Piacenza, Italy (cap. Piacenza), is an important strategic and commercial center, and has manufactures of silk, cottons, hats, and pottery. It is the seat of a bishop and has a cathedral in the Lombardo-Romanesque style (dating from the 12th century), with an immense crypt, a campanile 223 ft. high, and paintings by Carracci, Guercino, and others. The church of San Sisto (1499) formerly contained the famous Sistine Madonna by Raphael. There is a public library possessing a valuable collection of manuscripts. Founded as a Roman colony at the same time as Cremona, Placentia, as it was then called, was destroyed by the Gauls in 200 B. C. The French, under Louis XII., took it, but it was recaptured by Pope Julius II.; and it remained a

papal domain until 1545, when Paul III. gave it to his son, Peter Farnese. It formed part of the duchy of Parma until incorporated in 1860 with the kingdom of Italy.

Piatra, Roumania (Moldavia), has a large trade in timber.

Pietermaritzburg, cap. Natal, Africa, has government buildings, a town hall, hospital, and asylum, and is the see of an Anglican bishop.

Pilsen, Austria (Bohemia), is famous for its beer, though it also produces sugar, clothing, porcelain and pottery, machinery, furniture, hardware, spirits, leather, sulphuric acid, and paper. Coal, iron, and alum are mined. The 13th century church of St. Bartholomew, the town hall (1555), and the museums are the chief public buildings.

Pine Bluff, Arkansas (c. h. Jefferson Co.), is an important trade center in an agricultural region in which cotton is one of the chief crops. It has cotton-gins, cotton compresses, cottonseed-oil mills, foundry, furniture factory, sheet iron works, railroad shops, etc. The annual state fair is held here. It is the seat of the state colored normal college and other educational institutions.

Pingyang, Korea, is an ancient city, and was the scene of battles between Japan and China in 1592 and 1894, China winning the former, Japan the latter.

Pinsk, Russia (Minsk), has potteries, tanneries, breweries, oil and soap works, and leather manufactories.

Piotrkof, Russia (Poland), is one of Poland's oldest towns, and has remains of a royal mediæval castle.

Piqua, Ohio (Miami Co.), has large linseed-oil works, strawboard mills, woollen mills, furniture factory, rolling-mills, iron-works, and sheet steel mills.

Piræus, Greece (Attica), is the chief port of entry of Greece, and exports olives, olive-oil, and marble. It has an arsenal, and some cotton-mills. It was founded by Themistocles and Pericles; was destroyed by Sulla in 86 B. C.; and was rebuilt in the 19th century. It was in ancient times connected with Athens by the "Long Walls."

Pirmasens, Germany (Bavaria), manufactures boots and shoes and leather goods.

Pisa, Italy (cap. Pisa), has cotton manufactures and exports olive-oil and marble. It is the seat of an archbishop and has a magnificent cathedral in the Gothic style, commenced in 1063, and completed in 1118. In front of the cathedral stands the circular marble baptistery (12th century), adorned with numerous columns and a dome nearly 200 ft. high. The Campo Santo, or cemetery, dates from the 13th century, and contains numerous interesting monuments, sarcophagi, friezes, etc. The famous leaning tower, a campanile, built entirely of marble, 178½ ft. high, was commenced in 1174 and completed in 1350. Amongst secular buildings are the university, founded in 1343; the Palazzo del Commune; the municipal museum; and the Academy of Arts. Pisa is an ancient Etruscan town, and became subject to Rome in the 2nd century B. C. Between the 11th and 14th centuries it was a powerful commercial republic, the rival of Genoa and Venice. After its great naval defeat by Genoa in 1284, Pisa began to decline until at the beginning of the 15th century it became subject to Florence.

Pistoja, Italy (Florence), is the seat of a bishop, and its 12th century cathedral is rich in works of art. Among other famous churches and secular buildings are the Madonna dell' Umiltà, San Giovanni, San Domenico, the Palazzo del Commune (1294), and the Ospedale del Ceppo (1277). Pistoja is famous for its gun-barrels and pistols, the latter of which are said to have derived their name from the town. It was here that Catiline fell in the battle of 62 B. C.; while during the middle ages the feuds between the Bianchi and Neri frequently devastated the town.

Pittsburg, Kansas (Crawford Co.), is a trade center in a productive agricultural region and has extensive railroad shops, foundries and machine-shops, packing houses, planing-mills, rolling-mills, glass factories, brick plants, etc. In the vicinity are coal-mines, and a state

manual training normal school is located here. It was settled in 1876.

Pittsburg, Pennsylvania (c. h. Allegheny Co.), is the second city of Pennsylvania, and one of the chief industrial centers of the United States, situated at the junction of the Monongahela and Allegheny rivers, which here unite to form the Ohio. The sister city of Allegheny, situated on the north bank of the Allegheny and extending down to the Ohio, was incorporated with Pittsburg in 1907, and is now known as the North Side. The great basis of the prosperity of the "Iron City" has been the fact that it stands in the center of one of the richest coal districts of the globe, and of the chief supply of natural gas, the use of which as a fuel gave a great impetus to its manufacturing industries, although the gas is now mainly used for domestic purposes. The staple manufactures of Pittsburg are iron, steel, and glass. The Pittsburg district produces 65 per cent. of all the coke manufactured in the United States, and the output of bituminous coal is enormous. Pig iron, structural iron and steel, steel rails, armor-plate for battleships, electrical machinery, air-brakes, locomotives, steam-engines, and other machinery are extensively manufactured. The largest manufactory of corks in the world is located here. It also has one of the most famous establishments for the manufacture of instruments of precision. It is engaged in the manufacture of salt, fire brick, building brick, stoves, brass ware, bronze ware, white lead, pottery, leather, paper, and lumber. It is also the center of large and important chemical industries and is heavily engaged in the production of food products, meats, preserves, and pickles. The Western University of Pennsylvania, founded in 1787, has about 1,000 students in its various departments of law, literature, engineering, medicine, dental surgery, and pharmacy. Theological seminaries of the Presbyterian, United Presbyterian, and the Reformed Presbyterian churches are located here. The Pennsylvania College for Women is an important institution, and there is a large number of denominational schools and private academies. The University of Pittsburg has about 1,000 students. The Carnegie Institute, with an extensive gallery of art, one of the largest museums in America, and an exceptional and perfect music hall, has, in recent years, achieved a world-wide reputation as a center of artistic and scientific activity. Connected with the institute is a school for industrial training, for which Andrew Carnegie has given an endowment of $2,000,000. The Carnegie Free Library of Pittsburg, founded by the princely benefactor whose name it bears, is the greatest establishment of its kind in America, contains upwards of 300,000 volumes, and maintains branch libraries and numerous agencies for the free distribution of literature within Greater Pittsburg. Pittsburg occupies the site of the French Fort Duquesne, erected in 1754 and abandoned on the advance of Gen. Forbes in 1758. Its place was taken by the English Fort Pitt, and the laying out of the town of Pittsburg may be dated from about 1765.

Pittsfield, Massachusetts (c. h. Berkshire Co.), noted for its beautiful homes, is situated in the center of the Berkshire Hills. The chief industrial establishments are foundries and machine-shops, electrical works, paper-mills, etc. The Berkshire Athenæum, the Crane Art Museum, and a public library with 50,000 volumes, are among the institutions of note. It was settled in 1743.

Pittston, Pennsylvania (Luzerne Co.), is in the midst of the anthracite coal region, and its chief industries are connected with the mining and shipping of coal. It has large flour and lumber mills, knitting mills, foundries, machine-shops, brick and terra-cotta works, stove and engine works, breweries, etc. It was settled in 1770.

Plainfield, New Jersey (Union Co.), is a residential suburb of New York, and has a number of manufacturing establishments, the chief of which are tool works, machine-shops, kid glove factories, silk and cotton mills, aluminum works, etc. It was settled in 1684.

Plattsburg, New York (c. h. Clinton Co.), is a port of entry on Lake Champlain and the center of a region in which there are many famous summer resorts. The chief manufacturing establishments are lumber mills, sewing-machine works, flour-mills, pulp-mills, etc. It is the seat of a state normal and training school, and has numerous other educational institutions. Adjoining the village on the south is a United States military reservation. It was settled in 1784. The first naval battle between American and British forces took place on October 11, 1776, off Valcour Island, a short distance from Plattsburg.

Plauen, Germany (Kg. of Saxony), has extensive lace-making, embroidery, and white cotton goods industries.

Ployesti, Roumania (Wallachia), has a considerable trade in wool, and large petroleum refineries.

Plymouth, England (Devon), is a strongly fortified port, and one of the chief naval stations of the country, with large naval and military establishments. It has an extensive trade, and exports lead, tin, copper, fish, and machinery. Plymouth was the starting-point of the expedition against the Armada in 1588, and the last European point touched by the Mayflower in 1620. It was unsuccessfully besieged by the Royalists in the civil war.

Plymouth, Massachusetts (c. h. Plymouth Co.), is of abiding interest as the landing-place of the Pilgrim Fathers, and the site of the first settlement in New England. The Plymouth Rock, on which the Pilgrims landed from the Mayflower in December, 1620, is carefully preserved and is now covered by a granite canopy. Other places of interest in the town are Pilgrim Hall, where books, pictures, and other valuable relics of early times are kept, and Cole's Hill and Burial Hill, where many of the first settlers are buried. Plymouth also has the national monument to the Pilgrims, dedicated in August, 1889. Its manufactures include duck, cordage, wire nails, and electrical supplies; its fisheries are of considerable importance. On account of its beautiful situation and historic interest, Plymouth is a popular summer and tourist resort.

Plymouth, Pennsylvania (Luzerne Co.), is situated in the anthracite coal region, and its chief industries are the mining and shipping of coal. It also manufactures drills and other mining machinery. It was settled in 1768.

Point de Galle, Ceylon, was for a long time the chief port of the island, but since the transference of the shipping traffic to Colombo, trade has much declined, although cocoanut, plumbago, and tea are still exported.

Poitiers, France (cap. Vienne), is the seat of a bishop, and has a cathedral founded in 1161 by Eleanor of Aquitaine, queen of Henry II. of England. Other noteworthy buildings are the church of St. Hilaire le Grand, in which rest the remains of the famous bishop of Poitiers (320-68), and the church of St. Radegonde, long a place of pilgrimage. In the neighborhood Clovis defeated Alaric, king of the Visigoths, in 507, and in 1356 the Black Prince defeated the French.

Pola, Austria (Coastland), is the principal naval station of Austria, with a snug, strongly fortified natural harbor. There are many Roman and mediæval remains, and a modern cathedral. Pola was taken in 1148 by Venice, and repeatedly destroyed during the contests between the Venetians and the Genoese.

Poltava, Russia (cap. Poltava), has breweries, brick works, tanneries, flour-mills, and manufactories of cloth, flannel, soap, candles, tobacco, carts, and carriages. Poltava is the seat of a Greek Orthodox bishop, and contains a cathedral (1710) and a military school. Two monuments (besides the cathedral) commemorate the victory of July 9, 1709, when Russia, under Peter the Great, definitely worsted Sweden under Charles XII.

Ponce, Porto Rico (cap. Ponce), is the second largest city of Porto Rico, and the first in commercial importance. Coffee is its most important export. It also exports sugar, rum, molasses, and tobacco. At its port, Playa Ponce (2½ miles), are a large harbor, accommodating vessels drawing 25 ft., and the custom-house. Ponce has a cathedral, hospitals, theatres, fine residences, and gardens, and in the vicinity are many extensive plantations.

Pondichery, India (Madras), is the chief settlement

of the French in India. It has no harbor, and is dependent on artesian wells for its water-supply. The chief industry is weaving and dyeing. The French first settled here in 1674; the town was taken by the Dutch in 1693, and by the British in 1761, 1778, 1793, and 1803. It was finally restored to the French in 1815.

Pontiac, Michigan (c. h. Oakland Co.), has large carriage manufactories, machine-shops, flour and lumber mills, gasoline engine works, etc. The Eastern Michigan Asylum for the Insane, with 1,200 patients, is located here. Pontiac was settled in 1818.

Pontypridd, Wales (Glamorgan), has mines of iron and coal, foundries, and chain and cable works.

Poona, India (Bombay), is the headquarters of the Bombay army, and the seat, during the rainy season, of the government of the presidency. Gold, silver, and brass ware, ivory-carving, paper-making, and the modelling of small clay figures are its chief industries. It has two arts colleges and a college of science. Poona is the center of Brahmanical influence in Western India.

Portage la Prairie, Manitoba, Canada, is a railroad and trade center in a rich agricultural section. It has numerous grain elevators, threshing machine and farm implement works, oatmeal mills, etc.

Port Arthur, China (Manchuria), is a fortified port on Kwangtung peninsula. Port Arthur was taken by the Japanese in 1894 during their war with China, and in 1898 it was leased by China to the Russians, who strongly fortified it, and made it the terminus of the Siberian Railway. Here the Japanese commenced war with Russia in February, 1904, by a torpedo attack, and early in January, 1905, the town capitulated to the Japanese after a siege of 7 months.

Port Arthur, Ontario, Canada, is a port on Thunder Bay, Lake Superior, with a large transhipment trade. Its industries include a blast furnace, boiler works, stock yards, elevators, etc.

Port au Prince, cap. Haiti, is a port at the inner extremity of Gulf of Gonaive and is the seat of the bulk of the foreign trade of Haiti. Exports coffee, cocoa, mahogany, etc. It is partially fortified, and is an archbishop's see.

Port Elizabeth, Cape Colony, Africa, is a port on Algoa Bay, and the third city of the colony. Exports wool, skins, and hides.

Port Huron, Michigan (c. h. St. Clair Co.), is a port of entry, the southern outlet of Lake Huron, and is a favorite summer resort. The chief industries are shipbuilding works, engine and thresher works, fiber works, etc. A tunnel under the Saint Clair River, over a mile long, connects this city with Sarnia, Ontario. It was settled in 1790.

Portland, Maine (c. h. Cumberland Co.), is the largest city of Maine, and an important seaport and railroad terminus, which is noted for its exports of grain, fruit, and live stock. Its industries include boot and shoe factories, rolling-mills, foundries, machine-shops, locomotive works, engine and boiler works, petroleum refineries, lumber mills, etc. Fishing and the shell-fish industry are extensively pursued. A United States marine hospital and a state hospital are located here. Its harbor is deep, and well protected by 6 forts. Casco Bay, an admirable yachting water, is crowded with pretty wooded islands, many of which are favorite summer resorts. The Wadsworth-Longfellow Mansion, erected by Gen. Peleg Wadsworth, the poet's grandfather, in 1785-6, is of note and, in an addition built in 1903, contains the library of the Maine Historical Society. The town was founded in 1633. It suffered severely at the hands of the French and the Indians in 1775; and was almost totally destroyed by the Indians. After the war it was rebuilt.

Portland, Oregon (c. h. Multnomah Co.), is the largest city of the state, a port of entry with an extensive coastwise and transpacific trade, and an important railroad terminus, located on the banks of the Willamette River, 120 miles from the Pacific Ocean. Its exports are chiefly wheat, flour, wool, fish, and timber. The country surrounding Portland is rich in timber and minerals, and a vast agricultural region is tributary to this rapidly growing metropolis. Its principal manufactures are lumber, flour, furniture, woollen goods, cordage, clothing, canned goods, wagons, and carriages. Excellent water-power for manufacturing purposes is obtained from the Willamette Falls, Oregon City, 12 miles distant. Among the educational institutions are the Portland University and the medical and law schools of the State University. Portland ranks among the wealthiest cities of its size in the country. It was settled in 1843 and its growth since has been rapid and uninterrupted.

Port Louis, cap. Mauritius, is a port and coaling station, and the chief commercial center of the colony. In 1810 it was taken by the British.

Porto Alegre, Brasil (cap. Rio Grande do Sul), is a port and the most important city of southern Brazil. It exports tobacco, Paraguay tea, hides, and cattle.

Port of Spain, cap. Trinidad, is a port and coaling station, and exports cocoa, sugar, and asphalt. It contains a royal college, governor's house, and botanical gardens.

Porto Novo, cap. Dahomey, Africa, has considerable trade in palm-oil, kola-nuts, etc.

Porto Praya, Cape Verde Islands, exports coffee.

Port Said, Egypt, is a port and coaling station at the Mediterranean entrance to the Suez Canal. Exports cotton and oil-seeds.

Portsmouth, England (Hampshire), is a fortified port, coaling station, and the chief naval station of England. It has the greatest arsenal, and is the most strongly fortified place in the kingdom. The church of St. Thomas Becket is notable. Portsmouth rose to importance in the 13th century, and was strongly fortified in the 16th century.

Portsmouth, New Hampshire (Rockingham Co.), is the only seaport of the state, with some coasting trade, principally in coal, but very little foreign commerce. It has cotton-mills, marble works, boot and shoe factories and some ship-building. It contains many colonial houses of historical interest, chief among which are the Wentworth house and the Langdon house. The Athenæum contains a library of 25,000 volumes. The United States Navy Yard, which is officially known as the Portsmouth Navy Yard, is situated on an island named Piscataqua, within the limits of the township of Kittery, Maine. It has a large dry dock, and, in the days of wooden ships, many were built here. The treaty of peace between Russia and Japan was signed here on September 5, 1905. Portsmouth was first settled in 1623 and was under the jurisdiction of Massachusetts until 1679, when New Hampshire became a separate colony. It was incorporated as a town in 1653, when it received the name of Portsmouth, and became a city in 1849. It was the capital of the state for some time and is now, alternately with Concord, the seat of the sessions of the United States district court of New Hampshire.

Portsmouth, Ohio (c. h. Scioto Co.), is a commercial center of an extensive mining and manufacturing section, situated at the confluence of the Scioto and Ohio rivers, and at the southern terminus of the Ohio canal. In the vicinity are valuable deposits of fire-clay. The chief manufactures are foundry and machine-shop products, paving and building brick, cars, lumber products, stoves, rectified spirits, carriages, furniture, etc. In the vicinity are interesting remains of the Mound Builders. It was settled in 1803.

Portsmouth, Virginia (c. h. Norfolk Co.), is an important trade and railroad center opposite Norfolk, with which it is connected by ferry. It has a good harbor and considerable coastwise trade. It exports large quantities of lumber, cotton products, naval stores, pig iron, fruit, and vegetables. It has a United States navy yard, which has a plant for constructing steel ships, and there are two dry docks. It has also a United States navy hospital. It was settled in 1752.

Posen, Germany (cap. Posen), is the most important fortress on Germany's eastern frontier. Agricultural implements, machinery, liqueurs, beer, and cigars are the principal products. Posen was a flourishing member of the Hanseatic League during the middle ages.

Potosi, Bolivia, was long famous for the silver mines

in the neighboring mountain (Cerro de Potosi), where silver was discovered in 1545, but the production has greatly decreased of late. The tin mines are, however, still extensively worked. The town has a university, a Bolivian mint, and a metallurgical museum.

Potsdam, Germany (Prussia), contains the palace of Sans-Souci and the castle of Babelsberg, the former built by the Great Elector in 1660. Frederick the Great greatly embellished the town and its surroundings. To the Friedenskirche is attached a mausoleum, in which lie the Emperor Frederick and his consort, the Princess Royal of Great Britain. Cottons, woollens, tobacco, and sugar are manufactured.

Pottstown, Pennsylvania (Montgomery Co.), is the trade center of a fertile agricultural region. There is a considerable amount of mineral wealth in the vicinity. It has large manufactories of iron and steel, rolling-mills, blast-furnaces, steel-mills, bridge works, boiler works, carriage works, automobile factories, etc.

Pottsville, Pennsylvania (c. h. Schuylkill Co.), is one of the most important mining centers and shipping points in the Schuylkill anthracite coal region. Has steel mills, blast-furnaces, rolling-mills, and foundries. Here anthracite coal was first successfully used in smelting iron ore. It was settled in 1800.

Poughkeepsie, New York (c. h. Dutchess Co.), is an important commercial center on the east bank of the Hudson River, which is here spanned by a cantilever bridge 1½ miles long and 200 ft. above high water. Its chief industrial establishments are foundry and machine-shops, lumber mills, mowing-machine works, flour-mills, etc. Two miles north of the city is the Hudson River State Hospital for the Insane. Two miles east of the city is Vassar College, with about 1,000 students, the first and perhaps the best known of the American colleges for women, founded in 1861 by Matthew Vassar.

Prague, Austria (cap. Bohemia), is the commercial and manufacturing center of Bohemia, and has engineering and iron-works, and manufactures of chemicals, cement, earthenware, porcelain, furniture, cottons, carpets, paper, and hats, as well as breweries, distilleries, flour and lumber mills, and printing establishments. Prague contains many historic buildings. The stone Charles's Bridge (1357-1503), adorned with statues and groups, is an object of pilgrimage to Czechs, because of the martyrdom there of John of Nepomuk (1383). The Teyn church (1360-1460) played a prominent part in the religious troubles of Bohemia, and contains the tomb of the Danish astronomer, Tycho Brahé (1546-1601). The Clementinum contains the archiepiscopal seminary, the university library, the astronomical observatory, and part of the university. The last named, founded in 1348, was in 1882 divided into a German and a Czech section. Other buildings of note are the cathedral (begun 1344), and the Bohemian parliament palaces. Prague contains two polytechnic schools, an academy of music, an academy of fine arts, the Bohemian National Museum, and several big monasteries. It is the seat of a Roman Catholic archbishop. Founded apparently about the 8th century, Prague was developed in the 13th and 14th centuries. It became the center of the Hussites, having been taken by them in 1424. The Thirty Years' War commenced there in 1618. It was taken by the Imperialists in 1620, the Saxons in 1631, Wallenstein in 1632, the Swedes in 1648, the French and Bavarian troops in 1741, the Imperialists again in 1743, and Frederick the Great in 1744. Near it, May 6, 1757, the Prussians under Frederick the Great defeated the Austrians under Charles of Lorraine. It was taken by the Prussians in 1866.

Prato, Italy (Florence), has a cathedral dating from the 12th century and containing a Madonna by Andrea della Robbia (1489). Another noted church is that of the Madonna delle Carceri. It has manufactures of straw, woollen and cotton goods, and near by are quarries of serpentine.

Pressburg, Hungary, is a strongly fortified town, and the trading center of a fertile wheat and wine growing district. It has manufactures of furniture, musical instruments, gloves, silks, flour, spirits, and wine. It is the seat of a Roman Catholic bishop, and has a cathe-

dral, founded in the 11th century, where the kings of Hungary were crowned. On the capture of Buda by the Turks in 1541, Pressburg was declared the capital of Hungary, which, with a short interval, it remained until 1784.

Preston, England (Lancashire), is a port 12 miles from the mouth of Ribble River, and a center of the cotton-spinning industry. It has, besides, ship-building, iron and brass foundries, machine and electrical engineering works, and other industrial establishments.

Pretoria, cap. Transvaal, Africa, was laid out in 1855, and named after Pretorius, first president of the South African Republic. It became the seat of government in 1863. In 1900 it was entered by Lord Roberts. The Parliament House is the principal building.

Prince Albert, Saskatchewan, Canada, is an important trade and railroad center with extensive lumber interests, flour-mills, etc.

Prince Rupert, British Columbia, Canada, is the Pacific terminus of the Grand Trunk Pacific railway, with extensive harbor facilities. It has sawmills and other lumber industries.

Prizrend, Turkey (Kossovo), is famous for its manufacture of weapons; there are also steel, glass, pottery, and saddlery industries. It is the seat of a Roman Catholic and of a Greek Orthodox archbishop, and has a Greek Orthodox seminary for priests.

Prossnitz, Austria (Moravia), has manufactures of clothing, shoes, agricultural machines, and beer.

Providence, cap. Rhode Island (c. h. Providence Co.), is the largest city of the state, and a port of entry on the Providence River, a tidal arm of the Narragansett Bay, 25 miles from the Atlantic Ocean. It is the second city of New England, and one of the great industrial centers of the United States, noted for the variety of its manufactured products and the greatness of its industries. It has extensive manufactories of jewelry, silverware, cotton goods, gold and bronze ware, woollen goods, steam-engines, fine tubes and machinery, screws, locomotives, leather belting, rubber goods, boilers, stoves, as well as scores of other manufacturing plants. Brown University, founded in 1764, with upwards of 1,000 students, a state normal school, Rhode Island School of Design, Williams College, and other educational institutions are located here. The Athenaeum contains a library of 70,000 volumes and some interesting portraits. The city was founded by Roger Williams in 1636 after his expulsion from Massachusetts.

Przemysl, Austria (Galicia), is a strongly fortified town, and trades in wood, leather, corn, and linen. It is the seat of a Roman Catholic and of a Greek Orthodox bishop, and has two cathedrals.

Pskof, Russia (cap. Pskof), has tanneries, distilleries, saw mills, flour-mills, manufactures of tobacco, cordage, flax materials, and sail-cloth. Pskof is the seat of a Greek Orthodox archbishop, and has a cathedral (1689-98) in the Russo-Byzantine style. From 1348 to 1510, when it was finally conquered by Moscow, Pskof formed an autonomous city-state, closely similar to Novgorod.

Puebla, Mexico (cap. Puebla), is one of the oldest cities of the republic, and contains a fine cathedral, a state college, episcopal palace, and the Palafoxiana library. It has cotton and woollen mills, iron and bronze works, and potteries.

Pueblo, Colorado (c. h. Pueblo Co.), is an active commercial and industrial city and an important railway and mining center. In the vicinity are extensive coal and oil fields, also deposits of limestone and iron ore. The city is noted for its smelting and steel works, with many thousand employees. It also has large railroad car shops. There is a state mineral palace, with a complete collection of the minerals of Colorado. A temporary Mormon settlement was located here in 1846 and a trading post was established in 1850. In 1854 the Ute Indians massacred the residents of the post, and, in 1859, the present city was laid out.

Puerto Cabello, Venezuela (Carabobo), is a seaport with modern docks and an export trade in coffee, cocoa, hides, and cattle. It has handsome buildings, including the finest custom-house in Venezuela.

Puerto Plata, Santo Domingo, West Indies, is a

port with good anchorage, and has considerable trade in tobacco.

Punta Arenas, Chile (cap. Magellanes), is the southernmost town in the world, but only as far south as Dublin in Ireland is north. It has good wharves and warehouses and a growing trade in cattle, sheep, and furs. It is a Chilean naval station and a coaling station for the ships of all nations.

Quebec, cap. Quebec, Canada, is a port and coaling station on the St. Lawrence River, 400 miles from its mouth; the third city of Canada; and the most important military position in the Dominion. It is picturesquely situated, being largely built on the precipitous flanks of a bold headland, which is crowned by an ancient citadel. It is an important railroad and steamship terminal, and the largest ocean steamers can reach the city, but the port is usually closed by ice from the middle of December to April. It exports grain, cattle, and lumber, and has manufactories of iron, leather, and cotton. Quebec is the seat of a Roman Catholic cardinal archbishop and of an Anglican bishop, and has Roman Catholic and Protestant cathedrals. The fine Parliament house (1887) is also worth mentioning. Among the educational institutions are Laval University and Morrin College. The site of Quebec was visited by Cartier in 1535; the city was founded by the French under Champlain in 1608; taken by the British in 1629 and restored in 1632; unsuccessfully attacked by the British in 1690; besieged by the British under Wolfe in 1759, and taken after the battle of Quebec in Sept., 1759; ceded to Great Britain in 1763; and unsuccessfully attacked by the Americans under Montgomery in 1775. In the battle of Quebec, on the Plains of Abraham, near Quebec, Sept. 13, 1759, a victory was gained by the British under Wolfe over the French under Montcalm. It resulted in the fall of Quebec, and ultimately in the loss of Canada by the French.

Querétaro, Mexico (cap. Querétaro), was the scene of many patriotic episodes in Mexican history. The movement for independence began here, and here also the Emperor Maximilian was shot in 1867. Cotton and woollen goods are manufactured and opals are found in the vicinity.

Quezaltenango, Guatemala, Central America, has manufactures of cotton, linen, and woollen fabrics. In the vicinity are numerous interesting antiquities.

Quincy, Illinois (c. h. Adams Co.), is a trade center in a productive agricultural region, with extensive commercial and manufacturing interests. It has large machine-shops, stove factories, wire fence and brass casting works, lumber mills, canning factories, engine and plow works, flour-mills, etc. Among the educational institutions are a Roman Catholic college, opened in 1860, a Roman Catholic institute, with nearly 1,000 students, and a Methodist Episcopal school for boys. A soldiers' and sailors' home and a home for the aged and infirm are located here. It was settled in 1822.

Quincy, Massachusetts (Norfolk Co.), is a residential suburb of Boston, with many places of historic interest and some buildings of Revolutionary days. It has extensive quarrying industries, a large shipbuilding plant, and numerous educational institutions. It was settled in 1625 and came into great prominence in the last quarter of the 19th century, on account of an educational movement under the leadership of Francis Wayland Parker. It was the birthplace of John Adams, John Quincy Adams, and John Hancock.

Quito, cap. Ecuador, has manufactories of coarse cotton and woollen goods, blankets, carpets, leather, lace, etc. The most important buildings are the churches; others are the government house, the archbishop's palace, the municipal buildings, the astronomical observatory, the university, and the institute of sciences.

Racine, Wisconsin (c. h. Racine Co.), is a commercial and industrial center in a productive agricultural region, with a good harbor on Lake Michigan. It has extensive manufactories of agricultural implements,

carriages, wagons, automobiles, bicycles, copper goods, iron specialties, boots and shoes, etc. A Protestant Episcopal college, founded in 1852, and other educational institutions are located here. It was settled in 1834.

Radcliffe, England (Lancashire), has manufactures of cottons, chemicals, iron, and machinery, as well as coal-mines.

Radom, Russia (Poland), is the center of the Polish leather industry, and has sawmills, breweries, and tanneries, and a trade in cattle.

Ragusa, Italy (Syracuse), trades in oil, wine and cheese, and manufactures cotton and woollen goods. In the neighborhood are rock caverns and grottoes.

Raleigh, cap. North Carolina (c. h. Wake Co.), has, in addition to the various departments of the state government, state institutions for the insane, the blind, the deaf and dumb, also a state penitentiary. A national cemetery and a federal cemetery are located here. Among the educational institutions are a state college of agricultural and mechanical arts, with 600 students, a Baptist university, a Presbyterian institute, etc. Cotton and tobacco products are the leading articles of commerce. The largest manufacturing establishments are flour-mills, phosphate works, cotton-mills, and cottonseed-oil mills. The site of the present city was selected in 1792 by the legislature for the location of the state capital, and, in the same year, the city was founded and named in honor of Sir Walter Raleigh. In 1794 the legislature held its first session here. The city was occupied by Gen. Sherman during a part of 1865.

Rampur, India (United Provinces), is the capital of a native state of the same name, and has manufactures of damask, pottery, and jewelry.

Ramsgate, England (Kent), is a popular seaside resort with very bracing air, and has a handsome esplanade and a marine drive. Queen Victoria, before her accession to the throne, often resided here. It contains a Jewish theological college (1867), which has a valuable collection of Hebrew manuscripts.

Randers, Denmark (Jutland), has manufactures of iron, machinery, gloves, and watches. It possesses an interesting church dating from the 14th century.

Rangoon, India (cap. Lower Burma), is the chief port and coaling station of Burma, situated on a branch of the Irawadi River, 21 miles from the sea. It exports teak, petroleum, rice, and spices. Rangoon came into British possession in 1852, and since then it has been transformed into a prosperous modern mercantile city.

Ratibor, Germany (Prussia), has railway shops and manufactories of glass, iron, and steel.

Ratisbon, Germany (Bavaria), is exceptionally rich in mediæval remains and works of art. The cathedral of St. Peter was commenced in 1275. It has manufactures of tobacco, machinery, pencils, and soap. Six miles below the town, above the Danube, stands the Walhalla, or hall of fame for distinguished Germans. The town was founded by Tiberius, and was known to the Romans as Castra Regina. It was the capital of the Eastern Franks in the 9th century. Near the cathedral is the hall in which the German Diets held their meetings from 1645 to 1806. Numerous ecclesiastical councils have been held here.

Ravenna, Italy (cap. Ravenna), has numerous churches and monasteries, notably the cathedral or Basilica Ursiana, San Giovanni in Fonte, San Francesco (with Dante's tomb, 1482), Santo Spirito, and San Vitale. There is an Academy of Fine Arts and a famous library. It is the seat of an archbishop. The principal trade is in wine, silks, musical instruments, glass, soap, and starch. It was in old times a seaport and Augustus made it the headquarters of his Adriatic fleet. Under the succeeding emperors it became one of the chief cities of Italy. It was because of its strength as a fortress (owing to the marshes) that in 404 A. D., the emperor Honorius took up his abode there. From that time until about 750 Ravenna was considered the capital of Italy.

Rawalpindi, India (Punjab), has a military can-

tonment and an arsenal. It was the scene of the surrender of the Sikhs after the battle of Gujarat (1849).

Rawtenstall, England (Lancashire), has cotton, woollen, slipper, and carpet-felt factories, coal-mines, and stone-quarries.

Reading, England (cap. Berkshire), has considerable trade, and manufactures of biscuits, iron, ale, etc. It contains ruins of a Benedictine abbey, founded in 1121. It was the headquarters of the Danes in their inroad on Wessex in 871, and the scene of one of their defeats; was burned by the Danes in 1006; and was taken by the Parliamentarians under the Earl of Essex in 1643.

Reading, Pennsylvania (c. h. Berks Co.), is a trade and manufacturing center in an agricultural region, in which there is great mineral wealth. The chief industries are connected with iron and steel productions, but the coal-mines and limestone quarries contribute much to the industrial wealth of the city. It has large railroad shops, machine-shops, foundries, potteries, paper and wood pulp mills, carriage works, and iron and steel works. It was first settled by the Germans, but later a number of English colonies settled here and named the place after Reading, England. It was laid out in 1748.

Recklinghausen, Germany (Prussia), has manufactures of damask, tobacco, cabinet wares, etc. Coal is mined here.

Reggio, Italy (cap. Reggio di Calabria), is the seat of an archbishop, and has a fine cathedral. It manufactures perfumes, silks, and terra-cotta. Reggio, the ancient Rhegium, was founded by Calcidians about 720 B. C. It was partially destroyed by an earthquake (91 B. C.). Having become a Roman municipium (88 B. C.), it was made the headquarters of the Roman fleet and army by Octavius in the war with Sextus Pompeius. It was annexed to the kingdom of Naples early in the 16th century. Enormous damage was done by earthquakes in 1783, 1894, and 1908.

Reggio nell' Emilia, Italy (cap. Reggio nell' Emilia), is the seat of a bishop. Among the numerous churches the most picturesque is that of the Madonna della Ghiara, constructed in 1597. Reggio is the birthplace of the poet Ariosto and of the astronomer Secchi.

Regina, cap. Saskatchewan, Canada, is a railroad and trade center in the midst of a rich agricultural district. It has considerable wholesale trade in agricultural implements and merchandise. It is the seat of a collegiate institute and a normal school. A Dominion land-office is located here.

Reichenberg, Austria (Bohemia), is a center of cloth industry, and woollens, carpets, machinery, and leather are manufactured. The industrial Museum for Northern Bohemia is located here.

Remscheid, Germany (Prussia), is the center of the German cutlery trade, and does an enormous export business.

Rennes, France (cap. Ille-et-Vilaine), is an important railway center, and has manufactures of sail-cloth, table linen, leather, and agricultural implements. The ancient capital of Brittany, it was almost destroyed by fire in 1720. The principal edifice is the stately law-courts (1618), the parliament house of the states of Brittany. Rennes is the seat of an archbishop. The town hall contains an exceptionally fine picture-gallery. It is the birthplace of General Boulanger (1837-91), and here, in 1899, Dreyfus was tried for the second time.

Reno, Nevada (c. h. Washoe Co.), is a railroad and trade center in an agricultural region, largely under irrigation. It has large reduction works and manufactories of machinery, brick, etc. It is the seat of the state university.

Rensselaer, New York (Rensselaer Co.), is a residential and manufacturing suburb of Albany, and has felt factories, tannery, aniline works, etc.

Resht, Persia, has a trade in raw silk and cocoons, and exports rice, fruit, tobacco, textiles, cotton dyes, and wood.

Reus, Spain (Tarragona), has silk, calico, soap, and leather manufactories, and iron-smelting works.

Reval, Russia (cap. Esthonia), is a port and coaling station on the Gulf of Finland. It has a 13th century Danish castle, now the governor's residence, and other noteworthy buildings. The chief manufactures are cotton, paper, leather, hosiery, liquors, and ices. The chief exports are cheese, flax, oats, rye, linseed, joiners' materials, skins, pigs' bristles, game, hides, leather, and spirit of wine. Reval was founded in 1219-28 as a Danish town, and from 1238 it was a Hanseatic trade center. In 1346 it was sold by Denmark to the Teutonic Knights; in 1561 it became Swedish; in 1710 it was captured by Peter the Great, and definitely became Russian in 1721.

Revere, Massachusetts (Suffolk Co.), is a residential suburb of Boston and a summer resort. It has a state bath-house built and maintained by the state. It was settled in 1627.

Rheims, France (Marne), is a strongly fortified town, and an important railway center. It lies in a sheep-feeding district; hence its large manufacture of woollens, especially merino, and mixed silk and wool fabrics. It is also the headquarters for the production of champagne. Rheims is the seat of an archbishop, is renowned for its historic associations, and, above all, for its cathedral, begun in 1212, one of the finest specimens of Gothic architecture in Europe. Other notable structures are the archbishop's palace; the Romanesque church of St. Rémy or Remigius (1160-80), which contains the saint's tomb; and the Porta Martis, one of the old Roman gates of the city. Christianity became established here in the 4th century, and under Rémy, who was its bishop (459-533), Rheims first asserted its claim to be the coronation place of the kings of France—a privilege abolished in 1830. It is the birthplace of Colbert, the famous finance minister of Louis XIV.

Rheydt, Germany (Prussia), has manufactures of silks and velvets, cotton goods, machinery, and colored paper.

Rhondda, Wales (Glamorgan), is a center of coal-mining and iron industry.

Richmond, England (Surrey), is a suburb of London, and was for centuries a favorite royal residence; here died Edward III. and Queen Elizabeth. The Old Park on the north includes Kew Gardens; on the south is the New Park. White Lodge, in the park, was long the residence of the Duke and Duchess of Teck, and here was born Prince Edward of Wales (1894).

Richmond, Indiana (c. h. Wayne Co.), is the commercial and industrial center of a productive agricultural region. The chief manufactures are agricultural implements, carriages, milling machinery, threshing-machines, clothing, flour, and dairy products. A Friends' college and a Roman Catholic academy are located here; as well as a state insane hospital. It was settled in 1816 by Friends from North Carolina.

Richmond, cap. Virginia (c. h. Henrico Co.), is the largest city of the state, and a thriving commercial and industrial center, situated at the head of navigation on the James River. The chief industries of the city are connected with the preparation of tobacco, there being over fifty establishments for this purpose. It also has large foundries and machine-shops, locomotive works, extensive fertilizer works, carriage works, lumber mills, railroad car shops, etc. It has an extensive wholesale trade with all parts of the South. The James River furnishes water-power, which is used largely for manufacturing, also for propelling the street cars and lighting the city. It was settled in 1737 on the site of the home of the famous Indian chief, Powhatan, and had still only a few hundred inhabitants when made capital of the state in 1779. In 1861, it became the seat of government for the seceding states, and the capture of Richmond became ultimately the chief objective point of the Union troops. It was defended with great stubbornness by the Confederates, who threw up strong lines of earthworks all around it. On April 2, 1865, the city was evacuated, and a large part of it was destroyed. All traces of this devastation have been removed, and the city is now in thriving condition.

Riga, Russia (cap. Livonia), is a port and coaling station on the Duna, 5 miles from the Gulf of Riga, and an important industrial center. The chief exports are flax, timber, eggs, grain, hides and skins, oil-cake, and linseed. The harbor is ice-free from April to November. Riga has machine works, breweries, distilleries, and sawmills. The cathedral (with one of the largest organs in the world) and the castle are notable. It is the seat of a Greek Orthodox archbishop and of a Roman Catholic bishop.

Rimini, Italy (Forli), has iron-works and silk-mills, and a trade in silks and sulphur. The church of St. Francesco, now the cathedral, built in the 14th century, assumed its present form in 1450. The municipal buildings contain a famous collection of paintings. Among the noteworthy monuments is a bronze statue of Pope Paul V., and a memorial, dating from the 15th century, of the crossing of the Rubicon by Cæsar. The library was founded in 1617. Among the Roman antiquities is the Porta Romana, a triumphal arch. The Palazzo Ruffo was the scene of the murder (1285) of Francesca da Rimini, immortalized by Dante. Rimini, the ancient Ariminum, was made a bishopric in 260, and the celebrated council of Arians and Athanasians was held here in 359. The Malatesta family occupied the town in the 13th century, until it was ceded to Venice (1503).

Rio de Janeiro, cap. Brazil (Rio de Janeiro), is the largest city of Brazil, its chief commercial center, and a port and coaling station. It has under construction one of the finest systems of docks in South America. The Cattete palace is the residence of the president; the Palacete Itamaraty, official residence of the minister of foreign affairs, has been the scene of many historical events; and the government printing-office is one of the finest edifices in the city. The botanical garden, cathedral, university, national museum, and the observatory on Morro do Castello should be mentioned. There are also a conservatory of music, and military and naval academies. The climate is unhealthy, yellow fever and tuberculosis being very prevalent. Coffee forms two-thirds of the total exports, though gold, manganese, and hides are also exported. In 1762 it was made the capital of the state of Brazil, to which Maranhao (northern Brazil) was attached in 1774. It was the residence of the Portuguese court 1808-21, and became the capital of the empire of Brazil in 1823. Until 1834 it was also the capital of the province of Rio de Janeiro. The revolution of 1889 occurred here, and in 1893 the city was bombarded during the naval rebellion.

Rio Grande do Sul, Brasil (Rio Grande do Sul), has a good port and a large trade in hides, tobacco, and dried meats. It contains breweries and tanneries. The city is the location of a Protestant Episcopal missionary bishopric and has a theological seminary.

Riverside, California (c. h. Riverside Co.), is a rich residential city and winter resort in the heart of the orange growing region. Sherman Institute, a famous Indian school, is located here.

Rixdorf, Germany (Prussia), is a suburb of Berlin and has manufactories of linoleum, woollen and leather goods.

Roanne, France (Loire), is at the head of navigation of the Loire, and has a large transit trade, handling the manufactures of Lyons, the iron and coal of St. Etienne, and Levant wares. It has cotton manufactures, and is an entrepôt for agricultural produce.

Roanoke, Virginia (Roanoke Co.), is the trade center of a rich agricultural and mining section. It has extensive iron manufacturing interests, rolling-mills, locomotive and car shops, iron furnaces, bridge works, etc. The Virginia College for young ladies is located here.

Rochdale, England (Lancashire), manufactures woollens, cottons, velvet and plush; there are also large iron-works. The church of St. Chad (12th century) retains portions of ancient architecture. The town hall is a handsome modern structure, and a bronze statue commemorates John Bright, long closely associated with the town.

Rochefort, France (Charente-Inférieure), is a fortified port and naval arsenal. There are ship-building,

cannon foundries, and similar military industries. The naval hospital is one of the best in Europe. Rochefort's commerce consists in salt, timber, oysters, salt fish, and colonial wares. It was here that Napoleon surrendered to Captain Maitland of the Bellerophon, July, 1815.

Rochester, England (Kent), has agricultural implement and cement works. The bishopric was founded early in the 7th century. The cathedral dates chiefly from the 11th and 12th centuries. The church of St. Margaret, rebuilt 1829, dates from the 11th century. A museum is installed in Eastgate House, 16th century. Other buildings are the 17th century Guildhall and Watt's Charity House (1579).

Rochester, New York (c. h. Monroe Co.), is the third city of the state, and one of the most important commercial and industrial centers of the country. The Genesee River, which runs through the center of the city, has here, in the upper and lower falls, a descent of 176 ft. Among the chief industrial interests are the manufactures of clothing, and of boots and shoes. Rochester is the home of the camera, and practically all of the film cameras that are made in the world are manufactured here, as well as a majority of all the plate cameras; the city leads in other photographic apparatus, also in the making of optical instruments. It has extensive fruit canning works. The nursery interests in and around the city are very extensive, and several large seed houses are located here. The University of Rochester, founded in 1850, has a faculty of 23 instructors, with upwards of 300 students. The Rochester Athenæum and Mechanics' Institute gives instruction in practical arts and sciences to about 4,000 students and ranks high among the technical trade schools of the country; there are numerous other educational institutions. A state hospital for the insane is located here. The city was settled in 1812. On March 17, 1865, a great flood damaged property to the extent of over $1,000,000 and on February 26, 1904, a fire devastated a large portion of the dry goods district, causing a loss of over $3,000,000.

Rockford, Illinois (c. h. Winnebago Co.), is one of the most important industrial centers of northern Illinois. The chief manufactures are agricultural implements, furniture, silver-plated ware, watches, flour, and grape sugar. It is the seat of Rockford College for women. It was settled in 1834.

Rock Island, Illinois (c. h. Rock Island Co.), is the seat of a government arsenal and armory located on an island in the Mississippi River. It has extensive manufactories of agricultural implements, lumber mills, and a number of smaller industries. A Lutheran college and several other educational institutions are located here. It was settled in 1726.

Rome, cap. Italy (cap. Rome), is the capital of the ancient world, the center of the Roman Catholic Church, and the most famous of all cities. The city is distinguished by its vast ruins, its many historic gates, and its monuments, ancient and modern. Modern Rome is enclosed by a circle of forts, 30 miles in extent. Among the existing remains of the ancient city, the Forum, Colosseum, Forum of Trajan, Cloaca Maxima, catacombs, Pantheon, column of Aurelius, theater of Marcellus, pyramid of Cestius, arches of Constantine, Titus, and Septimius Severus, baths of Titus and Caracalla, ruins on the Palatine, basilica of Constantine, temples of Concord, Fortune, Saturn, and Neptune, palace of Caligula, mausoleum of Hadrian, and column of Trajan are notable. Rome contains about 300 churches, many of them interesting from their historical associations or from the archæological or artistic treasures preserved in them. At present St. Peter's is the largest church in the world, occupying 18,000 sq. yds.; the extreme height is 435 ft. Under the high altar is the tomb of the apostle. North of St. Peter's is the Vatican Palace, the largest in the world, covering 13½ acres, with over a thousand halls, chapels, and rooms. Originally built by Symmachus, it has been enlarged and embellished by successive popes, especially by Nicholas V. The famous Sistine Chapel, with frescoes by Michael Angelo, dates from Sixtus IV. (1481). The state apartments and loggie were adorned by Raphael with frescoes

by order of Julius II. and Leo X. A picture gallery, founded by Pius VII. with the pictures restored by France in 1815, a collection of marble and bronze statuary, and a marvellous library containing 34,000 MSS. founded by Nicholas V., make the Vatican an unrivalled shrine of art and literature. A law of 1871 recognised the Vatican, with the Lateran and the Castello Gandolfo, as being outside civic jurisdiction. Another large palace is the Quirinal, the residence of the King of Italy. Rome contains many picture galleries and museums, filled with priceless art treasures, and several libraries with valuable collections of books and manuscripts. Many of the palaces and villas of the old Roman nobility have also collections of pictures and sculptures. Chief among them are the Borghese Palace, the Colonna, Doria, Farnese, Corsini, and Barberini palaces; and the Villa Borghese, Albani, Ludovisi, Medici, etc. The history of Rome was for centuries the history of the civilised world; and even after it ceased to be the capital of the empire, it was the center of civilization, and the most interesting and influential city of Christendom. Rome was the capital of a kingdom which gradually grew till the foundation of the republic in 509 B. C. The republic steadily extended, and after wars with Æquians, Volscians, Latins, Samnites, Sabines, Tarentines, etc. Rome became mistress of Italy by the middle of the 3rd century B. C. Then came the wars with Carthaginians and Macedonians, with Jugurtha and Mithridates, and Rome made herself the mistress of southern Europe, northern Africa and western Asia. First Gaul, and then southern Britain were conquered by Julius Cæsar (51 B. C.). The republic was overthrown, and Augustus, the first emperor, was at peace with all the world soon after the birth of Christ. The empire was extended to Germany and Dacia, in Parthia and Asia; but in the 3rd century A. D. the northern nations—especially the Goths and kindred tribes—began to do more than hold their own, and the empire contracted on the north. The seat of the empire was removed from Rome to Byzantium, or Constantinople, by Constantine in 330 A. D., and in 364 the empire was divided into an eastern and a western empire, Rome remaining capital of the western half. Ere long Rome was taken and retaken by the barbarians (410,476), and, retaken again (553) by Belisarius, was made dependent on Constantinople. But as capital of the popes, new glories were in store for her; Charlemagne and Otho of Germany were crowned emperors of the West at Rome, and the city became the independent capital of the increasing papal dominions, or States of the Church, the Romagna, Bologna, and Perugia being conquered by Pope Julius II. in 1503. Rome remained the mother city of Christendom, and continued to flourish in spite of the temporary sojourn of the popes at Avignon and the short-lived republic of Rienzi (1347). Again in 1798 the French proclaimed Rome a republic, and in 1808 the city became part of the French kingdom of Italy. It was restored to the popes in 1814, who, save during the troubles of 1848-9, retained it as capital of the States of the Church, under French protection; but in 1860 the papal states revolted to Sardinia, and in 1870 Rome became part of the Italian kingdom and the national capital.

Rome, New York (Oneida Co.), is a trade and manufacturing city in a section devoted to farming and dairying. Its chief industrial establishments are brass and copper mills, locomotive works, knitting mills, and canneries. A state asylum for incurable insane is located here. It was settled in 1760.

Rosario, Argentina (Santa Fe), is a port and coaling station on the Parana River 240 miles from its mouth, the second city of Argentina, and an important railway center. It exports wheat, linseed-oil, wool, hides, etc. It is the seat of a national college.

Rostock, Germany (Mecklenburg-Schwerin), manufactures machinery, ships, chocolate, chemicals, furniture, and beer. The university, founded in 1419, is attended by some 540 students.

Rostof, Russia (Yaroslaf), is an iron-manufacturing center, and makes candles, wax, tallow, linen, vinegar, soap, leather, white lead, treacle, and sweetmeats;

has also tanneries, breweries, distilleries, and brick works. The church of the Assumption was founded in 1230, and has a famous belfry built about 1590; the Spasso-Yakovlevskii monastery has a renowned treasury. It is a center of pilgrimage, and the seat of a Greek Orthodox bishop.

Rotherham, England (York), has steel and iron-works, glass manufactures, and coal-mines.

Rotterdam, Netherlands (South Holland), is the chief port and coaling station of Netherlands and its second city in size. About 50 per cent. of the total imports by sea into the Netherlands, and almost as large a proportion of the sea-borne exports, pass through Rotterdam, and in addition, 80 per cent. of the Rhine traffic. Grain, timber, dyewoods, metals, hardware, petroleum, drugs and chemicals, rice, coffee, tobacco, and palm kernels are the principal exports. The chief industrial establishments are machine and engineering shops, sugar, tobacco, and metal factories, and distilleries. There is also ship-building. It is one of the chief emigration ports of the continent. The principal church (15th century) possesses a famous organ. Rotterdam has a well-known picture gallery, the Museum Boymans (1864-7), with works by Rembrandt, Cuyp, Bol, Eeckhout Hobbema, and others. There are also a national museum, a marine museum, an academy of the fine arts, and a music school. Erasmus, the poet Tollens, the statesman Van Hogendorp, Grinling Gibbons, and James, Duke of Monmouth, were natives of the town. Rotterdam was burned in 1563, and was taken by the Spaniards in 1572. It developed rapidly in the 19th century.

Roubaix, France (Nord), is the first woollen manufacturing town of France, and possesses a famous "conditioning house" for testing wool and woollens.

Rouen, France (cap. Seine-Inférieure), is a port on the Seine, an important railroad center, and the chief cotton manufacturing city of France. It also has dyeworks and manufactures of linen, wool, silks, hosiery, chemicals, confectionery, and paper, and engineering and ship-building works. Rouen is the seat of an archbishop, and has some of the grandest churches in France. Chief of these are the cathedral dating from the 13th century onwards, which once possessed the heart of Cœur de Lion (now transferred to the museum of antiquities); St. Maclou (15th century), in the florid style; St. Ouen (14th to 15th century), one of the most delicate and graceful of all Gothic churches; and St. Gervais, one of the oldest churches in France. Other interesting structures are the law-courts (15th century): the gate-house (1517); the famous Hôtel de Bourgthéroulde (1506); the belfry (1389) where the curfew still rings daily at 9 P. M.; the public library (140,000 vols.); and the museum of antiquities, which contains the finest painted glass in Europe. It is the birthplace of La Salle, the first explorer of the Mississippi River.

Roulers, Belgium (West Flanders), has manufactures of linen and cotton goods, lace, and silk ribbons.

Rowley Regis, England (Stafford), has iron-works, collieries, potteries, and manufactures of anchors, chains, rivets, nails, and gun-barrels.

Rugby, England (Warwick), is the seat of a famous public school, founded in 1567, which has had many renowned principals. The town has large engineering works, and important cattle markets.

Rustchuk, Bulgaria, is a port on the Danube, and has manufactures of tobacco, cigars, pottery, and soap. It is the seat of a Greek Orthodox archbishop and of an Armenian bishop.

Rutland, Vermont (c. h. Rutland Co.), is well known for its marble quarries, and the industries connected with the shipment of marble; the quarries are now in the suburban towns, but the marble works, which have been of considerable importance since 1830, continue here as formerly. Iron ore, fire-clay, and slate are found in the vicinity. The city also has extensive manufactories of machinery for quarrying and channeling marble, engine and boiler works, a scale factory, lumber mills, brick-yards, etc. It was settled in 1770.

Ryasan, Russia (cap. Ryasan), is the seat of a Greek Orthodox archbishop, and its cathedral (1690,

rebuilt 1770) is one of the most striking in Russia. There are breweries, distilleries, and manufactories of candles, malt, cotton stuffs, and machinery.

Sacramento, cap. California (c. h. Sacramento Co.), is a port of entry and an important commercial, railway, and industrial center. Its industries include flour-mills, lumber-mills, railroad car shops, and manufactories for saddlery, harness, furniture, etc. The State Capitol building, completed in 1869, cost $2,500,000, and houses the state library of over 150,000 volumes. The Crocker Art Gallery contains pictures and California minerals, and has a school of art. In Fort Sutter Park is a reproduction of the fort established here in 1840, the first white settlement having been in 1839 by Capt. John A. Sutter. On November 3, 1852, a conflagration destroyed property exceeding $5,000,000 in value.

Saga, Japan (cap. Saga), is a port and commercial center, with trade in coal, tobacco, etc.

Sagar, India (Central Provinces), exports salt, sugar, and cloth. It has a fort and an arsenal.

Saginaw, Michigan (c. h. Saginaw Co.), is an important manufacturing, jobbing, and railroad center on the Saginaw River, 16 miles from Saginaw Bay. It manufactures a great variety of lumber products, furniture, pianos, iron and steel products, glass, salt, beet sugar, flour, etc. Bituminous coal is found in the vicinity. The Michigan Employment Institution for the Blind, a manual training school, and a German Lutheran theological seminary are located here. It was settled in 1815.

Saharanpur, India (United Provinces), is a trade center. It has an old Rohilla fort, a handsome mosque, and botanical gardens.

Saigon, cap. French Indo-China, is a port and coaling station on the Donoi River, 42 miles from the sea. It has a government palace, law-courts, cathedral, hospitals, botanical and zoological gardens, rice-mills, sawmills, varnish, and soap factories.

Saint Catharines, Ontario, Canada, is a railroad and trade center in an agricultural and fruit growing section, with paper mills, flour mills, canneries, etc. It is the seat of Bishop Ridley College.

Saint Denis, France (Seine), is a suburb of Paris, and has a trade in corn, wine, wood, wool, and manufactures of machinery, rolling-stock, small steamers, chemicals, and soap. The abbey church of St. Denis, the patron saint of France, has been the burying-place of most of its kings since the time of Dagobert (638). Dagobert's church was rebuilt (1137-44) by Abbot Suger, and the building endured the caprices of many restorers down to the time of Napoleon III. Amongst the finest of the memorials are the tomb of Louis XII. (1498-1515); the cenotaph of Francis I. (1515-47); the tomb of Henry II. (1547-59); the monument to Francis II., and, in the 11th century crypt, the coffins of the last Bourbon sovereigns. In 1793 the contents of fifty-one tombs were flung into a heap and buried in a trench.

Saint Denis, cap. Reunion Island, is a port, with an exposed roadstead, and is the residence of the French governor of the island. It has a cathedral, and a botanical garden.

Saint Etienne, France (cap. Loire), is built on a coal-field and on an iron field, and can draw supplies of raw silk from the Rhone valley. The total annual value of ribbons made here and in the vicinity is about $20,-000,000. The National Arms Factory turns out most of the small arms for the French army; and the general iron and steel manufactures include iron-mongery of all kinds, iron and steel plates, machinery castings, gun armor, and iron masts. Glass bottles are made.

Saint Gallen, Switzerland (cap. St. Gallen), is the center of the muslin, embroidery, and cotton trade. It grew up round the abbey founded in the 7th century by the Irish monk Gallus, and later became one of the principal literary centers of Europe. It was secularised in 1798, but still has a cathedral; the MSS. in its library are considered to form one of the most precious collections in Europe.

Saint Gilles, Belgium (Brabant), is a suburb of Brussels. Its industries include chemicals, perfumery,

lace, flax-spinning, marble-dressing, and market-gardening.

Saint Helens, England (Lancashire), has manufactures of glass and chemicals, and coal and fire-clay are worked.

Saint Henri, Quebec, Canada, is a suburb of Montreal.

Saint Hyacinthe, Quebec, Canada, is a trade center in a prosperous agricultural region. Its industries include boots and shoes, spinning wheels, organs, threshing machines, etc. It is the seat of a Dominican college and convent.

Saint John, New Brunswick, Canada, is a port and coaling station on the Bay of Fundy, and the commercial metropolis of the province. It exports lumber, grain, and meat. It has large sawmills, extensive fisheries, and manufactories of iron castings, engines, machinery, tools, cotton goods, woollen goods, boots and shoes, pulp, etc. It was settled chiefly by American loyalists at the close of the Revolution; was chartered as a city in 1785, and was partly destroyed by fire in 1877.

Saint Johns, cap. Newfoundland, is a port and coaling station. Its harbor is one of the best on the Atlantic coast, and was formerly strongly fortified, but the fortifications have been allowed to fall into ruins. It has two cathedrals, the Roman Catholic being spared by the fire of 1892, which consumed more than half the city. The trade consists chiefly of supplies for the fisheries, and during the spring season the harbor is thronged with vessels. Settlement was made here by Sir Humphrey Gilbert in 1582.

Saint Joseph, Missouri (c. h. Buchanan Co.), is the third city of Missouri, and an important railroad, commercial, and industrial center on the Missouri River. It has many large wholesale establishments engaged in all branches of trade, while its stock-yards, slaughtering and packing industries are very extensive. It is also a great horse and mule market. It is located in one of the finest and wealthiest agricultural sections of the United States. A trading post was established here in 1826, which grew into a town, incorporated in 1845, and then into a city, incorporated in 1851. Its growth has been steady and, in recent years, very rapid.

Saint Louis, Missouri (c. h. Saint Louis City Co.), is the largest city of Missouri, and the principal commercial and banking center of the southwest. It is located on the west bank of the Mississippi River, 20 miles below the mouth of the Missouri River, and is a port of entry, and one of the most important railroad centers of the country, with a large river trade. Its position in the center of the great Mississippi Valley gives it an immense trade, among the staples of which are breadstuffs, packed meats, tobacco, live stock, timber, grain, wool, furs, etc. In manufactures, Saint Louis ranks fourth among the American cities, and high among the cities of the world in tobacco-making. It produces immense quantities of beer, flour, boots and shoes, hardware, stoves, railway cars, woodenwares, chemicals, drugs, etc. It has a horse and mule market which is one of the largest in the world. Washington University, noted for the breadth of its charter, includes an ordinary undergraduate department, schools of engineering, fine arts, law, medicine, dentistry, and botany, a manual training school, and schools for boys and girls. It is attended by about 2,000 university students and by about 1,000 others. Its group of buildings in the Tudor-Gothic style, is built of red Missouri granite. The School and Museum of Fine Arts is housed in the building originally erected as the Fine Arts Building of the Louisiana Purchase Exposition and contains a large collection of casts, sculptures, paintings, etc. A Roman Catholic university, founded in 1829, a College of Christian Brothers, and a state normal school are among the other educational institutions. The public library has over 200,000 volumes. The Missouri Botanical Garden contains 75 acres and is excellently equipped. The parks of Saint Louis have an area of over 2,300 acres, the largest of which is the Forest Park containing 1,370 acres. Jefferson Barracks, a United States military post, is 12 miles south of the city. Saint Louis was founded by the French in 1764 and still bears traces of its French origin in the names of some of its streets and

leading families. In 1770, the Spanish authority was established here; in 1803, when the population of the city was still below 1,000, with the rest of the territory then known as Louisiana, it became a portion of the United States. It was incorporated as a city in 1822. On May 27th, 1896, Saint Louis was visited by a terrific tornado, which destroyed 300 lives and property to the value of $10,000,000. The floods of 1903 raised the river 38 ft. and did great damage in its suburb, East Saint Louis. In 1904, Saint Louis was the scene of the Louisiana Purchase Exposition, held to commemorate the centenary of the purchase of Louisiana from France.

Saint Louis, cap. Senegal, Africa, is a seaport, and the terminus of caravan routes from the Sahara. It is one of the finest towns in West Africa, and has a cathedral, a governor's palace, and a normal training college.

Saint Nazaire, France (Loire-Inférieure), is a port on the Loire estuary. The chief exports include brandy, poultry, eggs, vegetables, butter, sardines, glass, and dyes. The main industry is iron ship-building.

Saint Nicolas, Belgium (East Flanders), has manufactures of textiles and other industries.

Saint Ouen, France (Seine), is a residential suburb of Paris.

Saint Paul, cap. Minnesota (c. h. Ramsey Co.), is situated at the head of navigation on the Mississippi River and ranks with its twin city of Minneapolis as the leading commercial, banking, and industrial center of the Northwest. It is a great railway center, and has a large wholesale trade by rail and river. The State Capitol, erected in 1898-1906, at a cost of $4,500,000, is a large and handsome edifice of granite and Georgia marble, and one of the most impressive and successful state capitol buildings in the United States. It has mural paintings by prominent American artists, and contains a state law library of 25,000 volumes and a state historical library of over 75,000 volumes. The Agricultural School of the University of Minnesota is located here. Another notable public building is the Auditorium, with a hall for meetings, conventions, etc., having a seating capacity of 10,000. The wholesale traffic of the city has grown to remarkable dimensions, extending over the entire western portion of the continent. It has a large trade in tea, furs, provisions, and general merchandise. It has extensive manufactures of furniture, stoves, lumber products, etc. Fort Snelling, a United States military post, lies on the west bank of the Mississippi, 6 miles southwest of the city. The first white settler, a Canadian voyager, built a house here in 1838, and in 1841 the place received its name from a chapel dedicated to Saint Paul by a French priest. In 1854, when it received a city charter, it contained 3,000 inhabitants, and since then its growth has been very rapid.

Saint Petersburg, cap. Russia (esp. Saint Petersburg), is a port and coaling station at the head of the Gulf of Finland. The Winter Palace, the center of court life in the capital, is an immense quadrilateral, built in 1732-62 (after the plans of Rastrelli), burnt in 1837, and rebuilt in 1837-8. It contains many magnificent state rooms. The other imperial and princely palaces include the Old and New Mikhailovskii residences, built in 1797-1801 and in 1819-25 by Paul I. and Alexander I. respectively; the Taurida palace, built by Catherine II. for Potemkin; the Anitchkof palace, a favorite residence of Alexander III.; the Marble or Orlof palace, another of the creations of Catherine II.; and the palaces of the Grand Dukes Vladimir and Michael. Several of these contain valuable art collections, though many of their treasures have been concentrated at the Winter Palace and the Hermitage. The last named (founded in 1765, rebuilt and greatly enlarged in 1840-52) is the Louvre of Saint Petersburg; it contains some of the finest artistic and archæological collections in the world, including Greek antiquities from the Black Sea coasts. Nowhere else is there such a wealth of Rembrandt's masterpieces; only in Madrid is there a greater treasure of the best works of Velasquez and Murillo; and no continental gallery has so good a collection of English art. The imperial, or National Library ranks after those of Paris and London; it possesses 1,500,000 printed volumes and 27,-100 manuscripts. Among other libraries and collections

are those of the Academy of Sciences (the chief learned society of the empire); of the Asiatic Museum; of the Academy of Fine Arts; of the National Museum of Alexander III., especially representative of Russian painting; of the university (which was founded in 1819, and has 2,000 students), of the School of Mines, of the Museum of Artillery, of the Semenof Gallery in Vasili Ostrof, of the Imperial Geographical Society, and of the Historical, Archæological, and Economical Societies. Besides the university, Saint Petersburg possesses a school of mines, with a mineralogical collection unsurpassed in Europe; a technological institute, a school of civil engineering, a forestry school, the military academy of the general staff, academies of artillery, military law, medicine, and engineering, of Greek Orthodox and Roman Catholic theology, of fine arts, and of music, as well as a naval academy. The great departmental state buildings include the Admiralty, founded in 1704 by Peter the Great, and containing the archives and official rooms of the ministry of marine and the naval museum; the general staff buildings containing the military archives of Russia, and the official quarters of the ministries of war, finance, and foreign affairs; the buildings of the governing Senate and the Holy Synod. Saint Petersburg also possesses thirteen cathedral churches, of which those of St. Isaac the Dalmatian, of the Kazan Mother of God, of St. Alexander Nevskii, of St. Peter and St. Paul, of the Resurrection, and of the Emperor Alexander II. are the most noteworthy. The greater industrial establishments of Saint Petersburg consist chiefly of metal works, iron-foundries, sugar-refineries, distilleries, breweries, ship- and boat-building yards, printing works, and manufactories of tobacco, crystal and glass, silk, cotton and cloth stuffs, leather, cordage, carts and carriages, gold embroidery, pottery and porcelain, and machinery. The chief exports are grain, linseed, oil-cake, flax, petroleum, wool, beer, eggs, and timber. Founded in 1703 by Peter the Great, the city was in 1712 formally created the imperial capital.

Saint Quentin, France (Aisne), has manufactories of cotton, woollens, silk, sugar, and machinery. Of great architectural interest are the City Hall (1331-1509), and a church (12th to 15th century), in the purest Gothic style. Near Saint Quentin were fought two great battles—the first in 1557, when the French were utterly defeated by a Spanish army; the second in 1871, when the Germans destroyed a French army which was attempting to relieve Paris.

Saint Thomas, Ontario, Canada, is a railroad and trade center, with car shops, planing-mills, flour-mills, etc.

Sakai, Japan (Osaka), is a port on Osaka Bay, and has manufactures of cotton goods, cutlery, bricks, and saké.

Salem, India (Madras), has industries of weaving, carpet manufactures and cutlery.

Salem, Massachusetts (c. h. Essex Co.), is known as the mother city of Massachusetts. It has a good harbor and considerable manufacturing interests, also a large cotton-mill, tanneries, a lead factory, chemical works, cordage works, and shoe factories. Essex Institute, containing an interesting collection of historical paintings, portraits, and relics; a library of nearly 400,000 volumes, and the Peabody Academy of Sciences, containing the East India Marine Museum, with important Japanese and other ethnological, maritime and natural history collections, are of note. Naumkeag, on the site of Salem, was first visited by Roger Conant, one of the Cape Ann immigrants, in 1626, and a permanent settlement was made here by Gov. Endicott two years later. Gov. Winthrop landed here in 1630, and for a time Salem was the capital of Massachusetts. In 1692 Salem was the scene of the extraordinary witchcraft delusion. The legislature of Massachusetts met at Salem in 1774, the last time under the English Crown, and issued a call for a Continental Congress. Privateersmen from Salem were very active during the war. After the war, Salem engaged in the East India trade, and many of its citizens attained great wealth and influence. Indeed, it is said that about 1810 a Salem merchant was the largest ship-owner in the world. Among the famous natives of Salem are Nathaniel Hawthorne, who was Surveyor of the Port

here from 1846 to 1850; W. H. Prescott, the historian; Maria S. Cummins, author of "The Lamplighter"; Pierce, the mathematician; and W. W. Story, the sculptor.

Salem, cap. Oregon (c. h. Marion Co.), is the center of the most important hop-growing region in the world. It has flour-mills, woollen mills, etc. The capitol is an imposing building, and state institutions for deaf and dumb, blind, and insane, and a penitentiary are located here. It is the seat of a Methodist Episcopal university.

Salerno, Italy (cap. Salerno), is the seat of an archbishop, and has a cathedral dating from the 11th century. It was famous in the middle ages for its medical school.

Salford, England (Lancashire), is a suburb of Manchester, with extensive manufacturing interests.

Salisbury, England (Wiltshire), has a famous cathedral, founded in 1220, which is a beautiful example of early English architecture. The cloisters, built in the 13th century, are the most perfect in England, and form a quadrangle 181 ft. square. The spire, 404 ft., was added in the 14th century, and is the loftiest in England. Within the close are the chapter-house, the library, the episcopal palace, and the deanery. Salisbury contains three other ancient churches, and several medieval buildings.

Saloniki, Turkey (cap. Saloniki), is a port and coaling station at the head of the Gulf of Saloniki, and the second city in size and commercial importance of Turkey in Europe. Of particular interest are the antiquities, notably its ancient churches and a Roman triumphal arch. The city is overlooked by an ancient citadel. Its industries include cotton-mills, soap-works, and silk manufactories. It exports chiefly corn, tobacco, opium, cotton, cocoons, wool, and hides. Founded 315 B. C., it became the capital of Macedonia. Under the name of Thessalonica it was visited by St. Paul. In 904 the Saracens sold 22,000 of the inhabitants as slaves. It was wrested from Venice by the Turks in 1430.

Saltillo, Mexico (cap. Coahuila), contains a cathedral, has cotton-mills, and is noted for its woollen shawls.

Salt Lake City, cap. Utah (c. h. Salt Lake Co.), is the metropolis of the state, an important railroad and mining center, and the center of trade for the vast "intermountain" region. Great smelters for the reduction of ores are located in and near the city, and it has large beet-sugar factories and manufactories of iron products, machinery, etc. Fort Douglas, a United States military post, is situated on a plateau 500 ft. above and about 3 miles from the heart of the city. It is the headquarters of the Church of Jesus Christ of Latter Day Saints (Mormons), and the home of the president and many of the high officers of the church. The Temple, Tabernacle, and Assembly Hall occupy a square in the central portion of the city. The Tabernacle, built in 1864-7, is a huge and extraordinary structure, in the shape of an oval or ellipse, 250 ft. long, 150 ft. wide, and 70 ft. high. It is surmounted by a wooden roof with iron shingles, resembling the shell of a turtle or the inverted hull of a ship, supported by 44 sandstone pillars. The Temple is a large and handsome building of granite, erected in 1853-93 at a cost of over $4,000,000. It is 186 ft. long from east to west and 99 ft. wide. At each end are 3 pointed towers, the loftiest of which, in the center of the eastern or principal façade, is 210 ft. high and is surmounted by a colossal gilded figure of the Angel Moroni. The Assembly Hall, to the south of the Tabernacle, is a granite building with accommodation for 3,000 people, intended for divine service. Salt Lake City was founded in 1847 by the Mormons, under Brigham Young, who had been driven from Nauvoo the previous year and had made a long and perilous journey across the Indian-haunted plains. The district was then a barren and unpromising desert, but the industrious Mormons set to work at once to plough and plant, and began that system of irrigation, which has drawn out the latent capabilities of the soil and made the Utah valleys among the most productive regions in the country.

Salzburg, Austria (cap. Salzburg), is a town noted for its picturesque situation, and one of the principal touring resorts of Austria. It is also an important

railway center. It is the birthplace of Mozart, and of Hans Makart, the painter.

Samara, Russia (cap. Samara), is a port on the Volga, and an important railway center with large trade in grain. Its industries include tanneries, flour-mills, breweries, brick-works, iron and bell metal foundries. It is the seat of an Orthodox bishop, and has two old cathedrals, built in 1685 and 1730-5, and one completed in 1894; also a museum of natural history and archæology.

Samarang, Java, is a fortified town, and one of the most important commercial centers of the island. Has no port, but vessels a mile out at sea take away sugar, coffee, tobacco, and indigo.

Samarkand, Russia (Russian Turkestan), has an active commerce and manufactures of cotton, silk, etc. Among the objects of interest are the grave of Timur, the citadel, three colleges, and neighboring ruins. The ancient city was destroyed by Alexander the Great. In the middle ages Samarkand was a large and flourishing city, renowned as a seat of learning. It was taken and destroyed by Jenghis Khan in 1219; became the capital of Timur in 1369; and was annexed to Russia in 1868.

San Angelo, Texas (c. h. Tom Green Co.), is an important trade center in an agricultural region, largely devoted to dry farming.

San Antonio, Texas (c. h. Bexar Co.), is one of the largest cities of Texas, and an important railroad and trade center. It has large flouring-mills, cotton-presses, cement works, oil-mills, broom works, etc., and is a prominent cotton, wool, horse, mule, and cattle market, as well as the financial center of a very extensive stock-raising section. It is surrounded by an agricultural region supplied with water from artesian wells. In and near the city are numerous medicinal springs. Fort Sam Houston, an important military post and headquarters of the department of Texas, is located here. There is also a United States arsenal for storage of ordnance supplies and the manufacturing and repairing of equipment. It is frequented as a winter resort, and is noted for its old Spanish mansions near the city, having been settled by the Spaniards about 1690. The Church of the Mission del Alamo, which seems to have derived its name from having been built in a grove of alamo or cottonwood trees, is a low and strong structure of adobe, with very thick walls, built in 1744, and preserved as a national monument. The most salient event in San Antonio's history is the "Fall of the Alamo" in 1836. Texas had determined to resist certain obnoxious laws imposed by Mexico, and the latter sent an army under Santa Ana to reduce the rebels. The advance-guard of 4,000 men reached San Antonio on Feb. 22nd, and found the fortified Church of the Alamo garrisoned by a body of 145 Americans (afterwards joined by 25 or 30 more), under Travis, Bowie, and Davy Crockett, who refused to surrender. After a siege of 12 days, the church was finally carried by assault (March 6th) and all the survivors of the gallant little band of defenders were put to the sword.

San Diego, California (c. h. San Diego Co.), is a port of entry and an important health resort and trading center on San Diego Bay, with one of the best harbors on the California coast. It is noted for its climate, which is mild and equable. It is the commercial center of a region in which are produced large quantities of nuts, fruit, and honey, and has flour and planing-mills, machine-shops, wagon works, etc. It is a naval coaling station, and has a United States garrison. Coronado Beach, on a small peninsula opposite the city, is a favorite resort. The Mission of San Diego, 6½ miles from the city, was the first settlement made by white men in California. The bay was discovered in 1542.

Sandusky, Ohio (c. h. Erie Co.), is a port of entry on Sandusky Bay, and has one of the best landlocked harbors on Lake Erie. It is a trade and manufacturing center in a fertile agricultural region, of which the chief productions are grapes, peaches, apples, and grain. Fishing is an important industry of the city, and a large amount of wine is shipped. The chief industrial establishments are factories, in which are made baskets and crates for

the shipment of grapes and peaches; cooperages, in which are made casks for the shipment of wine; machine-shops; engine, boiler, threshing-machine, and cement works; flour-mills; and furniture factories. There are large coal and lumber yards. Considerable coal is brought here for manufacture and reshipment, and the ship-building industry is also important. The United States hatchery at Put-in-Bay encourages the fish industry. A state sailors' and soldiers' home, with accommodations for 1,600 persons, is located here.

San Francisco, California (c. h. San Francisco Co.), is the largest city in California, also the largest on the western coast of America or in the United States west of the Mississippi River. It is a port of entry situated on a peninsula between the Bay of San Francisco and the Pacific Ocean, with a sheltered harbor about 90 miles long and from 5 to 15 miles wide, having excellent anchorage ground and an abundance of deep water. It is extensively engaged in the manufacture of mining machinery, the canning of fruits and vegetables, ship-building, etc., and the principal industrial establishments are sugar and molasses refineries, slaughtering and meat-packing houses, leather manufactories, foundries, machine-shops, smelting works, etc. It is an important railroad terminus and has a large wholesale trade in merchandise of every description. Its foreign commerce is very extensive, and among the chief exports are gold and silver, wine, fruit, wool, oil, lumber, flour, and breadstuffs. The principal imports are tea from China and Japan. A United States branch mint, ranking second in the country, is located here. A large part of the city was destroyed by earthquake and fire in 1906, but the business district has been rebuilt in an improved manner, and is now noted for its modern commercial buildings, magnificent hotels, etc. One of the most interesting historical relics of San Francisco is the old Mission Dolores, dating from about 1778; the old church is built of adobe and is adjoined by a tangled and neglected little churchyard. Among the educational institutions of the city may be mentioned the Cooper Medical College, the Medical Department of the University of California, the Cogswell Polytechnic School, and the California School of Mechanical Arts. The Presidia, or government military reservation, has an area of 1,500 acres and stretches along the Golden Gate for about 4 miles. Golden Gate Park, having an area of 1,013 acres, contains a museum, Japanese tea-garden, etc. The city was founded in 1835. In 1848, the year of the discovery of gold in California, its population was about 500; in 1850 it was about 25,000, and, in each decade, it has seen an extraordinary increase.

San José, California (c. h. Santa Clara Co.), is situated in a fertile agricultural region, the beautiful Santa Clara Valley, famed for the variety and amount of its fruit. It has large canneries, packing-houses, and shipping industries connected with the fruit industry, also flour-mills, lumber mills, grain elevators, woollen factories, basket and box factories, machine-shops, etc. It has a considerable wholesale trade. Among the educational institutions are a state normal school, a Methodist Episcopal university, a Roman Catholic college, etc. On account of its equable climate, it is much frequented by tourists, and is a delightful residential city. The first settlement was made here about 1770 and the pueblo of San José was established in 1782. It was the state capital under the first constitution of California.

San José, cap. Costa Rica, is a well-built city, which contains a cathedral, a university, a national museum, and one of the finest opera-houses in Central America. The city was founded about 1738, and has been the capital, except for short intervals, since 1824.

San Juan, Porto Rico (cap. San Juan), is a fortified port and the largest city of Porto Rico. The harbor is excellent, affording anchorage for the largest ships,— the only good harbor in Porto Rico and one of the best in the West Indies. The entrance, however, is very narrow, and difficult in stormy weather. The chief exports are tobacco, sugar, and coffee. It contains a number of notable public buildings, including the bishop's palace, a large cathedral, the old government building, the city hall, the Casa Blanca, the custom-

house, a military hospital, and an arsenal. A new capitol is being built. San Juan was founded in 1511 and was the seat of the Spanish provincial government. It was bombarded by the U. S. fleet under Admiral Sampson in May, 1898.

San Juan del Norte, Nicaragua, Central America, is the chief port of Nicaragua on the Caribbean Sea. Exports mahogany, rubber, bananas, hides, and indigo. Steamers run from here up the San Juan River into Lake Nicaragua.

San Luis Potosi, Mexico (cap. San Luis Potosi), is an important railroad center, with a large trade in cattle, hides, and wool. The famous silver mines worked in the neighborhood since the end of the 16th century are now almost abandoned. It has handsome public buildings including a cathedral, a university, and a city hall.

San Miguel, Salvador, Central America, is the center of a fertile agricultural region, and has a trade in indigo.

San Salvador, cap. Salvador, Central America, has more than once been destroyed by earthquakes. It contains a university, a cathedral, a national museum and library, and an observatory. It has an active trade in agricultural products, particularly indigo.

San Sebastian, Spain (cap. Guipuscoa), is the most fashionable seaside resort of Spain and one of the most picturesque places in Europe, with a magnificent beach and splendid parks. It is the seat of a Roman Catholic bishop. It has flour- and saw-mills, iron foundries, and manufactures of cloth and hats, and exports lead, copper, iron, and cement.

Santa Ana, Salvador, Central America, is the center of a sugar-growing district, which has also rich deposits of iron, copper, silver, and zinc.

Santa Barbara, California (c. h. Santa Barbara Co.), is one of the most attractive winter resorts of California, noted for the beauty of its surroundings, the luxuriance of its roses and other flowers, and its excellent bathing beach. The old mission, founded by Padre Junipero Serra in 1786, is one of its attractions.

Santa Cruz, Canary Island (cap. Tenerife), is a port and coaling station. Exports wine, almonds, cochineal, sugar, and agricultural products.

Santa Fé, Argentina (cap. Santa Fé), is a port on the Paraná River and an important railway center, with large exports of cattle, wool, and lumber. It has a cathedral, a normal school, and a seminary for priests.

Santander, Spain (cap. Santander), is a port on the Bay of Biscay with a large foreign trade, chiefly in iron ore. Fishing is carried on and tobacco is manufactured. It is the seat of an archbishop.

Santiago, cap. Chile (cap. Santiago), contains a Congress house which is one of the handsomest buildings of the kind in South America; the government house, used as a mint under Spanish rule, is an imposing edifice. Other important buildings are the cathedral, university, archiepiscopal palace, the municipal offices, the national museum, and observatory. The city was damaged by earthquake on Aug. 16, 1906, but has been largely rebuilt.

Santiago, Spain (Coruna), is the seat of an archbishop, and is famous for its superb Romanesque cathedral, the shrine of St. James (Sant' Iago) the Great, patron saint of Spain. It was formerly a great place of pilgrimage. A decaying university still exists.

Santiago de Cuba, Cuba (cap. Oriente), is a port and coaling station, with a magnificent landlocked harbor, and is the second city of the island in size and importance. It is situated in the center of a rich mineral district, and exports iron ore, manganese, and copper, as well as sugar, tobacco, and coffee. In the Spanish-American War of 1898, the Spanish fleet under Cervera was blockaded inside the harbor by an American fleet under Sampson, and when the Spanish fleet attempted to escape, it was destroyed outside the harbor entrance, July 3, 1898.

Santo Domingo, cap. Santo Domingo, West Indies, is a port, and perhaps the most perfect specimen of the 16th century Spanish city in America. In the old cathedral are buried many of the notable characters

here from 1846 to 1850; W. H. Prescott, the historian; Maria S. Cummins, author of "The Lamplighter"; Pierce, the mathematician; and W. W. Story, the sculptor.

Salem, cap. Oregon (c. h. Marion Co.), is the center of the most important hop-growing region in the world. It has flour-mills, woollen mills, etc. The capitol is an imposing building, and state institutions for deaf and dumb, blind, and insane, and a penitentiary are located here. It is the seat of a Methodist Episcopal university.

Salerno, Italy (cap. Salerno), is the seat of an archbishop, and has a cathedral dating from the 11th century. It was famous in the middle ages for its medical school.

Salford, England (Lancashire), is a suburb of Manchester, with extensive manufacturing interests.

Salisbury, England (Wiltshire), has a famous cathedral, founded in 1220, which is a beautiful example of early English architecture. The cloisters, built in the 13th century, are the most perfect in England, and form a quadrangle 181 ft. square. The spire, 404 ft., was added in the 14th century, and is the loftiest in England. Within the close are the chapter-house, the library, the episcopal palace, and the deanery. Salisbury contains three other ancient churches, and several mediæval buildings.

Saloniki, Turkey (cap. Saloniki), is a port and coaling station at the head of the Gulf of Saloniki, and the second city in size and commercial importance of Turkey in Europe. Of particular interest are the antiquities, notably its ancient churches and a Roman triumphal arch. The city is overlooked by an ancient citadel. Its industries include cotton-mills, soap-works, and silk manufactories. It exports chiefly corn, tobacco, opium, cotton, cocoons, wool, and hides. Founded 315 B. C., it became the capital of Macedonia. Under the name of Thessalonica it was visited by St. Paul. In 904 the Saracens sold 22,000 of the inhabitants as slaves. It was wrested from Venice by the Turks in 1430.

Saltillo, Mexico (cap. Coahuila), contains a cathedral, has cotton-mills, and is noted for its woollen shawls.

Salt Lake City, cap. Utah (c. h. Salt Lake Co.), is the metropolis of the state, an important railroad and mining center, and the center of trade for the vast "intermountain" region. Great smelters for the reduction of ores are located in and near the city, and it has large beet-sugar factories and manufactories of iron products, machinery, etc. Fort Douglas, a United States military post, is situated on a plateau 500 ft. above and about 3 miles from the heart of the city. It is the headquarters of the Church of Jesus Christ of Latter Day Saints (Mormons), and the home of the president and many of the high officers of the church. The Temple, Tabernacle, and Assembly Hall occupy a square in the central portion of the city. The Tabernacle, built in 1864-7, is a huge and extraordinary structure, in the shape of an oval or ellipse, 250 ft. long, 150 ft. wide, and 70 ft. high. It is surmounted by a wooden roof with iron shingles, resembling the shell of a turtle or the inverted hull of a ship, supported by 44 sandstone pillars. The Temple is a large and handsome building of granite, erected in 1853-93 at a cost of over $4,000,000. It is 186 ft. long from east to west and 99 ft. wide. At each end are 3 pointed towers, the loftiest of which, in the center of the eastern or principal façade, is 210 ft. high and is surmounted by a colossal gilded figure of the Angel Moroni. The Assembly Hall, to the south of the Tabernacle, is a granite building with accommodation for 3,000 people, intended for divine service. Salt Lake City was founded in 1847 by the Mormons, under Brigham Young, who had been driven from Nauvoo the previous year and had made a long and perilous journey across the Indian-haunted plains. The district was then a barren and unpromising desert, but the industrious Mormons set to work at once to plough and plant, and began that system of irrigation, which has drawn out the latent capabilities of the soil and made the Utah valleys among the most productive regions in the country.

Salzburg, Austria (cap. Salzburg), is a town noted for its picturesque situation, and one of the principal touring resorts of Austria. It is also an important

railway center. It is the birthplace of Mozart, and of Hans Makart, the painter.

Samara, Russia (cap. Samara), is a port on the Volga, and an important railway center with large trade in grain. Its industries include tanneries, flour-mills, breweries, brick-works, iron and bell metal foundries. It is the seat of an Orthodox bishop, and has two old cathedrals, built in 1685 and 1730-5, and one completed in 1894; also a museum of natural history and archæology.

Samarang, Java, is a fortified town, and one of the most important commercial centers of the island. Has no port, but vessels a mile out at sea take away sugar, coffee, tobacco, and indigo.

Samarkand, Russia (Russian Turkestan), has an active commerce and manufactures of cotton, silk, etc. Among the objects of interest are the grave of Timur, the citadel, three colleges, and neighboring ruins. The ancient city was destroyed by Alexander the Great. In the middle ages Samarkand was a large and flourishing city, renowned as a seat of learning. It was taken and destroyed by Jenghiz Khan in 1219; became the capital of Timur in 1369; and was annexed to Russia in 1868.

San Angelo, Texas (c. h. Tom Green Co.), is an important trade center in an agricultural region, largely devoted to dry farming.

San Antonio, Texas (c. h. Bexar Co.), is one of the largest cities of Texas, and an important railroad and trade center. It has large flouring-mills, cotton-presses, cement works, oil-mills, broom works, etc., and is a prominent cotton, wool, horse, mule, and cattle market, as well as the financial center of a very extensive stock-raising section. It is surrounded by an agricultural region supplied with water from artesian wells. In and near the city are numerous medicinal springs. Fort Sam Houston, an important military post and headquarters of the department of Texas, is located here. There is also a United States arsenal for storage of ordnance supplies and the manufacturing and repairing of equipment. It is frequented as a winter resort, and is noted for its old Spanish mansions near the city, having been settled by the Spaniards about 1600. The Church of the Mission del Alamo, which seems to have derived its name from having been built in a grove of alamo or cottonwood trees, is a low and strong structure of adobe, with very thick walls, built in 1744, and preserved as a national monument. The most salient event in San Antonio's history is the "Fall of the Alamo" in 1836. Texas had determined to resist certain obnoxious laws imposed by Mexico, and the latter sent an army under Santa Ana to reduce the rebels. The advance-guard of 4,000 men reached San Antonio on Feb. 22nd, and found the fortified Church of the Alamo garrisoned by a body of 145 Americans (afterwards joined by 25 or 30 more), under Travis, Bowie, and Davy Crockett, who refused to surrender. After a siege of 12 days, the church was finally carried by assault (March 6th) and all the survivors of the gallant little band of defenders were put to the sword.

San Diego, California (c. h. San Diego Co.), is a port of entry and an important health resort and trading center on San Diego Bay, with one of the best harbors on the California coast. It is noted for its climate, which is mild and equable. It is the commercial center of a region in which are produced large quantities of nuts, fruit, and honey, and has flour and planing-mills, machine-shops, wagon works, etc. It is a naval coaling station, and has a United States garrison. Coronado Beach, on a small peninsula opposite the city, is a favorite resort. The Mission of San Diego, 6½ miles from the city, was the first settlement made by white men in California. The bay was discovered in 1542.

Sandusky, Ohio (c. h. Erie Co.), is a port of entry on Sandusky Bay, and has one of the best landlocked harbors on Lake Erie. It is a trade and manufacturing center in a fertile agricultural region, of which the chief productions are grapes, peaches, apples, and grain. Fishing is an important industry of the city, and a large amount of wine is shipped. The chief industrial establishments are factories, in which are made baskets and crates for

the shipment of grapes and peaches; cooperages, in which are made casks for the shipment of wine; machine-shops; engine, boiler, threshing-machine, and cement works; flour-mills; and furniture factories. There are large coal and lumber yards. Considerable coal is brought here for manufacture and reshipment, and the ship-building industry is also important. The United States hatchery at Put-in-Bay encourages the fish industry. A state sailors' and soldiers' home, with accommodations for 1,600 persons, is located here.

San Francisco, California (c. h. San Francisco Co.), is the largest city in California, also the largest on the western coast of America or in the United States west of the Mississippi River. It is a port of entry situated on a peninsula between the Bay of San Francisco and the Pacific Ocean, with a sheltered harbor about 90 miles long and from 5 to 15 miles wide, having excellent anchorage ground and an abundance of deep water. It is extensively engaged in the manufacture of mining machinery, the canning of fruits and vegetables, ship-building, etc., and the principal industrial establishments are sugar and molasses refineries, slaughtering and meat-packing houses, leather manufactories, foundries, machine-shops, smelting works, etc. It is an important railroad terminus and has a large wholesale trade in merchandise of every description. Its foreign commerce is very extensive, and among the chief exports are gold and silver, wine, fruit, wool, oil, lumber, flour, and breadstuffs. The principal imports are tea from China and Japan. A United States branch mint, ranking second in the country, is located here. A large part of the city was destroyed by earthquake and fire in 1906, but the business district has been rebuilt in an improved manner, and is now noted for its modern commercial buildings, magnificent hotels, etc. One of the most interesting historical relics of San Francisco is the old Mission Dolores, dating from about 1776; the old church is built of adobe and is adjoined by a tangled and neglected little churchyard. Among the educational institutions of the city may be mentioned the Cooper Medical College, the Medical Department of the University of California, the Cogswell Polytechnic School, and the California School of Mechanical Arts. The Presidio, or government military reservation, has an area of 1,500 acres and stretches along the Golden Gate for about 4 miles. Golden Gate Park, having an area of 1,013 acres, contains a museum, Japanese tea-garden, etc. The city was founded in 1835. In 1848, the year of the discovery of gold in California, its population was about 500; in 1850 it was about 25,000; and, in each decade, it has seen an extraordinary increase.

San José, California (c. h. Santa Clara Co.), is situated in a fertile agricultural region, the beautiful Santa Clara Valley, famed for the variety and amount of its fruit. It has large canneries, packing-houses, and shipping industries connected with the fruit industry, also flour-mills, lumber mills, grain elevators, woollen factories, basket and box factories, machine-shops, etc. It has a considerable wholesale trade. Among the educational institutions are a state normal school, a Methodist Episcopal university, a Roman Catholic college, etc. On account of its equable climate, it is much frequented by tourists, and is a delightful residential city. The first settlement was made here about 1770 and the pueblo of San José was established in 1782. It was the state capital under the first constitution of California.

San José, cap. Costa Rica, is a well-built city, which contains a cathedral, a university, a national museum, and one of the finest opera-houses in Central America. The city was founded about 1738, and has been the capital, except for short intervals, since 1824.

San Juan, Porto Rico (cap. San Juan), is a fortified port and the largest city of Porto Rico. The harbor is excellent, affording anchorage for the largest ships,—the only good harbor in Porto Rico and one of the best in the West Indies. The entrance, however, is very narrow, and difficult in stormy weather. The chief exports are tobacco, sugar, and coffee. It contains a number of notable public buildings, including the bishop's palace, a large cathedral, the old government building, the city hall, the Casa Blanca, the custom-

house, a military hospital, and an arsenal. A new capitol is being built. San Juan was founded in 1511 and was the seat of the Spanish provincial government. It was bombarded by the U. S. fleet under Admiral Sampson in May, 1898.

San Juan del Norte, Nicaragua, Central America, is the chief port of Nicaragua on the Caribbean Sea. Exports mahogany, rubber, bananas, hides, and indigo. Steamers run from here up the San Juan River into Lake Nicaragua.

San Luis Potosi, Mexico (cap. San Luis Potosi), is an important railroad center, with a large trade in cattle, hides, and wool. The famous silver mines worked in the neighborhood since the end of the 16th century are now almost abandoned. It has handsome public buildings including a cathedral, a university, and a city hall.

San Miguel, Salvador, Central America, is the center of a fertile agricultural region, and has a trade in indigo.

San Salvador, cap. Salvador, Central America, has more than once been destroyed by earthquakes. It contains a university, a cathedral, a national museum and library, and an observatory. It has an active trade in agricultural products, particularly indigo.

San Sebastian, Spain (cap. Guipuscoa), is the most fashionable seaside resort of Spain and one of the most picturesque places in Europe, with a magnificent beach and splendid parks. It is the seat of a Roman Catholic bishop. It has flour- and saw-mills, iron foundries, and manufactures of cloth and hats, and exports lead, copper, iron, and cement.

Santa Ana, Salvador, Central America, is the center of a sugar-growing district, which has also rich deposits of iron, copper, silver, and zinc.

Santa Barbara, California (c. h. Santa Barbara Co.), is one of the most attractive winter resorts of California, noted for the beauty of its surroundings, the luxuriance of its roses and other flowers, and its excellent bathing beach. The old mission, founded by Padre Junipero Serra in 1786, is one of its attractions.

Santa Cruz, Canary Island (cap. Tenerife), is a port and coaling station. Exports wine, almonds, cochineal, sugar, and agricultural products.

Santa Fé, Argentina (cap. Santa Fé), is a port on the Paraná River and an important railway center, with large exports of cattle, wool, and lumber. It has a cathedral, a normal school, and a seminary for priests.

Santander, Spain (cap. Santander), is a port on the Bay of Biscay with a large foreign trade, chiefly in iron ore. Fishing is carried on and tobacco is manufactured. It is the seat of an archbishop.

Santiago, cap. Chile (cap. Santiago), contains a Congress house which is one of the handsomest buildings of the kind in South America; the government house, used as a mint under Spanish rule, is an imposing edifice. Other important buildings are the cathedral, university, archiepiscopal palace, the municipal offices, the national museum, and observatory. The city was damaged by earthquake on Aug. 16, 1906, but has been largely rebuilt.

Santiago, Spain (Coruna), is the seat of an archbishop, and is famous for its superb Romanesque cathedral, the shrine of St. James (Sant' Iago) the Great, patron saint of Spain. It was formerly a great place of pilgrimage. A decaying university still exists.

Santiago de Cuba, Cuba (cap. Oriente), is a port and coaling station, with a magnificent landlocked harbor, and is the second city of the island in size and importance. It is situated in the center of a rich mineral district, and exports iron ore, manganese, and copper, as well as sugar, tobacco, and coffee. In the Spanish-American War of 1898, the Spanish fleet under Cervera was blockaded inside the harbor by an American fleet under Sampson, and when the Spanish fleet attempted to escape, it was destroyed outside the harbor entrance, July 3, 1898.

Santo Domingo, cap. Santo Domingo, West Indies, is a port, and perhaps the most perfect specimen of the 16th century Spanish city in America. In the old cathedral are buried many of the notable characters

of early American history, including the family of Columbus. Another landmark is the castle of Columbus. It is the seat of a Roman Catholic archbishop, and of a university, founded in 1730. It exports sugar and coffee.

Santos, Brasil (São Paulo), is a port and coaling station, whose chief export is coffee. It claims to be the greatest coffee shipping port in the world. The church of Our Lady of Montserrat, on a hill overlooking the town, is one of the oldest shrines in Brasil.

Sao Paulo, Brasil (cap. São Paulo), is the second city of Brasil, the center of its coffee-growing region, and an important railroad and industrial point. It is a modern, handsome city, and among its fine buildings are the cathedral, the treasury, the government building, and the Ypiranga Palace, a magnificent structure erected to commemorate the Declaration of Independence. It is the seat of a Roman Catholic bishop. Sao Paulo was founded by Jesuits in 1554 as a mission station.

Sapporo, Japan (Hokkaido), is the seat of an agricultural college, and has a museum and a botanical garden.

Saragossa, Spain (cap. Saragossa), has manufactures of silk and woollen goods. Among its noteworthy features are the cathedral, university (1474), leaning clock tower (Torre Nueva), and the seven-arched bridge (1437) over the Ebro. After being captured from the Moors, Saragossa became the capital of Aragon in 1118. The French attacked it in 1808 and 1809; Agustina, the "Maid of Saragossa," aided in the defense.

Sarajevo, Austria (Bosnia), has manufactures of tobacco, brass, copper, iron goods, pottery, and silk. It is the seat of a Roman Catholic bishop and has a cathedral.

Saratof, Russia (cap. Saratof), is a trading center for east Russia, and has flour-mills, oil-presses, tobacco factories, distilleries, and iron-works. A society of naturalists maintains a biological station.

Saratoga Springs, New York (Saratoga Co.), is a noted watering-place famous for its saline mineral springs. The waters are cathartic, tonic, alterative, and diuretic. On account of its spacious hotels, which can accommodate about 20,000 visitors, and convention halls, one of the latter seating 5,000, it is a favorite place for political conventions. The first hotel was erected here in 1774, and thousands visit this village to use the waters, making it one of the most popular summer resorts.

Sarnia, Ontario, Canada, is a port at the mouth of St. Clair River. It has lumber mills, engine works, salt works, etc.

Sasebo, Japan (Nagasaki), is a naval port.

Saskatoon, Saskatchewan, Canada, is an important trade center and the seat of the projected Provincial University. It has wood working plants, brick works, etc.

Sassari, Italy (cap. Sassari), is the seat of an archbishop and has a university, founded in 1677.

Sault Sainte Marie, Michigan (c. h. Chippewa Co.), is a port of entry near the outlet of Lake Superior on the Sault Sainte Marie ship-canal, which connects Lakes Superior and Huron, and is an important railway center. The international railroad bridge, across Saint Mary's River, connects the city with the village of the same name in Canada. The principal industrial establishments are lumber mills, machine-shops, foundries, planing-mills, flour-mills, brick works, and shingle-mills. Enormous water-power has been developed from the rapids of the Saint Mary's River. Fort Brady, a United States military post, is near by. The place was settled in 1662 by Père Marquette, and constitutes the first permanent settlement within the present limits of Michigan. It was incorporated as a city in 1887.

Sault Sainte Marie, Ontario, Canada, has extensive steel rail works, iron works, pulp and paper mills, a blast furnace, etc.

Savannah, Georgia (c. h. Chatham Co.), is the second city of Georgia, a port of entry on the Savannah River, 18 miles from its mouth, and one of the most important commercial and shipping cities of the Southern States.

It has an extensive coastwise and export trade, the chief articles being cotton, lumber, rice, and naval stores. Its industries include fertilising plants, cotton-seed-oil mills, iron foundries, sawmills, planing-mills, rice and flour-mills, cotton-mills, machine-shops, ice factories, etc. The Telfair Academy of Arts and Sciences is located here; it contains a collection of casts, paintings, and various objects of art and historical interest. Savannah was settled in 1733 by Gen. James Oglethorpe, the founder of the youngest of the thirteen original states. In 1778, this city was captured by the British, who repulsed a Franco-American attempt to take it the following year. The port of Savannah was closed to commerce by the Federal fleet from 1861 to 1865, and Sherman occupied the city in December, 1864, at the end of his triumphant "March through Georgia." Since the war its progress has been rapid.

Savona, Italy (Genoa), is a fortified port, with manufactures of cloth, silk, paper, glass, and soap, which is supposed to have been invented here. It is the seat of a bishop, and has a cathedral dating from the 17th century.

Scarborough, England (York), is a much-frequented seaside resort with a picturesque situation, and extensive promenades and gardens.

Schaerbeek, Belgium (Brabant), is a suburb of Brussels.

Schenectady, New York (c. h. Schenectady Co.), is an important industrial center in a fertile agricultural region, having extensive manufactures of locomotives, electrical construction works, foundries and machine-shops, broom and brush factories, etc. It was settled in 1662.

Schiedam, Netherlands (South Holland), is noted for its production of gin and other liquors.

Schoneberg, Germany (Prussia), is a suburb of Berlin.

Schweidnitz, Germany (Silesia), manufactures woollen goods, leather, needles, gloves, agricultural implements, and machinery. Its beer has been famous since the 16th century.

Schwerin, Germany (cap. Mecklenburg-Schwerin), has manufactures of carriages, colors, varnish, soap, musical instruments, furniture, and bricks. It contains two palaces, a 14th-15th century cathedral, museums, and picture-galleries.

Scranton, Pennsylvania (c. h. Lackawanna Co.), is the third city of Pennsylvania, and an important railroad and industrial center in the chief anthracite coal district of the United States. There are many collieries within its limits. It has extensive iron industries, manufactories of silk and silk goods, mining machinery, etc. A famous correspondence school is located here, and the city has numerous other educational institutions. It was settled in 1788.

Seattle, Washington (c. h. King Co.), is the largest city of Washington, and an important port on Puget Sound. It has an extensive coastwise and export trade, with regular steamers to Alaskan and Asiatic ports. It is the chief entrepôt of the Alaskan gold-fields, and a large amount of gold-dust is brought to the United States Assay Office here. It is the principal commercial and industrial center of a large region containing extensive natural resources, the most important of which are lumber, fish, mines, and agricultural products. The chief exports are coal, timber, hops, and fish. The most important industrial establishments are shipyards, lumber and shingle mills, flour-mills, iron-works, rolling-mills, foundries, and wood-working establishments. The power used in a great number of the mills and factories is electricity, generated by the falls in the rivers flowing down from the Cascade Mountains. The United States Navy Yard of Puget Sound is located near by at Bremerton, on Port Orchard Bay, across Puget Sound. At Magnolia Bluff is a United States Army Post. The State University, with 1,500 students, is located here; its grounds furnished the site of the Alaska-Yukon-Pacific Exposition of 1909. Among the other educational institutions are Seattle Seminary and a Roman Catholic college. The public library contains over 70,000 volumes. The city was founded in 1852 and

named after an Indian chief. Almost all of the business quarter was destroyed by fire in 1889, but it has been rebuilt in a more substantial manner.

Sedalia, Missouri (c. h. Pettis Co.), is a trade center in an agricultural and coal-mining region. In the vicinity are deposits of zinc, iron, lead, and fire-clay, also beds of emery and large deposits of limestone. The chief industrial establishments are railroad shops, car shops, foundries, machine-shops, agricultural implement works, woollen mills, etc.

Selma, Alabama (c. h. Dallas Co.), is an industrial and trade center in a rich cotton-growing section. It has cotton-gins, cottonseed-oil mills, lumber mills, fertiliser works, yarn mills, etc.

Semipalatinsk, Russia in Asia (cap. Semipalatinsk), has tanneries and distilleries, and manufactures flour, soap, and candles.

Sendai, Japan (cap. Miyagi), has a trade in fish and salt.

Seoul, cap. Korea, has a Roman Catholic cathedral, a bell-tower (1468), and a marble pagoda. Since 1904 the city has been under the control of the Japanese.

Seraing, Belgium (Liège), has large engineering and machinery works.

Seres, Turkey (Saloniki), is the center of the Macedonian cotton industry and exports cotton, tobacco, hides, and cocoons. It is the see of an archbishop of the Greek Church.

Sergievsk, Russia (Moscow), has a convent with fabulous treasures, which attracts hosts of pilgrims.

Serpukhof, Russia (Moscow), has a cathedral dating from 1710 and a variety of manufactures.

Sevastopol, Russia (Taurida), is the chief Russian naval port of the Black Sea, and a summer resort. It has extensive dockyards and naval arsenals, two navigation schools, a naval hospital, and a biological station. On Oct. 5, 1854, the English and allied armies invested Sevastopol, and, after one of the most famous sieges of modern history, gained possession on Sept. 8, 1855. In 1890 the commercial port was transferred to Theodosia.

Seville, Spain (cap. Seville), was the seat of a Roman colony (Hispalis), and was a Moorish kingdom from 712 until its conquest by Ferdinand III. in 1248. Its exquisite Hispano-Gothic cathedral, commenced in 1403 and finished in 1519, is one of the finest churches in the world; its Mudejar tower (Giralda) especially merits attention. This was begun in 1000 by the Moors, but its upper portion has been to some extent Christianised. It stands 350 ft. high. The Moorish palace, or Alcazar, and the Charity Hospital, with pictures by Murillo, a native of the city, deserve attention. There are also an art museum (with works by Murillo), a university, and an academy of fine arts. Seville is an archiepiscopal see. The city, once the center of the commerce with America, is still a busy and prosperous port, shipping wine, oil, fruit, lead, and cork, and having manufactures of ceramics, leather, tobacco, and machinery.

Shahjahanpur, India (United Provinces), manufactures sugar.

Shamokin, Pennsylvania (Northumberland Co.), is an industrial and commercial center in the anthracite coal region. The chief industrial establishments are foundries, machine-shops, and knitting, planing, and flour-mills; a large number of the inhabitants are employed in the mines in the vicinity. It was settled in 1835.

Shanghai, China (Kiang Su), is a treaty port and coaling station. Its facilities for distribution and the lack of deep water at the treaty ports in the north, have made it the entrepôt of all foreign trade north of Fuchau, and the greatest foreign market of the empire. Miles of wharves and five large dockyards hardly suffice for the traffic. It exports principally silk, cotton, tea, rice, hides, wool, and cereals. Cotton-spinning, silk-winding, ship-building, and the manufacture of matches, rice, paper, ice, and furniture are the chief industries. Shanghai was taken by the British in 1842, and was opened to foreign trade in the same year.

Shashi, China (Hupeh), is a treaty port on the Yangtze River. It is an important mart for cotton stuffs.

Sheboygan, Wisconsin (c. h. Sheboygan Co.), is a commercial and industrial center in a large agricultural region, and a port on Lake Michigan at the mouth of the Sheboygan River. It has large chair, furniture, boot and shoe factories, a shipyard, lime-kilns, boiler works, etc., and a large trade in cheese.

Sheffield, England (York), is the chief seat of English cutlery manufacture. Among the articles manufactured are knives, scissors, razors, tools of all kinds, rails, armor-plates, castings, surgical instruments, machinery, silver-plate, axles, etc. Its cutlery has been celebrated from early times. Mary Queen of Scots was confined in the castle between 1570 and 1584.

Shenandoah, Pennsylvania (Schuylkill Co.), is a commercial and industrial center in a region rich in anthracite coal. Several of the largest collieries in the country are located here. The chief industries of the place are connected with the mining and shipping of coal. There are machine-shops, mining-tool works, foundries, etc. It was settled in about 1850.

Sherbrooke, Ontario, Canada, has manufactures of paper, and of woollen and worsted goods.

Sheridan, Wyoming (c. h. Sheridan Co.), is the largest city of northern Wyoming, and has considerable trade in wool and lumber. Coal is mined near by.

Sherman, Texas (c. h. Grayson Co.), is a trade center in the Red River Valley, with an extensive trade in cotton and grain products. It has a cotton-gin, several large cottonseed-oil mills, flour-mills, lumber mills, furniture factories, etc. Large coal-mines are situated near by in Oklahoma. A Methodist Episcopal college and several other educational institutions are located here.

Shidzuoka, Japan (cap. Shidzuoka), has manufactures of lacquer-ware and baskets.

Shimonoseki, Japan (Yamaguchi), is a fortified seaport on the Inland Sea, near its west entrance, and has been open to foreign trade since 1890. In 1864 it was bombarded by a combined fleet of British, American, Dutch, and French vessels as a protest against the Japanese firing on foreign vessels; the Japanese paid an indemnity in 1875. The treaty signed at Shimonoseki in April, 1895, put an end to the Japanese-Chinese war.

Shiraz, Persia, exports tobacco and inlaid woodwork, and is famous for its wine.

Sholapur, India (Bombay), has manufactures of silk and cotton.

Shreveport, Louisiana (c. h. Caddo Parish), is the commercial and industrial center of the northwestern part of Louisiana, and has an extensive wholesale trade. It is situated in a rich agricultural region in which cotton is the chief product. There are several cotton-presses, cottonseed-oil mills, railroad shops, lumber- and stock-yards, etc. Its railroad facilities and opportunities for river transportation give it especial advantage as a shipping point and a jobbing place. It was incorporated as a city in 1839. During the Civil War, from the time Baton Rouge was captured by the Federal forces until the close of the war, Shreveport was the capital of the state.

Shusha, Russia (Elisavetpol), has cotton, wool, and silk mills, tanneries, dye-works, and soap factories.

Sialkot, India (Punjab), manufactures paper and cloth.

Siangtan, China (Hunan), is a great trading center.

Sibonga, Philippine islands (Cebu), is a trade center with extensive fisheries.

Siena, Italy (cap. Siena), has several very interesting public buildings, and is exceptionally rich in works of art. It is the seat of a bishop, and its cathedral, built in the 13th and 14th centuries, is a splendid example of Gothic architecture. The university was founded in 1203. The Piazza del Campo, celebrated by Dante in his Purgatorio, contains the Loggie di San Paolo, the seat of a commercial tribunal in the middle ages. Siena was the center of an art school, whose chief representatives were Buoninsegna, Sodoma, and Perussi.

Simbirsk, Russia (Simbirsk), is a river port, and has breweries, tanneries, distilleries, brick works, flour-mills, iron and copper foundries, and manufactories of candles,

soap, malt, and potash, as well as fisheries. It is the seat of an Orthodox bishop and contains two cathedrals.

Simferopol, Russia (cap. Taurida), is the seat of a Greek Orthodox bishop, and has a fine cathedral, in the Venetian style, and a museum of antiquities and natural history. There are vineyards, nursery-gardens, extensive orchards and manufactures of soap, candles, and tobacco.

Singan, China (cap. Shensi), is an important trade center.

Singapore, Straits Settlements, is a port and coaling station on the direct route between India and China, and has become the chief commercial emporium of southeastern Asia, being at the same time a free port. The chief exports are tin, gambier, sago, tapioca, sugar, pepper, tortoise-shell, mother-of-pearl, gutta-percha, india-rubber, nutmegs and mace, camphor, gum elastic, coffee, tobacco, rice, hides, sapan-wood, stick-lac, and rattans, the trade being mostly transport trade. Singapore manufactures tapioca, sago, gambier, white pepper, tools, implements, vehicles, and furniture, and has ship-building, machine-shops, and engineering works.

Sioux City, Iowa (c. h. Woodbury Co.), is the commercial and industrial center of northwestern Iowa, northeastern Nebraska, and the southeastern part of South Dakota. It has extensive meat-packing houses, flour-mills, broom and furniture factories, stove and engine works, agricultural implement works, carriage and wagon works, tile works, creameries, etc. There are also large grain elevators, stock- and lumber-yards, and a number of wholesale houses. It was settled in 1849 by traders, and, for several years, was the place where gold seekers, going to the Black Hills, were fitted with all necessary material for gold-hunting and camp life.

Sioux Falls, South Dakota (c. h. Minnehaha Co.), is a trade center in an agricultural and stock-raising region. In the vicinity are extensive quarries of marble, jasper, and other stones. It is situated on the Sioux River, which here has a fall of 100 ft. in a half mile, forming a series of cascades that furnish abundant water-power. It has large breweries, a basket factory, stone-quarries, flour-mills, marble-works, etc. Its trade in agricultural implements is quite large. A state penitentiary and a state children's home are located here. It was settled in 1856.

Siut, Egypt, is noted for its pipe bowls. A dam across the Nile is located here.

Sivas, Turkey in Asia, has manufactures of coarse woollen goods. Silver, lead, copper, iron, coal, and salt are abundant in the vicinity.

Skutari, Turkey, has wool weaving manufactories, ship-building, and exports wool, corn, sumach, and hides.

Slivno, Bulgaria (Eastern Roumelia), has distilleries, printing works, etc.

Smethwick, England (Stafford), has iron, glass, and chemical works.

Smichow, Austria (Bohemia), is a suburb of Prague.

Smolensk, Russia (cap. Smolensk), has tanneries, brick-fields, oil works, breweries, distilleries, flour-mills, potteries, limekilns, and manufactures of soap, tallow, linen, and leather. Smolensk is one of the oldest East Slav cities; it existed before the Rurik immigration of 862, and was the capital of the Krivichi. It was an appanage of the Kiev grand princes from 882 to 1054; from 1054 to 1395 it was governed by its own princes. From 1610 to 1654 it was held by Poland, and was acquired by Russia in 1686. Here took place some of the fiercest fighting in the Napoleonic invasion of 1812.

Smyrna, Turkey in Asia, is the principal port of Asia Minor and a coaling station. Its principal industry is the manufacture of carpets, other branches being connected with machine-shops, iron-foundries, silks, cottons, woollens, and pottery. The exports consist chiefly of carpets, figs, tobacco, valonia, raisins, and silk. It is the seat of archbishops of the Greek Orthodox, Roman Catholic, and Armenian Churches. Smyrna is a very ancient city, and it claimed to be the birthplace of Homer. It was one of the seven cities addressed by John in the Revelation. It was destroyed by an earthquake in 178

A. D., and was restored by Marcus Aurelius; was occupied by the Knights of St. John in the 14th century; and was sacked by Timur in 1402. From 1424 it has been under Turkish rule.

Sofia, cap. Bulgaria, exports hides, corn, linen, cloth, and silk. It is the seat of a Greek Orthodox archbishop, and of a university. It has hot mineral springs. A famous church council was held here in 343 A. D. It fell into the hands of the Turks in 1382, and the Russians occupied it in 1878.

Solingen, Germany (Prussia), is the center of an important steel manufacturing district, turning out cutlery especially.

Somerville, Massachusetts (Middlesex Co.), is a residential suburb of Boston, and, in its industrial life, is identified with Boston in many ways. It was settled in 1629 and was a part of Charlestown until 1842, when it was set off as an independent town. In 1872 it was incorporated as a city.

Southampton, England (Hampshire), is a port, a coaling station, the chief British military transport port, and the regular port of call for several important mail steamship lines. Exports cotton manufactures, apparel, leather, etc., and has yacht-building and marine engineering works. Among the most important old buildings are St. Mary's (rebuilt 1884), St. Michael's (in part Norman), Holy Rood (ancient tower and spire), the 12th century chapel of God's House, now used for French Protestant service and King John's palace (12th century). A considerable portion of the old town walls, erected by Richard II., still exist, including parts of towers and three gateways. One, Bargate, a massive structure in two stages, has a large chamber used as the Guildhall, and contains ancient paintings of the local legendary hero, Sir Bevis of Hampton, and of the giant Ascupart. During the middle ages the port had great trade with Venice; its modern prosperity dates from the opening of the new docks (1842).

South Bend, Indiana (c. h. Saint Joseph Co.), is the industrial center for a large portion of northern Indiana, and has extensive manufactures of automobiles, wagons, agricultural implements, woollen goods, paper, sewing-machine parts, furniture, boilers, electrical appliances, cutlery, harness, steel castings, indurated fiber goods, lumber, rubber stamps, etc. It is situated in a fertile agricultural region in which stock-raising receives considerable attention. Just outside the limits of the city, in the village of Notre Dame, are the large Roman Catholic schools,—the University of Notre Dame for men, and Saint Mary's College for women.

South Bethlehem, Pennsylvania (Northampton Co.), is a commercial and industrial center in a section rich in iron ore deposits and coal. The chief industrial establishments are iron and steel works, machine-shops and foundries, zinc and brass works, wood-working factories, knitting mills, silk-mills, and furniture works. A large quantity of armor-plates, shafts for marine engines, and heavy ordnance is manufactured here. It is the seat of Lehigh University (700 students), which ranks high for its work in engineering, physics, biology, and metallurgy. It was settled in 1741.

Southbridge, Massachusetts (Worcester Co.), is an industrial and commercial center, with optical works, shuttle works, woollen and cotton-mills, knife mills, etc. It was settled about 1730.

South Omaha, Nebraska (Douglas Co.), is the third city of Nebraska, and an industrial suburb of Omaha, with large stock-yards and stock markets. It is one of the largest slaughtering and meat-packing centers in the United States.

Southport, England (Lancashire), is a watering-place, with a handsome esplanade and marine drive. A marine park with lake (44 acres) fronts the shore. There are botanical gardens, a winter garden, and a zoological park. The public buildings include town hall, Cambridge Hall, Atkinson Art Gallery and Free Library, and market hall, and there are several hydropathic establishments.

South Shields, England (Durham), is a port on the Tyne, with manufactures of glass and chemicals, ship-

building, iron, rope, paint and varnish works, and a great trade in coal. The parish church of St. Hilda, rebuilt about 1811, preserves its ancient tower; the town hall dates from 1768.

Spandau, Germany (Prussia), is a fortified town, 8 miles northwest of Berlin. Here is the Julius tower, containing, since 1874, the sum of $30,000,000 in coin for immediate use in case of war, the money being part of the war indemnity paid by France after the Franco-German war.

Spartanburg, South Carolina (c. h. Spartanburg Co.), is a trade center in an agricultural region, in which cotton is the principal product. It has cotton-mills, and nearby are large limestone quarries and iron-mines. It is the seat of Wofford College, founded in 1853.

Spezia, Italy (Genoa), is a strongly fortified port on the Gulf of Genoa, and the chief naval station and arsenal of Italy, with large ship-building yards and docks. It is a favorite summer resort, and in the bay Shelley lost his life.

Spokane, Washington (c. h. Spokane Co.), is an important railroad, jobbing, and manufacturing center, and the metropolis of the so-called "Inland Empire." It has extensive lumber mills and manufactories of lumber products, sewer and water-pipe works, flour-mills, flour mill machinery works, brick and terra-cotta works, foundries and machine-shops, iron works, furniture factories, etc. It is in the center of an important mining district. Spokane River, which here has a fall of 132 ft., furnishes abundant water-power, made available largely by the generation of electricity. Spokane is the commercial center of the interior Columbia River basin, which is rich in mines, lumber, and agricultural lands. It has a large trade in wheat, horticultural products, and an extensive distributing trade in merchandise throughout eastern Washington, Idaho, Montana, and British Columbia. Gonzaga College and several other institutions of prominence are located here. The first settlement was made in 1873, and the rapid growth of the city began with the entrance of the Northern Pacific R.R. in 1881, which has continued to this day. On August 4, 1889, the business section of the city was almost wholly destroyed by fire, this was quickly rebuilt, and the city is now one of the most progressive and solidly built among the larger interior cities of the country.

Springfield, cap. Illinois (c. h. Sangamon Co.), is an important trade and manufacturing center in the great corn belt, and has extensive coal-mining, farming, horse and cattle breeding, and large manufacturing interests. Among the more important industries are watch factories, textile works, woollen mills, wood-working mills, machine-shops, boiler and car works, flour-mills, etc. The Illinois State Fair is permanently located here. Among the institutions of learning are an Ursuline Convent, a Lutheran College, etc. Springfield was first settled in 1819. It was the home of Abraham Lincoln, who is buried in Oak Ridge Cemetery.

Springfield, Massachusetts (c. h. Hampden Co.), is an important banking, manufacturing, and residential city in the center of one of the most populous and prosperous sections of New England. It has extensive manufactories of firearms, automobiles, railroad cars and supplies, envelopes, stationery, cotton goods, skates, printed books, and electrical supplies. The United States armory and arsenal are located here, and about 1,500 men are employed in making arms for the government. The French-American College, opened in 1885, and the International Y. M. C. A. Training School are located here. The public library contains over 150,000 volumes. The art and science museums, which are departments of the city library, are, perhaps, unrivaled in any city of similar size in the country. Springfield was first settled in 1636.

Springfield, Missouri (c. h. Greene Co.), is a trade center in a region known for its valuable deposits of zinc and lead. The chief industrial establishments are railroad shops, iron-works, machine-shops, wagon and carriage factories, furniture factories, etc. It has an extensive trade in manufactures, also in lead and zinc, and is the jobbing center of a large part of the southern

counties of Missouri. A state normal school, a Roman Catholic academy, and a Congregational college are located here.

Springfield, Ohio (c. h. Clark Co.), is a railroad and manufacturing center in a rich agricultural region. It has extensive manufactories of agricultural implements, gas- and steam-engines, piano plates, machinery in vast variety, tools, iron, steel, and brass goods, etc. In the vicinity are the state fraternal homes of the Free Masons, the Independent Order of Odd Fellows, and the Knights of Pythias. The Wittenberg College, founded in 1845 by the Lutheran church, is located here.

Srinagar, India (cap. Kashmir), was formerly a shawl-weaving center, but is now engaged chiefly in silver-working, carpet-weaving, and the manufacture of paper and leather.

Stamford, Connecticut (Fairfield Co.), is a manufacturing and residential suburb of New York. It has extensive manufactories of locks, typewriters, woollen goods, lumber, hats, shoes, machine-shop products, pottery, chemicals, etc. Several private schools of prominence are located here. It was settled in 1641.

Stanislau, Austria (Galicia), has railway workshops, tanneries, flour-mills, dye-works, and tile works.

Stargard, Germany (Prussia), has a cathedral dating from the 14th century. It manufactures machinery, and woollen and cotton goods.

Stavanger, Norway (cap. Stavanger), is a seaport with large fishing establishments. It has a cathedral in the Norman style, built in the 13th century and restored in 1866.

Stavropol, Russia (Stavropol), has flour-mills and gardening. It is the seat of a Greek Orthodox bishop.

Steelton, Pennsylvania (Dauphin Co.), is a manufacturing suburb of Harrisburg. The chief industrial establishments are steel bridge and construction works, blast-furnaces, rail and blooming-mills.

Stettin, Germany (Prussia), is a port and coaling station on the Oder, 36 miles from its mouth. It manufactures cement, clothing, machinery, sugar, chemicals, porcelain, and paper, and is one of the most important ship-building centers in Germany. It has churches of the 12th and 14th centuries, and a royal palace (1575).

Steubenville, Ohio (c. h. Jefferson Co.), is a trade center in an agricultural, coal, and natural gas region. Natural gas is near by, and is brought to the city by pipes and pumping. Coal surrounds the city and is mined extensively. Valuable clay deposits are in the vicinity and petroleum wells are within a few miles. Large quarries, which contain excellent building-stone, contribute largely to the prosperity of the city. Among the manufacturing establishments are iron and steel works, tin plate and clay works, potteries, glass factories, blast-furnaces, tube works, and nail factories. The city was settled in 1797.

Stillwater, Minnesota (c. h. Washington Co.), is a trade center in an agricultural region, and is extensively engaged in manufacturing, the chief industrial establishments being flour and feed mills, grain elevators, lumber mills, foundries, machine-shops, and furniture factories. The city has a large trade in logs, lumber and lumber products, wheat, flour, and live stock. It was settled in 1840.

Stockholm, Sweden (cap. Stockholm), is a port and coaling station at the outlet of Lake Malar, the chief naval station of the country, and the principal emporium for the commerce of central and northern Sweden. It is noted for its location and environs, and is remarkable for its numerous beautiful promenades and delightful views, especially on the Djurgard, with its outdoor museum illustrative of the national life, and the Scandinavian northern museum. The principal buildings are the Riddarholms Church, a large mediæval structure mostly in the Renaissance style, which is the Westminster Abbey of Sweden, being the burial-place of the kings and distinguished men of the country; the Royal Palace, a vast square in the Italian Style, built 1697-1753; the National Museum; the Northern Museum; and the Royal Library. Stockholm is the seat of the Swedish academy, the academy of sciences, the academy of belles-lettres, history and antiquities, and of a uni-

versity. The industries include sugar-refineries, tobacco, silk, stearin, and tallow factories, linen and cotton weaving and spinning, also iron-foundries. Stockholm has a small export trade in iron, timber, and tar, but a large import trade. The harbor is closed by ice from three to five months every year. In the environs are several very beautiful royal châteaux in picturesque parks.

Stockport, England (Cheshire), has cotton manufacture, hat-making, iron-founding, machinery, and brewing.

Stockton, California (c. h. San Joaquin Co.), is the metropolis and chief distributing center of the San Joaquin Valley, and is located at the head of the all-year navigation of the San Joaquin River system, being connected with the river by a navigable channel 2½ miles in length. Stockton has a large trade in fruits, grain, hay, and other agricultural products. Among the important industrial establishments are flour-mills, agricultural implement factories, window-glass factories, woollen mills, coal briquette factory, machine works, and canneries. A state hospital for the insane is located here. The growth of the city dates from the discovery of gold in 1848.

Stockton-on-Tees, England (Durham), is a port on the Tees River, 5 miles from its estuary. It has blast-furnaces and rolling-mills, and its extensive industries include the construction of marine engines, bridge-building, ship-building, and railway material, and glass-bottle works. A castle was built here soon after the conquest; it was taken by the parliamentarians in 1644.

Stoke-upon-Trent, England (Stafford), is famous as the chief center of the porcelain and earthenware manufacture, with which the names of Wedgwood, Minton, and Copeland are associated. The Minton memorial building contains a free library and museum, and schools of science and art, and in the church is a monument to Wedgwood (died 1793).

Stolp, Germany (Prussia), has manufactures of amber goods, linen, and alcohol.

Stralsund, Germany (Prussia), has manufactures of playing-cards, machinery, electric lamps, sugar, and starch. It has a town hall dating from the 14th century, a high school from 1560, and an arsenal. It is notable for its successful resistance to the siege commanded by Wallenstein, in 1628.

Strassburg, Germany (cap. Alsace-Lorraine), is a fortified town and has many handsome public buildings erected mostly since 1875, namely, the university, imperial palace, library, and the provincial government building. It is the seat of a bishop, and its famous minster was founded before 1015, and completed in 1439. The principal manufactures are cigars, leather, surgical instruments, and machinery. Here the Emperor Julian defeated the Alemanni in 357. The town was seized by Louis XIV. in 1681, and capitulated to the Germans in 1870.

Stratford, England (Essex), is a suburb of London, and has chemical works, distilleries, and various manufactories.

Stratford, Ontario, Canada, is a trade center in a rich agricultural district and has locomotive repair shops, a fence factory, agricultural implement works, bridge and iron works, etc.

Stratford-on-Avon, England (Warwick), is famous as the birthplace of Shakespeare. The house in which the poet was born is preserved, as are also the site of New Place, where the poet spent his later years, and the cottage of Anne Hathaway, the poet's wife. The Shakespeare Memorial is a handsome building, erected in 1879 on the bank of the Avon, and contains a theatre, a library with a rare collection of Shakespeareana, and a picture gallery. The parish church of Holy Trinity, dating from the 13th century, contains the tomb of Shakespeare. Another interesting building is Harvard House (16th century), the birthplace of the mother of John Harvard, the founder of Harvard University.

Streator, Illinois (Lasalle Co.), is a trade center in a fertile agricultural region in which there is considerable coal, fire-clay, and building stone. The manufacturing establishments include brick and tile works, glass factories, flour-mills, machine-shops, and foundries. The trade is chiefly in manufactures, coal, grain, live-stock, and dairy products.

Stretford, England (Lancashire), is an active center of the cotton industry.

Stryj, Austria (Galicia), has tanneries and manufactories of matches.

Stuttgart, Germany (cap. Württemberg), is one of the chief publishing centers in Germany, and manufactures furniture, pianos, chemicals, colors, chocolate, carriages, and leather. The old castle, now used as government offices, was built in 1553-70; the other notable buildings include the royal palace (1746-1807), polytechnic, conservatory of music, library, national industrial museum, and picture gallery. It is the birthplace of Hegel, Hauff, and Schwab. The counts of Württemberg took up their residence here in 1316.

Suchau, China (Kiang Su), is situated in the heart of the silk district, and is one of the wealthiest cities in China. It was almost destroyed in the Taiping rebellion, but has since largely recovered. It was opened to foreign trade in 1896.

Sucre, cap. Bolivia (Chuquisaca), is the residence of the archbishop of Bolivia, and contains a university. It was founded in 1539 by a companion of Pizarro, and was called Chuquisaca till the declaration of independence. It exports chiefly goods in transit.

Suez, Egypt, is a port and coaling station at the southern end of the Suez Canal. Once a flourishing emporium for Oriental trade, it fell into decay and until the opening of the Suez Canal, was a wretched village. The modern town is built on a desert peninsula. It exports chiefly goods in transit.

Sumy, Russia (Kharkof), has breweries, distilleries, tanneries, brick-fields, tallow foundries, and candle manufactories.

Sunderland, England (Durham), is a port and coaling station at the mouth of the Wear River, and one of the foremost steel ship-building centers in the United Kingdom. It has besides engineering, anchor, rope and chain cable works, glass-works, forges, and other industries. The parish church of St. Peter was founded in the 7th century by Biscop, in connection with the monastery where Bede became a student; St. Michael's (originally 10th century), contains some 13th century work. The town hall is a handsome modern structure.

Superior, Wisconsin (c. h. Douglas Co.), is a railroad and trade center, sharing with Duluth the commercial advantage of being the extreme western port of the Great Lakes system of the United States. It has three connecting harbors well sheltered and deep, making a combined length of 13 miles, with an extreme width of 3 miles. The principal manufacturing establishments are flour and lumber mills, iron- and steel-works, bag factories, cooperage, chair factory, wagon and carriage works, while lath, shingles, and other lumber products are manufactured quite extensively. It also has shipyards, coal docks of great capacity, large grain elevators, a large dry dock, and a number of wholesale merchandise establishments. The Finnish University and a state normal school are located here.

Surabaya, Java, is a port with the best harbor in Java, and is the naval and military headquarters of the Dutch East Indies. It has an arsenal, ship-building yards, and exports rice, cotton, coffee, and sugar.

Surakarta, Java, has an active trade.

Surat, India (Bombay), has manufactories of cotton and silk.

Swansea, Wales (Glamorgan), is a port and coaling station, the chief seat of the tin-plate manufacture, and one of the most important copper smelting and refining towns in the world. Extensive coal mines are near by. It is the seat of an Anglican bishop.

Swatow, China (Kwangtung), is a treaty port at the mouth of the Kan River, opened to foreign trade in 1869. It has large exports of sugar, tea, paper, tobacco, beans, and bean-cake.

Swindon, England (Wiltshire), has limestone quarries, and large locomotive and railway car works.

Sydney, cap. New South Wales, Australia, is a port and coaling station on Port Jackson, 4 miles from the

Pacific Ocean. Exports wool, silver ore, gold, wheat, meat, etc. The industries include coach factories, foundries, engineering works, and cloth mills. Sydney has large and beautiful parks, and among its public buildings are St. Andrew's Cathedral, a fine Gothic edifice; St. Mary's Cathedral (Roman Catholic); the town hall, of vast size and possessing a very fine organ; the university, a fine edifice in the Gothic style; the Government House, in the Tudor style; and the Museum.

Sydney, Nova Scotia, Canada, is a seaport, coaling station, and mining town on Cape Breton Island. It has large steel works, iron furnaces, coal mining interests, chemical works, etc.

Syracuse, Italy (cap. Syracuse), was in antiquity a famous city of Sicily, founded by Corinthian settlers about 733 B. C. The colonists seem to have occupied the little island of Ortygia, stretching southeast from the shore; but later the city extended to the mainland. The seat successively of "tyranny" and democracy, Syracuse was involved in a great struggle with Athens (415-414 B. C.) and the celebrated siege in which it came off victorious, Dionysius' fierce war with Carthage (397 B. C.) raised the renown of Syracuse still higher. In 212 B. C. the city was conquered by the Romans after a two years' siege, it having sided with the Carthaginians. Under the Romans Syracuse slowly declined, though with its handsome public buildings and its artistic and intellectual culture, it always continued to be the first city of Sicily. It was captured and burned by the Saracens in 878 A. D., and after that sank into complete decay. The modern city (Siracusa) is confined to the original limits, Ortygia, which, however, is no longer an island, but a peninsula. It has a cathedral (the ancient temple of Minerva), a museum of antiquities, a public library, the fountain of Arethusa (its waters mingled with sea-water since the earthquake of 1170), and remains of temples, aqueducts, the citadel Euryalus, a Greek theatre, a Roman amphitheatre, and quarries, besides ancient Christian catacombs. It carries on a brisk trade in fruits, oil, and wine.

Syracuse, New York (c. h. Onondaga Co.), is, in point of population, the fourth city of the state, and the commercial metropolis of central New York. It is an important railroad and manufacturing center, and has a large wholesale trade in merchandise. It has extensive manufactories of typewriters, automobiles, iron and steel products, etc. The salt industry, formerly important here, is now a relatively small item in the business of the city, but the water from the salt springs in the marshes bordering on Onondaga Lake, has become both indispensable and profitable, through combination with limestone in a chemical process, for the production of sodium nitrate and its by-products. Syracuse University, comprising the departments of liberal arts, fine arts, law, medicine, and applied science, with an attendance of over 3,200 students, a faculty of nearly 220, and a library of 90,000 volumes, was chartered in 1870 under the auspices of the Methodist Episcopal church; it is among the largest educational institutions of the United States and maintains a high standard of scholarship. The annual State Fair is permanently located here.

Syzran, Russia (Simbirsk), is the seat of a Greek Orthodox bishop, and has a cathedral dating from the 18th century. There are tanneries, iron and steel foundries distilleries, breweries, dye-works, brick works, and manufactures of agricultural machinery.

Szegedin, Hungary, has manufactures of soap and cloth, ship-building yards, and is the trade center of a large and fertile agricultural region. It suffered terribly from floods in 1879, when about 2,000 people lost their lives.

Szekesfejervar, Hungary, manufactures knives and soap and has an extensive trade in wine.

Szentes, Hungary, has a large trade in agricultural produce.

Tabriz, Persia, is the most important commercial center of northwestern Persia, and exports dry fruit, raisins, cotton, and carpets. The principal features of the town are the blue mosque and a famous tower built by Greeks.

Tacoma, Washington (c. h. Pierce Co.), is an important port and railroad terminus at the head of navigation on Puget Sound. It has a harbor, with warehouses, elevators, and extensive railroad trackage, extending for many miles along its deep water front, affording excellent facilities for its enormous distributing trade in wheat, lumber, coal, and merchandise. This city is tributary to one of the richest farming sections of the Northwest. It has extensive manufactories of lumber and lumber products, car shops, rolling-mills, shipyards, packing-houses, furniture factories, smelting works, flour-mills, and iron-foundries. Puget Sound University (Methodist), Pacific University (Lutheran), Whitworth College (Presbyterian), and numerous other educational institutions are located here. Tacoma was laid out in 1868 and was incorporated as a city in 1883.

Taganrog, Russia (Don Cossacks), is a port on the Gulf of Taganrog, with large exports of grain. There are tanneries, macaroni and tobacco manufactories, tallow foundries, and fisheries. It is the seat of a Greek Orthodox bishop.

Taiwan, Japan (cap. Formosa), is one of the chief ports of the island, with exports of tea, sugar, rice, and jute.

Takamatsu, Japan (cap. Kagawa), is a port on the Inland Sea and an important trade center.

Takaoka, Japan (Toyama), has dye-works and manufactures of cotton and silk.

Takasaki, Japan (Gumma), has an important trade in cotton manufactures.

Talca, Chile (cap. Talca), is an important trade center.

Tambof, Russia (cap. Tambof), has manufactures of tallow, soap, candles, woollens, sail-cloth, tobacco, oil, and tiles, also iron-foundries, distilleries, breweries and alum works. It is the seat of a Greek Orthodox archbishop.

Tammerfors, Russia (Finland), has paper, cotton, linen, and woollen mills.

Tampa, Florida (c. h. Hillsboro Co.), is an important railroad terminus and port of entry on Tampa Bay, 29 miles from the Gulf of Mexico. It has the best harbor on the western coast of Florida, and regular steamer connections with all of the large Gulf ports, also with several of the Atlantic ports. In the vicinity are large mines of phosphate. Lumber and naval stores are largely produced and shipped. The chief manufacturing industry is connected with the tobacco product, the supplies coming largely from Cuba; the imports of this product largely exceed those of any other United States port except New York. It is a famous winter resort, and has several large hotels. Tampa is the historical landing-place of Narvaez and De Soto, leading ill-fated early Spanish expeditions. Its first settlement began with the establishment of the United States military post of Fort Brooke during the war with the Seminole Indians.

Tampico, Mexico (Tamaulipas), has an excellent harbor and is a modern, progressive shipping and manufacturing city. It exports precious metals, coffee, goatskins and fiber.

Tamsui, Japan (Formosa), is a treaty port.

Tananarivo, cap. Madagascar, Africa, is the chief town of Madagascar, and was for over a century the capital of the Hova chiefs.

Tangier, Morocco, is a fortified seaport and health resort on the Strait of Gibraltar. It is the chief commercial city of Morocco, and the diplomatic headquarters. Here are a great mosque, a castle, the sultan's palace, and the governor's residence. The chief exports are cattle, eggs, slippers, goatskins, and woollens. It was ceded to England in 1662 as part of the dowry of Catherine of Braganza, wife of Charles II.; but in 1684 it was given up to Morocco.

Tanjore, India (Madras), has a Hindu temple, a dismantled fort, and a palace, and is noted for its silks, carpets, jewels, and inlaid metal-work.

Tanta, Egypt, is noted for its fairs.

Taranto, Italy (Lecce), is a fortified seaport, and has a naval arsenal and two harbors. Oyster and mussel fisheries are the chief industries, with manufacture of barrels, soap, oil, velvets, and cottons. It is the seat

of a Roman Catholic bishop, and the cathedral and the episcopal palace are among the principal buildings. The city was famous in antiquity and known as Tarentum.

Tarbes, France (cap. Hautes-Pyrénées), has manufactures of coarse woollen goods, machinery, cannon, and paper. It has a cathedral dating from the 12th century.

Tarnopol, Austria (Galicia), has brewing, distilling, and corn-milling industries.

Tarnow, Austria (Galicia), has manufactures of agricultural implements and glass.

Tarragona, Spain (cap. Tarragona), is an ancient Græco-Roman city, once the capital of Roman Eastern Spain (Tarraco). It has important ruins—cyclopean walls, palace of Augustus, and a fine aqueduct. It is one of the best ports on the Mediterranean, with large exports of wine and fruit, and a busy weaving industry.

Tashkend, Russia (cap. Russian Turkestan), has an astronomical observatory. Silk, leather, and metal goods are manufactured.

Taunton, Massachusetts (c. h. Bristol Co.), is an important cotton manufacturing and trade center. It has manufactures of cutlery, tools, nails, jewelry, machinery, silverware, stoves, etc. A state insane hospital, with over 1,000 patients, is located here. It was settled in 1638.

Tegucigalpa, cap. Honduras, Central America, has a cathedral and a university. In the vicinity are gold, silver, and copper mines.

Teheran, cap. Persia, is the center of some caravan trade, but has little commercial importance. About 12 miles from it is the royal mosque of Shah-Abdul-Azim, where Nasr-ed-Din Shah was assassinated in 1896.

Temesvar, Hungary, is a fortified town, and has manufactures of tobacco, cloth, paper, leather, and oil. It is the seat of a Roman Catholic and of a Greek Orthodox bishop. Its castle was founded by Hunyady in 1442, and its Roman Catholic cathedral by Maria Theresa in 1736-57. Held by the Turks from 1552 to 1716, Temesvar was retaken by Prince Eugene.

Teplitz, Austria (Bohemia), is a watering-place, with alkaline-saline mineral springs (73.5° to 98.5° F.,) which were known to the ancient Germans. In the vicinity are lignite mines. It has manufactures of pottery, chemicals, cotton, lace, machinery, and furniture.

Teramo, Italy (cap. Teramo), has silk-spinning and manufactures of pottery and leather. Its cathedral dates from the 14th century.

Terre Haute, Indiana (c. h. Vigo Co.), is a railroad and trade center in a rich agricultural region, with extensive coal-fields near by. It has a large distributing trade in coal, and the manufacturing industries embrace rolling-mills, foundries, flour-mills, car works, railroad shops, glass factories, etc. In the vicinity are large deposits of shale and clay, and a number of clay plants are in operation near by. A state normal school and several other educational institutions are located here. It is one of the oldest settlements in the state, and was laid out as a city in 1816.

Theodosia, Russia (Taurida), is a port on the Black Sea. It has a college, and a museum rich in local antiquities. As "Kaffa of the Genoese," it was the most famous mediæval port on the Black Sea, but it had a long and prosperous existence as a Milesian colony in pre-Christian times.

Thorn, Germany (Prussia), is a fortified town, and has manufactures of machinery, soap, and tobacco, and iron-foundries. It was founded by the Teutonic Knights, in the 13th century, and was a member of the Hanseatic League. It was ceded by the Knights to Poland in 1466, and remained in that country's possession until 1793. It is the birthplace of the astronomer Copernicus.

Three Rivers, Quebec, Canada, is a port of entry on the St. Lawrence River. It has a Roman Catholic cathedral, convents and a college. Its industries include several lumber mills, foundries, cotton mills, etc.

Tientsin, China (Chili), is a treaty port on the Huen River, 30 miles from its mouth. It is the port of supply for the whole of China north of the Yangtze River. Skins, wool, tea, wood, coal, and gold are the chief exports. The treaty of 1858; the occupation by the allies in the

winter of 1860-1; the massacre of Roman Catholics and other foreigners, June 21, 1870; the bombardment of foreign settlements, June 17 to July 14, 1900; the capture of the city by the allied forces, July 15, 1900; and the administration of Tientsin by foreign provisional government until Aug. 15, 1902, are the most notable events of recent years.

Tiffin, Ohio (c. h. Seneca Co.), is a commercial and industrial center in an agricultural region. In the vicinity are deposits of clay and glass sand. The manufacturing establishments are glass-works, potteries, woollen mills, iron-works, and furniture factories. Heidelberg University, opened in 1850 and having an attendance of over 400 students, and Ursuline College are located here. It was settled in 1817.

Tiflis, Russia (cap. Tiflis), is a commercial center, exporting silk, carpets, cotton, tobacco, and silver ornaments in filigree and enamel work. It has a museum and a botanical garden.

Tilburg, Netherlands (North Brabant), has manufactures of woollens, cloth, and leather.

Tilsit, Germany (Prussia), has glass, soap, and oil works, engineering works, iron-foundries, distilleries, and tanneries. Here was signed, between Alexander I. and Napoleon, on July 7, 1807, the treaty of Tilsit.

Tipton, England (Stafford), has coal-mining and manufactures of heavy iron goods.

Tiraspol, Russia (Cherson), is famous for its horticulture and has manufactures of tobacco and candles, tallow foundries, and flour-mills.

Tiumen, Siberia (Tobolsk), has manufactures of leather, carpets, soap, and pottery.

Tlemcen, Algeria, has manufactures of arms, morocco leather, and carpets.

Tobolsk, Siberia (cap. Tobolsk), has a kremlin, cathedral, monument to Yermak, also residences of the governor, and of the archbishop of Tobolsk and Siberia. Tobolsk was founded in 1587.

Tokat, Turkey in Asia, has manufactures of copper ware and leather. It was the scene of an Armenian massacre in November, 1895.

Tokushima, Japan (cap. Tokushima), is a trade center.

Tokyo, cap. Japan (Tokyo), contains the Mikado's palace, the government offices, the imperial university, and diplomatic residences. Large vessels lie 5 or 6 miles from the city, and the trade is conducted at Yokohama, 18 miles distant by railway. Its importance dates from 1590, when it fell into the hands of Iyeyasu, who, in 1603, became Shogun and made it his capital.

Toledo, Ohio (c. h. Lucas Co.), is an important railroad and commercial center on the Maumee River, 6 miles from Lake Erie, with a large transhipment trade in coal, grain, iron ore, timber, etc. The location of Toledo at the western end of Lake Erie makes it a most convenient shipping point for a large portion of the "winter wheat belt," and it is one of the most important primary grain markets of the United States, after Chicago. It has many large grain elevators. Toledo is also the leading clover-seed market of the world, and its quotations govern the prices of clover-seed for the United States. It is also one of the largest shipping ports in the world for soft coal, which comes by rail from the mines of West Virginia, southern and eastern Ohio, and Pennsylvania, and is transported by water to all ports on the upper lakes. Among its leading manufacturing establishments are steel-mills, blast-furnaces, glass-works, iron-works, malleable iron works, automobile works, flour-mills, spice-mills, scale works, distilleries, petroleum refineries, etc. It was settled in 1817.

Toledo, Spain (cap. Toledo), is a romantic Moorish walled city situated on the Tagus. It was the capital of Gothic Spain, and of an independent kingdom under the Moors. It was captured by Alfonso VI., who made it the capital of Castile in 1085. Its superb Gothic cathedral is only rivalled in Spain by that of Burgos, and its Moorish and Mudejar remains are of the highest interest. Its swords were once famous throughout the world, and a large trade in steel cutlery is still carried on, as well as some cloth weaving.

Toluca, Mexico (cap. Mexico), has cotton-mills and breweries.

Tomsk, Siberia (cap. Tomsk), has a university, opened in 1888, and other educational institutions. It has considerable trade.

Tonopah, Nevada (c. h. Nye Co.), is an important gold-mining center. It has considerable trade in merchandise with outlying mining camps.

Topeka, cap. Kansas (c. h. Shawnee Co.), is an important railroad and commercial center with coal-mines and stone-quarries in the vicinity. It has large railroad shops, several flour-mills, foundries, machine-shops, lumber mills, boiler factories, etc. The city is an important center of wholesale trade. The State Capitol is a handsome stone edifice, and there are also located here the state asylum for the insane, a state reform school, etc. It is the seat of Washburn College, Kansas Medical College, and several other educational institutions. It was settled in 1854.

Toronto, cap. Ontario, Canada, is a port on Lake Ontario, and, in size, is the second city of Canada. It has many public buildings of architectural merit. It is the seat of the University of Toronto, Trinity College, and several other educational institutions of note. Toronto is the see of an Anglican bishop and of a Roman Catholic archbishop. It has a public library, as well as the university and the legislative libraries. There are foundries, agricultural implement works, machine-shops, breweries, a very large distillery, tanneries, soap-works, and canning and packing establishments. The city was founded (as York) in 1794 by Governor Simcoe, and it has been the capital of the province since 1797.

Torquay, England (Devon), is a seaside resort. Some remains exist of Tor Abbey, and about a mile from the center of the town is Kent's Hole, a noted bone cavern.

Torre del Greco, Italy (Naples), is a seaport, a fishing town, and seaside resort at the base of Mt. Vesuvius. It has frequently suffered from the eruptions of Vesuvius. Coral working is the leading industry.

Torreon, Mexico (Coahuila), is an industrial and commercial center in a cotton growing section.

Tortosa, Spain (Tarragona), has a magnificent Gothic cathedral dating from the 14th century.

Totonicapam, Guatemala, Central America, has manufactures of textiles, pottery, and woodenware.

Tottenham, England (Middlesex), is a suburb of London.

Toulon, France (Var), is the chief naval port of France on the Mediterranean. It is strongly fortified, and has a government arsenal, nautical museum, etc. There is a dock area of 740 acres, with 80 acres of deep water floating docks, which will hold the largest ships. Toulon was the Greek Telonion and the Roman Telo Martius. In the 17th century its arsenal and dockyards were begun by Vauban. Both were destroyed by the British in 1793.

Toulouse, France (cap. Haute-Garonne), has manufactures of silks, woollens, leather, copper, and cannon. Among its interesting buildings are the Romanesque church of St. Sernin (Saturnin), consecrated 1096, and the cathedral of St. Etienne (13th century). The museum contains the ivory horn of the renowned paladin Roland. Toulouse is the seat of a Catholic archbishop, and has a university. Between 920 and 1271, it was the capital of the county of Toulouse, and became a center of Provencal troubadours. It was reunited to France under Philip the Bold (1271), and was the scene of Huguenot massacres in 1562 and 1572, as well as of Albigensian persecutions. Here, in 1814, Wellington defeated Soult.

Tourcoing, France (Nord), has important manufactures of woollens, also of cottons, velvets, and carpets.

Tournay, Belgium (Hinnaut), has manufactures of hosiery, carpets, and woollen and cotton goods. It possesses a magnificent cathedral, which contains pictures by Rubens.

Tours, France (cap. Indre-et-Loire), has manufactures of steel and iron goods, glass and earthenware, and chemicals. Its 12th century cathedral contains a beautiful monument to the children of Charles VIII.

The ruins of the castle Plessis-les-Tours are in the vicinity. In 732, Charles Martel defeated the Saracens near Tours.

Toyama, Japan (cap. Toyama), is an important commercial center.

Trapani, Italy (cap. Trapani), is a seaport with exports of olive-oil, salt, salted fish, wine, and building stone, and manufactures of coral, alabaster, and mother-of-pearl goods.

Traverse City, Michigan (c. h. Grand Traverse Co.), is a trade center in an agricultural and fruit growing region. It has an excellent harbor, and is especially noted for the manufacture of oval wood dishes and corn-starch. It also manufactures fruit baskets, farming implements, flour, leather, foundry and machine-shop products, etc. The Northern Michigan Insane Asylum is located here. It was settled in 1852.

Trebizond, Turkey in Asia, is a fortified port on the Black Sea, with exports of cattle, hazel-nuts, tobacco, etc. Originally a Greek colony (c. 600 B. C.) from Sinope, Trebizond in 1204 became the capital of the empire of Trebizond under Alexius Comnenus; in 1461 it was captured by Mohammed II. of Turkey. In October, 1895, the town was the scene of Armenian massacres.

Trenton, cap. New Jersey (c. h. Mercer Co.), is located at the head of tide-water navigation of the Delaware River, and is an important industrial and railroad center. The State Capitol is a handsome edifice overlooking the river. A state insane asylum and a penitentiary are also located here. It has extensive iron-works, wire-works, machine-shops, potteries, rubber works, furniture works, watch works, oilcloth factories, etc. Trenton has a number of noted educational institutions, among which are the Model schools, the state school for deaf mutes, an industrial school for girls, a Roman Catholic College, etc. On Dec. 26, 1776, Washington crossed the Delaware here, and surprised and routed the Hessians under Colonel Rahl, following up this success by the battle of Jan. 2, 1777, in which he maintained his ground against Lord Cornwallis.

Treves, Germany (Prussia), has manufactures of woollens, cottons, linen, and leather, but is especially noted for Roman remains. These include the basilica, or palace of Constantine, now used as a Protestant church; the Porta Nigra, a fortified gate dating from the 4th century; an amphitheatre, capable of accommodating 30,000 spectators; the imperial palace; baths, in capital preservation; and the piers of the river bridge. In the cathedral, built on the site of a 4th century basilica, are preserved a nail reputed to be from the cross, and the "Holy Coat," brought to Treves by the Empress Helena, and exhibited as lately as 1891 to 2,000,000 pilgrims. Other buildings are the Provincial Museum, the church of Our Lady, and the Municipal Library, containing very valuable manuscripts. Originally the capital of the Treviri, Treves became a Roman colony, and during the 3rd and 4th centuries was a favorite residence of the Roman emperors. From the beginning of the 9th century till 1786, when the electorate was transferred to Coblenz, Treves was an archbishopric of very great importance.

Treviso, Italy (cap. Treviso), has manufactures of iron goods, pottery, silks, and woollens. It is the seat of a bishop, and among its famous buildings are its 15th century cathedral (containing Titian's "Annunciation") and the Gothic church of S. Niccolo.

Trichinopoli, India (Madras), has manufactures of cigars, hardware, and gold and silver jewelry. It is a mission center for both Roman Catholics and Protestants, and has two colleges affiliated to the University of Madras.

Trient, Austria (Tyrol), is a fortified town, and has silk weaving, and manufactures of pottery and salami (sausage). In the neighborhood are vineyards, and quarries of gypsum and marble. It is the seat of a bishop. Here met between 1545 and 1563 the famous Council of Trent.

Trieste, Austria (cap. Coastland), is the chief port of Austria and a coaling station. It is the naval arsenal and storehouse for the Austrian navy. Ship-build-

ing is carried on, and marine steam-engines, anchors, ropes, soap, leather, and furniture are manufactured. Its trade in grain, wine, and oil is considerable, more than half being with the Levant and the Far East.

Trikhala, Greece (cap. Trikhala), is situated in a fruit-growing district, and has a trade in grain and tobacco, as well as manufactures of cotton and woollen goods and leather. It is the seat of a Greek Orthodox archbishop. It was once famous for its temple of Æsculapius.

Tripoli, cap. Tripoli, Africa, is a port on the Mediterranean Sea. It is a typical Moorish city, with high walls, beautiful gardens, many mosques, and several large churches.

Trivandrum, India (Madras), is the capital of the native state of Travancore, and has the palaces of the maharajah, a large temple of Vishnu, a mint, and the maharajah's college.

Trondhjem, Norway (cap. Sondre Trondhjem), is a port on Trondhjem Fiord. The Anglo-Norman cathedral of St. Olaf, the most interesting ecclesiastical edifice of Norway, dates from the 13th century, but all that remains of the original building, since the fire of 1530, is the richly adorned Late Gothic choir. The work of restoration was begun in 1861. The chief exports are wood and woodenware, copper, herrings, blubber, timber, and cellulose. Trondhjem is the seat of a bishop, and has an Academy of Sciences. The city, originally called Nidaros, was founded in 997 by Olaf Tryggvason, and was formerly the capital of Norway. It was in the middle ages a pilgrim resort (shrine of St. Olaf), and is now the crowning-place of the kings of Norway. King Haakon VII. and his queen were crowned here in 1906.

Troy, New York (c. h. Rensselaer Co.), is located at the head of tide-water navigation on the Hudson River, and is an important railroad and manufacturing center. The chief manufactures are collars and cuffs, laundry machinery, stoves and other iron products, electrical machinery, Bessemer steel, mechanical and mining engineering machinery, bells, paints, brushes, horseshoes, etc. The Emma Willard Seminary, the Rensselaer Polytechnic Institute, a celebrated school, and other educational institutions are located here.

Troyes, France (cap. Aube), has manufactures of hosiery. It is said to have given the name to troy weight. The cathedral has a particularly fine Flamboyant western front, and the old abbey of St. Lupus contains the town library. Here was signed in 1420 a peace between Henry V. of England and Charles VI. of France.

Tsaritsyn, Russia (Saratof), is an important river port on the Volga. It has tanneries, breweries, brick works, manufactories of hydromel, mustard, conserves, and machinery; but fisheries and market-gardening are the chief industries.

Tsu, Japan (cap. Miye), is a port on Owari Bay. It has several fine temples.

Tucson, Arizona (c. h. Pima Co.), is the largest city in Arizona. It has extensive mining and cattle interests, railroad shops, and considerable wholesale trade. The University of Arizona is located here.

Tucuman, Argentina (cap. Tucuman), has a national college, normal school, and other schools. It has an extensive trade in oxen and mules, and is the center of the sugar plantation region. There are large groves of oranges and lemons in the vicinity and the city is considered one of the most picturesque points in the republic. It is the seat of a Roman Catholic bishop. The Argentine declaration of independence was signed here on July 9, 1816.

Tula, Russia (cap. Tula), is the seat of a bishop, and has a kremlin of the 16th century containing two cathedrals. There are iron-foundries, tanneries, breweries, distilleries, sugar-refineries, and manufactures of arms, machinery, samovars, as well as of tiles, bricks, tobacco, confectionery, candles, and brushes.

Tunbridge Wells, England (Kent), is a watering-place. The springs (chalybeate) were discovered in the reign of James I., and were much frequented in the 18th century

Tunis, cap. Tunis, Africa, is a port and coaling station. Several of the mosques are magnificently decorated, as is the bey's palace. The citadel contains a fine collection of antiquities.

Turin, Italy (cap. Turin), is an important railway and commercial center, the Mount Cenis and the Simplon tunnels favoring its transit trade. Electricity is used in most of the factories. Silk, woollen goods, paper, cotton, linen, hats, jewelry, liqueurs, and leather are manufactured. Wine and silk are exported. Among the older buildings are the Palazzo Madama; the Palazzo Carignano (now a natural history museum); the cathedral, built in the 7th century and rebuilt in the 15th century in the Renaissance style; the royal palace; the modern buildings include the Mole Antonelliana (containing the Risorgimento Italiano Museum). Picture galleries and museums are numerous and valuable —Royal Albertine Academy, Royal Pinacoteca, Museum of Antiquities, the Civic Museum. The university, founded in 1804, has over 2,600 students; and there are a school of engineering, a school of veterinary medicine, a seminary for priests, and a musical academy. Turin came under the rule of the dukes of Savoy in the 11th century. During the 16th and 17th centuries it was alternately under France and Savoy. It was the capital of the kingdom of Sardinia till 1860, when it became the capital of Italy, remaining so till 1865.

Tver, Russia (cap. Tver), is the seat of a Greek Orthodox archbishop, and has a kremlin; a cathedral (1682), containing tombs of princes of Tver; the church of the Trinity (1684); and an imperial palace, erected by Catherine II., with museum of ethnology and archæology. It manufactures cotton, sail-cloth and other cloths, candles, linen, cordage, hats, bells, oils, and earthenware; has iron-foundries, breweries, distilleries, and dye-works, and is a prominent river port. In 1246 it became the capital of an independent principality.

Tynemouth, England (cap. Northumberland), is a port at the mouth of the Tyne River. It has shipbuilding and industries connected with shipping.

Udaipur, India (Rajputana), is the capital of a native state of the same name, and has considerable trade.

Udine, Italy (cap. Udine), has manufactures of silks, velvets, cottons, and leather. The castle, once the residence of the patriarchs of Aquileia, is now used as barracks. Among its other buildings are its Romanesque cathedral, the archiepiscopal palace, and the municipal buildings.

Ufa, Russia (cap. Ufa), is a river port, and has tanneries, tallow foundries, breweries, distilleries, brick works, sawmills, manufactories of soap, candles, gingerbread, and pottery. It is the seat of a Greek Orthodox bishop and has a cathedral.

Ujpest, Hungary, is a suburb of Budapest.

Ulm, Germany (Württemberg), is a fortified town and river port on the Danube. Tobacco, cement, machinery, and linen and cotton goods are manufactured. Ornamental pipe bowls and Ulm bread are characteristic products. The Late Gothic cathedral (1377) is second in Germany to that of Cologne, and its tower, completed in 1890, is 528 ft. high. Ulm was an important trade center in the late middle ages. Here, in 1805, the Austrian general Mack capitulated with about 30,000 men to the French.

Uman, Russia (Kief), has an agricultural college, and manufactures of tobacco, malt, candles, bricks, vinegar, also flour-mills and iron-foundries.

Union, New Jersey (Union Co.), is a residential suburb of Hoboken.

Upsala, Sweden (cap. Upsala), is the seat of an archbishop, and has a cathedral (built 1230-1435 and restored 1885-93), with the monuments of many kings and famous men. The university, founded in 1477, has the largest library in Sweden (300,000 vols. and 12,000 MSS.), and is attended by 1,500 students. Upsala has chemical factories, breweries, and brick works.

Uralsk, Siberia (cap. Uralsk), has manufactures of candles, soap, glue, and sheepskin coats, as well as fisheries, tanneries, distilleries, breweries, and brickworks. It has two cathedrals; one Orthodox, founded

in 1837; the other, belonging to the Old Believers, founded at the beginning of the 18th century.

Urmia, Persia, is noted as a center of missionary activity, and is the seat of a seminary for priests, and a college. It is a summer resort of the Persian nobility. It exports raisins and molasses.

Uskup, Turkey (Kossovo), is an important trade center and has leather, dyeing, and weaving industries.

Utica, New York (c. h. Oneida Co.), is an important railroad and commercial center in a prosperous agricultural and industrial region. The chief manufactures are hot-air furnaces, hosiery and print goods, men's clothing, machine-shop products, steam fitting and heating apparatus, tobacco products, lumber products, etc. It is a distributing center for an extensive region, and is an important cheese market; large quantities of flowers, especially roses, are shipped to the larger cities. A state hospital, the Masonic Home of New York State, and other important institutions are located here.

Utrecht, Netherlands (cap. Utrecht), has been since 1723 the headquarters of the Jansenists and the seat of their archbishop. Its university, founded in 1636, has 850 students. Textiles, machinery, chemicals, tobacco, bricks, and beer are the chief manufacturing products. Here was formed in 1579 the union of the seven Protestant provinces—the union out of which grew the nation of the Netherlands; and here in 1713 was signed the treaty of peace which ended the war of the Spanish Succession. Pope Adrian VI. was born at Utrecht in 1459.

Utsunomiya, Japan (cap. Tochigi), is an important trade center.

Valence, France (cap. Drôme), has manufactures of glass, silk, hosiery, and the printing of handkerchiefs. Its Romanesque cathedral was consecrated by Pope Urban II. in 1095.

Valencia, Spain (cap. Valencia), is the third city of Spain. Its cathedral, though much spoiled by restoration, has fine paintings of the Valencian school; and many pictures by Ribalta are in other churches and public buildings. Its university is one of the most famous, as the art school is one of the most important in Spain. Tobacco, silk, fans, and gloves are manufactured.

Valencia, Venezuela, is an important trading center situated in a rich agricultural region, and has an extensive trade in coffee, sugar, rum, cattle, and hides. It contains a cathedral built in 1813, a university, and a national college.

Valenciennes, France (Nord), has manufactures of coarse lace, hosiery, cambric, glass, and iron goods. The manufacture of Valenciennes lace has been discontinued for 50 years. The artist Watteau was born here.

Valetta, cap. Malta, is one of the most important British naval and coaling stations in the Mediterranean. The town with its two harbors and four dry docks, is defended by forts, partly constructed in 1530 by the Knights of St. John. Among the many fine buildings are the cathedral, the palace of the grandmasters of the Knights of St. John, and the university.

Valladolid, Spain (cap. Valladolid), was the capital of Spain before Madrid. It contains the royal palace of Philip III. and the palace of Cardinal Mendoza, now a museum. Columbus died and Cervantes lived here.

Valparaiso, Chile (cap. Valparaiso), is the largest town on the Pacific coast of South America and the chief port and coaling station of Chile. It has a dangerous open harbor which is being improved, and there are two floating docks, and a government mole with hydraulic machinery. It is fortified and has a naval school. It has locomotive and ship-building works, and exports copper, hides, wheat, and wool. The earthquake of August 16, 1906, destroyed many important buildings, but the city has been rebuilt.

Vancouver, British Columbia, Canada, is a port and coaling station on an arm of the Strait of Georgia and the largest city of British Columbia. It is an important railway terminus and port of departure for Alaskan, Asiatic, and Australian steamers. Exports

fish, lumber, gold, etc., and has railroad shops, ship-building yards, sugar-refineries, canneries, breweries, distilleries, cooperages, and manufactories of machinery, furniture, glass, etc. Vancouver College, affiliated with McGill University, is located here.

Varna, Bulgaria, is a fortified port on the Black Sea, and exports cattle, butter, hides, and grain. It is the seat of a Greek Orthodox bishop. Here, on Nov. 10, 1444, Ladislaus, king of Hungary and Poland, was defeated and slain by the Turks.

Vellore, India (Madras), is a military station, and has an interesting old fort.

Venice, Italy (cap. Venice), is a fortified port and coaling station on the Gulf of Venice, ranking next to Genoa in importance. Exports textiles, grain, hemp, and glass and enamel ware. The foundations of the city rest on piles driven into the mud. Glass is one of its most famous manufactures; and beads, all sorts of Venetian ware, mirrors, mosaics, electric lamps, candelabra, tapestry, brocades, lace, velvets, silks, machinery and chemicals are also manufactured. Ship-building and the making of torpedoes are carried on. Among the many fine churches, the basilica of St. Mark is marked out as one of the three finest in Italy; others are the churches of SS. Giovanne Paolo (the burial-place of the doges), Santa Maria Gloriosa dei Frari, Santa Maria della Salute, San Giorgio degli Schiavoni, San Salvatore, and San Pietro di Castello. Other places and objects of great interest are the Academy of Fine Arts (containing Venetian canvases from the earliest times), the municipal museum, the quay Riva degli Schiavoni, the Rialto, the Bridge of Sighs, and the granite column, surmounted by the Winged Lion of St. Mark. The famous Campanile collapsed on July 14, 1902; it is now in process of reconstruction. After the fall of Napoleon, Venice was ceded to Austria; but the government was so oppressive that she revolted, and under the leadership (1848-9) of Daniele Manin she drove out the Austrians and joined in the struggle for Italian independence. Even after the failure of Charles Albert's expeditions into Lombardy, she refused to yield, and was only reduced, after a year's siege, by famine and cholera. When Lombardy was freed in 1859, Venice remained under the Austrian yoke, and it was not until 1866 that she was united to the Italian kingdom.

Vera Cruz, Mexico (Vera Cruz), is the chief port of Mexico, with exceptional facilities for a large and growing trade. Its exports consist principally of ores, coffee, tobacco, hides, dyewoods, and drugs. It was founded at the close of the 16th century, and was taken by General Scott in 1847.

Vercelli, Italy (Novara), is an important railroad center and has manufactories of cotton and woollen goods, silks, and machinery. The church of St. Andrea dates from the 13th century, and there is a museum of Roman antiquities.

Verona, Italy (cap. Verona), is a fortified city, and has manufactures of nails, pianos, organs, cotton and silk goods, paper, and flour, and trade in fruit, wines, and marble. The Gothic cathedral dates from the 12th century, and contains a fine Assumption by Titian; the very fine Romanesque church of San Zeno Maggiore stands in the west corner of the city. Verona was originally a Celtic foundation, but became a Roman colony with the title of Augusta. In the later empire it was often a residence of the court. It possesses a Roman amphitheater, also a few remains of a theater, and of an ancient gate, and wall of Gallienus, erected in 265 A. D.

Verviers, Belgium (Liège), has long been noted for woollens, and has large manufactures of woollen yarn.

Viborg, Russia (Finland), is a fortified town and naval station, with manufactures of machinery and soap, and an export trade in timber, iron, paper, tar, tallow, fish, and butter. It is the seat of a Greek Orthodox archbishop, and has a fine cathedral. It has a naval school, and Russian, Swedish, and Finnish colleges.

Vicenza, Italy (Vicenza), has manufactures of silk, woollen and cotton goods, and pottery. Its Gothic cathedral dates from the 13th century.

Vichy, France (Allier), is a famous watering-place,

which has mineral springs known since the Roman times. Large quantities of Vichy water are shipped annually to all parts of the world.

Vicksburg, Mississippi (c. h. Warren Co.), is the largest city of the state, and a trade center in an agricultural region, of which cotton is the principal product. It is situated on the Mississippi River and has excellent railroad facilities. The chief manufacturing establishments are cottonseed-oil mills, planing mills, railroad shops, foundries, and machine-shops. A Roman Catholic college for boys and a colored college are located here. The National Cemetery above the city contains over 16,000 graves. The name of Vicksburg is well known from its prominence in the Civil War, when, as the key of the Mississippi, it was strongly fortified and garrisoned by the Confederates. After baffling Farragut and Sherman in 1862, it was finally captured by Grant in 1863, in a campaign which cost him 9,000 men.

Victoria, Brasil (cap. Espirito Santo), exports coffee, sugar and rice.

Victoria, cap. British Columbia, Canada, is a port and coaling station on Juan de Fuca Strait, and a tourist resort. It has one of the best harbors on the Pacific coast, and is a port of call for transpacific steamships running to Vancouver, Seattle, and Tacoma. It has extensive sawmills, ship-building and iron-working plants, fish canneries, and chemical and other factories. It exports lumber, fish, and agricultural produce. Its leading architectural features are the Parliament buildings, government house, the Canadian Pacific hotel, post-office, and churches. The climate is mild, the rainfall small, and it is noted for its natural beauty. There is a Canadian garrison in the city.

Vienna, Austria (cap. Lower Austria), is the chief commercial and industrial city of the country, and an important railway and waterway center. It has extensive manufactures of iron and steel goods, silk, cotton, furniture, fancy goods, leather, and bronze, as well as breweries and distilleries; and has a large trade in grain, cattle, wood, and manufactured articles. Vienna is one of the handsomest of European cities, and although an ancient place, is of modern aspect, having been almost completely rebuilt in the last 50 years. It is the seat of a Roman Catholic archbishop and possesses the famous cathedral of St. Stephen (1300-1510). Among the principal buildings are the Hofburg, or Imperial Palace, containing many historic relics, such as the coronation regalia of Charlemagne, the famous diamond of Charles the Bold of Burgundy, etc.; the magnificent Parliament House; the City Hall; the Opera House; the Museum of Natural History; and the magnificent Picture Gallery, rich in canvases by Dürer, Rubens, and the Venetian masters. It is the seat of a famous university, founded in 1365; of an Academy of Sciences; of a great number of technical and special schools; and of two rich libraries. Vienna became the capital of the Hapsburg dominions in 1276; was occupied by the French in 1805 and in 1809; and was the scene of a revolutionary outbreak in 1848.

Villa Rica, Paraguay, is the center of an agricultural region, producing tobacco, cotton, and oranges.

Vilna, Russia (cap. Vilna), is a river port, and manufactures buttons, gloves, tobacco, confectionery, pencils, boots and shoes, brushes, hats, and artificial flowers. It is the seat of a Greek Orthodox archbishop and of a Roman Catholic bishop, and possesses two fine cathedrals.

Vincennes, France (Seine), is a suburb of Paris, on the border of the Bois de Vincennes, and has manufactures of chemicals and cartridges. The keep of its castle was the prison of Henry IV., Condé, Diderot, Mirabeau, and others.

Vincennes, Indiana (c. h. Knox Co.), is a trade center in a fertile agricultural region, and has manufactures of flour, lumber products, sewer pipe, iron and tin products, agricultural implements, machine-shop products, etc. Vincennes University, founded in 1806, is located here. It is the oldest place in the state. In 1702 the French built here a fort, and, for several years, it was the seat of the empire of France in the Ohio Valley. In 1763 the British obtained possession

of the place, and in 1779 it was captured by the Virginians, who held it until 1783, when it was ceded to the United States. In 1787 the first court was held, and in 1800 the Indiana Territory was made the capital. In 1813 the territorial capital was removed to Corydon.

Vinnitsa, Russia (Podolia), has manufactures of tobacco, soap, candles, carts and carriages.

Vitebsk, Russia (cap. Vitebsk), is a river port and has dye-works, and manufactures of candles, tobacco, varnish, woollen and linen cloth. It is the seat of a Greek Orthodox bishop, and has the cathedral of St. Nicholas (1664), and the old churches of St. Elias (1643) a fine example of Old Russian style, the Assumption (1777), and St. Antony (1731).

Vitoria, Spain (cap. Alava), has manufactures of leather, soap, mirrors and picture frames. The cathedral dates from the 12th century. On June 21, 1813, Wellington defeated here the French under Joseph Bonaparte and Jourdain.

Vladikavkaz, Russia (cap. Terek), is an important military station and has considerable trade.

Vladimir, Russia (cap. Vladimir), is a river port and has dye-works, cotton-mills, manufactures of malt, and tobacco, also gardening and fruit-raising, especially cherries. It is the seat of a Greek Orthodox archbishop, and has a cathedral built in the 12th century, and restored in the 19th century. Founded in 1116 by Vladimir Monomakh, Vladimir became the capital of a line of grand princes, who replaced those of Kief in power among Russian rulers.

Vladivostok, Siberia (Coast Province), is a fortified port, the chief Russian naval station of Asia, and the terminus of the Trans-Siberian Railway. Exports skins, furs, edible seaweed, etc. Its harbor is one of the most beautiful in the world. The climate is severe, the mean temperature for January being 6°F., while for August (the hottest month) it is 70°. In the Russo-Japanese war it escaped attack by the Japanese, but suffered from naval mutiny and unrest in the Russian disturbances of 1905-6.

Vologda, Russia (cap. Vologda), is a river port, and possesses some ancient churches, including the cathedral of St. Sophia (1568).

Volsk, Russia (Saratof), has iron-works, tanneries, and flour-mills.

Voronezh, Russia (cap. Voronezh), has manufactures of soap, candles, bricks, bells, tobacco, woollen and linen cloth, and vitriol. It is the seat of a Greek Orthodox bishop, and has a fine cathedral. It possesses a palace of Peter the Great, and has a technical school and a military college.

Waco, Texas (c. h. McLennan Co.), is a railroad and trade center in a rich agricultural region, of which grain and cotton are the chief products. The chief manufactures are cotton goods, clothing, wagons and carriages, saddlery and harness foundry and machine-shop products. It is the principal interior cotton market of the United States. Among the educational institutions are a Baptist college, opened in 1845, a Christian college, opened in 1873, a Roman Catholic college, opened in 1900, and an African Methodist Episcopal college, opened in 1881.

Wakamatsu, Japan (Fukushima), is noted for its manufactures of lacquer ware.

Wakayama, Japan (cap Wakayama), has an important trade in cotton.

Wakefield, England (York), manufactures worsted yarn and cocoa-matting. The battle of Wakefield was fought in 1460, when the Yorkists were defeated by the Lancastrians.

Wakefield, Massachusetts (Middlesex Co.), is an industrial center in an agricultural region. It has iron and brass factories, rattan goods works, shoe factories, machine-shops, etc.

Wallasey, England (Cheshire), is a suburb of Birkenhead.

Walla Walla, Washington (c. h. Walla Walla Co.), is a trade center in a fertile agricultural region known as the Walla Walla Valley, and forms a part of the "Inland Empire." It is one of the most important commercial cities of the eastern part of the state, and has extensive

manufactories of farming implements and machinery, flour-mills, lumber mills, etc. A United States land-office is located here. It is the seat of Whitman College, founded in 1859.

Walsall, England (Stafford), has iron and brass founding, tanning, currying, the manufacture of saddlery and harness, and extensive collieries.

Waltham, Massachusetts (Middlesex Co.), is an industrial center celebrated for its watchmaking works, which are said to be the largest of the kind in the world. It also has large cotton-mills, saddlery and harness works, foundry and machine-shops, wagon and carriage factories, emery-wheel works, lumber mills, furniture factories, men's clothing factories, etc. Among the educational institutions are a state school for the feeble-minded, and a normal training school.

Walthamstow, England (Essex), is a suburb of London.

Wandsbek, Germany (Prussia), is a suburb of Hamburg, and has manufactures of tobacco, etc.

Warrington, England (Lancashire), has iron manufactures.

Warsaw, Russia (cap. Poland), was formerly the capital of the kingdom of Poland. The palaces are now put to municipal use; the Casimir Palace is occupied by the university, which is attended by over 1,500 students. Iron and steel goods, machinery, carriages, plated goods, and boots and shoes are manufactured. Trade is carried on in grain, leather, sugar, and coal; the wool and hop fairs are important. Warsaw was the residence of the dukes of Mazovia till 1526. In 1609 it superseded Krakow as the capital of Poland. It was taken by the Swedes in 1655 and 1656, and was captured by the Russians in 1764 and 1794. In 1807 it was made capital of the duchy of Warsaw, and in 1813 came entirely under Russian rule.

Washington, District of Columbia, is the capital of the United States, lies on the left bank of the Potomac, 156 miles from the Chesapeake Bay, and is, in many respects, one of the most beautiful cities of the United States. The present site of the national capital was selected in 1790, mainly through the agency of George Washington; and the Federal District of Columbia, 100 square miles in area, was set apart for this purpose, on territory ceded by Maryland and Virginia. The Virginia portion of the district was, however, retroceded in 1846, and the present area of the District of Columbia is 69 square miles. The commerce and manufactures of Washington are relatively unimportant, and its prosperity depends on its position as the seat of Congress and the Government Offices. There are upwards of 50,000 army and navy officers and civil servants in Washington, and these, with their families, make up a large proportion of the population. It is emphatically the scientific center of the country and its many scientific societies have thousands of members. The sobriquet of "City of Magnificent Distances," applied to Washington when its framework seemed unnecessarily large for its growth, is still deserved, perhaps, for the width of its streets and the spaciousness of its parks and squares. The Capitol dominates the entire city with its soaring dome and ranks among the most beautiful buildings in the world. It is 751 ft. in length, from 121 to 324 ft. wide, and consists of a main edifice of sandstone, painted white, and of two wings of white marble. It covers an area of 3½ acres. The main building with its original low-crowned dome, was completed in 1827; the wings of the new iron dome were added in 1851-65. It contains the Hall of Representatives, Senate Chamber, and the Supreme Court Room. It stands in a park of about 50 acres in extent and near by are the Senate and House of Representatives buildings, two white marble edifices in classic style, completed in 1908, containing offices for the senators and representatives. The Library of Congress, an enormous structure of the Italian Renaissance style, 470 ft. long and 340 ft. wide, erected in 1888-97, at a cost of over $6,000,000, is rich in sculptural adornments and its interior is sumptuously decorated with paintings by 50 American artists. The library contains upwards of 2,000,000 volumes, over 100,000 manuscripts, and more than 100,000 maps, etc.

The Treasury Department, the oldest of the governmental buildings, is constructed entirely of freestone and granite in the Ionic style. The city Post-Office and Post-Office Department building, constructed entirely of granite, occupies an entire block, and is one of the handsomest structures of the modern styles of architecture. The State, War, and Navy Departments occupy a large structure built in the Renaissance style and cover more than 4¼ acres. The building commonly known as the Patent Office, but which is really the office of the Secretary of the Interior Department, is a large structure of white marble in the pure Doric style. The old building of the Post-Office Department, a structure built of marble in the Corinthian style, covering one block, is now occupied by the General Land-Office and Indian Bureau. The Pension Bureau is housed in a brick building in which are held many of the great functions known as Inaugural Balls. The home of the President is an artistically plain building of freestone painted white, from which it derives the name, "White House." The offices of the White House are contained in a low white building at the western extremity of a long colonnade extending from the White House. Directly south of the White House is the Washington Monument, a marble shaft 555 ft. in height, to the building of which almost every country on the face of the earth contributed a stone. The Washington Public Library, a beautiful building, the gift of Andrew Carnegie, was opened in 1902. The Smithsonian Institution, a building of brownstone in the ancient Gothic style of architecture, was built by money bequeathed by James Smithson, an Englishman. The policy of the institution is to encourage research, and it has been the chief promoter of scientific investigation of the climate, products, and antiquities of the United States. It possesses a library of 250,000 volumes and issues publications of great scientific value. The National Museum, which is under the direction of the Smithsonian Institution, contains valuable and excellently arranged collections of natural history, anthropology and geology, derived mainly through the scientific investigations of the United States government, and is housed in a large and dignified building of white granite, which has been designed to harmonize with the older public buildings of Washington built on classic lines. The Army Medical Museum contains a pathological collection, a collection of army and navy supplies, and a library of over 200,000 volumes. The Department of Agriculture occupies a new white marble and brick building, of which the wings were completed in 1908, and includes a herbarium and conservatories of economic plants; the grounds in front of it are devoted to an arboretum arranged by families. The Bureau of Engraving and Printing, where all of the paper money, stamps, and bonds are made, is near by. The Governmental Printing Office, the largest and most complete printing establishment in the world, is housed in a huge brick structure. Other large buildings are the Naval Observatory, the Naval Museum of Hygiene, and the Coast Survey. The Union Railway Station, completed in 1908 at a cost of $4,000,000, is undoubtedly one of the most successful buildings of the kind in the country. It is constructed of white marble, and is 630 ft. long and 210 ft. wide. The Corcoran Gallery of Art occupies a handsome white marble structure erected in 1894-7 and contains a large and valuable collection of casts, sculptures, paintings, and other objects of art. It has a school of art with over 300 pupils. Washington is the seat of many universities, colleges, etc. The Catholic University of America, founded in 1885, is situated in one of the northern suburbs and is one of the foremost institutions in the country for the study of advanced philosophy. Georgetown University, founded in 1789, and conducted by the Jesuit order of the Roman Catholic church, has all the departments of a modern university. Howard University, established in 1867, and supported almost entirely by the government, furnishes the negro with opportunities for an education. The George Washington University is another well-equipped university, having legal, medical, dental, collegiate, and scientific departments; it also has a school of diplomacy and jurispru-

dence in connection with its law school and a modern, well-equipped hospital in connection with its medical department. Trinity College, a Roman Catholic institution, designed for the higher education of women, is conducted by the Sisters of Notre Dame de Namur. The Columbia Institution for the deaf and dumb, known also as Gallaudet College, founded in 1857, is one of the foremost institutions of the kind in the country. The American University, organized by the Methodist church, has a great future. There is a fine zoological garden in the environs of Washington.

Waterbury, Connecticut (c. h New Haven Co.), is a commercial and industrial center, with extensive manufactures of brass ware products, brass castings, buttons, men's clothing, stamped ware, watches, etc. It has many educational institutions, and a public library containing upwards of 100,000 volumes. It was settled in 1677.

Waterford, Ireland (cap. Waterford), has Protestant and Roman Catholic cathedrals. Fragments of the city walls remain, including an 11th century tower. The harbor is formed chiefly by the estuary of the Suir and Barrow. Butter and bacon are the principal exports.

Waterloo, Belgium (Brabant), is a village, 11 miles south of Brussels, which gave its name to the decisive battle fought near it on June 18, 1815, between Napoleon and the allied armies under Wellington and Blücher, when the former was defeated. The result was the final overthrow of Napoleon, and his exile to St. Helena.

Waterloo, Iowa (c. h. Blackhawk Co.), is a trade center in an agricultural and stock-raising region. The chief manufacturing establishments are foundries, machine-shops, gas-engine works, automobile works, etc. It was settled in 1845.

Watertown, Massachusetts (Middlesex Co.), is an industrial center with woollen mills, hosiery works, starch factory, needle factory, etc. A United States arsenal is located here. In this city is Mount Auburn Cemetery, one of the most famous in the country.

Watertown, New York (c. h. Jefferson Co.), is the commercial and industrial center of a rich agricultural region. It has an extensive wholesale trade in all kinds of merchandise, and is the distributing place for a large number of towns and villages. It has large paper-mills, carriage works, vise works, portable steam-engine works, silk-mills, machinery works, brass works, etc. A state armory is located here. It was settled in 1800.

Watervliet, New York (Albany Co.), is an industrial suburb of Albany and Troy, at the head of navigation of the Hudson River. It has large manufactories of woollen goods, bells, iron products, lumber products, street cars, scales, etc. A United States arsenal, one of the largest plants for the construction of siege ordnance and field and coast defense, is located here. The industrial growth of the place has been closely connected with the work of the government arsenal.

Waukesha, Wisconsin (c. h. Waukesha Co.), is a famous health resort with several mineral springs, the waters of which possess valuable medicinal properties. The shipment of mineral waters is one of the principal industries, and it has also iron-works, steel bridge works, and a canning factory. It is the seat of Carrol College, and of the State industrial school for boys. It was settled in 1836 and incorporated in 1848.

Wausau, Wisconsin (c. h. Marathon Co.), is a trade center in a region having extensive lumbering interests. The chief manufacturing establishments are sawmills, sash and door factories, box factories, machine-shops, flour-mills, paper-mills, etc. It has a training school for teachers and a county agricultural college. It was settled in 1845.

Webster, Massachusetts (Worcester Co.), is a trade center in an agricultural region, and has considerable industrial interests connected with farm products. It has cotton and woollen mills, shoe factories, and machine-shops.

Weihaiwei, China (Shantung), is a British naval and coaling station, leased to England since 1898, which commands the entrance to the Gulf of Pechili. In the Chino-Japanese war of 1895 the Japanese captured the town and destroyed the Chinese fleet, which had

taken refuge in the harbor. It is a favorite summer resort for Europeans.

Weimar, Germany (Saxe-Weimar), is a seat of the book trade, and has manufactures of cloth and leather. In the 18th and 19th centuries it was a great literary center, having associated with it Goethe, Schiller, Herder, Wieland, Immermann, and others. Among the numerous buildings are the ducal palace, the 15th century town church, containing Cranach's "Crucifixion," and the Liszt Museum.

Weissenfels, Germany (Saxony), has manufactures of boots and paper, and coal-mines.

Wellington, cap. New Zealand, is a port and coaling station on Port Nicholson. Among the chief public buildings are Government House, Houses of Parliament, government buildings, general post-office, government life insurance offices, town hall, and Harbor Board offices. The chief industrial establishments are foundries, refrigerating works, woollen mills, soap and candle works, wax-match factory, pottery works, boot factories and rope works.

Wenchau, China (Chehkiang), is a treaty port. Formerly the center of Japanese trade with China, it is now no longer of commercial importance.

West Bay City, Michigan (Bay Co.), is an industrial suburb of Bay City and has a number of industries connected with lumber products. The chief manufacturing establishments are lumber and planing-mills, sugar-beet factories, sash and door factories, salt works, etc. It also has shipyards, foundries, and machine-shops.

West Bromwich, England (Stafford), is noted for hardware manufactures and has forges and foundries. Coal is mined.

Westfield, Massachusetts (Hampden Co.), is the center of an agricultural region and has extensive manufacturing interests. The chief manufacturing establishments are machine-shops, paper-mills, thread mills, whip factories, and cigar factories. The principal educational institution is a state normal school.

West Ham, England (Essex), is a suburb of London, and has railway works, and manufactures of iron, india-rubber, and gutta-percha; also ship-building yards, sugar-refineries, and soap and jute works.

West Hartlepool, England (Durham), has iron and steel works, electrical works, iron ship-building, and a large trade in coal.

West Hoboken, New Jersey (Hudson Co.), is a residential suburb of New York. St. Michael's Monastery (Passionist Fathers), with which is connected a theological school, is located here.

Weymouth, Massachusetts (Norfolk Co.), is a trade center with manufactories of boots and shoes, hammocks, fireworks, and machinery. It also has car repair shops and extensive wool scouring yards. It was settled in 1623.

Wheeling, West Virginia (c. h. Ohio Co.), is the largest city of West Virginia and an important railroad and trade center. The city has steamer connections with all Ohio River ports and does an extensive wholesale trade. In the vicinity are coal-fields of great importance. The chief manufactures are iron and steel products, pottery, glass, tobacco products, leather, lumber products, canned goods, wagons and carriages, etc. It was the first town founded on the Ohio River. The first settlement was made in 1769. In 1861 it was made the capital of the "restored government of Virginia" by the people of Virginia who were opposed to secession. In 1863, it was the meeting place of the convention that formed the state of West Virginia. It was the capital of the state from 1863-70 and again from 1875-85.

White Plains, New York (c. h. Westchester Co.), is a residential suburb of New York, situated in a beautifully developed section. Bloomingdale Asylum for the insane is nearby. During the Revolutionary War, it was the scene of many battles and skirmishes; the Battle of White Plains, which resulted in the defeat of Washington by Lord Howe, was fought here on Oct. 28, 1776.

Wichita, Kansas (c. h. Sedgwick Co.), is an important commercial and industrial center, situated in a fertile agricultural region in southern Kansas. The

chief products of this section are wheat, corn, hogs, and cattle. The principal manufactures of the city are confectionery, flour, saddlery and harness, and men's clothing. Large meat-packing houses and a number of wholesale and jobbing houses are located here. A Friends' University, founded in 1898, a Congregational college, opened in 1892, and a Roman Catholic college, opened in 1900, are among the institutions of learning. It was settled in 1869.

Wichita Falls, Texas (c. h. Wichita Co.), is a railroad and trade center in a rich agricultural region.

Wiener Neustadt, Austria (Lower Austria), manufactures locomotives, machinery, wire, bells, pottery, ribbons, etc. It was founded in 1192.

Wiesbaden, Germany (Prussia), is a famous watering-place. Of its 20 hot springs, which were known to the Romans, the principal is the "Kochbrunnen" ("Boiling-spring"; 156°F.). The saline hot springs containing silica and iron, are efficacious in gout, rheumatism, scrofula, and other skin diseases and nervous affections. Though the public gaming-tables were abolished in 1872, the number of visitors annually is about 60,000.

Wigan, England (Lancashire), has cotton factories, oil works, foundries, and railway-wagon works, and trade in coal from collieries in the district.

Wiju, Korea, is the chief city of northern Korea, and has a large trade with China.

Wilhelmshaven, Germany (Prussia), is the principal naval station of Germany on the North Sea, and has a commodious harbor, large docks, extensive arsenal, ship-building yards, and all the latest equipments of a modern naval station.

Wilkes-Barre, Pennsylvania (c. h. Luzerne Co.), is the center of one of the richest anthracite coal regions in the country, and the mining and shipping of coal is the most important industry. It has large manufactories of silk, lace, axles, iron and steel, wire rope, etc., also large railroad shops. It was settled in 1772.

Wilkinsburg, Pennsylvania (Allegheny Co.), is a residential suburb of Pittsburg.

Willesden, England (Middlesex), is a suburb of London and a great railway junction.

Williamsport, Pennsylvania (c. h. Lycoming Co.), is a trade center in an agricultural and mining region. The chief manufacturing establishments are steel-works, lumber mills, furniture and rubber goods factories, woodworking machinery works, silk-mills, etc. The city has an extensive trade in lumber products and coal. It was settled in 1779.

Wilmington, Delaware (c. h. Newcastle Co.), is the largest city of Delaware, as well as its chief industrial and trade center. It has large car shops, steel-works, machine-shops, paper-mills, cotton-mills, iron-works and furniture factories. Other industries of importance are ship- and boat-building, meat-packing, pulp manufacturing, etc. A few miles outside of the city is a large powder manufacturing establishment. The old Swedes' Church, built of brick in 1698, is said to be the oldest building in the United States that has been in continuous use as a church since its erection. Among its educational institutions may be mentioned a state industrial school for girls, and a reform school for boys. Wilmington was settled in 1638.

Wilmington, North Carolina (c. h. New Hanover Co.), is the largest city of North Carolina and a port of entry on the Cape Fear River. It is an important commercial center and has an extensive trade in cotton, rice, turpentine, vegetables, lumber, and naval stores. It has large manufactories of lumber products, naval stores, cotton products, etc. A United States marine hospital is located here. It was settled in 1730.

Wimbledon, England (Surrey), is a residential suburb of London and has an ancient earthwork, traditionally ascribed to Cæsar.

Winchester, England (Hampshire), is an ancient city, the Caer Gwent of the Britons and the Venta Belgarum of the Romans. A residence of the Saxon kings, it became after the conquest the capital of the country. It possesses a famous cathedral founded in the 11th century, the transepts, crypt, and part of the nave being

from that period. It is the largest cathedral in England, 560 ft. in length, and has a magnificent interior of various periods of architecture. Among many ancient monuments and memorials to Saxon and Danish kings and Norman princes are those of Hardicanute and William Rufus. It has, besides, several ancient churches, and a hospital founded in the 12th century, which contains a handsome Norman church. It is the seat of Winchester College founded in 1387.

Windsor, England (Berkshire), is an ancient town, and contains Windsor Castle, the chief royal palace of England, whose buildings and immediate grounds cover 12 acres. Not far from the castle is the Great Park with an area of about 1,800 acres, which contains the palace and mausoleum of Frogmore, where the remains of Queen Victoria and of the Prince Consort rest. Near by is Eton, the location of one of the greatest public schools in England.

Windsor, Ontario, Canada, is a residential suburb of Detroit, and has considerable manufacturing and wholesale trade.

Winnipeg, cap. Manitoba, Canada, is the largest city of Manitoba and western Canada, and has a large transit trade in grain and general merchandise. It is the largest grain market in the British Empire and an important railroad center, with the shops of three transcontinental railway lines. In 1871 it had a population of 241, and was one of the Hudson Bay Co.'s trading posts, known as Fort Garry. Since its choice as the capital of Manitoba its growth has been very rapid, and its trade has increased faster even than its population. Winnipeg is the seat of the University of Manitoba, partly made up of Manitoba College (Presbyterian), St. John's Episcopal College, and the Manitoba Medical College; and of St. Boniface College (Roman Catholic), located at St. Boniface, on the opposite side of the Red River.

Winona, Minnesota (c. h. Winona Co.), is a commercial and industrial center on the Mississippi River, situated in a rich agricultural and stock-raising region. It has large manufactories of patent medicines, flour, lumber, agricultural implements, wagons, etc. There are large railroad shops. The city was settled in 1851.

Winston-Salem, North Carolina (c. h. Forsyth Co.), is the commercial center of a fertile agricultural region, especially noted for its tobacco. Tobacco manufacture is the leading industry, but there are also chemical works, rolling-mills, cotton-mills, etc.

Winterthur, Switzerland (Zürich), is a great railway center, and possesses important factories, particularly of locomotives and other machinery.

Witten, Germany (Prussia), has iron and steel industries, glass-works, and manufactures of chemicals and leather.

Woburn, Massachusetts (Middlesex Co.), is a residential suburb of Boston, and has manufactures of leather, chemicals, machinery, glue, and foundry products. The city was settled in 1640.

Wolverhampton, England (Stafford), is a great center of the iron industry; motor cars and cycles, japanned and tin wares, are also made, and the manufacture of electrical machinery and plants is another important industry. Coal and iron are mined.

Wood Green, England (Middlesex), is a suburb of London.

Woonsocket, Rhode Island (Providence Co.), is an industrial center, with large manufactures of cotton goods, foundry and machine-shop products, worsted goods, hosiery and knit goods, woollen goods, furniture, etc.

Worcester, England (cap. Worcester), contains a cathedral, dating in part from the 11th century, in which are tombs of St. Oswald, Wulfstan, and King John. There are several ancient churches, an 11th century commandery, and timbered houses. Its manufactures include gloves, porcelain, carriages, railway signals, sauce, vinegar, and British wines.

Worcester, Massachusetts (c. h. Worcester Co.), is the second city of Massachusetts, and is noted for the number and variety of its manufacturing establishments. It has a large number of foundries and machine-

shops, boot and shoe factories, envelope works, wireworks, woollen goods mills, worsted goods mills, lumber mills, clothing factories, etc. A state armory and a state asylum are located here. An art museum contains a good collection of casts, valuable paintings, engravings, etc. Adjacent to the court-house is the American Antiquarian Society, one of the leading learned bodies of America, founded in 1812, possessing a library of over 100,000 volumes. Among the educational institutions are Clark University, a postgraduate school for original research, and a state normal school. Worcester was settled in 1673.

Worms, Germany (Hesse), has manufactures of leather, cloth, tobacco, machinery, and chemicals, and produces the Liebfrauenmilch wine. The massive Romanesque cathedral, with two cupolas and four towers, was founded in the 8th, rebuilt in the 11th and 12th centuries, and carefully restored in the last quarter of the 19th century. The synagogue (11th century) is one of the oldest in Germany. The town house was restored in 1885. Worms is one of the oldest cities of Germany; in it is laid the scene of the Nibelungenlied. It was occupied by the Romans, destroyed by Attila, and afterwards rebuilt by Clovis. It was frequently the residence of Charlemagne and his Carlovingian successors, and was erected into a free imperial city by the Emperor Henry V. The most famous diet held here was that in 1521, at which Luther defended himself before Charles V.

Wuchang, China (cap. Hupeh), has a mint and cotton cloth mills.

Wuchau, China (Kwangsi), is a treaty port and has considerable trade.

Wuhu, China (Nganwhei), is a treaty port on the Yangtse River. It exports grain to the south, and has a large trade in coal, tea, and silk.

Wursburg, Germany (Bavaria), has manufactures of tobacco and cigars, musical instruments, furniture, machinery, and railway carriages, also breweries, and carries on a trade in wine. It is the seat of a Roman Catholic bishop, and its Romanesque cathedral was founded in the 9th century, enlarged in the 11th century, and restored in 1882-3. Among its other churches are Saint Burkard, the university and seminary churches, the Gothic chapel of St. Mary, and the Neumünster church. It possesses a university with a celebrated faculty of medicine.

Yakoba, Northern Nigeria, Africa, has considerable trade and some manufactures of cotton.

Yamagata, Japan (cap. Yamagata), is a trade center.

Yarkand, China (cap. Eastern Turkestan), is the chief trading center with north India across the Karakorum Pass.

Yarmouth, England (Norfolk), is a port and popular seaside resort, with a marine parade 3½ miles long, beautiful large gardens, and popular amusement parks. It is the center of the herring and other fisheries. Shipbuilding is carried on.

Yaroslaf, Russia (cap. Yaroslaf), has manufactures of cotton, tobacco, and flour. It is the seat of a Greek Orthodox archbishop, and its cathedral dates from the early part of the 13th century. Yaroslaf was the capital of an independent principality from 1126 to 1471.

Yeisk, Russia (Kuban), is a port on the Gulf of Taganrog and exports grain.

Yezd, Persia, has manufactures of silk, cotton, pottery, and felt. It is almost the only seat in Persia of the Gebrs, or Parsis, followers of Zoroaster.

Yochau, China (Hunan), is an important trade center.

Yokkaichi, Japan (Miye), has manufactures of silk, paper, and porcelain.

Yokohama, Japan (cap. Kanagawa), is the chief port of Japan, and a coaling station. Exports silk and silk goods, tea, metals, etc. It has a number of notable or even imposing public buildings (prefectural buildings, custom-house, post-office, court-house, railway depot), and in its massive piers and quays, docks, club-houses, churches of various denominations, etc., wears a semi-European aspect.

Yokosuka, Japan (Kanagawa), is an important naval station, with dock yards, arsenals, and large facilities for ship-building.

Yonkers, New York (Westchester Co.), is a residential suburb of New York, and has manufactures of carpets, foundry and machine-shop products, patent medicines, furniture, etc. It also has large elevator works and ship-building yards. It was settled in 1650. The Manor-House, built in 1682, is now the city hall.

York, England (cap. York), is the seat of an archbishop, and its cathedral church of St. Peter was founded in the 7th century. The present building, a magnificent cruciform structure, exhibits various styles of architecture, having been begun by Archbishop Roger (1171), and continued by his successors till 1472. It retains a Norman crypt, and possibly some fragments of Saxon work, and among several tombs of the archbishops is one of Walter de Grey (13th century). There are several other ancient churches. The castle at York was founded by William the Conqueror, but the present buildings are of various later dates. Its manufactures include glass, confectionery, leather, artificial manure, railway cars, iron-foundries, etc.

York, Pennsylvania (c. h. York Co.), is a trade and industrial center in a productive farming section. Its leading industries are foundry and machine-shops, agricultural implement works, carriage and wagon works, flour- and grist-mills, furniture factories, etc. It has an extensive trade in agricultural products and general merchandise. The city was settled in 1735.

Youngstown, Ohio (c. h. Mahoning Co.), is an important commercial and railroad center, with many large manufactories. It has extensive iron-works, steel and rail mills, blast-furnaces, boiler works, car shops, engine-works, stove works, and also large oilcloth and rubber works.

Zaandam, Netherlands (North Holland), has oil-mills, sawmills, and factories for paper, cement, and colors. Here Peter the Great of Russia (1697) learned ship-building. Zaandam was formerly the chief Dutch port for the Greenland whale fisheries.

Zacatecas, Mexico (cap. Zacatecas), is an important mining center, and its chief industry is the reduction of silver ore. Pottery is also made here. It contains a cathedral, a mint, and an institute of sciences.

Zagazig, Egypt, is a grain and cotton center.

Zanesville, Ohio (c. h. Muskingum Co.), is a trade center in a fertile agricultural region, and near the city are extensive deposits of coal, clay, and limestone. The manufactories are varied and consist of the largest tile works in the world, potteries, agricultural implement works, bent-wood works, glass works, leather tanneries, woollen mills, etc. It was settled in 1799.

Zanzibar, cap. Zanzibar, Africa, is a port on a low-lying peninsula, projecting from the west coast. The buildings include the sultan's palace, a French hospital, the English hospital, an English church, a Hindu temple, the old fort, and a government ship-repairing factory. It was formerly a center of the slave trade.

Zeitz, Germany (Prussia), has manufactures of cotton, wool, machinery, pianos, and furniture.

Zhitomir, Russia (cap. Volhynia), has manufactures of tobacco, spirits, soap, etc. It is an educational center, and the seat of a Greek archbishop and a Catholic bishop.

Zittau, Germany (Saxony), has manufactures of damask and woollen goods. Lignite is mined in the vicinity.

Zurich, Switzerland (cap. Zurich), is the most populous and most important town in Switzerland, built on the Limmat, as it issues from the Lake Zurich. Of recent years it has vastly increased in size and splendor. It is the center of the Swiss silk industry, while much cotton and machinery are also manufactured. Zurich was the scene of the labors of Zwingli, the reformer.

Zwickau, Germany (Saxony), manufactures chemicals, glass, earthenware, and machinery. Coal is mined. The church of St. Mary dates from 1451. Schumann, the musical composer, was born here.

Zwolle, Netherlands (cap. Overyssel), has several interesting buildings dating from the 15th to the 17th centuries. Thomas à Kempis, the author of "De Imitatione Christi," spent here the last sixty-four years of his life (1407-71).

STATISTICS OF THE WORLD

Area..................................196,900,000 sq. miles
Area of land..........................52,372,000 sq. miles
Area of fertile land..................28,270,000 sq. miles
Population............................1,610,000,000 souls
Age of earth..........................40 to 100 million years
Distance from Sun.....................93,000,000 miles
Equatorial diameter...................7,926.5 miles
Equatorial circumference..............24,902 miles
Polar diameter........................7,809.5 miles
Density compared with water...........5.55

Continental Area and Population

Continent	Area in sq. miles	Population	
		Number	Per sq. mi.
Africa........	11,515,000	130,000,000	11.
America, N.....	9,323,000	115,000,000	12.
America, S.....	6,882,000	45,000,000	6.
Asia........	17,057,000	900,000,000	53.
Australasia.....	3,456,000	5,500,000	1.
Europe........	3,879,000	414,500,000	107.
Polar Regions.....	253,000	....	....

Approximate Area of the Largest Countries

1. British Empire........................11,450,000 sq. miles
2. Russia (and possessions)..............8,500,000 sq. miles
3. France (and possessions).............4,330,000 sq. miles
4. China (and possessions)..............4,300,000 sq. miles
5. United States (and possessions)......3,750,000 sq. miles

Ocean Depth and Area

The oceans, including the inland seas connected with them, cover about 144,500,000 sq. miles, or 73.39% of the total surface.

Ocean	Greatest Depth	Area
Atlantic...............	27,366 feet	34,000,000 sq. miles
Pacific................	31,000 feet	71,000,000 sq. miles
Indian.................	18,582 feet	28,000,000 sq. miles
Arctic.................	9,000 feet	4,000,000 sq. miles
Antarctic..............	25,200 feet	7,500,000 sq. miles

The greatest ocean depth is approximately the same as the greatest land height, but the average depth of the sea floor is about 12,000 feet, while the average land height above the sea is only 2,300 feet. The maximum depth in the Pacific is near the island of Mindanao and that in the Atlantic off the coast of Porto Rico.

Population by Races

Division	Color	Section	Est. Population
Mongolian...	Yellow.......	Asia.............	685,000,000
Caucasian...	White........	Europe & America..	650,000,000
Negro........	Black.......	Africa..........	160,000,000
Semitic.......	White.......	North Africa.......	50,000,000
Malayan......	Brown.......	Australasia, inc.	
		Malay Peninsula.	50,000,000
Indian........	Red..........	America, N. and S...	15,000,000

Population by Religion

Christian..539,000,000
Catholic......................................270,000,000
Protestant....................................174,000,000
Greek Church..................................110,000,000
Abbysinian Church.............................3,000,000
Armenian Church...............................1,700,000
Others..300,000
Hindooism..206,000,000
Mohammedanism..243,000,000
Buddhism...450,000,000
Polytheism...142,000,000
Jewish Religion..10,000,000

Of the total population of the world, about one-third are Christians.

Proportion of Women

In Europe, there are 1,000 men to 1,027 women; in Africa, 1,000 men to 1,045 women; in America, 1,000 men to 964 women; in Asia, 1,000 men to 961 women; in Australia, 1,008 men to 937 women.

The highest proportion of women is found in Uganda, where there are 1,467 to every 1,000 men. The lowest proportion is in Alaska and the Malay States, where there are in the former 391 and in the latter 389 to every 1,000 men.

Largest Cities in the World

City	Country	Population
London...........	England.............	4,523,000
Greater London..........	England.............	7,252,963
New York...........	United States.......	4,766,883
Paris................	France..............	2,888,110
Tokyo...............	Japan...............	2,186,079
Chicago.............	United States.......	2,185,283
Berlin..............	Germany.............	2,071,257
Vienna..............	Austria.............	2,031,468
St. Petersburg......	Russia..............	1,907,708
Philadelphia........	United States.......	1,549,008
Moscow..............	Russia..............	1,481,200
Constantinople......	Turkey..............	1,125,000
Osaka...............	Japan...............	995,945
Canton..............	China...............	1,250,000
Buenos Aires........	Argentina...........	1,333,532
Calcutta............	India...............	1,222,313
Peking..............	China...............	1,077,209

Principal Languages of the World

	Tongues		Tongues
English..........	150,000,000	Spanish........	55,000,000
German..........	120,000,000	Italian........	40,000,000
Russian.........	90,000,000	Portuguese.....	30,000,000
French..........	60,000,000		

There are approximately 3,500 languages or dialects in the world. The English has grown faster than any other language during the past century.

Highest Structures in the World

	Feet
Eiffel Tower, Paris..................................	984
Woolworth Building, New York City...................	750
Metropolitan Building, New York City................	700
Singer Building, New York City	612
Municipal Building, New York City...................	560
Washington Monument, Washington, D. C...............	555
City Hall, Philadelphia, Pa.........................	549
Bankers' Trust Co. Building, New York City..........	539
Cologne Cathedral, Cologne, Germany.................	501
Pyramid of Cheops, Memphis, Egypt...................	450
St. Peter's, Rome, Italy	391
Park Row Building, New York City....................	386
St. Paul's, London, England	366
Times Building, New York City	363
Milan Cathedral, Milan, Italy.......................	360

World's Postal Statistics
(From Annual Reports)

Country	Letters and Post Cards	No. of Post Offices
United States...................	8,000,000,000	59,239
United Kingdom..................	4,000,000,000	24,000
Germany.........................	4,000,000,000	50,500
Austria-Hungary.................	1,500,000,000	16,000
France..........................	1,350,000,000	13,000
Japan...........................	1,200,000,000	7,000
Russia..........................	1,000,000,000	15,000
India...........................	759,000,000	19,500
Canada..........................	500,000,000	12,800
Italy...........................	400,000,000	10,000
Argentina.......................	350,000,000	3,000
Australia.......................	320,000,000	7,750
Switzerland.....................	270,000,000	4,100

World's Crops

The latest statistics of total crops raised per year on the earth are:

Corn......	3,672,636,000 bu.	Barley.....	1,385,245,000 bu.
Wheat....	3,626,336,000 bu.	Oats......	4,410,686,000 bu.
Rye......	1,675,898,000 bu.	Cotton....	19,171,000 bales
Tobacco...	2,595,247,000 lbs.	Potatoes...	5,523,864,000 bu.
	Sugar......	35,498,809,920 lbs.	

Height of Famous Mountain Peaks

ASIA	Everest.......................	29,002 feet
	Godwin-Austen................	28,865 feet
	Kinchinginga.................	28,156 feet
	Bride Peak...................	25,100 feet
SOUTH AMERICA	Aconcagua...................	23,080 feet
	Mercidario...................	22,315 feet
	Tupunagato..................	21,550 feet
	Illampu.....................	21,490 feet
	Cotopaxi....................	19,612 feet
NORTH AMERICA	McKinley....................	20,498 feet
	Logan.......................	19,539 feet
	Rainier......................	14,526 feet
	Whitney.....................	14,499 feet
EUROPE	Elbruz......................	18,347 feet
	Mont Blanc..................	15,781 feet
	Matterhorn..................	14,709 feet
	Jungfrau....................	13,761 feet

World's Longest Rivers

River	Outlet	Length in Miles
Amazon................	Atlantic..............	4,200
Nile...................	Mediterranean........	3,800
Yenisei................	Arctic Sea............	3,200
Yangtse...............	North Pacific.........	3,300
Missouri & Mississippi.	Gulf of Mexico.......	3,200
Ob....................	Arctic Sea............	3,200
Congo.................	Atlantic..............	2,900
Niger.................	Gulf of Guinea.......	2,700
Amur.................	Sea of Okhotsk.......	2,700
Hoang................	Yellow Sea...........	2,500
Volga.................	Caspian Sea..........	2,300
Mackenzie.............	Beaufort Sea.........	2,400
La Plata..............	South Atlantic........	2,300
Yukon................	Behring Sea..........	2,000
St. Lawrence..........	Gulf of St. Lawrence.	2,300

Famous Waterfalls of the World

Name and Location	Height in Feet
Gavarnie, France...........................	1,385
Grand, Labrador...........................	2,000
Minnehaha, Minnesota......................	50
Missouri, Montana..........................	90
Montmorenci, Quebec.......................	265
Multnomah, Oregon.........................	850
Murchison, Africa...........................	120
Niagara, New York-Ontario..................	164
Rjukan, Norway............................	780
Schaffhausen, Switzerland...................	100
Skjaeggedalsfos, Norway....................	530
Shoshone, Idaho...........................	210
Staubbach, Switzerland......................	1,000
Stirling, New Zealand.......................	500
Sutherland, New Zealand....................	1,904
Takkakaw, British Columbia.................	1,200
Twin, Idaho...............................	180
Yellowstone (upper), Montana...............	110
Yellowstone (lower), Montana...............	310
Ygnassu, Brazil............................	210
Yosemite (upper), California................	1,436
Yosemite (middle), California...............	626
Yosemite (lower), California................	400
Vettis, Norway.............................	950
Victoria, Africa............................	400
Voringfos, Norway..........................	600

World's Greatest Lakes

Chad.......	50,000 sq. miles	Huron......	23,800 sq. miles
Superior.....	31,200 sq. miles	Michigan...	22,450 sq. miles
Vict. Hyanza.	26,500 sq. miles		

World's Largest Islands

Greenland..	750,000 sq. miles	Madagascar.	228,000 sq. miles
New Guinea.	330,000 sq. miles	Sumatra....	160,000 sq. miles
Borneo.....	280,000 sq. miles	Gr. Britain..	88,000 sq. miles

WORLD'S LARGEST ISLANDS—Continued.

Australia, of course, is the largest island of all, with an area of nearly 3,000,000 sq. miles, but Australia is now generally considered to be a continent.

Railways of the World *

Old World	Miles	New World	Miles
Europe.............	207,488	North America......	283,511
Asia................	63,341	South America......	43,638
Africa..............	22,905	Australasia.........	19,275
	293,734		346,424

A total of 640,158 miles for the whole globe, which is a gain of 11,460 miles in one year, of which increase 6,221 miles were in the Old World and 8,239 in the New.

In 1849 there were only 4,772 miles of railroad. In the last 10 years, 149,092 miles have been built. Thirty per cent. of the railroads are government owned or controlled—107,746 miles in Europe, 36,365 in Asia, and 18,036 miles out of the 19,275 miles in Australasia.

Miles of Telegraph Line

Country	Miles	Country	Miles
United States.......	225,000	Argentina..........	38,000
Russia.............	126,559	Mexico............	22,000
Germany...........	142,020	Brazil.............	36,199
India..............	74,828	Italy..............	35,000
France.............	113,583	Turkey............	28,000
Austria-Hungary....	44,100	Spain.............	26,441
United Kingdom....	61,296	China.............	29,327
Canada............	42,312		
Australia..........	44,100	The World....	1,355,624

In proportion to the number of people, New Zealand and Australia have the most telegraph lines.

OCEAN CABLES

Ownership	Miles	Ownership	Miles
British.............	160,000	German............	10,000
United States.......	50,000	Other Nations......	12,000
French.............	25,000		
Danish.............	11,000	Total...........	268,000

The first ocean cable was laid across the English Channel between Dover and Calais in 1850.

Telephones in the World

Country	Miles of Wire	Stations
United States.......................	17,017,393	7,659,475
German Empire.....................	3,121,000	1,006,800
Great Britain......................	2,047,680	639,900
Canada...........................	568,746	284,373
France............................	830,520	230,700
Sweden...........................	223,200	186,000
Russia and Finland.................	318,500	172,900
Japan............................	222,098	117,394
Austria...........................	302,000	111,880
Denmark..........................	229,600	95,700
Switzerland........................	210,600	78,000
Italy..............................	144,600	72,500
Norway...........................	119,400	54,300
Spain.............................	55,660	25,300
Elsewhere.........................	1,233,370	500,965
Total..........................	26,644,367	11,235,987

World's Longest Bridges

Name	Location	Length Miles Yds.	
Tay................	Scotland..............	2	73
Ohio...............	Illinois...............	2	—
Forth..............	Scotland..............	1	1,005
Missouri...........	Missouri..............	1	784
Queensborough.....	New York............	1	1,107
Williamsburgh......	New York............	1	213
Victoria...........	New York............	1	520
Manhattan.........	Montreal.............	1	413
Susquehanna.......	Maryland............	1	345
Brooklyn..........	New York............	1	78

Plans of the proposed bridge across San Francisco Bay to Oakland have been filed with the Board of Supervisors in San Francisco. It will be 3 miles long. The total suspension will be 5,946 yards. The estimated cost is $26,000,000.

* From the *Archiv für Eisenbahnwesen.*

COLONIES AND DEPENDENCIES OF THE PRINCIPAL COUNTRIES OF THE WORLD

British Empire

Europe

	Area, sq. mi.	Population
Isle of Man	220	52,034
Channel Islands	70	96,900
Malta and Gozo	118	228,534
Gibraltar	1.8	19,586

Asia

	Area, sq. mi.	Population
British India (including ten great and four small provinces)	1,097,821	244,267,542
Feudatory or Native States under control of India.		
Haiderábád	82,698	13,374,676
Baroda	8,099	2,032,798
Mysore	29,444	5,806,193
Kashmir	80,900	3,158,126
Rájputána	127,541	10,530,432
Central India	78,774	9,356,980
Bombay States	65,761	7,411,675
Madras States	9,969	4,811,841
Central Provinces States	31,188	2,117,002
Bengal States	32,773	4,538,161
U. P. States	5,079	832,036
Punjab States	36,532	4,212,794
Baluchistan	86,511	396,432
Eastern Bengal and Assam	15,986	575,835
Sikkim	2,818	88,169
N. W. Frontier	......	1,622,094
Total Native States	691,253	70,864,995
Ceylon	25,332	4,109,054
Straits Settlements (inc. Singapore)	1,660	714,069
Federated Malay States	27,506	1,036,999
Hong Kong and Territory	405	366,145
Weihaiwei	285	147,177
North Borneo (British)	31,100	208,183
Brunei	4,000	30,000
Sarawak	42,000	500,000
Laccadive Islands	......	10,274
Other British Malay States (5)	23,450	925,652
Aden, Perim, Sokotara, and Kuria Muria Islands	10,382	58,165
Bahrein Islands	546	90,000
*Cyprus	3,584	274,108

Africa

	Area, sq. mi.	Population
Union of South Africa:		
Cape Colony	277,000	2,564,965
Natal	35,400	1,194,043
Transvaal	110,400	1,686,212
Orange Free State	50,400	528,174
Basutoland	11,716	405,903
Bechuanaland Protectorate	275,000	125,350
Swaziland	6,536	99,959
Rhodesia	439,575	1,770,871
British West Africa:		
Gambia and Protectorate	4,504	146,100
Gold Coast and Protectorate	80,000	1,502,898
Sierra Leone	25,000	1,403,132
Northern Nigeria	255,700	9,269,000
Southern Nigeria and Protectorate	79,880	7,855,749
British East Africa:		
East Africa Protectorate	250,000	4,038,000
Uganda Protectorate	117,681	2,843,325
Zanzibar Protectorate	1,020	198,914
Nyasaland Protectorate	40,000	970,430
British Somaliland	68,000	302,859
*Egypt	400,000	11,189,978
Sudan	984,520	3,000,000
Mauritius and Dep	720	377,083
Seychelles and Dep	160	26,000
Ascension	34	186
St. Helena	47	3,520
Tristan da Cunha	......	75

*See Turkey.

British Empire—continued

America

	Area, sq. mi.	Population
Dominion of Canada (10 provinces and Northwest Ter.)	3,729,665	7,204,838
Newfoundland	42,734	238,670
Labrador	120,000	3,949
British West Indies:		
Jamaica	4,424	831,383
Bahamas	4,403	55,944
Leeward Islands	701	127,189
Windward Islands	516	157,264
Barbados	166	171,982
Trinidad and Tobago	1,868	330,074
British Guiana	90,300	296,000
British Honduras	8,600	37,479
Bermuda	19.3	18,994
Falkland Islands (inc. S. Georgia)	7,500	3,275

Oceania

	Area, sq. mi.	Population
Australia (6 provinces)	2,974,581	4,568,707
New Zealand	104,751	1,008,468
Fiji Islands	7,435	139,541
Papua (British New Guinea)	90,540	272,000
Tonga, Solomon and Gilbert Is.	390	30,045
Pacific Islands unattached	......	......

Belgium

	Area, sq. mi.	Population
Belgian Kongo (Africa)	909,654	15,000,000

China

	Area, sq. mi.	Population
Manchuria (Asia)	363,610	20,000,000
Tibet (Asia)	463,200	6,500,000
Sin-Kiang (Asia) Consisting of Chinese Turkestan, Kulja, Zungaria and Outer Kansu	550,340	1,200,000
Mongolia (Asia)	1,367,600	2,600,000

Denmark

	Area, sq. mi.	Population
Iceland (Europe)	39,756	85,188
Greenland (N. America)	46,740	13,517
Danish West Indies (N. America) (St. Croix, St. Thomas and St. John)	138	27,086

France

Asia

	Area, sq. mi.	Population
French India	196	282,386
Pondichéry	......	184,754
Karikal	......	56,579
Chandernagar	......	25,293
Mahé	......	10,729
Yanaon	......	5,033
French Indo-China:		
Anam	52,100	5,554,822
Cambodia	45,000	1,634,252
Cochin China	20,000	3,050,785
Tonkin	46,400	6,119,720
Laos	98,000	640,877
Kwangchauwan	190	150,000

Africa

	Area, sq. mi.	Population
Algeria	343,500	5,231,850
Morocco	305,548	6,000,000
French West Africa		
Sénégal	74,000	1,172,096
Upper Sénégal and Niger	72,000	4,471,031
Guinea	95,000	1,498,000
Ivory Coast	130,000	1,132,812
Dahomey	......	825,950
Mauretania	344,967	223,000
Tunis	45,779	1,923,217
French Kongo	669,280	5,000,000

France—continued

	Area, sq. mi.	Population
French Sahara...................	1,544,000	800,000
Réunion.....................	9.0	173,822
Madagascar...................	228,000	3,104,881
Mayotta and Comoro Islands......	760	10 ,833
French Somali Coast.............	5,790	208,000

America

St. Pierre and Miquelon..........	93	4,209
Guadeloupe and Dep.............	688	212,430
Martinique...................	385	181,004
French Guiana.................	30,500	49,009

Oceania

New Caledonia and Dep..........	8,548	57,208
*New Hebrides.................	1,000	50,000
French Establishments consisting of Society Islands and other smaller groups..............	1,520	30,563

*Jointly administered by France and Great Britain.

German Empire

Africa

Togo........................	33,700	1,000,372
Kamerun.....................	191,130	2,303,200
German So: West Africa..........	322,450	83,000
German East Africa.............	384,180	10,032,000

Asia

Kiauchau....................	200	168,900

Oceania

German New Guinea Protectorate:

Kaiser Wilhelm's Land..........	70,000	301,700
Bismarck Archipelago..........	20,000	188,400
Eastern Caroline Islands......		43,000
Western Caroline Islands, inc. } Pelew and Marianne Islands	810	17,000
Solomon Islands..............	4,200	....
Marshall Islands.............	150	15,000
Samoan Islands:		
Savaii.......................	660 }	35,500
Upolu......................	340 }	

Italy

Erytrea (Africa)................	45,800	450,000
Somaliland (Africa).............	139,430	400,000
Tripoli (Africa)................	406,000	529,176
Tienttsin (China)................	18	17,000

Japan

Asia

Chosen or Korea................	86,000	13,461,299
Formosa.....................	13,458	3,443,679
Sakhalin.....................	12,500	43,273
Kwantung....................	1,256	446,714
Hokoto or Pescadores (12 Islands).	50	....

Netherlands

	Area, sq. mi.	Population
Dutch East Indies (Oceania):		
Java and Madura..............	50,554	30,09 ,008
Island of Sumatra..............	162,608	4,029,300
Riau Lingaa Archipelago........	16,301	112.216
Banca	4,446	115,18
Billiton	1,863	36,858
Dutch Borneo	212,737	1,233,655
Island of Celebes.............	71,470	851,905
Molucca Islands	43,864	407,906
Timor Archipelago.............	17,698	308,660
Bali and Lombok..............	4,065	523,535
Dutch New Guinea............	151,789	200,000
Dutch West Indies (America):		
Dutch Guiana	46,060	86,233
Curaçao (6 Islands)............	403	54,469

Portugal

Africa

Cape Verde Islands..............	1,480	142,552
Portuguese Guinea.............	13,940	820,000
Islands of St. Thomé and Principe..	360	42.103
Angola.......................	484,800	4,119,000
Mozambique..................	293,400	3,120,000

Asia

Goa (India)...................	1,469	475,513
Damao, Diu (India).............	169	56,285
Macao, Taipa Cologne (China)....	4	63,991
Portuguese Timor (India).........	7,330	300,000

Russia

Asia

Bokhara......................	83,000	250,000
Khiva.......................	24,000	800,000

Spain

Rio de Oro and Adrar (Africa)....	73,000	12,000
Spanish Guinea or Rio Muni (Africa)	12,000	200,000
Fernando Po., Annabon, Corisco, Great Elobey, Little Elobey (Africa).................	814	23,844
‡Canary Islands................	2,807	358,564

Turkey

*Egypt (Africa)................	400,000	11,139,978
†Crete (Europe)................	3,365	312,151
*Cyprus (Asia)................	3,584	274,108
†Samos (Asia).................	180	53,400

United States

Philippine Islands..............	127,853	8,368,427
Hawaii......................	6,449	196,227
Porto Rico...................	3,606	1,135,783
Guam.......................	210	12,517
Panama Canal Zone.............	448	62,810
Tutuila Group.................	77	6,800
Midway, Wake and other islands in the Pacific...................		

*A tributary state but under British protection. See British Empire.

†A tributary state but having autonomous government. Crete will probably become a part of Greece as soon as the Balkan-Turko treaty is signed.

‡The Canary Islands are considered part of Spain, not a possession.

THE OLDEST ORDER

What is the oldest order in existence? The claim is made for that of the Holy Sepulchre, the grand officership of which has just been conferred by the Pope on a member of the Irish Nationalist party, Sir Thomas Grattan Esmond. It appears that no date or the name of a founder can be assigned to the Order of the Holy Sepulchre, though there is a legendary tradition that traces its origin to the time of Charlemagne. In the middle of the last century, however, when the Latin Patriarchate of Jerusalem was re-established, the office of grand master of the order was transferred to it by Pope Pius the Ninth, who many years later, in 1868, created by statute three ranks of the order—the Grand Cross, Commander, and Knight. The costume is a white cloak, with the Cross of Jerusalem in red enamel. The Pope himself is grand master of the order.

UNDER WATER 6 1-2 MINUTES RECORD

When a man named Enocs remained under water for four minutes and forty-six seconds in March, 1896, the swimming world was astonished. A Frenchman named Poulyuen tried to break this record in 1907. He remained under water for four minutes and thirty-one seconds and then came up more dead than alive. On Nov. 2, 1912, Poulyuen succeeded in breaking the record by remaining under for six minutes and thirty seconds.

Newfoundland is a land of lakes. So numerous are they that it is estimated they cover about one-third of the total area of the island. There are 687 named lakes, and 30,000 known ones without names. The island has about 4,000 miles of sea coast.

HOMESTEADS IN CANADA

WHO IS ELIGIBLE.—A homestead (160 acres) may be taken up by any person who is the sole head of a family or by any male eighteen years of age or over who is a British subject or who declares his intention to become a British subject.

A widow having minor children of her own, dependent upon her for support, is permitted to make homestead entry as the sole head of a family.

ACQUIRING HOMESTEAD.—To acquire a homestead an applicant must make entry in person, either at the Dominion Lands Office for the district in which the land applied for is situated, or at a sub-agency authorized to transact business in such district. At the time of entry a fee of $10 must be paid.

APPLICATION FOR PATENT.—When a homesteader has completed his residence and cultivation duties, he makes his application for patent before the Agent of Dominion Lands for the district in which the homestead is situated, or before a sub-agent authorized to deal with lands in such district. If the duties have been satisfactorily performed patent issues to the homesteader shortly after without any further action on his part and the land thus becomes his absolute property.

RESIDENCE.—To earn patent for homestead, a person must reside in a habitable house upon the land for six months during each of three years. Such residence, however, need not be commenced before six months after the date on which entry for the land was secured.

IMPROVEMENT DUTIES.—Before being eligible to apply for patent, a homesteader must break (plough up) thirty acres of the homestead, of which twenty acres must be cropped. It is also required that a reasonable proportion of this cultivation must be done during each homestead year. Before being eli-

gible to apply for patent, the homesteader must have a house on the homestead worth at least $300.

PRE-EMPTIONS.—In certain districts in Southern Alberta and Saskatchewan, an additional 160 acres may be purchased under certain residence and improvement conditions by a person who has secured a homestead but who has not previously obtained a pre-emption under any Dominion Lands Act. Usually entry for homestead and pre-emption is made at the same time.

ENTRY.—As in the case of homesteads, entry must be made in person before the Agent of Dominion Lands in whose district the land is situated, or before a sub-agent authorized to deal with lands in such district. An entry fee of $10 must be paid at the time of entry. Only a person with a homestead entry may enter for a pre-emption.

RESIDENCE DUTIES.—In addition to the six months' residence in each of three years required in connection with homestead, a person who has entered for both homestead and pre-emption must put in six months' residence in each of three other years to secure patent for both. This residence may be put in on either homestead or pre-emption and must be in a habitable house.

IMPROVEMENT DUTIES.—The cultivation required in connection with a homestead and pre-emption is eighty acres. This may be done on either the homestead or pre-emption or part of it on each. A reasonable proportion of such cultivation must be done each year.

PAYMENT.—Payment for a pre-emption must be made at the rate of $3.00 per acre as follows:

One-third of the purchase price at the end of three years from date of entry. Balance in five equal annual installments with interest at 5 per cent. at the end of each year from the date of the pre-emption entry.

CANADIAN PROVINCES BY DISTRICTS AND SUB-DISTRICTS
AREA AND POPULATION

ALBERTA

DISTRICT	SQ. M.	1911	1901
Calgary....	5,736.750	60,502	8,362
Edmonton..	121,929.509	57,045	12,823
MacLeod..	9,407.241	34,504	7,856
Medicine Hat	25,619.493	70,606	10,804
Red Deer....	21,839.823	61,372	10,314
Strathcona..	10,750.242	49,473	12,345
Victoria....	57,641.942	41,161	10,518

BRITISH COLUMBIA

DISTRICT	SQ. M.	1911	1901
Comox-Atlin	143,251.38	42,263	21,457
Kootenay...	27,016.28	50,772	31,962
Nanaimo...	2,717.00	31,822	22,293
New Westminster...	4,844.50	55,679	23,976
Vancouver city......	652.00	123,902	28,895
Victoria city.	2.96	31,660	20,919
Yale and Cariboo...	174,932.08	56,382	29,155

MANITOBA

DISTRICT	SQ. M.	1911	1901
Brandon....	2,913.91	39,734	25,047
Dauphin...	20,614.451	44,000	22,631
Lisgar......	1,708.120	23,501	24,736
Macdonald..	3,738.460	35,841	23,866
Marquette..	5,209.201	33,598	20,431
Portage la Prairie...	2,741.537	27,950	23,483

MANITOBA Cont'd.

DISTRICT	SQ. M.	1911	1901
Provencher..	5,641.60	40,693	24,434
Selkirk.....	17,881.517	53,091	24,021
Souris.....	3,858.194	29,049	24,222
Winnipeg city.....	19.922	128,157	42,340

NEW BRUNSWICK

DISTRICT	SQ. M.	1911	1901
Carleton....	1,310.60	21,446	21,621
Charlotte...	1,283.40	21,147	22,415
Gloucester.	1,869.81	32,662	27,936
Kent.......	1,778.02	24,376	23,958
Kings and Albert....	2,101.73	30,285	32,580
Northumberland...	4,740.60	31,194	28,543
Restigouche.	3,269.68	15,687	10,586
St. John city and county	615.88	53,572	51,759
Sunbury and Queens...	2,529.28	17,116	16,906
Victoria and Madawaska	3,364.92	28,222	21,136
Westmorland	1,442.18	44,621	42,060
York.......	3,605.26	31,561	31,620

NOVA SCOTIA

DISTRICT	SQ. M.	1911	1901
Annapolis...	1,323.88	18,581	18,842
Antigonish..	556.00	11,962	13,617

NOVA SCOTIA Cont'd.

DISTRICT	SQ. M.	1911	1901
Cape Breton N. & Victoria.....	1,355.10	29,888	24,650
Cape Breton S.....	721.90	53,352	35,087
Colchester..	1,451.00	23,664	24,900
Cumberland.	1,683.00	40,543	36,168
Digby......	1,000.00	20,167	20,322
Guysborough..	4,656.00	17,048	18,320
Halifax city & county.	2,123.38	80,257	74,662
Hants.....	1,229.00	19,703	20,056
Inverness..	1,408.75	25,571	24,353
Kings.....	864.00	21,780	21,937
Lunenburg.	1,202.00	33,260	32,389
Pictou.....	1,124.00	35,858	33,459
Richmond.	489.00	13,273	13,515
Shelburne & Queens...	2,022.48	24,211	24,428
Yarmouth..	858.76	23,220	22,869

ONTARIO

DISTRICT	SQ. M.	1911	1901
Algoma E...	49,114.64	44,628	25,211
Algoma W-O.....	22,263.11	28,752	17,894
Brant	334.23	19,259	18,273
Brantford..	86.86	26,617	19,867
Brockville..	274.87	18,531	18,721
Bruce N.....	950.95	23,783	27,424
Bruce S.....	699.46	26,249	31,596

ONTARIO Cont'd.

DISTRICT	SQ. M.	1911	1901
Carleton....	650.87	28,406	24,380
Dufferin....	556.64	17,740	21,036
Dundas....	383.12	18,165	19,757
Durham....	628.98	26,411	27,570
Elgin E....	362.52	17,597	17,901
Elgin W-O.	357.58	26,715	25,645
Essex N....	239.27	38,006	28,789
Essex S....	467.53	29,541	29,955
Frontenac...	1,595.91	21,944	24,746
Glengarry...	477.59	21,259	22,131
Grenville...	462.83	17,545	21,021
Grey E....	688.06	19,650	23,661
Grey N....	448.12	26,991	24,874
Grey S....	571.70	19,250	21,053
Haldimand.	488.13	21,562	21,233
Halton....	362.68	22,208	19,545
Hamilton E.	2.69	39,793	24,000
Hamilton W -O....	3.54	37,279	28,634
Hastings E..	1,291.41	24,978	27,943
Hastings W -O....	1,031.57	30,825	31,348
Huron E....	428.24	16,289	19,227
Huron S....	466.46	19,508	22,881
Huron W- O....	400.71	17,186	19,712
Kent E....	414.59	23,698	25,328
Kent W-O..	503.07	32,297	31,866
Kingston ..	3.54	20,660	19,788
Lambton E..	548.66	22,223	26,919
Lambton W -O....	575.57	29,109	29,723
Lanark N...	566.64	14,624	17,236
Lanark S...	571.35	19,751	19,996
Leeds....	624.81	18,222	19,254
Lennox and Addington	1,169.77	20,386	23,346
Lincoln....	332.41	35,429	30,552
London....	6.85	46,300	37,976
Middlesex E.	413.62	20,814	20,228
Middlesex N	436.46	13,737	16,419
Middlesex W -O....	379.68	16,214	18,079
Muskoka...	1,585.38	21,233	20,971
Nipissing..	31,573.07	74,130	28,309
Norfolk....	634.26	27,110	29,147
Northumber- land E....	438.65	19,927	20,495
Northumber- land W-O.	265.64	12,965	13,055
Ontario N..	504.82	17,141	18,390
Ontario S...	347.69	23,865	22,018
Ottawa city pt	4.75	73,193	57,640
Oxford N..	410.56	25,077	25,644
Oxford S...	353.99	22,294	22,760
Parry Sound	3,928.29	26,547	24,936
Peel....	468.51	22,102	21,475
Perth N....	429.77	30,235	29,256
Perth S....	409.81	18,947	20,615
Peterbor- ough E...	891.38	15,499	16,291
Peterbor- ough W-O	553.81	26,151	20,704
Prescott....	494.29	26,968	27,035
Prince Ed- ward....	390.40	17,150	17,864
Renfrew N..	1,057.81	23,617	24,556
Renfrew S..	1,644.95	27,852	27,676
Russell....	698.68	39,434	35,166
Simcoe E...	529.39	35,294	29,845
Simcoe N..	574.88	24,699	26,071
Simcoe S...	558.61	25,060	26,399
Stormont...	412.33	24,775	27,042
Thunder Bay & Rainy R.	72,578.38	67,249	28,987
Toronto Centre	1.02	53,125	43,861

ONTARIO Cont'd.

DISTRICT	SQ. M.	1911	1901
Toronto E..	3.43	68,912	40,194
Toronto N..	2.92	56,469	40,886
Toronto S..	4.56	43,956	38,108
Toronto W- O....	4.91	105,291	44,991
Victoria....	2,834.23	36,499	38,511
Waterloo N.	273.20	33,619	27,124
Waterloo S..	242.63	28,988	25,470
Welland....	387.27	42,163	31,588
Wellington N	580.46	22,292	26,120
Wellington S	438.88	33,200	29,526
Wentworth..	451.97	34,634	26,818
York Centre	333.73	26,048	21,505
York N....	430.56	22,415	22,419
York S....	188.98	68,018	20,699

PRINCE EDWARD ISLAND

DISTRICT	SQ. M.	1911	1901
Kings....	641.18	22,636	24,725
Prince....	778.23	32,779	35,400
Queens....	764.95	38,313	43,134

QUEBEC

DISTRICT	SQ. M.	1911	1901
Argenteuil..	783.367	16,766	16,407
Bagot....	346.141	18,206	18,181
Beauce....	1,891.040	5,399	43,129
Beauharnois	147.039	20,802	21,732
Bellechasse..	652.640	21,141	18,706
Berthier..	2,192.748	19,872	19,980
Bonaventure	3,463.61	28,110	24,495
Brome....	488.159	13,216	13,397
Chambly & Vercheres.	337.000	28,715	24,318
Champlain..	9,926.95	42,758	32,015
Charlevoix..	2,273.491	20,637	19,334
Chateauguay	265.279	13,322	13,583
Chicoutimi & Saguenay.	137,179.74	63,341	48,291
Compton..	1,439.041	29,630	26,460
Deux-Mon- tagnes....	279.258	13,868	14,438
Dorchester..	941.600	25,096	21,007
Drummond & Arthabaska.	1,197.82	41,590	38,999
Gaspe....	4,551.470	35,001	30,683
Hochelaga..	2.788	75,049	56,919
Huntingdon.	361.250	13,240	13,979
Jacques- Cartier..	115.317	65,023	26,168
Joliette....	3,013.50	23,911	22,255
Kamouraska	1,037.509	20,888	19,099
Labelle....	3,837.58	40,351	32,901
Laprairie & Napierville	319.200	19,335	19,633
L'Assomp-.. tion....	246.647	15,164	13,995
Laval....	148.886	29,977	19,743
Levis....	271.838	28,913	26,210
L'Islet....	772.806	16,435	14,439
Lotbiniere..	726.401	22,158	20,039
Maisonneuve	9.903	170,978	65,178
Maskinonge.	2,940.000	16,509	15,813
Megantic..	780.163	31,314	23,878
Missisquoi..	375.219	17,466	17,339
Montcalm..	4,201.750	13,862	13,001
Montmagny.	630.134	17,356	14,757

QUEBEC Cont'd.

DISTRICT	SQ. M.	1911	1901
Montmor- ency....	2,136.959	13,215	12,311
Montreal— Ste. Anne.	1.295	21,676	23,368
Montreal— St. Antoine	1.659	48,638	47,653
Montreal— St. Jacques	.684	44,057	42,618
Montreal— St. Laurent	.850	55,860	48,808
Montreal— Ste. Marie	.970	54,910	40,631
Nicolet....	626.078	30,055	27,209
Pontiac....	19,917.341	29,416	25,722
Portneuf...	1,488.800	30,529	27,159
Quebec Centre	1.288	21,143	20,366
Quebec East.	.928	47,429	39,325
Quebec W-O	.786	9,618	9,149
Quebec co...	2,728.498	25,844	22,101
Richelieu...	221.253	20,686	19,518
Richmond & Wolfe....	1,224.320	39,491	34,137
Rimouski...	5,585.106	51,490	40,157
Rouville...	242.977	13,131	13,407
St. Hyacinthe	277.611	22,342	21,543
St. Jean & Iberville..	403.022	21,882	20,679
Shefford....	567.200	23,976	23,628
Sherbrooke..	237.599	23,211	18,426
Soulanges...	136.111	9,400	9,928
Stanstead...	432.478	20,765	18,998
Temiscouata	1,806.189	36,430	29,185
Terrebonne.	781.822	29,018	26,816
Trois-Rivieres & St.Mau- rice....	2,568.050	36,153	29,311
Vaudreuil..	200.647	11,039	10,445
Wright....	2,427.67	48,332	42,830
Yamaska...	364.966	19,511	20,564

SASKATCHEWAN

DISTRICT	SQ. M.	1911	1901
Assiniboia...	7,505.491	42,556	9,332
Battleford..	54,515.615	47,075	6,171
Humboldt..	1,702.920	52,195	2,166
Mackenzie..	8,445.060	40,558	13,537
Moosejaw..	33,850.306	87,725	5,761
Prince Albert...	103,262.192	36,319	12,795
Qu'Appelle..	5,359.320	35,608	17,178
Regina....	7,930.773	70,556	7,703
Saltcoats...	4,199.430	28,695	9,479
Saskatoon..	6,610.890	51,145	7,157

YUKON

DISTRICT	SQ. M.	1911	1901
Yukon....	207.076	8,512	27,219

NORTHWEST TERRITORIES

DISTRICT	SQ. M.	1911	1901
Northwest Territories	1,921.685	17,196	20,129

INDEX
CITIES AND TOWNS OF CANADA
LATEST POPULATION FIGURES

Capitals of provinces are printed in capital letters. The index references enclosed in parentheses, indicate location of city or village on the accompanying maps; those marked with an asterisk (*) refer to map of Ontario, Northern Part.

ALBERTA

Alix, (H8) 267	Champion, (H10) 230	Grassy Lake, (K11) 247	Medicine Hat,	Redcliff, (L10) 220
Athabaska Landing,	Claresholm, (H10) 809	Grouard, (E5) 447	(I.10) Est. 17,000	Red Deer, (H8) 2,118
(J6) 227	Coalhurst, (J11) 450	Hardieville, (J11) 330	Milk River, (J11) 230	St. Albert, (H7) 614
Banff, (F9) 937	Cochrane, (G9) 395	Hardisty, (K8) 351	Monarch, (H11) 200	Sedgewick, (K8) 331
Bankhead, (F9) 694	Coleman, (G11) 1,557	High River, (H10) 1,182	Morinville, (H7) 385	Stafford Village, (J11) 985
Barnwell, (J11) 225	Coronation, (K8) 1,200	Hillcrest Mines,	Nanton, (H10) 571	Stavely, (H10) 245
Bassano, (J10) 540	Crossfield, (G9) 262	(G11) 481	North Edmonton,	Stettler, (J8) 1,444
Bawlf, (J8) 270	Daysland, (J8) 349	Hill Spring, (H11) 275	(H7) 404	Stirling, (J11) 514
Bellevue, (G11) 463	Diamond City, (J11) 510	Innisfail, (H8) 602	North Red Deer,	Stony Plain, (G7) 305
Blairmore, (G11) 1,137	Didsbury, (H9) 726	Irricana, (H9) 400	(H8) 304	Strathcona, (H7) 5,579
Bow Island, (K11) 307	EDMONTON,	Irvine, (L10) 372	Ogden, (H10) 530	Strathmore, (H9) 531
Brooks Station, (K10) 486	(H7) Est. 68,500	Kimball, (H11) 265	Okotoks, (G10) 516	Taber, (K11) 1,400
Burdett, (K11) 230	Edson, (E7) 497	Lacombe, (H8) 1,029	Olds, (H9) 917	Tofield, (J7) 586
Calgary, (G10) Est. 84,000	Exshaw, (F9) 250	Leduc, (H7) 523	Ouellettville, (J10) 220	Trochu, (H9) 353
Camrose, (J7) 1,586	Fitzhugh, (C8) 225	Lethbridge	Passburg, (G11) 305	Vegreville, (K7) 1,029
Canmore, (F9) 754	Fort Saskatchewan,	(H11) Est. 11,000	Pincher Creek,	Vermilion, (L7) 625
Cardiff, (H7) 246	(J7) 782	Lignite, (H8) 330	(H11) 1,027	Wainwright, (L8) 788
Cardston, (H11) 1,207	Frank, (G11) 806	Lille, (G11) 303	Ponoka, (H8) 642	Walsh, (L11) 200
Carmangay, (J10) 286	Gadsby, (J8) 213	Lloydminster, (M7) 222	Provost, (L8) 329	Warner, (J11) 321
Carstairs, (H9) 270	Gleichen, (J10) 583	Macleod, (H11) 1,844	Queenstown, (J10) 666	Wetaskiwin, (H7) 2,411
Castor, (J8) 1,659	Granum, (H11) 250	Magrath, (J11) 995	Raymond, (J11) 1,465	

BRITISH COLUMBIA

Abbotsford 300	Courtenay, (F13) 200	Kamloops, (L12) 3,772	Nelson, (O13) 4,476	Princeton, (L13) 400
Agassiz, (K13) 330	Cranbrook, (R13) 3,090	Kaslo, (P13) 722	New Denver, (O13) 300	Revelstoke, (O12) 3,017
Ainsworth, (O13) 320	Creston, (P13) 300	Kelowna, (M13) 1,663	New Michel, (S13) 900	Rivers Inlet, (D11) 200
Alberni, (G13) 891	Cumberland, (F13) 1,237	Kincolith, (B7) 200	New Westminster	Rossland, (O13) 2,826
Aldergrove, (J13) 225	Duncans Station,	Kitchener, (P13) 235	(H13) 13,199	Salmon Arm, (M12) 400
Armstrong, (M12) 810	(H14) 600	Kyuquot, (D12) 350	Nicola, (L12) 200	Sechelt, (H13) 400
Arrowhead, (N12) 560	Earls Road 300	Ladner, (H13) 800	North Bend, (K13) 200	South Vancouver,
Ashcroft, (K12) 535	Edmonds 225	Ladysmith, (H14) 3,295	North Saanich, (H14) 200	(H13) 16,126
Atlin, (U3) 325	Ebolt, (N13) 325	Langley Fort, (J13) 250	North Vancouver,	Steveston, (H13) 1,100
Barkerville, (K9) 320	Enderby, (M12) 855	Langley Prairie, (J13) 320	(J13) 8,196	Stewart, (B7) 800
Beaver Lake, (K10) 300	Erie, (O13) 220	Lillooet, (K12) 300	Peachland, (M13) 325	Summerland, (M13) 835
Brechin 250	Esquimalt, (H14) 4,001	Lulu Island, (H13) 300	Penticton, (M13) 750	Telegraph Creek,
Camborne, (O12) 300	Fairview, (M13) 330	Lytton, (K12) 300	Phoenix, (N13) 662	(V5) 250
Central Park 500	Ferguson, (O12) 235	Marysville, (R13) 300	Pilot Bay, (P13) 200	Trail, (O13) 1,460
Chemainus, (H14) 550	Fernie, (R13) 3,146	Mayne, (H14) 200	Point Grey, (H13) 4,320	Trout Lake, (O12) 325
Chilliwack, (K13) 1,657	Fort Steele, (R13) 276	Merritt, (L12) 703	Port Alberni 700	Vancouver, (H13) 100,401
Clayoquot, (E13) 300	Golden, (O11) 932	Michel, (S13) 850	Port Essington, (B8) 350	Vernon, (M12) 2,671
Clo-oose, (G14) 200	Grand Forks, (N13) 1,577	Midway, (N13) 250	Port Hammond,	VICTORIA, (H14) 31,660
Cloverdale, (J13) 220	Greenwood, (N13) 778	Mission City, (J13) 500	(J13) 300	Wardner, (R13) 220
Cobble Hill, (H14) 285	Hazelton, (D7) 550	Moyie, (R13) 560	Port Haney, (J13) 230	Wellington, (G13) 370
Comaplix, (O12) 275	Hedley, (L13) 320	Nakusp, (O12) 347	Port Moody, (J13) 550	Yale, (K13) 200
Corbin, (S13) 450	Hosmer, (R13) 2,019	Nanaimo, (G13) 8,168	Prince Rupert, (A8) 4,184	Ymir, (O13) 600

MANITOBA

Alexander, (B7) 300	Carman, (E7) 1,271	Gimli, (F6) 496	Lauder, (B7) 225	Newdale, (B6) 225
Altona, (E7) 450	Cartwright, (C7) 300	Gladstone, (D6) 782	Le Pas, (A3) 265	Ninga, (C7) 325
Arden, (C6) 225	Crystal City, (D7) 535	Glenboro, (C7) 560	Letellier, (E7) 284	Norway House, (F2) 1,150
Austin, (D7) 260	Cypress River, (D7) 305	Grand Rapids, (C3) 219	Macgregor, (D7) 550	Notre Dame De
Baldur, (C7) 415	Dauphin, (C5) 2,815	Grandview, (B5) 637	Manitou, (D7) 639	Lourdes, (D7) 100
Barrows, (A4) 245	Deloraine, (B7) 808	Gretna, (E7) 519	Melita, (A7) 690	Oak Lake, (B7) 449
Beausejour, (F6) 847	Dominion City, (E7) 225	Griswold, (B7) 325	Miami, (D7) 375	Oak River, (B6) 225
Belmont, (C7) 350	Dunrea, (C7) 200	Hamiota, (B6) 565	Minota, (A6) 325	Pierson, (A7) 230
Benito, (A5) 225	Elgin, (B7) 365	Hartney, (B7) 623	Minnedosa, (C6) 1,483	Pilot Mound, (D7) 457
Binscarth, (A6) 260	Elkhorn, (A7) 574	Headingly, (E7) 150	Morden, (D7) 1,130	Plumas, (C6) 255
Birtle, (B6) 437	Elm Creek, (E7) 265	Holland, (D7) 361	Morris, (E7) 598	Plum Coulee, (E7) 380
Boissevain, (C7) 918	Emerson, (E7) 1,043	Inkster, (E7) 225	Napinka, (B7) 326	Portage la Prairie,
Brandon, (B7) Est. 17,500	Garson, (F6) 220	Killarney, (C7) 1,010	Neepawa, (C6) 1,864	(D6) 5,892
Carberry, (C7) 878	Gilbert Plains, (B5) 542	La Riviere, (D7) 225	Nelson House, (D1) 468	Rapid City, (B6) 580

Index of Canada, Copyright, 1914, by C. S. Hammond & Co., N. Y.

Rathwell, (D7)_____ 230
Reston, (A7)_____ 416
Rivers, (B6)_____ 950
Roblin, (A5)_____ 350
Roland, (E7)_____ 433
Rosenfeld, (E7)_____ 225
Russell, (A6)_____ 562
St. Boniface, (F7)__7,483
St. Charles, (E7)___ 150

Sair'e Agathe, (E7) 150
Sainte Anne des
    Chenes, (F7)____ 325
St. George, (F6)____ 200
St. Jean Baptiste,
    (E7)_____1,174
St. Laurent, (D6)__ 581
St. Norbert, (F7)__1,448
St. Pierre-Jolys,(F7) 473

Selkirk, (F6)_____2,977
Shoal Lake, (B6) __ 591
Somerset, (D7)_____ 275
Souris, (B7)_____1,854
Stonewall, (E6)_____1,005
Stony Mountain,
    (E6)_____ 200
Strathclair Station,
    (B6)_____ 230

Swan Lake, (D7)___ 250
Swan River, (A4)___ 574
Teulon, (E6)_____ 200
Transcona, (F7)____ 300
Treherne, (D7)_____ 600
Tyndall, (F6)_____ 235
Virden, (A7)_____1,550
Waskada, (B7)_____ 350

Wawanesa, (C7)____ 375
Winkler, (E7)_____ 458
WINNIPEG,
    (E7)_____Est. 225,000
Winnipeg Beach,
    (F6)_____ 245
Winnipegosis, (B5). 518
York Factory, (G1) 291

NEW BRUNSWICK

Albert, (E3)_____ 350
Alma, (D3)_____ 250
Andover, (B2)_____ 289
Apohaqui, (D3)____ 300
Argyle, (B2)_____ 100
Aroostook Junction,
    (B2)_____ 500
Avondale, (B2)_____ 150
Baie Verte, (E3)___ 300
Baillie, (B3)_____ 600
Balmoral, (C1)_____ 600
Barachois, (E2)____ 300
Barnesville, (D3)__ 300
Bass River, (D2)___ 300
Bath, (B2)_____ 500
Bathurst, (D1)_____ 960
Bay du Vin, (D1)__ 300
Bayfield, (F2)_____ 150
Beaver Harbour,(B2) 270
Belledune, (D1)____ 400
Belledune River,(D1)200
Benton, (B3)_____ 200
Beresford, (D1)____ 100
Berrys Mills, (D2)_ 500
Blacks Harbor, (C3) 300
Blackville, (D2)____ 450
Blissfield, (C2)____ 250
Blissville, (C3)____ 300
Bloomfield, (B2)___ 400
Boiestown, (C2)____ 250
Bonney River, (C3) 200
Bristol, (B2)_____ 300
Buctouche, (E2)____ 500
Burnt Church, (D1) 100
Butternut Ridge,
    (D2)_____ 500
Calhoun, (E2)_____ 200
Campbellton, (C1).3,817
Campo Bello, (C4)_ 400
Canaan Station,(D2) 100
Canterbury Station,
    (B3)_____ 250
Cape Bald, (E2)___ 300
Cape Tormentine,
    (F2)_____ 100
Caraquet, (D1)____2,800
Caron Brook, (A1)_ 100

Centreville, (B2)___ 600
Charlo, (C1)_____ 100
Chatham, (D2)____4,666
Chelmsford, (D2)__ 100
Chipman, (D2)_____ 350
Clair, (A1)_____ 200
Clifton, (D1)_____ 200
Coal Branch, (D2)_ 130
Cocagne, (E2)_____ 900
Coldstream, (B2)__ 300
Coles Island, (D3)_ 300
College Bridge, (E3) 100
Connor, (A1)_____ 100
Cork, (C3)_____ 350
Cormierville, (E2)_ 500
Corn Hill, (D3)____ 500
Cross Creek, (C2)_ 100
Cumberland Bay,
    (D2)_____ 100
Dalhousie, (C1)___1,650
Dalhousie Junction,
    (C1)_____ 200
Debec, (B2)_____ 200
Deer Lake, (B3)___ 100
Derby, (D2)_____ 200
Doaktown, (C2)___ 400
Dorchester, (E3)__1,080
Douglas Harbor,
    (C3)_____ 200
Douglastown, (D1) 500
Drummond, (B1)__ 100
Dumbarton, (B3)__ 100
Durham, (C2)_____ 100
Edmundston, (A1).1,821
Eel River, (C1)____ 175
Elgin, (D3)_____ 250
Escuminac, (E1)___ 330
Fairville, (C3)____2,000
Five Fingers, (B1)_ 500
Flatlands, (C1)____ 150
Florenceville, (B2) 500
Fox Creek, (E2)___ 500
FREDERICTON,
    (C3)_____7,208
Fredericton Junc-
    tion, (C3)_____ 350
Gagetown, (C3)___ 233

Gardners Creek,(D3) 200
Gaspereaux, (C3).. 100
Gibson, (C3)_____ 800
Grande Anse, (D1)_ 700
Grand Falls, (B2)..1,280
Green River, (A1).. 200
Hampstead, (C3)__ 250
Hampton, (D3)____ 554
Harcourt, (D2)____ 400
Hartland, (B2)____ 844
Harvey, (E3)_____1,800
Harvey Station,(C3) 300
Head Of Tide, (C1) 100
Hillsborough, (E3)_ 911
Hopewell Cape, (E3) 300
Indiantown, (D2)__ 150
Inkerman, (E1)____ 500
Jacksonville, (B2)_ 100
Jacquet River, (D1) 400
Keswick, (C2)_____ 200
Kingsclear, (C3)__ 100
Kingston, (C3)____ 500
Kirkland, (B2)____ 200
Kouchibouguac,
    (E2)_____ 600
Lawrence, (B3)____ 100
Lepreau, (C3)_____ 300
Little Shippegan,
    (E1)_____ 250
Loggieville, (D2)__ 800
Lower Caraquet,
    (E1)_____1,500
Lower Prince Wil-
    liam, (B3)_____ 400
McAdam Junction,
    (B3)_____1,000
McLeods Mills,(D2) 250
Maces Bay, (C3) .. 100
Marysville, (C2)__1,837
Mascarene, (C3)__ 300
Memramcook, (E2) 300
Middle Southamp-
    ton, (B3)_____ 600
Milford, (C3)_____ 600
Millerton, (D2)____ 400
Millidgeville, (C3). 300
Millstream, (D3).. 400

Milltown, (B3)____1,904
Millville, (B2)_____ 300
Minto, (C2)_____ 100
Miscou Centre,
    (E1)_____ 100
Miscou Harbour,
    (E1)_____ 300
Moncton, (E2)___11,345
Moores Mills, (B3). 300
Musquash, (C3)___ 250
Neguac, (D1)_____ 150
Nelson, (D2)_____ 200
Newcastle, (D2)__2,945
New Denmark, (B2) 500
New Mills, (C1)___ 200
North Head, (C4)_ 100
Norton, (D3)_____ 500
Notre Dame, (E2). 400
Oak Bay, (B3)_____ 150
Oromocto, (C3)___ 250
Pennfield Centre,
    (C3)_____ 600
Penobsquis, (D3).. 300
Perth, (B2)_____ 200
Petersville Church,
    (C3)_____ 200
Petitcodiac, (D3)__ 750
Petit Rocher, (D1).. 250
Plaster Rock, (B2). 150
Pointe Du Chene,
    (E2)_____ 150
Port Elgin, (E2)___ 800
Prince William,(D3) 200
Quaco, (D3)_____ 900
Quispamsis, (D3)__ 100
Red Rapids, (B2)__ 200
Rexton, (E2)_____ 600
Richibucto, (E2)__ 871
Richibucto Village,
    (E2)_____ 300
Richmond Corner,
    (B2)_____ 275
Riley Brook, (B1)_ 100
Riordan, (D1)_____ 100
River Charlo, (C1) 350
Riverside, (E3)____ 250
Rogersville, (D2)_1,000

Rolling Dam, (B3). 500
Rothesay, (D3)____ 300
Round Hill, (C3)__ 300
Sackville, (E3)____2,039
Saint Andrews, (B3) 987
Saint Basil, (B1)__ 100
Sainte Croix, (B3). 300
Saint George, (C3). 988
Saint Jacques, (A1) 200
Saint John, (D3)_42,511
Saint Joseph, (E3). 400
Saint Leonards,(B1) 276
Saint Louis de Kent,
    (E2)_____ 400
Saint Martins, (D3)1,500
Saint Mary, (E2)__ 100
Saint Marys Ferry,
    (C3)_____ 900
Saint Paul, (D2)___ 600
Saint Stephen, (B3)2,836
Salisbury, (D2)____ 300
Second Falls, (C3). 200
Shediac, (E2)_____1,442
Sheffield, (C3)____ 350
Shippegan, (E1)___ 500
Southampton, (B3) 300
Springfield, (D3)__ 300
Springfield, (B2)__ 300
Stanley, (C3)_____ 100
Stonehaven, (D1).. 300
Sussex, (D3)_____1,906
Sussex Corner, (D3) 300
Tabusintac, (D1).. 600
Taymouth, (C2)___ 100
Tracadie, (E1)____1,500
Tracey Station,(C3) 250
Upham, (D3)_____ 200
Upper Gagetown,
    (C3)_____ 300
Waterford, (D3)__1,450
Waweig, (C3)_____ 100
Welsford, (C3)____ 200
West Branch St.
    Nicholas River,
    (E2)_____ 300
Woodstock, (B2)_3,856
Youngs Cove, (D3) 300

NEWFOUNDLAND

Alexander Bay, (L4) 100
Badger Brook, (K4) 40
Bay Of Islands,
    (H4)_____1,400
Bonaventure, (L5)_ 50
Bonavista, (L4)___2,000
Brigus, (L5)_____1,000

Carbonear, (L5)__3,000
Catalina, (L5)____1,000
Clarenville, (K5)__ 200
Codroy Pond, (H5) 500
Fortune, (J5)_____ 600
Gambo, (K4)_____ 600
Glenwood, (K4)___ 100

Grand Falls, (K4).. 500
Harbor Grace,(L5).5,000
Hearts Content,
    (L5)_____ 900
Keel, (L4)_____ 150
Lewisport, (K4)___ 100
Norris Arm, (K4).. 150

Placentia, (K5)___1,000
Port Basque, (H5)_ 800
Port Blandford,
    (K5)_____ 200
River Head, (J4)__ 150
St. Georges, (H5)_ 800

ST. JOHNS,(L5)_29,594
Shoal Harbor, (K5) 100
Stephenville, (H5)_ 100
Trinity, (L5)_____1,000
Twillingate, (K4)_3,500
Whitbourne, (L5).. 350

NOVA SCOTIA

AdvocateHarbor,(E3)850
Amherst, (E3)____8,973
Annapolis Royal,
    (D4)_____1,019
Antigonish, (H3)_1,787
Apple River, (E3).. 400
Arcadia, (C5)_____ 300
Argyle, (D5)_____ 125

Arichat, (H3)_____ 800
Ariasig, (G3)_____ 200
Athol, (E3)_____ 100
AtwoodsBrook, (D5) 200
Avonport, (E3)____ 220
Aylesford, (E3)___ 350
Baddeck, (J2)____1,650
Barrington, (D5).. 700

Baxters Harbor, (E3) 200
Bay St. Lawrence,
    (J1)_____ 400
Bear River (D4)..1,475
Bedford, (F4)_____ 600
Belleville, (D5)___ 300
Berwick (E3)_____1,000
Boylston, (H3)____ 300

Bridgeport, (K2)_1,200
Bridgetown, (D4)_ 996
Bridgewater, (E4).2,775
Brookfield, (F3)___ 400
Brooklyn, (E4)____ 300
Caledonia, (G3)__ 300
Caledonia, Queens,
    (D4)_____ 500

Cambridge Station,
    (E3)_____ 150
Canning, (E3)____1,500
Canso, (H3)_____1,617
Cape North, (J1).. 350
Cape Sable Island,
    (D5)_____ 200
Centreville (E3)__ 100

ONTARIO

Smiths Falls, (P3) _6,370
Smithville, (H6)____ 650
Snelgrove, (G5)____ 135
Sombra, (B7)_____ 500
Southampton, (D3)_1,685
South End, (H6)___ 125
South Gloucester, (Q2)_____ 120
South Indian, (Q2)_ 275
South Lancaster, (R2)_____ 130
South March, (P2)_ 115
South Mountain, (Q3)_____ 420
South River, (L20)_ *593
Southwold Station, (D7)_____ 140
South Woodslee, (A8)_____ 350
Sparta, (D7)_____ 425
Spencerville, (P3)_ 500
Sprague, (K17)____ *400
Springbrook, (L1)__ 225
Springfield, (E7)___ 454
Springford, (E7)___ 300
Springvale, (E7)___ 115
Spring Valley, (F4)_ 125
Spruce Dale, (H2)__ 150
Stamford, (H6)____ 310
Stanleyville, (O3)__ 100
Staples, (A8)_____ 308
Stayner, (F4)_____ 1,049
Steelton, (K16)___ *3,936
Stella, (N4)_____ 200
Stevensville, (B7)__ 350
Stirling, (L4)_____ 848
Stittville, (P2)____ 250
Stockdale, (L4)____ 110
Stony Creek, (G6)__ 530
Stony Point, (A8)__ 378
Stouffville, (H5)__1,014
Straffordville, (E7)_ 270
Stratford, (D6)__12,916

Strathcona, (N4)___ 370
Strathroy, (C7)___2,8.3
Streetsville, (G5)__ 543
Stromness, (G7)___ 100
Stroud, (G4)_____ 100
Sturgeon Bay, (G3)_ 175
Sturgeon Falls, (K20)_____*2,199
Sudbury, (K18)__*4,150
Sulphide, (M3)____ 125
Summerstown, (R2) 200
Summerstown Station, (R2)_____ 260
Sunbury, (O4)_____ 100
Sunderland, (H4)__ 650
Sundridge, (L20)__ *420
Sutton West, (H4)_ 753
Swansea, (H5)____ 500
Sydenham, (N4)___ 630
Talbotville Royal, (D7)_____ 100
Tamworth, (N4)___ 850
Tara, (D4)_____ 551
Tavistock, (E6)___ 981
Tecumseh, (A8)___ 300
Teeswater, (D4)___ 854
Teeterville, (E7)__ 200
Terra Cotta, (G5)_ 180
Thamesford, (L6)__ 500
Thamesville, (B7)__ 807
Thedford, (C6)____ 559
Thessalon, (K16)_*1,945
Thistletown, (G5)__ 100
Thornbrie, (H6)___ 550
Thornhill, (H5)___ 900
Thornton, (G4)___ 250
Thornbury, (F3)___ 793
Thorold, (B6)___2,273
Tilbury, (B8)___1,168
Tillsonburg, (E7)_2,758
Tincap, (P3)_____ 120
Tiverton, (C4)____ 340
Tobermory, (C2)__ 100

Todmorden, (H5)__ 400
Toledo, (O3)_____ 450
TORONTO, (H5)376,538
Tottenham, (G4)__ 517
Treadwell, (Q1)___ 130
Trenton, (L4)____3,988
Trout Creek, (K20) *400
Troy, (F6)_____ 110
Tupperville, (B7)__ 165
Tweed, (M4)____1,368
Tyrone, (J5)_____ 170
Udora, (H4)_____ 175
Uffington, (H3)___ 100
Underwood, (C4)__ 150
Union, (D7)_____ 250
Unionville, (H5)__ 470
Uptergrove, (H3)__ 120
Ursa, (K3)_____ 100
Utterson, (H2)____ 100
Uttoxeter, (C6)___ 115
Uxbridge, (H4)__1,433
Vanessa, (F7)_____ 220
Vankleek Hill, (R1)_1,577
Varna, (C5)_____ 135
Ventnor, (Q3)____ 150
Verner, (K19)____ *500
Vernon, (Q2)_____ 225
Verona, (N4)_____ 270
Victoria Harbour, (G3)_____1,616
Victoria Mines, (K18)_____ *600
Victoria Road, (H3) 260
Victoria Square, (H5)_____ 100
Vienna, (E7)_____ 332
Villanova, (F7)___ 120
Vineland, (H6)___ 230
Virgil, (H6)_____ 160
Vittoria, (F7)_____ 500
Vroomanton, (H4)_ 125
Waldemar, (F5)___ 175
Wales, (R2)_____ 100

Walkerton, (D4)_2,601
Walkerville, (A8)_3,302
Wallaceburg, (B7)_5,438
Wallacetown, (D7)_ 300
Walsh, (F7)_____ 120
Walsingham Centre, (E7)_____ 350
Walters Falls, (E4) 200
Walton, (D5)_____ 150
Wanstead, (B7)___ 110
Wardsville, (C7)__ 240
Warkworth, (L4)__ 700
Warren, (K19)___ *500
Warsaw, (K4)____ 175
Warwick, (C6)____ 165
Washago, (H3)___ 200
Waterdown, (G6)_ 756
Waterford, (F7)_1,083
Waterloo, (E6)__4,359
Watford, (C7)___1,092
Waubaushene, (G3) 800
Waverley, (G3)___ 120
Webbwood, (K18)_ *657
Welland, (H7)__5,318
Welland Port, (G6)_ 225
Wellesley, (E6)___ 650
Wellington, (M5)__ 785
Wellmans Corners, (L4)_____ 115
Wendover, (Q1)___ 220
Westboro, (P2)___ 550
Westbrook, (N4)__ 100
West Flamborough, (F6)_____ 325
West Gravenhurst, (H3)_____ 100
West Hill, (H5)___ 175
West Lorne, (C7)__ 740
Weston, (G5)___1,875
West Osgoode, (Q2) 130
Westover, (F6)___ 120
Westport, (O3)___ 803
Westwood, (K4)__ 135

Wheatley, (B8)___ 650
Whitby, (H5)___2,248
Whitechurch, (D5)_ 225
White Lake, (N2)__ 150
White River, (H14)_*500
Whitevale, (H5)___ 200
Whitney, (L21)__*2,000
Wiarton, (D3)__2,266
Wicklow, (L5)____ 350
Wikwemikong,(L18)*900
Wilberforce, (K2)__ 150
Wilkesport, (B7)__ 215
Williamsburg, (Q3)_ 350
Williamsford, (E4)_ 275
Williamstown, (R2) 640
Wilton, (N4)_____ 200
Winchester, (Q2)_1,143
Winchester Springs, (Q3)_____ 360
Windham Centre, (F7)_____ 265
Windsor, (A8)__17,829
Winger, (H7)_____ 120
Wingham, (D5)__2,238
Winona, (G6)____ 500
Winterbourne, (F5) 100
Wolfe Island, (O4)_ 550
Wolverton, (E6)__ 200
Woodbridge, (G5)_ 607
Woodham, (D6)___ 100
Woodslee, (A8)___ 100
Woodstock, (E6)_9,320
Woodville, (H4)__ 394
Wooler, (L4)_____ 325
Wroxeter, (D5)___ 366
Wyebridge, (G3)__ 275
Wyevale, (G3)____ 150
Wyoming, (B7)___ 569
Yarker, (N4)_____ 475
York, (G6)_____ 375
Youngs Point, (K4) 125
Zephyr, (H4)_____ 175
Zurich, (C6)_____ 700

PRINCE EDWARD ISLAND

Abrams Village, (E2)_____ 350
Alberton, (E2)____ 700
Bedeque, (F2)____ 300
Belfast, (G2)_____ 300
Breadalbane, (F2)_ 200
Caledonia, (G2)___ 200
Cape Traverse, (F2) 250
Cardigan, (G2)___ 250
Cavendish, (F2)___ 200
CHARLOTTETOWN, (F2)_____11,203

Coleman, (E2)____ 300
Conway Station, (F2)_____ 100
Crapaud, (F2)____ 400
Ellerslie, (F2)____ 300
Elmsdale, (E2)____ 200
Emerald, (F2)____ 150
Fairfield, (E2)____ 100
Flat River, (F3)___ 250
Freetown, (F2)___ 150
Gaspereaux, (G2)_ 100
Georgetown, (G2)_1,010

Hunter River, (F2)_ 250
Kensington, (F2)__ 500
Kildare, (E2)_____ 100
Mill River, (E2)__ 200
Miscouche, (E2)__ 100
Montague, (G2)_1,100
Montrose, (E2)___ 150
Mount Stewart, (G2) 300
Murray Harbor,(G3) 200
North Rustico (F2)_ 150
O'Leary Station,(E2) 300
Orwell, (G2)_____ 250

Pinette, (F2)_____ 400
Pisquid (G2)_____ 230
Richmond, (E2)__ 100
Rustico, (F2)_____ 150
St. Eleanors, (F2)_ 200
St. Louis, (E2)___ 200
St. Margarets, (G2) 150
St. Peters, (G2)__ 200
Souris, (G2)____1,089
Stanley Bridge, (F2)_____ 350

Suffolk Station, (F2)_____ 200
Summerside, (F2)_2,678
Tignish, (F2)____ 450
Tryon, (F2)_____ 250
Tyne Valley, (F2)_ 200
Valleyfield, (G2)__ 200
Victoria, (F2)____ 400
Wellington Station, (E2)_____ 150
Wood Island, (G3)_ 150

QUEBEC

Abbotsford, (J6)_1,000
Abenakis Springs, (J4)_____ 200
Abercorn, (J6)___ 250
Actonvale, (K5)_1,402
Adamsville, (J6)__ 375
Adstock, (M4)___ 350
Agnes, (N5)_____ 417
Amqui, (O17)___*1,070
Ancienne Lorette, (M3)_____1,100
Ange Gardien, (M4)_____1,450
Ange Gardien de Rouville, (J6)__ 400
Angers, (D5)_____ 450
Anjou, (H4)_____ 200
Armagh, (N3)___1,000
Arthabaska, (L4)_1,458
Arthurville, (N3)__ 300
Arundel, (E5)____ 250

Asbestos, (L5)__2,224
Ascot Corner, (L6)_ 770
Athelstan, (F6)__ 200
Ayers Cliff, (K6)_ 316
Aylmer (East), (C6)3,109
Aylwin, (B5)_____ 200
Baie St. Paul, (N2)_1,857
Baillargeon, (M3)_ 350
Barnston, (L6)___ 200
Batiscan, (K3)__1,100
Beaconsfield, (G6)_ 375
Beauce Junction, (M4)_____ 300
Beauceville Est, (N4)_____1,677
Beauceville West, (N4)_____ 300
Beauharnois, (G6)_2,015
Beaulac, (G4)____ 431
Beaulieu, (M3)___ 200
Beaumont, (N3)__ 350

Beauport, (M3)__3,000
Beaupre, (N2)____ 500
Beaurivage, (M4)_ 300
Becancour, (K4)__ 311
Bedford, (J6)___1,432
Beebe, (K6)_____ 808
Beebe Junction, (K6)_____ 200
Beliales Milis, (F4)_ 250
Belle Riviere, (F5)_ 400
Beloeil Station, (H5)_____ 800
Beloeil Village (H5)1,501
Berangar (J6)____ 200
Bergerville, (M3)_ 300
Bernier, (L4)_____ 628
Berthier, (H4)__1,350
Berthier (en bas', (N3)_____1,335
Berthier Junction (H4)_____ 225

Bienville, (M3)_1,004
Black Lake, (M4)_ ...
Blanche, (D5)____ 200
Bolduc, (N5)_____ 750
Bolton Centre, (K6)_____ 200
Bord a Plouffe, (G5)_____ 200
Bordeaux, (G5)__ 994
Boucherville, (H5)_1,097
Bouchette, (C4)__ 350
Breakeyville, (M3)_ 700
Brigham, (J6)____ 300
Brome, (K6)_____ 548
Bromptonville, (L6)1,239
Brookdale, (E5)__ 600
Broughton Station, (M4)_____ 250
Brownsburg, (F5)_ 525
Bryson, (A5)_____ 477
Buckingham, (D5)3,854

Buckland, (N3)___ 500
Bury, (M6)_____ 600
Calumet, (E5)____ 650
Calumet Island, (A5)_____ 400
Canrobert, (J6)__ 229
Capelton, (L6)___ 600
Cap Magdeleine, (K4)_____1,750
Cap St. Ignace, (O2)_____2,650
Cap St. Ignace Station, (O2)_____ 200
Cap sante, (L3)__ 700
Capucins, (O18)__ *350
Cartierville, (G6)_ 905
Cascades Point, (F6)_____ 200
Caughnawaga, (G6)2,000
Causapscal, (O18)*1,000
Cedars, (F6)_____ 500

PRINCE EDWARD ISLAND

QUEBEC

SASKATCHEWAN

POPULATION OF THE UNITED STATES

The thirteenth census of the United States was taken by the Bureau of the Census as of April 15, 1910. The total area enumerated includes continental United States, the territories of Alaska and Hawaii, and Porto Rico. The enumeration also includes persons stationed abroad in the military and naval service of the Government (including civilian employees, etc.), who were specially enumerated through the cooperation of the War and Navy Departments.

The following table gives the total population for the area enumerated in 1910. The corresponding census figures for 1900 are also given for purposes of comparison.

Area	1910	1900
The United States (total area of enumeration)..........	93,402,151	*77,256,636
Continental United States.......	91,972,266	75,994,575
Noncontiguous territory.........	1,429,885	1,262,055
Alaska.....................	64,356	63,592
Hawaii.....................	191,909	154,001
Porto Rico.................	1,118,012	†953,243
Military and naval service stationed abroad.............	55,608	91,219

*Includes 953,243 persons enumerated in Porto Rico in 1899.
†According to the census of Porto Rico taken in 1899 under the direction of the War Department.

The rate of increase from 1900 to 1910 was 20.9 per cent. for the total area of enumeration and 21 per cent. for continental United States. It should be noted that this table does not cover all the outlying possessions of the United States. Including the population of the Philippines and other possessions, the population living under the American flag is approximately as follows:

Population of the United States and possessions.................................	101,100,000
Enumerated at the census of 1910..............	93,402,151
Philippine Islands, 1903........................	*7,635,426
Guam, estimated.............................	9,000
Samoa, estimated............................	6,100
Panama Canal Zone, estimated................	50,000

*The census of 1911 for Philippines gives 8,368,427.
†The census of 1911 for Panama Canal Zone gives 154,255.

The population of continental United States is 91,972,266. Compared with the population of 75,994,575 in 1900 this represents an increase during the past decade of 15,977,691, or 21 per cent. The rate of increase was slightly greater than during the preceding decade, 1890–1900, when it was 20.7 per cent.

AREA AND POPULATION OF CONTINENTAL U. S. SINCE FIRST CENSUS

Census Year	Gross Area Square Miles	Population	Increase Over Preceding Census		Adjusted percentages Increase
			Number	Per cent.	
1910.....	3,026,789	91,972,266	15,977,691	21.0	21.0
1900.....	3,026,789	75,994,575	13,046,861	20.7	20.7
1890.....	3,026,789	62,947,714	12,791,931	25.5	24.9
1880.....	3,026,789	50,155,783	11,597,412	30.1	26.0
1870.....	3,026,789	38,558,371	7,115,050	22.6	26.6
1860.....	3,026,789	31,443,321	8,251,445	35.6	35.6
1850.....	2,997,119	23,191,876	6,122,423	35.9	35.9
1840.....	1,792,223	17,069,453	4,203,433	32.7	32.7
1830.....	1,792,223	12,866,020	3,227,567	33.5	33.5
1820.....	1,792,223	9,638,453	2,398,572	33.1	33.1
1810.....	1,720,122	7,239,881	1,931,398	36.4	36.4
1800.....	892,135	5,308,485	1,379,269	35.1	35.1
1790.....	892,135	3,929,214	....	....	....

RANK OF TWENTY-FIVE LARGEST CITIES

Cities	Population		
	1910	1900	1890
New York, N. Y.......	1 4,766,883	1 3,437,202	1 2,507,414
Chicago, Ill..........	2 2,185,283	2 1,698,575	2 1,099,850
Philadelphia, Pa.....	3 1,549,008	3 1,293,697	3 1,046,964
St. Louis, Mo........	4 687,029	4 575,238	4 451,770
Boston, Mass.........	5 670,585	5 560,892	5 448,477
Cleveland, O.........	6 560,663	7 381,768	9 261,353
Baltimore, Md........	7 558,485	6 508,957	6 434,439
Pittsburgh, Pa.......	8 533,905	11 321,616	12 238,617
Detroit, Mich........	9 465,766	13 285,704	14 205,876
Buffalo, N. Y........	10 423,715	8 352,387	10 255,664
San Francisco, Cal....	11 416,912	9 342,782	7 298,997
Milwaukee, Wis.......	12 373,857	14 285,315	15 204,468
Cincinnati, O........	13 363,591	10 325,902	8 296,908
Newark, N. J........	14 347,469	16 246,070	16 181,830
New Orleans, La......	15 339,075	12 287,104	11 242,039
Washington, D. C....	16 331,069	15 278,718	13 230,392
Los Angeles, Cal.....	17 319,198	36 102,479	56 50,395
Minneapolis, Minn....	18 301,408	19 202,718	17 164,738
Jersey City, N. J.....	19 267,779	17 206,433	18 163,003
Kansas City, Mo......	20 248,381	22 163,752	23 132,716
Seattle, Wash........	21 237,194	48 80,671	69 42,837
Indianapolis, Ind....	22 233,650	21 169,164	26 105,436
Providence, R. I......	23 224,326	20 175,597	24 132,146
Louisville, Ky.......	24 223,928	18 204,731	19 161,129
Rochester, N. Y......	25 218,149	24 162,608	21 133,896

CITIES OF FASTEST GROWTH, 1900 TO 1910

Rank.	City	Population 1910	Pr. ct. inc. 1900–1910
1.	Oklahoma City, Okla...........	64,205	539.7
2.	Muskogee, Okla................	25,278	494.2
3.	Birmingham, Ala..............	132,685	245.4
4.	Pasadena, Cal.................	30,291	232.2
5.	Los Angeles, Cal.............	319,198	211.5
6.	Berkeley, Cal.................	40,434	206.0
7.	Flint, Mich...................	38,550	194.2
8.	Seattle, Wash.................	237,194	194.0
9.	Spokane, Wash................	104,402	183.3
10.	Fort Worth, Texas.............	73,312	174.7
11.	Huntington, W. Va............	31,161	161.4
12.	El Paso, Tex..................	39,279	146.9
13.	Tampa, Fla...................	37,782	138.5
14.	Schenectady, N. Y.............	72,826	129.9
15.	Portland, Ore.................	207,214	129.2
16.	Oakland, Cal.................	150,174	124.3
17.	San Diego, Cal...............	39,578	123.6
18.	Tacoma, Wash................	83,743	122.0
19.	Dallas, Tex...................	92,104	116.0
20.	Wichita, Kan.................	52,450	112.6
21.	Waterloo, Iowa...............	26,693	112.2
22.	Jacksonville, Fla..............	57,699	103.0

AREA OF UNITED STATES

Accession	Gross area (sq. mi.)	Accession	Gross area (sq. mi.)
CONTINENTAL U. S.	3,026,789	OUTLYING POSSESSIONS	716,517
Area of U.S. in 1790[1]	892,135	Alaska, 1867.....	590,884
Louisiana Pur., 1803	827,987	Hawaii, 1898.....	6,449
Florida, 1819.......	58,666	Philippines, 1899 .	115,026
Territory gained by Treaty with Spain, 1819.................	13,435	Porto Rico, 1899..	3,435
		Guam, 1899......	210
Texas, 1845........	389,166	Samoa, 1900......	77
Oregon, 1846.......	286,541	Panama Canal Zone, 1904.....	436
Mex. Cession, 1848 .	529,189		
Gadsden Pur., 1853.	29,670		

[1] Includes the drainage basin of the Red River of the North, not a part of any acquisition, but previously considered a part of the Louisiana Purchase.

POPULATION OF STATES AND TERRITORIES

States and Capitals	Population			Per Cent of Increase		Density per sq. mile
	1910	1900	1890	1900 to 1910	f 1890 to 1900	1910
Alabama..........Montgomery....	2,138,093	1,828,697	1,513,017	16.9	20.9	41.7
Arizona...........Phoenix	204,354	122,931	59,620	66.2	106.2	1.8
Arkansas.........Little Rock.....	1,574,449	1,311,564	1,128,179	20.0	16.3	30.0
California........Sacramento.....	2,377,549	1,485,053	1,203,130	60.1	22.4	15.3
Colorado.........Denver..........	799,024	539,700	412,198	48.0	30.9	7.7
Connecticut.....Hartford........	1,114,756	908,420	746,258	22.7	21.7	231.3
Delaware........Dover..........	202,322	184,735	168,493	9.5	9.6	103.0
Dist. of Columbia..........	331,069	278,718	230,392	18.8	21.0	5,517.8
Florida...........Tallahassee.....	752,619	528,542	391,422	42.4	35.0	13.7
Georgia..........Atlanta.........	2,609,121	2,216,331	1,837,353	17.7	20.6	44.4
Idaho............Boise...........	325,594	161,772	84,385	101.3	91.7	3.9
Illinois...........Springfield.....	5,638,591	4,821,550	3,826,351	16.9	26.0	100.6
Indiana..........Indianapolis....	2,700,876	2,516,462	2,192,404	7.3	14.8	74.9
Iowa............Des Moines.....	2,224,771	2,231,853	1,911,896	a0.3	16.7	40.0
Kansas..........Topeka.........	1,690,949	1,470,495	1,427,096	15.0	3.0	20.7
Kentucky........Frankfort.......	2,289,905	2,147,174	1,858,635	6.6	15.5	57.0
Louisiana........Baton Rouge...	1,656,388	1,381,625	1,118,587	19.9	23.5	36.5
Maine...........Augusta........	742,371	694,466	661,086	6.9	5.0	24.8
Maryland........Annapolis......	1,295,346	1,188,044	1,042,390	9.0	14.0	130.3
Massachusetts.....Boston.........	3,366,416	2,805,346	2,238,943	20.0	25.3	418.8
Michigan.........Lansing........	2,810,173	2,420,982	2,093,889	16.1	15.6	48.9
Minnesota.......St. Paul........	2,075,708	1,731,394	1,301,826	18.5	34.5	25.7
Mississippi.......Jackson........	1,797,114	1,551,270	1,289,600	15.8	20.3	38.8
Missouri.........Jefferson City..	3,293,335	3,106,665	2,679,184	6.0	16.0	47.9
Montana........Helena.........	376,053	243,329	132,159	54.5	84.1	2.6
Nebraska.......Lincoln........	1,192,214	1,066,300	1,058,910	11.8	0.7	15.5
Nevada.........Carson City....	81,875	42,335	45,761	93.4	a7.5	.7
New Hampshire...Concord.......	430,572	411,588	376,530	4.6	9.3	47.7
New Jersey......Trenton........	2,537,167	1,883,669	1,444,933	34.7	30.4	337.7
New Mexico.....Santa Fe.......	327,301	195,310	153,593	67.6	27.2	2.7
New York.......Albany........	9,113,614	7,268,894	5,997,853	25.4	21.2	191.2
North Carolina.....Raleigh........	2,206,287	1,893,810	1,617,947	16.5	17.1	45.3
North Dakota.....Bismarck.......	577,056	319,146	182,719	80.8	74.7	8.2
Ohio............Columbus......	4,767,121	4,157,545	3,672,316	14.7	13.2	117.0
Oklahoma.......Oklahoma City..	1,657,155	b790,391	b258,657	b109.7	b205.6	23.9
Oregon..........Salem..........	672,765	413,536	313,767	62.7	31.8	7.0
Pennsylvania.....Harrisburg.....	7,665,111	6,302,115	5,258,014	21.6	19.9	171.0
Rhode Island.....Providence.....	542,610	428,556	345,506	26.6	24.0	508.5
South Carolina.....Columbia.......	1,515,400	1,340,316	1,151,149	13.1	16.4	49.7
South Dakota.....Pierre..........	583,888	401,570	328,808	45.4	22.1	7.6
Tennessee........Nashville.......	2,184,789	2,020,616	1,767,518	8.1	14.3	52.4
Texas............Austin.........	3,896,542	3,048,710	2,235,523	27.8	36.4	14.8
Utah............Salt Lake City..	373,351	276,749	207,905	34.9	33.1	4.5
Vermont.........Montpelier.....	355,956	343,641	332,422	3.6	3.4	39.0
Virginia..........Richmond......	2,061,612	1,854,184	1,655,980	11.2	12.0	51.2
Washington......Olympia........	1,141,990	518,103	349,390	120.4	48.3	17.1
West Virginia.....Charleston......	1,221,119	958,800	762,794	27.4	25.7	50.8
Wisconsin........Madison........	2,333,860	2,069,042	1,686,880	12.8	22.7	42.2
Wyoming.........Cheyenne......	145,965	92,531	60,705	57.7	52.4	1.5
Continental U. S................	91,972,266	75,994,575	e62,622,250	21.0	20.7	....
Alaska..........Juneau.........	64,356	63,592	32,052	1.2	98.4	.1
Hawaii..........Honolulu........	191,909	154,001	89,990	24.6	71.1	29.8
Porto Rico.....San Juan.......	1,118,012	c 953,243	....	....	....	325.5
Military and Naval............	55,608	91,219				
U. S. including dependencies named above......................	93,402,151	76,303,387	e 63,069,756	20.9	(d)	30.9

(a) Decrease. (b) For purposes of comparison the 1900 population figures of Oklahoma and Indian Territory are combined. (c) 1899. (d) In the last line of this table the 1900 and 1890 population figures do not include Porto Rico. (e) Includes population (325,464) of Indian Territory and Indian reservations specially enumerated in 1890 but not included in the general report on population in 1890 (f) The percentages in this column are figured on the basis of the actual county totals in each State exclusive of Indian reservations.

NUMBER OF MALES OF VOTING AGE

The 1910 census for continental United States gives the males of voting age as 26,999,151, and constituting 29.4 per cent. of the entire population (91,972,266). They are divided as follows: Native whites of native parentage, 13,211,731, or 48.9 per cent.; native whites of foreign or mixed parentage, 4,498,966, or 16.7 per cent.; foreign-born whites, naturalized, 3,035,333, or 11.2 per cent.; foreign-born whites, not naturalized, 3,611,273, or 13.4 per cent.; negroes, 2,459,327, or 9.1 per cent.

POPULATION, 1910, BY SEX, RACE AND NATIVITY

State	Sex		Race, Nativity, and Parentage							
	Male	Female	Native white of native parentage	Native white of foreign parentage	Foreign white	Negro	Indian	Chinese	Japanese	All other
Alabama	1,074,209	1,063,884	1,177,457	32,438	18,946	908,275	909	61	3	4
Arizona	118,582	85,772	82,480	42,175	46,844	2,067	29,201	1,236	351	......
Arkansas	810,025	764,424	1,077,509	36,608	16,913	442,891	460	59	8	1
California	1,322,973	1,054,576	1,106,533	635,970	517,319	21,645	16,371	36,197	41,324	2,190
Colorado	430,697	368,327	475,136	181,432	126,971	11,453	1,482	360	2,190	......
Connecticut	563,641	551,115	395,649	374,546	328,737	15,174	152	427	71	......
Delaware	103,435	98,887	127,809	25,873	17,421	31,181	5	29	4	......
Dist. of Col.	158,050	173,019	166,711	45,066	24,351	94,446	68	369	47	11
Florida	394,166	358,453	373,967	35,828	33,851	308,669	74	184	45	1
Georgia	1,305,019	1,304,102	1,391,058	25,677	15,081	1,176,987	95	219	4	......
Idaho	185,546	140,048	203,604	75,254	40,444	646	3,488	838	1,308	12
Illinois	2,911,653	2,726,938	2,600,565	1,724,489	1,201,928	109,041	188	2,104	276	......
Indiana	1,383,299	1,317,577	2,140,168	350,747	159,118	60,280	279	249	35	......
Iowa	1,148,171	1,076,600	1,303,526	632,182	273,388	15,078	471	93	30	3
Kansas	885,912	805,037	1,207,087	292,077	134,719	54,504	2,444	15	103	......
Kentucky	1,161,709	1,128,196	1,863,157	124,775	40,023	261,656	234	50	10	......
Louisiana	835,275	821,113	776,569	112,728	51,828	713,874	780	493	31	85
Maine	377,053	365,118	494,918	135,188	109,911	1,364	892	90	8	......
Maryland	644,225	651,121	766,628	191,841	104,176	232,249	55	374	23	......
Massachusetts	1,655,226	1,711,190	1,104,361	1,370,793	1,050,899	38,042	688	2,493	140	......
Michigan	1,454,534	1,355,639	1,224,841	965,217	595,200	17,115	7,519	239	40	2
Minnesota	1,108,511	967,197	575,081	941,415	542,857	7,084	9,053	250	66	2
Mississippi	905,761	891,153	757,233	19,495	9,391	1,009,487	1,253	249	2	4
Missouri	1,687,838	1,605,497	2,387,909	518,141	228,695	157,452	313	532	91	2
Montana	226,866	149,187	162,129	106,811	91,647	1,834	10,745	1,276	1,593	18
Nebraska	627,782	564,432	642,075	362,153	175,883	7,689	3,502	109	574	29
Nevada	52,551	29,524	35,413	20,956	18,102	513	5,240	900	839	12
New Hampshire	216,290	214,282	230,231	103,118	96,560	564	34	64	1	......
New Jersey	1,286,463	1,230,704	1,009,009	777,859	658,159	89,760	168	1,109	203	......
New Mexico	175,245	152,056	255,609	26,431	22,662	1,628	20,573	246	252	......
New York	4,584,581	4,529,033	3,230,154	3,007,507	2,729,260	134,181	6,046	5,235	1,217	14
North Carolina	1,098,417	1,107,616	1,485,705	8,855	5,953	697,843	7,851	78	2	......
North Dakota	317,554	259,502	162,461	251,256	156,138	617	6,486	39	59	......
Ohio	2,434,765	2,312,156	3,033,215	1,024,177	597,255	111,443	127	574	70	......
Oklahoma	881,573	775,582	1,310,403	94,044	40,088	137,612	74,825	137	46	......
Oregon	384,255	288,510	416,851	135,241	103,002	1,519	5,090	7,359	3,418	285
Pennsylvania	3,942,137	3,712,974	4,222,016	1,806,392	1,438,752	193,908	1,503	1,749	189	2
Rhode Island	270,359	272,251	159,821	194,646	178,031	9,529	284	266	33	......
South Carolina	731,842	763,558	661,970	11,138	6,054	835,843	331	56	8	......
South Dakota	317,101	266,787	245,665	217,478	100,628	817	19,137	120	43	......
Tennessee	1,103,491	1,081,298	1,654,606	38,367	18,460	473,088	216	43	8	1
Texas	2,017,612	1,878,930	2,602,958	361,926	240,012	690,020	702	575	341	8
Utah	196,857	176,394	171,671	131,527	63,404	1,143	3,123	373	2,105	5
Vermont	182,568	173,188	229,182	75,055	49,861	1,621	26	8	3	......
Virginia	1,035,348	1,036,264	1,325,238	37,943	26,628	671,096	539	154	14	......
Washington	658,650	483,340	585,401	282,529	241,227	6,058	10,997	2,706	12,886	186
West Virginia	644,044	577,075	1,042,107	57,638	57,072	64,173	36	90	3	......
Wisconsin	1,208,541	1,125,119	763,224	1,044,764	512,569	2,900	10,142	224	34	3
Wyoming	91,666	54,299	80,711	32,497	27,165	2,235	1,486	244	1,971	56
Total	47,332,122	44,640,144	49,488,441	18,900,663	13,343,583	9,828,294	265,683	70,944	71,722	2,936

GROWTH OF JEWISH POPULATION

According to a recent report of the Hebrew Sheltering and Immigrant Aid Society 1,496 of Jewish immigrants arriving at the port of New York in 1911 were on their way to points west of the Mississippi. It is to be observed, however, that no less than 475 gave their destination as Missouri—meaning, no doubt, chiefly St. Louis. The entire number of those who were going West, however, was only 4.25 per cent. of the total Jewish immigration of the year.

The Society has published a chart showing facts with regard to the Jewish population of 81 cities in 24 States. The growth of population in ten years in these 81 cities varied from 100 to 200 per cent. The increase of Jewish population in that period was anywhere from 200 to 800 per cent. Sixty-one cities showed marked increase of Jewish residents, only 26 showed no notable gain.

The table showed that while Duluth had gained 300 per cent., Atlanta 100 per cent., Sioux City 700 per cent., Seattle 400 per cent., Wichita, Kan., 500 per cent., Grand Forks, N. D., 800 per cent.; Portland, Ore, 200 per cent.; Richmond, Va., had gained only 30 per cent., and Charleston, S. C., only 50 per cent.

STATES BY COUNTIES

ALABAMA

COUNTY	1910	1900	1890
Autauga	20,038	17,915	13,330
Baldwin	18,178	13,194	8,941
Barbour	32,728	35,152	34,898
Bibb	22,791	18,498	13,824
Blount	21,456	23,119	21,927
Bullock	30,196	31,944	27,063
Butler	29,030	25,761	21,641
Calhoun	39,115	34,874	33,835
Chambers	36,056	32,554	26,319
Cherokee	20,226	21,096	20,459
Chilton	23,187	16,522	14,549
Choctaw	18,483	18,136	17,526
Clarke	30,987	27,790	22,624
Clay	21,006	17,099	15,765
Cleburne	13,385	13,206	13,218
Coffee	26,119	20,972	12,170
Colbert	24,802	22,341	20,189
Conecuh	21,433	17,514	14,594
Coosa	16,634	16,144	15,906
Covington	32,124	15,346	7,536
Crenshaw	23,313	19,668	15,425
Cullman	28,321	17,849	13,439
Dale	21,873	21,189	17,225
Dallas	53,401	54,657	49,350
Dekalb	28,261	23,558	21,106
Elmore	28,245	26,099	21,732
Escambia	18,889	11,320	8,666
Etowah	39,109	27,361	21,926
Fayette	16,248	14,132	12,823
Franklin	19,369	16,511	10,681
Geneva	26,230	19,096	10,690
Greene	22,717	24,182	22,007
Hale	27,883	31,011	27,501
Henry	20,943	36,147	24,847
Houston	32,414	......	......
Jackson	32,918	30,508	28,026
Jefferson	226,476	140,420	88,501
Lamar	17,487	16,084	14,187
Lauderdale	30,936	26,559	23,739
Lawrence	21,984	20,124	20,725
Lee	32,867	31,826	28,694
Limestone	26,880	22,387	21,201
Lowndes	31,894	35,651	31,550
Macon	26,049	23,126	18,439
Madison	47,041	43,702	38,119
Marengo	39,923	38,315	33,095
Marion	17,495	14,494	11,347
Marshall	28,553	23,289	18,935
Mobile	80,854	62,740	51,587
Monroe	27,155	23,666	18,990
Montgomery	82,178	72,047	56,172
Morgan	33,781	28,820	24,089
Perry	31,222	31,783	29,332
Pickens	25,055	24,402	22,470
Pike	30,815	29,172	24,423
Randolph	24,659	21,647	17,219
Russell	25,937	27,083	24,093
St. Clair	20,715	19,425	17,353
Shelby	26,949	23,684	20,886
Sumter	28,699	32,710	29,574
Talladega	37,921	35,773	29,346
Tallapoosa	31,034	29,675	25,460
Tuscaloosa	47,559	36,147	30,352
Walker	37,013	25,162	16,078
Washington	14,454	11,134	7,935
Wilcox	33,810	35,631	30,816
Winston	12,855	9,554	6,552

The State 2,138,093 1,828,697 1,513,017

ARIZONA

COUNTY	1910	1900	1890
Apache	9,196	8,297	4,281
Cochise	34,591	9,251	6,938
Coconino	8,130	5,514	......
Gila	16,780	4,973	2,021
Graham	23,547	14,162	5,670
Maricopa	34,488	20,457	10,986
Mohave	3,773	3,426	1,444
Navajo	11,491	8,829	......
Pima	22,818	14,089	12,673
Pinal	9,045	7,779	4,251
Santa Cruz	6,766	4,545	......
Yavapai	15,996	13,799	8,685
Yuma	7,733	4,145	2,671
San Carlos Indian Res.	...	a3,065	......

The Territory, 204,354 122,931 59,620

(a) In Gila, Graham, and Navajo counties, but the population in each county was not separately returned in 1900.

ARKANSAS

COUNTY	1910	1900	1890
Arkansas	16,103	12,973	11,432
Ashley	25,268	19,734	13,295
Baxter	10,389	9,298	8,527
Benton	33,389	31,611	27,716
Boone	14,318	16,396	15,816
Bradley	14,518	9,651	7,972
Calhoun	9,894	8,539	7,267
Carroll	16,829	18,848	17,288
Chicot	21,987	14,528	11,419
Clark	23,686	21,289	20,997
Clay	23,690	15,886	12,200
Cleburne	11,903	9,628	7,884
Cleveland	13,481	11,620	11,362
Columbia	23,820	22,077	19,893
Conway	22,729	19,772	19,459
Craighead	27,627	19,505	12,025
Crawford	23,942	21,270	21,714
Crittenden	22,447	14,529	13,940
Cross	14,042	11,051	7,693
Dallas	12,621	11,518	9,296
Desha	15,274	11,511	10,324
Drew	21,960	19,451	17,352
Faulkner	23,708	20,780	18,342
Franklin	20,638	17,395	19,934
Fulton	12,193	12,917	10,984
Garland	27,271	18,773	15,328
Grant	9,425	7,671	7,786
Greene	23,852	16,979	12,908
Hempstead	28,285	24,101	22,796
Hot Spring	15,022	12,748	11,603
Howard	16,898	14,076	13,789
Independence	24,776	22,557	21,961
Izard	14,561	13,506	13,038
Jackson	23,501	18,383	15,179
Jefferson	52,734	40,972	40,881
Johnson	19,698	17,448	16,758
Lafayette	13,741	10,594	7,700
Lawrence	20,001	16,491	12,984
Lee	24,252	19,409	18,886
Lincoln	15,118	13,389	10,255
Little River	13,597	13,731	8,903
Logan	26,350	20,563	20,774
Lonoke	27,983	22,544	19,263
Madison	16,056	19,864	17,402

ARKANSAS Cont'd

COUNTY	1910	1900	1890
Marion	10,203	11,377	10,390
Miller	19,555	17,558	14,714
Mississippi	30,468	16,384	11,635
Monroe	19,907	16,816	15,336
Montgomery	12,455	9,444	7,923
Nevada	19,344	16,609	14,832
Newton	10,612	12,538	9,950
Ouachita	21,774	20,892	17,033
Perry	9,402	7,294	5,538
Phillips	33,535	26,561	25,341
Pike	12,565	10,301	8,537
Poinsett	12,791	7,025	4,272
Polk	17,216	18,352	9,283
Pope	24,527	21,715	19,458
Prairie	13,853	11,875	11,374
Pulaski	86,751	63,179	47,329
Randolph	18,987	17,156	14,485
St. Francis	22,548	17,157	13,543
Saline	16,657	13,122	11,311
Scott	14,302	13,183	12,635
Searcy	14,825	11,988	9,664
Sebastian	52,278	36,935	33,200
Sevier	16,616	16,339	10,072
Sharp	11,688	12,199	10,418
Stone	8,946	8,100	7,043
Union	30,723	22,495	14,977
Van Buren	13,509	11,220	8,567
Washington	33,889	34,256	32,024
White	28,574	24,864	22,946
Woodruff	20,049	16,304	14,009
Yell	26,323	22,750	18,015

The State 1,574,449 1,311,564 1,128,179

CALIFORNIA

COUNTY	1910	1900	1890
Alameda	246,131	130,197	93,864
Alpine	309	509	667
Amador	9,086	11,116	10,320
Butte	27,301	17,117	17,939
Calaveras	9,171	11,200	8,882
Colusa	7,732	7,364	14,640
Contra Costa	31,674	18,046	13,515
Del Norte	2,417	2,408	2,592
Eldorado	7,492	8,986	9,232
Fresno	75,657	37,862	32,026
Glenn	7,172	5,150	......
Humboldt	33,857	27,104	23,469
Imperial	13,591		
Inyo	6,974	4,377	3,544
Kern	37,715	16,480	9,808
Kings	16,230	9,871	......
Lake	5,526	6,017	7,101
Lassen	4,802	4,511	4,239
Los Angeles	504,131	170,298	101,454
Madera	8,368	6,364	......
Marin	25,114	15,702	13,072
Mariposa	3,956	4,720	3,787
Mendocino	23,929	20,465	17,612
Merced	15,148	9,215	8,085
Modoc	6,191	5,076	4,986
Mono	2,042	2,167	2,002
Monterey	24,146	19,380	18,637
Napa	19,800	16,451	16,411
Nevada	14,955	17,789	17,369

CALIFORNIA Cont'd.

COUNTY	1910	1900	1890
Orange	34,436	19,696	13,589
Placer	18,237	15,786	15,101
Plumas	5,259	4,657	4,933
Riverside	34,696	17,897	.....
Sacramento	67,806	45,915	40,339
San Benito	8,041	6,633	6,412
San Bernardino	56,706	27,929	25,497
San Diego	61,665	35,090	34,987
San Francisco	416,912	342,782	298,997
San Joaquin	50,731	35,452	28,629
San Luis Obispo	19,383	16,637	16,072
San Mateo	26,585	12,094	10,087
Santa Barbara	27,738	18,934	15,754
Santa Clara	83,539	60,216	48,005
Santa Cruz	26,140	21,512	19,270
Shasta	18,920	17,318	12,133
Sierra	4,098	4,017	5,051
Siskiyou	18,801	16,962	12,163
Solano	27,559	24,143	20,946
Sonoma	48,394	38,480	32,721
Stanislaus	22,522	9,550	10,040
Sutter	6,328	5,886	5,469
Tehama	11,401	10,996	9,916
Trinity	3,301	4,383	3,719
Tulare	35,440	18,375	24,574
Tuolumne	9,979	11,166	6,082
Ventura	18,347	14,367	10,071
Yolo	13,926	13,618	12,684
Yuba	10,042	8,620	9,636

The State 2,377,549 1,485,053 1,208,130

COLORADO

COUNTY	1910	1900	1890
Adams	8,892	.....	.....
Arapahoe	10,263	153,017	132,135
Archuleta	3,302	2,117	826
Baca	2,516	759	1,479
Bent	5,043	3,049	1,313
Boulder	30,330	21,544	14,082
Chaffee	7,622	7,085	6,612
Cheyenne	3,687	501	534
Clear Creek	5,001	7,082	7,184
Conejos	11,285	8,794	7,193
Costilla	5,498	4,632	3,491
Custer	1,947	2,937	2,970
Delta	13,688	5,487	2,534
Denver	213,381	.....	.....
Dolores	642	1,134	1,498
Douglas	3,192	3,120	3,006
Eagle	2,985	3,008	3,725
El Paso	43,321	31,602	21,239
Elbert	5,331	3,101	1,856
Fremont	18,181	15,636	9,156
Garfield	10,144	5,835	4,478
Gilpin	4,131	6,690	5,867
Grand	1,862	741	604
Gunnison	5,897	5,331	4,359
Hinsdale	646	1,609	862
Huerfano	13,320	8,395	6,882
Jackson	1,013	.....	.....
Jefferson	14,231	9,306	8,450
Kiowa	2,899	701	1,243
Kit Carson	7,483	1,580	2,472
La Plata	10,812	7,016	5,509
Lake	10,600	18,054	14,663
Larimer	25,270	12,168	9,712
Las Animas	33,643	21,842	17,208
Lincoln	5,917	926	689
Logan	9,549	3,292	3,070
Mesa	22,197	9,267	4,260
Mineral	1,239	1,913	.....
Montezuma	5,029	3,058	1,529

COLORADO Cont'd.

COUNTY	1910	1900	1890
Montrose	10,291	4,535	3,980
Morgan	9,577	3,268	1,601
Otero	20,201	11,522	4,192
Ouray	3,514	4,731	6,510
Park	2,492	2,998	3,548
Phillips	3,179	1,583	2,642
Pitkin	4,566	7,020	8,929
Prowers	9,520	3,766	1,969
Pueblo	52,223	34,448	31,491
Rio Blanco	2,332	1,690	1,200
Rio Grande	6,563	4,080	3,451
Routt	7,561	3,661	2,369
Saguache	4,160	3,853	3,313
San Juan	3,063	2,342	1,572
San Miguel	4,700	5,379	2,909
Sedgwick	3,061	971	1,293
Summit	2,003	2,744	1,906
Teller	14,351	29,002	.....
Washington	6,002	1,241	2,301
Weld	39,177	16,808	11,736
Yuma	8,499	1,729	2,596

The State 799,024 539,700 412,198

CONNECTICUT

COUNTY	1910	1900	1890
Fairfield	245,322	184,203	150,081
Hartford	250,182	195,480	147,180
Litchfield	70,260	63,672	53,542
Middlesex	45,637	41,760	39,524
New Haven	337,282	269,163	209,058
New London	91,253	82,758	76,634
Tolland	26,459	24,523	25,081
Windham	48,361	46,861	45,158

The State 1,114,756 908,420 746,258

DELAWARE

COUNTY	1910	1900	1890
Kent	32,721	32,762	32,664
Newcastle	123,188	109,697	97,182
Sussex	46,413	42,276	38,647

The State 202,322 184,735 168,493

DISTRICT OF COLUMBIA

1910	1900	1890
331,069	278,718	230,392

FLORIDA

COUNTY	1910	1900	1890
Alachua	34,305	32,245	22,934
Baker	4,805	4,516	3,333
Bradford	14,090	10,295	7,516
Brevard	4,717	5,158	3,401
Calhoun	7,465	5,132	1,681
Citrus	6,731	5,391	2,394
Clay	6,116	5,635	5,154
Columbia	17,689	17,094	12,877
Dade	11,933	4,955	861
De Soto	14,200	8,047	4,944
Duval	75,163	39,733	26,800
Escambia	36,549	28,313	20,188
Franklin	5,201	4,890	3,308

FLORIDA Cont'd.

COUNTY	1910	1900	1890
Gadsden	22,198	15,294	11,894
Hamilton	11,825	11,881	8,507
Hernando	4,997	3,638	2,476
Hillsboro	78,374	36,013	14,941
Holmes	11,557	7,762	4,336
Jackson	29,821	23,377	17,544
Jefferson	17,210	16,195	15,757
Lafayette	6,710	4,987	3,686
Lake	9,509	7,467	8,034
Lee	6,294	3,071	1,414
Leon	19,427	19,887	17,752
Levy	10,361	8,603	6,586
Liberty	4,700	2,956	1,452
Madison	16,919	15,446	14,316
Manatee	9,550	4,663	2,895
Marion	26,941	24,403	20,796
Monroe	21,563	18,006	18,786
Nassau	10,525	9,654	8,294
Orange	19,107	11,374	12,584
Osceola	5,507	3,444	3,133
Palm Beach	5,577	.....	.....
Pasco	7,502	6,054	4,249
Polk	24,148	12,472	7,905
Putnam	13,096	11,641	11,186
St. John	13,208	9,165	8,712
St. Lucie	4,075	.....	.....
Santa Rosa	14,097	10,293	7,961
Sumter	6,696	6,187	5,363
Suwanee	18,603	14,554	10,524
Taylor	7,103	3,999	2,122
Volusia	16,510	10,003	8,467
Wakulla	4,802	5,149	3,117
Walton	16,460	9,346	4,816
Washington	16,403	10,154	6,426

The State 752,619 528,542 391,422

GEORGIA

COUNTY	1910	1900	1890
Appling	12,318	12,336	8,676
Baker	7,973	6,704	6,144
Baldwin	18,354	17,768	14,608
Banks	11,244	10,545	8,562
Bartow	25,388	20,823	20,616
Ben Hill	11,863	.....	.....
Berrien	22,772	19,440	10,694
Bibb	56,646	50,473	42,370
Brooks	23,832	18,606	13,979
Bryan	6,702	6,122	5,520
Bulloch	26,464	21,377	13,712
Burke	27,268	30,165	28,501
Butts	13,624	12,805	10,565
Calhoun	11,334	9,274	8,438
Camden	7,690	7,669	6,178
Campbell	10,874	9,518	9,115
Carroll	30,855	26,576	22,301
Catoosa	7,184	5,823	5,431
Charlton	4,722	3,592	3,335
Chatham	79,690	71,239	57,740
Chattahoochee	5,586	5,790	4,902
Chattooga	13,608	12,952	11,202
Cherokee	16,661	15,243	15,412
Clarke	23,273	17,708	15,186
Clay	8,960	8,568	7,817
Clayton	10,453	9,598	8,295
Clinch	8,424	8,732	6,652
Cobb	28,397	24,664	22,286
Coffee	21,953	16,169	10,483
Colquitt	19,789	13,636	4,794
Columbia	12,328	10,653	11,281
Coweta	28,800	24,980	22,354
Crawford	8,310	10,368	9,315
Crisp	16,423	.....	.....
Dade	4,139	4,578	5,707

GEORGIA Cont'd.

COUNTY	1910	1900	1890
Dawson	4,686	5,442	5,612
Decatur	29,045	29,454	19,949
Dekalb	27,881	21,112	17,189
Dodge	20,127	13,975	11,452
Dooly	20,554	26,567	18,146
Dougherty	16,035	13,679	12,206
Douglas	8,953	8,745	7,794
Early	18,122	14,828	9,792
Echols	3,309	3,209	3,079
Effingham	9,971	8,334	5,599
Elbert	24,125	19,729	15,376
Emanuel	25,140	21,279	14,703
Fannin	12,574	11,214	8,724
Fayette	10,966	10,114	8,728
Floyd	36,736	33,113	28,391
Forsyth	11,940	11,550	11,155
Franklin	17,894	17,700	14,670
Fulton	177,733	117,363	84,655
Gilmer	9,237	10,198	9,074
Glascock	4,669	4,516	3,720
Glynn	15,720	14,317	13,420
Gordon	15,861	14,119	12,758
Grady	18,457	......	......
Greene	18,512	16,542	17,051
Gwinnett	28,824	25,585	19,899
Habersham	10,134	13,604	11,573
Hall	25,730	20,752	18,047
Hancock	19,189	18,277	17,149
Haralson	13,514	11,922	11,316
Harris	17,886	18,009	16,797
Hart	16,216	14,492	10,887
Heard	11,189	11,177	9,557
Henry	19,927	18,602	16,220
Houston	23,609	22,641	21,613
Irwin	10,461	13,645	6,316
Jackson	30,169	24,039	19,176
Jasper	16,552	15,033	13,879
Jeff Davis	6,050		
Jefferson	21,379	18,212	17,213
Jenkins	11,520		
Johnson	12,897	11,409	6,129
Jones	13,103	13,358	12,709
Laurens	35,501	25,908	13,747
Lee	11,679	10,344	9,074
Liberty	12,924	13,093	12,887
Lincoln	8,714	7,156	6,146
Lowndes	24,436	20,036	15,102
Lumpkin	5,444	7,433	6,867
McDuffie	10,325	9,804	8,789
McIntosh	6,442	6,537	6,470
Macon	15,016	14,093	13,183
Madison	16,851	13,224	11,024
Marion	9,147	10,080	7,728
Meriwether	25,180	23,339	20,740
Miller	7,986	6,319	4,275
Milton	7,239	6,763	6,208
Mitchell	22,114	14,767	10,906
Monroe	20,450	20,682	19,137
Montgomery	19,638	16,359	9,248
Morgan	19,717	15,813	16,041
Murray	9,763	8,623	8,461
Muscogee	36,227	29,836	27,761
Newton	18,449	16,734	14,310
Oconee	11,104	8,602	7,713
Oglethorpe	18,680	17,881	16,951
Paulding	14,124	12,969	11,948
Pickens	9,041	8,641	8,182
Pierce	10,749	8,100	6,379
Pike	19,495	18,761	16,300
Polk	20,203	17,856	14,945
Pulaski	22,835	18,489	16,559
Putnam	13,876	13,436	14,842
Quitman	4,594	4,701	4,471
Rabun	5,562	6,285	5,606
Randolph	18,841	16,847	15,267
Richmond	58,886	53,735	45,194
Rockdale	8,916	7,515	6,813

GEORGIA Cont'd.

COUNTY	1910	1900	1890
Schley	5,213	5,499	5,443
Screven	20,202	19,252	14,424
Spalding	19,741	17,619	13,117
Stephens	9,728		
Stewart	13,437	15,856	15,682
Sumter	29,092	26,212	22,107
Talbot	11,696	12,197	13,258
Taliaferro	8,766	7,912	7,291
Tattnall	18,569	20,419	10,253
Taylor	10,839	9,846	8,666
Telfair	13,288	10,083	5,477
Terrell	22,003	19,023	14,503
Thomas	29,071	31,076	26,154
Tift	11,487	......	......
Toombs	11,206	......	......
Towns	3,932	4,748	4,064
Troup	26,228	24,002	20,723
Turner	10,075	......	......
Twiggs	10,736	8,716	8,195
Union	6,918	8,481	7,749
Upson	12,757	13,670	12,188
Walker	18,692	15,661	13,282
Walton	25,393	20,942	17,467
Ware	22,957	13,761	8,811
Warren	11,860	11,463	10,957
Washington	28,174	28,227	25,237
Wayne	13,069	9,449	7,485
Webster	6,151	6,618	5,695
White	5,110	5,912	6,151
Whitfield	15,934	14,509	12,916
Wilcox	13,486	11,097	7,980
Wilkes	23,441	20,866	18,081
Wilkinson	10,078	11,440	10,781
Worth	19,147	18,664	10,048

The State 2,609,121 2,216,331 1,837,353

IDAHO

COUNTY	1910	1900	1890
Ada	29,088	11,559	8,368
Bannock	19,242	11,702	......
Bear Lake	7,729	7,051	6,057
Bingham	23,306	10,447	13,575
Blaine	8,387	4,900	......
Boise	5,250	4,174	3,342
Bonner	13,588	......	......
Canyon	25,323	7,497	......
Cassia	7,197	3,951	3,143
Custer	3,001	2,049	2,176
Elmore	4,785	2,286	1,870
Fremont	24,606	12,821	......
Idaho	12,384	9,121	2,955
Kootenai	22,747	10,216	4,108
Latah	18,818	13,451	9,173
Lemhi	4,786	3,446	1,915
Lincoln	12,676	1,784	......
Nez Perce	24,860	13,748	2,847
Oneida	15,170	8,933	6,819
Owyhee	4,044	3,804	2,021
Shoshone	13,963	11,950	5,382
Twin Falls	13,543	......	......
Washington	11,101	6,882	3,836

The State.... 325,594 161,772 a84,385

(a) Includes population of Alturas and Logan counties (population 2,629 and 4,169, respectively, in 1890) absorbed by other counties since 1890.

ILLINOIS

COUNTY	1910	1900	1890
Adams	64,588	67,058	61,888
Alexander	22,741	19,384	16,563
Bond	17,075	16,078	14,550
Boone	15,481	15,791	12,203
Brown	10,397	11,557	11,951
Bureau	43,975	41,112	35,014
Calhoun	8,610	8,917	7,652
Carroll	18,035	18,963	18,320
Cass	17,372	17,222	15,963
Champaign	51,829	47,622	42,159
Christian	34,594	32,790	30,531
Clark	23,517	24,033	21,899
Clay	18,661	19,553	16,772
Clinton	22,832	19,824	17,411
Coles	34,517	34,146	30,093
Cook	2,405,233	1,838,735	1,191,922
Crawford	26,281	19,240	17,283
Cumberland	14,281	16,124	15,443
Dekalb	33,457	31,756	27,066
Dewitt	18,906	18,972	17,011
Douglas	19,591	19,097	17,669
Dupage	33,432	28,196	22,551
Edgar	27,336	28,273	26,787
Edwards	10,049	10,345	9,444
Effingham	20,055	20,465	19,358
Fayette	28,075	28,065	23,367
Ford	17,096	18,359	17,035
Franklin	25,943	19,675	17,138
Fulton	49,549	46,201	43,110
Gallatin	14,628	15,836	14,935
Greene	22,363	23,402	23,791
Grundy	24,162	24,136	21,024
Hamilton	18,227	20,197	17,800
Hancock	30,638	32,215	31,907
Hardin	7,015	7,448	7,234
Henderson	9,724	10,836	9,876
Henry	41,736	40,049	33,338
Iroquois	35,543	38,014	35,167
Jackson	35,143	33,871	27,809
Jasper	18,157	20,160	18,188
Jefferson	29,111	28,133	22,590
Jersey	13,954	14,612	14,810
Jo Daviess	22,657	24,533	25,101
Johnson	14,331	15,667	15,013
Kane	91,862	78,792	65,061
Kankakee	40,752	37,154	28,732
Kendall	10,777	11,467	12,106
Knox	46,159	43,612	38,752
Lake	55,058	34,504	24,235
Lasalle	90,132	87,776	80,798
Lawrence	22,661	16,523	14,693
Lee	27,750	29,894	26,187
Livingston	40,465	42,035	38,455
Logan	30,216	28,680	25,489
McDonough	26,887	28,412	27,467
McHenry	32,509	29,759	26,114
McLean	68,008	67,843	63,036
Macon	54,186	44,003	38,083
Macoupin	50,685	42,256	40,3o0
Madison	89,847	64,694	51,535
Marion	35,094	30,446	24,341
Marshall	15,679	16,370	13,o53
Mason	17,377	17,491	16,0o7
Massac	14,200	13,110	11,313
Menard	12,796	14,336	13,120
Mercer	19,723	20,945	18,5x5
Monroe	13,508	13,847	12,948
Montgomery	35,311	30,836	30,003
Morgan	34,420	35,006	32,636
Moultrie	14,630	15,224	14,481
Ogle	27,864	29,129	28,710
Peoria	100,255	88,608	70,378
Perry	22,088	19,830	17,529
Platt	16,376	17,706	17,062
Pike	28,622	31,595	31,000
Pope	11,215	13,585	14,016
Pulaski	15,650	14,554	11,355

ILLINOIS Cont'd.

COUNTY	1910	1900	1890
Putnam	7,561	4,746	4,730
Randolph	29,120	28,001	25,049
Richland	15,970	16,391	15,019
Rock Island	70,404	55,249	41,917
St. Clair	119,870	86,685	66,571
Saline	30,204	21,685	19,342
Sangamon	91,024	71,593	61,195
Schuyler	14,852	16,129	16,013
Scott	10,067	10,455	10,304
Shelby	31,693	32,126	31,191
Stark	10,098	10,186	9,982
Stephenson	36,821	34,933	31,338
Tazewell	34,027	33,221	29,556
Union	21,856	22,610	21,549
Vermilion	77,996	65,635	49,905
Wabash	14,913	12,583	11,866
Warren	23,313	23,163	21,281
Washington	18,759	19,526	19,262
Wayne	25,697	27,626	23,806
White	23,052	25,386	25,005
Whiteside	34,507	34,710	30,854
Will	84,371	74,764	62,007
Williamson	45,098	27,796	22,226
Winnebago	63,153	47,845	39,938
Woodford	20,506	21,822	21,429
The State	5,638,591	4,821,550	3,826,351

INDIANA

COUNTY	1910	1900	1890
Adams	21,840	22,232	20,181
Allen	93,386	77,270	66,689
Bartholomew	24,813	24,594	23,867
Benton	12,688	13,123	11,903
Blackford	15,820	17,213	10,461
Boone	24,673	26,321	26,572
Brown	7,975	9,727	10,308
Carroll	17,970	19,953	20,021
Cass	36,368	34,545	31,152
Clark	30,260	31,835	30,259
Clay	32,535	34,285	30,536
Clinton	26,674	28,202	27,370
Crawford	12,057	13,476	13,941
Daviess	27,747	29,914	26,227
Dearborn	21,396	22,194	23,364
Decatur	18,793	19,518	19,277
Dekalb	25,054	25,711	24,307
Delaware	51,414	49,624	30,131
Dubois	19,843	20,357	20,253
Elkhart	49,008	45,052	39,201
Fayette	14,415	13,495	12,630
Floyd	30,293	30,118	29,458
Fountain	20,439	21,446	19,558
Franklin	15,335	16,388	18,366
Fulton	16,879	17,453	16,746
Gibson	30,137	30,099	24,920
Grant	51,426	54,693	31,493
Greene	36,873	28,530	24,379
Hamilton	27,026	29,914	26,123
Hancock	19,030	19,189	17,829
Harrison	20,232	21,702	20,786
Hendricks	20,840	21,292	21,498
Henry	29,758	25,088	23,879
Howard	33,177	28,575	26,186
Huntington	28,982	28,901	27,644
Jackson	24,727	26,633	24,139
Jasper	13,044	14,292	11,185
Jay	24,961	26,818	23,478
Jefferson	20,483	22,913	24,507
Jennings	14,203	15,757	14,608
Johnson	20,394	20,223	19,561
Knox	39,183	32,746	28,044
Kosciusko	27,936	29,109	28,645
Lagrange	15,148	15,284	15,615
Lake	82,864	37,892	23,886

INDIANA Cont'd.

COUNTY	1910	1900	1890
Laporte	45,797	38,386	34,445
Lawrence	30,625	25,729	19,792
Madison	65,224	70,470	36,487
Marion	263,661	197,227	141,156
Marshall	24,175	25,119	23,818
Martin	12,950	14,711	13,973
Miami	29,350	28,344	25,823
Monroe	23,426	20,873	17,673
Montgomery	29,296	29,388	28,025
Morgan	21,182	20,457	18,643
Newton	10,504	10,448	8,803
Noble	24,009	23,533	23,359
Ohio	4,329	4,724	4,955
Orange	17,192	16,854	14,678
Owen	14,053	15,149	15,040
Parke	22,214	23,000	20,296
Perry	18,078	18,778	18,240
Pike	19,684	20,486	18,544
Porter	20,540	19,175	18,052
Posey	21,670	22,333	21,529
Pulaski	13,312	14,033	11,233
Putnam	20,520	21,478	22,335
Randolph	29,013	28,653	28,085
Ripley	19,452	19,881	19,350
Rush	19,349	20,148	19,034
St. Joseph	84,312	58,881	42,457
Scott	8,323	8,307	7,833
Shelby	26,802	26,491	25,454
Spencer	20,676	22,407	22,060
Starke	10,567	10,431	7,339
Steuben	14,274	15,219	14,478
Sullivan	32,439	26,005	21,877
Switzerland	9,914	11,840	12,514
Tippecanoe	40,063	38,659	35,078
Tipton	17,459	19,116	18,157
Union	6,260	6,748	7,006
Vanderburg	77,438	71,769	59,809
Vermilion	18,865	15,252	13,154
Vigo	87,930	62,035	50,195
Wabash	26,926	28,235	27,126
Warren	10,899	11,371	10,955
Warrick	21,911	22,329	21,161
Washington	17,445	19,409	18,619
Wayne	43,757	38,970	37,628
Wells	22,418	23,449	21,514
White	17,602	19,138	15,671
Whitley	16,892	17,328	17,768
The State	2,700,876	2,516,462	2,192,404

IOWA

COUNTY	1910	1900	1890
Adair	14,420	16,192	14,534
Adams	10,998	13,601	12,292
Allamakee	17,328	18,711	17,907
Appanoose	28,701	25,927	18,961
Audubon	12,671	13,626	12,412
Benton	23,156	25,177	24,178
Blackhawk	44,865	32,399	24,219
Bremer	15,843	16,305	14,630
Boone	27,626	28,200	23,772
Buchanan	19,748	21,427	18,997
Buena Vista	15,981	16,975	13,548
Butler	17,119	17,955	15,463
Calhoun	17,090	18,569	13,107
Carroll	20,117	20,319	18,828
Cass	19,047	21,274	19,645
Cedar	17,765	19,371	18,253
Cerro Gordo	25,011	20,672	14,864
Cherokee	16,741	16,570	15,659
Chickasaw	15,375	17,037	15,019
Clarke	10,736	12,440	11,332
Clay	12,766	13,401	9,309
Clayton	25,576	27,750	26,733
Clinton	45,394	43,832	41,199
Crawford	20,041	21,685	18,894

IOWA Cont'd.

COUNTY	1910	1900	1890
Dallas	23,628	23,058	20,479
Davis	13,315	15,620	15,258
Decatur	16,347	18,115	15,643
Delaware	17,888	19,185	17,349
Des Moines	36,145	35,989	35,324
Dickinson	8,137	7,995	4,328
Dubuque	57,450	56,403	49,848
Emmet	9,816	9,936	4,274
Fayette	27,919	29,845	23,141
Floyd	17,119	17,754	15,424
Franklin	14,780	14,996	12,871
Fremont	15,623	18,546	16,842
Greene	16,023	17,820	15,797
Grundy	13,574	13,757	13,215
Guthrie	17,374	18,729	17,380
Hamilton	19,242	19,514	15,319
Hancock	12,731	13,752	7,621
Hardin	20,921	22,794	19,003
Harrison	23,162	25,597	21,356
Henry	18,640	20,022	18,895
Howard	12,920	14,512	11,182
Humboldt	12,182	12,667	9,836
Ida	11,296	12,327	10,705
Iowa	18,409	19,544	18,270
Jackson	21,258	23,615	22,771
Jasper	27,034	26,976	24,943
Jefferson	15,951	17,437	15,184
Johnson	25,914	24,817	23,082
Jones	19,050	21,954	20,233
Keokuk	21,160	24,979	23,862
Kossuth	21,971	22,720	13,120
Lee	36,702	39,719	37,715
Linn	60,720	55,392	45,303
Louisa	12,855	13,516	11,873
Lucas	13,462	16,126	14,563
Lyon	14,624	13,165	8,680
Madison	15,621	17,710	15,977
Mahaska	29,860	34,273	28,805
Marion	22,995	24,159	23,058
Marshall	30,279	29,991	25,842
Mills	15,811	16,764	14,548
Mitchell	13,435	14,916	13,299
Monona	16,633	17,980	14,515
Monroe	25,429	17,985	13,666
Montgomery	16,604	17,803	15,848
Muscatine	29,505	28,242	24,504
O'Brien	17,262	16,985	13,060
Osceola	8,956	8,725	5,574
Page	24,002	24,187	21,348
Palo Alto	13,845	14,354	9,318
Plymouth	23,129	22,209	19,568
Pocahontas	14,808	15,339	9,553
Polk	110,438	82,624	65,410
Pottawattamie	55,832	54,336	47,430
Poweshiek	19,589	19,414	18,394
Ringgold	12,904	15,325	13,556
Sac	16,555	17,639	14,522
Scott	60,000	51,558	43,164
Shelby	16,552	17,932	17,611
Sioux	25,248	23,337	18,370
Story	24,083	23,159	18,127
Tama	22,156	24,585	21,651
Taylor	16,312	18,784	16,384
Union	16,616	19,928	16,900
Van Buren	15,020	17,354	16,253
Wapello	37,743	35,426	30,426
Warren	18,194	20,376	18,269
Washington	19,925	20,718	18,468
Wayne	16,184	17,491	15,670
Webster	34,629	31,757	21,582
Winnebago	11,914	12,725	7,325
Winneshiek	21,729	23,731	22,528
Woodbury	67,616	54,610	55,632
Worth	9,950	10,887	9,247
Wright	17,951	18,227	12,057
The State	2,224,771	2,231,853	1,911,896

KANSAS

COUNTY	1910	1900	1890
Allen	27,640	19,507	13,509
Anderson	13,829	13,938	14,203
Atchison	28,107	28,606	26,758
Barber	9,916	6,594	7,973
Barton	17,876	13,784	13,172
Bourbon	24,007	24,712	28,575
Brown	21,314	22,369	20,319
Butler	23,059	23,363	24,055
Chase	7,527	8,246	8,233
Chautauqua	11,429	11,804	12,297
Cherokee	38,162	42,694	27,770
Cheyenne	4,248	2,640	4,401
Clark	4,093	1,701	2,357
Clay	15,251	15,833	16,146
Cloud	18,388	18,071	19,295
Coffey	15,205	16,643	15,856
Comanche	3,281	1,619	2,549
Cowley	31,790	30,156	34,478
Crawford	51,178	38,809	30,286
Decatur	8,976	9,234	8,414
Dickinson	24,361	21,816	22,273
Doniphan	14,422	15,079	13,535
Douglas	24,724	25,096	23,961
Edwards	7,033	3,682	3,600
Elk	10,128	11,443	12,216
Ellis	12,110	8,626	7,942
Ellsworth	10,444	9,626	9,272
Finney (a)	6,908	3,469	3,350
Ford	11,393	5,497	5,308
Franklin	20,884	21,354	20,279
Geary	12,681	10,744	10,423
Gove	6,044	2,441	2,994
Graham	8,700	5,173	5,029
Grant	1,087	422	1,308
Gray	3,121	1,264	2,415
Greeley	1,335	493	1,264
Greenwood	16,060	16,196	16,309
Hamilton	3,360	1,426	2,027
Harper	14,748	10,310	13,266
Harvey	19,200	17,591	17,601
Haskell	993	457	1,077
Hodgeman	2,930	2,032	2,395
Jackson	16,861	17,117	14,626
Jefferson	15,826	17,533	16,620
Jewell	18,148	19,420	19,349
Johnson	18,288	18,104	17,385
Kearny	3,206	1,107	1,571
Kingman	13,386	10,663	11,823
Kiowa	6,174	2,365	2,873
Labette	31,423	27,387	27,586
Lane	2,603	1,563	2,060
Leavenworth	41,207	40,940	38,485
Lincoln	10,142	9,886	9,709
Linn	14,735	16,689	17,215
Logan	4,240	1,962	3,384
Lyon	24,927	25,074	23,196
McPherson	21,521	21,421	21,614
Marion	22,415	20,676	20,539
Marshall	23,880	24,355	23,912
Meade	5,055	1,581	2,542
Miami	20,030	21,641	19,614
Mitchell	14,089	14,647	15,037
Montgomery	49,474	29,039	23,104
Morris	12,397	11,967	11,381
Morton	1,333	304	724
Nemaha	19,072	20,376	19,249
Neosho	23,754	19,254	18,561
Ness	5,883	4,535	4,944
Norton	11,614	11,325	10,617
Osage	19,905	23,659	25,062
Osborne	12,827	11,844	12,083
Ottawa	11,811	11,182	12,581
Pawnee	8,859	5,084	5,204
Phillips	14,150	14,442	13,661
Pottawatomie	17,522	18,470	17,722
Pratt	11,156	7,085	8,118
Rawlins	6,380	5,241	6,756

KANSAS Cont'd.

COUNTY	1910	1900	1890
Reno	37,853	29,027	27,079
Republic	17,447	18,248	19,002
Rice	15,106	14,745	14,451
Riley	15,783	13,828	13,183
Rooks	11,282	7,960	8,018
Rush	7,826	6,134	5,204
Russell	10,800	8,489	7,333
Saline	20,338	17,076	17,442
Scott	3,047	1,098	1,262
Sedgwick	73,095	44,037	43,626
Seward	4,091	822	1,503
Shawnee	61,874	53,727	49,172
Sheridan	5,651	3,819	3,733
Sherman	4,549	3,341	5,261
Smith	15,365	16,384	15,613
Stafford	12,510	9,829	8,520
Stanton	1,034	327	1,031
Stevens	2,453	620	1,418
Sumner	30,654	25,631	30,271
Thomas	5,455	4,112	5,538
Trego	5,398	2,722	2,535
Wabaunsee	12,721	12,813	11,720
Wallace	2,759	1,178	2,468
Washington	20,229	21,963	22,894
Wichita	2,006	1,197	1,827
Wilson	19,810	15,621	15,286
Woodson	9,450	10,022	9,021
Wyandotte	100,068	73,227	54,407

The State 1,690,949 1,470,495 1,427,096

(a) Garfield county (population 881 in 1890) annexed to Finney county in 1893.

KENTUCKY

COUNTY	1910	1900	1890
Adair	16,503	14,888	13,721
Allen	14,882	14,657	13,692
Anderson	10,146	10,051	10,610
Ballard	12,690	10,761	8,390
Barren	25,293	23,197	21,490
Bath	13,988	14,734	12,813
Bell	28,447	15,701	10,312
Boone	9,420	11,170	12,246
Bourbon	17,462	18,069	16,976
Boyd	23,444	18,834	14,033
Boyle	14,668	13,817	12,948
Bracken	10,308	12,137	12,369
Breathitt	17,540	14,322	8,705
Breckinridge	21,034	20,534	18,976
Bullitt	9,487	9,602	8,291
Butler	15,805	15,896	13,956
Caldwell	14,063	14,510	13,186
Calloway	19,867	17,633	14,675
Campbell	59,369	54,223	44,208
Carlisle	9,048	10,195	7,612
Carroll	8,110	9,825	9,266
Carter	21,966	20,228	17,204
Casey	15,479	15,144	11,848
Christian	38,845	37,962	34,118
Clark	17,987	16,694	15,434
Clay	17,789	15,364	12,447
Clinton	8,153	7,871	7,047
Crittenden	13,296	15,191	13,119
Cumberland	9,846	8,962	8,452
Daviess	41,020	38,667	33,120
Edmonson	10,469	10,080	8,005
Elliott	9,814	10,387	9,214
Estill	12,273	11,669	10,836
Fayette	47,715	42,071	35,698
Fleming	16,066	17,074	16,078
Floyd	18,623	15,552	11,256
Franklin	21,135	20,852	21,267

KENTUCKY Cont'd.

COUNTY	1910	1900	1890
Fulton	14,114	11,546	10,005
Gallatin	4,697	5,163	4,611
Garrard	11,894	12,042	11,138
Grant	10,581	13,239	12,671
Graves	33,539	33,204	28,534
Grayson	19,958	19,878	18,688
Green	11,871	12,255	11,463
Greenup	18,475	15,432	11,911
Hancock	8,512	8,914	9,214
Hardin	22,696	22,937	21,304
Harlan	10,566	9,838	6,197
Harrison	16,873	18,570	16,914
Hart	18,173	18,390	16,439
Henderson	29,352	32,907	29,536
Henry	13,716	14,620	14,164
Hickman	11,750	11,745	11,637
Hopkins	34,291	30,995	23,505
Jackson	10,734	10,561	8,261
Jefferson	262,920	232,549	188,598
Jessamine	12,613	11,925	11,248
Johnson	17,482	13,730	11,027
Kenton	70,355	63,591	54,161
Knott	10,791	8,704	5,438
Knox	22,116	17,372	13,762
Larue	10,701	10,764	9,433
Laurel	19,872	17,592	13,747
Lawrence	20,067	19,612	17,702
Lee	9,531	7,988	6,205
Leslie	8,976	6,753	3,964
Letcher	10,623	9,172	6,920
Lewis	16,887	17,868	14,803
Lincoln	17,897	17,059	15,962
Livingston	10,627	11,354	9,474
Logan	24,977	25,994	23,812
Lyon	9,423	9,319	7,628
McCracken	35,064	28,733	21,051
McLean	13,241	12,448	9,887
Madison	26,951	25,607	24,348
Magoffin	13,654	12,006	9,196
Marion	16,330	16,290	15,648
Marshall	15,771	13,692	11,287
Martin	7,291	5,780	4,209
Mason	18,611	20,446	20,773
Meade	9,783	10,533	9,484
Menifee	6,153	6,818	4,666
Mercer	14,063	14,426	15,034
Metcalfe	10,453	9,988	9,871
Monroe	13,663	13,053	10,989
Montgomery	12,868	12,834	12,367
Morgan	16,259	12,792	11,249
Muhlenberg	28,598	20,741	17,955
Nelson	16,830	16,587	16,417
Nicholas	10,601	11,952	10,764
Ohio	27,642	27,287	22,946
Oldham	7,248	7,078	6,754
Owen	14,248	17,553	17,676
Owsley	7,979	6,874	5,975
Pendleton	11,985	14,947	16,346
Perry	11,255	8,276	6,331
Pike	31,679	22,686	17,378
Powell	6,268	6,443	4,698
Pulaski	35,986	31,293	25,731
Robertson	4,121	4,900	4,684
Rockcastle	14,473	12,416	9,841
Rowan	9,438	8,277	6,129
Russell	10,861	9,695	8,136
Scott	16,956	18,076	16,546
Shelby	18,041	18,340	16,521
Simpson	11,460	11,624	10,878
Spencer	7,567	7,406	6,760
Taylor	11,961	11,075	9,353
Todd	16,488	17,371	16,814
Trigg	14,539	14,073	13,902
Trimble	6,512	7,272	7,140
Union	19,886	21,326	18,729
Warren	30,579	29,970	30,158
Washington	13,940	14,182	13,622

KENTUCKY Cont'd.

COUNTY	1910	1900	1890
Wayne	17,518	14,892	12,852
Webster	20,974	20,097	17,196
Whitley	31,982	25,015	17,590
Wolfe	9,864	8,764	7,180
Woodford	12,571	13,134	12,380

The State 2,289,905 2,147,174 1,858,635

LOUISIANA

COUNTY	1910	1900	1890
Acadia	31,847	23,483	13,231
Ascension	23,887	24,142	19,545
Assumption	24,128	21,620	19,629
Avoyelles	34,102	29,701	25,112
Bienville	21,776	17,588	14,108
Bossier	21,738	24,153	20,330
Caddo	58,200	44,499	31,555
Calcasieu	62,767	30,428	20,176
Caldwell	8,593	6,917	5,814
Cameron	4,288	3,952	2,828
Catahoula	10,415	16,351	12,002
Claiborne	25,050	23,029	23,312
Concordia	14,278	13,559	14,871
De Soto	27,689	25,063	19,860
East Baton Rouge	34,580	31,153	25,922
East Carroll	11,637	11,373	12,362
East Feliciana	20,055	20,443	17,903
Franklin	11,989	8,890	6,900
Grant	15,958	12,902	8,270
Iberia	31,262	29,015	20,997
Iberville	30,954	27,006	21,848
Jackson	13,818	9,119	7,453
Jefferson	18,247	15,321	13,221
La Salle	9,402		
Lafayette	28,733	22,825	15,966
Lafourche	33,111	28,882	22,095
Lincoln	18,485	15,898	14,753
Livingston	10,627	8,100	5,769
Madison	10,676	12,322	14,135
Morehouse	18,786	16,634	16,786
Natchitoches	36,455	33,216	25,836
Orleans	339,075	287,104	242,039
Ouachita	25,830	20,947	17,985
Plaquemines	12,524	13,039	12,541
Pointe Coupee	25,289	25,777	19,613
Rapides	44,545	39,578	27,642
Red River	11,402	11,548	11,318
Richland	15,769	11,116	10,230
Sabine	19,874	15,421	9,390
St. Bernard	5,277	5,031	4,326
St. Charles	11,207	9,072	7,737
St. Helena	9,172	8,479	8,062
St. James	23,009	20,197	15,715
St. John the Baptist	14,338	12,330	11,359
St. Landry	66,661	52,906	40,250
St. Martin	23,070	18,940	14,834
St. Mary	39,368	34,145	22,416
St. Tammany	18,917	13,335	10,160
Tangipahoa	29,160	17,625	12,655
Tensas	17,060	19,070	16,647
Terrebonne	28,320	24,464	20,167
Union	20,451	18,520	17,304
Vermilion	26,390	20,705	14,234
Vernon	17,384	10,327	5,903
Washington	18,886	9,628	6,700
Webster	19,186	15,125	12,466
W. Baton Rouge	12,636	10,285	8,363
West Carroll	6,249	3,685	3,748
West Feliciana	13,449	15,994	15,062
Winn	18,357	9,648	7,082

The State 1,656,388 1,381,625 1,118,587

MAINE

COUNTY	1910	1900	1890
Androscoggin	59,822	54,242	48,968
Aroostook	74,664	60,744	49,589
Cumberland	112,014	100,689	90,949
Franklin	19,119	18,444	17,053
Hancock	35,575	37,241	37,312
Kennebec	62,863	59,117	57,012
Knox	28,981	30,406	31,473
Lincoln	18,216	19,669	21,996
Oxford	36,256	32,238	30,586
Penobscot	85,285	76,246	72,865
Piscataquis	19,887	16,949	16,134
Sagadahoc	18,574	20,330	19,452
Somerset	36,301	33,849	32,627
Waldo	23,383	24,185	27,759
Washington	42,905	45,232	44,482
York	68,526	64,885	62,829

The State 742,371 694,466 661,086

MARYLAND

COUNTY	1910	1900	1890
Allegany	62,411	53,694	41,571
Anne Arundel	39,553	39,620	34,094
Baltimore	122,399	90,755	72,909
Baltimore city	558,485	508,957	434,439
Calvert	10,325	10,223	9,860
Caroline	19,216	16,248	13,903
Carroll	33,934	33,860	32,376
Cecil	23,759	24,662	25,851
Charles	16,386	17,662	15,191
Dorchester	28,669	27,962	24,843
Frederick	52,673	51,920	49,512
Garrett	20,105	17,701	14,213
Harford	27,965	28,269	28,993
Howard	16,106	16,715	16,269
Kent	16,957	18,786	17,471
Montgomery	32,089	30,451	27,185
Prince Georges	36,147	29,898	26,080
Queen Annes	16,839	18,364	18,461
St. Marys	17,030	17,182	15,819
Somerset	26,455	25,923	24,155
Talbot	19,620	20,342	19,736
Washington	48,671	45,133	39,782
Wicomico	26,815	22,852	19,930
Worcester	21,841	20,865	19,747

The State 1,294,450 1,188,044 1,042,390

MASSACHUSETTS

COUNTY	1910	1900	1890
Barnstable	27,542	27,826	29,172
Berkshire	105,259	95,667	81,108
Bristol	318,573	252,029	186,465
Dukes	4,504	4,561	4,369
Essex	436,477	357,030	299,995
Franklin	43,600	41,209	38,610
Hampden	231,369	175,603	135,713
Hampshire	63,327	58,820	51,859
Middlesex	669,915	565,696	431,167
Nantucket	2,962	3,006	3,268
Norfolk	187,506	151,539	118,950
Plymouth	144,337	113,985	92,700
Suffolk	731,388	611,417	484,780
Worcester	399,657	346,958	280,787

The State 3,366,416 2,805,346 2,238,943

MICHIGAN

COUNTY	1910	1900	1890
Alcona	5,703	5,691	5,409
Alger	7,675	5,868	1,238
Allegan	39,819	38,812	38,961
Alpena	19,965	18,254	15,581
Antrim	15,692	16,568	10,413
Arenac	9,640	9,821	5,683
Baraga	6,127	4,320	3,036
Barry	22,633	22,514	23,783
Bay	68,238	62,378	56,412
Benzie	10,638	9,685	5,237
Berrien	53,622	49,165	41,285
Branch	25,605	27,811	26,791
Calhoun	56,638	49,315	43,501
Cass	20,624	20,876	20,953
Charlevoix	19,157	13,956	9,686
Cheboygan	17,872	15,516	11,986
Chippewa	24,472	21,338	12,019
Clare	9,240	8,360	7,558
Clinton	23,129	25,136	26,509
Crawford	3,934	2,943	2,962
Delta	30,108	23,881	15,330
Dickinson	20,524	17,890	(b)
Eaton	30,499	31,668	32,094
Emmet	18,561	15,931	8,756
Genesee	64,555	41,804	39,430
Gladwin	8,413	6,564	4,208
Gogebic	23,333	16,738	13,166
Grand Traverse	23,784	20,479	13,355
Gratiot	28,820	29,889	28,668
Hillsdale	29,673	29,865	30,660
Houghton	88,098	66,063	35,389
Huron	34,758	34,162	28,545
Ingham	53,310	39,818	37,666
Ionia	33,550	34,329	32,801
Iosco	9,753	10,246	15,224
Iron	15,164	8,990	4,432
Isabella	23,029	22,784	18,784
Jackson	53,426	48,222	45,031
Kalamazoo	60,427	44,310	39,273
Kalkaska	8,097	7,133	5,160
Kent	159,145	129,714	109,922
Keweenaw	7,156	3,217	2,894
Lake	4,939	4,957	6,505
Lapeer	26,033	27,641	29,213
Leelanau	10,608	10,556	7,944
Lenawee	47,907	48,406	48,448
Livingston	17,736	19,664	20,858
Luce	4,004	2,983	2,455
Mackinac	9,249	7,703	7,830
Macomb	32,606	33,244	31,813
Manistee	26,688	27,856	24,230
Marquette	46,739	41,239	39,521
Mason	21,832	18,885	16,385
Mecosta	19,466	20,693	19,697
Menominee	25,648	27,046	33,639
Midland	14,005	14,439	40,657
Missaukee	10,606	9,308	5,048
Monroe	32,917	32,754	32,337
Montcalm	32,069	32,754	32,637
Montmorency	3,755	3,234	1,487
Muskegon	40,577	37,036	40,013
Newaygo	19,220	17,673	20,476
Oakland	49,576	44,792	41,245
Oceana	18,379	16,644	15,698
Ogemaw	8,907	7,765	5,583
Ontonagon	8,650	6,197	3,756
Osceola	17,889	17,859	14,630
Oscoda	2,027	1,468	1,904
Otsego	6,552	6,175	4,272
Ottawa	45,301	39,667	35,358
Presque Isle	11,249	8,821	4,687
Roscommon	2,274	1,787	2,033
Saginaw	89,290	81,222	82,273
St. Clair	52,341	55,228	52,105
St. Joseph	25,499	23,889	25,356
Sanilac	33,930	35,055	32,589
Schoolcraft	8,681	7,889	5,818

MICHIGAN Cont'd.

COUNTY	1910	1900	1890
Shiawassee	33,246	33,866	30,952
Tuscola	34,913	35,890	32,508
Van Buren	33,185	33,274	30,541
Washtenaw	44,714	47,761	42,210
Wayne	531,590	348,793	257,114
Wexford	20,769	16,845	11,278

The State 2,810,173 2,420,982 2,093,889

(a) Includes population (860) of Manitou annexed to Charlevoix and Leelanau in 1896, and (135) of Isle Royal annexed to Keweenaw in 1897.

(b) Organized from parts of Iron, Marquette and Menominee in 1891.

MINNESOTA

COUNTY	1910	1900	1890
Aitkin	10,371	6,743	2,462
Anoka	12,493	11,313	9,884
Becker (a)	18,840	14,375	9,401
Beltrami	19,337	11,030	312
Benton	11,615	9,912	6,284
Bigstone	9,367	8,731	5,722
Blue Earth	29,337	32,263	29,210
Brown	20,134	19,787	15,817
Carlton	17,559	10,017	5,272
Carver	17,455	17,544	16,532
Cass	11,620	7,777	1,247
Chippewa	13,458	12,499	8,555
Chisago	13,537	13,248	10,359
Clay	19,640	17,942	11,517
Clearwater (a)	6,870		
Cook	1,336	810	98
Cottonwood	12,651	12,069	7,412
Crow Wing	16,861	14,250	8,852
Dakota	25,171	21,733	20,240
Dodge	12,094	13,340	10,864
Douglas	17,669	17,964	14,606
Faribault	19,949	22,055	16,708
Fillmore	25,680	28,238	25,966
Freeborn	22,282	21,838	17,962
Goodhue	31,637	31,137	28,806
Grant	9,114	8,935	6,875
Hennepin	333,480	228,340	185,294
Houston	14,297	15,400	14,653
Hubbard	9,831	6,578	1,412
Isanti	12,615	11,675	7,607
Itasca	17,208	4,573	743
Jackson	14,491	14,793	8,924
Kanabec	6,461	4,614	1,579
Kandiyohi	18,969	18,416	13,997
Kittson	9,669	7,889	5,387
Koochiching	6,431		
Lac qui Parle	15,435	14,289	10,382
Lake	8,011	4,654	1,299
Le Sueur	18,609	20,234	19,057
Lincoln	9,874	8,966	5,691
Lyon	15,722	14,591	9,501
McLeod	18,691	19,595	17,026
Mahnomen (a)	3,249		
Marshall	16,338	15,698	9,130
Martin	17,518	16,936	9,403
Meeker	17,022	17,753	15,456
Mille Lacs	10,705	8,066	2,845
Morrison	24,053	22,891	13,325
Mower	22,640	22,335	18,019
Murray	11,755	11,911	6,692
Nicollet	14,125	14,774	13,382
Nobles	15,210	14,932	7,958
Norman	13,446	15,045	10,618
Olmsted	22,497	23,119	19,806
Otter Tail	46,036	45,375	34,232

MINNESOTA Cont'd.

COUNTY	1910	1900	1890
Pine	15,878	11,546	4,052
Pipestone	9,553	9,264	5,132
Polk	36,001	35,429	30,192
Pope	12,746	12,577	10,032
Ramsey	223,675	170,554	139,796
Red Lake	15,940	12,195	
Redwood	18,425	17,261	9,386
Renville	23,123	23,693	17,099
Rice	25,911	26,080	23,968
Rock	10,222	9,668	6,817
Roseau	11,338	6,994	
St. Louis	163,274	82,932	44,862
Scott	14,888	15,147	13,831
Sheburne	8,136	7,281	5,908
Sibley	15,540	16,862	15,199
Stearns	47,733	44,464	34,844
Steele	16,146	16,524	13,232
Stevens	8,293	8,721	5,251
Swift	12,949	13,503	10,161
Todd	23,407	22,214	12,930
Traverse	8,049	7,573	4,516
Wabasha	18,554	18,924	16,972
Wadena	8,652	7,921	4,053
Waseca	13,466	14,760	13,313
Washington	26,013	27,808	25,992
Watonwan	11,382	11,496	7,746
Wilkin	9,063	8,080	4,346
Winona	33,398	35,686	33,797
Wright	28,082	29,157	24,164
Yellow Medicine	15,406	14,602	9,854

The State 2,075,708 1,751,394 1,301,826

(a) Includes population of part of White Earth Indian Reservation (population 3,486 in 1910) in Becker, Clearwater, and Mahnomen counties.

MISSISSIPPI

COUNTY	1910	1900	1890
Adams	25,265	30,111	26,031
Alcorn	18,159	14,987	13,115
Amite	22,954	20,708	18,198
Attala	28,851	26,248	22,213
Benton	10,245	10,510	10,585
Bolivar	48,905	35,427	29,980
Calhoun	17,726	16,512	14,688
Carroll	23,139	22,116	18,773
Chickasaw	22,846	19,892	19,891
Choctaw	14,357	13,036	10,847
Claiborne	17,403	20,787	14,516
Clarke	21,630	17,741	15,826
Clay	20,203	19,563	18,607
Coahoma	34,217	26,293	18,342
Copiah	35,914	34,395	30,233
Covington	16,909	13,076	8,299
De Soto	23,130	24,751	24,183
Forrest	20,722		
Franklin	15,193	13,678	10,424
George	6,599		
Greene	6,050	6,795	3,906
Grenada	15,727	14,112	14,974
Hancock	11,207	11,886	8,318
Harrison	34,658	21,002	12,481
Hinds	63,726	52,577	39,279
Holmes	39,088	36,828	30,970
Issaquena	10,560	10,400	12,318
Itawamba	14,526	13,544	11,708
Jackson	15,451	16,513	11,251
Jasper	18,498	15,394	14,785
Jefferson	18,221	21,292	18,947
Jefferson Davis	12,860		
Jones	29,885	17,846	8,333
Kemper	20,348	20,492	17,961

MISSISSIPPI Cont'd.

COUNTY	1910	1900	1890
Lafayette	21,883	22,110	20,553
Lamar	11,741		
Lauderdale	46,919	38,150	29,661
Lawrence	13,080	15,103	12,318
Leake	18,298	17,360	14,803
Lee	28,894	21,956	20,040
Leflore	36,290	23,834	16,869
Lincoln	28,597	21,552	17,912
Lowndes	30,703	29,095	27,047
Madison	33,505	32,493	27,321
Marion	15,599	13,501	9,532
Marshall	26,796	27,674	26,043
Monroe	35,178	31,216	30,730
Montgomery	17,706	16,536	14,459
Neshoba	17,980	12,726	11,146
Newton	23,085	19,708	16,625
Noxubee	28,503	30,846	27,338
Oktibbeha	19,676	20,183	17,694
Panola	31,274	29,027	26,977
Pearl River	10,593	6,697	2,957
Perry	7,685	14,682	6,494
Pike	37,212	27,545	21,203
Pontotoc	19,688	18,274	14,940
Prentiss	16,931	15,788	13,679
Quitman	11,593	5,435	3,286
Rankin	23,944	20,955	17,922
Scott	16,723	14,316	11,740
Sharkey	15,694	12,178	8,382
Simpson	17,201	12,800	10,138
Smith	16,603	13,055	10,635
Sunflower	28,787	16,084	9,384
Tallahatchie	29,078	19,600	14,361
Tate	19,714	20,618	19,253
Tippah	14,631	12,983	12,951
Tishomingo	13,067	10,124	9,302
Tunica	18,646	16,479	12,158
Union	18,997	16,522	15,606
Warren	37,488	40,912	33,164
Washington	48,933	49,216	40,414
Wayne	14,709	12,539	9,817
Webster	14,853	13,619	12,060
Wilkinson	18,075	21,453	17,592
Winston	17,139	14,124	12,089
Yalobusha	21,519	19,742	16,629
Yazoo	46,672	43,948	36,394

The State 1,797,114 1,551,270 1,289,600

MISSOURI

COUNTY	1910	1900	1890
Adair	22,700	21,728	17,417
Andrew	15,282	17,332	16,000
Atchison	13,604	16,501	15,533
Audrain	21,687	21,160	22,074
Barry	23,869	25,532	22,943
Barton	16,747	18,253	18,504
Bates	25,869	30,141	32,223
Benton	14,881	16,556	14,973
Bollinger	14,576	14,650	13,121
Boone	30,533	28,642	26,043
Buchanan	93,020	121,838	70,100
Butler	20,624	16,769	10,164
Caldwell	14,605	16,656	15,152
Callaway	24,400	25,984	25,131
Camden	11,582	13,113	10,040
Cape Girardeau	27,621	24,315	22,060
Carroll	23,098	26,455	25,742
Carter	5,504	6,706	4,659
Cass	22,973	23,636	23,301
Cedar	16,080	16,923	15,620
Chariton	23,503	26,826	26,254
Christian	15,832	16,939	14,017
Clark	12,811	15,383	15,126
Clay	20,302	18,903	19,856

MISSOURI Cont'd.

COUNTY	1910	1900	1890
Clinton	15,297	17,363	17,138
Cole	21,957	20,578	17,281
Cooper	20,311	22,532	22,707
Crawford	13,576	12,959	11,961
Dade	15,613	18,125	17,526
Dallas	13,181	13,903	12,647
Daviess	17,605	21,325	20,456
DeKalb	12,531	14,418	14,539
Dent	13,245	12,986	12,149
Douglas	16,664	16,802	14,111
Dunklin	30,328	21,706	15,085
Franklin	29,830	30,581	28,056
Gasconade	12,847	12,298	11,706
Gentry	16,820	20,554	19,018
Greene	63,831	52,713	48,616
Grundy	16,744	17,832	17,876
Harrison	20,466	24,398	21·033
Henry	27,242	28,054	28.235
Hickory	8,741	9,985	9,453
Holt	14,539	17,083	15,469
Howard	15,653	18,337	17,371
Howell	21,065	21,834	18,618
Iron	8,563	8,716	9,119
Jackson	283,522	195,193	160,510
Jasper	89,673	84,018	50,500
Jefferson	27,878	25,712	22,484
Johnson	26,297	27,843	28,132
Knox	12,403	13,479	13,501
Laclede	17,363	16,523	14,701
Lafayette	30,154	31,679	30,184
Lawrence	26,583	31,662	26,228
Lewis	15,514	16,724	15,935
Lincoln	17,033	18,352	18,346
Linn	25,253	25,503	24,121
Livingston	19,453	22,302	20,668
McDonald	13,539	13,574	11,283
Macon	30,868	33,018	30,575
Madison	11,273	9,975	9,268
Maries	10,088	9,616	8,600
Marion	30,572	26,331	26,233
Mercer	12,335	14,706	14,581
Miller	16,717	15,187	14,162
Mississippi	14,557	11,837	10,134
Moniteau	14,375	15,931	15,630
Monroe	18,304	19,716	20,790
Montgomery	15,604	16,571	16,850
Morgan	12,863	12,175	12,311
New Madrid	19,488	11,280	9,317
Newton	27,136	27,001	22,108
Nodaway	28,833	32,938	30,914
Oregon	14,681	13,906	10,467
Osage	14,283	14,096	13,080
Ozark	11,926	12,145	9,795
Pemiscot	19,559	12,115	5,975
Perry	14,898	15,134	13,237
Pettis	33,913	32,438	31,151
Phelps	15,796	14,194	12,636
Pike	22,556	25,744	26,321
Platte	14,429	16,193	16,248
Polk	21,561	23,255	20,339
Pulaski	11,438	10,394	9,387
Putnam	14,308	16,688	15,365
Ralls	12,913	12,287	12,294
Randolph	26,182	24,442	24,893
Ray	21,451	24,805	24,215
Reynolds	9,592	8,161	6,803
Ripley	13,099	13,186	8,512
St. Charles	24,695	24,474	22,977
St. Clair	16,412	17,907	16,747
St. Francois	35,738	24,051	17,347
St. Louis	82,417	50,040	36,307
St. Louis city	687,029	575,238	451,770
Ste. Genevieve	10,607	10,359	9,883
Saline	29,448	33,703	33,762
Schuyler	9,062	10,840	11,249
Scotland	11,869	13,232	12,674
Scott	22,372	13,092	11,228

MISSOURI Cont'd.

COUNTY	1910	1900	1890
Shannon	11,443	11,247	8,898
Shelby	14,864	16,167	15,642
Stoddard	27,807	24,669	17,327
Stone	11,559	9,892	7,090
Sullivan	18,598	20,282	19,000
Taney	9,134	10,127	7,973
Texas	21,458	22,192	19,406
Vernon	28,827	31,619	31,505
Warren	9,123	9,919	9,913
Washington	13,378	14,263	13,153
Wayne	15,181	15,309	11,927
Webster	17,377	16,640	15,177
Worth	8,007	9,832	8,738
Wright	18,315	17,519	14,484

The State 3,293,335 3,106,665 2,679,184

MONTANA

COUNTY	1910	1900	1890
Beaverhead	6,446	5,615	4,655
Broadwater	3,491	2,641	......
Carbon	13,962	7,533	......
Cascade	28,833	25,777	8,755
Chouteau	17,191	10,966	4,741
Custer	14,123	7,891	5,308
Dawson	12,725	2,443	2,056
Deer Lodge	12,988	17,393	15,155
Fergus	17,385	6,937	3,514
Flathead	18,785	9,375	......
Gallatin	14,079	9,553	6,246
Granite	2,942	4,328	......
Jefferson	5,601	5,330	6,026
Lewis and Clark	21,853	19,171	19,145
Lincoln	3,638	......	......
Madison	7,229	7,695	4,692
Meagher	4,190	2,526	4,749
Missoula	23,596	13,964	14,427
Park	10,731	7,341	6,881
Powell	5,904	......	......
Ravalli	11,666	7,822	......
Rosebud	7,985	......	......
Sanders	3,713	......	......
Silver Bow	56,848	47,635	23,744
Sweet Grass	4,029	3,086	......
Teton	9,546	5,080	......
Valley	13,630	4,355	......
Yellowstone	22,944	6,212	2,065

The State.... 376,053 a243,329 132,159

(a) Includes population of Crow Indian Reservation (2,660) in Rosebud and Yellowstone counties.

NEBRASKA

COUNTY	1910	1900	1890
Adams	20,900	18,840	24,303
Antelope	14,003	11,344	10,399
Banner	1,444	1,114	2,435
Blaine	1,672	603	1,146
Boone	13,145	11,689	8,683
Boxbutte	6,131	5,572	5,494
Boyd	8,826	7,332	695
Brown	6,083	3,470	4,359
Buffalo	21,907	20,254	22,162
Burt	12,726	13,040	11,069
Butler	15,403	15,703	15,454
Cass	19,786	21,330	24,080
Cedar	15,191	12,467	7,028
Chase	3,613	2,559	4,807
Cherry	10,414	6,541	6,428
Cheyenne	4,551	5,570	5,693
Clay	15,729	15,735	16,310
Colfax	11,610	11,211	10,453
Cuming	13,782	14,584	12,265

NEBRASKA Cont'd.

COUNTY	1910	1900	1890
Custer	25,668	19,758	21,677
Dakota	6,564	6,286	5,386
Dawes	8,254	6,215	9,722
Dawson	15,961	12,214	10,129
Deuel	1,786	2,630	2,893
Dixon	11,477	10,535	8,084
Dodge	22,145	22,298	19,260
Douglas	168,546	140,590	158,008
Dundy	4,098	2,434	4,012
Fillmore	14,674	15,087	16,022
Franklin	10,303	9,455	7,693
Frontier	8,572	8,781	8,497
Furnas	12,083	12,373	9,840
Gage	30,325	30,051	36,344
Garden	3,538	......	......
Garfield	3,417	2,127	1,659
Gosper	4,933	5,301	4,816
Grant	1,097	763	458
Greeley	8,047	5,691	4,869
Hall	20,361	17,206	16,513
Hamilton	13,459	13,330	14,096
Harlan	9,578	9,370	8,158
Hayes	3,011	2,708	3,953
Hitchcock	5,415	4,409	5,799
Holt	15,545	12,224	13,672
Hooker	981	432	426
Howard	10,783	10,343	9,430
Jefferson	16,852	15,196	14,850
Johnson	10,187	11,197	10,333
Kearney	9,106	9,866	9,061
Keith	3,692	1,951	2,556
Keyapaha	3,452	3,076	3,920
Kimball	1,942	758	959
Knox	18,358	14,343	8,582
Lancaster	73,793	64,835	76,395
Lincoln	15,684	11,416	10,441
Logan	1,521	960	1,378
Loup	2,188	1,305	1,662
McPherson (a)	2,470	517	401
Madison	19,101	16,976	13,669
Merrick	10,379	9,255	8,758
Morrill	4,584	......	......
Nance	8,926	8,222	5,773
Nemaha	13,095	14,952	12,930
Nuckolls	13,019	12,414	11,417
Otoe	19,323	22,288	25,403
Pawnee	10,582	11,770	10,340
Perkins	2,570	1,702	4,364
Phelps	10,451	10,772	9,869
Pierce	10,122	8,445	4664
Platte	19,006	17,747	15,437
Polk	10,521	10,542	10,817
Redwillow	11,056	9,604	8,837
Richardson	17,448	19,614	17,574
Rock	3,627	2,809	3,083
Saline	17,866	18,252	20,097
Sarpy	9,274	9,080	6,875
Saunders	21,179	22,085	21,577
Scotts Bluff	8,355	2,552	1,888
Seward	15,895	15,690	16,140
Sheridan	7,328	6,033	8,687
Sherman	8,278	6,550	6,399
Sioux	5,599	2,055	2,452
Stanton	7,542	6,959	4,619
Thayer	14,775	14,325	12,738
Thomas	1,191	628	517
Thurston	8,704	6,517	3,176
Valley	9,480	7,339	7,092
Washington	12,738	13,086	11,869
Wayne	10,397	9,862	6,169
Webster	12,008	11,619	11,210
Wheeler	2,292	1,362	1,683
York	18,721	18,205	17,279

The State 1,192,214 1,066,300 1,058,910

(a) Arthur county (population 91 in 1890) annexed to McPherson county since 1890.

NEVADA

COUNTY	1910	1900	1890
Churchill.......	2,811	830	703
Clark..........	3,321	......	......
Douglas........	1,895	1,534	1,551
Elko..........	8,133	5,688	4,794
Esmeralda......	9,695	1,972	2,148
Eureka........	1,830	1,954	3,275
Humboldt......	6,825	4,463	3,434
Lander.........	1,786	1,534	2,266
Lincoln........	3,489	3,284	2,466
Lyon.	3,568	2,268	1,987
Nye.	7,513	1,140	1,290
Ormsby........	3,089	2,893	4,883
Storey.	3,045	3,673	8,806
Washoe........	17,434	9,141	6,437
White Pine.....	7,441	1,961	1,721
The State.....	81,875	42,335	45,761

NEW HAMPSHIRE

COUNTY	1910	1900	1890
Belknap........	21,309	19,526	20,321
Carroll........	16,316	16,895	18.124
Cheshire.......	30,659	31,321	29,579
Coos..........	30,753	29,468	23,211
Grafton........	41,652	40,844	37,217
Hillsboro.......	126,072	112,640	93,247
Merrimack.....	53,335	52,430	49,435
Rockingham....	52,188	51,118	49,650
Strafford......	38,951	39,337	38,442
Sullivan.......	19,337	18,009	17,304
The State.....	430,572	411,588	376,530

NEW JERSEY

COUNTY	1910	1900	1890
Atlantic........	71,894	46,402	28,036
Bergen........	138,002	78,441	47,226
Burlington....	66,565	58,241	58,528
Camden.......	142,029	107,643	87,687
Cape May.....	19,745	13,201	11,268
Cumberland....	55,153	51,193	45,438
Essex.........	512,886	359,053	256,098
Gloucester.....	37,368	31,905	28,649
Hudson.......	537,231	386,048	275,126
Hunterdon....	33,569	34,507	35,355
Mercer........	125,657	95,365	79,978
Middlesex.....	114,426	79,762	61,754
Monmouth.....	94,734	82,057	69,128
Morris........	74,704	65,156	54,101
Ocean.........	21,318	19,747	15,974
Passaic.......	215,902	155,202	105,046
Salem.........	26,999	25,530	25,151
Somerset......	38,820	32,948	28,311
Sussex........	26,781	24,134	22,259
Union........	140,197	99,353	72,467
Warren........	43,187	37,781	36,553
The State	2,537,167	1,883,669	1,444,933

NEW MEXICO

COUNTY	1910	1900	1890
Bernalillo.......	23,606	28,630	20,913
Chaves........	16,850	4,773	......
Colfax.........	16,460	10,150	7,974
Curry.........	11,443	......	......
Dona Ana......	12,893	10,187	9,191
Eddy..........	12,400	3,229	......
Grant.........	14,813	12,883	9,657
Guadalupe.....	10,927	5,429	......

NEW MEXICO Cont'd

COUNTY	1910	1900	1890
Lincoln.........	7,822	4,953	7,081
Luna..........	3,913	......	......
McKinley......	12,963	......	......
Mora..........	12,611	10,304	10,618
Otero.........	7,069	4,791	......
Quay..........	14,912	......	......
Rio Arriba.....	16,719	13,777	11,534
Roosevelt......	12,064	......	......
San Juan......	8,504	4,828	1,890
San Miguel....	22,930	22,053	24,204
Sandoval......	8,579	......	......
Santa Fe......	14,770	14,658	13,562
Sierra.........	3,536	3,158	3,630
Socorro........	14,761	12,195	9,595
Taos..........	12,008	10,889	9,868
Torrance.......	10,119	......	......
Union.........	11,404	4,528	......
Valencia.......	13,320	13,895	13,876
The State.....	327,396	195,310	153,593

NEW YORK

COUNTY	1910	1900	1890
Albany.........	173,666	165,571	164,555
Allegany.......	41,412	41,501	43,240
Broome........	78,809	69,149	62,973
Cattaraugus....	65,919	65,643	60,866
Cayuga........	67,106	66,234	65,302
Chautauqua....	105,126	88,314	75,202
Chemung......	54,662	54,063	48,265
Chenango.....	35,575	36,568	37,776
Clinton........	48,230	47,430	46,437
Columbia......	43,658	43,211	46,172
Cortland.......	29,249	27,576	28,657
Delaware......	45,575	46,413	45,496
Dutchess......	87,661	81,670	77,879
Erie..........	528,985	433,686	322,981
Essex.........	33,458	30,707	33,052
Franklin.......	45,717	42,853	38,110
Fulton........	44,534	42,842	37,650
Genesee.......	37,615	34,561	33,265
Greene........	30,214	31,478	31,598
Hamilton......	4,373	4,947	4,762
Herkimer......	56,356	51,049	45,608
Jefferson......	80,297	76,748	68,806
Kings.........	1,634,351	1,166,582	838,547
Lewis.........	24,849	27,427	29,806
Livingston.....	38,037	37,059	37,801
Madison.......	39,289	40,545	42,892
Monroe........	283,212	217,854	189,586
Montgomery ...	57,567	47,488	45,699
Nassau........	83,930	55,448	......
New York....	2,762,522	2,050,600	1,515,301
Niagara........	92,036	74,961	62,491
Oneida........	154,157	132,800	122,922
Onondaga.....	200,298	168,735	146,247
Ontario.......	52,286	49,605	48,453
Orange........	115,751	103,859	97,859
Orleans.......	32,000	30,164	30,803
Oswego........	71,664	70,881	71,883
Otsego........	47,216	48,939	50,861
Putnam........	14,665	13,787	14,849
Queens.......	284,041	152,999	128,059
Rensselaer.....	122,276	121,697	124,511
Richmond.....	85,969	67,021	51,693
Rockland......	46,873	38,298	35,162
St. Lawrence...	89 005	89'083	85,048
Saratoga.......	61,917	61,089	57,663
Schenectady....	88,235	46,852	29,797
Schoharie......	23,855	26,854	29,164
Schuyler.......	14,004	15,811	16,711
Seneca........	26,972	28,114	28,227
Steuben.......	83,362	82,822	81,473
Suffolk........	96,138	77,582	62,491

NEW YORK Cont'd.

COUNTY	1910	1900	1890
Sullivan........	33,808	32,306	31,031
Tioga.........	25,624	27,951	29,935
Tompkins......	33,647	33,830	32,923
Ulster.........	91,769	88,422	87,062
Warren........	32,223	29,943	27,866
Washington....	47,778	45,624	45,690
Wayne........	50,179	48,660	49,729
Westchester....	283,055	184,257	146,772
Wyoming......	31,880	30,413	31,193
Yates.........	18,642	20,318	21,001
The State	9,113,614	7,268,894	5,997,853

NORTH CAROLINA

COUNTY	1910	1900	1890
Alamance.......	28,712	25,665	18,271
Alexander......	11,592	10,960	9,430
Alleghany......	7,745	7,759	6,523
Anson.........	25,465	21,870	20,027
Ashe..........	19,074	19,581	15,628
Beaufort.......	30,877	26,404	21,072
Bertie.........	23,039	20,538	19,176
Bladen........	18,006	17,677	16,763
Brunswick.....	14,432	12,657	10,900
Buncombe.....	49,798	44,288	35,266
Burke.........	21,408	17,699	14,939
Cabarrus......	26,240	22,456	18,142
Caldwell.......	20,579	15,694	12,298
Camden.......	5,640	5,474	5,667
Carteret.......	13,776	11,811	10,825
Caswell........	14,858	15,028	16,028
Catawba.......	27,918	22,133	18,689
Chatham.......	22,635	23,912	25,413
Cherokee......	14,136	11,860	9,976
Chowan.......	11,303	10,258	9,167
Clay..........	3,909	4,532	4,197
Cleveland......	29,494	25,078	20,394
Columbus.....	28,020	21,274	17,856
Craven........	25,594	24,160	20,533
Cumberland....	35,284	29,249	27,321
Currituck......	7,693	6,529	6,747
Dare..........	4,841	4,757	3,768
Davidson......	29,404	23,403	21,702
Davie.........	13,394	12,115	11,621
Duplin........	25,442	22,405	18,690
Durham.......	35,276	26,233	18,041
Edgecombe....	32,010	26,591	24,113
Forsyth.......	47,311	35,261	28,434
Franklin.......	24,692	25,116	21,090
Gaston........	37,063	27,903	17,764
Gates.........	10,455	10,413	10 252
Graham........	4,749	4,343	3,313
Granville......	25,102	23,263	24,484
Greene........	13,083	12,038	10,039
Guilford.......	60,497	39,074	28,052
Halifax........	37,646	30,793	28,908
Harnett.......	22,174	15,988	13,700
Haywood......	21,020	16,222	13,346
Henderson.....	16,262	14,104	12,589
Hertford.......	15,436	14,294	13,851
Hyde..........	8,840	9,278	8,903
Iredell........	34,315	29,064	25,462
Jackson.......	12,998	11,853	9,512
Johnston......	41,401	32,250	27,239
Jones.........	8,721	8,226	7,403
Lee..........	11,376	......	......
Lenoir........	22,769	18,639	14,879
Lincoln.......	17,132	15,498	12,586
McDowell.....	13,538	12,567	10,939
Macon........	12,191	12,104	10,102
Madison.......	20,132	20,644	17,805
Martin........	17,797	15,383	15,221
Mecklenburg...	67,031	55,268	42,673
Mitchell........	17,245	15,221	12,807

NORTH CAROLINA Cont'd.

COUNTY	1910	1900	1890
Montgomery....	14,967	14,197	11,239
Moore............	17,010	23,622	20,479
Nash.............	33,727	25,478	20,707
New Hanover....	32,037	25,785	24,026
Northampton....	22,323	21,150	21,242
Onslow..........	14,125	11,940	10,303
Orange..........	15,064	14,690	14,948
Pamlico.........	9,966	8,045	7,146
Pasquotank.....	16,693	13,660	10,748
Pender..........	15,471	13,381	12,514
Perquimans.....	11,054	10,091	9,293
Person..........	17,356	16,685	15,151
Pitt.............	36,340	30,889	25,519
Polk............	7,640	7,004	5,902
Randolph........	29,491	28,232	25,195
Richmond........	19,673	15,855	23,948
Robeson.........	51,945	40,371	31,483
Rockingham.....	36,442	33,163	25,363
Rowan...........	37,521	31,066	24,123
Rutherford......	28,385	25,101	18,770
Sampson.........	29,982	26,380	25,096
Scotland........	15,363	12,553	......
Stanly...........	19,909	15,220	12,136
Stokes..........	20,151	19,866	17,199
Surry...........	29,705	25,515	19,281
Swain...........	10,403	8,401	6,577
Transylvania....	7,191	6,620	5,881
Tyrrell.........	5,219	4,980	4,225
Union...........	33,277	27,156	21,259
Vance...........	19,425	16,684	17,581
Wake...........	63,229	54,626	49,207
Warren..........	20,266	19,151	19,360
Washington......	11,062	10,608	10,200
Watauga.........	13,556	13,417	10,611
Wayne..........	35,698	31,356	26,100
Wilkes..........	30,282	26,872	22,675
Wilson..........	28,269	23,596	18,644
Yadkin..........	15,428	14,083	13,790
Yancey..........	12,072	11,464	9,490

The State 2,206,287 1,893,810 1,617,947

NORTH DAKOTA

COUNTY	1910	1900	1890
Adams..........	5,407	......	......
Barnes..........	18,066	13,159	7,045
Benson..........	12,681	8,320	2,460
Billings........	10,186	975	176
Bottineau.......	17,295	7,532	2,893
Bowman.........	4,668	......	......
Burleigh........	13,087	6,081	4,247
Cass............	33,935	28,625	19,613
Cavalier........	15,659	12,580	6,471
Dickey..........	9,839	6,061	5,573
Dunn...........	5,302	......	159
Eddy............	4,800	3,330	1,377
Emmons.........	9,796	4,349	1,971
Foster..........	5,313	3,770	1,210
Grand Forks....	27,888	24,459	18,357
Griggs..........	6,274	4,744	2,817
Hettinger.......	6,557	......	81
Kidder..........	5,962	1,754	1,211
Lamoure........	10,724	6,048	3,187
Logan...........	6,168	1,625	597
McHenry........	17,627	5,253	1,584
McIntosh.......	7,251	4,818	3,248
McKenzie.......	5,720	......	3
McLean.........	14,578	4,791	893
Mercer.........	4,665	1,778	428
Mountrail......	8,491	......	122
Morton.........	25,289	8,069	4,728
Nelson.........	10,140	7,316	4,293
Oliver..........	3,577	990	464
Pembina........	14,749	17,869	14,334

NORTH DAKOTA Cont'd.

COUNTY	1910	1900	1890
Pierce.........	9,740	4,765	905
Ramsey.........	15,199	9,198	4,418
Ransom.........	10,345	6,919	5,393
Richland........	19,659	17,387	10,751
Rolette.........	9,558	7,995	2,427
Sargent.........	9,202	6,039	5,076
Sheridan........	8,103	...	5
Stark...........	12,504	7,621	2,328
Steele..........	7,616	5,888	3,777
Stutsman........	18,189	9,143	5,266
Towner..........	8,963	6,491	1,450
Traill..........	12,545	13,107	10,217
Walsh..........	19,491	20,288	16,587
Ward...........	42,185	7,961	1,681
Wells..........	11,814	8,310	1,212
Williams........	20,249	1,530	875

The State 577,056 a319,146 b182,719

(a) Includes population (2,208) of Standing Rock Indian reservation, not returned by counties in 1900.

(b) Includes population (809) of Fort Yates and Standing Rock Indian Agency, and of Church, Renville, Stevens, and Williams counties.

OHIO

COUNTY	1910	1900	1890
Adams..........	24,755	26,328	26,093
Allen...........	56,580	47,976	40,644
Ashland........	22,975	21,184	22,223
Ashtabula......	59,547	51,448	43,655
Athens.........	47,798	38,730	35,194
Auglaize........	31,246	31,192	28,100
Belmont........	76,856	60,875	57,413
Brown..........	24,832	28,237	29,899
Butler..........	70,271	56,870	48,597
Carroll.........	15,761	16,811	17,566
Champaign......	26,351	26,642	26,980
Clark...........	66,435	58,939	52,277
Clermont.......	29,551	31,610	33,553
Clinton.........	23,680	24,202	24,240
Columbiana.....	76,619	68,590	59,029
Coshocton......	30,121	29,337	26,703
Crawford.......	34,036	33,915	31,927
Cuyahoga.......	637,425	439,120	309,970
Darke..........	42,933	42,532	42,961
Defiance........	24,498	26,387	25,760
Delaware.......	27,182	26,401	27,189
Erie............	38,327	37,650	35,462
Fairfield.......	39,201	34,259	33,939
Fayette.........	21,744	21,725	22,309
Franklin........	221,567	164,460	124,087
Fulton.........	23,914	22,801	22,023
Gallia..........	25,745	27,918	27,005
Geauga.........	14,670	14,744	13,489
Greene..........	29,733	31,613	29,820
Guernsey.......	42,716	34,425	28,645
Hamilton.......	460,732	409,479	374,573
Hancock........	37,860	41,993	42,563
Hardin.........	30,407	31,187	28,939
Harrison........	19,076	20,486	20,830
Henry..........	25,119	27,282	25,080
Highland........	28,711	30,982	29,048
Hocking........	23,650	24,398	22,658
Holmes.........	17,909	19,511	21,139
Huron..........	34,206	32,330	31,949
Jackson........	30,791	34,248	28,408
Jefferson.......	65,423	44,357	39,415
Knox...........	30,181	27,768	27,600
Lake...........	22,927	21,680	18,235
Lawrence.......	39,488	39,534	39,556
Licking........	55,590	47,070	43,279

OHIO Cont'd.

COUNTY	1910	1900	1890
Logan..........	30,084	30,420	27,386
Lorain..........	76,037	54,857	40,295
Lucas..........	192,728	153,559	102,296
Madison........	19,902	20,590	20,057
Mahoning.......	116,151	70,134	55,979
Marion.........	33,971	28,678	24,727
Medina.........	23,598	21,958	21,742
Meigs..........	25,594	28,620	29,813
Mercer.........	27,536	28,021	27,220
Miami..........	45,047	43,105	39,754
Monroe.........	24,244	27,031	25,175
Montgomery.....	163,763	130,146	100,852
Morgan.........	16,097	17,905	19,143
Morrow.........	16,815	17,879	18,120
Muskingum......	57,488	53,185	51,210
Noble..........	18,601	19,466	20,753
Ottawa.........	22,360	22,213	21,974
Paulding.......	22,730	27,528	25,932
Perry..........	35,396	31,841	31,151
Pickaway.......	26,158	27,016	26,959
Pike...........	15,723	18,172	17,482
Portage........	30,307	29,246	27,868
Preble.........	23,834	23,713	23,421
Putnam.........	29,972	32,525	30,188
Richland........	47,667	44,289	38,072
Ross...........	40,069	40,940	39,454
Sandusky.......	35,171	34,311	30,617
Scioto.........	48,463	40,981	35,377
Seneca.........	42,421	41,163	40,869
Shelby.........	24,663	24,625	24,707
Stark..........	122,987	94,747	84,170
Summit.........	108,253	71,715	54,089
Trumbull.......	52,766	46,591	42,373
Tuscarawas.....	57,035	53,751	46,618
Union..........	21,871	22,342	22,860
Van Wert.......	29,119	30,394	29,671
Vinton.........	13,096	15,330	16,045
Warren.........	24,497	25,584	25,468
Washington.....	45,422	48,245	42,380
Wayne.........	38,058	37,870	39,005
Williams.......	25,198	24,953	24,897
Wood..........	46,330	51,555	44,392
Wyandot........	20,760	21,125	21,722

The State 4,767,121 4,157,545 3,672,316

OKLAHOMA

COUNTY	1910	1907
Adair..........	10,535	9,115
Alfalfa.........	18,138	16,070
Atoka..........	13,808	12,113
Beaver.........	13,631	13,364
Beckham........	19,699	17,758
Blaine.........	17,960	17,227
Bryan..........	29,854	27,865
Caddo..........	35,685	30,241
Canadian.......	23,501	20,110
Carter..........	25,358	26,402
Cherokee.......	16,778	14,274
Choctaw........	21,862	17,340
Cimarron.......	4,553	5,927
Cleveland......	18,843	18,460
Coal...........	15,817	15,585
Comanche......	41,489	31,738
Craig..........	17,404	14,955
Creek..........	26,223	18,365
Custer.........	23,231	18,478
Delaware.......	11,469	9,876
Dewey..........	14,132	13,329
Ellis..........	15,375	13,978
Garfield........	33,050	28,300
Garvin.........	26,545	22,787
Grady..........	30,309	23,420
Grant..........	18,760	17,638

OKLAHOMA Cont'd.

COUNTY	1910	1907
Greer	16,449	23,624
Harmon	11,328	......
Harper	8,189	8,089
Haskell	18,875	16,865
Hughes	24,040	19,945
Jackson	23,737	17,087
Jefferson	17,430	13,439
Johnston	16,734	18,672
Kay	26,999	24,757
Kingfisher	18,825	18,010
Kiowa	27,526	22,247
Latimer	11,321	9,340
Le Flore	29,127	24,678
Lincoln	34,779	37,293
Logan	31,740	30,711
Love	10,236	11,134
McClain	15,659	12,888
McCurtain	20,681	13,198
McIntosh	20,961	17,975
Major	15,248	14,307
Marshall	11,619	13,144
Mayes	13,596	11,064
Murray	12,744	11,948
Muskogee	52,743	37,467
Noble	14,945	14,198
Nowata	14,223	10,453
Okfuskee	19,995	15,595
Oklahoma	85,232	55,849
Okmulgee	21,115	14,362
Osage	20,101	15,332
Ottawa	15,713	12,827
Pawnee	17,332	17,112
Payne	23,735	22,022
Pittsburg	47,650	37,677
Pontotoc	24,331	23,057
Pottawatomie	43,595	43,272
Pushmataha	10,118	8,295
Roger Mills	12,861	13,239
Rogers	17,736	15,485
Seminole	19,964	14,687
Sequoyah	25,005	22,499
Stephens	22,252	20,148
Texas	14,249	16,448
Tillman	18,650	12,869
Tulsa	34,995	21,693
Wagoner	22,086	19,529
Washington	17,484	12,813
Washita	25,034	22,007
Woods	17,567	15,517
Woodward	16,592	14,595

The State 1,657,155 1,414,177

OREGON

COUNTY	1910	1900	1890
Baker	18,076	15,597	6,764
Benton	10,663	6,706	8,650
Clackamas	29,931	19,658	15,233
Clatsop	16,106	12,765	10,016
Columbia	10,580	6,237	5,191
Coos	17,959	10,324	8,874
Crook	9,315	3,964	3,244
Curry	2,044	1,868	1,709
Douglas	19,674	14,565	11,864
Gilliam	3,701	3,201	3,600
Grant	5,607	5,948	5,080
Harney	4,059	2,598	2,559
Hood River	8,016	......	......
Jackson	25,756	13,698	11,455
Josephine	9,567	7,517	4,878
Klamath	8,554	3,970	2,444
Lake	4,658	2,847	2,604
Lane	33,783	19,604	15,198
Lincoln	5,587	3,575	......

OREGON Cont'd.

COUNTY	1910	1000	1890
Linn	22,662	18,603	16,265
Malheur	8,601	4,203	2,601
Marion	39,780	27,713	22,934
Morrow	4,357	4,151	4,205
Multnomah	226,261	103,167	74,884
Polk	13,469	9,923	7,858
Sherman	4,242	3,477	1,792
Tillamook	6,266	4,471	2,932
Umatilla	20,309	18,049	13,381
Union	16,191	16,070	12,044
Wallowa	8,364	5,538	3,661
Wasco	16,336	13,199	9,183
Washington	21,522	14,467	11,972
Wheeler	2,484	2,443	......
Yamhill	18,285	13,420	10,692

The State.... 672,765 413,536 313,767

PENNSYLVANIA

COUNTY	1910	1900	1890
Adams	34,319	34,496	33,486
Allegheny	1,018,463	775,058	551,959
Armstrong	67,880	52,551	46,747
Beaver	78,353	56,432	50,077
Bedford	38,879	39,468	38,644
Berks	183,222	159,615	137,327
Blair	108,858	85,099	70,866
Bradford	54,526	59,403	59,233
Bucks	76,530	71,190	70,615
Butler	72,689	56,962	55,339
Cambria	166,131	104,837	66,375
Cameron	7,644	7,048	7,238
Carbon	52,846	44,510	38,624
Center	43,424	42,894	43,269
Chester	109,213	95,695	89,377
Clarion	36,638	34,283	36,802
Clearfield	93,768	80,614	69,565
Clinton	31,545	29,197	28,685
Columbia	48,467	39,896	36,832
Crawford	61,565	63,643	65,324
Cumberland	54,479	50,344	47,271
Dauphin	136,152	114,443	96,977
Delaware	117,906	94,762	74,683
Elk	35,871	32,903	22,239
Erie	115,517	98,473	86,074
Fayette	167,449	110,412	80,006
Forest	9,435	11,039	8,482
Franklin	59,775	54,902	51,433
Fulton	9,703	9,924	10,137
Greene	28,882	28,281	28,935
Huntingdon	38,304	34,650	35,751
Indiana	66,210	42,556	42,175
Jefferson	63,090	59,113	44,005
Juniata	15,013	16,054	16,655
Lackawanna	259,570	193,831	142,088
Lancaster	167,029	159,241	149,095
Lawrence	70,032	57,042	37,517
Lebanon	59,565	53,827	48,131
Lehigh	118,832	93,893	76,631
Luzerne	343,186	257,121	201,203
Lycoming	80,813	75,663	70,579
McKean	47,868	51,343	46,863
Mercer	77,699	57,387	55,744
Mifflin	27,785	23,160	19,996
Monroe	22,941	21,161	20,111
Montgomery	169,590	138,995	123,290
Montour	14,868	15,526	15,645
Northampton	127,667	99,687	84,220
Northumberland	111,420	90,911	74,698
Perry	24,136	26,263	26,276
Philadelphia	1,549,008	1,293,697	1,046,964
Pike	8,033	8,766	9,412
Potter	29,729	30,621	22,778
Schuylkill	207,894	172,927	154,163

PENNSYLVANIA Cont'd.

COUNTY	1910	1900	1890
Snyder	16,800	17,304	17,651
Somerset	67,717	49,461	37,317
Sullivan	11,293	12,134	11,620
Susquehanna	37,746	40,043	40,093
Tioga	42,829	49,086	52,313
Union	16,249	17,592	17,820
Venango	56,359	49,648	46,640
Warren	39,573	38,946	37,585
Washington	143,680	92,181	71,155
Wayne	29,236	30,171	31,010
Westmoreland	231,304	160,175	112,819
Wyoming	15,509	17,152	15,891
York	136,405	116,413	99,489

The State 7,665,111 6,302,115 5,258,014

RHODE ISLAND

COUNTY	1910	1900	1890
Bristol	17,602	13,144	11,428
Kent	36,378	29,976	26,754
Newport	39,335	32,599	28,552
Providence	424,417	328,683	255,123
Washington	24,942	24,154	23,649

The State 542,610 428,556 345,506

SOUTH CAROLINA

COUNTY	1910	1900	1890
Abbeville	34,804	33,400	46,854
Aiken	41,849	39,032	31,822
Anderson	69,568	55,728	43,696
Bamberg	18,544	17,296	......
Barnwell	34,209	35,504	44,613
Beaufort	30,355	35,495	34,119
Berkeley	23,487	30,454	55,428
Calhoun	16,634	......	......
Charleston	88,594	88,006	59,903
Cherokee	26,179	21,359	......
Chester	29,425	28,616	26,660
Chesterfield	26,301	20,401	18,468
Clarendon	32,188	28,184	23,233
Colleton	35,390	33,452	40,293
Darlington	36,027	32,388	29,134
Dillon	22,615	......	......
Dorchester	17,891	16,294	......
Edgefield	28,281	25,478	49,259
Fairfield	29,442	29,425	28,599
Florence	35,671	28,474	25,027
Georgetown	22,270	22,846	20,857
Greenville	68,377	53,490	44,310
Greenwood	34,225	28,343	......
Hampton	25,126	23,738	20,544
Horry	26,995	23,364	19,256
Kershaw	27,094	24,696	22,361
Lancaster	26,650	24,311	20,761
Laurens	41,550	37,382	31,610
Lee	23,318	......	......
Lexington	32,040	27,264	22,181
Marion	20,596	35,181	29,976
Marlboro	31,189	27,639	23,500
Newberry	34,586	30,182	26,434
Oconee	27,337	23,634	18,687
Orangeburg	55,893	59,663	49,393
Pickens	25,422	19,375	16,389
Richland	55,143	45,589	36,821
Saluda	20,943	18,966	......
Spartanburg	83,465	65,560	55,385
Sumter	38,472	51,237	43,605
Union	29,911	25,501	25,363
Williamsburg	37,626	31,685	27,777
York	47,718	41,684	38,831

The State 1,515,400 1,340,316 1,151,149

SOUTH DAKOTA

COUNTY	1910	1900	1890
Armstrong......	647	8	34
Aurora.........	6,143	4,011	5,045
Beadle.........	15,776	8,081	9,586
Bonhomme......	11,061	10,379	9,057
Brookings......	14,178	12,561	10,132
Brown.........	25,867	15,286	16,855
Brule..........	6,451	5,401	6,737
Buffalo........	1,589	1,790	993
Butte..........	4,993	2,907	1,068
Campbell.......	5,244	4,527	3,510
Charles Mix....	14,899	8,498	4,178
Clark..........	10,901	6,942	6,728
Clay..........	8,711	9,316	7,509
Codington......	14,092	8,770	7,037
Corson........	2,929	......	......
Custer.........	4,458	2,728	4,891
Davison........	11,625	7,483	5,449
Day...........	14,372	12,254	9,168
Deuel..........	7,768	6,656	4,574
Dewey.........	1,145	......	......
Douglas........	6,400	5,012	4,600
Edmunds.......	7,654	4,916	4,399
Fall River......	7,763	3,541	4,478
Faulk.........	6,716	3,547	4,062
Grant.........	10,303	9,103	6,814
Gregory........	13,061	2,211	483
Hamlin........	7,475	5,945	4,625
Hand..........	7,870	4,525	6,546
Hanson........	6,237	4,947	4,267
Harding........	4,228	......	167
Hughes........	6,271	3,684	5,044
Hutchinson....	12,319	11,897	10,469
Hyde..........	3,307	1,492	1,860
Jerauld........	5,120	2,798	3,605
Kingsbury......	12,560	9,866	8,562
Lake..........	10,711	9,137	7,508
Lawrence.......	19,694	17,897	11,673
Lincoln........	12,712	12,161	9,143
Lyman.........	10,848	2,632	437
McCook........	9,589	8,689	6,448
McPherson.....	6,791	6,327	5,940
Marshall.......	8,021	5,942	4,544
Meade.........	12,640	4,907	4,712
Miner.........	7,661	5,864	5,165
Minnehaha.....	29,631	23,926	21,879
Moody.........	8,695	8,326	5,941
Pennington.....	12,453	5,610	7,050
Perkins........	11,348	......	......
Potter.........	4,466	2,988	2,910
Roberts........	14,897	12,216	1,997
Sanborn........	6,607	4,464	4,610
Schnasse.......	292	......	......
Spink.........	15,981	9,487	10,581
Stanley........	14,975	1,341	1,207
Sterling........	252	......	96
Sully..........	2,462	1,715	2,412
Tripp..........	8,323	......	......
Turner.........	13,840	13,175	10,256
Union.........	10,676	11,153	9,130
Walworth.......	6,488	3,839	2,153
Yankton.......	13,135	12,649	10,444
Pine Ridge Indian reservation (c).	6,607	6,827	......
Rosebud Indian reservation (c).	3,960	5,201	......
The State	583,888	a401,570	b328,808

(a) Includes population (4,015) of Cheyenne and Standing Rock Indian reservations not returned by counties in 1900.

(b) Includes population (40) of Washington county in 1890.

(c) Includes unorganized counties (Bennett, Mellette, Shannon, Todd, Washa-

SOUTH DAKOTA Cont'd.

baugh, and Washington) for which population was not separately returned in 1910; part of Rosebud Indian reservation attached to Gregory and Tripp counties since 1900

TENNESSEE

COUNTY	1910	1900	1890
Anderson.......	17,717	17,634	15,128
Bedford........	22,667	23,845	24,739
Benton.........	12,452	11,888	11,230
Bledsoe........	6,329	6,626	6,134
Blount.........	20,809	19,206	17,589
Bradley........	16,336	15,759	13,607
Campbell.......	27,387	17,317	13,486
Cannon........	10,825	12,121	12,197
Carroll........	23,971	24,250	23,630
Carter.........	19,838	16,688	13,389
Cheatham......	10,540	10,112	8,845
Chester........	9,090	9,896	9,069
Claiborne......	23,504	20,696	15,103
Clay..........	9,009	8,421	7,260
Cocke.........	19,399	19,153	16,523
Coffee.........	15,625	15,574	13,827
Crockett.......	16,076	15,867	15,146
Cumberland....	9,327	8,311	5,376
Davidson.......	149,478	122,815	108,174
Decatur........	10,093	10,439	8,995
Dekalb........	15,434	16,460	15,650
Dickson........	19,955	18,635	13,645
Dyer..........	27,721	23,776	19,878
Fayette........	30,257	29,701	28,878
Fentress.......	7,446	6,106	5,226
Franklin.......	20,491	20,392	18,929
Gibson.........	41,630	39,408	35,859
Giles..........	32,629	33,035	34,957
Grainger.......	13,888	13,512	13,196
Greene.........	31,083	30,596	26,614
Grundy........	8,322	7,802	6,345
Hamblen.......	13,650	12,728	11,418
Hamilton......	89,267	61,695	53,482
Hancock.......	10,778	11,147	10,342
Hardeman.....	23,011	22,976	21,029
Hardin........	17,521	19,246	17,698
Hawkins.......	23,587	24,267	22,246
Haywood.......	25,910	25,189	23,558
Henderson.....	17,030	18,117	16,336
Henry.........	25,434	24,208	21,070
Hickman.......	16,527	16,367	14,499
Houston........	6,224	6,476	5,390
Humphreys.....	13,908	13,398	11,720
Jackson........	15,036	15,039	13,325
James.........	5,210	5,407	4,903
Jefferson.......	17,755	18,590	16,478
Johnson........	13,191	10,589	8,858
Knox..........	94,187	74,302	59,557
Lake..........	8,704	7,368	5,304
Lauderdale	21,105	21,971	18,756
Lawrence	17,569	15,402	12,286
Lewis.........	6,033	4,455	2,555
Lincoln........	25,908	26,304	27,382
Loudon........	13,612	10,838	9,273
McMinn........	21,046	19,163	17,890
McNairy.......	16,356	17,760	15,510
Macon.........	14,559	12,881	10,878
Madison.......	39,357	36,333	30,497
Marion........	18,820	17,281	15,411
Marshall.......	16,872	18,763	18,906
Maury.........	40,456	42,703	38,112
Meigs.........	6,131	7,491	6,930
Monroe........	20,716	18,585	15,329
Montgomery...	33,672	36,017	29,697
Moore.........	4,800	5,706	5,975
Morgan........	11,458	9,587	7,639
Obion.........	29,946	28,286	27,273
Overton........	15,854	13,353	12,039
Perry.........	8,815	8,800	7,785

TENNESSEE Cont'd.

COUNTY	1910	1900	1890
Pickett........	5,087	5,366	4,736
Polk..........	14,116	11,357	8,361
Putnam........	20,023	16,890	13,683
Rhea..........	15,410	14,318	12,647
Roane.........	22,860	22,738	17,418
Robertson......	25,466	25,029	20,078
Rutherford.....	33,199	33,543	35,097
Scott..........	12,947	11,077	9,794
Sequatchie.....	4,202	3,326	3,027
Sevier.........	22,296	22,021	18,761
Shelby.........	191,439	153,557	112,740
Smith.........	18,548	19,026	18,404
Stewart........	14,860	15,224	12,193
Sullivan........	28,120	24,935	20,879
Sumner........	25,621	26,072	23,668
Tipton.........	29,459	29,273	24,271
Trousdale......	5,874	6,004	5,850
Unicoi.........	7,201	5,851	4,619
Union.........	11,414	12,894	11,459
Van Buren.....	2,784	3,126	2,863
Warren........	16,534	16,410	14,413
Washington....	28,968	22,604	20,354
Wayne.........	12,062	12,936	11,471
Weakley.......	31,929	32,546	28,955
White.........	15,420	14,157	12,348
Williamson.....	24,213	26,429	26,321
Wilson........	25,394	27,078	27,148
The State	2,184,789	2,020,616	1,767,518

TEXAS

COUNTY	1910	1900	1890
Anderson.......	29,650	28,015	20,923
Andrews........	975	87	24
Angelina.......	17,705	13,481	6,306
Aransas........	2,106	1,716	1,824
Archer.........	6,525	2,508	2,101
Armstrong......	2,682	1,205	944
Atascosa.......	10,004	7,143	6,459
Austin.........	17,699	20,676	17,859
Bailey.........	312	4	......
Bandera.......	4,921	5,332	3,795
Bastrop........	25,344	26,845	20,736
Baylor.........	8,411	3,052	2,595
Bee...........	12,090	7,720	3,720
Bell..........	49,186	45,535	33,377
Bexar.........	119,676	69,422	49,266
Blanco.........	4,311	4,703	4,649
Borden........	1,386	776	222
Bosque........	19,013	17,390	14,224
Bowie.........	34,827	26,676	20,267
Brazoria.......	13,299	14,861	11,506
Brazos........	18,919	18,859	16,650
Brewster.......	5,220	2,356	1,033
Briscoe........	2,162	1,253	......
Brown.........	22,935	16,019	11,421
Burleson.......	18,687	18,367	13,001
Burnet........	10,755	10,528	10,747
Caldwell.......	24,237	21,765	15,769
Calhoun........	3,635	2,395	815
Callahan.......	12,973	8,768	5,457
Cameron.......	27,158	16,095	14,424
Camp.........	9,551	9,146	6,624
Carson.........	2,127	469	356
Cass..........	27,587	22,841	22,554
Castro.........	1,850	400	9
Chambers......	4,234	3,046	2,241
Cherokee......	29,038	25,154	22,975
Childress......	9,538	2,138	1,175
Clay..........	17,043	9,231	7,503
Cochran........	65	25	......
Coke..........	6,412	3,430	2,059
Coleman.......	22,618	10,077	6,112
Collin.........	49,021	50,087	36,736

TEXAS Cont'd.

COUNTY	1910	1900	1890
Collingsworth...	5,224	1,233	357
Colorado.........	18,897	22,203	19,512
Comal............	8,434	7,008	6,398
Comanche........	27,186	23,009	15,608
Concho..........	6,654	1,427	1,065
Cooke...........	26,603	27,494	24,696
Coryell..........	21,703	21,308	16,873
Cottle...........	4,396	1,002	240
Crane...........	331	51	15
Crockett.........	1,296	1,591	194
Crosby..........	1,765	788	346
Dallam..........	4,001	146	112
Dallas..........	135,748	82,726	67,042
Dawson.........	2,320	37	29
De Witt.........	23,501	21,311	14,307
Deaf Smith.....	3,942	843	179
Delta...........	14,566	15,249	9,117
Denton..........	31,258	28,318	21,289
Dickens.........	3,092	1,151	295
Dimmit.........	3,460	1,106	1,049
Donley.........	5,284	2,756	1,056
Duval..........	8,964	8,483	7,598
Eastland........	23,421	17,971	10,373
Ector..........	1,178	381	224
Edwards........	3,768	3,108	1,970
Ellis...........	53,629	50,059	31,774
El Paso.........	52,599	24,886	15,678
Erath..........	32,095	29,966	21,594
Falls..........	35,649	33,342	20,706
Fannin.........	44,801	51,793	38,709
Fayette.........	29,796	36,542	31,481
Fisher..........	12,596	3,708	2,996
Floyd..........	4,638	2,020	529
Foard..........	5,726	1,568	
Fort Bend......	18,168	16,538	10,586
Franklin........	9,331	8,674	6,481
Freestone.......	20,557	18,910	15,987
Frio...........	8,895	4,200	3,112
Gaines..........	1,255	55	68
Galveston.......	44,479	44,116	31,476
Garza..........	1,995	185	14
Gillespie.......	9,447	8,229	7,056
Glasscock.......	1,143	286	208
Goliad..........	9,909	8,310	5,910
Gonzales........	28,055	28,882	18,016
Gray...........	3,405	480	203
Grayson.........	65,996	63,661	53,211
Gregg..........	14,140	12,343	9,402
Grimes.........	21,205	26,106	21,312
Guadalupe.......	24,913	21,385	15,217
Hale...........	7,566	1,680	721
Hall...........	8,279	1,670	703
Hamilton........	15,315	13,520	9,313
Hansford........	935	167	133
Hardeman.......	11,213	3,634	3,904
Hardin..........	12,947	5,049	3,956
Harris..........	115,693	63,786	37,249
Harrison........	37,243	31,878	26,721
Hartley........	1,298	377	252
Haskell.........	16,249	2,637	1,665
Hays...........	15,518	14,142	11,352
Hemphill.......	3,170	815	519
Henderson......	20,131	19,970	12,285
Hidalgo........	13,728	6,837	6,534
Hill...........	46,760	41,355	27,583
Hockley........	137	44	
Hood..........	10,008	9,146	7,614
Hopkins........	31,038	27,950	20,572
Houston........	29,564	25,452	19,360
Howard.........	8,881	2,528	1,210
Hunt...........	48,116	47,295	31,885
Hutchinson.....	892	303	58
Irion..........	1,283	848	870
Jack...........	11,817	10,224	9,740
Jackson.........	6,471	6,094	3,281
Jasper..........	14,000	7,138	5,592
Jeff Davis......	1,678	1,150	1,394

TEXAS Cont'd.

COUNTY	1910	1900	1890
Jefferson.......	38,182	14,239	5,857
Johnson........	34,460	33,819	22,313
Jones..........	24,299	7,053	3,797
Karnes.........	14,942	8,681	3,637
Kaufman........	35,323	33,376	21,598
Kendall........	4,517	4,103	3,826
Kent..........	2,655	899	324
Kerr...........	5,505	4,980	4,462
Kimble.........	3,261	2,503	2,243
King...........	810	490	173
Kinney.........	3,401	2,447	3,781
Knox...........	9,625	2,322	1,134
La Salle........	4,747	2,303	2,139
Lamar..........	46,544	48,627	37,302
Lamb..........	540	31	4
Lampasas.......	9,532	8,625	7,584
Lavaca.........	26,418	28,121	21,887
Lee...........	13,132	14,595	11,952
Leon..........	16,583	18,072	13,841
Liberty........	10,686	8,102	4,230
Limestone......	34,621	32,573	21,678
Lipscomb.......	2,634	790	632
Live Oak.......	3,442	2,268	2,055
Llano..........	6,520	7,301	6,772
Loving.........	249	33	3
Lubbock........	3,624	293	33
Lynn..........	1,713	17	24
McCulloch......	13,405	3,960	3,217
McLennan.......	73,250	59,772	39,204
McMullen.......	1,091	1,024	1,038
Madison........	10,318	10,432	8,512
Marion.........	10,472	10,754	10,862
Martin.........	1,549	332	264
Mason..........	5,683	5,573	5,180
Matagorda......	13,594	6,097	3,985
Maverick.......	5,151	4,066	3,698
Medina.........	13,415	7,783	5,730
Menard.........	2,707	2,011	1,215
Midland........	3,464	1,741	1,033
Milam..........	36,780	39,666	24,773
Mills..........	9,694	7,851	5,493
Mitchell........	8,956	2,855	2,059
Montague.......	25,123	24,800	18,863
Montgomery.....	15,679	17,067	11,765
Moore..........	561	209	15
Morris.........	10,439	8,220	6,580
Motley.........	2,396	1,257	139
Nacogdoches....	27,406	24,663	15,984
Navarro........	47,070	43,374	26,373
Newton.........	10,850	7,282	4,650
Nolan..........	11,999	2,611	1,573
Nueces.........	21,955	10,439	8,093
Ochiltree.......	1,602	267	198
Oldham.........	812	349	270
Orange.........	9,528	5,905	4,770
Palo Pinto......	19,506	12,291	8,320
Panola.........	20,424	21,404	14,328
Parker.........	26,331	25,823	21,682
Parmer.........	1,555	34	7
Pecos..........	2,071	2,360	1,326
Polk...........	17,459	14,447	10,332
Potter.........	12,424	1,820	849
Presidio........	5,218	3,673	1,698
Rains..........	6,787	6,127	3,909
Randall........	3,312	963	187
Reagan.........	392		
Red River......	28,564	29,893	21,452
Reeves.........	4,392	1,847	1,247
Refugio........	2,814	1,641	1,239
Roberts........	950	620	326
Robertson......	27,454	31,480	26,506
Rockwall.......	8,072	8,531	5,972
Runnels........	20,858	5,379	3,193
Rusk..........	26,946	26,099	18,559
Sabine.........	8,582	6,394	4,969
San Augustine...	11,264	8,434	6,688
San Jacinto.....	9,542	10,277	7,360

TEXAS Cont'd.

COUNTY	1910	1900	1890
San Patricio.....	7,307	2,372	1,312
San Saba.......	11,245	7,569	6,641
Schleicher......	1,893	515	155
Scurry.........	10,924	4,158	1,415
Shackelford.....	4,201	2,461	2,012
Shelby.........	26,423	20,452	14,365
Sherman........	1,376	104	34
Smith.........	41,746	37,370	28,324
Somervell.......	3,931	3,498	3,419
Starr..........	13,151	11,469	10,749
Stephens.......	7,980	6,466	4,926
Sterling........	1,493	1,127	......
Stonewall.......	5,320	2,183	1,024
Sutton.........	1,569	1,727	658
Swisher........	4,012	1,227	100
Tarrant........	108,572	52,376	41,142
Taylor.........	26,293	10,499	6,957
Terrell.........	1,430	......	......
Terry..........	1,474	48	21
Throckmorton...	4,563	1,750	902
Titus..........	16,422	12,292	8,190
Tom Green......	17,882	6,804	5,152
Travis.........	55,620	47,386	36,322
Trinity........	12,768	10,976	7,648
Tyler..........	10,250	11,899	10,877
Upshur.........	19,960	16,266	12,695
Upton..........	501	48	52
Uvalde.........	11,233	4,647	3,804
Val Verde......	8,613	5,263	2,874
Van Zandt......	25,651	25,481	16,225
Victoria........	14,990	13,678	8,737
Walker.........	16,061	15,813	12,874
Waller.........	12,138	14,246	10,888
Ward..........	2,389	1,451	77
Washington.....	25,561	32,931	29,161
Webb..........	22,503	21,851	17,586
Wharton........	21,123	16,942	7,584
Wheeler........	5,258	636	778
Wichita........	16,094	5,806	4,831
Wilbarger......	12,000	5,759	7,092
Williamson.....	42,228	38,072	25,909
Wilson.........	17,066	13,961	10,655
Winkler........	442	60	18
Wise..........	26,450	27,116	24,134
Wood..........	23,417	21,048	13,932
Yoakum.........	602	26	4
Young.........	13,657	6,540	5,049
Zapata.........	3,809	4,760	3,562
Zavalla........	1,889	792	1,097

The State 3,896,542 3,048,710 2,235,523

UTAH

COUNTY	1910	1900	1890
Beaver.........	4,717	3,613	3,340
Boxelder.......	13,894	10,009	7,642
Cache..........	23,062	18,139	15,509
Carbon.........	8,624	5,004	......
Davis..........	10,191	7,996	6,751
Emery.........	6,750	4,657	5,076
Garfield........	3,660	3,400	2,457
Grand..........	1,595	1,149	541
Iron...........	3,933	3,546	2,683
Juab..........	10,702	10,082	5,582
Kane..........	1,652	1,811	1,685
Millard.........	6,118	5,678	4,033
Morgan.........	2,467	2,045	1,780
Piute..........	1,734	1,954	2,842
Rich..........	1,883	1,946	1,527
Salt Lake......	131,426	77,725	58,457
San Juan.......	2,377	1,023	365
Sanpete........	16,704	16,313	13,146
Sevier..........	9,775	8,451	6,199

UTAH Cont'd.

COUNTY	1910	1900	1890
Summit	8,200	9,439	7,733
Tooele	7,924	7,361	3,700
Uinta	7,050	6,458	2,762
Utah	37,942	32,456	23,768
Wasatch	8,920	4,736	3,595
Washington	5,123	4,612	4,009
Wayne	1,749	1,907	......
Weber	35,179	25,239	22,723
The State	373,351	276,749	207,905

VERMONT

COUNTY	1910	1900	1890
Addison	20,010	21,912	22,277
Bennington	21,378	21,705	20,448
Caledonia	26,031	24,381	23,436
Chittenden	42,447	39,600	35,389
Essex	7,384	8,056	9,511
Franklin	29,866	30,198	29,755
Grand Isle	3,761	4,462	3,843
Lamoille	12,585	12,289	12,831
Orange	18,703	19,313	19,575
Orleans	23,337	22,024	22,101
Rutland	48,139	44,209	45,397
Washington	41,702	36,607	29,606
Windham	26,932	26,660	26,547
Windsor	33,681	32,225	31,706
The State	355,956	343,641	332,422

VIRGINIA

COUNTY	1910	1900	1890
Accomac	36,650	32,570	27,277
Albemarle	29,871	28,473	32,379
Alexandria	10,231	6,430	8,593
Alleghany	14,173	16,330	9,283
Amelia	8,720	9,037	9,068
Amherst	18,932	17,864	17,551
Appomattox	8,904	9,662	9,589
Augusta	32,445	32,370	37,005
Bath	6,538	5,595	4,587
Bedford	29,549	30,356	31,213
Bland	5,154	5,497	5,129
Botetourt	17,727	17,161	14,854
Brunswick	19,244	18,217	17,245
Buchanan	12,334	9,692	5,867
Buckingham	15,204	15,266	14,383
Campbell	23,043	23,256	41,087
Caroline	16,596	16,709	16,681
Carroll	21,116	19,303	15,497
Charles City	5,253	5,040	5,066
Charlotte	15,785	15,343	15,077
Chesterfield	21,299	18,804	26,211
Clarke	7,468	7,927	8,071
Craig	4,711	4,293	3,835
Culpeper	13,472	14,123	13,233
Cumberland	9,195	8,996	9,482
Dickenson	9,199	7,747	5,077
Dinwiddie	15,442	15,374	13,515
Elizabeth City	21,225	19,460	16,168
Essex	9,105	9,701	10,047
Fairfax	20,536	18,580	16,655
Fauquier	22,526	23,374	22,590
Floyd	14,092	15,388	14,405
Fluvanna	8,323	9,050	9,508
Franklin	26,480	25,953	24,985
Frederick	12,787	13,239	17,880
Giles	11,623	10,793	9,090
Gloucester	12,477	12,832	11,653
Goochland	9,237	9,519	9,958
Grayson	19,856	16,853	14,394

VIRGINIA Cont'd.

COUNTY	1910	1900	1890
Greene	6,937	6,214	5,622
Greenesville	11,890	9,758	8,230
Halifax	40,044	37,197	34,424
Hanover	17,200	17,618	17,402
Henrico	23,437	30,062	103,394
Henry	18,459	19,265	18,208
Highland	5,317	5,647	5,352
Isle of Wight	14,929	13,102	11,313
James City	3,624	3,688	5,643
King and Queen	9,576	9,265	9,669
King George	6,378	6,918	6,641
King William	8,547	8,380	9,605
Lancaster	9,752	8,949	7,191
Lee	23,840	19,856	18,216
Loudoun	21,167	21,948	23,274
Louisa	16,578	16,517	16,997
Lunenburg	12,780	11,705	11,372
Madison	10,055	10,216	10,225
Mathews	8,922	8,239	7,584
Mecklenburg	28,956	26,551	25,359
Middlesex	8,852	8,220	7,458
Montgomery	17,268	15,852	17,742
Nansemond	26,886	23,078	19,692
Nelson	16,821	16,075	15,336
New Kent	4,682	4,865	5,511
Norfolk	52,744	50,780	77,038
Northampton	16,672	13,770	10,313
Northumberland	10,777	9,846	7,885
Nottoway	13,462	12,366	11,582
Orange	13,486	12,571	12,814
Page	14,147	13,794	13,092
Patrick	17,195	15,403	14,147
Pittsylvania	50,709	46,894	59,941
Powhatan	6,099	6,824	6,791
Prince Edward	14,266	15,045	14,694
Prince George	7,848	7,752	7,872
Prince William	12,026	11,112	9,805
Princess Anne	11,526	11,192	9,510
Pulaski	17,246	14,609	12,790
Rappahannock	8,044	8,843	8,678
Richmond	7,415	7,088	7,146
Roanoke	19,623	15,837	30,101
Rockbridge	21,171	21,799	23,062
Rockingham	34,903	33,527	31,299
Russell	23,474	18,031	16,126
Scott	23,814	22,694	21,694
Shenandoah	20,942	20,253	19,671
Smyth	20,326	17,121	13,360
Southampton	26,302	22,848	20,078
Spotsylvania	9,935	9,239	14,233
Stafford	8,070	8,097	7,362
Surry	9,715	8,469	8,256
Sussex	13,664	12,082	11,100
Tazewell	24,946	23,384	19,899
Warren	8,589	8,837	8,280
Warwick	6,041	4,888	6,650
Washington	32,830	28,995	29,020
Westmoreland	9,313	9,243	8,399
Wise	34,162	19,653	9,345
Wythe	20,372	20,437	18,019
York	7,757	7,482	7,596
The State	2,061,612	1,854,184	1,655,980

WASHINGTON

	1910	1900	1890
Adams	10,920	4,840	2,098
Asotin	5,831	3,366	1,580
Benton	7,937		
Chehalis	35,590	15,124	9,249
Chelan	15,104	3,931	
Clallam	6,755	5,603	2,771
Clarke	26,115	13,419	11,709
Columbia	7,042	7,128	6,709

WASHINGTON Cont'd.

COUNTY	1910	1900	1890
Cowlitz	12,561	7,877	5,917
Douglas	9,227	4,926	3,161
Ferry	4,800	4,562	......
Franklin	5,153	486	696
Garfield	4,199	3,918	3,897
Grant	8,698		
Island	4,704	1,870	1,787
Jefferson	8,337	5,712	8,368
King	284,638	110,053	63,989
Kitsap	17,647	6,767	4,624
Kittitas	18,561	9,704	8,777
Klickitat	10,180	6,407	5,167
Lewis	32,127	15,157	11,499
Lincoln	17,539	11,969	9,312
Mason	5,156	3,810	2,826
Okanogan	12,887	4,689	1,467
Pacific	12,532	5,983	4,358
Pierce	120,812	55,515	50,940
San Juan	3,603	2,928	2,072
Skagit	29,241	14,272	8,747
Skamania	2,887	1,688	774
Snohomish	59,209	23,950	8,514
Spokane	139,404	57,542	37,487
Stevens	25,297	10,543	4,341
Thurston	17,581	9,927	9,675
Wahkiakum	3,285	2,819	2,526
Walla Walla	31,931	18,680	12,224
Whatcom	49,511	24,116	18,591
Whitman	33,280	25,360	19,109
Yakima	41,709	13,462	4,429
The State	1,141,990	518,103	349,390

WEST VIRGINIA

COUNTY	1910	1900	1890
Barbour	15,858	14,198	12,702
Berkeley	21,999	19,469	18,702
Boone	10,331	8,194	6,885
Braxton	23,023	18,904	13,928
Brooke	11,098	7,219	6,660
Cabell	46,685	29,252	23,595
Calhoun	11,258	10,266	8,155
Clay	10,233	8,248	4,659
Doddridge	12,672	13,689	12,183
Fayette	51,903	31,987	20,542
Gilmer	11,379	11,762	9,746
Grant	7,838	7,275	6,802
Greenbrier	24,833	20,683	18,034
Hampshire	11,694	11,806	11,419
Hancock	10,465	6,693	6,414
Hardy	9,163	8,449	7,567
Harrison	48,381	27,690	21,919
Jackson	20,956	22,987	19,021
Jefferson	15,889	15,935	15,553
Kanawha	81,457	54,696	42,756
Lewis	18,281	16,980	15,895
Lincoln	20,491	15,434	11,246
Logan	14,476	6,955	11,101
McDowell	47,856	18,747	7,300
Marion	42,794	32,430	20,721
Marshall	32,388	26,444	20,735
Mason	23,019	24,142	22,863
Mercer	38,371	23,023	16,002
Mineral	16,674	12,883	12,085
Mingo	19,431	11,359	......
Monongalia	24,334	19,049	15,705
Monroe	13,055	13,130	12,429
Morgan	7,848	7,294	6,744
Nicholas	17,699	11,403	9,309
Ohio	57,572	48,024	41,557
Pendleton	9,349	9,167	8,711
Pleasants	8,074	9,345	7,539
Pocahontas	14,740	8,572	6,814
Preston	26,341	22,727	20,355

WEST VIRGINIA Cont'd.

COUNTY	1910	1900	1890
Putnam	18,587	17,330	14,342
Raleigh	25,633	12,436	9,597
Randolph	26,028	17,670	11,633
Ritchie	17,875	18,901	16,621
Roane	21,543	19,852	15,303
Summers	18,420	16,265	13,117
Taylor	16,554	14,978	12,147
Tucker	18,675	13,433	6,459
Tyler	16,211	18,252	11,962
Upshur	16,629	14,696	12,714
Wayne	24,081	23,619	18,652
Webster	9,680	8,862	4,783
Wetzel	23,855	22,880	16,841
Wirt	9,047	10,284	9,411
Wood	38,001	34,452	28,612
Wyoming	10,392	8,380	6,247

The State....1,221,119 958,800 762,794

WISCONSIN

COUNTY	1910	1900	1890
Adams	8,604	9,141	6,889
Ashland	21,965	20,176	20,063
Barron	29,114	23,677	15,416
Bayfield	15,987	14,392	7,390
Brown	54,098	46,359	39,164
Buffalo	16,006	16,765	15,997
Burnett	9,026	7,478	4,393
Calumet	16,701	17,078	16,639
Chippewa	32,103	33,037	25,143
Clark	30,074	25,848	17,708
Columbia	31,129	31,121	28,350
Crawford	16,288	17,286	15,987
Dane	77,435	69,435	59,578
Dodge	47,436	46,631	44,984
Door	18,711	17,583	15,682
Douglas	47,422	36,335	13,468

WISCONSIN Cont'd.

COUNTY	1910	1900	1890
Dunn	25,260	25,043	22,664
Eau Claire	32,721	31,692	30,673
Florence	3,381	3,197	2,604
Fond du Lac	51,610	47,589	44,088
Forest	6,782	1,396	1,012
Grant	39,007	38,881	36,651
Green	21,641	22,719	22,732
Green Lake	15,491	15,797	15,363
Iowa	22,497	23,114	22,117
Iron	8,306	6,616	......
Jackson	17,075	17,466	15,797
Jefferson	34,306	34,789	33,530
Juneau	19,569	20,629	17,121
Kenosha	32,929	21,707	15,581
Kewaunee	16,784	17,212	16,153
La Crosse	43,996	42,997	38,801
Lafayette	20,075	20,959	20,265
Langlade	17,062	12,553	9,465
Lincoln	19,064	16,269	12,008
Manitowoc	44,978	42,261	37,831
Marathon	55,054	43,256	30,369
Marinette	33,812	30,822	20,304
Marquette	10,741	10,509	9,676
Milwaukee	433,187	330,017	236,101
Monroe	28,881	28,103	23,211
Oconto	25,657	20,874	15,009
Oneida	11,433	8,875	5,010
Outagamie	49,102	46,247	38,690
Ozaukee	17,123	16,363	14,943
Pepin	7,577	7,905	6,932
Pierce	22,079	23,943	20,385
Polk	21,367	17,801	12,968
Portage	30,945	29,483	24,798
Price	13,795	9,106	5,258
Racine	57,424	45,644	36,268
Richland	18,809	19,483	19,121
Rock	55,538	51,203	43,220
Rusk	11,160		
St. Croix	25,910	26,830	23,139
Sauk	32,869	33,006	30,575

WISCONSIN Cont'd.

COUNTY	1910	1900	1890
Sawyer	6,227	3,593	1,977
Shawano	31,884	27,475	19,236
Sheboygan	54,888	50,345	42,489
Taylor	13,641	11,262	6,731
Trempealeau	22,928	23,114	18,920
Vernon	28,116	28,351	25,111
Vilas	6,019	4,929	......
Walworth	29,614	29,259	27,860
Washburn	8,196	5,521	2,926
Washington	23,784	23,589	22,751
Waukesha	37,100	35,229	33,270
Waupaca	32,782	31,615	26,794
Waushara	18,886	15,972	13,507
Winnebago	62,116	58,225	50,097
Wood	30,583	25,865	18,127

The State 2,333,860 2,069,042 1,686,880

WYOMING

COUNTY	1910	1900	1890
Albany	11,574	13,084	8,865
Bighorn	8,886	4,328	......
Carbon	11,282	9,589	6,857
Converse	6,294	3,337	2,738
Crook	6,492	3,137	2,338
Fremont	11,822	5,357	2,463
Johnson	3,453	2,361	2,357
Laramie	26,127	20,181	16,777
Natrona	4,766	1,785	1,094
Park	4,909		
Sheridan	16,324	5,122	1,972
Sweetwater	11,575	8,455	4,941
Uinta	16,982	12,223	7,414
Weston	4,960	3,203	2,422
National Park reservation	519	369	467

The State.....145,965 92,531 60,705

INDEX
CITIES AND TOWNS OF THE UNITED STATES
1910 CENSUS

In this compilation the official population figures, as determined by the Thirteenth (1910) Census of the United States, are given for all cities, villages and boroughs seperately enumerated by the government officials; of the places not seperately enumerated by the Census Bureau, and for which there are no government figures obtainable, recent estimates, supplied by local officials and by other reliable authorities, are given. Capitals of states are printed in capital letters, and places having one hundred or more inhabitants are named. The index references enclosed in parentheses, indicate location of city or village on the accompanying maps.

Some comparatively unimportant places have been omitted from the maps to avoid crowding and consequent indistinctness in the engraving; such omitted towns will be be found in the list, and their situation on the map, in each case, may readily be determined by means of the index-reference letters and numbers,

ALABAMA

Abanda, (H4)...... 100	Bellwood, (G8)..... 201	Catherine, (C6)..... 200
Abbeville, (H7)...1,141	Benton, (E6)....... 600	Cecil, (F6)......... 100
Abbott, (D7) 200	Berry (C3)........ 372	Cedar Bluff, (G2).. 230
Abercrombie, (D5).. 350	Bessemer, (D4)..10,864	Cedarcove, (D4)... 400
Abernant, (D4).... 300	Bethany, (B4)..... 100	Cedarville, (C5)... 250
Acton, (E4).......1,500	Beulah, (H5)...... 150	Center, (G2)...... 256
Ada, (F6)........ 100	Bezar, (B2)....... 300	Centerville, (D5).. 730
Adamsville, (E3).. 649	Billingsley, (E5).... 256	Central, (F5)...... 100
Addison, (D2)..... 100	Birmingham,	Central Mills, (D6). 200
Adger, (D4).....1,200	(E4)........132,685	Ceylon, (D8)...... 100
Akron, (C5)....... 100	Bissell, (H2)....... 150	Chambers, (F6)... 250
Alabama City, (G2)4,313	Black, (G8)........ 485	Chatom, (B8)..... 100
Alameda, (C7)..... 200	Blackwell, (D3).... 200	Chelsea, (E4)..... 100
Alamuchee, (B6)... 100	Bladon Springs,(B7). 430	Chepultepec, (E3). 100
Alberta, (D6)...... 150	Blocton, (D4)....2,800	Cherokee, (C1).... 269
Albertville, (F2)...1,544	Blossburg, (E3)...2,000	Chesterfield, (G2).. 100
Aldrich, (E4)...... 450	Blount Sprs. (E3).. 300	Childersburg, (F4). 449
Alexander City,(G5)1,710	Bleasantville, (E2).. 287	Chinagrove, (C6).. 150
Alexis, (G2)....... 150	Blue Springs, (G7). 117	Choccolocco, (G3).. 250
Aliceville, (B4).... 640	Bluffton, (G2).... 300	Chunchula, (B9)... 100
Allenton, (D7)..... 200	Boaz, (F2).......1,010	Citronelle, (B8)... 935
Almond, (H4)..... 100	Boligee, (B5)...... 200	Claiborne, (C7)... 100
Alpine, (F4)....... 100	Bolling, (E7)...... 350	Clanton, (E5)....1,123
Altoona, (F2).....1,071	Bon Air, (F4)..... 250	Clayton, (H7)....1,130
America, (D3)..... 100	Bon Secour, (C10).. 350	Cleveland, (E2)... 150
Andalusia, (E8)...2,480	Bradleyton, (F7)... 150	Clinton, (C5)..... 100
Anderson, (D1).... 230	Brachead, (D4).... 250	Clio, (G7)........ 580
Anniston, (G3)..12,794	Brantley, (F7)..... 803	Cloverdale, (C2)... 150
Arab, (E2)........ 150	Brewton, (D8)...2,185	Coalburg, (E3).... 250
Ariton, (G7)...... 431	Bridgeport, (G1)..2,125	Coal City, (F3).... 685
Arkadelphia, (E3).. 210	Bridgeton, (E4).... 100	Coal Creek, (D3).. 300
Arlington, (C6).... 100	Brierfield, (E4)...2,100	Coaldale, (E3).... 500
Ashford, (H8)..... 479	Brighton, (E4)....1,502	Coaling, (D4)..... 250
Ashland, (G4)....1,062	Brilliant, (C3).... 700	Coalmont, (E4).... 100
Ashville, (F3)..... 278	Brompton, (F3)... 250	Coal Valley, (D3).. 400
Athens, (E1)....1,715	Brooklyn, (E8).... 350	Coatopa, (B6)..... 100
Atmore, (C8).....1,060	Brooks, (F8)...... 100	Coatsbend, (G2)... 100
Attalla, (F2).....2,513	Brookside, (E3).... 623	Cobb, (G3)....... 100
Aubrey, (D4)...... 300	Brooksville, (F2)... 100	Cochrane, (B4).... 200
Auburn, (G5)....1,408	Brookwood, (D4)..1,550	Coden, (B10)..... 250
Austinville, (E2)... 671	Browns, (E6)...... 200	Coffee Sprs. (G8). 503
Autaugaville, (E6). 313	Brownsboro, (E1).. 100	Coffeeville, (B7)... 100
Avondale, (E3)...4,500	Brundidge, (G7)... 815	Colbran, (G2)..... 100
Baltzell, (C5)...... 100	Buenavista, (D7)... 200	Collinsville, (G3).. 673
Bangor, (E3)...... 200	Burleson, (B2).... 100	Collirene, (E6).... 150
Banks, (G7)....... 307	Burnsville, (E6)... 200	Columbia, (H8)...1,122
Bankston, (C3).... 100	Butler, (B6)...... 250	Columbiana, (E4)..1,079
Barlow Bend, (C7). 200	Cahaba, (D6)..... 200	Comer, (H6)...... 200
Barton, (C1)...... 100	Calebee, (G6)..... 250	Cooper, (E5)...... 100
Batesville, (H6)... 143	Caledonia, (D7)... 200	Coosada Sta., (F6). 150
Battelle, (G1)....1,600	Calera, (E4)...... 754	Copeland, (B8)... 300
Bay Minette, (C9). 749	Calumet, (D3)..... 350	Cordova, (D3)...1,747
Bear Creek, (C2).. 600	Camden, (D7)..... 648	Corona, (E3)....1,500
Beatrice, (D7)..... 345	Campbell, (C7).... 100	Cortelyou, (B8)... 200
Belcher, (H7)..... 200	Camphill, (G5).... 896	Cottondale, (D4).. 520
Belgreen, (C2).... 300	Capitol Heights,(F6) 403	Cottonwood, (H8). 500
Belknap, (D6)..... 100	Carbonhill, (D3)..1,627	Courtland, (D1)... 478
Bellamy, (B6)..... 500	Cardiff, (E3)...... 426	Covington, (E3)... 500
Belle Ellen, (D4).. 400	Carlisle, (F2)...... 100	Cowarts, (H8).... 150
Belle Mina, (E1)... 100	Carlowville, (D6).. 200	Cragford, (G4).... 150
Belle Sumter, (D4). 100	Carney, (C9)...... 150	Crawford, (H6)... 200
Belleville, (D8).... 170	Carrollton, (B4)... 444	Crews Depot, (B3). 150
Bells Landing, (D7). 100	Castleberry, (D8)... 225	Crichton, (B9).... 100

Cropwell, (F3)..... 110	Elvira, (E4)........ 250	
Crossville, (G2)... 250	Elyton, (E3).....1,000	
Cuba, (B6)....... 650	Emauhee, (F4).... 300	
Cullman, (E2)...2,130	Enon, (H6)....... 300	
Culpepper, (C7)... 100	Ensley, (E3).....4,850	
Curls Station, (B6).. 200	Enterprise, (G8)..2,322	
Curtiston, (F2).... 200	Epes, (B5)........ 374	
Cusseta, (H5)..... 150	Equality, (F5)..... 200	
Cypress, (C5)..... 150	Escatawpa, (B8)... 130	
Dadeville, (G5)...1,193	Estaboga, (F3)..... 94	
Daleva, (G8)..... 200	Eufaula, (H7)....4,259	
Daleville, (G8).... 500	Eunola, (G8)...... 321	
Damascus, (F8)... 100	Euphronia, (F6)... 100	
Danville, (D2).... 220	Eureka, (F3)...... 130	
Daphne, (C9)..... 700	Eutaw, (C5).....1,001	
Davis Creek, (C3).. 100	Evergreen, (D8)...1,582	
Daviston, (G4).... 127	Exsho, (B8)...... 100	
Dawson, (G2)..... 130	Ezra, (D4)........ 250	
Dayton, (C6)..... 382	Fackler, (G1)..... 150	
De Armanville, (G3) 130	Fairhope, (C10)... 590	
Deatsville, (F5)... 194	Fair View, (E1)... 237	
Decatur, (D1)...4,228	Falkville, (E2).... 335	
Deer Creek, (D3).. 279	Farmersville, (E6). 100	
Deerpark, (B8).... 120	Fatama, (D7)..... 100	
Delmar, (C2)..... 100	Faunsdale, (C6)... 352	
Demopolis, (C6)..2,417	Fayette, (C3)..... 636	
Deposit, (F1)..... 100	Fayetteville, (F4). 190	
Detroit, (B2)..... 160	Fernbank, (B3)... 180	
Dixons Mills, (C6).. 200	Finchburg, (D7)... 200	
Dolive, (C9)...... 150	Fisk, (E1)........ 300	
Dolomite, (D4)..1,000	Fitzpatrick, (G6).. 398	
Dora, (D3)....... 916	Fivepoints, (H4)... 200	
Doran, (C8)...... 150	Flat Creek, (D3)... 600	
Dothan, (H8)....7,016	Flatwood, (D6).... 100	
Double Sprs. (D2). 200	Flint, (E2)........ 197	
Douglas, (G8).... 100	Flomaton, (D9)... 539	
Downs, (G6)..... 100	Florala, (F8).....2,439	
Dozier, (F7)...... 268	Florence, (C1)....6,689	
Dravo, (G5)...... 400	Flossy, (D1)...... 180	
Drifton, (D3)..... 100	Foley, (C10)...... 200	
Dudleyville, (G5).. 100	Forest Home, (E7). 300	
Duke, (G3)....... 100	Forkland, (C5).... 100	
Duncanville, (D4).. 250	Forney, (H2)..... 120	
Dunham, (E7).... 300	Fort Davis, (G6)... 200	
Dunnavant, (E3).. 100	Fort Deposit, (E7). 893	
Eastaboga, (F3)... 500	Fort Payne, (G2)..1,317	
EastBirmingham,(E3)500	Fosters, (C4)..... 110	
East Lake, (E3)... 4,100	Franklin, (D7).... 150	
East Tallassee, (G5)2,000	Fredonia, (H5).... 250	
East Thomas, (E3).1,500	Fruitdale, (B8).... 250	
Eclectic, (G5)..... 315	Fruithurst, (G3)... 257	
Eden, (F3)....... 165	Fulton, (C7)...... 518	
Edwardsville, (G3).. 393	Furman, (E6)..... 125	
Effie, (E8)........ 300	Gadsden, (F3)..10,557	
Effort, (F7)....... 100	Gainestown, (C8).. 100	
Elamville, (G7).... 100	Gainesville, (B5).. 532	
Elba, (F8).......1,079	Gallion, (C6)..... 200	
Elbert, (C4)...... 200	Gantt, (E8)....... 300	
Eldridge, (C3).... 120	Gantts Quarry, (F4) 300	
Eliza, (G1)....... 150	Garden City, (E2).. 250	
Elk, (D2)......... 100	Garland, (E7)..... 250	
Elkmont, (E1).... 188	Gasque, (C10).... 130	
Elmore, (F5)...... 300	Gaston, (B6)...... 100	

INDEX

CITIES AND TOWNS OF THE UNITED STATES

1910 CENSUS

In this compilation the official population figures, as determined by the Thirteenth (1910) Census of the United States, are given for all cities, villages and boroughs seperately enumerated by the government officials; of the places not seperately enumerated by the Census Bureau, and for which there are no government figures obtainable, recent estimates, supplied by local officials and by other reliable authorities, are given. Capitals of states are printed in capital letters, and places having one hundred or more inhabitants are named. The index references enclosed in parentheses, indicate location of city or village on the accompanying maps.

Some comparatively unimportant places have been omitted from the maps to avoid crowding and consequent indistinctness in the engraving; such omitted towns will be be found in the list. and their situation on the map, in each case, may readily be determined by means of the index-reference letters and numbers.

ALABAMA

Abanda, (H4)...... 100	Bellwood, (G8)..... 201	Catherine, (C6).... 200	Cropwell, (F3)...... 110	Elvira, (E4)........ 250
Abbeville, (H7)...1,141	Benton, (E6)....... 600	Cecil, (F6)........ 100	Crossville, (G2)... 250	Elyton, (E3)......1,000
Abbott, (D7)...... 200	Berry (C3)......... 372	Cedar Bluff, (G2).. 230	Cuba, (B6)......... 650	Emauhee, (F4)..... 300
Abercrombie, (D5). 350	Bessemer, (D4)..10,864	Cedarcove, (D4)... 400	Cullman, (E2).....2,130	Eoon, (H6)........ 300
Abernant, (D4).... 300	Bethany, (B4)...... 100	Cedarville, (C5)... 250	Culpepper, (C7).... 100	Ensley, (E3)......4,850
Acton, (E4).......1,500	Beulah, (H5)....... 150	Center, (G2)....... 256	Curls Station, (B6). 200	Enterprise, (G8)..2,322
Ada, (F6)......... 100	Bexar, (B2)........ 300	Centerville, (D5).. 730	Curtiston, (F2).... 200	Epes, (B5)......... 374
Adamsville, (E3)... 649	Billingsley, (E5)... 256	Central, (F5)...... 100	Cusseta, (H5)...... 150	Equality, (F5)..... 200
Addison, (D2)..... 100	Birmingham,	Central Mills, (D6). 200	Cypress, (C5)...... 150	Escatawpa, (B8)... 130
Adger, (D4).......1,200	(E4)..........132,685	Ceylon, (D8)....... 100	Dadeville, (G5)...1,193	Estaboga, (F3)..... 94
Akron, (C5)....... 100	Bissell, (H2)....... 150	Chambers, (F6).... 250	Daleva, (G8)....... 200	Eufaula, (H7).....4,259
Alabama City, (G2)4,313	Black, (G8)........ 485	Chaton, (B6)....... 100	Daleville, (G8).... 500	Eunola, (G8)....... 321
Alameda, (C7).... 200	Blackwell, (D3).... 200	Chelsea, (E4)...... 100	Damascus, (F8).... 150	Euphronia, (F6)... 100
Alamuchee, (B6)... 100	Bladon Springs, (B7). 430	Chepultepec, (E3).. 200	Danville, (D2)..... 220	Eureka, (F3)....... 130
Alberta, (D6)..... 150	Blocton, (D4).....2,800	Cherokee, (C1).... 269	Daphne, (C9)...... 700	Eutaw, (C5).......1,001
Albertville, (F2)..1,544	Blossburg, (E3)...2,600	Chesterfield, (G2). 100	Davis Creek, (C3). 100	Evergreen, (D8)..1,582
Aldrich, (E4)...... 450	Blount Spra, (E3). 300	Childersburg, (F4). 449	Daviston, (G4)..... 127	Exzho, (B8)....... 100
Alexander City,(G5)4,270	Blountsville, (E3).. 287	Chinagrove, (C6).. 150	Dawson, (G2)...... 130	Ezra, (D4)........ 250
Alexis, (G2)....... 150	Blue Springs, (G7). 117	Choccolocco, (G3). 250	Dayton, (C5)...... 382	Fackler, (G1)...... 150
Aliceville, (B4).... 640	Bluffton, (G2)..... 300	Chuachula, (B9)... 100	De Armanville, (G3) 130	Fairhope, (C10)... 590
Allenton, (D7).... 200	Boaz, (F2).......1,010	Citronelle, (B8)... 935	Deatsville, (F5)... 194	Fair View, (E1)... 237
Almond, (H4)..... 100	Boligee, (B5)...... 200	Claiborne, (C7).... 100	Decatur, (D3).....4,228	Falkville, (E2)..... 335
Alpine, (F4)...... 100	Bolling, (E7)...... 350	Clanton, (E5).....1,123	Deer Creek, (D3).. 279	Farmersville, (E6). 100
Altoona, (F2).....1,071	Bon Air, (F4)..... 250	Clayton, (H7)....1,130	Deerpark, (B8).... 120	Fatama, (D7)...... 100
America, (D3)..... 100	Bon Secour, (C10). 350	Cleveland, (E2)... 150	Delmar, (C2)...... 100	Faunsdale, (C6)... 352
Andalusia, (E8)..2,480	Bradleyton, (F7)... 150	Clinton, (C5)...... 100	Demopolis, (C5)..2,417	Fayette, (C3)...... 636
Anderson, (D1)... 230	Braehead, (D4).... 250	Clio, (G7)......... 580	Deposit, (D1)..... 100	Fayetteville, (F4). 190
Anniston, (G3)..12,794	Brantley, (F7)..... 803	Cloverdale, (C2)... 150	Detroit, (B2)...... 160	Fernbank, (B3).... 180
Arab, (E2)........ 150	Brewton, (D8)....2,185	Coalburg, (E3).... 250	Dixons Mills, (C6). 200	Finchburg, (D7)... 200
Ariton, (G7)...... 431	Bridgeport, (G1)..2,125	Coal City, (F3)... 685	Dolive, (C9)...... 150	Fisk, (E1)......... 300
Arkadelphia, (E3). 210	Bridgeton, (E4)... 100	Coal Creek, (D3).. 300	Dolomite, (D4)...1,000	Fitzpatrick, (G6).. 398
Arlington, (C6)... 100	Brierfield, (E4)..2,100	Coaldale, (E3)..... 500	Dora, (D3)........ 916	Fivepoints, (H4).. 200
Ashford, (H8)..... 479	Brighton, (E4)...1,502	Coaling, (D4)..... 250	Doran, (C8)....... 150	Flat Creek, (D3).. 600
Ashland, (G4)....1,062	Brilliant, (C3).... 700	Coalmont, (E4).... 100	Dothan, (H8).....7,016	Flatwood, (D6).... 100
Ashville, (F3)..... 278	Brompton, (F3).... 250	Coal Valley, (D3).. 400	Double Sprs, (D2). 200	Flint, (E2)........ 197
Athens, (E1).....1,715	Brooklyn, (E8).... 350	Coatopa, (B6)..... 100	Douglas, (G8)..... 100	Flomaton, (D9)... 539
Atmore, (C8).....1,060	Brooks, (F8)...... 100	Coatsbend, (G2)... 100	Downs, (G6)....... 100	Florala, (F8).....3,439
Attalla, (F2).....2,513	Brookside, (E3)... 623	Cobb, (G3)........ 100	Dozier, (F7)....... 288	Florence, (C1)....6,689
Aubrey, (D4)...... 300	Brooksville, (F2).. 100	Cochrane, (B4).... 200	Dravo, (G4)....... 400	Flowy, (D1)....... 180
Auburn, (G5).....1,408	Brookwood, (D4).1,550	Coden, (B10)...... 250	Drifton, (D3)..... 100	Foley, (C10)...... 200
Austinville, (E2).. 671	Browns, (D6)...... 200	Coffee Sprs, (G8). 503	Dudleyville, (G5).. 100	Forest Home, (E7). 300
Autaugaville, (E6). 313	Brownsboro, (E1).. 100	Coffeeville, (B7).. 100	Duke, (G3)........ 150	Forkland, (C5).... 100
Avondale, (E3)...4,500	Brundidge, (G7)... 815	Coffman, (G2).... 100	Duncanville, (D4). 250	Forney, (H2)...... 120
Baitzell, (C5)..... 100	Buenavista, (D7).. 200	Collinsville, (G2). 673	Dunham, (E7)..... 300	Fort Davis, (G6).. 200
Bangor, (E3)...... 200	Burleson, (B2).... 100	Collirene, (E6).... 150	Dunnavant, (E3).. 100	Fort Deposit, (E7). 893
Banks, (G7)....... 307	Burnsville, (E6)... 200	Columbia, (H8)...1,122	Eastaboga, (F3)... 500	Fort Payne, (G2)..1,317
Bankston, (C3).... 100	Butler, (B6)....... 250	Columbiana, (E4)..1,079	EastBirmingham,(E3)500	Fosters, (C4)...... 110
Barlow Bend, (C7). 200	Cahaba, (D6)..... 200	Comer, (H6)...... 200	East Lake, (E3)...4,100	Franklin, (D7).... 150
Barton, (C1)...... 100	Calebee, (G6)..... 250	Cooper, (E5)...... 100	East Tallassee, (G5)2,000	Fredonia, (H5)... 250
Batesville, (H6)... 143	Caledonia, (D7)... 200	Coosada Sta., (F6).. 150	East Thomas, (E3). 1,500	Fruitdale, (B8)... 250
Battelle, (G1)....1,600	Calera, (E4)....... 754	Copeland, (B8).... 300	Eclectic, (G5)..... 315	Fruithurst, (G3)... 257
Bay Minette, (C9). 749	Calumet, (D3)..... 350	Cordova, (D3).....1,747	Eden, (F3)........ 165	Fulton, (C7)....... 518
Bear Creek, (C2).. 600	Camden, (D7)..... 648	Corona, (D3)......1,500	Edwardsville, (G3).. 395	Furman, (E6)...... 125
Beatrice, (D7).... 345	Campbell, (C7).... 100	Cortelyou, (H8)... 200	Effie, (E8)........ 300	Gadsden, (F3)...10,557
Belcher, (H7)..... 200	Camphill, (G5).... 896	Cottondale, (D4).. 320	Effort, (G7)....... 100	Gainestown, (C8). 100
Belgreen, (C2).... 300	Capitol Heights,(F6) 403	Cottonwood, (H8).. 350	Elamville, (G7).... 200	Gainesville, (B5).. 532
Belknap, (D6)..... 100	Carbonhill, (D3)..1,627	Courtland, (D1)... 478	Elba, (F8).......1,079	Gallion, (C6)...... 200
Bellamy, (B6)..... 500	Cardiff, (E3)...... 426	Covington, (E3)... 500	Elbert, (C4)....... 200	Gantt, (E8)........ 300
Belle Ellen, (D4).. 400	Carlisle, (F3)..... 100	Cowarts, (H8)..... 150	Eldridge, (C3)..... 120	Gantts Quarry, (F4) 300
Belle Mina, (E1).. 100	Carlowville, (D6).. 200	Cragford, (G4).... 150	Elias, (G1)........ 150	Garden City, (E2).. 250
Belle Sumter, (D4). 100	Carney, (C9)...... 150	Crawford, (H6).... 200	Elk, (D2)......... 100	Garland, (E7)..... 250
Belleville, (D8)... 170	Carrollton, (B4)... 444	Crews Depot, (B3). 150	Elkmont, (E1)..... 188	Gasque, (C10)..... 130
Bells Landing, (D7). 100	Castleberry, (D8).. 225	Crichton, (B9).... 100	Elmore, (F5)...... 300	Gaston, (B6)...... 100

Sayreton, (E3)...... 600	
Schuster, (D7)..... 100	
Scotia, (F6)....... 150	
Scotland, (D7)..... 100	
Scottsboro, (G1)...1,019	
Scotts Station, (D5). 200	
Scyrene, (C7)...... 200	
Seale, (H6)....... 312	
Searight, (F8)..... 120	
Searles, (D4)..... 700	
Section, (G1)...... 250	
Seddon, (F3)...... 133	
Sellers, (F6)...... 100	
Sellersville, (G6).... 100	
Selma, (E6).....13,649	
Seloca, (E2)...... 300	
Seman, (F3)....... 100	
Seminole, (C10).... 200	
Shady Grove, (F7). 500	
Sheffield, (C1)....4,865	
Shelby, (E4)...... 750	
Shiloh, (C6)...... 100	
Short Creek, (D3).. 150	
Shorter, (G6)...... 500	
Shorterville, (H7).. 100	
Shottsville, (B2)... 100	
Sistrunk, (F5)..... 100	
Six Mile, (D4)..... 100	
Sligo, (F2)....... 100	
Slocomb, (G8)..... 896	
Sloss, (E4)....... 320	
Smith Hill, (D4)... 422	
Snowdoun, (F6)... 200	
Snow Hill, (E7)... 500	

Society Hill, (G6)... 100	
Somerville, (E2)... 300	
South, (E8)....... 100	
Speigner, (F5)..... 100	
Spring Garden, (G3)........... 300	
Springhill, (B9)..... 200	
Springvalley, (C1)... 100	
Springville, (E3)... 350	
Stafford, (B4)..... 650	
Stamp, (G1)...... 200	
Standing Rock, (H4)........... 300	
Stanton, (E5)..... 400	
Steele, (F3)....... 100	
Stevenson, (G1)... 574	
Stewart,(C5)...... 300	
Stewartsville, (F4).. 150	
Stockdale, (F4).... 100	
Stocks Mill, (G2)... 300	
Stockton, (C9)...1,400	
Stouts M'tn, (E2)... 500	
Strata, (F6)....... 100	
Stroud, (H4)...... 100	
Sturkie, (G5)..... 100	
Stutts, (C1)...... 100	
Styx, (C9)....... 150	
Suggsville, (C7).... 340	
Sulligent, (B3)..... 619	
Sullingint, (C8)..... 150	
Sulphur Sprs., (G1). 100	
Summerdale, (C10)........... 100	
Summerfield, (D6).. 450	

Summit, (E2)..... 150	
Sumterville, (B5)... 300	
Sunny South, (C7)... 150	
Superior, (E4)..... 130	
Suspension, (G6)... 100	
Sycamore, (F4)... 500	
Sykes Mills, (F5)... 100	
Sylacauga, (F4)...1,456	
Talladega, (F4)...5 854	
Talassee, (G6)...1,347	
Taylorsville, (E1)... 200	
Tecumseh, (H2)... 200	
Tennille, (G7)..... 100	
Tensaw, (C9)..... 100	
Texas, (C3)....... 300	
Tharin, (F6)...... 100	
Theodore, (B9).... 300	
Thomas, (E3)...1,500	
Thomaston, (C6)... 200	
Thomasville, (C7)..1,181	
Thompson, (G6)... 263	
Thorsby, (E5)..... 900	
Three Notch, (G6).. 180	
Tilden, (D6)...... 230	
Toulminville, (B9).. 150	
Towncreek, (D1).. 344	
Townly, (D3)..... 235	
Triana, (E1)...... 200	
Trinity, (D1)...... 198	
Troy, (G7)......4,961	
Trussville, (E3)... 900	
Tunnel Springs, (D7)........... 150	
Turnbull, (D7)..... 150	

Turner, (F3)....... 100	
Tuscaloosa, (C4)..8,407	
Tuscumbia, (C1)..3,324	
Tuskegee, (G6)...2,803	
Tuskegee Institute, (G6)........... 100	
Union, (C5)...... 100	
Union Grove, (F2).. 100	
Union Sprs, (G6)...4,055	
Uniontown, (C6)...1,836	
University, (C4).... 150	
Valley Head, (G1).. 300	
Vandorn, (C6)..... 100	
Verbena, (F5)..... 400	
Vernon, (B3)..... 423	
Victoria, (G7)..... 100	
Vida, (E5)....... 150	
Vienna, (B4)...... 79	
Vina, (B2)....... 110	
Vincent, (F4)..... 995	
Vinegar Bend, (A8). 540	
Vinemont, (E2).... 100	
Wadley, (G4)..... 426	
Wadsworth, (F5)... 300	
Walker Springs, (C7)........... 100	
Wallace, (D8)..... 200	
Walnut Grove,(F1). 204	
Warrenton, (F2)... 130	
Warrior, (E3)..... 660	
Warriorstand, (G6). 400	
Warsaw, (B5)..... 150	
Waterloo, (B1)..... 435	
Water Valley, (B7).. 100	

Watkins, (F3)...... 200	
Watson, (E3)......1,000	
Waverly, (G5)..... 170	
Wawbeek, (D8)... 200	
Weaver, (G3)..... 100	
Webb, (H8)....... 256	
Wedgworth, (C5)... 100	
Wedowee, (H4).... 435	
Woogufka, (F4).... 200	
Wesobulga, (G4)... 150	
West Blocton, (D4). 892	
West End, (E4)...2,500	
Wetumpka, (F5)...1,103	
Whatley, (C7)..... 187	
Whistler, (B9)....3,500	
Whiteoak Springs, (H7)........... 249	
Whiteplains, (G3).. 250	
Whitfield, (B6)..... 200	
Whitney, (F3)..... 100	
Wilmer, (B9)...... 500	
Wilsonville, (F4)... 933	
Winfield, (C3)..... 419	
Womack Hill, (B7).. 100	
Woodlawn, (E3)..3,200	
Woodstock, (D4)... 500	
Woodville, (F1)... 800	
Woodward, (D4)..1,000	
Wyeth City, (F2)... 400	
Wylam, (E4)....3,000	
Yellow Pine, (B8).. 500	
Yolande, (C4)....1,000	
York, (B6)....... 710	
Yucca, (G1)...... 150	

ARIZONA

Agua Caliente, (B5) 100	
Alhambra, (C5)... 200	
Arivaca, (D7)..... 150	
Arlington, (C5)... 100	
Ashfork, (C3)..... 200	
Bellevue, (E5)..... 100	
Benson, (E7)...... 900	
Bigbug, (C4)...... 100	
Bisbee, (F7)..... 9,019	
Blackwater, (D5).. 250	
Bonita, (F6)....... 100	
Bowie, (F6)....... 200	
Brownell, (C6)..... 200	
Bryce, (F6)....... 100	
Buckeye, (C5)..... 400	
Campverde, (D4).. 300	
Canille, (E7)...... 200	
Casa Blanca, (D5). 1,200	
Casagrande, (D6).. 350	
Cavecreek, (D5)... 100	
Central, (E7)...... 300	
Cerro Colorado, (D7)........... 200	
Chiricahua, (F7)... 200	
Chloride, (A3)..... 470	
Christmas, (E5)... 200	
Clifton, (F5)....4,874	
Cochise, (F6)..... 100	
Cochran, (D5)..... 100	

Columbia, (C4).... 100	
Concho, (F4)...... 260	
Congress, (B4).... 1,000	
Constellation, (C4) 250	
Copper Creek, (E6) 100	
Cottonwood, (C4). 100	
Courtland, (F7)... 500	
Crittenden, (E7)... 170	
Crowley, (E5)..... 150	
Crown King, (C4). 200	
Dome, (A6)....... 100	
Don Luis, (E7)... 250	
Douglas, (F7)...6,437	
Dudleyville, (E6).. 150	
Duncan, (F6)...1,200	
Duquesne, (E7)... 150	
Eden, (F6)....... 500	
Escuela, (E6)..... 200	
Fairbank, (E7)..... 180	
Fairview, (F6)..... 200	
Flagstaff, (D3)...1,633	
Florence, (D5)..... 807	
Fort Apache, (F5). 300	
Fort Huachuca, (E7) 400	
Fort Thomas, (E5) 150	
Franklin, (F6)..... 100	
Geronimo, (F5)... 100	
Gila Bend, (C6)... 200	
Gleeson, (F7)..... 500	

Glendale, (C5).... 250	
Globe, (E5)......7,083	
Grand Canyon, (C2)........... 250	
Greer, (F4)....... 100	
Groom Creek, (C4) 300	
Hackberry, (B3)... 100	
Hamburg, (E7)... 150	
Harshaw, (E7).... 100	
Helvetia, (E7)..... 100	
Holbrook, (E4)... 450	
Hubbard, (F6)..... 100	
Humboldt, (C4)... 100	
Jerome, (C4).....2,393	
Junction, (C4)..... 100	
Kelvin, (E5)...... 350	
Kingman, (A3)...1,000	
Lehi, (D5)....... 300	
Liberty, (C5)...... 150	
Lochiel, (E7)...... 100	
Lowell, (F7).....2,500	
McCabe, (C4)..... 300	
Mammoth, (E6)... 400	
Maricopa, (D5)... 100	
Matthews, (F6)... 200	
Maxton, (C4)..... 200	
Mesa, (D5).....1,692	
Metcalf, (F5)...2,500	
Miami, (E5)...... 300	

Middlemarch, (E7) 100	
Mohave City, (A3). 180	
Morenci, (F5)...5,010	
Mowry, (E7)...... 500	
Naco, (F7)....... 200	
Nogales, (E7)... 3,514	
Octave, (C4)..... 500	
Oro Blanco, (E7).. 100	
Owens, (E4)...... 150	
Paradise, (F7)..... 100	
Parker, (A4)...... 400	
Patagonia, (E7)... 100	
Payson, (D4)..... 170	
Pearce, (F7)...... 700	
PHOENIX, (C5). 11,134	
Pima, (F6)....... 500	
Pine, (D4)....... 100	
Pinedale, (E4)..... 100	
Pirtleville, (F7)... 1,500	
Prescott, (C4)...5,092	
Quartzsite, (A5)... 300	
Ray, (E5)........ 100	
Rice, (E5)....... 300	
Roosevelt, (D5)... 800	
Rosemont, (E7)... 150	
Sacaton, (D5)..... 250	
Safford, (F6)...... 929	
Saint David, (E7).. 500	
Saint Johns, (F4).. 1,250	

San Carlos, (E5)... 5,000	
Sasco, (D6)....... 300	
Show Low, (F4).... 150	
Silverbell, (D6)... 700	
Silverking, (D5)... 100	
Snowflake, (E4)... 500	
Solomonsville, (F6) 750	
Springerville, (F4). 600	
Supai, (C3)....... 100	
Superior, (D5)..... 200	
Taylor, (E4)...... 100	
Tempe, (C5)..... 1,473	
Thatcher, (F6)..... 904	
Tombstone, (F7).. 1,582	
Toreva, (E3)...... 550	
Troy, (E5)....... 200	
Tubac, (D7)...... 100	
Tucson, (E6)...13,193	
Twin Buttes, (D7). 300	
Walker, (C4)...... 150	
Warren, (F7)...... 180	
Wellton, (B6)..... 200	
Wenden, (B5)..... 160	
Whitehills, (A3)... 100	
Wickenburg, (C5). 570	
Willcox, (F6)...... 500	
Williams, (C3)...1,267	
Winslow, (E3)...2,381	
Yuma, (A6)......2,914	

ARKANSAS

Abbott, (A3)...... 300	
Actus, (A3)....... 150	
Ada, (C3)........ 100	
Adona, (C3)...... 180	
Ain, (C4)........ 100	
Alco, (C3)....... 200	
Alexander, (C4)... 141	
Aficia, (D3)....... 168	
Alix, (B3)........ 500	
Alleene, (A5)...... 130	
Allis, (D5)........ 100	
Alma, (A3)....... 565	
Almond, (D3)..... 300	

Almyra, (D4)..... 252	
Alpena Pass, (B2). 300	
Alpine, (B4)...... 100	
Altheimer, (D4)... 500	
Altus, (B3)....... 659	
Aly, (B4)........ 200	
Amity, (B4)....... 813	
Amos, (C2)....... 150	
Antimony, (A4)... 150	
Antoine, (B4)..... 300	
Aplin, (C4)....... 200	
Appleton, (C3).... 300	
Archillion, (E3).... 100	

Ard, (B3)........ 150	
Arden, (A5)...... 400	
Argenta, (C4)... 11,138	
Arkadelphia, (B4). 2,745	
Arkana, (C2)..... 200	
Arkansas City,(D5) 1,485	
Arkansas Post, (D4) 100	
Arkinda, (A5)..... 160	
Armada, (A3)..... 300	
Armorel, (F2)..... 300	
Ashdown, (A5)... 1,247	
Ash Flat, (D2)..... 150	
Ashton, (D5)...... 200	

Ashvale, (D4)..... 100	
Askew, (E4)...... 31	
Athelstan, (E3).... 100	
Athens, (A4)..... 115	
Atkins, (C3)..... 1,258	
Atlanta, (C5)..... 150	
Augusta, (D3)... 1,520	
Aurora, (B3)..... 100	
Austin, (C3)...... 177	
Auvergne, (D3)... 150	
Ava, (B4)........ 100	
Avoca, (A2)...... 100	
Avon, (A4)....... 300	

Balboa, (D2)...... 200	
Bald Knob, (D3).. 617	
Banks, (C5)....... 200	
Banner, (D3)...... 100	
Barber, (B3)...... 100	
Bardstown, (E3)... 200	
Barfield, (F3)..... 100	
Barling, (A3)..... 100	
Barren Fork, (D3). 350	
Barton, (E4)...... 250	
Bates, (A4)....... 272	
Batesville, (D3).. 3,399	
Bauxite, (C4)..... 300	

CALIFORNIA

COLORADO

La Junta, (L?).... 4,154
Lake City, (D6).... 405
Lakeside, (F3)..... 103
Lamar, (No?..... 2,977
La Plata, (C8)..... 100
Laporte, (H?)..... 120
La Salle, (J2?..... 150
Las Animas, (M6). 2,008
Lasauses, (G8)..... 160
Lavalley, (H8)..... 190
La Veta, (H7)..... 691
Lawrence, (H5)..... 62
Lawson, (G4)..... 100
Leadville, (F4)... 7,508
Lime, (F7)..... 100
Limon, (L4)..... 534
Littleton, (J4)... 1,373
Loma, (A4)..... 800
Longmont, (J2).. 4,256
Loretto, (H3)..... 170
Louisville, (J1)... 1,706
Louviers, (H3)..... 300
Loveland, (H2).. 3,651
Lujane, (C6)..... 200
Lyons, (H3)..... 632
Magnolia, (H3)..... 100
Maitland, (J?)..... 350
Malvern, (J4)..... 150
Manassa, (G8)..... 788
Mancos, (B8)..... 567
Manitou, (J4)... 1,357
Manzanola, (L6).. 428
Marble, (D4)..... 782
Mead, (J2)..... 114
Meeker, (C2)..... 807
Merino, (M2)..... 100
Mesa, (B4)..... 100
Milliken, (J2)..... 100
Minturn, (F5)..... 241

Mirage, (G6)..... 120
Moffat, (G6)..... 100
Molina, (B4)..... 100
Montclair, (J3)..... 410
Monte Vista, (F7). 2,544
Montezuma, (G3).. 134
Montrose, (C6)... 3,254
Monument, (J4)... 149
Morrison, (H3)..... 251
Mosca, (G7)..... 150
Mountain View (J5) 390
Mount Morrison,
   (H3)..... 350
Nederland, (H3)..... 446
Nevadaville, (G3). 367
Newcastle, (C3)..... 493
Newett, (G5)..... 100
Niwot, (H2)..... 100
North Creede, (E7) 200
North Longmont,
   (H2)..... 260
Norwood, (B6)..... 212
Nunn, (J1)..... 143
Oak Creek, (E2).. 222
Ohio, (E5)..... 153
Olathe, (C5)..... 458
Ophir, (C7)..... 124
Ordway, (L6)..... 705
Ortiz, (F8)..... 500
Osier, (F8)..... 130
Ouray, (C6)..... 1,644
Overland, (H3)..... 250
Overton, (J6)..... 180
Oxford, (C8)..... 500
Pagosa Jc, (G8)..... 100
Pagosa Sprs, (E8).. 669
Paisaje, (G8)..... 200
Palisades, (B4)..... 900
Palmer, (J4)..... 150

Palmer Lake, (J4). 163
Paonia, (C5)..... 1,007
Parker, (J3)..... 100
Parlin, (E5)..... 100
Perigo, (G3)..... 200
Perins, (C8)..... 150
Petersburg, (J3)..... 100
Piedmont, (J8)..... 100
Pierce, (J1)..... 350
Pine, (H4)..... 100
Pinon, (J6)..... 100
Pitkin, (E5)..... 250
Platteville, (J2)..... 430
PonchoSprings,(F6) 43
Portland, (H6)..... 600
Primero, (J8)..... 2,000
Prospect Heights,
   (H6)..... 157
Pueblo, (J6)..... 44,395
Radiant, (H6)..... 200
Ramah, (K4)..... 200
Redcliff, (F4)..... 383
Red Mountain,(C7) 26
Rico, (B7)..... 368
Ridgway, (C6)..... 376
Rifle, (C3)..... 698
Robinson, (F4)..... 78
Rockvale, (H6).. 1,413
Rockyford, (L6). 3,230
Rollinsville, (H3).. 160
Romeo, (F8)..... 200
Rosemont, (J5)..... 500
Rosita, (H6)..... 42
Roswell, (J5)..... 350
Roubideau, (B5)..... 200
Rouse, (J8)..... 550
Rugby, (J8)..... 250
Russell Gulch,
   (H3)..... 700

Rye, (J?)..... 250
Saguache, (C6)..... 704
Saint Elmo, (F3)..... 46
Salida, (G5)..... 4,425
Salina, (H2)..... 100
San Bernardo, (C7) 100
Sanford, (G8)..... 564
San Luis, (G8)..... 900
San Rafael, (F5)..... 261
Sargents (E6)..... 100
Sawpit, (C7)..... 121
Sedalia, (J4)..... 100
Sedgwick, (N1)..... 200
Segundo (J8)..... 1,600
Seibert, (N4)..... 130
Severance, (J1)..... 150
Sheridan, (H3)..... 498
Silt, (C3)..... 100
Silver Cliff, (H6).. 250
Silver Plume, (G3). 460
Silverton, (C7).. 2,153
Simla, (K4)..... 100
Smuggler, (C7)..... 150
Sneffels, (C6)..... 150
Snyder, (L2)..... 120
Somerset, (D5)..... 420
Sopris, (J8)..... 300
South Canon, (H6). 1,321
South Canon, (D3) 200
South Fork, (E7).. 150
Springfield, (N8)..... 100
Starkville, (K8)... 1,000
Steamboat Springs,
   (E2)..... 1,227
Sterling, (M1)... 3,044
Sugar City, (L6)..... 808
Sulphur Sprs,(F2).. 182
Sunshine, (H2)..... 180
Superior, (H3)..... 349

Swandyke, (G3)... 100
Swink, (L?)..... 310
Tarryall, (H4)..... 100
Teller, (E7)..... 100
Telluride, (C7)... 1,756
Tercio, (H8)..... 500
Thomasville, (E4). 200
Timnath, (J1)..... 150
Tioga, (J7)..... 250
Tolland, (G3)..... 100
Trinidad, (K8).. 10,204
Turret, (G5)..... 200
Twin Lakes, (F4).. 130
Valverde, (J3)..... 800
Victor, (H5)..... 3,162
Villagrove, (F6)..... 130
Walden, (F1)..... 162
Waldorf, (G3)..... 100
Walsenburg, (J7).. 2,423
Ward, (G2)..... 129
Weldona, (L2)..... 100
Wellington, (J1)..... 459
Westcliffe, (H6).. 232
Westcreek, (H4)..... 34
Weston, (J8)..... 600
Wetmore, (H6)..... 100
Whitehorn, (G5)..... 500
Whitewater, (A5).. 150
Wiley, (N6)..... 197
Williamsburg,
   (H6)..... 556
Windsor, (J2)..... 935
Wolcott, (E3)..... 100
Woodland Park,
   (J4)..... 163
Wootton, (K8)..... 100
Wray, (O2)..... 1,000
Yampa, (E2)..... 332
Yuma, (N2)..... 333

CONNECTICUT

Abington, (H5).... 223
Adams, (C7)..... 100
Addison, (F6)..... 300
Allingtown, (E7)..... 300
Andover, (G6)..... 300
Ansonia, (E7)... 15,152
Ashford, (H5)..... 300
Aspetuck, (C8)..... 150
Avon, (E5)..... 1,100
Bakersville, (D5)..... 200
Ballouville, (J5)..... 250
Baltic, (H6)..... 750
Bantam, (D6)..... 500
Barkhamsted, (F5). 150
Beacon Falls, (D7). 850
Bean Hill, (H6)..... 450
Berlin, (F6)..... 900
Bethany, (D7)..... 400
Bethel, (C7)..... 3,041
Bethlehem, (D6)..... 400
Black Hall, (G7)..... 220
Bloomfield, (F5)... 1,650
Boardman, (C6)..... 200
Bolton, (G6)..... 300
Botsford, (C7)..... 700
Bozrah, (H6)..... 400
Bozrahville, (H6)..... 200
Branchville, (C7)..... 175
Branford, (E7)... 2,560
Bridgeport, (D8) 102,054
Bridgewater, (C6). 500
Bristol, (F6)..... 9,527
Broad Brook, (F5). 1,400
Brookfield, (C7)..... 550
Brookfield Center,
   (C7)..... 350
Brooklyn, (J5)... 1,750
Brooksvale, (E7)..... 100
Buckland, (F5)..... 300
Burlington, (E5).. 1,125
Burnside, (F5)..... 800

Burrville, (D5)..... 150
Campville, (D6)..... 100
Canaan, (C4)..... 100
Cannon Station,
   (C8)..... 200
Canterbury, (H6).. 300
Canton Center, (E5) 200
Center Brook, (G7) 175
Center Groton (H7) 500
Central Village,(J6) 1,000
Chapinville, (C4).. 200
Chaplin, (H5)..... 300
Cheshire, (E7)... 1,500
Chester, (G7)..... 1,300
Chesterfield, (H7).. 270
Chestnut Hill, (G6) 100
Clarks Falls, (J7).. 280
Clinton, (F7)..... 1,200
Clintonville, (E7).. 100
Cobalt, (F6)..... 320
Colchester, (G6).. 978
Colebrook, (D5)..... 275
Colebrook River,(D5)200
Collinsville, (E5). 2,000
Columbia, (G6)..... 450
Comstock's Bridge,
   (G6)..... 200
Cornwall, (C5)..... 100
Cornwall Bridge,(C5)100
Cornwall Hollow,
   (C5)..... 100
Coscob, (B8)..... 500
Coventry, (G5)..... 100
Cranbury, (C8)..... 500
Cromwell, (F6).. 2,000
Danbury, (C7).. 20,234
Danielson, (J5).. 2,934
Darien, (C8)..... 2,400
Dayville, (J5)..... 400
Deep River, (G7). 1,480
Derby, (D7)..... 8,991

Durham, (F7)..... 500
Durham Center, (F7) 390
Eagleville, (H5)..... 300
East Berlin, (F6).. 700
East Canaan, (C4). 500
Eastford, (H5)..... 180
East Glastonbury,
   (F6)..... 350
East Granby, (E5) 450
East Haddam, (G7) 1,350
East Hampton,
   (G6)..... 1,400
East Hartford (F5) 5,500
East Hartford Mea-
   dow, (F5)..... 700
East Hartland, (E5) 300
Easthaven, (E7).. 1,250
East Killingly, (J5) 100
East Litchfield, (D5) 100
East Lyme, (H7).. 500
East Morris, (D6).. 100
East Norwalk, (C8) 3,600
Easton, (C7)..... 300
East Port Chester,
   (B8)..... 2,000
East River, (F7)..... 300
East Thompson,
   (J5)..... 300
East Wallingford,
   (F7)..... 150
East Willington, (H5) 150
East Windsor, (F5) 240
East Windsor Hill,
   (F5)..... 350
East Woodstock,
   (H5)..... 280
Ekonk, (J6)..... 130
Ellington, (G5)... 1,900
Elliott, (H5)..... 125
Elmwood, (F6)..... 300
Enfield, (F5)..... 850

Essex, (G7)..... 2,100
Fairfield, (C8)... 3,100
Falls Village, (C5). 600
Farmington, (E6). 897
Fitchville, (H6)..... 150
Forestville, (E6).. 3,400
Gales Ferry, (H7).. 125
Gardner Lake, (H7) 130
Gaylordsville, (C6) 275
Georgetown, (C7). 230
Gildersleeve, (F6). 900
Gilead, (G6)..... 350
Glasgo, (J6)..... 700
Glastonbury, (F6). 1,800
Glenbrook, (B8)..... 450
Glenville, (B8)..... 800
Goshen, (D5)..... 350
Granby, (E5)..... 360
Greenfield Hill, (C8) 900
Greens Farms, (C8) 125
Greenwich, (B8).. 3,886
Griswold, (J6)..... 400
Grosvenor Dale,
   (J5)..... 800
Groton, (H7)..... 1,895
Grovebeach, (F7).. 100
Guilford, (F7)..... 1,608
Gurleyville, (H5).. 230
Haddam, (F7)..... 400
Haddam Neck, (G6) 100
Hadlyme, (G7)..... 250
Hallville, (H6)..... 250
Hamburg, (G7)..... 260
Hamden, (E7)..... 4,300
Hampton, (H5)..... 480
Hanover, (H6)..... 400
HARTFORD,
   (F5)..... 98,915
Hartland, (F5)..... 190
Harwinton, (D5).. 1,300
Hawleyville, (C7).. 600

Hazardville, (F5).. 1,200
Hebron, (G6)..... 420
Higganum. (F7). 1,000
Highland Park,
   (G5)..... 200
High Ridge, (B8). 300
Highwood, (E7)..... 150
Hockanum, (F6)..... 250
Hopewell, (F6)..... 500
Hop River, (G6)..... 100
Hotchkissville, (C6) 250
Huntington, (D7).. 1,000
Ivoryton, (G7)..... 350
Jewett City, (J6).. 3,023
Kensington, (E6).. 1,950
Kent, (C6)..... 400
Kent Furnace, (C6) 180
Kibbe, (G4)..... 180
Killingly, (J5)..... 400
Killingworth, (F7). 575
Lakeville, (C5)..... 750
Laurelglen, (J7)..... 75
Lebanon, (H6)..... 1,100
Ledyard, (J7)..... 850
Leete Island, (F5). 150
Leonard Bridge,
   (G6)..... 100
Liberty Hill, (H6). 125
Lime Rock, (C5)..... 450
Litchfield, (D6)..... 903
Long Hill, (C7)..... 400
Longridge, (B8)..... 430
Lyme, (H7)..... 680
Madison, (F7)..... 850
Manchester, (G5).. 3,600
Manchester Green,
   (G5)..... 300
Mansfield, (G5)..... 200
Mansfield Center,
   (H5)..... 350

Mansfield Depot,
(G5)............ 250
Marble Dale, (C6).. 200
Marion, (E6)...... 280
Marlboro, (G6).... 260
Massapeag, (H7).. 100
Mechanicsville, (J5) 600
Melrose, (F5)...... 130
Meriden, (E6)...27,265
Merrow, (G5)...... 100
Mianus, (B8)...... 500
Middlebury, (D6).. 750
Middlefield, (F6).. 650
Middle Haddam,
(F6)............ 550
Middletown, (F6). 11,851
Milford, (D8)..... 4,000
Mildale, (E6)...... 350
Millington, (G7)... 100
Mill Plain, (B7)... 350
Millstone, (H7)... 180
Milton, (C5)...... 100
Minortown, (D6).. 160
Mohegan, (H7).... 220
Monroe, (D7)..... 275
Montowese, (E7).. 300
Montville, (H7)... 1,000
Moodus, (G7)..... 750
Moosup, (J6)..... 2,300
Morris, (D6)...... 400
Mount Carmel, (E7) 500
Mount Carmel Cen-
ter, (E7)........ 400
Mystic, (J7)..... 3,900
Naugatuck, (D7)..12,722
Neraug, (E5)...... 100
New Boston, (J4).. 100
New Britain, (E6) 43,916
New Canaan, (B8). 1,672
New Fairfield, (B7) 100
New Hartford, (D5)1,500
New Haven, (E7) 133,605
Newington, (F6)... 400
Newington Jc, (F6) 200
New London, (H7) 19,659
New Milford, (C6). 4,100
New Preston, (C6). 450
Newtown, (C7)... 434
Niantic, (H7)..... 1,200
Nichols, (D8)..... 300
Noank, (J7)...... 1,100
Norfolk, (D5)..... 1,350
Noroton, (B8)..... 600
Noroton Heights, (B8) 500
North Ashford, (H5) 140
North Branford,(F7) 370
North Canton, (E5) 200
Northfield, (D6)... 600
Northford, (E7)... 370

North Franklin, (H6) 300
North Granby, (E4) 420
North Grosvenor
Dale, (J5)...... 2,500
North Guilford, (F7) 500
North Haven, (E7) 1,700
North Kent, (C5).. 150
North Lyme, (G7). 150
North Madison,
(F7)............ 300
North Ridgefield,
(B7)............ 140
North Stamford, (B8) 550
North Sterling, (J5) 130
North Stonington,
(J7)............ 480
Northville, (C6)... 275
North Westchester,
(G6)............ 250
North Wilton, (C8) 150
North Windham,
(H6)............ 200
North Woodbury,
(D6)............ 350
North Woodstock,
(H5)............ 280
Norwalk, (C8)... 6,954
Norwich, (H6).. 20,367
Norwichtown, (H6) 1,400
Oakdale, (H7).... 330
Oakville, (D6).... 600
Occum, (H6)..... 225
Old Lyme, (G7)... 730
Old Mystic, (J7)... 700
Old Saybrook, (G7) 675
Oneco, (J6)...... 410
Orange, (D7).... 1,500
Orehill, (C5)..... 120
Oronoque, (D8)... 100
Oxford, (D7)..... 400
Packer, (J6)...... 200
Pendleton Hill, (J6) 100
Pequabuck, (E6).. 350
Phoenixville, (H5). 100
Pine Meadow, (D5) 150
Pineorchard, (E7). 500
Plainfield, (J6)... 1,200
Plainville, (E6)... 2,500
Plantsville, (E6).. 1,800
Plattsville, (C8)... 310
Pleasant Valley, (E5) 300
Plymouth, (D6)... 2,400
Pomfret, (J5)..... 1,100
Pomfret Center, (H5) 300
Poquetanuck, (H7) 500
Poquonock, (F5).. 900
Poquonock Bridge,
(H7)............ 250
Portland, (F6)... 2,300

Preston, (J6) 500
Putnam, (J5)..... 6,637
Quaker Hill, (H7). 180
Quinebaug, (J4)... 300
Rainbow, (F5).... 300
Redding, (C7).... 600
Redding Ridge,(C7) 240
Reynolds Bridge,(D6)200
Ridgebury, (C7)... 250
Ridgefield, (B7)... 1,114
Riverbank, (B8)... 100
Riverside, (B8)... 250
Riverton, (E5).... 200
Rockfall, (F6)..... 200
Rockville, (G5)... 7,977
Rockyhill, (F6)... 1,050
Roundhill, (B8).. 1,000
Rowayton, (C8)... 1,150
Roxbury, (C6).... 300
Roxbury Falls, (C6) 125
Roxbury Station,
(C6)............ 300
Salem, (G7)...... 350
Salisbury, (C5)... 2,250
Sandy Hook, (C7). 1,600
Sanford, (C7).... 340
Saugatuck, (C8)... 600
Saybrook, (G7)... 700
Scitico, (F5)..... 540
Scotland, (H6).... 400
Seymour, (D7)... 4,000
Shailerville, (F7).. 100
Shaker Station, (F4) 120
Sharon, (C5)..... 1,500
Sharon Valley, (C5) 300
Shelton, (D7).... 4,807
Sherman, (C6).... 400
Short Beach, (E7). 300
Silverlane, (F5)... 250
Silver Mine, (C8). 400
Simsbury, (E5)... 1,600
Somers, (E5)..... 480
Somerville, (G5).. 880
Sound Beach, (B8). 1,000
South Britain, (C7) 450
Southbury, (D7).. 400
South Cheshire, (E7) 200
South Coventry, (G5) 950
Southford, (D7)... 200
South Glastonbury,
(F6)............ 1,300
Southington, (E6). 3,714
South Killingly, (J5) 150
South Lyme, (G7). 175
South Manchester,
(G6)............ 9,000
South Meriden, (E6) 880
South Norwalk,(C8)8,968
Southport, (C8)... 1,200

South Wethersfield,
(F6)............ 175
South Willington,
(G5)............ 250
South Wilton, (C8) 210
South Windham,
(H6)............ 400
South Windsor,(F5)1,075
South Woodstock,
(J5)............ 280
Springdale, (B8)... 540
Stafford, (G5).... 900
Stafford Springs,
(G5)........... 3,059
Staffordville, (G4). 450
Stamford, (B8). 25,138
Stanwich, (B8).... 300
Stepney, (C7).... 300
Stepney Depot, (D7) 225
Sterling, (J6)..... 450
Stevenson, (D7)... 150
Stonington, (J7).. 2,083
Stony Creek, (E7). 1,200
Storrs, (G5)...... 200
Stratford, (D8)... 4,000
Suffield, (F5).... 2,800
Taftville, (H6)... 3,800
Talcottville, (G5).. 475
Tariffville, (E5)... 300
Terryville, (D6)... 2,400
Thomaston, (D6). 3,200
Thompson, (J5)... 625
Thompsonville,
(F4)........... 6,000
Tolland, (G5)..... 100
Torrington, (D5). 15,483
Tracy, (E6)....... 170
Trumbull, (D8)... 900
Turnerville, (G6).. 150
Tyler City, (D7)... 150
Tylerville, (G7)... 100
Uncasville, (H7)... 670
Union, (H5)...... 100
Union City, (D7). 3,300
Unionville, (E5).. 2,200
Vernon, (G3)..... 325
Vernon Center, (G5) 150
Versailles, (H6)... 190
Vinton Mills, (F5). 100
Voluntown, (J6)... 100
Wallingford, (E7). 8,690
Wapping, (F5).... 550
Warehouse Point,
(F5)........... 1,250
Warren, (C6)..... 380
Warrenville, (H5). 100
Washington, (C6). 480
Washington Depot,
(C6)............ 440

Waterbury, (D6). 73,141
Waterford, (H7)... 2,550
Watertown, (D6). 3,400
Waterville, (D6). 3,000
West Mystic, (J6). 400
Weatogue, (E5)... 200
West Ashford, (H5) 100
Westbrook, (G7)... 700
West Cheshire, (E6) 300
Westchester, (G6). 250
West Cornwall, (C5) 350
Westfield, (H5)... 100
West Goshen, (D5) 175
West Granby, (E5) 320
West Hartford,(E5)3,800
West Hartland, (E4) 120
West Haven, (E7). 8,543
Westminster, (H6) 200
West Mystic, (J7). 100
West Norfolk, (D5) 100
West Norwalk, (C8) 500
Weston, (C8)..... 480
Westport, (C8)... 3,200
West Redding, (C7) 180
West Simsbury,(E5) 200
West Stafford, (G5) 240
West Suffield, (F5) 820
West Thompson,
(J5)........... 150
West Torrington,
(D5)........... 150
West Willington,
(G5)........... 150
West Woodstock,
(H5)........... 280
Wethersfield, (F6). 2,750
Whitneyville, (E7). 300
Willimantic, (H6) 11,230
Willington, (G5)... 260
Wilsonville, (J4)... 220
Wilton, (C8)..... 420
Winchester Center,
(D5)........... 400
Windham, (H6)... 600
Windsor, (F5)... 2,600
Windsor Locks,
(F5)........... 3,500
Windsorville, (F5). 250
Winnipauk, (C8).. 600
Winsted, (D5).... 7,754
Wolcott, (E6)..... 200
Woodbridge, (E7). 800
Woodbury, (D6). 1,000
Woodmont, (D8).. 194
Woodstock, (J5)... 280
Woodstock Valley,
(H5)........... 280
Yalesville, (E7)... 1,500
Yantic, (H6)...... 600

DELAWARE

Ashland, (G1)..... 160
Bayard, (H3)..... 220
Beaver Valley, (G1) 290
Bellevue, (G1).... 250
Bethamy Bch, (H3) 56
Bethel, (G3)...... 370
Blackbird, (G2)... 100
Blades, (G3)...... 500
Bowers, (H2)..... 212
Brandywine Springs,
(G1)........... 100
Bridgeville,(G3)... 939
Camden,...(H2)... 553
Centerville, (G1).. 200
Cheswold, (G2)... 223
Christiana, (G1)... 400
Claymont, (H1)... 400
Clayton, (G2)..... 764
Concord, (G3).... 350
Coolspring, (H3).. 150
Dagsboro, (H3)... 176

Delaware City,(G1) 1,132
Delmar, (G4)..... 530
DOVER, (H2)... 3,720
Edgemoor, (H1)... 500
Ellendale, (H3)... 216
Elsmere, (G1).... 374
Farmington, (G3). 255
Farnhurst, (G1)... 330
Faulkland, (G1)... 250
Felton, (G2)..... 451
Frankford, (H3)... 395
Frederica, (H2)... 659
Georgetown, (H3). 1,609
Glasgow, (G1).... 100
Greenwood, (G3).. 362
Grubbs, (H1)..... 100
Harbeson, (H3)... 150
Harrington, (G3).. 1,500
Hartly, (G2)..... 200
Henry Clay Factory,
(G1)........... 1,000

Hickman, (G3)... 300
Hockessin, (G1)... 400
Hollyoak, (H1).... 220
Houston Sta, (H3). 150
Kenton, (G2)..... 209
Kirkwood, (G1)... 120
Laurel, (G3)..... 2,166
Lebanon, (H2).... 200
Leipsic, (G2)..... 271
Lewes, (H3)..... 2,158
Lincoln, (H3)..... 400
Little Creek, (H2). 285
Magnolia, (H2)... 210
Marshallton, (G1). 430
Middletown, (G2). 1,399
Midway, (H3).... 100
Milford, (H3).... 2,603
Millsboro, (H3)... 451
Millville, (H3).... 193
Milton, (H3).... 1,038
Montchanin, (G1). 130

Mount Cuba, (G1). 100
Mount Pleasant,
(G1)........... 200
Nassau, (H3)..... 150
Newark, (G1).... 1,913
Newcastle, (G1).. 3,351
Newport, (G1).... 722
Oakel, (G3)...... 100
Ocean View, (H3). 302
Odessa, (G2)..... 585
Omar, (H3)...... 100
Port Penn, (G1)... 299
Redlion, (G1)..... 100
Rehoboth, (H3)... 327
Risingsun, (G2)... 300
Rockland, (G1)... 400
Roxana, (H3)..... 155
Saint Georges, (G1) 264
Seaford, (G3).... 2,108
Selbyville, (H4)... 342
Smyrna, (G2).... 1,843

Stanton, (G1)..... 300
Stockley, (H3).... 170
Summit Bridge,
(H5)........... 130
Taylors Bridge, (G2) 250
Townsend, (G2)... 494
Viola, (G2)...... 250
Whitesville, (G4).. 100
Willowgrove, (G2). 200
Wilmington, (G1) 87,411
Winterthur, (H1).. 220
Wooddale, (G1)... 200
Woodland, (G3)... 270
Woodside, (G2)... 300
Wyoming, (G2)... 517
Yorklyn, (G1)..... 330

DISTRICT OF
COLUMBIA
WASHINGTON,
(D3)........ 331,069

CONNECTICUT

Mansfield Depot, (G5).......... 250
Marble Dale, (C6). 200
Marion, (E6)...... 280
Marlboro, (G6).... 260
Massapeag, (H7).. 100
Mechanicsville, (J5) 600
Melrose, (F5)..... 130
Meriden, (E6)... 27,265
Merrow, (G5)..... 100
Mianus, (B8)..... 500
Middlebury, (D6). 750
Middlefield, (F6).. 650
Middle Haddam, (F6).......... 550
Middletown, (F6). 11,851
Milford, (D8).... 4,000
Mildale, (E6)..... 350
Millington, (G7).. 100
Mill Plain, (B7)... 350
Millstone, (H7)... 180
Milton, (C5)..... 100
Minortown, (D6).. 160
Mohegan, (H7)... 220
Monroe, (D7)..... 275
Montowese, (E7).. 300
Montville, (H7)... 1,000
Moodus, (G7)..... 750
Moosup, (J6)..... 2,300
Morris, (D6)...... 400
Mount Carmel, (E7) 500
Mount Carmel Center, (E7)........ 400
Mystic, (J7)...... 3,900
Naugatuck, (D7)..12,722
Neraug, (E5)...... 100
New Boston, (J4).. 100
New Britain, (E6) 43,916
New Canaan, (B8). 1,672
New Fairfield, (B7) 100
New Hartford, (D5)1,500
New Haven, (E7) 133,605
Newington, (F6).. 400
Newington Jc, (F6) 200
New London, (H7) 19,659
New Milford, (C6). 4,100
New Preston, (C6). 450
Newtown, (C7).... 434
Niantic, (H7)..... 1,200
Nichols, (D8)..... 300
Noank, (J7)....... 1,100
Norfolk, (D5)..... 1,350
Noroton, (B8)..... 600
Noroton Heights,(B8) 500
North Ashford, (H5) 140
North Branford,(F7) 370
North Canton, (E5) 200
Northfield, (D6)... 600
Northford, (E7)... 370

North Franklin, (H6) 300
North Granby, (E4) 420
North Grosvenor Dale, (J5)...... 2,500
North Guilford, (F7) 500
North Haven, (E7) 1,700
North Kent, (C5).. 150
North Lyme, (G7). 150
North Madison, (F7).......... 300
North Ridgefield, (B7).......... 140
North Stamford, (B8) 550
North Sterling, (J5) 130
North Stonington, (J7)........... 480
Northville, (C6)... 275
North Westchester, (G6).......... 250
North Wilton, (C8) 150
North Windham, (H6).......... 200
North Woodbury, (D6).......... 350
North Woodstock, (H5).......... 280
Norwalk, (C8).... 6,954
Norwich, (H6).. 20,367
Norwichtown, (H6) 1,400
Oakdale, (H7).... 330
Oakville, (D6).... 600
Occum, (H6)..... 225
Old Lyme, (G7)... 730
Old Mystic, (J7)... 700
Old Saybrook, (G7) 675
Oneco, (J6)....... 410
Orange, (D7)..... 1,500
Orehill, (C5)..... 120
Oronoque, (D8)... 100
Oxford, (D7)..... 400
Packer, (J6)...... 200
Pendleton Hill, (J6) 100
Pequabuck, (E6).. 350
Phoenixville, (H5). 100
Pine Meadow, (D5) 150
Pineorchard, (E7). 500
Plainfield, (J6).... 1,200
Plainville, (E6)... 2,500
Plantsville, (E6).. 1,800
Plattsville, (C8)... 310
Pleasant Valley, (E5) 300
Plymouth, (D6)... 2,400
Pomfret, (J5)..... 1,100
Pomfret Center, (H5) 300
Poquetanuck, (H7) 500
Poquonock, (F5).. 900
Poquonock Bridge, (H7)........... 250
Portland, (F6).... 2,300

Preston, (J6)..... 500
Putnam, (J5)..... 6,637
Quaker Hill, (H7). 180
Quinebaug, (J4)... 300
Rainbow, (F5).... 300
Redding, (C7).... 600
Redding Ridge,(C7) 240
Reynolds Bridge,(D6)200
Ridgebury, (C7)... 250
Ridgefield, (B7)... 1,114
Riverbank, (B8)... 100
Riverside, (B8).... 250
Riverton, (E5).... 200
Rockfall, (F6).... 200
Rockville, (G5)... 7,977
Rockyhill, (F6)... 1,050
Roundhill, (B8)... 1,000
Rowayton, (C8)... 1,150
Roxbury, (C6).... 300
Roxbury Falls, (C6) 125
Roxbury Station, (C6).......... 300
Salem, (G7)...... 350
Salisbury, (C5)... 2,250
Sandy Hook, (C7). 1,600
Sanford, (C7)..... 340
Saugatuck, (C8)... 600
Saybrook, (G7)... 700
Scitico, (F5)...... 540
Scotland, (H6).... 400
Seymour, (D7).... 4,000
Shailerville, (F7).. 100
Shaker Station, (F4) 120
Sharon, (C5)..... 1,500
Sharon Valley, (C5) 300
Shelton, (D7).... 4,807
Sherman, (C6).... 400
Short Beach, (E7). 300
Silverlane, (F5)... 250
Silver Mine, (C8). 400
Simsbury, (E5)... 1,600
Somers, (G5)..... 480
Somerville, (G5).. 880
Sound Beach, (B8). 1,000
South Britain, (C7) 450
Southbury, (D7)... 400
South Cheshire, (E7) 200
South Coventry, (G5) 950
Southford, (D7)... 200
South Glastonbury, (F6).......... 1,100
Southington, (E6). 3,714
South Killingly, (J5) 150
South Lyme, (G7). 175
South Manchester, (G6).......... 9,000
South Meriden,(E6) 700
South Norwalk, (C8)8,968
Southport, (C8)... 1,200

South Wethersfield, (F6).......... 175
South Willington, (G5).......... 250
South Wilton, (C8) 210
South Windham, (H6).......... 400
South Windsor,(F5) 1,075
South Woodstock, (J5).......... 280
Springdale, (B8)... 540
Stafford, (F5).... 900
Stafford Springs, (G5).......... 3,059
Staffordville, (G4). 450
Stamford, (B8).. 25,138
Stanwich, (B8)... 300
Stepney, (C7)..... 300
Stepney Depot, (D7) 225
Sterling, (J6)..... 450
Stevenson, (D7)... 150
Stonington, (J7)... 2,083
Stony Creek, (E7). 1,200
Storrs, (G5)...... 200
Stratford, (D8)... 4,900
Suffield, (F5)..... 2,800
Talcottville, (G5).. 475
Tariffville, (E5)... 300
Terryville, (D6)... 2,400
Thomaston, (D6).. 3,200
Thompson, (J5)... 625
Thompsonville, (F5).......... 6,000
Tolland, (G5)..... 1,000
Torrington, (D5).. 15,483
Tracy, (E6)....... 170
Trumbull, (D8)... 900
Turnerville, (G6).. 150
Tyler City, (D7)... 150
Tylerville, (G7)... 100
Uncasville, (H7)... 670
Union, (H3)...... 100
Union City, (D7).. 3,300
Unionville, (E5)... 2,200
Vernon, (G5)..... 325
Vernon Center, (G5) 150
Versailles, (H6)... 190
Vinton Mills, (F5). 100
Voluntown, (J6)... 600
Wallingford, (E7). 8,690
Wapping, (F5).... 550
Warehouse Point, (F5).......... 1,250
Warren, (C6)..... 380
Warrenville, (H5). 100
Washington, (C6).. 480
Washington Depot, (C6).......... 440

Waterbury, (D6). 73,141
Waterford, (H7).. 2,550
Watertown, (D6).. 3,400
Waterville, (D6). 3,000
Wauregan, (J6).... 400
Westogue, (E5).... 200
West Ashford, (H5) 100
Westbrook, (G7).. 700
West Cheshire, (E6) 300
Westchester, (G6). 250
West Cornwall, (C5) 350
Westford, (H5).... 100
West Goshen, (D5) 175
West Granby, (E5) 320
West Hartford,(E5) 8,800
West Hartland,(E4) 120
West Haven, (E7). 8,543
Westminster, (H6) 200
West Mystic, (J7). 100
West Norfolk, (D5) 100
West Norwalk, (C8) 500
Weston, (C8)..... 480
Westport, (C8)... 3,200
West Redding, (C7) 180
West Simsbury, (E5) 200
West Stafford, (G5) 240
West Suffield, (F5). 820
West Thompson, (J5).......... 150
West Torrington, (D5).......... 150
West Willington, (G5).......... 150
West Woodstock, (H5).......... 280
Wethersfield, (F5). 2,750
Whitneyville, (E7). 300
Willimantic, (H6) 11,230
Willington, (G5)... 260
Wilsonville, (J4)... 220
Wilton, (C8)...... 420
Winchester Center, (D5).......... 400
Windham, (H6)... 600
Windsor, (F5).... 2,600
Windsor Locks, (F5).......... 3,500
Windsorville, (F5). 250
Winnipauk, (C8).. 600
Winsted, (D5).... 7,754
Wolcott, (E6)..... 200
Woodbridge, (E7). 800
Woodbury, (D6).. 1,000
Woodmont, (D8).. 194
Woodstock Valley, (H5).......... 280
Yalesville, (E7)... 1,500
Yantic, (H6)...... 600

DELAWARE

Ashland, (G1)..... 160
Bayard, (H3)...... 220
Beaver Valley, (G1) 290
Bellevue, (G1)..... 250
Bethamy Bch, (H3) 56
Bethel, (G3)...... 370
Blackbird, (G2)... 100
Blades, (G3)...... 500
Bowers, (H2)..... 212
Brandywine Springs, (G1).......... 100
Bridgeville,(G3).. 939
Camden, ..(H2)... 553
Centerville, (G1)... 200
Cheswold, (G2).... 223
Christiana, (G1)... 400
Claymont, (H1).... 400
Clayton, (G2)..... 764
Concord, (G3)..... 350
Coolspring, (H3)... 150
Dagsboro, (H3).... 176

Delaware City,(G1) 1,132
Delmar, (G4)..... 530
DOVER, (H2).... 3,720
Edgemoor, (H1)... 500
Ellendale, (H3).... 216
Elsmere, (G1)..... 374
Farmington, (G3). 255
Farnhurst, (G1)... 330
Faulkland, (G1)... 250
Felton, (G2)...... 451
Frankford, (H3)... 395
Frederica, (H2)... 659
Georgetown, (H3). 1,609
Glasgow, (G1)..... 100
Greenwood, (G3).. 362
Grubbs, (H1)..... 100
Harbeson, (H3)... 150
Harrington, (G3).. 1,500
Hartly, (G2)...... 200
Henry Clay Factory, (G1).......... 1,000

Hickman, (G3).... 300
Hockessin, (G1)... 400
Hollyoak, (H1).... 220
Houston Sta, (H3). 150
Kenton, (G2)..... 209
Kirkwood, (G1)... 120
Laurel, (G3)...... 2,166
Lebanon, (H2).... 200
Leipsic, (G2)..... 271
Lewes, (H3)...... 2,158
Lincoln, (H3)..... 400
Little Creek, (H2). 285
Magnolia, (H2)... 210
Marshallton, (G1). 430
Middletown, (G2). 1,399
Midway, (H3)..... 100
Milford, (H3).... 2,603
Millsboro, (H3)... 451
Millville, (H3).... 193
Milton, (H3).... 1,038
Montchanin, (G1). 130

Mount Cuba, (G1). 100
Mount Pleasant, (G1).......... 200
Nassau, (H3)..... 150
Newark, (H3).... 1,913
Newcastle, (G1)... 3,351
Newport, (G1).... 722
Oakel, (G3)...... 100
Ocean View, (H3). 302
Odessa, (G2)..... 585
Omar, (H3)...... 100
Port Penn, (G1)... 299
Redlion, (G1)..... 100
Rehoboth, (H3)... 327
Risingsun, (G2)... 300
Rockland, (G1).... 100
Roxana, (H3)..... 155
Saint Georges, (G1) 264
Seaford, (G3).... 2,108
Selbyville, (H4)... 342
Smyrna, (G2).... 1,843

Stanton, (G1)..... 300
Stockley, (H3).... 170
Summit Bridge, (G1).......... 130
Taylors Bridge, (G2) 250
Townsend, (G2)... 494
Viola, (G2)....... 250
Whitesville, (G4).. 100
Willowgrove, (G2). 200
Wilmington, (G1) 87,411
Winterthur, (G1).. 220
Wooddale, (G1)... 200
Woodland, (G3)... 270
Woodside, (G2)... 300
Wyoming, (G2).... 517
Yorklyn, (G1)..... 330

DISTRICT OF COLUMBIA

WASHINGTON, (D3)........ 331,069

FLORIDA

New River, (D2)...	130	Panasoffkee, (D3).	200	Rio, (F4).........	100	South Jacksonville,		Watertown, (D1)..	250
New Smyrna, (F2).	1,121	Paola, (E3).......	100	River Jc, (B1)....	600	(E1)..........	1,147	Wauchula, (E4)..	1,099
Newtown, (D2)...	200	Parish, (D4)......	100	Rochelle, (D2)...	200	South Lake Weir,		Waukeenah, (C3)..	300
Niceville, (B7) ...	150	Pauway, (E3)......	200	Rockledge, (F3)..	100	(D3)...:......	100	Wausau, (A1).....	400
Nichols, (D4).....	300	Pedro, (D3).......	150	Rockwell, (D2)...	100	South Port, (A1)..	200	Webster, (D3).....	301
Nixon, (A1).......	180	Pensacola, (B7).	22,982	Rosalie, (E4).....	100	Spring Garden,(E2)	150	Weirsdale, (F3)....	150
Nocatee, (E4).....	250	Perry, (C1)......	1,012	Rosewood, (D2)...	250	Springhill, (B1)..	200	Welaka, (E2).....	294
Noma, (A1).......	816	Picolata, (E2).....	200	Sagano, (D3).....	150	Starke, (D2).....	1,135	Welcome, (D4)	100
Norwalk, (E2).....	120	Piedmont, (E3)....	100	Saint Andrew, (A1)	675	Steinhatchee, (C2).	150	Millborn, (D1)...	247
Oakhill, (F3).....	100	Pierce, (E4)......	200	Saint Augustine,		Stuart, (F5)......	275	West Bay, (C7)...	100
Oakland, (E3).....	211	Pierson, (E2).....	250	(E2).........	5,494	Summerfield, (D2).	230	Westlake, (C1)...	200
O'Brien, (D1).....	280	Pine, (D2)	120	Saint Catherine,		Summit, (E2)	120	West Palm Beach,	
Ocala, (D2)......	4,370	Pine Barren, (A6).	300	(D3).........	100	Sumterville, (D3).	200	(F5)........	1,743
Ocoee, (E3).......	130	Pinecastle, (E3)..	100	Saint Cloud, (E3)..	900	Sorrento, (E3)	150	West Tampa,	
Ojus, (F6)........	150	Pinehurst, (E1)...	150	Saint Francis, (E2)	50	Sutherland, (D3)..	300	(D4)........	8,258
Okahumpka, (E3).	200	Pinemount, (D1)..	275	Saint Joseph, (D3).	150	Suwanee, (C1)....	125	Westville, (C6)....	650
Olive, (A7)......	100	Pinetta, (C1).....	225	Saint Leo, (D3)....	100	Sycamore, (D4)...	150	Wewahitchka, (A1)	250
Olney, (F4)......	150	Plant City, (D3).	2,481	Saint Marks, (B1).	200	Switzwater, (F4)..	150	White City, (F4)...	350
Olustee, (D1).....	300	Planter, (F6).....	300	Saint Nicholas, (E1)	350	TALLAHASSEE,		White Springs,	
Orangebend, (E3).	75	Platt, (D4).......	100	Saint Petersburg,		(B1)........	5,018	(D1)........	1,177
Orange City, (E3).	490	Point Washington,		(D4).........	4,127	Tampa, (D4)...	37,782	Whitfield, (B7)...	250
Orange Heights,		(B7)........	300	Sampson, (D2)....	100	Tangerine, (E3) ..	100	Wildwood, (D3)..	329
(D2).........	150	Pomona, (E2).....	301	San Antonio, (D3).	131	Tarpon Spra,(D3).	2,212	Willeford, (D2)...	100
Orangepark, (E1).	372	Pompano (F5)...	269	Sanborn, (B1)....	150	Tavares, (E3)	175	Williston, (D2)...	371
Orange Sprs, (E1).	200	Ponce De Leon,		Sanderson, (D1)...	150	Taylor, (B1).....	100	Wilmarth, (C1)....	70
Orient, (D4)......	225	(C6).........	200	Sanford, (E3)....	3,570	Terrell, (F3).....	150	Windsor, (D2).....	300
Orlando, (E3)....	3,894	Ponce Park, (F2).	150	Sanibel, (D5).....	100	Thonotosassa, (D3)	300	Winter Garden,	
Ormond, (E2).....	780	Portland, (B6)....	75	San Mateo, (E3)..	110	Titusville, (F3) ..	868	(E3)........	351
Osteen, (E3)......	300	Port Orange, (E2).	200	Sarasota, (D4)....	840	Tooganka, (A1) ..	200	Winterhaven,	
Oviedo, (E3)......	250	Port Tampa, (D4).	100	Seabreeze, (F2)...	308	Trenton (D2).....	304	(E3)........	600
Oxford, (D3).....	330	Port Tampa City,		Sebastian, (F4)...	220	Tulliy, (D3)......	290	Winter Park,	
Pablo Beach, (E1).	249	(D4).........	1,343	Seffner, (D3).....	300	Turners (E3).....	100	(E3)........	570
Paisley, (E3).....	80	Potolo, (C6)......	50	Seville, (E2).....	200	Tyler, (E4)......	100	Wiscon, (D3).....	100
Palatka, (E2).....	3,779	Princeton, (F6)...	100	Shady Grove, (C1).	100	Unadilla, (D4)...	283	Woodville, (B1)..	180
Palatka Heights,		Providence, (D2).	300	Silas, (C6).......	100	Venice, (D4).....	100	Yalaha, (E3).....	100
(E2)........	367	Punta Gorda, (E5).	1,012	Silver Spring, (E2).	100	Venus, (E4)......	100	Youkon, (E1).....	150
Palm Beach, (F5)..	403	Putnam Hall, (E2).	100	Smithcreek, (B1)..	150	Vernon, (A1)....	100	Yulee, (E1)......	250
Palmetto, (D4)...	773	Quincy, (B1)....	3,204	Sneads, (B1).....	506	Vicksburg, (A1)..	150	Zellwood, (E3)....	100
Panama City, (E1)	422	Reddick, (D2).....	498	Socrum, (E3).....	160	Wall, (D3).......	540	Zephyrhills, (D3)..	300
Panama Park,		Redland, (F6).....	130	Sopchoppy, (B1)..	192	Waldo, (E2)......	150	Zolfo, (E4)......	171
(E1).........	100	Redlevel, (D2).....	150	Sorrento, (E3).....		Warrington, (B7)..	1,400	Zuber, (D2)......	100

GEORGIA

Abbeville, (E8)...	1,201	Atkinson, (H9)....	200	Bethlehem, (D4)...	209	Bristol, (G9).....	198	Carsonville, (C6)..	100
Acree, (D9)......	200	ATLANTA,		Between, (D4).....	104	Bronwood, (C8)..	465	Cartecay, (C2)....	150
Acworth, (B3).....	1,043	(C4).......	154,839	Beverly, (F3).....	14	Brooklet, (H7)....	361	Cartersville, (B3).	4,067
Ada, (D8)........	100	Atlanta Heights,		Bibb City, (B7)...	463	Brooks, (C5).....	200	Cass Station, (B3).	150
Adabelle, (H7)....	100	(C4)........	100	Big Creek, (C3)...	200	Broxton, (F8)....	1,040	Cassville, (B3)...	500
Adairsville, (B3)..	751	Attapulgus, (B10).	360	Bingen, (B10).....	200	Brunswick, (J9)..	10,182	Cavespring, (A3)..	805
Adel, (E9).......	1,092	Atwater, (C6).....	59	Birmingham, (C3).	100	Buchanan, (A4)..	462	Cecil, (E9)......	354
Adrian, (F6).....	816	Auburn, (D3).....	217	Bishop, (E4).....	268	Buckhead, (E4)..	384	Cedartown, (A4)..	3,551
Aikenton, (D5)...	99	Augusta, (H5)....	41,040	Blackshear, (G9)..	1,235	Buenavista, (C7).	1,016	Cement, (B3).....	100
Alley, (F7)......	306	Auraria, (C3).....	150	Blairsville, (D2)...	203	Buford, (D3).....	1,683	Center, (E3).....	208
Aimar, (H8)......	100	Austell, (B4)....	755	Blakely, (A9).....	1,838	Bullochville, (B6)..	204	Chattahatchee(A5)	119
Ainslie, (E7).....	200	Autreyville, (D9)..	200	Blanton, (E10)....	100	Burroughs, (J8)...	130	Chalybeate, (B6)..	147
Alamo, (F7)......	249	Avalon, (E3).....	60	Bloomingdale, (J7)	100	Burtsboro, (C2)...	250	Chamblee, (C4)...	129
Albany, (C8).....	8,190	Avera, (G5)......	228	Blue Ridge, (C2)..	898	Burwell, (A4)....	150	Charing, (C7).....	100
Alexanderville,		Babcock, (B9).....	402	Bluffton, (B8)....	325	Bushnell, (F8) ...	173	Chatsworth, (B2)..	314
(F10).......	200	Baconton, (C8)...	391	Blythe, (G5).....	100	Butler, (C6).....	705	Chattahoochee,(C4)	1,000
Aline, (G7)......	100	Baden, (D10).....	300	Bochee, (B3).....	300	Butts, (G6)......	100	Chauncey, (E7)...	350
Allapaha, (E9)...	532	Bainbridge, (B10).	4,217	Bogart, (D4).....	257	Byromville, (D7)..	300	Cherokee, (B3)...	100
Allatoona, (B3)...	100	Bairdstown, (E4)..	-100	Bold Springs, (D4)	100	Byron, (D6)......	300	Chester, (E7).....	278
Allentown, (E6)...	150	Baldwin, (D3).....	280	Bolingbroke, (D6).	144	Cadwell, (E7).....	154	Chestlehurst, (C5).	150
Alma, (G8)......	458	Ball Ground, (C3).	448	Bolton, (C4).....	150	Cairo, (C10).....	1,505	Chestnut Mountain,	
Alpharetta, (C3)..	356	Bamboo, (H9).....	300	Bolway, (A3).....	100	Calhoun, (B3).....	1,652	(D3)........	100
Altamaha, (G8)...	100	Banning, (A4).....	470	Boston, (D10).....	1,130	Camak, (F5).....	241	Chickamauga, (A2)	312
Alto, (D2).......	109	Bannockburn, (E9)	350	Bostwick, (D4)....	333	Camilla, (C9).....	1,827	China Hill, (E8)...	180
Ambrose, (F8).....	100	Barnesville, (C5).	3,068	Bowdon, (A4).....	541	Campania, (G5)..	300	Chipley, (B6).....	742
Americus, (C7)...	8,063	Barnett, (F5).....	450	Bowersville, (E3)..	398	Campbellton, (B4).	110	Clarkesville, (D2)..	528
Amsterdam, (C10).	250	Barney, (D10)....	303	Bowman, (E3).....	738	Campton, (D4)...	145	Clarkston, (C4)....	349
Andersonville, (C7)	174	Bartow, (F6).....	384	Boxspring, (B6)...	100	Canoe Station, (G7)	100	Claxton, (H7).....	1,008
Apalachee, (D4)...	481	Barwick, (D10)....	381	Boyd, (J6).......	100	Canon, (E3).....	728	Clayton, (E2).....	541
Appling, (G5).....	200	Batson, (E7).....	100	Boykin, (B9).....	64	Canton, (B3)....	2,002	Clelland, (E9)....	150
Arabi, (D8)......	433	Battle Hill, (C4)..	300	Brag, (H7).......	100	Capel, (E10)....	100	Cleveland, (D2)..	750
Aragon, (A3).....	1,200	Baxley, (G8).....	831	Braganza, (G9)...	100	Capitola, (D6)...	100	Clifton, (J8).....	800
Arcadia, (J8).....	150	Beach, (F9)......	358	Braswell, (B3)...	95	Carlsworth, (D2).	100	Climax, (C10)....	328
Argyle, (F9).....	280	Beards Creek, (H8)	150	Bremen, (A4).....	890	Carl, (D3).......	166	Clinton, (D6).....	350
Arlington, (B9)...	1,308	Beaumont, (A2)..	150	Brentwood, (G8)..	100	Carlton, (E3)....	325	Clyde, (J7)......	100
Armena, (C8).....	162	Belfast, (J8).....	300	Brewton, (F6)....	214	Cornesville, (E3).	322	Clyo, (J7).......	200
Ashburn, (D8)....	2,214	Bellton, (D3).....	193	Bridgeboro, (D9)..	350	Cusidine, (B7)...	150	Coal Mountain,	
Atco, (B3).......	200	Bellville, (G7)....	400	Bridgetown, (E9)..	130	Carrollton, (A4).	3,297	(C3)........	100
Athens, (E4).....	14,913	Bemiss, (E10)....	100	Brinson, (B10)....	707	Carrs Station, (E5)	100	Cobbtown, (G7)...	254

FLORIDA

GEORGIA

Morgan, (B9)..... 302
Morganton, (C2).. 195
Morrison, (H7).... 100
Morrow, (C4)..... 250
Morven, (D10).... 383
Moultrie. (D9).. 3,349
Mountain City, (E2) 158
Mountain Scene,
(D2)........... 250
Mount Airy, (E2).. 256
Mount Vernon, (F7) 605
Mountville, (B5).. 226
Moye, (B8)........ 100
Musella, (C6)...... 100
Myers, (H7)...... 100
Myrtle, (D7)...... 100
Mystic, (E8)...... 140
Nacoochee, (D2).. 200
Nahunta, (G9).... 150
Nashville, (E9)... 990
Naylor, (E10).... 538
Needham, (G9).... 150
Nellieville, (G5)... 503
Nellwood, (H7).... 100
Nelson, (C3)...... 550
Newborn, (D4).... 475
Newell, (H10).... 200
New England City,
(A1).......... 139
New Holland, (D3) 2,000
Newnan, (B5).. 5,548
Newton, (C9)..... 364
Nichols, (F9)..... 720
Nicholson, (E3).... 167
Nielly, (F8)....... 200
Nile, (D10)....... 200
Nona, (E5)........ 350
Norcross, (C4).... 968
Norman Park, (D9) 648
Norristown, (F7).. 87
North Rome, (A3). 300
Norwood, (F5).... 340
Nuberg, (F3)...... 100
Nunez, (G7)...... 174
Nye, (F5)......... 300
Oakfield, (D8).... 276
Oakhurst, (C4)... 233
Oakland, (B5).... 100
Oak Park, (G7)... 144
Oakwood, (D3)... 110
Ochlochnee, (C10). 350
Ocilla, (E8)..... 2,017
Odessadale, (B5).. 161
Odum, (G8)...... 258
Offerman, (C9)... 483
Ogeechee, (H6)... 130
Oglethorpe, (C7).. 924
Ohoopee, (G7)... 101
Ola, (C3)........ 150
Oliver, (J6)...... 243
Olympia, (E10)... 200
Omaha, (B7)..... 209
Omega, (D9).... 274
Orange, (C3)..... 100
Orchard Hill, (C5). 100
Orland, (F7)..... 457
Orsman, (A3).... 100
Osierfield, (E8)... 300
Ousley, (E10).... 100
Owensbyville, (A5) 150
Owens Ferry, (H10) 200
Oxford, (D5)..... 655
Ozell, (D10)..... 200
Palmetto, (B4).... 922
Parish, (H7)...... 100
Parrott, (B8)..... 360
Patten, (D10).... 32

Patterson, (G9).... 264
Pavo, (D10)....... 572
Payne, (B3)....... 100
Pearson, (F9)..... 558
Pelham, (C9).... 1,880
Pembroke, (H7)... 467
Pendergrass, (D3). 239
Penfield, (E4).... 475
Pepperton, (D5)... 454
Perkins, (H6).... 230
Perry, (D7)...... 649
Pettyville, (G9)... 150
Pfeiffer, (J6)..... 150
Phinizy, (G4)..... 100
Pidcock, (D10).... 310
Pinebloom, (F9)... 200
Pinegrove, (G8)... 150
Pinehurst, (D7)... 451
Pineview, (E7)... 708
Pinia, (D8)...... 100
Pitts, (D8)....... 279
Plains, (C7)...... 400
Plainville, (B3)... 148
Pleasant Hill, (C6). 110
Point Peter, (E4).. 150
Pooler, (J7)...... 337
Popes Ferry, (D6). 100
Poplar, (C6)...... 100
Portal, (H7)...... 200
Porterdale, (D4).. 1,000
Potterville, (C7).. 300
Poulan, (D9)..... 652
Powder Spgs, (B4). 315
Powelton, (F5)... 100
Powersville, (D6).. 100
Prattsburg, (C6).. 100
Preston, (B7).... 259
Pretoria, (C8).... 369
Pride, (F5)....... 100
Primrose, (B5)... 57
Princeton, (E4)... 63
Princeton, (D4)... 250
Pulaski, (H7)..... 207
Quitman, (D10).. 3,915
Raccoon Mills, (A3) 113
Raleigh, (B6).... 150
Raymond, (B5)... 150
Rays Mill, (E9).. 300
Rebecca, (D8).... 252
Rebie, (E7)...... 200
Redan, (C4)..... 120
Redoak, (B4).... 100
Register, (H7).... 300
Reidsville, (G7)... 454
Remerton, (E10).. 200
Rentz, (F7)...... 275
Resaca, (B2)..... 112
Reynolds, (J6).... 150
Reynolds, (C6)... 521
Rhine, (E8)...... 321
Richland, (B7)... 1,250
Richwood, (D7).. 300
Riddleville, (F6).. 140
Ridgeville, (J9)... 300
Rincon, (J7)..... 100
Ringgold, (A2).... 398
Rising Fawn, (A1). 225
Ritch, (G8)...... 210
Riverdale, (C4)... 139
Riverside, (J9).... 170
Roberta, (C6).... 227
Rochelle, (E8).... 860
Rockingham, (G9). 100
Rockledge (F7)... 152
Rockmart, (B1).. 1,034
Rocky Ford, (H6). 385
Rocky Mount, (B5) 61

Rogers, (G6)..... 300
Rolston, (C3).... 100
Rome, (A3)..... 12,099
Roopville, (A5)... 173
Rosewood, (A3)... 100
Rossville, (A1)... 1,059
Roswell, (C3).... 1,158
Roundoak, (D5)... 100
Roy, (C3)....... 300
Royston, (E3)... 1,422
Ruckersville, (F3). 88
Rural Vale, (B2).. 100
Ruskin, (C9)..... 100
Russell, (D3).... 120
Rutledge, (D4)... 696
Saint Charles, (B5) 91
Saint Clair, (G3).. 200
Saint George, (G10) 272
Saint Marks, (B3). 54
Saint Marys, (H10) 691
Saint Simons Mills,
(J9)........... 230
Sale City, (C9)... 402
Sandersville, (F6). 2,641
Sargent, (B5).... 200
Sasser, (C8)..... 441
Satilla Bluff,
(H10).......... 300
Savannah, (J7).. 65,064
Scarboro, (H6)... 150
Scotland, (F8).... 150
Scott, (F7)...... 212
Scottdale, (C4)... 400
Scottsboro, (E6).. 150
Screven, (H9).... 276
Seabrook, (J8).... 100
Seney, (A3)...... 120
Senoia, (B5).... 1,111
Sessoms, (G8).... 100
Seville, (D8)..... 193
Shadydale, (D5).. 344
Sharon, (F5)..... 246
Sharpe, (A2)..... 100
Sharpsburg, (B5). 166
Shaw, (A2)...... 100
Shearwood, (H7). 270
Shellman, (B8)... 985
Sheltonville, (C3). 150
Shiloh, (B6)..... 250
Shingler, (D9)... 100
Siloam, (E5)..... 300
Slate, (E3)...... 500
Smarrs, (D6)..... 300
Smiley, (H8)..... 100
Smithonia, (E4).. 266
Smithville, (C8).. 574
Smyrna, (B4).... 599
Snapping Shoals,
(D4).......... 250
Snider, (C2)..... 100
Snow, (D7)...... 300
Social Circle, (D5). 1,590
Somoraville, (B3). 100
Soperton, (F7)... 469
Southwell, (H7).. 100
Spann, (F6)..... 100
Sparks, (E9)..... 842
Sparta, (F5).... 1,715
Spread, (G5)..... 370
Springfield, (J7).. 504
Springhaven, (F7). 100
Springplace, (B2). 242
Springvale, (B8).. 162
Starrsville, (D4).. 400
Statenville, (E10). 150
Statesboro, (H7).. 2,529
Statham, (D4).... 621

Staunton, (E9)... 100
Stellaville, (G5)... 149
Stephens, (E4).... 100
Sterling, (H9).... 171
Stevens Pottery,
(E6)........... 200
Stilesboro, (B3)... 200
Stillmore, (G7)... 645
Stillwell, (J7).... 100
Stilson, (H7)..... 150
Stockbridge, (C4). 200
Stockton, (E10)... 200
Stone Mountain,
(C4).......... 1,062
Stonewall, (B4)... 170
Sugar Valley, (A2). 197
Summertown, (G6) 125
Summerville, (A2). 657
Summerville, (G5). 4,361
Summit, (G6).... 566
Sumner, (D9).... 336
Sunnyside, (C5)... 200
Surrency, (G8)... 259
Suwanee, (C3)... 250
Swainsboro, (G6).. 1,313
Swords, (E4).... 107
Sycamore, (D8)... 296
Sylvania, (H6)... 1,403
Sylvester, (D8)... 1,447
Talbotton, (C6)... 1,081
Talking Rock, (B3) 108
Tallapoosa, (A4).. 2,117
Tallulah Falls, (E2) 85
Talmo, (D3)..... 150
Tarboro, (E9).... 150
Tarrytown, (G7).. 236
Tarver, (F10).... 150
Tate, (C3)....... 500
Taylors Creek, (H8) 140
Taylorsville, (B3). 197
Tazewell, (C7)... 150
Temple, (A4).... 711
Tempy, (D9).... 100
Tennga, (B2).... 145
Tennille, (F6).... 1,622
Texas, (A5)..... 150
Thebes, (J8)..... 300
Thelma, (F10)... 200
The Rock, (C6)... 138
Thomaston, (C6).. 1,545
Thomasville,
(C10)......... 6,727
Thomson, (G5)... 2,151
Thunderbolt, (J7). 750
Tifton, (D9).... 2,381
Tiger, (E2)...... 125
Tignall, (F4).... 320
Tilton, (B2)..... 142
Toccoa, (E2).... 3,120
Toomsboro, (E6). 404
Towns, (F8)..... 150
Traders Hill,
(G10).......... 100
Trenton, (A1).... 302
Trion, (A2)..... 1,721
Tunnelhill, (B2)... 295
Turin, (B5)...... 263
Turman, (B9).... 100
Tybee, (J7)...... 786
Tyrone, (F4)..... 100
Tyty, (D8)...... 276
Unadilla, (D7)... 1,003
Union City, (B4).. 534
Unionpoint, (E4).. 1,363
Upton, (F9)..... 200
Valambrosa, (E7). 100
Valdosta, (E10).. 7,656

Van Wert, (A4)... 139
Varnells Station,
(B2).......... 100
Vidalia, (D7).... 1,776
Vidette, (G5)..... 75
Vienna, (D7).... 1,564
Villanow, (A2)... 100
Villa Rica, (B4)... 855
Vinings, (C4).... 250
Vivian, (J6)..... 100
Waco, (A4)..... 326
Wadley, (G6).... 872
Walden, (D6).... 140
Waleska, (B3).... 243
Walker, (G5)..... 100
Walthourville,
(H8).......... 100
Waresboro, (F9).. 149
Waring, (E5)..... 100
Warm Springs, (B6) 100
Warrenton, (F5).. 1,368
Warthen, (F5)... 151
Warwick, (D8)... 226
Washington, (F4). 3,065
Wassaw, (J7).... 543
Watkinsville, (E4). 483
Waverly Hall, (c6) 300
Waycross, (G9).. 14,485
Waynesboro, (G5). 2,729
Warnesville, (H9). 420
Waymmanville,
(C6).......... 320
Wayside, (D5).... 100
Wars Station, (J8). 100
West Buford, (D3) 250
Weston, (B7).... 319
Westpoint, (A6).. 1,906
Whaley, (F5).... 100
Whigham, (C10).. 627
Whitehall, (E4)... 230
Whiteplains, (E5). 407
Whitesburg, (B4). 312
White Sulphur Spgs,
(B6).......... 50
Whitesville, (A6).. 100
Wilcox, (F8)..... 130
Willacoochee,
(E9).......... 960
Willett, (H7).... 100
Williamson, (C5). 179
Wilmot, (A2).... 100
Wilsonville, (F9).. 100
Winder, (D4)... 2,443
Winokur, (H7)... 204
Winston, (B4).... 168
Winterville, (E4).. 465
Withers, (F10)... 100
Woodbine, (H10). 155
Woodbury, (B5).. 917
Woodcliff, (H6)... 100
Woodland, (C6).. 189
Woodstock, (B3).. 442
Woodville, (E4)... 250
Wooster, (B5).... 100
Worth, (D8)..... 169
Worthville, (D5).. 150
Wray, (E8)...... 400
Wrayswood, (E4). 100
Wrens, (G5)..... 616
Wrightsville, (F6). 1,389
Yatesville, (C6)... 366
Yorkville, (B4)... 120
Young Harris, (D2) 283
Zaidee, (G7)..... 200
Zebulon, (C5).... 602
Zeigler, (H6)..... 100

IDAHO

Ahsahka, (B3)..... 100
Albion, (E7)...... 392
Almo, (E7)....... 300
Alpha, (C5)....... 100
American Falls, (F7) 953
Ammon, (G6)..... 214
Arbon, (F7)...... 150
Archer, (G6)..... 250
Arco, (E6)....... 322
Artesian City, (D7) 100
Ashton, (G5)..... 502
Athol, (B2)...... 281
Atlanta, (C6)..... 300
Avon, (B3)...... 100
Badger, (G6)..... 100
Bancroft, (G7).... 200
Basalt, (F6)...... 200
Basin, (E7)...... 250
Bates, (G6)...... 100
Bayhorse, (D5)... 150

IDAHO

ILLINOIS

Thisbe (F11)..... 717	Victoria (D3)..... 334	Westville (J5)..... 2,607
Thomas (E2)..... 290	Vienna (G11)..... 1,124	West York (J7)..... 300
Thomasboro (H5), 321	Villa Grove (H6).. 1,823	Wetaug (F11)..... 218
Thompsonville, (G10)..... 573	Viraridge (F11)..... 700	Wethersfield (E3). 1,593
Thornton (D2)..... 487	Viola (C3)..... 760	Wheaton (H2).. 3,423
Thornton (J2)..... 1,030	Virden (E7)..... 4,000	Wheeler (H7)..... 255
Tice (K3)..... 200	Virginia (D6)..... 1,501	Wheeling (J1)..... 260
Tilden (E9)..... 774	Volo (H1)..... 180	Whitash (G10)..... 353
Tilton (J5)..... 710	Wadsworth (J1)..... 150	White City (E7) .. 421
Time (C4)..... 158	Waggoner (E7)..... 270	White Heath (H5). 200
Tinewell (C5)..... 230	Waldron (J3)..... 261	White Hall (D7).. 2,854
Tinley Park (J2)..... 309	Walnut (E2)..... 763	Whiterock (F1)..... 100
Tioga (B5)..... 300	Walnut Hill (F8).. 160	Wichert (J3)..... 130
Tiskilwa (E3)..... 857	Walpole (G10)..... 130	Wilmington (H3). 1,450
Toluca Point (G6)..... 100	Walshville (E7)..... 169	Williamsfield (D4) 480
Toledo (H7)..... 900	Waltonville (F9).. 250	Williamson (E8).. 648
Tidono (H6)..... 760	Wanlock (C3)..... 150	Williamsville (F6). 600
Tolono (F4)..... 2,407	Wapella (G5)..... 498	Willisville (E10).. 1,082
Tonica (F3)..... 483	Warren (E1)..... 1,331	Willow Hill (J8).. 444
Topeka (F5)..... 130	Warrensburg (F6). 504	Willow Springs (J2) 150
Torino (D3)..... 514	Warrenville (H2).. 100	Wilmette (J1).. 4,943
Toulon (K3)..... 1,208	Warsaw (B5)..... 2,254	Wilmington (D7). 204
Towanda (G4)..... 404	Wasco (H2)..... 150	Winchester (D6).. 1,639
Tower Hill (G7)..... 1,040	Washburn (F4)..... 777	Windsor (G7)..... 987
Tremont (F4)..... 782	Washington (F4).. 1,530	Winfield (H4)..... 150
Trenton (E8)..... 1,694	Wasson (G10)..... 350	Wing (H4)..... 120
Trilla (H7)..... 260	Wataga (D3)..... 444	Winnebago (F1).. 415
Trimble (J7)..... 130	Waterloo (D9)..... 2,091	Winnetka (J1).. 3,168
Triumph (G3)..... 150	Waterman (G2)..... 398	Winslow (E1)..... 426
Trivoli (E4)..... 150	Watertown (D2).. 525	Winthrop Harbor, (J1)..... 439
Troy (E3)..... 1,447	Wateka (J4)..... 2,476	Wireton Park (J2) 150
Troy Grove (F3).. 289	Watson (G7)..... 330	Witt (F7)..... 2,170
Tunnel Hill (G10). 150	Wauconda (H1)..... 368	Woburn (F8)..... 250
Tuscola (H6)..... 2,453	Waukegan (J1).. 16,069	Wolrab Mills (H10) 140
Ullin (F15)..... 670	Waverly (E6)..... 1,538	Womac (E7)..... 170
Union (G1)..... 432	Wayne (H2)..... 150	Woodbine (D1)..... 100
Union Hill (H3)..... 250	Wayne City (G9). 620	Woodburn (D7).. 175
Unionville (G11)..... 150	Waynesville (F5).. 546	Woodhull (D3)..... 692
Unity (F14)..... 150	Webster (H5)..... 100	Woodland (J4)..... 295
Upper Alton (D8). 2,918	Wedron (G3)..... 200	Woodlawn (F9).. 315
Urbana (H5)..... 8,245	Weldon (G5)..... 521	Woodriver (D8)..... 84
Ursa (B6)..... 170	Wellington (J4)..... 300	Woodson (D6)..... 257
Utica (F3)..... 1,500	Wenona (F3)..... 1,442	Woodstock (H1).. 4,331
Valier (F9)..... 200	West Brooklyn (F2) 266	Woodworth (J4).. 150
Valley (C6)..... 800	West Chicago (H2) 2,378	Woosung (E2)..... 100
Vandalia (F8)..... 2,974	West Dundee (H1) 1,380	Worden (E8)..... 1,082
Van Orin (F2)..... 190	Western Springs, (J2)..... 905	Worth (J2)..... 200
Varna (F3)..... 400	Westervelt (G7).. •180	Wrights (D7)..... 200
Venedy (F9)..... 160	Westfield (J7)..... 927	Wyanet (E3)..... 872
Venice (D8)..... 3,718	West Frankfort, (G10)..... 2,111	Wyoming (E3).. 1,506
Vera (E7)..... 150	West Hammond, (J2)..... 4,948	Yale (H7)..... 120
Vergennes (F10). 342	West Jersey (E3).. 100	Yates City (D4).. 596
Vermilion (J6)..... 287	West Liberty (H8) 220	York (J7)..... 169
Vermilion Grove (J6)..... 200	Weston (G4)..... 260	Yorkville (H2)..... 431
Vermont (D5)..... 1,118	West Point (B5).. 292	Youngstown (C4). 100
Vernon (F8)..... 333	West Salem (H8). 725	Zeigler (F10)..... 500
Verona (G3)..... 300	West Union (J7).. 400	Zion City (J1).. 4,789
Versailles (C6)..... 557		

INDIANA

IOWA

KANSAS

Maize, (L6)....... 200	Newton, (L5).... 7,862	Pleasanton, (Q5).. 1,373	Scranton, (O4).... 770	Udall, (L7)........ 330
Manchester, (L3).. 250	New Ulysses, (C6). 200	Plevna, (J6)...... 150	Sedan, (N7)..... 1,211	Uniontown, (P6).. 256
Manhattan, (M3). 5,722	Nickerson, (J8) .. 1,195	Plymouth, (N5).... 100	Sedgwick, (L6).... 626	Urbana, (P6)..... 100
Mankato, (J2).... 1,155	Niles, (L4)....... 160	Pomona, (P4)..... 523	Selden, (D2)..... 297	Utica, (E4)....... 300
Maple City (M7).. 100	Niotaze, (N7).... 317	Portis, (H2)...... 304	Seneca, (N2).... 1,806	Valeda, (P7)...... 130
Maplehill, (N3)... 217	Norcatur, (E2).... 482	Potter, (P3)...... 150	Severance, (P2)... 383	Valley Center, (L6) 381
Mapleton, (Q6)... 230	Northbranch, (J2) 150	Potwin, (M6)..... 249	Severy, (N6)..... 608	Valley Falls, (O3). 1,129
Marienthal, (C4). 150	Norton, (F2)..... 1,787	Powhattan, (O2).. 216	Seward, (H5)..... 110	Vance, (Q3)...... 100
Marion, (M5)... 1,841	Nortonville, (P3) . 638	Prairie View, (G2). 191	Sharon, (J7)..... 350	Vassar, (O4)...... 130
Marquette, (K4).. 715	Norway, (K2).... 100	Pratt, (H6)..... 3,302	Sharon Springs, (B4) 440	Vermillion, (N2).. 366
Marysville, (M2). 2,260	Norwich, (K7) ... 392	Prescott, (Q5)... 255	Shawnee, (Q3).... 450	Vernon, (O6)..... 110
Matfield Green,	Oakhill, (L3)..... 200	Preston, (H6)..... 278	Sherman, (P7)... 150	Vesper, (J3)...... 100
(N5)........ 220	Oakland, (O3).... 1,465	Pretty Prairie, (K6) 327	Sherwin Jc, (P7).. 100	Victoria, (G4)... 1,500
Mayetta, (O3).... 337	Oakley, (D3)..... 681	Princeton, (P5)... 270	Shorey, (O3)..... 130	Vining, (L2)...... 191
Mayfield, (K7)... 200	Oak Valley, (N7).. 210	Protection, (F7) .. 390	Silverdale, (M7).. 110	Viola, (K6)....... 156
Meade, (E7)...... 664	Oberlin, (D2).... 1,157	Purcell, (P2)..... 100	Silver Lake, (O3).. 260	Virgil, (N6)...... 150
Medicine Lodge,	Offerle, (F6)..... 100	Queneno, (O4).... 556	Simpson, (K3).... 211	Viets, (N2)....... 200
(H7)........ 1,229	Ogden, (M3)..... 230	Quincy, (N6)..... 200	Skiddy, (M4)..... 100	Wabaunsee, (N3).. 260
Melrose, (Q7).... 200	Oketo, (M2)...... 253	Quindaro, (Q3)... 300	Skidmore, (Q7)... 250	Wakarusa, (O4)... 100
Melvern, (O4).... 505	Olathe, (Q4).... 3,272	Quinter, (E3)..... 450	Smith Center, (H2) 1,292	Wakeeney, (F4).. 883
Menlo, (D3)...... 100	Olivet, (O5)...... 150	Radley, (Q6)..... 500	Smolan, (K4)..... 130	Wakefield, (L3)... 514
Meriden, (O3).... 467	Olmitz, (H4)..... 150	Rago, (J7)....... 110	Soldier, (O2)..... 358	Waldo, (H3)...... 200
Merriam, (Q3).... 150	Olpe, (N5)....... 215	Ramona, (L4)..... 265	Solomon, (L4).... 949	Waldron, (J7)..... 262
Michigan Valley, (O4)100	Olsburg, (M3).... 230	Randall, (J2)..... 325	Somerset, (Q4)... 130	Walnut, (P6)..... 639
Midway, (Q7).... 500	Onaga, (N3)...... 759	Randolph, (M3)... 455	South Haven, (L7). 483	Walton, (L5)...... 357
Milan, (K7)...... 350	Oneida, (O2)..... 211	Ranson, (P4)..... 204	South Hutchinson,	Wanego, (N3).... 1,714
Mildred, (P6)..... 375	Opolis, (Q7)..... 350	Ransomville, (P5). 200	(K5).......... 387	Washington, (M2). 1,547
Milford, (M3)..... 200	Oronoque, (F2)... 100	Rantoul, (P4)..... 200	South Park, (Q3).. 150	Waterville, (M2).. 704
Milton, (K7)...... 150	Osage City, (O4).. 2,432	Raymond, (J5).... 150	Sparks, (P2)..... 130	Wathena, (P2).... 777
Miltonvale, (K3).. 829	Osawatomie, (Q5). 4,046	Reading, (N4).... 289	Spearville, (F6)... 576	Wauneta, (N7).... 150
Mineral, (Q7).... 1,770	Osborne, (H3).... 1,566	Redfield, (Q6).... 232	Speed, (G2)...... 100	Waverly, (O5).... 751
Minneapolis, (K3). 1,895	Oskaloosa, (P3) .. 851	Reece, (N6)...... 200	Spivey, (J7)...... 252	Wayne, (L2)...... 120
Minneola, (F7)... 348	Oswego, (P7).... 2,713	Reno, (P5)....... 120	Spring Hill, (Q4).. 605	Webber, (K2)..... 300
Moline, (N7)..... 808	Otego, (J2)....... 200	Republic, (K2).... 450	Stafford, (H6)... 1,927	Webster, (G3).... 200
Monmouth, (Q7).. 150	Otis, (G4)....... 300	Reserve, (O2)..... 300	Stanley, (Q4)..... 200	Weir, (Q7)...... 2,289
Montana, (P7) ... 180	Ottawa, (P4).... 7,650	Rexford, (D3)..... 300	Stanton, (Q5).... 100	Welda, (P5)...... 250
Mont Ida, (P5) ... 150	Ottumwa, (O5)... 130	Richfield, (B4).... 53	Stark, (P6)....... 191	Wellington, (L7).. 7,034
Montrose, (J2).... 120	Overbrook, (O4).. 580	Richland, (O4).... 250	Sterling, (J5).... 2,133	Wellsford, (G6)... 100
Monument, (C3) . 300	Overland Park, (Q4) 150	Richmond, (P5) .. 425	Stilwell, (Q4)..... 300	Wellsville, (P4)... 648
Moran, (P6)...... 559	Oxford, (L7)...... 624	Riley, (M3)....... 343	Stippville, (Q7)... 250	West Mineral, (P7) 1,700
Morehead, (P7) .. 150	Ozawkie, (P3).... 290	Robinson, (P2) ... 492	Stockdale, (N3)... 110	Westmoreland, (N3) 484
Morganville, (L3).. 285	Padonia, (O2)..... 100	Rock, (M7)....... 140	Stockholm, (B4)... 100	Westphalia, (P5).. 500
Morland, (E3).... 237	Palacky, (J4)..... 100	Rockcreek, (O3)... 100	Stockton, (G3)... 1,317	Wetmore, (O2).... 483
Morrill, (O2)..... 398	Palco, (F3)....... 279	Rome, (L7)....... 110	Stone City, (Q7).. 500	Wheaton, (N2).... 250
Morrowville, (L2). 200	Palmer, (L2)..... 300	Rosalia, (M6).... 100	Strawn, (O5)..... 150	White Church, (Q3) 100
Mound City, (Q5) 698	Paola, (Q4)..... 3,207	Rosedale, (Q3)... 5,960	Strong, (M5)..... 762	White City, (M4).. 506
Moundridge, (K5). 626	Paradise, (H3) ... 150	Roseland, (Q7)... 396	Summerfield, (N2) 554	Whitecloud, (P2) .. 1,119
Mound Valley, (P7). 956	Parker, (Q5)..... 398	Rossville, (N3)... 672	Sun City, (H7)... 130	Whitewater, (L6).. 518
Mount Hope, (K6). 519	Parkerville, (M4).. 157	Rozel, (G5)....... 160	Sycamore, (O7)... 140	Whiting, (O2)..... 426
Mulberry, (Q6)... 997	Parsons, (P7)... 12,463	Rushcenter, (G5).. 200	Sylvan Grove, (J4) 464	Wichita, (L6).... 52,450
Mullinville, (G6)... 289	Partridge, (J6)... 246	Russell, (H4).... 1,692	Sylvia, (J6)...... 634	Wilder, (Q3)...... 100
Mulvane, (L7).... 1,084	Pawnee Rock, (G5) 458	Russell Springs,	Syracuse, (B6).... 1,126	Willard, (O3)..... 200
Munden, (L2) 275	Pawnee Station,	(C4)......... 82	Tablemound, (O7) 200	Williamsburg, (P5) 399
Munjor, (G4)..... 110	(Q6).......... 100	Sabetha, (O2).... 1,768	Talmage, (L3).... 150	Williamstown, (P3) 100
Murdock, (K6).... 150	Paxico, (N3)..... 240	Saffordville, (N5).. 250	Talmo, (K2)...... 200	Willis, (O2)...... 188
Muscotah, (P2) ... 491	Peabody, (L5).... 1,416	Saint Francis, (B2) 492	Tampa, (L4)...... 256	Wilmore, (G7).... 125
Narka, (L2)...... 278	Peck, (L7)....... 150	Saint George, (N3). 230	Tecumseh, (O3)... 300	Wilmot, (M7)..... 100
Nashville, (J7) ... 130	Penalosa, (J6).... 130	Saint John, (H5).. 1,785	Tescott, (K3)..... 421	Wilsey, (M4)..... 400
National Military	Peoria, (P4)...... 160	Saint Joseph, (L2). 100	Thayer, (P7)..... 542	Wilson, (J4)...... 981
Home (P3).... 2,500	Perry, (P3)...... 400	Saint Marks, (K6). 100	Tipton, (J3)...... 200	Winchester, (P3).. 456
Natoma, (H3).... 407	Perth, (L7)...... 200	Saint Marys, (N3). 1,397	Toledo, (N5)..... 50	Windom, (K5).... 176
Navarre, (L4).... 130	Peru, (N7)....... 575	Saint Paul, (P6) .. 927	Tonganoxie, (P3). 1,018	Winfield, (M7).. 6,700
Neal, (N6)....... 180	Peterton, (O4).... 200	Saint Peter, (E3).. 150	TOPEKA, (O3).. 43,684	Winona, (C3)..... 150
Nelson, (Q7)..... 100	Petrolia, (O6).... 100	Salina, (K4)..... 9,688	Toronto, (O6).... 627	Wolcott, (Q3)..... 150
Neodesha, (O7) .. 2,872	Pfeifer, (G4)..... 250	Santa Fe, (D6).... 200	Towanda, (M6)... 275	Woodbine, (M4).. 250
Neosho Falls, (O6). 571	Phillipsburg, (G2). 1,302	Savonburg, (P6) .. 257	Tradingpost, (Q5). 130	Woodruff, (G2)... 200
Neosho Rapids, (O5) 256	Piedmont, (N6)... 200	Sawyer, (H6)..... 180	Traer, (D2)...... 100	Woodston, (G3)... 299
Ness City, (F5) ... 712	Pierceville, (D6)... 130	Saxman, (J5)..... 125	Tribune, (B5)..... 158	Xenia, (P6)...... 100
Netawaka, (O2)... 250	Piper, (Q3)....... 100	Scammon, (Q7).. 2,233	Troy, (P2)....... 940	Yale, (Q7)....... 800
New Albany, (O6). 213	Piqua, (O6)...... 200	Scandia, (K2).... 579	Turck, (Q7)...... 110	Yates Center, (O6) 2,024
New Cambria, (K4) 200	Pittsburg, (Q7).. 14,755	Schoenchen, (G3). 200	Turner, (Q3)..... 230	Yocemento, (G4).. 200
New Lancaster, (Q5) 100	Plains, (D7)...... 333	Scott, (D5)...... 918	Turon, (J6)...... 572	Zenda, (J7)....... 200
New Salem, (M7).. 100	Plainville, (G3)... 1,090	Scottsville, (J2)... 248	Tyro, (O7)....... 603	Zurich, (G5)...... 100

KENTUCKY

Adairville, (E4).... 683	Allen Springs, (E4). 100	Ammie, (H3)...... 110	Ashbyburg, (D3)... 150	Augusta, (H2).... 1,787
Adams, (J2)...... 150	Allensville, (E4)... 436	Amos, (E4)....... 400	Ashcamp, (J3).... 100	Backbone, (J2)... 100
Addison, (E3)..... 100	Almo, (C4)....... 130	Anchorage, (F2)... 384	Ashland, (J2).... 8,688	Bagdad, (F2)..... 184
Adolphus, (E4).... 100	Alphoretta, (J3)... 110	Anthoston, (D3)... 100	Athens, (G3)..... 197	Bandana, (C3).... 337
Afton, (J2)....... 130	Alpine, (G4)...... 300	Arlington, (C4)... 555	Athertonville,	Barbourville, (H4). 1,633
Akersville, (F4)... 130	Alton, (G2)...... 150	Arnoldton, (D3)... 100	(F3).......... 300	Bardstown, (F3)... 2,126
Albany, (F4)...... 579	Alvaton, (E4)..... 200	Artemus, (H4)... 170	Athol, (H3)...... 100	Bardstown Jc, (F3) 100
Alexandria, (G1).. 353	Ambrose, (G3).... 150	Asbury, (G3)..... 200	Auburn, (E4) 631	Bardwell, (C4) ... 1,087

Sorgho, (D3)...... 130
So. Carrollton, (D3). 365
So. Elkhorn, (G2)... 100
Southgate, (G2).... 627
So. Park, (F2)...... 100
So. Portsmouth,
    (J2)........... 660
Sparks Quarry, (G3) 100
Sparta, (G2)....... 107
Spottsville, (D3).... 448
Springfield, (F3)...1,329
Springlake, (G2).... 100
Spring Lick, (E3).... 190
Spurlington, (F3)... 130
Stamping Ground,
    (G2)........... 381
Stanford, (G3)....1,532
Stanton, (H3)...... 278
Stearns, (G4)...... 100
Stephens, (H2)..... 130
Stephensburg, (E3).. 130
Stephensport, (E2).. 205
Stepstone, (H2)..... 150
Stewartsville, (G2).. 200
Stinson, (J2)....... 200
Stithton, (F3)...... 300
Stonewall, (G2)..... 100
Stowers, (E4)...... 120
Sturgis, (D3)......1,467

Sulphur, (F2)...... 255
Summer Shade,
    (F4).......... 250
Summersville, (F3).. 320
Sweeden, (E3)...... 100
Switzer, (G2)...... 380
Talcum, (H3)....... 200
Taylor Mines, (E3). 430
Taylorsport, (G2).. 150
Taylorsville, (F2).. 622
Templer, (G3)...... 200
Texas, (F3)........ 130
Thealka, (J3)...... 100
The Ridge, (H2)... 150
Thor, (H2)......... 100
Thornton, (J3)..... 100
Tiline, (C3)....... 100
Tilton, (H2)....... 113
Tolesboro, (H2).... 400
Tola, (C3)......... 180
Tompkinsville, (F4). 639
Torchlight, (J2).... 200
Towers, (G2)....... 100
Travellers Rest, (H3) 200
Trenton, (D4)...... 653
Troy, (G3)......... 100
Turners Station,
    (F2).......... 115
Turnersville, (G3).. 200

Tyler, (C3)........ 500
Tyner, (H3)........ 130
Tyrone, (G2)....... 544
Union, (G2)........ 280
Uniontown, (C3)...1,356
Upton, (F3)........ 141
Urban, (H3)........ 400
Utica, (D3)........ 300
Valley Station, (F2). 210
Valleyview, (G3)... 630
Vanceburg, (H2)...1,145
Vanderburg, (D3)... 150
Van Lear, (J3)..... 100
Verona, (G2)....... 200
Versailles, (G2)....2,268
Vine Grove, (E3)... 570
Virgie, (J3)........ 130
Visalia, (G2)....... 250
Viva, (G3)......... 250
Wabd, (G3)......... 180
Waco, (G3)......... 210
Waddy, (F2)........ 254
Wakefield, (F3).... 100
Wallace, (G2)...... 130
Wallins Creek, (H4). 150
Wallonia, (D4)..... 100
Walleend, (H4)..... 300
Walnut Grove, (H2) 174
Walton, (G2)....... 650

Warfield, (J3)...... 100
Warsaw, (G2)...... 900
Washington, (H2)... 433
Wasioto, (H4)..... 300
Waterford, (F2).... 210
Watervalley, (C4).. 228
Waverly, (D3)..... 311
Webbville, (J2).... 200
Wellsburg, (G2)... 150
Wentz, (H3)....... 200
W.Covington,(G1). 1,751
W. Irvine, (G3).... 150
W. Liberty, (H3)... 442
W. Louisville, (D3). 192
Weston, (C3)...... 140
West Point, (F3)... 782
Westport, (F2).... 300
Wheatcroft, (D3).. 490
Whitehouse, (J3)... 150
White Mills, (F3).. 100
Whiteoak, (H3).... 130
Whiteplains, (D3).. 281
Whitepost, (J3).... 100
Whitesburg, (J3)... 321
White Sulphur,
    (G3).......... 100
Whitesville, (E3)... 452
Whitley, (G4)..... 157
Wickliffe, (C4).... 989

Wildie, (G3)....... 100
Willard, (J2)....... 177
Williamsburg, (G4) 2,004
Williamsport, (J3). . 200
Williamstown, (G2). 800
Willisburg, (F3).... 150
Wilmore, (G3).....1,000
Wilsonville, (F2)... 150
Wilton, (H4)....... 200
Winchester, (G3)...7,156
Wingo, (C4)....... 404
Wisemantown, (H3) 100.
Wolf, (H2)........ 300
Wolf Creek, (E2)... 150
Woodbine, (G4).... 100
Woodburn, (E4).... 217
Woodbury, (E3).... 173
Woods, (J3)........ 100
Woodville, (C3).... 250
Worley, (G4)...... 150
Worthville, (F2)... 326
Wurtland, (J2).... 190
Xena, (H3)........ 130
Yabo, (G3)........ 100
Yale, (H2)......... 200
Yelvington, (E3)... 400
Yosemite, (G3)..... 98
Yost, (D3)........ 150
Zion, (D3)........ 224

LOUISIANA

Abbeville, (F7).....2,907
Abita Springs, (L6). 365
Acy, (J6).......... 500
Adeline, (G7)...... 800
Albemarle, (H7)...1,200
Alberta, (D2)...... 500
Alden Bridge, (C1). 500
Alexandria, (F4)..11,213
Allemands, (K7)... 500
Alto, (G2)......... 150
Alton, (L6)........ 100
Ama, (K7)......... 800
Amelia, (H7)...... 400
Amesville, (K7).... 200
Amite, (K5).......1,677
Anchor, (H5)...... 300
Andrew, (F6)...... 150
Angie, (L5)....... 346
Ansley, (E2)....... 600
Antrim, (C1)...... 350
Arabi, (L7)........ 250
Arbroth, (H5)..... 300
Arcadia, (E1)......1,079
Arcola, (K5)....... 100
Ariel, (J7)......... 100
Arnaudville, (G6).. 279
Ashland, (D2)..... 200
Athens, (D1)...... 514
Atkins, (D2)....... 300
Atlanta, (E3)...... 311
Avery Island, (G7)..200
Avoca, (H7)....... 200
Ayers, (D4)....... 250
Baldwin, (H7).....1,000
Bancruft, (C5)..... 200
Barataria, (K7).... 300
Barham, (D4)..... 250
Baskin, (G2)...... 150
Bastrop, (G1)..... 854
Batchelor, (G5)... 200
BATON ROUGE,
    (H6)........ 14,897
Bayou Barbary, (J6) 150
Bayou Chicot, (F5). 150
Bayou Goula, (H6).1,000
Bayou Sara, (H5).. 630
Baywood, (J5)..... 300
Belair, (L7)....... 600
Belcher, (C1)..... 200
Bell City, (E6)..... 330
Belle Alliance,(H6).1,000

Belledean, (F4).... 400
Belle Helene, (J6).. 300
Belle Rose, (H6)... 500
Benson, (C3)...... 320
Bentley, (E4)...... 200
Benton, (C1)...... 318
Bermuda, (F3).... 100
Bernice, (E1)...... 781
Bertrandville, (L7).. 300
Berwick, (H7).....2,183
Bethany, (B2)..... 250
Bienville, (E2)..... 606
Bigcane, (G5)..... 230
Blairstown, (J5)... 100
Blanchard, (C1).... 200
Bogalusa, (L5)....1,850
Boleyn, (D3)...... 350
Bolinger, (C1)..... 300
Bonami, (D5)..... 300
Bonita, (G1)...... 273
Bordelonville, (G4). 200
Bossier, (C1)...... 775
Buurg, (J7)....... 500
Boutte, (K7)...... 300
Bowie, (J7).......1,200
Boyce, (E4)....... 865
Breaux Bridge,(G6)1,319
Brockdale, (K4)... 100
Broussard, (G6)... 499
Brunett, (H1)..... 100
Brusly Landing,
    (H6).......... 390
Bryceland, (E2)... 250
Buckeye, (F4)..... 100
Bunkie, (F5)......1,765
Buras, (L8)........ 500
Burnside, (J6)..... 100
Burton, (J7)...... 400
Cades, (G6)....... 200
Calhoun, (F1)..... 260
Cameron, (D7).... 200
Campti, (D3)..... 664
Canton, (E5)...... 100
Carencro, (F6)..... 609
Carson, (D5)...... 500
Carville, (H6)..... 300
Caspiana, (C2).... 250
Castor, (D2)...... 150
Cataro, (F5)...... 200
Cecilia, (G6)...... 130
Centerville, (H7).. . 500

Central, (J6)...... 200
Chacahoula, (J7)... 100
Chalmette, (L7).... 100
Charenton, (G7)... 400
Chatham, (F2).... 181
Chauvin, (J8)..... 320
Cheneyville, (F4).. 498
Cheniere, (F1)..... 230
Choudrant, (E1)... 200
Church Point, (F6). 481
Cinclare, (H6)..... 400
Clarks, (F2)....... 750
Clinton, (J5)...... 918
Clio, (J6)......... 220
Cloutierville, (E3).. 300
Colfax, (E3)......1,049
Collinston, (G1)... 333
Columbia, (F2).... 500
Convent, (J6)..... 500
Converse, (C3).... 200
Cooper, (D4)..... 300
Cottonport, (F5)... 866
Cotton Valley, (C1). 750
Coushatta, (D2)... 564
Covington, (K6)...2,601
Crawford, (G7).... 250
Crescent, (H6)..... 150
Crowley, (F6).....5,099
Cut Off, (K7)..... 200
Cypremort, (G7)... 200
Daisy, (L8)....... 100
Darrow, (H6)..... 500
Delcambre, (F7)... 308
Delhi, (G2)....... 685
Delta, (J2)........ 200
Denham Springs,(J5) 574
Denson, (J6)...... 100
De Quincy, (D6)... 715
De Ridder, (D6)...2,100
Diamond, (L7).... 300
Dido, (E5)........ 200
Dime, (L7)....... 200
Dodson, (E2)..... 845
Donaldsonville, (H6)4,090
Donner, (J7)...... 200
Dorcyville, (H6)... 200
Doss, (G1)....... 100
Doyline, (D1)..... 150
Dry Prong, (E3)... 100
Dubach, (E1)..... 714

Dubberly, (D1).... 200
Dubuisson, (F5)... 250
Dunbar, (L6)...... 200
Duson, (F6)....... 120
Dutch Town, (J6)... 100
Duty, (G3)........ 100
Dykesville, (D1)... 100
East Point, (D2)... 100
Echo, (F4)........ 238
Edgard, (J7)...... 500
Edna, (F6)........ 200
Egan, (F6)........ 150
Elizabeth, (E5).... 200
Ellendale, (J7).... 100
Elton, (E6)....... 100
Empire, (L8)...... 270
Engelwood, (H2)... 100
Eola, (F5)......... 400
Erath, (F7)....... 575
Eros, (F2)........ 898
Esther, (F7)....... 100
Etherwood, (F6)... 544
Eunice, (F6).......1,684
Evangeline, (F6)... 400
Evans, (D5)....... 230
Evergreen, (F5)... 299
Fairmount, (E4)... 200
Farmerville, (F1).. 598
Ferriday, (G2).... 577
Fisher, (D4)...... 200
Flora, (K3)....... 100
Florence, (G3).... 110
Florien, (D4)..... 250
Floyd, (H1)....... 200
Folsom, (K5)...... 100
Fordoche, (G5)... 200
Fort Jesup, (D3)... 100
Fort Foster, (E7)... 100
Franklin, (G7)....3,857
Franklinton, (K5).. 814
Frierson, (C2)..... 300
Fullerton, (E4)....1,238
Fulton, (D6)....... 150
Gansville, (E2).... 100
Garden City, (H7).. 500
Garyville, (J6).....1,000
Gassler, (E6)...... 100
Geismar, (J6)..... 250
Genesee, (K5).... 600
Gheens, (K7)..... 500

Gibsland, (D1)....1,065
Gibson, (H7)..... 200
Gilbert, (G2)...... 230
Gilliam, (C1)...... 150
Gillis, (D6)....... 100
Girard, (G2)...... 100
Gladis, (K5)...... 500
Glencoe, (G7).... 100
Glenmora, (E5)... 100
Glenwild, (H7).... 500
Gloster, (C2)..... 150
Goldonna, (E2)... 150
Gonzales, (J6).... 150
Good Pine, (F3)... 500
Gordon, (D1)..... 100
Grace, (E3)....... 250
Gramercy, (J6).... 100
Grand Cane, (C2).. 485
Grand Coteau, (G6). 392
Grand Isle, (L8)... 150
Grappes Bluff, (D3). 300
Grayson, (F2)..... 200
Greensburg, (J5)... 268
Greenwood, (B2).. 250
Gretna, (K7).....3,700
Grosse Tete, (G6).. 500
Gueydan, (E7)....1,081
Gullett's Station,(J5) 200
Hackberry, (D7)... 200
Hackley, (K5)..... 100
Hadley, (C3)...... 100
Hahnville, (K7)... 300
Hale, (F1)........ 120
Hall City, (C5).... 800
Hammond, (K5)...2,942
Hard Times Landing
    (H2).......... 200
Harrisonburg, (G3). 561
Harvey, (K7)..... 340
Haughton, (D2)... 249
Hawthorn, (D4)... 200
Haynesville, (D1).. 663
Head of Island, (J6). 100
Hecker, (D6)..... 400
Hermitage, (H5)... 200
Hessmer, (F4).... 100
Hineston, (E4).... 150
Hobart, (J6)...... 130
Hodge, (E2)...... 300
Hohen Solms, (H6). 170
Holly, (C2)....... 400

MAINE

So. Thomaston, (E7) 500
So. Union, (E7)..... 100
So. Vassalboro, (C7). 500
So. Waldoboro, (E7). 150
So. Warren, (E7).... 100
So. Waterford, (B7). 225
Southwest Harbor,
 (G7)............ 850
So. Windham, (C8).1,200
So. Windsor, (D7).. 150
Springfield, (G3)... 300
Springvale, (B9)...2,000
Sprucehead, (E7)... 100
Staceyville, (F4)... 100
Standish, (B8)..... 800
Stark, (D6)........ 300
Steep Falls, (B8)... 400
Stetson, (E6)....... 350
Steuben, (H7) 760
Stickney Corner,
 (E7).............. 130
Stillwater, (F6).... 100
Stockholm, (G1)... 380
Stockton Springs,
 (F7)............. 700
Stonington, (F7)..1,900
Stow, (B7)......... 150
Stratton, (C5)..... 300
Strickland, (C7)... 340
Strong, (C6)....... 630
Stroudwater, (G8).. 500
Sullivan, (G6)..... 225
Sumner, (B7)...... 150
Sunset, (F7)....... 300
Sunshine, (F7)..... 150
Surry, (F7)........ 500
Swans Island, (G7).. 300
Swanville, (F7)..... 200
Sweden, (B7)...... 150
Temple, (C6)....... 200

Tenants Harbor,
 (E8)............1,700
The Forks, (D5).... 240
Thomaston, (E7)..2,100
Thorndike, (E6).... 500
Topsfield, (H5).... 150
Topsham, (C8)...1,700
Tremont, (G7)..... 300
Trenton, (G6)...... 150
Trevett, (D8)...... 100
Troy, (E6)......... 380
Turner, (C7)....... 950
Turner Center, (C7). 130
Union, (E7)........ 850
Unionville, (G6)... 130
Unity, (E6)........ 825
Upper Frenchville,
 (G1)............. 250
Upper Gloucester,
 (C8)............. 300
Upper Madawaska,
 (G1)............1,400
Upton, (B6)....... 150
Vanburen, (H1)..3,000
Vanceboro, (J4)... 500
Vassalboro, (D7)... 180
Venzie, (F6)....... 200
Vienna, (D7)....... 150
Vinalhaven, (F7)..2,200
Waite, (H5)........ 100
Waldo, (E7)........ 200
Waldoboro, (E7)..2,200
Waldo Station, (E7) 470
Wales, (C7)........ 270
Waltagrass, (G1).. 600
Walnut Hill, (C8).. 600
Walpole, (D8)..... 100
Waltham, (G6)..... 100
Warren, (E7)....1,400
Washburn, (G2)... 500

Washington, (E7).. 375
Waterboro, (B8)... 700
Waterford, (B7)... 210
Waterville, (D6).11,458
Wayne, (C7)....... 400
Webbs Mills, (B8). 200
Weeks Mills, (D7). 350
Welchville, (C7)... 180
Weld, (C6)......... 500
Wellington, (D5)... 200
Wells, (B9)......1,200
Wellsbranch, (B9).. 110
Wells Depot, (B9).. 150
Wesley, (H6)....... 100
W. Athens, (D6)... 150
W. Auburn, (C7)... 100
W. Baldwin, (B8)... 400
W. Bath, (D8)...... 100
W. Bethel, (B8).... 240
W. Boothbay Harbor,
 (D8)............. 100
W. Bowdoin, (C7).. 425
W. Brighton, (D6).. 200
Westbrook, (C8)..8,281
W. Brooklin, (F7).. 200
W. Brooksville, (F7) 200
W. Buxton, (B8)... 800
W. Charleston, (E5). 180
W. Cumberland,
 (C8)............. 240
W. Denmark, (B8).. 200
W. Dresden, (D7).. 220
W. Durham, (C8).. 250
W. Eden, (G7)..... 100
W. Enfield, (G5)... 100
W. Falmouth, (C8).. 780
W. Farmington,
 (C6)............. 390
Westfield, (G2)..... 500
W. Franklin, (G6)... 300

W. Gardiner, (D7).. 200
W. Garland, (E5)... 100
W. Gorham, (C8)... 320
W. Gouldsboro, (G7) 210
W. Gray, (C8)..... 100
W. Hampden, (F6).. 120
W. Hancock, (G6).. 100
W. Harpswell, (C8).. 120
W. Harrington, (H6) 100
W. Hollis, (B8).... 100
W. Jonesport, (H6).. 400
W. Kennebunk, (B9) 300
W. Lebanon, (B9)... 180
W. Leeds, (C7).... 100
W. Levant, (E6)... 140
W. Lisbec, (J6).... 100
W. Mills, (C6)..... 200
W. Minot, (C7).... 400
W. Mount Vernon,
 (D7)............. 100
W. Newfield, (B8).. 175
W. Weston, (H4)... 100
W. Palmyra, (E6).. 210
W. Paris, (B7)..... 700
W. Pembroke, (J6).. 600
W. Penobscot, (F6).. 100
W. Peru, (B7)...... 350
W. Poland, (C7)... 475
W. Pownal, (C8)... 200
W. Rockport, (E7).. 200
W. Scarboro, (C8).. 400
W. Searsmont, (E7). 100
W. Sedgwick, (F7).. 150
W. Sidney, (D7)... 280
W. Southport, (D8). 100
W. Sullivan, (G6).. 375
W. Sumner, (C7)... 250
W. Trement, (G7).. 100
W. Troy, (E6)..... 120

W. Waldoboro, (E7). 250
W. Warren, (E7).. 100
W. Washington, (E7) 100
W. Winterport, (F6) 130
W. Woolwich, (D8). 130
Whitefield, (D7)... 300
Whiterock, (C8)... 280
Whiting, (J6)...... 300
Whitneyville, (H6).. 150
Willard, (C8)...... 500
Willimantic, (E5).. 200
Wilsons Mills, (A6).. 100
Wilton, (C6).....1,000
Windham Center,
 (C8)............. 200
Windsor, (D7)..... 180
Windsorville, (D7). 150
Winn, (G5)........ 450
Winnecook, (E6).. 290
Winnegance, (D8).. 300
Winslow, (D6)...1,000
Windows Mills, (E7) 400
Winter Harbor, (G7) 210
Winterport, (F6)..1,300
Winthrop, (C7)...1,500
Winthrop Center,
 (D7)............. 150
Wiscasset, (D7)... 900
Woodland, (J5).... 600
Woodville, (F5).... 100
Woolwich, (D8).... 200
Wyman, (H6)...... 260
Wytopitlock, (G4).. 110
Yarmouth, (C8)... 700
Yarmouthville,
 (C8)............1,600
York Beach, (B9).. 200
York Corner, (B9).. 250
York Harbor, (B9) 1,200
York Village, (B9). 1,100

MARYLAND

Aberdeen, (F1)..... 616
Abingdon, (F2)..... 200
Accident, (B4)..... 200
Adamstown, (D2).. 209
Admiral, (E2)...... 120
Aireys, (F3)........ 100
Alberton, (E2)..... 750
Alesia, (E1)........ 100
Allegany, (C4)..... 100
Allen, (G4)........ 300
Andrews, (F4)..... 200
ANNAPOLIS
 (E3)............8,609
Annapolis Jc., (E2). 100
Antietam, (C2)..... 190
Aquasco, (E3)..... 350
Arlington, (E2)...1,000
Arundel on the Bay,
 (F3)............. 9
Ashland, (E1)...... 150
Atholton, (E2)..... 100
Baden, (E3)........ 150
Baldwin, (F1)...... 300
Baltimore, (E2)..558,485
Barclay, (G2)...... 150
Barnesville, (D2)... 200
Barton, (B4).....2,000
Bay View, (G1).... 290
Beachville, (F4).... 100
Beckleysville, (E1).. 150
Belair, (F1).......1,005
Belalton, (E4)..... 100
Belfast, (E1)....... 150
Bellevue, (F3)..... 150
Benedict, (E3)..... 110
Bencvola, (C1)..... 120
Bengies, (F2)...... 150
Bentley Springs,(E1) 150
Berlin, (H4).......1,317
Berwyn, (E3)...... 700

Bethesda, (D3)..... 250
Betterton, (F2)..... 308
Big Pool, (C1)..... 200
Bishop, (H4)....... 200
Bishops Head, (F4). 150
Bishopville, (H4).. 262
Bivalve, (C4)...... 500
Blackrock, (E2).... 180
Bladensburg, (E3). 460
Bloomington, (B4).. 372
Blue Mountain,(D1) 150
Bolivar, (C2)...... 100
Boonsboro, (C1)... 759
Boothbyhill, (F2)... 140
Bowens, (E3)...... 140
Bowie, (F3)....... 496
Boyds, (D2)....... 130
Bozman, (F3)...... 140
Bradshaw, (F2).... 500
Branchville, (E2)... 150
Brentedsville, (C1). 100
Brentwood, (E3)... 100
Bridgetown, (G2)... 19
Brighton, (D2).... 100
Brookeville, (B2).. 835
Brooklandville, (E2) 300
Brooklyn, (E2)...1,200
Brookview, (G3)... 100
Browningsville, (D2) 110
Bruceville, (D1)... 150
Brunswick, (C2)..3,721
Buckeystown, (D2). 420
Budds Creek, (E4).. 100
Burkittsville, (C2).. 228
Burnt Mills, (E2)... 100
Burrsville, (G3).... 130
Butler, (B2)....... 150
Calvary, (F2)...... 100
Calvert, (G1)...... 200
Cambridge, (F3)..6,407

Capitol Heights,
 (E3)............. 150
Cardiff, (F1)...... 500
Carlos, (B4)....... 670
Carlos Jc., (B4)... 500
Carroll, (E2).....2,000
Carrollton, (E1)... 150
Castleton, (F1).... 100
Catoctin, (D1).... 300
Catonsville, (E2)..4,500
Cavetown, (C1)... 150
Cearfoss, (C1)..... 100
Cecilton, (G2)..... 518
Cedar Grove, (D2).. 100
Centreville, (F2)..1,435
Chaptico, (F4).... 100
Charlestown, (G1).. 274
Chase, (F2)........ 300
Cherry Hill, (C1).. 350
Chesapeake Beach,
 (E3)............. 100
Chesapeake City,
 (G1)............1,016
Chesapeake Jc., (D2) 800
Chester, (F3)...... 350
Chestertown, (F2)..2,735
Chevy Chase, (D3) 250
Chewsville, (C1)... 130
Childs, (G1)....... 100
Chillum, (E3)...... 200
Choptank, (G3)... 230
Church Creek, (F3). 400
Church Hill, (G2).. 306
Churchton, (E2)... 500
Churchville, (F1)... 200
Clara, (G4)........ 100
Clarksburg, (D2)... 170
Clear Spring, (C1).. 521
Clinton, (E3)...... 100
Cockeysville, (E2)..1,500

Colgate, (E2)...... 100
College Park, (E3).. 130
Colora, (F1)....... 150
Conococheague, (C1) 100
Conowingo, (F1).. 300
Cordova, (F3)..... 110
Cornersville, (F3)... 100
Corriganville, (C4).. 200
Cowentown, (G1).. 190
Crapo, (F4)....... 150
Creagerstown, (D1). 140
Crellin, (A1)...... 200
Crisfield, (G5)....3,468
Croon, (E3)....... 100
Crumpton, (G2)... 228
Cub Hill, (E2)..... 150
Cumberland, (C4) 21,839
Damascus, (D2)... 170
Danes Quarter,(G4) 350
Dargan, (C2)...... 100
Darlington, (F1)... 205
Darnestown, (D2).. 100
Dawsonville, (D2).. 110
Deal Island, (G4)..2,000
Deer Park, (B4)... 988
Delmar, (G4)...... 959
Denton, (G3).....1,481
Detour, (D1)...... 140
Dickerson, (D2)... 230
Dickeyville, (E2)... 800
Dorsey, (E2)...... 180
Downsville, (C1)... 140
Dublin, (F1)...... 130
E. Brooklyn, (E2).. 850
E. New Market,(G3) 280
Easton, (F3).....3,083
Eastport, (F3)....1,500
Eccleston, (E2)... 350
Eckhart Mines,(C4) 1,700
Eden, (G4)........ 150

Edesville, (F2)..... 140
Edgemont, (D1)... 150
Edgewood, (F2)... 100
Eldorado, (G3).... 200
Elk Mills, (G1).... 300
Elk Ridge, (E2)..1,000
Elkton, (G1).....2,487
Ellerslie, (C4)..... 750
Ellicott City, (E2). 1,151
Ellsoll, (G4)...... 400
Emmitsburg, (D1) 1,054
Ewell, (F4)........ 730
Fairfield, (E2)..... 400
Fairlee, (F2)....... 120
Fairmount, (G4)... 300
Fairplay, (B1)..... 300
Fallston, (F1)..... 180
Farmington, (F1).. 100
Faulkner, (D3)..... 100
Federalsburg, (G3).1,050
Finksburg, (E1)... 150
Fishing Creek, (F4).. 650
Flint Stone, (A1)... 320
Fords Store, (D2).1,200
Forest Glen, (D2).. 100
Forest Hill, (F1)... 150
Forestville, (E3)... 200
Fork, (F2)........ 300
Fort Washington,
 (D2)............. 400
Four Locks, (C1)... 750
Franklin Mines,(D3) 200
Franklinville, (E2).. 100
Frederick, (D2)..10,411
Freedom, (E2)..... 100
Friendsville, (A4)... 466
Frizellburg, (D1).. 100
Frostburg, (B4)..6,028
Fruitland, (G4).... 350
Fulford, (F1)...... 100

MASSACHUSETTS

MICHIGAN

MINNESOTA

MISSISSIPPI

MISSOURI

Sappington, (M4)... 100
Sarcoxe, (D7).....1,311
Sargent, (J7)...... 160
Savannah, (C2) ...1,583
Saverton, (K2)..... 120
Sawyer, (N8)...... 100
Schell City, (D5)... 562
Scheve, (I 5) 50
Schlicht, (H6)..... 100
Scotland, (D7)..... 150
Sedalia, (F4).....17,822
Sedgewickvi'le, (N6) 130
Seligman, (D8) 400
Senath, (N8) 1,029
Seneca, (C8)....... 951
Seventysix, (N6)... 150
Seymour, (G7)..... 590
Shackelford, (I 3) .. 50
Shawneetown, (N6) 120
Shelbina, (H2) ...2,174
Shelbyville, (H2) .. 645
Sheldon, (D6)...... 528
Sheridan, (C1)..... 409
Shiley, (D3) 200
Sikeston, (N8), ...3,427
Silex, (K3)....... 276
Siloam Springs, (H8) 86
Simmons, (H7)..... 140
Skidmore, (B1)..... 562
Slater (F3).......3,258
Sligo, (K6)........ 300
Smithfield, (C7)... 500
Smithton, (F4).... 346
Smithville, (D3)... 680
So. Gifford, (G2).. 148
So. Gorin, (H3).... 746
So. Greenfield, (E7). 274
So Laneville, (F1) . 27
Southwest, (H8) .. 483
Sparta, (H7)...... 271

Speed, (G4) 190
Spickard, (E1) 638
Spoonerville, (M8).. 241
Sprague, (C5) 154
Spring City, (C8).. 150
Springfield, (F7) .35,201
Spring Fork, (F4).. 130
Spring Garden,(H3). 60
Sprott (M6)...... 150
Spurgeon, (D8).... 250
Stahl, (G1)....... 300
Stanberry (C1)...2,121
Steele, (N8)...... 500
Steelville, (K6)... 771
Steffenville, (J2).. 120
Stella, (C8)....... 180
Stephens Store,(H4) 150
Stewartsville, (D2). 543
Stockton, (E6).... 390
Stotesbury, (C6)... 159
Stotts City, (E7).. 518
Stoutland, (G6)... 250
Stoutsville, (J2)... 315
Stover, (F5) 386
Stratford, (H7) ... 500
Strasburg (H4)... 350
Strattmann, (M4).. 600
Sturgeon, (H3).... 663
Sue City, (H2).... 140
Sugar Creek, (D3).. 500
Sullivan, (K5)..... 934
Sulphur Spres.,(M8) 200
Summersville, (J5). 256
Sumner, (F2) 324
Swedeburg (H6).. 100
Sweetsprings (F4).1,122
Svente, (M6) 200
Syracuse, (G4).... 193
Tabersville, (E8) .. 140
Taft, (M6) 50

Tanesville, (G8).. 140
Taos, (H4)....... 140
Tarkio, (B1)....1,906
Tea, (K5)....... 50
Tebbetts, (J4).... 200
Tecumseh, (J6).... 150
Thayer, (K8).....1,613
Thomas Hill, (G3).. 100
Thomasville, (J8)... 100
Thornfield, (G6) .. 50
Tina, (F2)........ 304
Tin Lull, (E1)..... 100
Tipton, (G4) 1,273
Tracy, (C3) 176
Trask, (J4) 90
Trelmar, (K4)..... 43
Trenton, (E1).....5,656
Trimble, (D4)..... 222
Triplett (F2)..... 473
Tucker, (H4)....1,120
Tuscola, (K3)..... 280
Turney, (D2)..... 212
Tuscumbia, (H5).. 285
Tuxedo, (M1)..... 890
Tyler, (H8)....... 180
Ulman, (H5)...... 120
Union, (L5)....... 934
Union Star, (D2)... 388
Unionville, (N6)... 210
Univerville, (F1)..2,000
University, (M4) .2,417
Urbana, (F6).... 150
Urich, (D5)....... 484
Utica, (F1)....... 500
Valley Park (M4) .1,800
Vanburen (H3) ... 400
Vandalia (K1) ...1,525
Vandusar, (N8)... 388

Verona, (E8) 446
Versailles, (G5)..1,398
Viburnd (D3).... 110
Vichy, (J5)....... 200
Victoria (L5).... 190
Vienna, (H5) 300
Villaridge, (L5)... 150
Viola, (E8) 100
Vista, (E5) 100
Waco, (C7) 150
Wakenda (F1)... 279
Waldron, (C3).... 160
Walker, (D6) 364
Walnut, (C2) 160
Walnutgrove,(E7). 600
Wappapello, (M6).. 180
Warren, (J2) 100
Warrensburg,(F4).4,689
Warrenton (K4).. 795
Warsaw, (F5)..... 824
Washburn, (E8) .. 219
Washington, (K4) 3,670
Watson (J1)..... 245
Waverly, (F3).... 777
Wayland (J1).... 384
Waynesville (H6).. 257
Weatherby (D2).. 171
Wentleau, (J6).... 347
Webb City, (C7).11,817
Webster Groves,
(M4).......7,080
Wellington, (D8).. 558
Wellston, (M4)...7,312
Wellsville (J3)...1,194
Wentworth, (D2).. 154
Went ville, (L4)... 539
West Union, (M4).. 300
Westboro (B1)... 333
Westfork (K7).... 220
Westline, (C4) ... 120

Weston, (C3).....1,019
Westphalia, (J5).. 371
Westgate, (J5) .. 294
Westville, (G2).. 150
Wheatland, (F6) .. 400
Wheeling, (F2 ... 490
Whitecask, (G8).. 250
Whiteside, (K1) .. 129
Whitesville, (C2).. 200
Whitewater, (N8).. 150
Whitham (F2) ... 250
Whiting, (D8)..... 242
Willis, (M8)...... 200
Wilcox, (C1)...... 100
Wilhird, (E7)..... 160
Williamsburg, (J4). 140
Williamstown, (J1). 201
Willowsville, (L8). 417
Willowsprings,(H6).1,401
Wilton, (H4) 120
Winchester (K1).. 150
Windsom (M4)... 100
Windsor, (F4)....2,241
Winfield, (L4) 422
Winnigan, (G1)... 170
Winkler, (J6) 100
Winona, (K7) 444
Winston, (D2) ... 247
Wittenberg, (N8).. 87
Woodbridge, (G4).. 119
Woodland, (N5)... 160
Worland, (C7) ... 159
Worth, (D1)...... 170
Worthington (G1) 200
Wright City, (K4). 377
Wyaconda (J1)... 480
Wythe (D8)...... 200
Zalma, (M7)..... 200
Zincite (C7) 1,000

MONTANA

Albright, (F2)..... 110
Alder, (D4)....... 150
Aldridge, (F3).... 300
Anaconda, (C3)..10,134
Argo, (E3)........ 110
Arlee, (B2)....... 100
Armington, (F2)... 100
Armstead, (D5)... 300
Augusta, (D2)..... 240
Avon (D3)........ 100
Bainville, (M1)... 150
Baldhutte, (D3)... 150
Ballantine (H4)... 100
Bannack, (C4)... 150
Basin, (D3)....... 650
Bearcreek, (G4)... 302
Bearmouth, (C3).. 100
Belfry, (G4) 150
Belgrade (E4)..... 561
Belknap (A2)..... 100
Belt, (F2) 1,158
Bernice, (D8).... 100
Bigfork, (B1) 400
Big Sandy, (F1)... 390
Bigtimber, (F4)...1,022
Billings, (H4)...10,031
Blossburg, (D3)... 250
Bonner, (C4)..... 700
Boulder, (D3)....1,100
Bowler, (H4) 100
Bozeman, (E4)...5,107
Bridger, (G4).... 514
Bristun, (C4)..... 100
Broadview, (H4)... 200
Browning, (C1) .. 100
Butte, (D3) ...39,165
Cameron, (E4) .. 100
Canyon Ferry, (E3) 100
Carroll, (F3) 250
Cascade, (F2) ... 400
Castle, (F3)....... 100

Centerville (D4)..2,500
Chestnut, (E4) ... 150
Chinook, (G1)... 780
Chouteau, (D2)... 570
Clancey, (E3)..... 300
Coalville, (G4)... 200
Columbia Falls, (C1)..601
Columbus, (G4)... 521
Conrad, (F1)..... 888
Cooke, (E4)...... 100
Corbin, (D3)...... 150
Craig, (D2)....... 100
Crow Agency, (J4). 100
Culbertson, (M1).. 518
Cut Bank, (D1)... 250
Dagmar, (M1)... 200
Dayton, (B2)..... 100
Deborgia, (A2)... 150
Deer Lodge, (D3).2,570
Dillon (D4).....1,815
Divide, (D4)...... 100
Dodson, (H1).... 100
Drummond, (D3).. 350
Dupuyer (D1).... 380
East Helena, (E3).1,200
Eddy, (A2) 100
Ekalaka (M4)... 150
Electric (F4).... 400
Elkhorn (E4).... 150
Elkpark, (D3) ... 100
Elliston, (D3).... 200
Eureka, (B1)..... 604
Fairview, (M2)... 100
Fallon, (L4)..... 100
Family, (D1).... 110
Forsyth (K4)...1,398
Fort Benton, (F2) .1,904
Fort Keogh (L3).. 100
Fort Maginnis (G2) 100
Fort Shaw, (E2)... 380
Frenchtown, (B3).. 300

Fromberg, (H4)... 300
Gardiner, (F5).... 500
Garneill, (G3).... 150
Garnet, (C3)..... 250
Garrison, (D3)... 100
Gateway, (D1)... 150
Gilt Edge, (G2)... 300
Glasgow, (K1)...1,338
Glendive, (M2)...2,428
Gould, (D4)...... 100
Granite, (C3).... 350
Grant, (C4)...... 100
Grantsdale (B3).. 200
Great Falls, (E2).13,948
Hamilton, (B3)...2,246
Hardin, (J4) 250
Harlem, (H1).... 383
Harlowton, (G3).. 770
Harrison, (E4)... 150
Hassel, (E3)..... 100
Havre, (G1).....3,624
HELENA (E3)..12,515
Henderson, (A2).. 200
Heron, (A1)..... 150
Highwood, (F2)... 100
Hinsdale (H1)... 100
Hobson, (G2).... 200
Hoffman, (F4).... 200
Hogan (D2)..... 100
Huntley, (H4)... 250
Iron Mountain,(A2) 300
Ismay, (M4)..... 300
Jardine (F5)..... 110
Jefferson City, (E3) 100
Jennings, (A1)... 100
Jessup, (B1)..... 200
Jocko (B2)...... 550
Joliet (G4)...... 300
Judith Gap, (G3).. 200

Junction (J3)..... 100
Kalispell (B1)...5,549
Kendall (G2)....1,200
Lame Deer (K4)...1,550
Laurel (H4)..... 100
Laurin, (D4)..... 100
Lavina, (G3).... 100
Lewistown (G2)..2,992
Libby (A1)...... 100
Lima (D5)....... 350
Livingston, (F4)..5,359
Logan (E4)...... 150
Lothrop, (B4).... 250
Maiden (G2).... 100
Malta (H1)...... 431
Monmouth, (D4).. 100
Manhattan (E4).. 100
Marysville (D4)..2,000
Meaderville (D3).. 150
Meadow Creek, (B4) 100
Melrose (D4).... 200
Melstone (J3) ... 200
Miles City (L4)..4,697
Missoula (C4)...12,800
Monarch (E2)... 150
Mondak (M1)... 150
Moore (G2)...... 573
Musselshell, (J3)... 150
Neihart, (E2).... 268
New Chicago (C3). 200
Norton (A2)..... 100
Oldham (E1)..... 100
Para Inc (H2)... 100
Park City (H4)... 400
Philipsburg, (C3)..1,100
Plains, (B2)..... 450
Polson, (B2)..... 350
Pony, (E4)...... 300
Poplar (L1)..... 350
Raibersburg (F3).. 350
Rancher (J4)..... 130

Red Lodge (G4)..4,860
Riedel, (G4)..... 150
Romind, (D3).... 100
Rocker, (D4)..... 100
Roschud, (K4)... 250
Roundup, (H3)...1,513
Ruby (D4)...... 200
Ryan (J1)....... 200
Saint Peter, (D2).. 380
Saint Regis, (A2).. 350
Saltese (A2) 350
Sandcoulee (E2)..2,000
Shannon (F3).... 100
Shelby (E1)..... 200
Sheridan, (D4)... 399
Shirley, (L3).... 100
Sidney (M2)..... 300
Silesia (H4).... 110
Silverbow, (D4)... 100
Smelter, (E2).... 210
Snowshoe, (A1).. 100
Somers (B1).... 750
Stacey, (L4)..... 130
Stanford, (F2)... 200
Stevensville (C3).. 796
Stockett (E2)...1,400
Sturgis (A2)..... 100
Sun River, (E2).. 450
Superior, (B1)... 150
Sweetgrass (D1).. 200
Taft, (A2)....... 100
Terry (L4)...... 700
Thompson, (A2)... 500
Thompson Falls,
(A2)........ 325
Three Forks (E4).. 674
Teston, (E2)..... 150
Townsend (E3)... 759
Trident (E4).... 150
Trout Creek, (A2). 257
Troy, (A1)...... 200

Twin Bridges, (D4).	491	Walkerville, (D3)..	2,491	Whitefish, (B1).... 1,479
Twodot, (F3).....	200	Warmsprings,		Whitehall, (E4)... 417
Victor, (B3).....	374	(D3)...........	100	White Sulphur
Virginia City, (E4).	467	Washoe, (G4).....	300	Springs, (F3) 417

Wibaux, (M3).....	487	Windham, (F2)....	100
Wickes, (D3).....	200	Winston, (E3).....	150
Willow Creek,		Wisdom, (C4).....	250
(E4)..........	200	Zortman, (H2)....	350

NEBRASKA

Abie, (P5)........	210	Brule, (E5)........	120	Douglas, (Q6)..... 305
Adams, (P7)........	647	Bruning, (N7).....	353	Du Bois, (Q7)..... 339
Ainsworth, (J2).....	1,045	Bruno, (P5).......	245	Dunbar, (Q6)..... 216
Albion, (M4)......	1,584	Brunswick, (M3)...	278	Duncan, (N5)..... 100
Alda, (L6)........	100	Burchard, (Q7)....	315	Dundee, (R5)..... 1,023
Alexandria, (O7)...	447	Burkett, (M6).....	400	Dunning, (H4)..... 120
Allen, (P3)........	371	Burr, (Q6)........	113	Dwight, (O5)..... 184
Alliance, (C3).....	3,105	Burton, (J2).......	100	Eagle, (Q6)........ 360
Alma, (K7)........	1,066	Burwell, (K4).....	915	Eddyville, (J5).... 254
Alvo, (Q6)........	230	Butte, (L2).......	550	Edgar, (N7)....... 1,080
Amherst, (K6).....	256	Byron, (N7).......	184	Edison, (J7)....... 334
Anoka, (L2).......	145	Cairo, (L5).......	364	Elba, (L5)........ 302
Anselmo, (J4).....	351	Callaway, (J5).....	765	Elgin, (M4)....... 606
Ansley, (K5)......	700	Cambridge, (H7)...	1,029	Elk Creek, (Q7)... 240
Arapahoe, (J7)....	901	Campbell, (L7)....	573	Elkhorn, (Q5)..... 291
Arcadia, (K5).....	618	Carleton, (N7)....	410	Elm Creek, (K6)... 620
Arlington, (Q5)...	645	Carroll, (O3).....	382	Elmwood, (Q6)... 635
Armour, (Q7).....	100	Cedar Bluffs, (P5).	500	Elsie, (F6)........ 160
Arnold, (H5)......	231	Cedar Creek, (Q5).	180	Elwood, (J6)...... 464
Ashland, (Q5).....	1,379	Cedar Rapids, (M4)	576	Emerson, (P3)..... 838
Ashton, (L5)......	404	Center, (N2)......	119	Endicott, (O7)..... 204
Asylum, (M6).....	600	Central City, (M5).	2,428	Ericson, (L4)..... 100
Atkinson, (L2).....	810	Ceresco, (P5).....	276	Eustis, (H6)....... 403
Atlanta, (J7)......	200	Chadron, (C2).....	2,687	Ewing, (M3)...... 440
Auburn, (R7).....	2,729	Chapman, (M5)....	266	Exeter, (O6)...... 916
Aurora, (M6).....	2,630	Chappell, (D5)....	329	Fairbury, (O7)..... 5,294
Avoca, (Q6).......	249	Cheney, (P6)......	100	Fairfield, (M7).... 1,054
Axtell, (K7).......	394	Chester, (N7).....	560	Fairmont, (N6)... 921
Ayr, (M7)........	142	Clarks, (N5)......	605	Falls City, (R7)... 3,255
Bancroft, (P3)....	742	Clarkson, (O4)....	647	Farnam, (H6)..... 462
Barada, (R7)......	118	Clatonia, (P7)....	233	Farwell, (L5)..... 246
Barnston, (P7).....	228	Clay Center, (N6).	1,065	Filley, (P7)....... 194
Bartlett, (L4).....	100	Clearwater, (M3)..	414	Firth, (P6)....... 343
Bartley, (H7)......	511	Cody, (F2)........	185	Florence, (Q5)..... 1,526
Bassett, (K2).....	383	Coleridge, (O2)...	535	Fontanelle, (Q4)... 100
Battle Creek, (N4).	597	Collegeview, (P6).	1,508	Fort Calhoun, (Q5). 324
Bayard, (B4)......	261	Colon, (P5).......	160	Fort Crook, (R5).. 203
Bazile Mills, (N2)..	77	Columbus, (O5)...	5,014	Fort Robinson, (A2) 220
Beatrice, (P7).....	9,356	Comstock, (K4)...	323	Foster, (N3)...... 122
Beaver City, (J7)...	975	Concord, (P3).....	198	Franklin, (L7)..... 949
Beaver Crossing,		Cook, (Q7).......	387	Fremont, (P5)..... 8,718
(O6)...........	542	Cordova, (O6).....	201	Friend, (O6)...... 1,261
Bee, (O5)........	207	Corsica, (N4).....	90	Fullerton, (N5)... 1,638
Beemer, (P4).....	494	Cortland, (P7)....	364	Gandy, (H5)...... 200
Belden, (O3)......	247	Cotesfield, (L5)...	150	Garrison, (O5)..... 177
Belgrade, (M5)....	400	Cowles, (M7).....	190	Geneva, (N6)..... 1,741
Bellevue, (R5).....	596	Cozad, (J6).......	1,096	Genoa, (N5)....... 1,376
Bellwood, (O5)...	397	Crab Orchard, (P7).	274	Gering, (A4)...... 627
Belvidere, (N7)...	475	Craig, (Q4).......	339	Germantown, (P6). 275
Benedict, (N6)....	336	Crawford, (B2)....	1,323	Gibbon, (L6)...... 718
Benkelman, (F7)...	538	Creighton, (N3)...	1,373	Gilead, (O7)...... 181
Bennet, (P6)......	457	Creston, (O4).....	338	Giltner, (M6)..... 300
Bennington, (Q5)...	276	Crete, (O6).......	2,404	Gladstone, (O7)... 100
Benson, (Q5).....	3,170	Crofton, (N2).....	610	Glenville, (M6)... 304
Berlin, (Q6)......	196	Crookston, (G2)...	100	Goehner, (O6)..... 110
Bertrand, (J6).....	643	Culbertson, (G7)...	580	Gordon, (D2)..... 920
Berwyn, (J5)......	100	Curtis, (G6)......	613	Gothenburg, (H6). 1,730
Bethany, (P6).....	948	Dakota, (P3)......	474	Grafton, (N6)..... 353
Bigspring, (D5)...	140	Dalton, (C5)......	207	Grand Island, (M6) 10,326
Bladen, (L7)......	494	Danbury, (H7)....	268	Grant, (E6)....... 358
Blair, (Q4).......	2,584	Dannebrog, (M5)...	380	Greeley, (M4)..... 845
Bloomfield, (N2)...	1,264	Darr, (J6)........	32	Greenwood, (Q6). 387
Bloomington, (K7).	507	Davenport, (N7)...	484	Gresham, (O5)..... 344
Blue Hill, (M7)...	761	Davey, (P6).......	120	Gretna, (Q5)...... 484
Blue Springs, (P7).	712	David City, (O5)...	2,177	Gross, (L2)....... 111
Boelus, (L5)......	233	Dawson, (R7).....	340	Guide Rock, (M7). 690
Bostwick, (M7)....	150	Daykin, (O7)......	220	Hadar, (N3)....... 140
Bradshaw, (N6)...	359	Decatur, (Q4).....	782	Haigler, (E7)...... 205
Brady, (H5).......	308	Deshler, (N7).....	609	Hallam, (P6)...... 168
Brainard, (O5)....	465	Deweese, (M7)....	200	Hampton, (N6)... 383
Brewster, (J4).....	230	DeWitt, (P7).....	675	Harbine, (P7)..... 150
Bridgeport, (B4)...	541	Diller, (P7).......	506	Hardy, (N7)...... 496
Bristow, (L2)......	175	Dixon, (P3).......	217	Harrisburg, (A4)... 140
Brock, (R7).......	434	Dodge, (P4)......	661	Harrison, (A2)..... 186
Broken Bow, (J5)..	2,260	Doniphan, (M6)...	399	Hartington, (O2).. 1,413
Brownville, (R7)...	457	Dorchester, (O6)...	610	Harvard, (M6)..... 1,102

Hastings, (M6)....	9,338	Madrid, (E6)......	124
Havelock, (P6)...	2,680	Magnet, (N3).....	178
Hayes Center, (F6).	330	Malcolm, (P6).....	100
Hay Springs, (C2).	408	Malmo, (P5)......	214
Heartwell, (L6)...	100	Manley, (Q6).....	180
Hebron, (N7).....	1,778	Marion, (G7).....	180
Hemingford, (B3)..	272	Marquette, (N6)...	290
Henderson, (N6)...	391	Martinsburg, (P2).	291
Hendley, (J7).....	238	Mason City, (K5).	480
Herman, (Q4)....	345	Maxwell, (H5)...	289
Hershey, (G5)....	332	Maywood, (G6)...	443
Hickman, (P6)....	388	Mead, (Q5).......	330
Hildreth, (K7)....	450	Meadow Grove,	
Holbrook, (H7)....	414	(N3)..........	388
Holdrege, (K7)....	3,030	Memphis, (Q5)...	162
Holmesville, (P7)..	170	Merna, (J5).......	459
Holstein, (L7)....	323	Merriman, (E2)...	254
Homer, (P3)......	397	Milford, (O6).....	716
Hooper, (P4).....	741	Millard, (Q5).....	260
Hordville, (N5)....	130	Miller, (K6).......	330
Hoskins, (O3).....	262	Milligan, (O7)....	336
Howe, (R7).......	180	Minatare, (B4)....	338
Howell, (O4).....	800	Minden, (L7).....	1,559
Hubbard, (P3)....	150	Mitchell, (A4)....	640
Hubbell, (N7).....	295	Monowi, (M2)....	109
Humboldt, (R7)...	1,176	Monroe, (N5).....	282
Humphrey, (O4)..	868	Moorefield, (H6)..	180
Huntington, (M6)..	410	Morrill, (A4).....	346
Huntley, (K7).....	190	Morse Bluff, (P5)..	196
Hyannis, (E4)....	262	Mount Clare, (M7).	130
Imperial, (E6)....	402	Mullen, (F3)......	180
Inavale, (L7).....	130	Murdock, (Q6)...	222
Indianola, (H7)...	681	Murray, (R6).....	210
Inland, (M6)......	100	Naper, (K2)......	300
Inman, (L3)......	230	Naponee, (K7)...	195
Ithaca, (P5)......	171	Nebraska City, (R6)	5,488
Jackson, (P3).....	290	Nehawka, (Q6)...	200
Jansen, (O7)......	308	Neligh, (M3).....	1,566
Johnson, (R7).....	213	Nelson, (M7).....	978
Johnstown, (J2)...	200	Nemaha, (R7)....	325
Julian, (R6)......	130	Newcastle, (P2)...	436
Juniata, (L6).....	471	Newman Grove,	
Kearney, (K6).....	6,202	(N4)..........	850
Kenesaw, (L6)....	657	Newport, (K2)....	268
Kennard, (Q5).....	319	Nickerson, (P4)...	140
Kilgore, (G2).....	130	Niobrara, (N2)...	822
Kimball, (A5)....	454	Nora, (M7).......	100
Lanham, (P7).....	130	Norfolk, (N3).....	6,025
Laurel, (O3)......	514	Norman, (L7)....	130
Lawrence, (M7)...	475	North Bend, (P4)..	1,105
Lebanon, (H7)....	197	North Loup, (L5)..	519
Leigh, (O4).......	567	North Platte, (G5).	4,793
Leshara, (Q5).....	86	Oak, (N7)........	237
Lewellen, (D5)....	120	Oakdale, (N3)....	631
Lewiston, (Q7)....	130	Oakland, (Q4)...	1,073
Lexington, (J6)...	2,059	Oconto, (J5)......	245
Liberty, (P7)......	394	Octavia, (O5)....	150
LINCOLN, (P7)...	43,973	Odell, (P7).......	427
Lindsay, (N4).....	465	Ogallala, (E5)....	643
Linwood, (P5)....	329	Ohiowa, (O7)....	373
Litchfield, (K5)...	403	Omaha, (R5).....	124,096
Lodgepole, (C5)...	245	O'Neill, (L3).....	2,089
Long Pine, (J2)...	785	Ong, (N7)........	285
Loomis, (K7).....	284	Orchard, (M3)...	532
Loretto, (N4).....	100	Ord, (L4)........	1,960
Lorton, (Q6).....	115	Orleans, (K7).....	942
Louisville, (Q6)....	778	Osceola, (N5).....	1,105
Loup City, (K5)...	1,128	Oshkosh, (D5)....	180
Lushton, (N6).....	205	Osmond, (N3)....	567
Lynch, (M2)......	330	Overton, (J6).....	574
Lyons, (Q4)......	865	Oxford, (J7)......	593
McCook, (G7)....	3,765	Page, (M3).......	300
McCool Junction,		Palisade, (F7).....	380
(N6)..........	369	Palmer, (M5).....	373
McLean, (N3)....	130	Palmyra, (Q6)....	334
Madison, (N4)...	1,708	Panama, (P6).....	230

NEVADA

NEW HAMPSHIRE

NEW JERSEY

Athenia, (D2)...... 410
Atlantic City,(D5)46,150
Atlantic Highlands,
  (D3)............1,645
Atsion, (C4)........ 100
Auburn, (B4)....... 180
Audubon, (B4)....1,343
Aura, (B4)......... 130
Avalon, (C5)....... 230
Avenel, (D2)........ 200
Avon, (D3)......... 426
Baptistown, (C2)... 120
Bargaintown, (C5).. 100
Barnegat, (B4)..... 850
Barnegat City, (D4). 70
Barnsboro, (B4).... 200
Barrington, (B4)... 150
Basking Ridge, (C2). 640
Bayhead, (D3)...... 281
Bayonne, (D2)...55,545
Bayville, (D4)...... 200
Beach Haven, (D4).. 272
Beatyestown, (C2).. 170
Bedminster, (C2)... 130
Beemerville, (C1)... 250
Beesleys Point, (C5). 175
Belford, (D3)...... 250
Belleplain, (C5).... 100
Belleville, (D2)....8,500
Belmar, (E3).......1,433
Belvidere, (B2)....1,764
Bergenfield, (E2)...1,991
Berkeley Hts.,(D2). 300
Berlin, (C4)....... 800
Bernardsville, (C2).1,000
Beverly, (C3)......2,140
Billingsport, (B4) . 100
Birmingham, (C4)... 220
Bivalve, (B5)...... 220
Blackwells Mills,(C3) 210
Blackwood, (B4)... 600
Blairstown, (C2)... 800
Bloomfield, (D2) .15,070
Bloomingdale, (D1). 340
Bloomsbury, (B2).. 600
Blue Anchor, (C4)... 250
Bogota, (D2)......1,125
Boonton, (D2)....4,910
Bordentown, (C3)..4,250
Boundbrook, (C2)..3,970
Bradevelt, (D3)... 150
Bradley Beach, (E3)1,807
Branchville, (C1).. 643
Bridgeboro, (C3).. 230
Bridgeport, (D4)... 280
Bridgeport, (B4)... 640
Bridgeton, (B5)..14,209
Bridgeville, (B2)... 100
Brielle, (D3)....... 330
Brigantine, (D5).... 67
Broadway, (B2).... 200
Brookdale, (D2)... 250
Brookside, (C2).... 160
Browns Mills, (C4).. 300
Browntown, (D3)... 200
Budd Lake, (C2)... 120
Buddtown, (C4).... 250
Burleigh, (C5)..... 190
Burlington, (C3)..8,316
Burrsville, (D3).... 200
Butler, (D2)......2,765
Buttzville, (C2).... 310
Caldwell, (D2).....2,216
Califon, (C2)...... 500
Camden, (B4)....94,518
Campgaw, (D1)... 100
Canton, (B5)...... 190
Cape May, (C5)...2,471
Cape May Court
  House, (C5)...... 800
Cape May Point,
  (C5)............ 162
Carlstadt, (D2)....3,807

Carmel, (B5)...... 750
Carpenterasville, (B2) 210
Carteret, (D2).....3,500
Cassville, (D3)..... 220
Cedarbrook, (C4)... 430
Cedar Grove, (D2).. 670
Cedar Lake, (C4).. 110
Cedarville, (B5)...1,500
Centerton, (C4).... 100
Centerton, (B4).... 130
Changewater, (C2).. 210
Chapel Hill, (D3)... 200
Charlotteburg, (D1). 100
Chatham, (D2)....1,874
Chatsworth, (D4)... 150
Cheesequake, (D3).. 200
Chesilhurst, (C4)... 246
Chester, (C2)......1,100
Chesterfield, (C3).. 150
Chews, (B4)....... 440
Chrome, (D2)...... 100
Cinnaminson, (B4).. 150
Clarksboro, (B4)... 200
Clarksburg, (D3)... 250
Clayton, (B4).....1,926
Clementon, (C4).... 500
Clermont, (C5)..... 300
Cliffside, (D2).....3,394
Cliffwood, (D3).... 130
Clifton, (D2).....8,100
Clinton, (C2)...... 836
Closter, (E2).....1,483
Cokesbury, (C2) .. 100
Cold Spring, (C5).. 150
Colesville, (C1).... 200
Collingswood, (B4).4,795
Colonia, (D2)...... 100
Colts Neck, (D3)... 150
Columbia, (B2).... 200
Columbus, (C3).... 590
Cookstown, (C3)... 180
Copperhill, (C3)... 100
Coytesville, (E2)... 750
Cranbury, (D3).... 800
Cranford, (D2)....1,000
Creamridge, (D3).. 100
Cresskill, (E2)..... 550
Crosskeys, (B4)... 250
Crosswicks, (C3)... 400
Croton, (C2)...... 110
Daretown, (B4)... 150
Dayton, (C3)...... 390
Deal, (D3)........ 273
Deal Beach, (E3)... 180
Deerfield Street,(B4) 150
Delair, (B4)....... 230
Delanco, (C3)..... 750
Delawanna, (D2)... 710
Delaware, (B2).... 440
Delford, (D2)....1,005
Delmont, (C5)..... 175
Demarest, (E2).... 560
Dennisville, (C5)... 300
Denville, (D2)..... 630
Dias Creek, (C5)... 250
Dividing Creek, (B5) 500
Dorchester, (C5)... 100
Dorothy, (C5)..... 100
Dover, (C2)......7,468
Drakestown, (D2).. 200
Dumont, (E2).....1,783
Dundee Lake, (D2). 400
Dunellen, (D2)....1,990
Dutchneck, (C3)... 100
Eastcreek, (C5)... 100
E. Millstone, (C2).. 356
E. Newark, (D2)..3,163
E. Nutley, (D2)... 700
E. Orange, (D2)..34,371
E. Rutherford, (D2)4,275
Eatontown, (D3)...1,200
Edgewater, (E2)..2,655
Edgewater Park,(C3) 200

Egg Harbor City,
  (C4)...........2,181
Elberon, (E3)...... 100
Eldora, (C5)....... 100
Elizabeth, (D2)..73,409
Ellisburg, (B4).... 110
Ellisdale, (C3).... 275
Elm, (C4)......... 230
Elmer, (B4)......1,167
Elwood, (C4)...... 150
Emerson, (D2)..... 767
Englewood, (E2)..9,924
Englewood Cliffs,
  (E2)............ 410
English Creek, (C5). 300
Englishtown, (D3).. 468
Erma, (C5)........ 230
Essex Fells, (D2)... 442
Estelville, (C5).... 100
Everett, (C3)..... 100
Everittstown, (B2).. 150
Evesboro, (C4).... 100
Ewan, (B4)....... 250
Fairfield, (D2).... 150
Fairhaven, (D3),..1,100
Fairlawn, (D2).... 120
Fairmount, (C2)... 100
Fairton, (B5)...... 500
Fairview, (E2).....2,441
Fanwood, (D2).... 471
Far Hills, (C2)..... 300
Farmingdale, (D3).. 416
Fieldsboro, (C3)... 480
Finderne, (C2)..... 100
Finesville, (B2).... 150
Fishing Creek, (C5). 180
Flanders, (C2)..... 510
Flatbrookville, (C1). 100
Flemington, (C3)..2,693
Florence, (C3)....1,800
Florham Park, (D2). 558
Folsom, (C4)...... 232
Fords, (D2)....... 750
Forked River, (D4).. 500
Fort Lee, (E2)....4,472
Franklin Furnace,
  (C1)............1,800
Franklin Park, (C3).. 230
Franklinville, (B4).. 300
Freehold, (D3)....3,233
Frenchtown, (B2).. 984
Freneau, (D3)..... 200
Garfield, (D2)....10,213
Garwood, (D2)....1,118
Georgetown, (C3).. 100
German Valley, (C2) 530
Gibbsboro, (C4).... 300
Gibbstown, (B4)... 200
Glassboro, (B4)...2,100
Glendola, (D3).... 250
Glen Gardner, (C2).. 600
Glen Ridge, (D2)...3,260
Glen Rock, (D2)...1,055
Glenwood, (D1)... 300
Gloucester City,
  (B4)...........9,462
Goshen, (C5)..... 350
Grantwood, (E2)..1,600
Great Meadows, (C2) 150
Greenbrook, (C3)... 200
Greenpond, (C2)... 100
Green Village, (D2). 170
Greenwich, (B5)...1,000
Grenloch, (B4).... 200
Griggstown, (C3)... 330
Groveville, (C3)... 300
Guttenberg, (D2)..5,647
Hackensack, (D2). 14,050
Hackettstown, (C2).2,715
Haddonfield, (B4)..4,142
Haddon Hts., (B4). 1,452
Hainesburg, (B1)... 300
Hainesport, (C4)... 500

Hainesville, (C1)... 300
Haledon, (D2).....2,560
Haleyville, (B5).... 330
Hamburg, (C1).... 800
Hamilton, (D3).... 250
Hamilton Sq., (C3).. 430
Hammonton, (C4)..5,088
Hampton, (C2).... 914
Hancocks Bridge,
  (B4)........... 130
Hanover, (D2)..... 400
Hanover Neck, (D2). 220
Hardingville, (B4).. 100
Hardwick, (C1).... 100
Harlingen, (C3).... 200
Harmony, (B2).... 250
Harrington, (E2)... 150
Harrington Park,
  (E2)........... 377
Harrison, (D2)...14,498
Harrisonville, (B4).. 300
Hartford, (C4)..... 100
Harvey Cedars, (D4) 33
Hasbrouck Heights,
  (D2)...........2,155
Haskell, (D1)...... 200
Haworth, (D2).... 588
Hawthorne, (D2)..3,400
Hazen, (B2)....... 180
Hazlet, (D3)...... 200
Heislerville, (C5)... 420
Helmetta, (D3).... 661
Herbertsville, (D3).. 200
Hewitt, (D1)...... 210
Hibernia, (D2).....1,400
High Bridge, (C2)..1,545
Highland Park,
  (D3)...........1,517
Highlands, (E3)....1,386
Hightstown, (C3)..1,879
Highwood, (E2)... 400
Hillsdale, (D1).... 950
Hilton, (D2)...... 220
Hoboken, (D2)...70,324
Hohokus, (D1).... 488
Holland, (B2)..... 100
Holly Beach, (C5)..1,901
Holmdel, (D3)..... 850
Holmeson, (D3).... 100
Homestead, (D2).. 35
Hopatcong, (C2)... 146
Hope, (C2)....... 350
Hopewell, (C3)....1,073
Hornerstown, (D3).. 200
Hudson Heights,
  (D2)...........1,500
Huffville, (B4).... 200
Hyson, (D3)...... 150
Imlaystown, (D3)... 150
Iona, (B4)........ 100
Irvington, (D2)...11,877
Iselin, (D2)....... 200
Island Heights, (D4) 313
Jacksonville, (C3).. 100
Jacobstown, (C3).. 100
Jamesburg, (D3)...2,075
Janvier, (B4)..... 200
Jefferson, (B4).... 110
Jenkins, (D4)..... 100
Jersey City, (D2)267,779
Jobstown, (C3).... 200
Johnsonburg, (C2).. 100
Juliustown, (C3)...1,000
Keansburg, (D3)... 500
Keirry, (D2)....18,659
Kenilworth, (D2)... 779
Kenvil, (C2)...... 350
Keyport, (D3)....3,554
Kingsland, (D2)... 800
Kingston, (C3).... 300
Kingwood, (C3)... 100
Kinkora, (C3)..... 100
Kirkwood, (C4).... 160

Kresson, (C4)..... 305
La Fayette, (C1)... 400
Lake Como, (E3)... 230
Lakehurst, (D3)...1,000
Lakewood, (D3)...4,000
Lambertville, (C3)..4,657
Landing, (C2).... 120
Landisville, (C4)... 140
Lanoka, (D4)..... 200
Laurel Springs, (B4) 230
Lavallette, (D4)... 42
Lawnside, (B4).... 300
Lawrence St .. (C3). 250
Lawrenceville, (C3). 300
Layton, (C1)..... 320
Lebanon, (C2).... 360
Ledgewood, (C2).. 400
Leeds Point, (D5).. 300
Leesburg, (C5).... 250
Leonardo, (D3).... 200
Leonardville, (D3) . 150
Leonia, (E2).....1,486
Liberty Corner, (C2) 260
Lincoln, (C2)..... 200
Lincroft, (D3)..... 100
Linden, (D2)...... 610
Lindenwold, (C4)... 300
Linwood, (C5).... 602
Little Falls, (D2) .3,600
Little Ferry, (D2). 2,541
Little Silver, (D3).. 400
Littleton, (D2).... 350
Little York, (B2)... 150
Livingston, (D2)... 550
Locust, (D3)...... 150
Lodi, (D2)........4,138
Long Branch, (E3)13,298
Longport, (D5).... 118
Lower Bank, (C4).. 150
Lower Squankum,
  (D3)........... 250
Ludlow, (B2)..... 200
Lumberton, (C4)... 850
Lyndhurst, (D2)...1,350
Lyons Farms, (D2). 400
McAfee Valley, (D1) 200
McKee City, (C5).. 140
Madison, (D2)....4,658
Magnolia, (B4).... 400
Mahwah, (D1).... 550
Maine Avenue, (C5). 150
Malaga, (B4)..... 400
Manahawkin, (D4). 700
Manalapan, (D3).. 300
Manasquan, (D3)..1,582
Mantua, (B4)..... 750
Mapleshade, (C4).. 200
Maplewood, (D2).. 600
Marcella, (D2).... 130
Margate City, (D5). 129
Marlboro, (D3)... 400
Marlton, (C4)..... 700
Marmora, (C5)... 200
Martinsville, (C2).. 150
Masonville, (C4)... 150
Matawan, (D3)...1,646
Maurer, (D2)..... 520
Maurice River, (C5)1,100
Mauricetown, (B5). 560
Mayetta, (D4).... 100
Mays Landing (C5)1,800
Maywood, (D2).... 889
Medford, (C4)....1,129
Mendham, (C2)...1,129
Menlo Park, (D2) . 350
Merchantville, (B4)1,996
Metuchen, (D2)...2,138
Mickleton, (B4)... 130
Middlebush, (C3).. 130
Middletown, (D3).. 200
Middle Valley, (C2). 200
Midland Park, (D2)2,001
Midvale, (D1)..... 420

NEW MEXICO

NEW YORK

NORTH CAROLINA

Denhoff, (D5)..... 340
Deslacs, (C3)...... 160
Devils Lake, (F3).. 5,157
Dickey, (F6)...... 187
Dickinson, (B6)... 3,678
Dogden, (D4)...... 320
Donnybrook, (C3). 297
Douglas, (C4)...... 171
Doyon, (F3)........ 200
Drake, (D4)........ 348
Drayton, (G2)...... 587
Dresden, (F2)...... 220
Driscoll, (D6)...... 180
Dunseith, (D2)..... 478
Dwight, (H7)....... 150
Eckelson, (F6)...... 100
Eckman, (C3)....... 84
Edgeley, (F7)...... 749
Edinburg, (G2)..... 300
Edmore, (F3)....... 344
Edmunds, (E5)..... 120
Egeland, (E2)...... 266
Elbowoods, (B4).... 150
Ellendale, (F8).... 1,389
Emerado, (G4)..... 230
Enderlin, (G6)..... 1,540
Englevale, (G7)..... 100
Epping, (A3)....... 100
Esmond, (E3)....... 353
Fairdale, (F3)...... 140
Fairmount, (H7).... 387
Fargo, (H6)...... 14,331
Fessenden, (E4).... 713
Fingal, (G6)....... 360
Finley, (G4)....... 516
Flaxton, (B2)...... 301
Forbes, (F8)....... 221
Forest River, (G3). 233
Forman, (G7)...... 352
Fort Ransom, (G7) 100
Fort Yates, (D7)... 350
Fullerton, (F7).... 206
Gackle, (E6)....... 300
Gakhutt, (H7)..... 100
Galesburg, (G5).... 250
Gardena, (D2)..... 119
Gardner, (G5)..... 250
Garrison, (C4)..... 406
Geneseo, (G7)..... 200
Gilby, (G3)........ 320
Gladstone, (B6).... 240
Glenburn, (C2)..... 268
Glenullin, (C6)..... 921

Goodrich, (D4).... 410
Grafton, (G3)..... 2,229
Grand Forks, (G4) 12,478
Grand Harbor, (E3) 120
Grandin, (G5)..... 330
Grano, (C2)....... 150
Granville, (D3).... 455
Great Bend, (H7).. 191
Gwinner, (G7)..... 300
Hague, (E7)........ 183
Hamilton, (G2).... 213
Hampden, (F2).... 230
Hankinson, (H7)... 1503
Hannaford, (F5)... 340
Hannah, (F1)...... 550
Hansboro, (E2).... 320
Hartland, (C3)..... 100
Harvey, (D4).... 1,443
Hastings, (F6)..... 120
Hatton, (G4)...... 666
Havana, (G8)...... 387
Haynes, (B8)...... 100
Hazelton, (D6).... 150
Heaton, (E5)....... 190
Hebron, (B6)...... 597
Hensel, (G2)....... 120
Hettinger, (B7).... 766
Hillsboro, (G5)... 1,237
Hoople, (G2)...... 175
Hope, (G5)........ 909
Horace, (H6)...... 100
Hunter, (G5)...... 365
Inkster, (G3)...... 353
Jamestown, (F6).. 4,358
Joliette, (G2)...... 100
Jud, (F6).......... 100
Kathryn, (F6)..... 260
Kenmare, (B2)... 1,437
Kensal, (F5)....... 456
Kermit, (A2)...... 108
Kindred, (H6)..... 310
Knox, (E3)........ 330
Kramer, (D2)..... 181
Kulm, (F7)........ 645
Lakota, (F3)..... 1,023
Lamoure, (F7)..... 929
Langdon, (F2).... 1,214
Lankin, (G3)...... 341
Lansford, (C2).... 456
Larimore, (G4)... 1,224
Lawton, (F3)...... 160
Leal, (F5)......... 100
Leeds, (E3)....... 682

Lehigh, (B6)...... 100
Lehr, (E7)........ 182
Leonard. (G6)..... 190
Lidgerwood, (G7).. 1,019
Lignite, (B2)...... 150
Linton, (D7)...... 644
Lisbon, (G7)..... 1,758
Litchville, (F6).... 484
Lonetree, (C3)..... 190
Lucca, (G6)....... 100
Ludden, (F8)...... 109
McClusky, (D4)... 517
McHenry, (F4).... 398
McVille, (F4)..... 310
Maddock, (E4).... 378
Mandan, (D6)... 3,873
Manfred, (E4)..... 130
Mannhaven, (C5).. 120
Manning, (B5)..... 100
Manvel, (G3)..... 200
Mapleton, (H6)... 207
Marion, (F6)...... 120
Marmarth, (A7)... 790
Martin, (D4)...... 100
Max, (C4)......... 285
Maxbass, (C2).... 240
Mayville, (G4)... 1,070
Medina, (E6)...... 343
Medora, (A6)..... 100
Mekinock, (G3)... 200
Melville, (E5)..... 150
Mercer, (D4)...... 150
Merricourt, (F7)... 78
Michigan, (F4).... 449
Milnor, (G7)...... 641
Milton, (F2)...... 410
Minnewaukan, (E3) 510
Minot, (C3)...... 6,188
Minto, (G3)....... 701
Mohall, (C2)...... 493
Monango, (F7).... 238
Montpelier, (F6)... 100
Mooreton, (H7)... 160
Mott, (B7)........ 100
Mountain, (G2)... 150
Munich, (F2)...... 230
Mylo, (E2)........ 98
Napoleon, (E6).... 320
Neche, (G2)....... 528
Nekoma, (F2)..... 120
Newburg, (D2).... 102
New Rockford, (F4) 880
New Salem, (C6)... 621

Niagara, (G4)..... 157
Niobe, (B2)....... 200
Nome, (G6)....... 218
Noonan, (A2)..... 153
Norma, (C2)...... 100
North Minot, (C3). 432
Northwood, (G4).. 769
Norwich, (D3)..... 180
Oakes, (F7)..... 1,499
Oberon, (E4)...... 320
Olga, (F2)........ 200
Omemee, (D2).... 332
Oriska, (G6)...... 160
Osnabrock, (F2)... 253
Overly, (D2)...... 182
Page, (G5)........ 479
Palermo, (B3)..... 177
Park River, (G3).. 1,008
Pekin, (F4)....... 100
Pembina, (G2).... 717
Penn, (E3)........ 100
Perth, (E2)....... 221
Petersburg, (F4)... 353
Pingree, (F5)...... 250
Pisek, (G3)....... 312
Plaza, (C3)....... 224
Portal, (B2)...... 491
Portland, (G5).... 561
Powers Lake, (B2). 100
Ray, (A3)......... 436
Reeder, (B7)...... 198
Reynolds, (G4).... 412
Richardton, (B6).. 647
Rock Lake, (E2)... 194
Rolette, (E2)...... 408
Rolla, (E2)....... 587
Rugby, (E3)..... 1,630
Ruso, (D4)....... 141
Russell, (D2)..... 161
Rutland, (G7).... 224
Ryder, (C4)....... 338
Saint John, (E2)... 424
Saint Thomas, (G2) 513
Sanborn, (F6)..... 390
Sarles, (E2)....... 346
Sawyer, (C3)..... 327
Schafer, (A4)..... 150
Scranton, (A7).... 214
Sentinel Butte,
   (A6)........... 150
Sharon, (G4)..... 304
Sheldon, (G6)..... 358
Sherbrooke, (G5).. 100

Sherwood, (C2).... 328
Sheyenne, (E4).... 320
Sims, (C6)........ 96
Souris, (D2)...... 267
Spiritwood, (F6)... 250
Stanley, (B3)..... 518
Stanton, (C5)..... 100
Starkweather,
   (F2)........... 246
Steele, (E6)....... 500
Sterling, (D6)..... 100
Strasburg, (D7)... 273
Streeter, (E6)..... 100
Surrey, (C3)...... 120
Sweet Briar, (C6).. 100
Sykeston, (E5).... 276
Tagus, (C3)....... 105
Temvik, (D7)..... 100
Thompson. (G4)... 360
Thorne, (E2)...... 105
Tioga, (B3)....... 203
Tolley, (C2)...... 250
Tolna, (F4)....... 209
Tower City, (G6).. 452
Towner, (D3)..... 691
Turtle Lake, (D4).. 200
Underwood, (C5).. 422
Upham, (D2)..... 296
Valley City,
   (F6).......... 4,606
Velva, (D3)....... 837
Verona, (F7)...... 235
Wahpeton, (H7).. 2,467
Walcott, (G6)..... 250
Wales, (F2)....... 230
Walhalla, (G2).... 592
Walum, (F5)...... 100
Warwick, (F4).... 180
Washburn, (D5)... 657
Webster, (E3)..... 100
Westhope, (C2)... 592
Wheatland, (G6).. 520
Wheelock, (A3)... 100
White Earth,
   (B3)........... 264
Williston, (A3)... 3,124
Willow City, (D2).. 625
Wilton, (D5)..... 437
Wimbledon, (F5).. 571
Wishek, (E7)..... 432
Wyndmere, (G7).. 439
York, (E3)........ 300
Zeeland, (E8)..... 193

OHIO

Abbeyville, (N5)... 160
Aberdeen, (F15)... 568
Academia, (L8)..... 300
Ada, (E6).......... 2,465
Adams Mills, (N9).. 150
Adamsville, (N9)... 176
Adario, (K6)....... 100
Addison, (M14).... 220
Addyston (B13)... 1,543
Adelphi, (J12)..... 407
Adena, (R9)....... 570
Adrian, (G5)...... 210
Agosta, (G7)...... 504
Aid, (L15)........ 200
Aitch, (Q11)...... 110
Akron, (O5)..... 69,067
Albany, (M13)..... 546
Alcony, (D9)...... 130
Alexandersville, (C11)180
Alexandria, (K9)... 414
Alger, (E7)....... 730
Alikanna, (S8)..... 100
Allensville, (K13).. 220
Allentown (D6)... 110
Alliance, (Q6)... 15,083
Alpha, (D11)...... 180
Alton, (H10) 110

Alvada, (G5)...... 160
Alvordton, (C3)... 40?
Amanda, (K11)... 484
Amboy, (S2)...... 300
Amelia, (D13).... 417
Amesville, (N12)... 267
Amherst, (M4)... 2,106
Amsden, (G5)..... 150
Amsterdam, (R8).. 1,041
Andersonville, (J12). 200
Andover, (S3)..... 902
Anna, (D8)........ 460
Annapolis, (R8)... 200
Ansonia, (B9)..... 656
Antioch, (Q11).... 169
Antiquity, (N14)... 300
Antrim, (P9)...... 150
Antwerp, (A5).... 1,187
Apple Creek, (N6).. 400
Applegrove, (N14).. 100
Appleton, (K9).... 100
Arabia, (L15)..... 100
Arcadia, (G5)..... 380
Arcanum, (B10).. 1,361
Archbold, (C3)... 1,082
Arion, (H14)...... 200
Arlington, (F6)... 798

Arlington, (C10)... 200
Arlington Heights,
   (C13)........... 468
Arnettsville, (B10). 240
Arnheim, (E14).... 100
Arnold, (H9)...... 100
Asbury, (C13)..... 150
Ashland, (L6).... 6,795
Ashley, (J8)....... 706
Ashtabula, (R2)... 18,266
Ashville, (J11).... 972
Athalia, (L15)..... 226
Athens, (M12)... 5,463
Atlanta, (H11).... 140
Atlas, (Q10)...... 100
Attica, (J5)....... 719
Atwater, (Q5)..... 300
Atwater Center,
   (Q5)........... 100
Auburn, (Q4)..... 100
Augusta, (Q7)..... 300
Aultman, (P6)..... 250
Aurora, (P4)...... 600
Aurora Station, (P4) 600
Austinburg, (R2)... 300
Austintown, (R5)... 200
Ava, (O10)........ 500

Avery, (K4)....... 100
Avon, (M4)....... 750
Ayersville, (C5)... 100
Bachman, (C10)... 200
Bainbridge, (P4)... 100
Bainbridge, (G13).. 883
Bairdstown, (F5)... 240
Bakersville, (O8)... 290
Baltic, (O8)....... 377
Baltimore (K10).... 551
Baltimore, (C10)... 408
Bantam, (D13)..... 200
Barberton, (O6)... 9,410
Barlow, (O12)..... 150
Barnesville, (Q10). 4,233
Barnhill, (P8)..... 506
Barryville, (D6).... 200
Bartlett, (N12)..... 200
Barton, (R9)...... 800
Bascom, (G5)...... 380
Basil, (K10)....... 504
Batavia, (D13)... 1,034
Batesville, (Q10).... 282
Bath, (O5)........ 100
Batson, (B5)...... 130
Bay, (O4)......... 450
Bayard, (Q7)...... 140

Bays, (F4)........ 200
Beach City, (O7).. 671
Beach Park, (M4).. 200
Beallsville, (Q10).. 564
Beamsville, (B9)... 150
Beasleys Fork, (F15) 150
Beaver, (J13)..... 286
Beaverdam, (F6)... 455
Bedford, (O4).... 1,783
Beedle, (P8)...... 150
Belden, (M4)..... 150
Belfast, (F14)..... 200
Bellaire, (S8)... 12,946
Bellbrook, (D11)... 283
Belle Center, (F8).. 283
Bellefontaine, (F8). 8,238
Bellevalley, (O10).. 689
Bellevue, (H4).... 5,209
Bellpoint, (H9)..... 130
Bellville, (L7)... 1,056
Belmont, (Q9)..... 572
Beloit, (P5)....... 298
Belpre, (O13)..... 510
Belpre, (O13)... 1,249
Benton (H6)...... 120
Benton, (N7)...... 300
Benton Ridge, (E5). 352

NORTH DAKOTA

Denhoff, (D5)..... 340
Deslacs, (C3)..... 160
Devils Lake, (F3).. 5,157
Dickey, (F6)..... 187
Dickinson, (B6)... 3,678
Dogden, (D4)..... 320
Donnybrook, (C3). 297
Douglas, (C4)..... 171
Doyon, (F3)..... 200
Drake, (D4)..... 348
Drayton, (G2).... 587
Dresden, (F2)..... 220
Driscoll, (D6).... 180
Dunseith, (D2)... 478
Dwight, (H7)..... 150
Eckelson, (F6).... 100
Eckman, (C3)..... 84
Edgeley, (F7)..... 749
Edinburg, (G2)... 300
Edmore, (F3)..... 344
Edmunds, (E5).... 120
Egeland, (E2)..... 266
Elbowoods, (B4).. 150
Ellendale, (F8)... 1,389
Emerado, (G4)... 230
Enderlin, (G6).... 1,540
Englevale, (G7)... 100
Epping, (A3)..... 100
Esmond, (E3)..... 353
Fairdale, (F3)..... 140
Fairmount, (H7)... 387
Fargo, (H6)..... 14,331
Fessenden, (E4)... 713
Fingal, (G6)..... 360
Finley, (G4)..... 516
Flaxton, (B2)..... 301
Forbes, (F8)..... 221
Forest River, (G3). 233
Forman, (G7)..... 352
Fort Ransom, (G7) 100
Fort Yates, (D7)... 350
Fullerton, (F7).... 206
Gackle, (E6)..... 300
Galchutt, (H7).... 100
Galesburg, (G5)... 250
Gardena, (D2).... 119
Gardner, (G5).... 250
Garrison, (C4).... 406
Geneseo, (G7).... 200
Gilby, (G3)..... 320
Gladstone, (B6)... 240
Glenburn, (C2)... 268
Glenullin, (C6)... 921

Goodrich, (D4).... 410
Grafton, (G3)..... 2,229
Grand Forks, (G4) 12,478
Grand Harbor, (E3) 120
Grandin, (G5)..... 330
Grano, (C2)..... 150
Granville, (D3)... 455
Great Bend, (H7).. 191
Gwinner, (G7).... 300
Hague, (E7)..... 183
Hamilton, (G2)... 213
Hampden, (F2)... 230
Hankinson, (H7).. 1503
Hannaford, (F5)... 340
Hannah, (F1)..... 550
Hansboro, (E2)... 320
Hartland, (C3)... 100
Harvey, (D4)..... 1,443
Hastings, (F6).... 120
Hatton, (G4)..... 666
Havana, (G8)..... 387
Haynes, (B8)..... 100
Hazelton, (D6).... 150
Heaton, (E5)..... 190
Hebron, (B6)..... 597
Hensel, (G2)..... 120
Hettinger, (B7)... 766
Hillsboro, (G5)... 1,237
Hoople, (G2)..... 175
Hope, (G5)..... 909
Horace, (H6)..... 100
Hunter, (G5)..... 365
Inkster, (G3)..... 353
Jamestown, (F6).. 4,358
Joliette, (G2)..... 100
Jud, (F6)..... 100
Kathryn, (F6).... 260
Kenmare, (B2)... 1,437
Kensal, (F5)..... 456
Kermit, (A2)..... 108
Kindred, (H6).... 310
Knox, (E3)..... 330
Kramer, (D2)..... 181
Kulm, (F7)..... 645
Lakota, (F3)..... 1,023
Lamoure, (F7).... 929
Langdon, (F2)... 1,214
Lankin, (G3)..... 341
Lansford, (C2).... 456
Larimore, (G4)... 1,224
Lawton, (F3)..... 160
Leal, (F5)..... 100
Leeds, (E3)..... 682

Lehigh, (B6)..... 100
Lehr, (E7)..... 182
Leonard, (G6).... 190
Lidgerwood, (G7).. 1,019
Lignite, (B2)..... 150
Linton, (D7)..... 644
Lisbon, (G7)..... 1,758
Litchville, (F6)... 484
Lonetree, (C3).... 190
Lucca, (G6)..... 100
Ludden, (F8)..... 109
McClusky, (D4)... 517
McHenry, (F4).... 398
McVille, (F4)..... 310
Maddock, (E4).... 378
Mandan, (D6).... 3,873
Manfred, (E4).... 130
Mannhaven, (C5).. 120
Manning, (B5).... 100
Manvel, (G3)..... 200
Mapleton, (H6)... 207
Marion, (F6)..... 120
Marmarth, (A7)... 790
Martin, (D4)..... 100
Max, (C4)..... 285
Maxbass, (C2).... 240
Mayville, (G4)... 1,070
Medina, (E6)..... 343
Medora, (A6)..... 100
Mekinock, (G3)... 200
Melville, (E5).... 150
Mercer, (D4)..... 150
Merricourt, (F7)... 78
Michigan, (F4).... 449
Milnor, (G7)..... 641
Milton, (F2)..... 410
Minnewaukan, (E3) 510
Minot, (C3)..... 6,188
Minto, (G3)..... 701
Mohall, (C2)..... 493
Monango, (F7)... 238
Montpelier, (F6).. 100
Mooreton, (H7)... 160
Mott, (B7)..... 100
Mountain, (G2)... 150
Munich, (F2)..... 230
Mylo, (E2)..... 98
Napoleon, (E6)... 320
Neche, (G2)..... 528
Nekoma, (F2)..... 120
Newburg, (D2)... 102
New Rockford, (F4) 880
New Salem, (C6).. 621

Niagara, (G4).... 157
Niobe, (B2)..... 200
Nome, (G6)..... 218
Noonan, (A2)..... 153
Norma, (C2)..... 100
North Minot, (C3). 432
Northwood, (G4).. 769
Norwich, (D3).... 180
Oakes, (F7)..... 1,499
Oberon, (E4)..... 320
Olga, (F2)..... 200
Omemee, (D2)... 332
Oriska, (G6)..... 160
Osnabrock, (F2)... 253
Overly, (D2)..... 182
Page, (G5)..... 479
Palermo, (B3).... 177
Park River, (G3)... 1,008
Pekin, (F4)..... 100
Pembina, (G2)... 717
Penn, (E3)..... 100
Perth, (E2)..... 221
Petersburg, (F4)... 353
Pingree, (F5)..... 250
Pisek, (G3)..... 312
Plaza, (C3)..... 224
Portal, (B2)..... 491
Portland, (G5)... 561
Powers Lake, (B2). 100
Ray, (A3)..... 436
Reeder, (B7)..... 198
Reynolds, (G4)... 412
Richardton, (B6).. 647
Rock Lake, (E2)... 194
Rolette, (E2)..... 408
Rolla, (E2)..... 587
Rugby, (E3)..... 1,630
Ruso, (D4)..... 141
Russell, (D2)..... 161
Rutland, (G7).... 224
Ryder, (C4)..... 338
Saint John, (E2)... 424
Saint Thomas, (G2) 513
Sanborn, (F6).... 390
Sarles, (E2)..... 346
Sawyer, (C3)..... 327
Schafer, (A4)..... 150
Scranton, (A7).... 214
Sentinel Butte,
  (A6)..... 150
Sharon, (G4)..... 304
Sheldon, (G6).... 358
Sherbrooke, (G5).. 100

Sherwood, (C2).... 328
Sheyenne, (E4)... 320
Sims, (C6)..... 86
Souris, (D2)..... 267
Spiritwood, (F6)... 250
Stanley, (B3)..... 518
Stanton, (C5)..... 100
Starkweather,
  (F2)..... 246
Steele, (E6)..... 500
Sterling, (D6).... 100
Strasburg, (D7)... 273
Streeter, (E6)..... 100
Surrey, (C3)..... 100
Sweet Briar, (C6).. 100
Sykeston, (E5).... 276
Tagus, (C3)..... 105
Temvik, (D7)..... 100
Thompson, (G4)... 360
Thorne, (E2)..... 105
Tioga, (B3)..... 203
Tolley, (C2)..... 250
Tolna, (F4)..... 209
Tower City, (G6).. 452
Towner, (D3)..... 691
Turtle Lake, (D4).. 200
Underwood, (C5).. 422
Upham, (D2)..... 296
Valley City,
  (F6)..... 4,606
Velva, (D3)..... 837
Verona, (F7)..... 235
Wahpeton, (H7).. 2,467
Walcott, (G6).... 250
Wales, (F2)..... 230
Walhalla, (G2)... 592
Walum, (F5)..... 100
Warwick, (F4).... 180
Washburn, (D5)... 657
Webster, (F3)..... 100
Westhope, (C2)... 592
Wheatland, (G6).. 520
Wheelock, (A3)... 100
White Earth,
  (B3)..... 264
Williston, (A3)... 3,124
Willow City, (D2). 623
Wilton, (D5)..... 437
Wimbledon, (F5).. 571
Wishek, (E7)..... 432
Wyndmere, (G7).. 439
York, (E3)..... 300
Zeeland, (E8)..... 193

OHIO

Abbeyville, (N5)... 160
Aberdeen, (F15)... 568
Academia, (L8).... 300
Ada, (E6)..... 2,465
Adams Mills, (N9).. 150
Adamsville, (N9)... 176
Adario, (K6)..... 100
Addison, (M14)... 220
Addyston, (B13)... 1,543
Adelphi, (J12)..... 407
Adena, (R9)..... 570
Adrian, (G5)..... 210
Agosta, (G7)..... 304
Aid, (L15)..... 200
Aitch, (Q11)..... 110
Akron, (O5)..... 69,067
Albany, (M13)... 546
Alcony, (D9)..... 130
Alexandersville, (C11)180
Alexandria, (K9)... 414
Alger, (E7)..... 730
Alikanna, (S8).... 100
Allensville, (K13)... 220
Allentown, (D6)... 110
Alliance, (Q6)..... 15,083
Alpha, (D11)..... 180
Alton, (H10)..... 110

Alvada, (G5)..... 160
Alvordton, (C3)... 402
Amanda, (K11).... 484
Amboy, (S2)..... 300
Amelia, (D13).... 417
Amesville, (N12)... 267
Amherst, (M4)... 2,105
Amsden, (G5)..... 150
Amsterdam, (R8).. 1,041
Andersonville, (J12) 200
Andover, (S3)..... 902
Anna, (D8)..... 460
Annapolis, (R8)... 200
Ansonia, (B9)..... 656
Antioch, (Q11).... 169
Antiquity, (N14)... 300
Antrim, (P9)..... 150
Antwerp, (A5)... 1,187
Apple Creek, (N6). 400
Applegrove, (N14). 100
Appleton, (K9)... 100
Arabia, (L15)..... 100
Arcadia, (G5)..... 380
Arcanum, (B10)... 1,361
Archbold, (C3)... 1,082
Arion, (H14)..... 200
Arlington, (F6)... 798

Arlington, (C10)... 200
Arlington Heights,
  (C13)..... 468
Arnettsville, (B10). 240
Arnheim, (E14)... 100
Arnold, (H9)..... 100
Asbury, (C13)..... 150
Ashland, (L6)..... 6,795
Ashley, (J8)..... 704
Ashtabula, (R2)... 18,266
Ashville, (J11)..... 972
Athalia, (L15).... 226
Athens, (M12)... 5,463
Atlanta, (H11).... 140
Atlas, (Q10)..... 100
Attica, (J5)..... 719
Atwater, (Q5)..... 300
Atwater Center,
  (Q5)..... 100
Auburn, (Q4)..... 100
Augusta, (Q7)..... 300
Aultman, (P6)..... 250
Aurora, (P4)..... 690
Aurora Station, (P4) 600
Austinburg, (R2)... 300
Austintown, (R5)... 200
Ava, (O10)..... 500

Avery, (K4)..... 100
Avon, (M4)..... 750
Ayersville, (C5)... 100
Bachman, (C10)... 100
Bainbridge, (P4)... 100
Bainbridge, (G13).. 883
Bairdstown, (F5)... 240
Bakersville, (O8)... 290
Balin, (O8)..... 377
Baltimore, (K10)... 551
Baltimore, (C10)... 408
Bantam, (D13).... 200
Barberton, (O5)... 9,410
Barlow, (O12).... 158
Barnesville, (Q10).. 4,233
Barnhill, (P8)..... 506
Barrville, (O6)... 200
Bartlett, (N12).... 200
Barton, (R9)..... 800
Bascom, (G5)..... 380
Bath, (O5)..... 504
Batavia, (D13)... 1,034
Batesville, (Q10)... 282
Bath, (O5)..... 100
Batson, (H5)..... 130
Bay, (O4)..... 450
Bayard, (Q7)..... 140

Bays, (T4)..... 200
Beach City, (O7).. 671
Beech Park, M4... 200
Beallsville, (Q10).. 564
Bell Branch, (N4)... 150
Beaver, (J13)..... 286
Beaverdam, (F15).. 150
Beaver Creek, (P15) 150
Beaver Valley, (J13) 689
Bedford, (O4).... 8,238
Belle Center, (F8). 689
Bellaire, (S9)..... 5,209
Bellbrook, (C11)... 130
Bellefontaine, (F7). 1,056
Bellevue, (J4)..... 572
Belmore, (E5)..... 298
Beloit, (Q6)..... 510
Belpre, (N12)..... 1,249
Benton, (H6)..... 120
Benton, (N5)..... 300
Benton Ridge, (F5). 352

Uniontown, (Q9)... 120
Uniontown, (M10).. 210
Unionville, (R2)... 300
Unionville Center,
  (G9)............. 239
Uniopolis, (D7)... 245
Unity, (S6)...... 300
Upper Sandusky,
  (G6)............3,779
Urbana, (E9).....7,739
Utica, (L9).....1,729
Valley City, (N5).. 250
Vanburen, (F5)... 303
Vandalia, (D10)... 221
Vanlue, (G6)...... 400
Van Wert, (B6)...7,157
Vaughnsville, (D6).. 300
Venedocia, (C6)... 247
Venice, (J4)...... 100
Vermilion, (L4)...1,369
Verona, (C10)..... 250
Versailles, (B9)...1,580
Vickery, (J4)..... 200
Vienna, (F10)..... 368
Vienna, (S5)...... 150
Vigo, (J13)....... 200
Villa, (E10)...... 100
Vincent, (O12).... 250
Vinton, (L14)..... 324
Wabash, (R8)...... 100
Wabash, (A7)...... 300
Waco, (P7)........ 200
Wadsworth, (N5)..3,073
Wagram, (K10)..... 100
Wainwright, (P8).. 350
Wakeman, (L5)..... 930
Walbridge, (G3)... 500
Waldo, (H8)....... 319
Walhonding, (M8).. 110

Walkers, (S7)..... 200
Walnut Creek, (N7). 150
Walnutrun, (G10).. 210
Walton, (G12)..... 100
Wapakoneta, (D7).5,349
Warner, (P11)..... 200
Warnock, (R9)..... 160
Warren, (R5).....11,081
Warrensburg, (H8).. 100
Warrensville, (P4).. 150
Warrenton, (S9)... 300
Warsaw, (M8)...... 512
Warwick, (O6)..... 500
Washington Court
  House, (G11)....7,277
Washingtonville,
  (S6)............ 957
Waterford, (O11).. 300
Waterloo, (K15)... 150
Watertown, (O12).. 200
Waterville, (F3)... 834
Wauseon, (D3)...2,650
Waverly, (H13)...1,803
Wayland, (Q5)..... 100
Waynesburg, (Q7).. 760
Waynesfield, (E7).. 542
Waynesville, (D11). 705
WeaversCorners,(J5) 150
Weavers Station,
  (B9)............ 120
Webster, (B9)..... 250
Weldon, (S4)...... 200
Wellington, (N5)..2,131
Wellston, (K13)..6,875
Wellsville, (S7)..7,769
Welshfield, (Q4).. 950
Wengerlawn, (B10). 130
W. Alexandria,
  (B11)..........1,030

W. Andover, (S3)... 300
W. Austintown, (R5) 150
W. Bedford, (M8)... 130
Westboro, (E12)... 300
W. Brookfield, (O6). 300
W. Cairo, (D6).... 386
W. Carlisle, (M9).. 180
W.Carrol ton,(D11)1,285
W. Charleston,(D10).130
W. Chester, (C12).. 300
W. Clarksfield, (L5) 200
W. Dover, (N4).... 300
W. Elkton, (B11).. 230
Western Star, (O5).. 154
Westerville, (J9)..1,903
W. Farmington,(R4) 446
Westfield, (J8)..... 140
W. Independence,
  (F5)............ 130
W. Jefferson, (G10).1,043
W. Lafayette, (O8).. 840
W. Lancaster, (F11). 180
W. Lebanon, (O7).. 300
W. Leipsic, (D5)... 253
W. Liberty, (E9)...1,288
W. Lodi, (J5)..... 100
W. Manchester,
  (B10)........... 445
W. Mansfield, (F8).. 913
W. Mecca, (S4).... 200
W. Mentor, (P3)... 100
W. Middletown,
  (C11)........... 300
W. Millgrove, (F5).. 191
W. Milton, (C10)..1,207
Westminster, (E7).. 300
W. Newton, (E7)... 200
Weston, (E4)...... 913
W. Park, (N4)..... 300

W. Richfield, (O5).. 250
W. Rushville, (L10). 147
W. Salem, (M6)... 642
W. Sonora, (B10).. 200
W. Toledo, (F3)...2,500
W. Union, (F14)...1,080
W. Unity, (C3).... 980
Westview, (N4).... 600
W. Wheeling, (S9).. 250
Wetzel, (C6)...... 100
Weymouth, (N5)... 300
Wharton, (G6).... 485
Wheelersburg, (J15). 250
Wheldon, (K13)... 280
Whipple, (P11).... 150
White Cottage,
  (M10)........... 300
Whitehouse, (E3).. 506
Wick, (S3)........ 200
Wickliffe, (P3).... 500
Wilberforce, (E11). 300
Wilkesville, (L13).. 203
Wilkins, (L9)..... 100
Williamsburg, (D13) 948
Williams Center,
  (B4)............ 100
Williamsfield, (S3).. 200
Williamsport, (H11). 536
Williamstown, (F6). 100
Williston, (G3)... 100
Willoughby, (P3)..2,072
Willow, (O4)...... 230
Willshire, (A6)... 653
Wilmington, (E12).4,491
Wilmot, (O7)...... 258
Winchester, (F14).. 927
Windham, (Q5).... 261
Windsor, (R3)..... 700
Windsor Mills, (Q3). 150

Winesburg, (O7) .. 250
Winfield, (O7) 100
Winona, (R6)...... 180
Winterset, (P9)... 200
Wintersville, (S8)... 100
Withamsville,
  (C13)........... 390
Wolf Run, (R8).... 100
Wood, (B12)....... 130
Woodington, (B9).. 300
Woodland, (G8).... 120
Woodlawn, (B12).. 250
Woodsfield, (O11)..2,502
Woodstock, (F9)... 310
Woodville, (C4)... 807
Wooster, (N6).....6,136
Worthington, (H9).. 547
Wren, (A6)........ 277
Wrightsville,
  (F15)........... 130
Wyandot, (H7).... 160
Wynant, (C8)...... 100
Wyoming, (C13)..1,893
Xenia, (E11).....8,706
Yellow Bud, (J12).. 250
Yellow Springs,
  (E10)..........1,360
York, (G8)........ 100
Yorkshire, (B8)... 182
Yorkville, (S9).... 100
Young, (H14)..... 100
Youngstown,
  (S5)..........79,066
Zaleski, (L12)..... 476
Zanesfield, (F8)... 250
Zanesville, (M10).28,026
Zimmerman, (D11). 160
Zoar, (P7)........ 182
Zoar Station, (P7). . 200

OKLAHOMA

Ada, (G5)........4,349
Adair, (K2)....... 376
Adamson, (J5)..... 150
Addington, (E6)... 493
Afton, (L1)......1,279
Agra, (G3)........ 366
Akins, (L4)....... 150
Albany, (H7)...... 100
Albion, (K5)...... 200
Alderson, (J5)..... 786
Alex, (E5)........ 240
Alikchi, (K6)...... 75
Aline, (C2)....... 303
Allen, (H5)....... 645
Alma, (E6)........ 120
Altus, (B5)......4,821
Alva, (C1).......3,688
Amabala, (H4)..... 60
Ames, (D2)........ 450
Amorita, (D1)..... 100
Anadarko, (D4)...3,439
Antelope, (A3)..... 60
Antioch, (F5)..... 100
Antlers, (J6).....1,273
Apache, (D5)...... 950
Arapaho, (C3)..... 713
Arcadia, (F3)..... 275
Ardmore, (F6)....8,618
Ark, (F7)......... 200
Arlington, (G3)... 120
Arnett, (A2)...... 511
Asher, (G5)....... 381
Atoka, (H6)......1,968
Atwood, (H5)...... 100
Augusta, (C1)..... 150
Autwine, (F1)..... 150
Avard, (C1)....... 170
Avery, (G3)....... 240
Avoca, (G5)....... 150
Bacone, (K3)...... 100
Bailey, (E5)....... 110
Bartlesville, (H1)..6,181

Beaver, (C7)...... 326
Bebee, (G5)....... 120
Beggs, (H3)....... 855
Bellemont, (G4)... 120
Bennington, (J7)... 513
Berwyn, (F6)...... 378
Bessie, (B4)....... 420
Beulah, (B4)...... 265
Bigcabin, (K1).... 220
Bigheart, (H2).... 307
Billings, (F1)..... 524
Binger, (D4)...... 280
Bison, (E2)....... 375
Bixby, (J3)........ 384
Blackburn, (G2)... 335
Blackwell, (F1)...3,266
Blair, (B5)........ 503
Blanchard, (E4)... 629
Bliss, (F1)........ 100
Blocker, (J4)...... 131
Blue, (J6)......... 100
Bluejacket, (K1)... 508
Boise City, (A7)... 190
Bokchito, (H7).... 525
Bokoshe, (L4)..... 483
Boley, (G4)......1,334
Boswell, (J6)..... 828
Boynton, (J3)..... 679
Braden, (L4)...... 100
Bradley, (E5)..... 160
Braggs, (K3)...... 259
Braman, (F1)..... 339
Breckinridge, (E2). 150
Briartown, (K4)... 150
Bridgeport, (D3).. 428
Bristow, (H3)....1,667
Britton, (E3)..... 696
Broken Arrow, (J2).1,576
Bromide, (G6).... 200
Brookeen, (K4)... 200
Buck, (J5)........ 500
Buffalo, (A1)..... 282

Butler, (B3)...... 100
Byars, (F5)....... 487
Byron, (D1)....... 286
Cache, (C5)....... 317
Caddo, (H6)......1,143
Cade, (J6)........ 40
Calera, (H7)...... 575
Calumet, (D3).... 160
Calvin, (H5)...... 570
Cameron, (L4).... 206
Canadian, (J4).... 481
Caney, (H6)....... 295
Canton, (C2)...... 703
Canute, (B4)...... 250
Capitol Hill, (E4).. 1,400
Capron, (C1)...... 400
Carbon, (J5)...... 490
Carmen, (D1)..... 883
Carnegie, (C4).... 935
Carney, (G3)...... 260
Carrier, (D2)..... 250
Cashion, (E3)..... 289
Castle, (H3)...... 294
Catoosa, (J2)..... 401
Cement, (D5)..... 770
Center, (G5)...... 150
Centrahoma, (H5).. 600
Centralia, (K1)... 387
Ceres, (F2)....... 120
Cestos, (B2)...... 125
Chandler, (G3)...2,024
Chant, (K4)...... 882
Chattanooga, (C6) 471
Checotah, (K4)...1,683
Chelsea, (J1).....1,350
Cherokee, (D1)...2,016
Cheyenne, (A3)... 468
Chickasha, (E4)..10,320
Chilocco, (G1).... 300
Choctaw, (F4).... 242
Choska, (J3)...... 100
Choteau, (K2).... 483

Claremore, (J2)...2,866
Clarksville, (J3)... 388
Clayton, (K5)..... 100
Clearview, (H4)... 100
Cleo, (D2)........ 425
Cleveland, (H2)...1,310
Clinton, (C4).....2,781
Cloud Chief, (C4).. 175
Coalgate, (H6)...3,255
Colbert, (H7)..... 300
Collinsville, (J2)..1,324
Comanche, (E6)...1,301
Connerville, (G6).. 160
Cooperton, (C5)... 76
Copan, (J1)....... 307
Cordell, (C4).....1,950
Cornish, (E6)..... 489
Covington, (E2)... 183
Coweta, (J3).....1,187
Cowlington, (L4).. 378
Coyle, (F3)....... 413
Craig, (J5)....... 200
Crescent, (E3).... 903
Cropper, (E2)..... 120
Cross, (F1)....... 230
Crowder, (J4)..... 529
Cumberland, (G6). 450
Curtis, (B2)...... 150
Cushing, (G3)....1,072
Custer, (C3)...... 854
Damon, (C1)...... 146
Dale, (F4)........ 250
Davenport, (G3).. 394
Davidson, (B6)... 361
Davis, (F5)......1,416
Dawson, (J2)..... 300
Deer Creek, (E1).. 166
Delaware, (K1)... 662
Depew, (H3)...... 300
Devol, (C6)....... 440
Dewey, (J1)......1,344

Dibble, (E4)...... 190
Dill, (B4)........ 240
Dougherty, (F6)... 278
Douglas, (E2)..... 132
Dover, (E3)....... 375
Dow, (J5)......... 900
Doxey, (A4)....... 100
Drummond, (E2)... 425
Duke, (A5)........ 350
Duncan, (D6).....2,477
Durant, (H6).....5,330
Durwood, (F6).... 230
Dustin, (H4)...... 579
Eagle City, (C3)... 160
Earl, (G6)........ 230
Earlsboro, (G4)... 388
Eddy, (E1)........ 200
Edmond, (E3).....2,090
Eldorado, (A6).... 926
Elgin, (D5)....... 178
Elk City, (B4)....3,165
El Reno, (D3)....7,872
Emet, (G6)........ 500
Enid, (E2)......13,799
Enterprise, (K4)... 300
Erick, (A4)....... 915
Erin Springs, (E5). 210
Eufaula, (J4).....1,307
Evans, (J2)....... 100
Fairfax, (G1)..... 819
Fairland, (L1).... 569
Fairmont, (E2).... 210
Fairview, (C2)...2,020
Fallis, (F3)....... 248
Fame, (J4)........ 150
Fanshawe, (L5)... 130
Fargo, (A2)....... 341
Faxon, (C6)....... 215
Fay, (C3)......... 110
Featherston, (K4).. 100
Fitzhugh, (G5).... 175
Fletcher, (D5).... 374

OREGON

Acme, (B4)........ 150
Adams, (G2)...... 205
Airlie, (B3)....... 100
Albany, (C3)..... 4,275
Albee, (G3)....... 100
Alsea, (B3)....... 130
Amity, (B2)....... 407
Antelope, (E3)... 175
Arlington, (E2)... 317
Ashland, (C5)... 5,020
Ashwood, (D3).... 26
Astoria, (B1)... 9,599
Athena, (G2)..... 586
Aumsville, (C3)... 160
Aurora, (C2)..... 190
Austin, (G3)..... 144
Baker, (H3)..... 6,742
Ballston, (B2)... 100
Bandon, (A4)... 1,803
Banks, (B2)...... 300
Barlow, (C2)..... 69
Bay City, (B2)... 281
Beaver Hill, (A4). 149
Beaverton, (C2).. 386
Bend, (D3)....... 536
Bonanza, (D5)... 150
Boume, (G3)...... 77
Bridal Veil, (C2). 200
Brooks, (C2)..... 120
Brownsville, (C3). 919
Buena Vista, (B3). 100
Burns, (F4)...... 904
Butteville, (C2).. 49
Buxton, (B2)..... 200
Camas Valley, (B4) 100
Canby, (C2)...... 587
Canyon City, (G3). 364
Canyonville, (B5). 149
Carlton, (B2).... 386
Carson, (H3).... 200
Cascade Locks,(D2) 280
Cedar Mill, (C2).. 100
Centralpoint, (C5). 761
Champoeg, (C2).. 100
Chemawa, (C3)... 600
Clackamas, (C2).. 100
Clatskanie, (B1).. 747
Clatsop, (B1).... 220
Clifton, (B1).... 130
Cline Falls, (D3).. 30
Coburg, (B3).... 613
Condon, (E2)... 1,009
Copperfield, (J3).. 250
Coquille, (A4).. 1,398
Cornelius, (B2).. 459
Cornucopia, (H2).. 150
Corvallis, (B3).. 4,552
Cottagegrove, (B4).1,834
Cove, (H2)....... 433
Crawfordsville,(C3) 130
Creswell, (B4)... 367

Dallas, (B3)..... 2,124
Damascus, (C2)... 100
Dayton, (B2)..... 453
Dayville, (F3).... 150
Deerhorn, (C3)... 100
Deer Island, (C2).. 100
Deschutes, (D3)... 64
Detroit, (C3).... 100
Dilley, (B2)...... 100
Drain, (B4)...... 335
Drewsey, (G4).... 82
Dufur, (D2)...... 523
Dundee, (C2)..... 196
Durkee, (H3).... 100
Eagle Point, (C5).. 120
Eastside, (A4)... 252
Echo, (F2)....... 400
Eddyville, (B3)... 100
Elgin, (G2)..... 1,120
Elkhead, (B4).... 100
Elkton, (B4)..... 150
Empire, (A4)..... 147
Enterprise, (H2). 1,242
Estacada, (C2)... 405
Eugene, (B3)... 9,009
Fairview, (C2).... 204
Falls City, (B3).. 969
Flora, (H2)...... 100
Florence, (A4).... 311
Forestgrove, (B2). 1,772
Fort Klamath, (C5) 200
Fort Stevens, (A1). 100
Fossil, (E3)...... 421
Freewater, (G2)... 532
Galescreek, (B2).. 120
Gardiner, (A4)... 391
Garibaldi, (B2)... 200
Gaston, (B2)..... 150
Gates, (C3)...... 100
Gervais, (C2).... 276
Glencoe, (C2).... 150
Glendale, (B5)... 646
Goble, (C1)...... 160
Goldbeach, (A5)... 150
Goldhill, (B5).... 423
Granite, (G3).... 89
Grant, (E2)...... 210
Grants Pass, (B5). 3,897
Grass Valley, (E2). 342
Greenhorn, (G3)... 28
Gresham, (C2).... 540
Haines, (G3)..... 423
Halfway, (H3).... 186
Halsey, (B3)..... 337
Hamilton, (F3)... 100
Hammond, (A1)... 100
Hardman, (F2)... 200
Harney, (G4).... 58
Harrisburg, (B3).. 453
Helix, (G2)...... 109
Heppner, (F2).... 880

Hermiston, (F2)... 647
Hilgard, (G3).... 190
Hillsboro, (B2).. 2,016
Hillsdale, (C2)... 100
Hood River, (D2).. 2,331
Houlton, (C2).... 347
Hubbard, (C2)... 283
Huntington, (H3).. 680
Imbler, (G2)..... 300
Independence, (B3) 1,160
Ione, (F2)....... 239
Irrigon, (F2)..... 100
Island City, (H2).. 166
Jacksonville, (B5). 785
Jefferson, (B3)... 415
John Day, (G3)... 258
Jordan Valley, (J5) 150
Joseph, (H2).... 725
Junction City, (B3) 759
Kamela, (G3).... 100
Kent, (E2)....... 150
Kerby, (B5)...... 120
Kings Valley, (B3). 100
Klamath Falls,(D5) 2,758
Knappa, (B1).... 180
Lafayette, (B2)... 412
La Grande, (G2). 4,843
Laidlaw, (D3).... 49
Lakeview, (E5)... 1,253
Langlois, (A5).... 150
Latourell Falls,
 (C2)............ 100
Lebanon, (C3)... 1,820
Leland, (B5)..... 110
Lents, (C2)...... 500
Lexington, (F2)... 185
Linkville, (D5)... 350
Linnton, (C2)... 1,165
Lonerock, (F2)... 70
Longcreek, (G3).. 86
Lostine, (H2).... 230
Lyons, (C3)..... 150
Lytle, (D3)...... 162
McCoy, (B2).... 100
McEwen, (G3)... 110
McMinnville, (B2). 2,400
Madras, (D3).... 364
Marcola, (C3).... 200
Marion, (B3).... 120
Marshfield, (A4).. 2,980
Mayger, (B1).... 110
Mayville, (E2).... 150
Meacham, (G2)... 100
Medford, (C5)... 8,840
Mehama, (C3)... 100
Melrose, (B4).... 100
Merlin, (B5)..... 100
Merrill, (D5)..... 200
Mill City, (C3)... 500
Milton, (G2).... 1,280
Milwaukee, (C2).. 860

Mist, (B2)....... 250
Mitchell, (E3).... 210
Monmouth, (B3).. 493
Monroe, (B3).... 340
Montavilla, (C2).. 780
Monument, (F3).. 119
Mora, (E2)...... 378
Mosier, (D2)..... 150
Mount Angel, (C2) 545
Mount Tabor, (C2) 1,500
Myrtlecreek, (B4). 429
Myrtlepoint, (A4). 836
Needy, (C2)...... 100
Nehalem, (B2)... 119
New Astoria, (B1). 957
Newberg, (B2)... 2,260
New Era, (C2)... 200
New Pine Creek,
 (E5)............ 200
Newport, (A3).... 721
North Bend, (A4).. 2,078
North Powder, (G2) 455
North Yamhill, (B2) 450
Nyssa, (H4)..... 449
Oakland, (B4).... 467
Olney, (B1)...... 180
Ontario, (H3)... 1,248
Oregon City, (C2). 4,287
Orient, (C2)..... 300
Oswego, (C2)... 1,107
Paisley, (E5)..... 200
Parkersburg, (A4). 100
Parkplace, (C2)... 200
Pendleton, (G2). 4,460
Peoria, (B3)..... 100
Perry, (G2)...... 300
Perrydale, (B3)... 100
Philomath, (B3)... 505
Phoenix, (C5).... 250
Pilotrock, (G2)... 197
Pine, (H3)....... 150
Portland, (C2).. 207,214
Port Orford, (A5).. 100
Prairie City, (G3). 348
Prineville, (E3)... 1,042
Prosper, (A4).... 250
Rainier, (C1)... 1,359
Redland, (H3).... 334
Rickreall, (B3)... 130
Riddle, (B5)..... 187
Roseburg, (B4)... 4,738
Rosland, (D4).... 126
Rufus, (E2)...... 100
Saint Helen, (C2). 742
Saint Johns, (C2). 4,872
Saint Paul, (C2).. 103
SALEM, (C3)... 14,094
Scappoose, (C2).. 200
Scholls, (C2).... 150
Scio, (C3)....... 295
Scottsburg, (B4).. 200

Scotts Mills, (C2).. 250
Seaside, (B2)... 1,121
Shaniko, (E3).... 495
Shedds, (B3).... 120
Sheridan, (B2)... 1,021
Sherwood, (C2)... 115
Silets, (B3)..... 100
Silverlake, (E4)... 150
Silverton, (C3).. 1,588
Sodaville, (C3)... 110
Springfield, (C3). 1,838
Stanfield, (F2)... 318
Stayton, (C3).... 703
Sublimity, (C3)... 138
Summerville, (G2). 237
Sumpter, (G3).... 643
Sweet Home, (C3). 202
Tangent, (B3).... 130
The Dalles, (D2). 4,880
Thurston, (C3)... 100
Tillamook, (B2).. 1,352
Toledo, (B3).... 541
Troutdale, (C2)... 309
Tualitin, (C2).... 120
Turner, (B3).... 191
Tygh Valley,
 (D2)............ 150
Ukiah, (G2)..... 110
Umatilla, (F2)... 198
Union, (H2).... 1,483
Vale, (H4)...... 992
Vernonia, (B2)... 69
Waldo, (B5)..... 100
Wallowa, (H2)... 793
Wamic, (D2).... 100
Warren, (C2).... 100
Warrendale, (D2). 150
Warrenton, (A1).. 339
Wasco, (E2).... 386
Waterloo, (C3)... 83
Wedderburn, (A5). 100
Wendling, (C3)... 300
Westfall, (H4)... 140
Weston, (G2).... 499
Westport, (B1)... 150
West Portland,
 (C2)............ 100
West Seaside, (B2). 149
Whiteson, (B2)... 100
Whitney, (G3).... 55
Wilbur, (B4).... 150
Wilderville, (B5).. 100
Willamette, (C2).. 317
Willamina, (B2).. 376
Woodburn, (C2). 1,616
Woods, (A2).... 100
Woodville, (B5)... 110
Yamhill, (B2).... 325
Yankton, (C2)... 130
Yaquina, (A3).... 300
Yoncalla, (B4).... 233

PENNSYLVANIA

Aaronsburg, (G6).. 300
Abbottstown, (H10). 400
Abington, (L9)... 600
Abrams, (L9).... 150
Academia, (G7)... 240
Academy Corners,
 (G2)............ 150
Ackermanville,(L6). 350
Acosta, (C9).... 200
Adah, (B10)..... 500
Adamsburg, (B8).. 300
Adamstown, (J9).. 675
Adamsville, (A3).. 200
Addingham, (L10). 100
Addison, (C11).. 350
Adelaide, (B9)... 500
Admire, (H10).... 150

Adrian, (B6)..... 100
Adrian Mines (D5). 210
Airville, (J10).... 100
Airydale, (F8).... 100
Aitch, (E8)...... 200
Akron, (J9)..... 719
Alba, (H3)...... 150
Albion, (A2).... 1,534
Alburtis, (K7)... 900
Aldan, (L10).... 661
Alden Station, (K5). 500
Aldenville, (L3)... 120
Alderson, (J4).... 600
Aleppo, (A10).... 100
Alexandria, (E7)... 432
Aliquippa, (A7).. 1,743
Allemans, (E7)... 100

Allen, (G9)...... 350
Allenport, (B9).. 1,000
Allens Mills, (D5).. 203
Allensville, (F7)... 290
Allentown, (K7).. 51,913
Allenwood, (H5)... 310
Allison Park, (B7).. 350
Allport, (E6)..... 203
Almedia, (J5).... 200
Altoona, (E7)... 52,127
Alum Bank, (D9).. 320
Alverton, (B9)... 600
Ambersons Valley,
 (F9)............ 450
Ambler, (L9)... 2,649
Ambridge, (A7).. 5,205
Amity, (A9)...... 250

Amsbry, (D7)..... 260
Andalusia, (M9)... 500
Anita, (C5).... 2,750
Annville, (H8)... 2,400
Ansonville, (D6).. 450
Antes Fort, (G5).. 330
Antrim, (G3).... 1,500
Apollo, (B7)... 3,006
Applewold, (B6)... 300
Aquashicola, (K6).. 300
Aram, (G10).... 125
Ararat, (K2).... 500
Arcadia (D6).... 1,400
Archbald, (L4)... 7,194
Ardara, (B8)..... 300
Arden, (A9)..... 400
Ardmore, (L10). 3,700

Arendtsville, (G10).. 383
Argentine, (B5)... 500
Ariel, (L4)...... 200
Aristes, (J6)..... 400
Armagh, (C8).... 82
Armbrust, (B9)... 225
Arnold, (B7)... 1,818
Arnot, (G3).... 2,750
Arona, (B8).... 683
Arrow, (D9)..... 300
Arroyo, (D4).... 200
Artemas, (E11)... 100
Ashbourne, (L9).. 650
Asherton, (H6)... 200
Ashfield, (K6).... 100
Ashland, (J6)... 6,855
Ashley, (K5).... 5,601

Lower Providence,
(L9)............ 220
Lowhill, (K7)...... 140
Lowville, (B1)...... 130
Loyalhanna, (C8)... 600
Loyalsock, (H5).... 850
Loysburg, (E9)..... 310
Loysville, (G8)..... 360
Lucesco, (B7)...... 100
Lucinda, (C4)...... 300
Lucinda Furnace,
(C4............ 100
Ludlow, (D3)....... 700
Lumber City, (D6).. 363
Lumberville, (L8)... 250
Lundys Lane, (A2).. 400
Lungerville, (H4)... 140
Luthersburg, (D5).. 375
Luxor, (C8)........ 200
Luzerne, (K4).....5,426
Lycippus, (C9)..... 120
Lykens, (H7).....2,943
Lyndora, (B6)....2,000
Lynnport, (K7)..... 150
Lyon Station, (K8).. 540
McAdoo, (J6).....3,389
McAlevys Fort, (F7) 225
McAlisterville,(G7). 560
McCance, (C8)..... 350
McClellandtown,
(B10)........... 800
McClure, (G7)...... 350
McConnellsburg,
(E10)........... 579
McConnells Mills,
(A8)........... 180
McConnellstown,
(E8)........... 310
McCoysville, (F8)... 140
McCracken, (A10).. 100
McDonald, (A8)...2,543
McElhattan, (G5)... 250
McEwensville,(H5).. 209
McGees Mills, (D6). 250
McGovern, (A9).... 150
McGrann, (C6)..... 200
McGraw, (C3)...... 150
McKean, (A2)...... 350
McKeansburg, (K7). 320
McKee, (E8)....... 180
McKees Half Falls,
(H7)........... 250
McKeesport, (B8).42,694
McKees Rocks,
(A8)........14,702
McKinley, (L9)..... 600
McKnightstown,
(G10)........... 130
McLallan Corners,
(A2)........... 100
McMahan, (B9).... 813
McMurray, (A8)... 100
McSherrystown,
(G10)........1,724
McSparran, (J11).. 200
McVeytown, (F8).. 514
Macdonaldton,
(D10)........1,000
Mackeyville, (G5)... 170
Macungie, (K7).... 772
Madeline, (D6)..... 300
Madera, (F6).....1,200
Madison, (B9)...... 421
Madisonburg, (F6).. 170
Mahaffey, (D6)..... 754
Mahanoy City,
(J6)........15,936
Mahanoy Plane,
(J6)........2,000
Mahoning, (C6).... 200
Maiden Creek, (K8). 100
Maiden Creek Station,
(K8)........... 100
Mainesburg, (H2).. 250

Mainville, (J6)..... 625
Maitland, (G7)..... 150
Maltby, (K4).....1,500
Malvern, (L9).....1,125
Mammoth, (C9)..1,000
Manada Hill, (H8).. 150
Manchester, (H9).. 347
Manheim, (J9)...2,302
Manns Choice, (D9) 341
Manoa, (L10)...... 500
Manor, (B8)......1,039
Manorville, (C6)... 545
Mansfield, (G2)...1,645
Mapleton Depot,
(F8)........... 752
Mapletown, (B10). 130
Maplewood, (L4)... 200
Marburg, (H10)... 150
Marchand, (C6)... 110
Marcushook, (L10).1,573
Marguerite, (C9)... 500
Marianna, (A9)...1,363
Marienville, (C4)...1,300
Marietta, (H9)...2,079
Marion, (F10)..... 260
Marion Center, (C6) 366
Marion Hts., (J6)..1,362
Marlstown, (G8)... 250
Markes, (F10)..... 180
Markham, (K10).. 130
Marklesburg, (E8).. 211
Markleysburg,
(B11)........... 227
Marple, (L10)..... 110
Mars, (A7).......1,215
Marsh, (K9)....... 230
Marshalls Creek,
(L5)........... 230
Marshallton, (K10).. 500
Marshwood, (K4).. 350
Marsteller, (D7)...1,000
Martha Furnace,
(E6)........... 170
Martin, (B10)..... 800
Martindale, (J9)... 140
Martinsburg, (E8). 920
Martins Corner,(K9) 100
Martins Creek, (L6). 300
Marvindale, (D3)... 100
Marysville, (H8)..1,693
Masontown, (B10).. 890
Masten, (H3)...... 300
Mastersonville, (J9). 150
Masthope, (L3).... 170
Matamoras, (M4)..1,388
Mattawana, (F8)... 320
Mauch Chunk,
(K6)........3,952
Maud, (M9)....... 240
Mausdale, (H6).... 100
Maxatawny, (K7).. 200
Mayburg, (C3).... 200
Mayfield, (L3)....3,662
Maytown, (H9).... 725
Mazeppa, (G6).... 140
Meadowbrook, (J6). 250
Meadow Lands,(A9) 500
Meadville, (A3)...12,780
Mechanic Grove,
(J10)........... 150
Mechanicsburg,
(H9)........4,469
Mechanicsburg,(C7) 159
Mechanics Valley,
(L8)........... 130
Mechanicsville,(M8) 200
Media, (L10).....3,562
Medix Run, (E4)... 400
Mehoopany, (J3).. 600
Melrose Park, (L9).. 200
Mench, (F10)..... 150
Menges Mills, (H10) 120
Mercer, (A5).....2,026

Mercersburg, (F10).1,410
Meredith, (C6).... 100
Merion Sta., (L9)... 600
Merionville, (L9)... 150
Mertztown, (K8)... 700
Meshoppen, (J3)... 630
Messmore, (B10)... 200
Mexico, (G7)...... 220
Meyersdale, (D10)..3,741
Middleboro, (A2)... 207
Middleburg, (G6).. 521
Middlebury Center,
(G2)........... 130
Middlecreek, (G6).. 140
Middleport, (J7)...1,100
Middlesex, (C9)... 130
Middle Spring, (F9). 160
Middletown, (H9)..5,374
Midland, (A7).....1,244
Midway, (A8)..... 941
Mifflin, (G7)...... 850
Mifflinburg, (G6)..1,559
Mifflintown, (G7).. 954
Mifflinville, (J5)... 600
Milan, (H2)....... 130
Milanville, (L3).... 200
Mildred, (J4)...... 400
Milerun, (H6)..... 110
Milesburg, (F6)... 531
Milford, (M4)..... 872
Milford Square, (L8) 200
Milford Sta., (C10). 100
Millbourne, (L10).. 322
Mill City, (K4).... 350
Mill Creek, (F8)... 308
Millersburg, (H7)..2,394
Millersburg, (J8).. 600
Millers Sta., (A2)... 100
Millerstown, (B6).. 993
Millerstown, (G7).. 549
Millersville, (J10)..1,350
Millerton, (H2).... 450
Millgrove, (J6).... 400
Millhall, (F5).....1,043
Millheim, (F6).... 626
Millmont, (G6).... 190
Millport, (E2)..... 150
Millrift, (M4)..... 150
Mill Run, (C10)... 200
Mills, (F2)........ 400
Millsboro, (A10)...1,060
Millstone, (C4).... 200
Millvale, (B8)....7,861
Mill Village, (B2).. 290
Millville, (H5).... 611
Millway, (J9)..... 175
Millwood, (C8).... 400
Milroy, (F7).......1,500
Milton, (H5).....7,460
Milton Grove, (J9).. 200
Milwaukee, (K4).. 160
Mina, (E3)....... 300
Mineral Point, (D8). 250
Miners Mills, (K5).3,159
Minersville, (J7)...7,240
Mines, (E8)....... 200
Minooka, (K4)...3,000
Mitchells Mills, (D7) 200
Modena, (K10)... 130
Mohnton, (K8)...1,536
Mohrsville, (J8)... 240
Mollenauer, (A8)... 800
Monaca, (A7).....3,376
Monessen, (B9)..11,775
Monocacy, (K9)... 260
Monongahela,
(B9)........7,598
Monroe, (J3)..... 403
Monroe, (B5)..... 150
Monroeton, (T3)... 500
Mont Alto, (F10).. 675
Montandon, (H6).. 600
Mont Clare, (T9).. 710
Monterey, (K7)... 180

Montgomery, (H5).1,490
Montgomery Ferry,
(G7)........... 130
Montoursville,(G4) 1,904
Montrose, (B7).... 200
Montrose, (K2)...1,914
Moon Run, (A8)...1,000
Moores, (L10).....1,400
Moorheadville,
(B1)........... 300
Moosic, (K4).....3,964
Mooween, (C7).... 200
Morann, (E6).....1,000
Moravia, (A6)..... 100
Morea Colliery, (J6) 830
Moredale, (G9).... 150
Moreland, (H5)... 210
Morgan, (A8).....3,000
Morgans Hill, (L6).. 100
Morgans Sta., (B9). 250
Morgantown, (K9). 200
Morganza, (A8)...1,000
Morris, (G3)...... 700
Morrisdale,
(E6)........... 670
Morris Run, (H3)..1,600
Morrisville, (M9)...2,002
Morstein, (K9).... 150
Morton, (L10)...1,071
Mortonville, (K10).. 130
Morwood, (L8).... 225
Moscow, (K4)..... 650
Moselem, (K7).... 200
Mosgrove, (C6).... 200
Mosiertown, (A2).. 200
Mount Aetna, (J8).. 330
Mountaindale, (E7).. 350
Mountain Eagle,
(F6)........... 150
Mountainhome, (L5) 200
Mountaintop, (K5) 1,200
Mountainville, (L7). 600
Mount Alton, (D2). 300
Mount Bethel, (L6). 400
Mount Braddock,
(B10)........1,000
Mount Carmel, (J7). 335
Mount Carmel,
(J6)........17,532
Mount Holly Springs,
(G9)........1,272
Mount Hope, (J9).. 100
Mount Jackson,
(A6)........... 400
Mount Jewett,(D3) 1,771
Mount Joy, (H9)..2,166
Mount Lebanon,(A8) 500
Mount Morris, (A11) 382
Mount Nebo, (J10).. 120
Mount Oliver, (A8).4,241
Mount Penn, (K8). 785
Mt. Pleasant, (B9)..5,812
Mt. Pleasant Mills,
(G7)........... 200
Mount Pocono,(L5) 500
Mountrock, (G9).. 140
Mt. Sterling, (B10). 500
Mount Union, (F8) 3,338
Mountville, (J9)... 803
Mount Wolf, (H8).. 200
Mount Zion, (J8)... 300
Moyer, (B9)....... 600
Muddy Creek Forks,
(J10)........... 200
Mummasburg, (G10) 110
Muncy, (H5).....1,904
Muncy Valley, (H4). 300
Munhall, (B8)....5,185
Munson Sta., (E6). 500
Murdocksville, (A8) 125
Murrell, (J9)...... 250
Murrysville, (B8).. 250
Mustard, (B9)..... 300

Mutual, (B9)...... 800
Myerstown, (J8)..2,000
Myra, (D8)....... 400
Nansen, (D3)..... 300
Nanticoke, (K5)..18,877
Nantmeal Village,
(K9)........... 200
Nanty Glo, (D8)..1,500
Narberth, (L9)...1,790
Nasby, (B1)....... 60
Nashville, (H10)... 150
Natalie, (H6)...... 250
Natrona, (B7)....4,000
Nauvoo, (G3).... 200
Nazareth, (L7)...3,978
Nebraska, (C4).... 250
Neffs, (K7)....... 200
Neffs Mills, (E7).. 110
Neffsville, (J9).... 400
Negley, (B8)...... 125
Nelson, (G2)...... 448
Nescopeck, (J5)...1,578
Neshaminy, (L9).. 100
Neshannock, (A5).. 300
Neshannock Falls,
(A5)........... 150
Nesquehoning, (K6)2,000
Nettleton, (D7).... 250
Nevilton, (A7)..... 300
New Albany, (J3)... 413
New Alexandria,
(C8)........... 505
New Baltimore,
(D10)........... 177
New Bedford, (A5).. 200
New Berlin, (G6)... 527
New Berlinville,(K8) 300
Newberrytown,(H9) 300
New Bethlehem,
(C5)........1,625
New Bloomfield,
(G8)........... 762
New Boston, (J5).. 650
New Bridgeville,
(J10)........... 200
New Brighton,
(A6)........8,329
New Britain, (L8).. 120
New Buenavista,
(D9)........... 180
New Buffalo, (H8).. 135
Newburg, (D6).... 274
Newburg, (F9).... 264
Newcastle, (A5)..36,280
Newcastle, (J6).... 600
New Centerville,
(L9)........... 110
New Centerville,
(C10)........... 106
Newchester, (G10). 250
New Columbia,(H5) 400
New Columbus, (J5) 175
Newcomer, (B10).1,000
New Cumberland,
(H9)........1,472
New Danville, (J10) 120
New Derry, (C8)... 350
New Eagle, (B9)... 400
Newell, (B9)...... 400
New England, (B8). 450
New Enterprise,(E9) 325
New Era, (J3)..... 150
New Florence, (C8). 717
Newfoundland, (L4). 900
New Franklin, (F10) 130
New Freedom, (H11) 726
New Freeport, (A10) 200
New Galilee, (A6).. 453
New Geneva, (B10) 290
New Germantown,
(F8)........... 200
New Grenada, (F9). 100
New Hamburg, (A4) 100
New Hanover, (K8). 510

York Furnace, (J10). 100
Yorkhaven, (H9).... 793
York New Salem, (H10)........... 230

York Run, (B10).... 100
York Springs, (G9)............ 299
Youngs, (L6)....... 100

Youngstown, (C8)... 323
Youngsville, (C2).. 1,406
Youngwood, (B9)..1,881
Yukon, (B9).......1,000

Zehner, (K5)....... 200
Zelienople, (A6)....1,388
Zerbe, (J7)........ 300
Zieglerville, (L9).... 400

Zion, (F6)......... 130
Zion Grove, (J6).... 200
Zollarsville, (A10)... 100
Zora, (G11)........ 100

RHODE ISLAND

Abbottrun, (L5).... 250
Adamsville, (M6)... 300
Albion, (L5)....... 450
Allenton, (L6)..... 700
Alton, (K7)........ 100
Anthony, (K6).....1,500
Apponaug, (L6)....1,300
Arcadia, (K6)...... 100
Arctic, (K6).......3,100
Arnold Mills, (L5)- . 300
Ashaway, (J6).....1,200
Ashland, (K5)...... 280
Ashton, (L5)......1,400
Austin, (K6)....... 100
Barrington Center, (L6)............. 520
Block Island, (K8). 1,200
Bradford, (K7) ... 150
Bridgeton, (K5).... 600
Bristol, (L6) 8,450
Bristol Ferry, (L5).. 100
Canonchet, (J7)... 100
Carolina, (K7)..... 300
Centerdale, (K5)..1,800
Centerville, (K6)..2,800
Central Falls, (L5).22,754
Charlestown, (K7).. 240
Chepachet, (K5)..1,200
Clayville, (K5)..... 300

Conimicut, (L6).... 200
Coventry, (K6)..... 450
Cranston, (L5)...21,171
Crompton, (K6)...1,800
Cumberland Hill, (L5)............. 150
Davisville, (L6).... 580
Diamond Hill, (L5). 250
E. Greenwich, (L6).3,300
E. Providence, (L5)........... 13,500
E. Providence Center, (L5)............. 200
Escoheag, (K6).... 100
Esmond, (L5)...... 500
Exeter, (K6)....... 250
Forestdale, (K5)... 400
Foster, (K5)....... 350
Foster Center, (K5). 600
Georgiaville, (K5)..1,000
Glendale, (K5)..... 370
Gould, (K7)....... 100
Grants Mills, (L4).. 300
Greene, (J6)....... 460
Greenville, (K5)... 700
Greystone, (K5)... 800
Hamilton, (L6).... 600
Harrisville, (K5)..2,200
Hillsgrove, (L6) ..1,200

Hope, (K6).......1,250
Hope Valley, (K7).. 850
Hopkinton, (J7)... 250
Hughesdale, (K5).. 450
Island Park, (M6). 225
Jamestown, (L7)..1,100
Kent, (K3)........ 170
Kenyon, (K7)...... 150
Kingston, (K7).... 450
La Fayette, (L6)... 350
Lakewood, (L5).... 650
Liberty, (K6)...... 100
Limerock, (L5).... 250
Little Compton, (M6)900
Longmeadow, (L6).. 350
Lonsdale, (L5)....1,900
Lymansville, (L5).. 700
Manville, (K5)....2,100
Mapleville, (K5)... 610
Middletown, (L5)..1,600
Narragansett Pier, (L7)...........1,250
Nasonville, (K5)... 550
Natick, (K6)......5,000
Nayatt Pt., (L6)... 650
Newport, (L6)...27,149
Nooseneck Hill,(K6) 100
No. Scituate, (K5)..1,050
No. Tiverton, (M6).1,650

Norwood, (L5)..... 650
Oakland, (K5)..... 490
Oakland Beach,(L6). 500
Pascoag, (K5).....3,000
Pawtucket, (L5).51,622
Peace Dale, (L7)...1,550
Phenix, (K6)......2,400
Phillipsdale, (L5).. 700
Pt. Judith, (L7)... 200
Pontiac, (L6).....1,900
Portsmouth, (M6)..2,400
Potter Hill, (J7).... 250
PROVIDENCE, (L5)..........224,326
Prudence, (L6)..... 200
River Point, (K6)..3,500
Riverside, (L5)....2,000
Rockland, (K5).... 400
Rockville, (J6)..... 250
Rumford, (L5)..... 500
Saundertown, (K5). 200
Saylesville, (L5)... 800
Shannock, (K7)... 100
Shawomet, (L6)... 100
Slatersville, (K5) .1,850
Slocum, (K6)...... 250
S. s F at r, (K5)... 100
So. Portsmouth, (M6)........... 100

So. Scituate, (K5)... 710
Stillwater, (K5).... 150
Summit, (K6)..... 220
Tarkiln, (K5)...... 190
Thornton, (K5).... 700
Tiverton, (M6)....1,950
Tiverton Four Cor-
ners, (M6)...... 325
Usquepaugh, (K7).. 100
Valley Falls, (L5)..3,400
Wakefield, (L7)...2,750
Warren, (L6).....6,450
Warwick, (L6).... 350
Washington, (K6).. 450
Watch Hill, (J7)............. 100
W. Barrington, (L6). 7.3
Westerly, (J7).....8,100
W. Gloucester, (J5) . 100
W. Greenwich Cen-
ter, (K6)....... 200
W. Kingston, (K7).. 100
White Rock, (J7)... 150
Wickford, (L6)....1,050
Wood River Jc., (K7)............ 150
Woodville, (K7)... 200
Woonsocket, (L5).38,125
Wyoming, (K6).... 200

SOUTH CAROLINA

Abbeville, (C3) ...4,459
Adamsburg, (D2).. 100
Adamsrun, (G6).... 200
Aiken, (D4)......3,911
Alcolu, (G4)....... 458
Allen, (J4)........ 200
Allendale, (E5) ...1,453
Almeda, (F6)...... 500
Alston, (E3)....... 110
Anderson, (B2)...9,654
Antreville, (C3) ... 100
Appleton, (E5).... 156
Arcadia, (D2) 300
Ariel, (J3)........ 500
Arkwright, (D2)... 500
Autun, (B2)....... 150
Bamberg, (E5)....1,937
Barnwell, (E5) ...1,324
Batesburg, (E4) ...1,995
Batesville, (C2)... 150
Bath, (D5)........ 500
Beaufort, (F7)....2,486
Belton, (C2)......1,652
Bennettsville, (H2)2,646
Bessie, (C2)....... 250
Bethune, (G3).... 317
Bingham, (J3).... 196
Bishopville, (G3)..1,659
Blacksburg, (E1)..1,119
Blackstock, (E2)... 192
Blackville, (E5)...1,278
Blaney, (F3)...... 100
Blenheim, (H2).... 228
Bluffton, (F7)..... 577
Blythewood, (F3).. 100
Bolen, (F4)....... 160
Bonneau, (H5).... 100
Bordeaux, (C4).... 200
Bowman, (F5).... 327
Bradley, (C3)..... 279
Branchville, (F5)..1,471
Brandon, (C2).... 300
Brighton, (E6).... 190

Brightsville, (H2)... 100
Brunson, (E6)..... 610
Bucklick, (F3)..... 400
Buffalo, (D2)...... 200
Burton, (F7)...... 100
Cades, (H4)....... 100
Calhoun, (B2)..... 215
Calhoun Falls, (B3) 296
Camden, (F3)....3,569
Cameron, (F4).... 421
Campobello, (C1).. 255
Carlisle, (E2)..... 367
Cartersville, (H3).. 320
Cateechee, (B2)... 300
Cayce, (E4)....... 180
Cedarspring, (D2). 150
Central, (B2)..... 886
Chapin, (E3)...... 345
Chappells, (D3)... 150
Charleston, (H6)..58,833
Cheraw, (E2)....2,873
Cherokee, (D1)... 150
Cherokee Falls, (D1) 591
Chester, (E2).....4,754
Chesterfield, (G2).. 618
Clarks Hill, (C4).. 100
Clifton, (D2).....1,700
Clinton, (D3).....3,272
Clio, (H2)........ 780
Clover, (E1)......1,207
Cokesbury, (C3)... 756
Colemans, (D3)... 150
Colliers, (C4)..... 150
COLUMBIA, (F4)26,319
Congaree, (F4).... 100
Connors, (G5).... 100
Converse, (D2)...1,000
Conway, (J4).....1,228
Cope, (F5)........ 174
Cordesville, (H5).. 130
Coronaca, (C3).... 199
Cottageville, (G6). 418
Cowpens, (D2)...1,101

Crocketville, (E6).. 200
Cromer, (D3)..... 100
Cross Anchor, (D2) 200
Crosshill, (C3).... 558
Crosskeys, (D2)... 120
Dantzler, (F5).... 130
Darlington, (H3)..3,789
Denmark, (E5)...1,075
Dillon, (J3)......1,757
Donalds, (C3)..... 268
Dovesville, (H3)... 169
Drayton, (D2).... 200
Duewest, (C3)... 672
Dunbar, (H2)..... 150
Duncan, (C2)..... 190
Dunklin, (C2).... 200
Earlybranch, (F6)... 160
Easley, (B2).....2,981
Eastover, (F4) ... 217
Eau Claire, (F4)..1,234
Ebenezer, (H5)... 420
Ebenezer, (E2).... 190
Edgefield, (D4)...1,771
Edgmoor, (F2).... 113
Edisto Island, (G6) 620
Ehrhardt, (E5).... 315
Elko, (E5)........ 114
Ellenton, (D5).... 367
Elliott, (G3)...... 500
Elloree, (F4)...... 540
Enoree, (D2).....1,000
Enterprise, (G6)... 200
Estill, (E6)....... 460
Ethel, (G6)....... 100
Eutawville, (G5)... 405
Exchange, (G6)... 300
Fairfax, (E6)...... 499
Fairforest, (C2)... 300
Fairmont, (C2).... 100
Fair Play, (A2).... 100
Ferguson, (G5).... 100
Field, (H2)....... 100
Fingerville, (C1)... 300

Florence, (H3)....7,057
Foreston, (G4).... 115
York, (J3)........ 134
Yorksboro, (C2)... 360
Fort Lawn, (F2)... 204
Fort Mill, (F2)...1,616
Fort Motte, (F4).. 392
Fortner, (B1)..... 200
Fountain Inn, (C2) 979
Frogmore, (F7)... 100
Furman, (E6)..... 100
Gaffney, (D1).....4,767
Gaston, (E4)..... 150
Georgetown, (J3).5,530
Gilbert, (E4)..... 450
Glendale, (D2)... 800
Glenn Springs, (D2) 178
Godhold, (J4).... 100
Goldville, (D3).... 210
Golightly, (D2)... 300
Gourdin, (H3).... 200
Govan, (E5)...... 111
Gowensville, (C1). 100
Graham ville, (F7).. 300
Graniteville, (D4)2,500
Graycourt, (C2)... 284
Great Falls, (F2)... 200
Greeleyville, (G4).. 630
Greenpond, (F6).. 360
Greenville, (C2)..15,741
Greenwood, (C3)..6,614
Greer, (C2)......1,673
Grover, (F5)...... 67
Gurley, (K3)..... 100
Hagood, (F3)..... 300
Hamburg, (D5)... 490
Hamer, (J3)...... 150
Hampton, (E6)... 748
Hardeeville, (F7).. 650
Harleyville, (G5)... 190
Harmon, (D3).... 150
Harper, (H5)..... 100
Hartsville, (G3)...2,365

Haskell, (J4)...... 100
Heath Spring, (F2) 452
Hebron, (H4).... 100
Helena, (D3)..... 425
Hendersonville,(F6) 319
Hickory Grove,(E1) 283
Highland, (C1)... 200
Hodges, (C2)..... 266
Hollyhill, (G5).... 342
Honeapath, (C3)..1,763
Hunterville, (C2).. 250
Inman, (C1)...... 474
Iona, (E5)........ 207
Iva, (B3)......... 804
Jackson, (D5).... 150
Jacksonboro, (G6). 54
James Island, (H6) 250
Jenburg, (D3).... 100
Jefferson, (G2)... 300
Johns Island, (G6) 1,000
Johnston, (D4)... 943
Jonesville, (D2)... 965
Kathwood, (D5)... 200
Jordon, (K4)..... 300
Kathwood, (D5)... 200
Jordon, (K4)..... 300
Kelton, (C2)...... 92
Kemper, (J3)..... 63
Kershaw, (F2).... 682
Killian, (F3)...... 110
Kings Creek, (E1).. 100
Kingstree, (H4)...1,372
Kline, (E5)....... 199
Ladies Island, (F7). 100
Lake City, (H4)...1,074
Lamar, (G3)...... 592
Lancaster, (F2)...2,098
Lando, (E2)...... 200
Landrum, (C1)... 449
Lane, (H4)....... 100
Langley, (D5)....1,000
Latta, (J3).......1,358
Laurel, (J4)...... 200
Laurens, (C3)....4,818
Leeds, (E2)....... 350

York Furnace, (J10). 100
Yorkhaven, (H9).... 793
York New Salem,
(H10)........... 230

York Run, (B10)... 100
York Springs,
(G9)........... 299
Youngs, (L6)....... 100

Youngstown, (C8)... 323
Youngsville, (C2).. 1,406
Youngwood, (B9)..1,881
Yukon, (B9).......1,000

Zehner, (K5)........ 200
Zelienople, (A6)....1,388
Zerbe, (J7)......... 300
Zieglerville, (L9).... 400

Zion, (F6).......... 130
Zion Grove, (J6)... 200
Zollarsville, (A10).. 100
Zora, (G11)........ 100

RHODE ISLAND

Abbottrun, (L5).... 250
Adamsville, (M6)... 300
Albion, (L5)....... 450
Allenton, (L6)..... 700
Alton, (K7)....... 100
Anthony, (K6).....1,500
Appomaug, (L6)...1,300
Arcadia, (K6)..... 450
Arctic, (K6).....3,100
Arnold Mills, (L5).. 300
Ashaway, (J6)...1,200
Ashland, (K5)..... 280
Ashton, (L5).....1,400
Austin, (K6)....... 100
Barrington Center,
(L6)........... 520
Block Island, (K8). 1,200
Bradford, (K7)... 150
Bridgeton, (K5)... 600
Bristol, (L6).....8,450
Bristol Ferry, (L6).. 100
Canonchet, (J7)... 100
Carolina, (K7)... 300
Centerdale, (K5)..1,800
Centerville, (K6)..2,800
Central Falls,(L5). 22,754
Charlestown, (K7).. 240
Chepachet, (K5)..1,200
Clayville, (K5)..... 300

Conimicut, (L6).... 200
Coventry, (K6).... 450
Cranston, (L5)...21,171
Crompton, (K6)...1,800
Cumberland Hill,
(L5)........... 150
Davisville, (L6).... 580
Diamond Hill, (L5). 250
E. Greenwich, (L6).3,300
E. Providence,
(L5).......... 13,500
E. Providence Center,
(L5)........... 200
Esmond, (K6)..... 100
Esmond, (L5)..... 500
Exeter, (K6)..... 250
Forestdale, (K5)... 400
Foster, (K5)..... 350
Foster Center (K5). 600
Georgiaville, (K5)..1,000
Glendale, (K5)... 370
Gould, (K7)..... 100
Grants Mills, (L6).. 300
Greene, (J6)..... 460
Greenville, (K5)... 700
Greystone, (K5)... 800
Hamilton, (L6)... 600
Harrisville, (K5)..2,200
Hillsgrove, (L6).... 300

Hope, (K6).......1,250
Hope Valley, (K7).. 850
Hopkinton, (J7)... 250
Hughsdale, (K5).. 450
Island Park, (M6). 225
Jamestown, (L7)..1,100
Kent, (K5)...... 170
Kenyon, (K7)... 150
Kingston, (K7).. 450
La Fayette, (L6).. 350
Lakewood, (L5).. 650
Liberty, (K6)... 100
Limerock, (L5)... 230
Little Compton,(M6)900
Longmeadow, (L6). 350
Lonsdale, (L5)...1,800
Lymansville, (L5).. 700
Manville, (K5)...2,100
Mapleville, (K5).. 610
Middletown, (L6).1,600
Narragansett Pier,
(L7)..........1,250
Nasonville, (K5).. 550
Natick, (K6)....5,000
Naxatt Pt., (L6).. 650
Newport, (L6)..27,149
Nonquit Hill,(K6) 100
No. Scituate, (K5) 1,050
No. Tiverton, (M6)1,050

Norwood, (L5).... 650
Oakland, (K5).... 490
Oakland Beach,(L6) 500
Pascoag, (K5)....5,000
Pawtucket, (L5).. 51,622
Peace Dale, (L7)..1,550
Phenix, (K6).....2,400
Phillipsdale, (L5).. 700
Pt. Judith, (L7).. 200
Pontiac, (L6).....1,900
Portsmouth, (M6).2,400
Potter Hill, (J7).. 250
PROVIDENCE,
(L5).........224,326
Prudence, (L6)... 200
River Point, (K6).3,500
Riverside, (L5)...2,000
Rockland, (K5).. 400
Rockville, (J6)... 250
Rumford, (L5)... 500
Saunderstown, (L6). 200
Saylesville, (L5).. 800
Shannock, (K7)... 100
Shawomet, (L6).. 100
Slatersville, (K5)..1,850
Slocum, (K6)... 250
So. F st r, (K5).. 100
So. Portsmouth,
(M6)........... 100

So. Scituate, (K5).. 710
Stillwater, (K5)... 150
Summit, (K6).... 220
Tarkiln, (K5)... 190
Thornton, (K5)... 700
Tiverton, (M6)...1,950
Tiverton Four Corners, (M6)..... 325
Usquepaugh, (K7).. 100
Valley Falls, (L5)..2,400
Wakefield, (L7)..2,750
Warren, (L6).....6,450
Warwick, (L6).... 350
Washington, (K6).. 450
Watch Hill,
(J7)........... 100
W. Barrington, (L6). 7,.3
Westerly, (J7)..8,100
W. Gloucester, (J5). 100
W. Greenwich Center, (K6)...... 200
W. Kingston, (K7).. 100
White Rock, (J7).. 150
Wickford, (L6)...1,050
Wood River Jc.,
(K7).......... 150
Woodville, (K7)... 200
Woonsocket, (L5).38,125
Wyoming, (K6).... 200

SOUTH CAROLINA

Abbeville, (C3)...4,459
Adamsburg, (D2).. 100
Adamsrun, (G6)... 200
Aiken, (D4).....3,911
Alcolu, (G4)..... 458
Allen, (J4)...... 200
Allendale, (E5)...1,453
Almeda, (F6).... 500
Alston, (E3)..... 110
Anderson, (B2)...9,654
Antreville, (C3)... 100
Appleton, (E5)... 156
Arcadia, (D2)... 300
Ariel, (J3)...... 500
Arkwright, (D2).. 500
Autun, (B2)..... 150
Bamberg, (E5)...1,937
Barnwell, (E5)...1,324
Batesburg, (D4)..1,995
Batesville, (C2)... 150
Bath, (D5)...... 500
Beaufort, (F7)...2,486
Belton, (C2).....1,652
Bennettsville, (H2)2,646
Bessie, (C2)..... 250
Bethune, (G3)... 317
Bingham, (J3)... 196
Bishopville, (G3).1,659
Blacksburg, (C1).1,119
Blackstock, (E2).. 192
Blackville, (E5)..1,278
Blaney, (F3).... 100
Blenheim, (H2).. 228
Bluffton, (F7)... 577
Blythewood, (F3).. 100
Bolen, (F4)..... 160
Bonneau, (H5)... 100
Bordeaux, (C4)... 200
Bowman, (F5)... 327
Bradley, (C3)... 279
Branchville, (F5).1,471
Brandon, (C2)... 300
Brighton, (E6)... 190

Brightsville, (H2)... 100
Brunson, (E6)... 610
Bucklick, (F3)... 400
Buffalo, (D2)... 200
Burton, (F7).... 100
Cades, (H4)..... 100
Calhoun, (B2)... 215
Calhoun Falls, (B3) 296
Camden, (F3)...3,569
Cameron, (F4)... 421
Campobello, (C1). 255
Carlisle, (E2)... 300
Cartersville, (H3).. 320
Cateechee, (B2)... 367
Cayce, (E4)..... 180
Cedarspring, (D2). 150
Central, (B2)... 886
Chapin, (E2).... 345
Chappells, (D3).. 150
Charleston, (H6)..58,833
Cheraw, (H2)...2,873
Cherokee, (D1)... 150
Cherokee Falls, (D1) 591
Chester, (E2)...4,754
Chesterfield, (G2).. 618
Clarks Hill, (C4).. 100
Clifton, (D2)...1,700
Clinton, (D3)...3,272
Clio, (H2)...... 780
Clover, (E1)....1,207
Cokesbury, (C3).. 756
Colemans, (D3)... 150
Colliers, (C4)... 150
COLUMBIA, (F4)26,319
Congaree, (F4)... 100
Connors, (G5)... 100
Converse, (D1)..1,000
Conway, (J4)...1,228
Cope, (F5)..... 174
Cordesville, (H5).. 130
Coronaca, (C3)... 190
Cottageville, (G6). 418
Cowpens, (D2)..1,101

Crocketville, (E6).. 200
Cromer, (D3)... 100
Cross Anchor, (D2) 200
Crosshill, (C3)... 558
Crosskeys, (D2)... 120
Dantzler, (F5)... 130
Darlington, (H3)..3,789
Denmark, (E5)...1,075
Dillon, (J3).....1,757
Donalds, (C3)... 268
Dovesville, (H3)... 169
Drayton, (D2)... 200
Duewest, (C3)... 672
Dunbar, (H2)... 150
Duncan, (C2)... 190
Dunklin, (C2)... 200
Earlybranch, (F6). 160
Easley, (B2)....2,983
Eastover, (F4)... 237
Eau Claire, (F4)..1,234
Ebenezer, (H3)... 420
Ebenezer, (E2)... 190
Edgefield, (D4)..1,771
Edgmoor, (F2)... 113
Edisto Island, (G6) 620
Ehrhardt, (E5)... 315
Elko, (E5)..... 114
Ellenton, (D5)... 367
Elliott, (G3)... 500
Elloree, (F4)... 540
Enoree, (D2)...1,000
Enterprise, (G6).. 200
Estill, (E6)..... 460
Ethel, (G6)..... 100
Eutawville, (G5).. 405
Exchange, (G6)... 300
Fairfax, (E6)... 499
Fairforest, (C2)... 300
Fairmont, (C2)... 100
Fair Play, (A2)... 100
Ferguson, (G5)... 100
Field, (B2)..... 100
Fingerville, (C1).. 300

Florence, (H3)...7,057
Foreston, (G4)... 113
Fork, (J3)...... 134
Forksboals, (C2).. 360
Fort Lawn, (F2).. 204
Fort Mill, (F2)...1,616
Fort Motte, (F4).. 392
Fortner, (B1)... 200
Fountain Inn, (C2) 979
Frogmore, (F7)... 100
Furman, (E6)... 100
Gaffney, (D1)...4,767
Gaston, (E4)... 150
Georgetown, (J5)..5,530
Gilbert, (E4)... 450
Glendale, (D2)... 800
Glenn Springs,(D2) 178
Godbold, (J4)... 100
Goldville, (D3)... 210
Golightly, (D2)... 300
Gourdin, (H5)... 200
Govan, (E5)... 111
Gowensville, (C1). 100
Grahamville, (F7). 300
Grahamville, (D4).2,500
Graycourt, (C2)... 284
Great Falls, (F2).. 200
Greeleyville, (G4). 630
Greenpond, (F6).. 360
Greenville, (C2)..15,741
Greenwood, (C3).6,614
Greer, (C2).....1,673
Grover, (E5)... 67
Gurley, (K3)... 100
Hagood, (F3)... 300
Hamburg, (D5).. 490
Hamer, (J3)... 150
Hampton, (E6)... 748
Hardeeville, (F7).. 650
Harleyville, (G5).. 190
Harmon, (D3)... 150
Harper, (H5)... 100
Hartsville, (G3)..2,365

Haskell, (J4)..... 100
Heath Spring, (F2). 452
Hebron, (H4)... 100
Helena, (D3)... 425
Hendersonville, (F6) 319
Hickory Grove,(E1) 285
Highland, (C1)... 200
Hodges, (C3)... 266
Hollyhill, (G5)... 342
Honeapath, (C3)..1,763
Huntersville, (C2). 250
Inman, (C1)..... 474
Irmo, (E3)..... 267
Iva, (B3)..... 894
Jackson, (D5)... 150
Jacksonboro, (G6). 50
James Island, (H6) 250
Jedburg, (G5)... 110
Jefferson, (G2)... 390
Johns Island, (G6). 1,000
Johnston, (D4)... 943
Jonesville, (D2)... 969
Jordan, (G4)... 300
Kathwood, (D5).. 200
Kelton, (D2)... 92
Kemper, (J3)... 62
Kershaw, (F2)... 682
Killian, (F3)... 110
Kings Creek, (E1). 100
Kingstree, (H4)..1,372
Kline, (E5)..... 199
Ladies Island, (F7) 100
Lake City, (H4)..1,074
Lamar, (G3)... 592
Lancaster, (F2)..2,098
Lando, (E2)... 200
Landrum, (C1)... 449
Lane, (H4)..... 100
Langley, (D5)...1,000
Latta, (J3).....1,358
Laurel, (J4)... 200
Laurens, (C3)...4,818
Leeds, (E2)... 350

Leesville, (D4)....	980	Midland Park, (G6)	100	Paxville, (G4).....	175	Sandyrun, (F4)....	110	Trio, (H5).......	198
Lena, (E6).......	100	Midway, (F5).....	96	Peak, (E3).......	183	Santuck, (D2)....	100	Troy, (C4)......	233
Lenud, (H5)......	250	Millettville, (D5)..	130	Pelham, (C2).....	300	Scotia, (E6).....	189	Tucapau, (D2)...	800
Levys, (E7)......	200	Modoc, (C4).....	108	Pelzer, (C2).....	6,000	Scranton, (H4)...	308	Ulmers, (E5)....	190
Lewiedale, (E4)...	162	Monetta, (D4)....	122	Pendleton, (B2)...	822	Seiglingville, (E5).	113	Union, (D2)....	5,623
Lexington, (E4)...	709	Monks Corner,(G5)	232	Perry, (E4)......	179	Sellers, (H3)....	458	Vances, (G5).....	97
Liberty, (B2)....	1,058	Monticello, (E3)...	100	Pickens, (B2)....	897	Seneca, (B2)....	1,313	Varnville, (E6)...	542
Libertyhill, (F3)...	200	Montmorenci, (D4)	100	Piedmont, (B2)...	4,000	Shandon, (F4)...	795	Vaucluse, (D4)...	850
Lincolnville, (G5)..	341	Moultrieville, (H6)	1,011	Pinewood, (G4)...	424	Sharon, (E2)....	374	Verdery, (C3)....	230
Little Mountain,(E3)	440	Mount Carmel,		Pinopolis, (G5)...	200	Shirley, (E6)....	130	Wagener, (E4)...	362
Little River, (K4).	200	(C3),..........	264	Plumbranch, (C4).	145	Silver, (G4).....	240	Walhalla, (A2)...	1,595
Little Rock, (J3)..	99	Mount Croghan,		Pomaria, (E3)....	200	Silverstreet (D3)..	200	Walterboro, (F6).	1,677
Livingston, (E4)...	168	(G2)..........	100	Ponpon, (G6)....	300	Simpsonville, (C2).	521	Walton, (E3)....	100
Lockhart, (E2)....	2,000	Mount Pleasant,		Port Harreison, (J4)	150	Smoaks, (F5)....	200	Wando, (H6)....	250
Lodge, (F5)......	300	(H6)..........	1,346	Port Royal, (F7)..	363	Smyrna (H4).....	109	Ward, (D4).....	183
Lonestar, (F4)....	100	Mountville, (D3)..	150	Pregnall, (G5)....	100	Snelling, (E5)....	338	Warrenville, (D4).	320
Loris, (K3)......	229	Mullins, (J3)....	1,832	Princeton, (C3)...	182	Societyhill, (H2).	500	Wateree, (F4)....	100
Lowndesville, (B3).	350	Navy Yard, (H6)..	1,000	Prosperity, (D3)...	737	South Lynchburg,		Waterloo, (C3)...	191
Lowryville, (E2)...	343	Neeses, (E4).....	143	Quick, (H2).....	200	(G3)..........	280	Wattacoo, (B1)...	100
Lucknow, (G3)....	139	Newberry, (D3)...	5,028	Reedy River Fac-		Sparjun, (D2)....	500	Wedgefield, (G4)..	310
Lugoff, (F3).....	150	New Brookland,		tory, (C2).....	240	Spartanburg, (D2)	17,517	Wellford, (C2)...	370
Lumber, (H3).....	300	(E4)..........	900	Reevesville, (F5)..	205	Springfield, (E4)..	438	Wells, (G5).....	100
Luray, (E6)......	100	Newmarket, (C3)..	160	Reidville, (C2)....	177	Starr, (B3)......	200	Westminster, (A2).	1,576
Lydia, (G3)......	100	Newport, (E2)....	100	Rembert, (F3)....	100	Steedman, (E4)...	200	West Union, (B2)..	320
Lynch, (H4)......	100	New Prospect, (D1)	130	Richburg, (F2)...	245	Stillwood, (E6)...	200	Westville, (H3)...	848
Lynchburg, (G3)..	466	Newry, (B2).....	900	Ridgeland, (F7)...	330	Stokesbridge (H4).	100	Whitehall, (F6)...	307
McBee, (G3).....	187	Neyles, (F6).....	300	Ridgespring, (D4).	505	Summerton (G4)..	678	Whitepond, (D5)..	250
McClellanville,(H5)	700	Nichols, (J3).....	118	Ridgeville, (G5)...	400	Summerville, (G5)	2,355	White Rock, (E3)..	80
McColl, (H2)....	1,628	Ninety Six, (D3)..	758	Ridgeway, (F3)...	370	Summit (H4)....	87	Whitestone, (D2)..	105
McConnells, (E2)..	279	Norris, (B2).....	180	Robertsville, (E6).	100	Sumter, (G4)...	8,109	Whitmire, (D2)...	1,045
McCormick, (C4)..	613	North, (E4)......	561	Rockhill, (F2)...	7,216	Swansea, (F4)...	523	Wilkins, (F7)....	100
Macbeth, (H5)....	100	NorthAugusta,(D5)	1,136	Rockton, (E3)....	100	Sycamore, (E5)...	99	Wilksburg, (E2)...	300
Madden, (D3)....	150	Norway, (E4)....	315	Rowesville, (F5)..	508	Taft, (H5).......	150	Williams, (F5)....	100
Mallory, (H3)....	96	Olanta, (H4)....	230	Ruby, (G2)......	194	Tarboro, (E6)....	100	Williamston, (C2).	1,957
Manning, (G4)...	1,854	Olar, (E5).......	350	Ruffin, (F5).....	100	Tatum, (H2)....	225	Willington, (C4)...	365
Marietta, (B1)...	310	Oldpoint, (E2)....	250	Rural, (G3)......	400	Taylors, (C2)....	250	Williston, (E5)...	624
Marion, (J3)....	3,844	Orangeburg, (F4).	5,906	Saint Charles, (G3)	100	Ten Mile, (H6)...	150	Wilson, (G4)....	100
Marlboro, (H2)...	100	Oswego, (G4)....	100	Saint George, (F5)	957	Terry, (C2).....	150	Windsor, (D5)...	200
Marsbluff, (H3)...	450	Owings, (C2)....	170	Saint Helena Island,		Tillman, (E6)....	100	Winnsboro, (E3)..	1,754
Maryville, (H6)...	473	Pacolet, (D2)....	410	(F7)...........	100	Timmonsville, (H3)	1,708	Winona, (H3)....	100
May, (J3).......	200	Pageland, (G2)...	360	Saint Matthews,		Tindal, (G4).....	100	Woodford, (E4)...	190
Maybinton, (E3)..	100	Pages Mills, (G3).	157	(F4)..........	1,377	Tirzah, (E1).....	128	Woodruff, (D2)...	1,880
Mayesville, (G4)..	751	Palmetto, (H3)...	100	Saint Stephen, (H5)	408	Toby, (J3)......	100	Woodward, (E2)..	400
Meggett, (G6)...	1,000	Panola, (G4)....	100	Salem, (A2).....	139	Townville, (B2)...	255	Yemassee, (F6)...	250
Meriwether, (C4)..	100	Paris Island, (F7).	300	Salley, (E4)......	311	Travellers Rest,		Yonges Island (G6)	200
Merritts Bridge,		Parksville, (C4)...	197	Saluda, (D3)....	610	(C2)..........	60	Yorkville, (E2)...	2,326
(D4)..........	100	Patrick, (G2)....	98	Sampit, (H5)....	200	Trenton, (D4)....	257	Zion, (J3)......	195

SOUTH DAKOTA

Aberdeen, (F2)...	10,753	Brentford, (F2)...	150	Colome, (E4).....	280	Ethan, (F4)......	312	Harrold, (E3)....	230
Akaska, (D2)....	114	Bridgewater, (G4).	934	Colton, (H4).....	407	Eureka, (E2)....	961	Hartford, (G4)...	648
Albee, (F2).....	131	Bristol, (G2).....	444	Columbia, (F2)...	235	Fairfax, (E4)....	500	Hayti, (G3).....	180
Alcester, (H4)...	409	Britton, (G2)....	901	Conde, (F2).....	592	Fairview, (H4)...	107	Hazel, (G3).....	229
Alexandria, (G4)..	955	Broadland, (F3)...	150	Corona, (H2)....	210	Faulkton, (E2)...	802	Hecla, (F2).....	462
Alpena, (F3)....	417	Brookings, (H3)..	2,971	Corsica, (F4)....	286	Fedora, (G3)....	100	Henry, (G3).....	441
Altamont, (H3)...	110	Bruce, (H3).....	262	Cottonwood, (C4).	250	Ferney, (F2)....	250	Hermosa, (A4)...	114
Andover, (G2)....	446	Bryant, (G3)....	645	Crandall, (G2)....	100	Flandreau, (H3)..	1,484	Herreid, (D2)....	414
Ardmore, (A4)....	146	Buffalo, (A2)....	150	Cresbard, (F2)...	320	Florence, (G2)...	270	Herrick, (E4)....	412
Arlington, (G3)...	791	Buffalogap, (A4)..	280	Crocker, (G2)....	100	Forest City, (D2).	100	Hetland, (G3)...	223
Armour (F4)....	968	Burke, (E4).....	311	Custer, (A4).....	602	Fort Pierre, (D3)..	792	Highmore, (E3)..	1,084
Artas, (E2).....	220	Bushnell, (H3)...	200	Dallas, (E4)....	1,277	Frankfort, (F3)...	408	Hill City, (A4)...	271
Artesian, (F4)....	583	Butler, (G2).....	200	Davis, (H4).....	164	Frederick, (F2)...	433	Hitchcock, (F3)..	259
Ashton, (F3)....	430	Camp Crook, (A2).	120	Deadwood, (A3)..	3,653	Freeman, (G4)...	615	Holmquist, (G2)..	100
Athol, (F2).....	150	Canastota, (G4)..	409	Dell Rapids, (H4).	1,367	Fulton, (G4)....	180	Hosmer, (E2)....	217
Aurora, (H3)....	236	Canova, (G4)....	311	Delmont, (F4)...	369	Galena, (A3)....	109	Hot Springs, (A4).	2,140
Avon, (G4).....	451	Canton, (H4)...	2,103	Dempster, (G3)...	100	Gannvalley, (F3)..	100	Hoven, (E2).....	209
Baltic, (H4).....	278	Carlyle, (E2)....	100	De Smet, (G3)..	1,063	Garden City, (G3).	320	Howard, (G3)...	1,026
Bancroft, (G3)...	100	Carthage, (G4)...	504	Dixon, (E4).....	150	Garretson, (H4)..	668	Hudson, (H4)...	404
Bellefourche, (A3).	1,352	Castlewood, (G3).	594	Doland, (F3)....	581	Gary, (H3)......	477	Humboldt, (G4)..	520
Belvidere, (C4)...	150	Cavour, (G3)....	403	Dolton, (G4)....	180	Gayville, (G5)...	257	Hurley, (G4)....	506
Bemis, (H3).....	100	Centerville, (H4).	971	Draper, (D4)....	211	Geddes, (F4)....	701	Huron, (F3)....	5,791
Beresford, (H4)...	1,117	Central City, (A3).	296	East Sioux Falls,		Gettysburg, (E2)..	936	Iona, (E4)......	100
Bigstone City,		Chamberlain, (E4)	1,215	(H4)..........	268	Glenham, (D2)...	182	Ipswich, (E2)....	810
(H2)..........	551	Chancellor, (H4)..	160	Edgemont, (A4)..	816	Goodwin, (H3)...	145	Irene, (G4).....	263
Bijou Hills, (E4)..	100	Chelsea, (F2)....	100	Effington, (H2)...	46	Greenmont, (A3).	150	Iroquois, (G3)...	578
Bison, (B2).....	100	Chester, (G4)....	160	Egan, (H4).....	516	Greenway, (E2)..	150	Java, (E2)......	473
Black Hawk, (A3).	100	Cheyenne Agency,		Elkpoint. (H5)..	1,200	Greenwood, (F5).	300	Jefferson, (H5)...	407
Blunt, (D3).....	566	(D2)..........	100	Elkton, (H4)....	742	Gregory, (E4)...	1,142	Kadoka, (C4)....	222
Bonesteel, (E4)...	563	Claremont, (F2)...	294	Emery, (G4).....	446	Grenville, (G2)...	100	Kennebec, (E4)...	252
Bowdle, (E2)....	671	Clark, (G3).....	1,220	Englewood, (A3).	100	Groton, (F2)....	1,108	Keystone, (A4)...	320
Bradley, (G2)....	351	Clearlake, (H3)...	704	Erwin, (G3).....	230	Harrisburg, (H4).	164	Kimball, (F4)....	713
Brandt, (H3)....	158	Colman, (H4)....	362	Estelline, (H3)...	509	Harrison, (F4)...	120	Kranaburg, (H3)...	100

TENNESSEE

TEXAS

Houston Heights, (M8) 6,984
Howe, (L4) 541
Howland, (M4) 406
Hubbard, (L4) 1,843
Buckabay, (J5) 150
Bufsmith, M? 150
Hughes Springs, N4 850
Humble, (M7) 3,456
Hungerford, L4 150
Huntington, N4 500
Hunterville, M? ..... 2??
Hutchins, L5 350
Hutto, (K7) .....
Huxley, O4 150
Hylton, (G5) 350
Iago, (M?) 150
Imperial, M8 .....
Independence, L7 300
Indiancreek, J6 100
Indian Gap, J6 150
Industry, L4 500
Inez, (L9) 150
Iola, (L7) 300
Iowa Park, J4 6,3
Iredell, (K6) 750
Irene, L6 446
Iris, (M6) 250
Ironees, N6 150
Isdtas, (H10) 250
Italy, (L5) 1,149
Itasca, (K5) 1,256
Iuka, (J8) 50
Jacksboro, J4 1,480
Jacksonville, M6) 2,875
Jardin, (M4) 130
Jasper, (N7) 450
Jayton, (G4) 314
Jefferson, (N5) 2,515
Jennings, M4) 150
Jewett, L6 546
Josquin, (N6) 150
Johnson City, J7 350
Johnsons Station, (K5) 100
Jonah, (K7) 120
Jonesboro, (K6) 450
Josephine, L4 500
Josbus, (K5) 800
Josserand, (N6) 300
Jourdanton, (J8) 450
Juan Saenz, (K10) 100
Junction, (H7) 550
Juno, (F7) 100
Justin, (K4) 500
Karnack, (N5) 100
Karnes City, (K9) 700
Katemcy, (M7) 150
Katy, (M8) 200
Kaufman, (L5) 1,950
Keenan, (M7) 500
Keene, (K5) 400
Keller, (K5) 250
Kellyville, (N5) 150
Kelm, (L5) 100
Kelsey, (N5) 150
Keltys, (N6) 1,000
Kemah, (M8) 200
Kemp, (L5) 1,250
Kempner, (K6) 170
Kenedy, (K9) 1,147
Kennard, (M6) 600
Kennedale, (K5) 200
Kenneth, (M7) 1,000
Kenney, (L7) 150
Kentuckytown, (L4) 100
Kerens, (L5) 945
Kermit, (E6) 100
Kerrville, (H7) 1,843
Kildare, (N5) 260
Kilgore, (N5) 450
Killeen, (K6) 1,265
Kimball, (K5) 200

UTAH

Robinson, (B4)......	400	Sigurd, (B5)...... 150	Summit, (A6)...... 100	Tucker, (C4)...... 150	Wellington, (D4).. 358
Rockport, (C3)......	110	Silver City, (B6).. 800	Sunnyside, (D4).. 750	Uinta, (C2)...... 230	Wellsville, (C2).. 1,195
Rockville, (A6)......	230	Smithfield, (C2)..1,865	Sunshine, (B3)... 100	Union, (C3)...... 700	Wendover, (A3)... 100
Roy, (B2)...........	250	Snowville, (B2)... 150	Syracuse, (B2)... 500	Upton, (C3)...... 250	West Portage, (B2) 120
Saint George, (A6). 1,769		South Jordan,		Taylorsville, (C3).. 550	Utahn, (D3)...... 100	Westwater, (E4)... 100
Saint John, (B3)...	200	(C3)............ 700	Teasdale, (C3)... 150	Venice, (C5)...... 140	Wheeler, (A6)...... 120
Salem, (C3).........	693	Spanish Fork, (C3) 3,464	Theodore, (D3)... 300	Vermilion, (B5)... 110	Whiterocks, (E3)... 100
Salina, (C5)....... 1,082		Spring City, (C4).. 1,100	Thistle, (C3)...... 200	Vernal, (E3)...... 836	Willard, (C2)...... 577
SALT LAKE CITY,		Spring Glenn, (D4) 200	Thurber, (C5)...... 280	Vernon, (B3)...... 150	Wilson, (C2)...... 350
(B3)............92,777		Springville, (C3).. 3,356	Tooele, (B3)...... 2,753	Virgin, (A6)...... 100	Winterquarters,
Sandy, (C3)....... 1,037		Spry, (B6)........ 280	Topliff, (B3)...... 100	Wales, (C4)...... 294	(C4)............ 200
Santa Clara, (A6)..	300	Standrod, (A2).... 150	Toquerville, (A6).. 350	Wallsburg, (C3)... 550	Woodland, (C3)... 300
Santaquin, (C4)...	915	Stateline, (A5).... 100	Torrey, (C5)...... 100	Wanship, (C3)... 230	Woodruff, (C2)... 500
Scipio, (B4).......	546	Sterling, (C4)..... 300	Tremonton, (B2).. 303	Washakie, (B2)... 200	Woods Cross, (B3). 1,000
Scofield, (C4)......	750	Stockton, (B3)..... 258	Trenton, (B2)..... 300	Washington, (A6). 424	Woodside, (D4)... 300
Sevier, (B5)........	130	Sugarhouse, (C3)..1,500	Tropic, (B6)...... 358	Weber, (B2)...... 500	Yost, (A2)......... 100

VERMONT

Adamant, (C3)..... 150	Cavendish, (B5)... 575	East Johnson, (B2) 300	Hartland, (C4) 800	Miles Pond, (D3).. 100
Addison, (A3)..... 575	Center Rutland,	East Middlebury,	Hartland Four Cor-	Milton, (A2)...... 634
Albany, (C2)...... 530	(A4)............ 600	(A4)............ 350	ners, (C4)...... 150	Monkton, (A3)... 300
Alburg, (A2)...... 425	Central, (D2)..... 100	East Monkton,(A3) 100	Uinta, (C5)...... 100	Monkton Ridge,
Alburg Center, (A2) 200	Charlotte, (A3).... 1,000	East Montpelier,	Heartwellville, (A6) 200	(A3)............ 280
Alburg Springs,(A2) 425	Chelsea, (C4)..... 800	(C3)............ 500	Highgate, (B2)... 600	Montgomery, (B2) 850
Amsden, (B5)...... 190	Chester, (B5)..... 666	East Peacham, (C3) 200	Highgate Center,	Montgomery Center,
Andover, (B5)..... 250	Chester Depot,(B5) 700	East Poultney, (A4) 300	(B2)............ 600	(B2)............ 800
Arlington, (A5)... 400	ChimneyPoint,(A3) 100	EastRandolph, (B4) 200	Highgate Springs,	MONTPELIER,
Ascutneyville, (C5) 200	Chippenhook, (A4) 100	East Rochford, (B2) 150	(A2)............ 200	(C3)............ 7,856
Athens, (B5)...... 100	Chittenden, (B4).. 350	East Roxbury, (B3) 150	Hinesburg, (A3)... 242	Moretown, (B3)... 500
Bakersfield, (B2).. 600	Clarendon, (B4)... 100	East Rupert, (A5). 100	Holden, (B4)...... 175	Morgan, (D2)..... 200
Barnard, (B4)..... 475	Clarendon Springs,	East Ryegate, (C3) 200	Holland, (D2)..... 200	Morgan Center,
Barnet, (D3)...... 550	(B4)............ 180	East Thetford, (C4) 130	Hortonville, (A4).. 190	(D2)............ 110
Barre, (C3).......10,734	Colchester, (A2)... 300	East Wallingford,	Hubbardton, (A4).. 200	Morristown, (B2).. 200
Barton, (C2)...... 1,339	Concord, (D3)..... 339	(B5)............ 280	Huntington, (A3).. 330	Morrisville, (B2).. 1,445
Bartonsville, (B5). 200	Copperfield, (C4).. 100	Eden, (B2)........ 300	Huntington Center,	Moscow, (B3)..... 130
Basin Harbor, (A3) 100	Corinth, (C3)..... 420	Eden Mills, (B2).. 400	(B3)............ 300	Mount Holly, (B5) 275
Beecher Falls, (E2) 350	Cornwall, (A4)... 475	Elmore, (B2)...... 500	Hyde Park, (B2).. 423	Newark, (D2)..... 100
Beldens, (A3)..... 200	Coventry, (C2)... 550	Enosburg, (B2)... 100	Hydeville, (A4)... 700	Newbury, (C3)... 412
Bellows Falls, (B5). 4,883	Craftsbury, (C2).. 425	EnosburgFalls,(B2) 1,153	Ira, (A4)......... 225	Newbury Center,
Belvidere, (B2)... 280	Cuttingsville, (B5). 350	Essex, (A3)...... 210	Irasburg, (C2)... 500	(C3)............ 100
Belvidere Center,	Danby, (B5)...... 800	Essex Junction,(A3)1,245	Island Pond, (D2). 1,573	Newfane, (B6)..... 136
(B2)............ 100	Danby Four Cor-	Evansville, (C2)... 300	Isle Lamotte, (A2).. 410	New Haven, (A3).. 500
Bennington, (A6)..6,211	ners, (A5)...... 170	Evarts, (C4)...... 150	Jacksonville, (B6).. 212	New Haven Mills,
Bennington Center,	Danville, (C3)..... 830	Fairfax, (A2)...... 500	Jamaica, (B5)..... 475	(A3)............ 400
(A6)............ 42	Davis Bridge, (B6).. 100	Fairfield, (B2)..... 500	Jay, (C2)......... 200	Newport, (C2)..... 2,548
Benson, (A4)..... 800	Derby, (C2)....... 316	Fair Haven, (A4) .. 2,554	Jeffersonville, (B2). 650	Newport Center,
Berkshire, (B2)... 180	Derby Line, (C2).. 390	Fairlee, (C4)...... 390	Jericho, (B3)...... 1,000	(C2)............ 250
Berlin, (B3)....... 100	Dorset, (A5)...... 600	Fayston, (B3)..... 200	Jericho Center,(B3) 275	North Bennington,
Bethel, (B4)...... 1,500	Dummerston, (B6). 120	Felchville, (B5)... 400	Johnson, (B2)..... 651	(A6)............ 663
Binghamville, (B2) 100	Duxbury, (B3).... 400	Ferrisburg, (A3)... 550	Jonesville, (A3)... 200	North Calais, (B3). 150
Bloomfield, (D2).. 200	East Alburg, (A2).. 200	Fisk, (A2)........ 100	Lake, (D2)........ 130	North Clarendon,
Bolton, (B3)...... 210	EastArlington,(A5) 600	Fitzdale, (D3)..... 180	Lake Dunmore, (A4) 100	(B4)............ 200
Boltonville, (C3).. 100	East Barnard, (B4) 175	Fletcher, (B2)..... 500	Lanesboro, (C3)... 100	North Craftsbury,
Bomoseen, (A4)... 100	East Barnet, (C3). 270	Forestdale, (P4)... 260	Larrabees Point,	(C2)............ 425
Bondville, (B5)... 300	East Barre, (C3).. 900	Fort Ethan Allen,	(A4)............ 100	North Danville,
Bradford, (C4)... 631	East Berkshire, (B2) 750	(A2)............ 100	Leicester, (A4).... 100	(C3)............ 250
Braintree, (B4)... 450	East Bethel, (B4).. 220	Fowler, (A4)...... 200	Leicester Junction,	North Duxbury,
Branch, (C2)...... 100	East Braintree,(B3) 250	Franklin, (B2)..... 900	(A4)............ 325	(B3)............ 150
Brandon, (A4)....1,608	East Brookfield,	Gallup Mills, (D2). 100	Lemington, (D2).. 100	North Fairfax, (A2) 100
Brattleboro, (B6).. 6,517	(B3)............ 200	Gassetts, (B5)..... 100	Lincoln, (B3)..... 500	North Ferrisburg,
Bridgewater, (B4). 350	East Burke, (D2).. 400	Gaysville, (B4)... 550	Londonderry, (B5). 400	(A3)............ 850
Bridgewater Cor-	East Calais, (C3).. 600	Georgia, (A2)..... 550	Lowell, (C2)...... 500	Northfield, (B3)... 1,918
ners, (B4)...... 200	East Charleston,	Georgia Plain, (A2) 150	Lower Cabot, (C3). 250	Northfield Falls,
Bridport, (A4)..... 750	(D2)............ 350	Glen, (B4)........ 300	Lower Waterford,	(B3)............ 400
Bristol, (A3)...... 1,180	East Charlotte,(A3) 100	Glover, (C2)...... 780	(D3)............ 100	North Hero, (A2).. 400
Brookfield, (B3)... 500	EastClarendon,(B4) 200	Goshen, (A4)...... 100	Ludlow, (B5)...... 1,621	North Hyde Park,
Brookline, (B5)... 100	East Concord, (D3) 100	Grafton, (B5)..... 650	Lunenburg, (D3).. 450	(B2)............ 500
Brownington, (C2). 230	East Corinth, (C3). 420	Granby, (D2)...... 85	Lyndon, (D2)..... 960	North Montpelier,
Brownington Cen-	East Craftsbury,	Grand Isle, (A2)... 700	Lyndon Center, (C2) 259	(C3)............ 400
ter, (C2)........ 220	(C2)............ 100	Graniteville, (C3).. 400	Lyndonville, (D2).. 1,573	North Pomfret, (B4) 100
Brownsville, (C5).. 520	East Dorset, (B5).. 600	Granville, (B4).... 300	McIndoe Falls, (C3) 400	North Pownal, (A6) 600
Burke, (D2)....... 100	East Dover, (B6)... 100	Green River, (B6). 500	Manchester, (A5).. 478	North Randolph,
Burlington, (A3)..20,468	East Dummerston,	Greensboro, (C2).. 500	Manchester Center,	(B4)............ 180
Cabot, (C3)....... 227	(B6)............ 700	Greensboro Bend,	(A5)............ 700	North Sheldon, (B2) 110
Cadys Falls, (B2).. 120	East Fairfield, (B2) 1,000	(C2)............ 300	Manchester Depot,	North Sherburne,
Calais, (C3)....... 200	East Fletcher, (B2) 100	Groton, (C3)...... 820	(B5)............ 760	(B4)............ 100
Cambridge, (B2).. 595	East Franklin, (B1) 150	Guildhall, (D2) ... 290	Marlboro, (B6)... 150	North Shrewsbury,
Cambridge Junc-	East Georgia, (A2). 100	Guilford, (B6)..... 200	Marshfield, (C3)... 700	(B4)............ 350
tion, (B2)....... 120	East Granville,(B3) 100	Halifax, (B6)..... 225	Mechanicsville, (B5) 550	North Springfield,
Cambridgeport,	EastHardwick,(C2) 470	Hancock, (B4).... 150	Mendon, (B4)..... 100	(B5)............ 430
(B5)............ 100	East Haven, (D2).. 100	Hanksville, (B3)... 100	Middlebury, (A3)..1,866	North Thetford,
Canaan, (D2)..... 275	East Highgate, (B2) 300	Hardwick, (C3).... 2,094	Middlesex, (B3)... 680	(C4)............ 250
Castleton, (A4)...1,000	East Jamaica, (B5). 100	Hartford, (C4).... 500	Middletown Sprs,(A5)600	North Troy, (C2).. 771

VIRGINIA

WASHINGTON

WEST VIRGINIA

Place	Pop.
Lanark, (D5)	500
Landgraff, (D6)	300
Landisburg, (E5)	300
Laneville, (H3)	333
Lansing, (D4)	200
Lashmeet, (D6)	100
Launa, (C5)	300
Laurel Creek, (D5)	130
Laurel Point, (G1)	100
Lawson, (D3)	100
Lawton, (E3)	500
Layland, (E3)	500
Layopolis, (E3)	156
Lazearville, (K5)	500
Leander, (E4)	200
Leatherwood, (K5)	144
Leebell, (F3)	100
Leetown, (L2)	100
Leon, (C3)	240
Leslie, (K6)	130
Lester, (D5)	300
Letart, (C3)	100
Lewisburg, (E3)	803
Lewiston, (C4)	300
Liberty, (C3)	100
Littlebirch, (E3)	100
Littleton, (E1)	713
Lizemores, (D4)	500
Lock Seven, (C4)	170
Logan, (B5)	1,640
London, (D4)	200
Lone Oak, (C3)	100
Longacre, (D4)	230
Longrun, (E2)	100
Lookout, (E4)	240
Lost City, (J3)	200
Lost Creek, (F2)	150
Lost River, (J3)	100
Lough, (D2)	100
Lowell, (E5)	270
Lowgap, (C4)	120
Lubeck, (C2)	100
Lumberport, (F2)	180
McComas, (D6)	656
McDowell, (D6)	100
McMechen, (K6)	1,500
Mabie, (F3)	2,921
Mabscott, (D5)	300
Macdonald, (D5)	561
Mace, (F4)	1,151
Madeline, (G3)	100
Madison, (C4)	150
Mahan, (D4)	295
Malden, (C4)	200
Mamie, (D3)	360
Mammoth, (D4)	300
Mannington, (F1)	2,672
Maplewood, (E5)	200
Marion, (E1)	100
Marlinton, (F4)	1,045
Marmet, (C4)	650
Marshall, (C4)	100
Marshes, (D5)	200
Marting, (D4)	400
Martinsburg, (K2)	10,698
Marvel, (D4)	250
Mason, (B2)	784
Mason Town, (G1)	520
Masseyville, (D5)	300
Matewan, (B5)	588
Matoaka, (D6)	150
May, (G3)	100
Maybeury, (D6)	2,000
Maysel, (D4)	100
Maysville, (H2)	150
Meador, (B5)	190
Meadow Bluff, (E5)	100
Meadow Creek, (E5)	250
Meadowville, (G2)	130
Merideth, (B4)	150
Metz, (F1)	180
Middlebourne, (E1)	546
Midrifton, (F2)	500
Middleway, (K2)	550

Place	Pop.
Midkiff, (B4)	140
Mill Creek, (G3)	740
Millville, (L2)	150
Millwood, (C3)	350
Milton, (B4)	817
Minden, (D5)	100
Minerva, (B4)	400
Mink, (C4)	200
Minnora, (D3)	300
Moatsville, (G2)	150
Mohawk, (C5)	200
Mona, (G1)	230
Monarch, (D4)	150
Monongah, (F2)	2,084
Montana Mines, (F1)	450
Montes, (G5)	250
Montgomery, (D4)	1,888
Montrose, (G2)	112
Moore, (G2)	150
Moorefield, (J2)	646
Morgansville, (E2)	300
Morgantown, (G1)	9,150
Morlan, (G2)	200
Morocco, (F4)	150
Mossy, (D5)	150
Mound, (C4)	200
Moundsville, (K6)	8,918
Mount Carbon, (D4)	450
Mount Clare, (F2)	250
Mount Hope, (D5)	494
Mount Lookout, (E4)	150
Mount Nebo, (E4)	200
Mucklow, (D4)	400
Mullens, (D5)	150
Murraysville, (C2)	100
Nabob, (C4)	100
Naugatuck, (B5)	100
Nelson, (C4)	100
Neola, (F5)	180
Nestorville, (G2)	150
Nettie, (E4)	200
Newburg, (G2)	821
Newcreek, (J2)	150
New Cumberland, (K4)	1,807
Newell, (K4)	150
New England, (C2)	500
Newhaven, (B3)	750
New Martinsville, (E1)	2,176
Newport, (C2)	150
New Richmond, (E5)	150
New Thacker, (B5)	180
Newton, (D3)	150
Nicolette, (D2)	150
Nolan, (B5)	150
Normantown, (E3)	110
Northfork, (D6)	425
North Point Pleasant, (B3)	452
Northspring, (C5)	100
Nuttallburg, (E4)	400
Oak Hill, (D5)	764
Oakvale, (E6)	278
Occana, (C5)	120
Odell, (D4)	300
Ogden City, (D5)	210
Ohley, (D4)	150
Okeefe, (B5)	150
Okott, (B5)	150
Olmstead, (C6)	260
Ompe, (K2)	100
Ona, (B4)	100
Oneal, (D5)	800
Orlando, (E3)	250
Orleans Crossroads, (K1)	150
Oswald, (D5)	546
Packs Branch, (D5)	200
Page, (D4)	500
Palestine, (D2)	230

Place	Pop.
Palmer, (E3)	100
Panther, (C6)	360
Parkersburg, (C2)	17,842
Parsons, (G2)	1,780
Patterson, (K5)	713
Pattersons Creek, (J1)	120
Pawpaw, (K1)	125
Pear, (E5)	300
Pearl, (E4)	100
Pecks Run, (F2)	150
Pence Springs, (E3)	100
Pennsboro, (D5)	100
Pennsboro, (D2)	930
Penrith, (K5)	300
Peru, (H3)	100
Petersburg, (H2)	350
Peterstown, (E6)	257
Petroleum, (D2)	100
Peytona, (C4)	200
Philippi, (F2)	1,038
Pickens, (F3)	600
Piedmont, (J1)	2,054
Pinegrove, (E1)	474
Pineville, (D5)	334
Piney, (E1)	150
Pittman, (E5)	250
Pleasant Valley, (K6)	210
Pleasant Valley, (K5)	346
Pluto, (D5)	100
Plymouth, (C3)	300
Poca, (C4)	350
Pocatalico, (C3)	150
Poe, (E4)	100
Point Pleasant, (B3)	2,045
Pond Gap, (D4)	200
Posey, (D5)	100
Powellton, (D4)	413
Powhatan, (D6)	450
Pratt, (D4)	300
Preston, (B4)	170
Prestonia, (E3)	120
Price Hill, (D5)	300
Prichard, (A4)	130
Priestley, (C4)	140
Prince, (E4)	250
Princeton, (D6)	3,027
Procious, (D4)	200
Proctor, (E1)	250
Prosperity, (D5)	300
Prudence, (D5)	400
Pruntytown, (F2)	400
Pullman, (E2)	230
Putney, (D4)	400
Quaker, (B4)	110
Queen Shoals, (D4)	180
Queens Ridge, (B4)	130
Quick, (D4)	100
Quiet Dell, (F2)	100
Quincy, (D4)	150
Quinnimont, (E5)	400
Radnor, (B4)	100
Raleigh, (D5)	350
Ranger, (B4)	100
Rapp, (B5)	200
Raven Rock, (D2)	100
Ravenswood, (C3)	1,081
Reader, (E1)	100
Redhouse Shoals, (C3)	200
Red Jacket, (B5)	1,200
Red Rock, (F2)	100
Redstar, (D5)	500
Reedsville, (G1)	208
Reedy, (D3)	313
Replete, (F3)	100
Richwood, (F4)	3,061
Ridgeley, (J1)	500
Ridgeway, (K2)	100
Rio, (J2)	150
Ripley, (C3)	891

Place	Pop.
Rippon, (L2)	100
Riverside, (D4)	170
Rivesville, (F1)	190
Robertsburg, (C3)	130
Robson, (D4)	250
Rock, (D6)	500
Rockford, (F2)	150
Rockport, (C2)	180
Romines Mills, (F2)	100
Romney, (J2)	1,112
Ronceverte, (F5)	2,157
Ronda, (D4)	100
Rosedale, (E3)	100
Roseville, (D5)	100
Rosina, (D4)	148
Rowlesburg, (G2)	936
Rupert, (E5)	100
Rushrun, (D5)	300
Ruth, (C4)	150
Sabraton, (G1)	700
Saint Albans, (C4)	1,209
Saint George, (G2)	245
Saint Marys, (D2)	1,358
Salem, (E2)	2,169
Salmon, (B5)	300
Salt Sulphur Springs, (F5)	200
Sandfork, (E3)	200
Sandyville, (C3)	250
Sarfis, (F2)	300
Saxon, (D5)	150
Sayre, (C5)	200
Scarbro, (D5)	1,533
Scary, (C4)	100
Sedan, (J2)	130
Seebert, (F4)	190
Selbyville, (F3)	180
Seng, (C5)	100
Seth, (C4)	200
Sewell, (E4)	200
Shadyspring, (D5)	150
Sharon, (D4)	100
Shaw, (H2)	250
Shenandoah Junction, (L2)	300
Shepherdstown, (L2)	1,070
Sheppard, (B5)	150
Sheridan, (B4)	200
Sherrard, (K6)	360
Shiloh, (E2)	130
Shinnston, (F2)	1,224
Shirley, (E2)	180
Shortcreek, (K5)	100
Silver Hill, (E1)	100
Simpson, (F2)	230
Sinks Grove, (E5)	100
Sisto, (C4)	200
Sir Johns Run, (K1)	120
Sissonville, (C3)	130
Sistersville, (D1)	2,684
Skelton, (D5)	300
Slab Fork, (D5)	250
Sleepy Creek, (K1)	300
Smithfield, (E1)	165
Smithton, (E2)	100
Smithville, (D2)	110
Sophia, (D5)	200
South Buckhannon, (F3)	681
South Carbon, (D4)	200
South Charleston, (C4)	500
South Keyser, (J2)	692
Spanishburg, (D6)	100
Sparks, (E4)	100
Spaulding, (B5)	130
Spencer, (D3)	1,328
Spilman, (B3)	250
Sprague, (D5)	250
Springfield, (J2)	135

Place	Pop.
Spring Hill, (C4)	150
Stirlington, (D6)	100
Star Gem, (C6)	100
Star City, (G1)	318
Stanley Heights, (F3)	511
Steelton, (E1)	100
Stockerts, (F2)	100
Stonecliff, (E5)	600
Stotlers Cross Roads, (K2)	100
Stuart, (D4)	697
Sullivan, (D5)	100
Sulphur, (H2)	100
Summersville, (E4)	204
Summit Point, (L2)	450
Sun, (B3)	800
Surveyor, (D5)	150
Sutton, (E3)	1,121
Sweeneyburg, (D5)	150
Sweetsprings, (F5)	300
Sylvia, (D5)	400
Tacy, (F3)	100
Talcott, (E5)	100
Tanner, (E3)	600
Tenmile, (F3)	100
Terra Alta, (G2)	1,126
Thacker, (B5)	200
Thayer, (E3)	500
Thoburn, (H2)	100
Thomas, (H2)	2,354
Thurmond, (D5)	315
Tioga, (F4)	120
Tolgate, (G2)	120
Triadelphia, (K5)	261
Tribble, (C3)	100
Troy, (E2)	144
Tunnelton, (G2)	792
Twin Branch, (C6)	400
Tyrconnell, (F2)	300
Tyrone, (D4)	160
Unceda, (C5)	100
Ungers Store, (K2)	200
Union, (E5)	298
Uniontown, (E1)	240
Upper Potomac, (H2)	100
Valley Bend, (F3)	090
Van, Cleveville, (C2)	100
Vandalia, (F3)	100
Vanetta, (D4)	100
Vaughan, (D4)	100
Villa, (D4)	200
Vinton, (D4)	150
Virginville, (K5)	250
Virosa, (F2)	100
Vivian, (D6)	600
Volcano, (D2)	100
Wabash, (H2)	200
Wadestown, (F1)	150
Wadesville, (C2)	150
Wake Forest, (D4)	200
Wallace, (F2)	200
Walton, (D3)	200
Ward, (D4)	300
Warden, (D5)	150
Wardensville, (J2)	122
War Eagle, (C5)	1,300
Warrior, (D4)	150
Watoga, (F4)	300
Watson, (F2)	400
Waverly, (D2)	100
Wayne, (B4)	500
Walton, (F3)	200
Webster Springs, (F4)	500
Weirton, (L5)	400
Welch, (C6)	1,500
Wellsburg, (K5)	4,189
West Columbia, (B3)	250
West End, (G2)	150

WISCONSIN

Place	Pop.	Place	Pop.	Place	Pop.	Place	Pop.	Place	Pop.
Glidden, (E3)	600	Ives, (L10)	100	Lyndon Station,		Morse, (E3)	100	Oso, (B6)	100
Goodrich, (F5)	120	Ives Grove, (L10)	100	(F8)	275	Mosinee, (G6)	482	Ontario, (E8)	383
Gordon, (C3)	300	Jackson, (K9)	150	Lynxville, (D9)	274	Mountain, (K5)	150	Oostburg, (L8)	380
Gotham, (F9)	350	Jacksonport, (M5)	200	Lyons, (K10)	400	Mount Calvary,		Oregon, (H10)	712
Grafton, (L9)	818	Janesville, (J10)	13,894	McCord, (G4)	100	(K8)	350	Orfordville, (H10)	449
Grand Rapids, (G7)	6,521	Jefferson, (J10)	2,582	McDill, (G7)	100	Mount Hope, (E10)	300	Osceola, (A5)	634
Grandview, (D3)	400	Jennings, (H5)	150	McMillan, (F6)	130	Mount Horeb, (G9)	1,048	Oshkosh, (K7)	33,062
Granite Heights,		Johnsburg, (K8)	200	Macfarland, (H9)	400	Mount Sterling,		Osseo, (D6)	548
(H5)	100	Johnson Creek, (J9)	425	MADISON, (H9)	25,531	(E9)	300	Otjen, (L10)	300
Granton, (E6)	250	Johnstown, (J10)	180	Magnolia, (H10)	100	Mount Vernon,		Ottawa, (K10)	200
Grantsburg, (A4)	721	Johnstown Center,		Maiden Rock, (B6)	337	(G10)	100	Owen, (E6)	745
Gratiot, (F10)	364	(J10)	200	Manawa, (J7)	820	Mukwonago, (K10)	615	Oxford, (G8)	250
Greenbay, (L6)	25,236	Juda, (H10)	290	Manchester, (H8)	220	Muscoda, (F9)	798	Packwaukee, (G8)	250
Greenbush, (L8)	120	Junction, (G6)	200	Mankowish, (F3)	100	Myra, (K9)	100	Palmyra, (J10)	649
Green Lake, (J8)	563	Juneau, (J9)	1,003	Manitowoc, (L7)	13,027	Namur, (L6)	100	Psdf, (G10)	150
Greenleaf, (L7)	100	Kansasville, (K10)	300	Maplewood, (L6)	260	Nashotah, (K9)	200	Pardeeville, (H8)	987
Greenstreet, (L7)	300	Kaukauna, (K7)	4,717	Marathon, (G6)	656	Nashville, (J5)	200	Park Falls, (E4)	1,972
Greenwood, (E6)	665	Kekoskee, (J8)	190	Marblehead, (K8)	280	National Home,		Parnell, (K8)	100
Gresham, (J6)	350	Kelley, (H6)	200	Marcy, (K9)	130	(L9)	2,100	Parrish, (H5)	100
Hackley, (F3)	500	Kellnersville, (L7)	440	Marengo, (E3)	250	Naugart, (G5)	120	Paskin, (C5)	100
Hales Corners,		Kempster, (H5)	200	Maribel, (L7)	300	Necedah, (F8)	1,054	Patch Grove, (E10)	180
(L10)	200	Kendall, (F8)	477	Marinette, (L5)	14,610	Neenah, (K7)	5,734	Paynesville, (L10)	500
Hamburg, (G5)	350	Kennan, (E4)	184	Marion, (J7)	795	Neillsville, (E6)	1,957	Pelicanlake, (H4)	200
Hammond, (B6)	408	Kenosha, (L10)	21,371	Markesan, (J8)	692	Nekoosa, (G7)	1,570	Pembine, (K4)	300
Hancock, (G7)	510	Kewaskum, (K8)	625	Marquette, (H8)	320	Nelson, (C7)	290	Pence, (F3)	410
Hanover, (H10)	100	Kewaunee, (M7)	1,839	Marshall, (H9)	459	Nelsonville, (H7)	100	Pennington, (E4)	130
Hartford, (K9)	2,982	Kiel, (K8)	1,244	Marshfield, (F6)	5,783	Nenno, (K9)	150	Pensaukee, (L6)	250
Hartland, (K9)	728	Kilbourn, (G8)	1,170	Martell, (B6)	190	Neopit, (J6)	150	Pepin, (B7)	397
Hatley, (H6)	300	Kimberly, (K7)	200	Martintown, (G10)	200	Neosho, (K9)	304	Perkinstown, (E5)	300
Haugen, (C4)	280	Kingston, (H8)	210	Mason, (D3)	750	Neshkoro, (H8)	379	Perry, (G10)	300
Hawkins, (E4)	400	Knapp, (B6)	413	Masonville, (F6)	150	New Amsterdam,		Peshtigo, (L5)	1,975
Hawthorne, (C3)	150	Kohlsville, (K9)	100	Mather, (F7)	100	(D8)	150	Petersburg, (K9)	100
Hayton, (K7)	200	Lac du Flambeau,		Mattoon, (J6)	868	New Auburn, (D5)	364	Pewaukee, (K7)	749
Hayward, (D4)	4,000	(F3)	200	Mauston, (F8)	1,701	New Berlin, (K10)	380	Pheasant Branch,	
Hazel Green, (E10)	621	La Crosse, (D8)	30,417	Mayfield, (K9)	100	Newburg, (K9)	400	(G9)	100
Hazelhurst, (G4)	180	Ladysmith, (D5)	2,352	Mayville, (K9)	2,282	New Cassel, (K8)	300	Phillips, (F4)	1,948
Heart Prairie, (J10)	100	La Farge, (E8)	654	Mazomanie, (G9)	917	New Diggings,		Phlox, (J5)	250
Hebron, (J10)	140	La Grange, (J10)	100	Medford, (F5)	1,846	(F10)	350	Pickett, (J8)	150
Heineman, (G5)	200	Lake Beulah, (K10)	100	Medina, (J7)	200	New Fane, (K8)	150	Pigeon Falls, (D7)	130
Helenville, (J9)	350	Lake Geneva,		Meeme, (L8)	100	New Glarus, (G10)	708	Pine Grove, (L7)	200
Hemlock, (G6)	100	(K10)	3,079	Mellen, (E3)	1,833	New Holstein, (K8)	839	Pine River, (H7)	200
Herman, (K9)	130	Lakemills, (J9)	1,672	Melrose, (D7)	250	New Lisbon, (F8)	1,074	Pittsville, (F7)	450
Hersey, (B6)	200	Lake Nebagamon,		Menasha, (K7)	6,081	New London, (J7)	3,383	Plain, (G9)	140
Hewitt, (F6)	280	(C3)	483	Mendota, (H9)	100	New Munster,		Plainfield, (G7)	723
High Bridge, (E3)	250	Lamartine, (J8)	150	Menomonee Falls,		(K10)	200	Plainville, (G8)	900
Highcliff, (K7)	250	Lancaster, (E10)	2,329	(K9)	919	New Richmond,		Platteville, (E10)	4,452
Highland, (F9)	1,096	Laney, (K6)	100	Menomonie, (C6)	5,036	(B5)	1,988	Pleasant Prairie,	
Hika, (L8)	280	Langlade, (J5)	100	Mercer, (F3)	200	New Rome, (G7)	160	(L10)	300
Hilbert, (K7)	571	Lannon, (K9)	450	Merrill, (G5)	8,689	Newton, (D8)	100	Plover, (H7)	320
Hiles, (J4)	250	Laona, (J4)	700	Merrillan, (E7)	625	Niagara, (K4)	1,700	Plum City, (B6)	305
Hillsboro, (F8)	804	La Rue, (G9)	250	Merrimack, (G9)	312	Niles, (L7)	200	Plymouth, (L8)	3,094
Hillsdale, (C5)	100	Lavalle, (F8)	421	Merton, (K9)	220	Norrie, (H6)	200	Poisar, (J5)	150
Hines, (C3)	500	Leadmine, (F10)	200	Middleton, (G9)	679	North Andover,		Polonia, (H6)	200
Hingham, (L8)	300	Leeman, (K6)	100	Midway, (D8)	100	(E10)	100	Poplar, (C2)	100
Hixton, (E7)	290	Lemonweir, (F8)	100	Mifflin, (F10)	350	North Cape, (K10)	150	Portage, (H8)	5,440
Holcombe, (D5)	300	Lena, (L6)	350	Milan, (F6)	100	North Crandon,		Port Edwards, (G7)	758
Hollandale, (G10)	205	Leon, (E8)	100	Milford, (J9)	130	(J4)	500	Portland, (J9)	250
Holmen, (D8)	300	Le Roy, (J8)	100	Milladore, (G6)	330	Northeim, (L8)	100	Port Washington,	
Holy Cross, (L9)	130	Levis, (D6)	200	Millston, (E7)	150	North Fond du Lac,		(L9)	3,792
Honey Creek,		Lima Center, (J10)	180	Milltown, (B4)	250	(K8)	1,960	Portwing, (D2)	250
(K10)	300	Limeridge, (F9)	250	Millville, (E10)	100	North Freedom, (G9)	647	Potosi, (E10)	464
Horicon, (J9)	1,881	Lincoln, (L6)	100	Milton, (J10)	833	North Hudson,		Pound, (K5)	300
Hortonville, (J7)	861	Linden, (F10)	580	Milton Junction,		(A5)	100	Poynette, (H9)	656
Houlton, (A6)	400	Lindsey, (F6)	100	(J10)	900	North La Crosse,		Poy Sippi, (J7)	180
Howards Grove,		Little Chute, (K7)	1,354	Milwaukee, (L9)	373,857	(D8) Pop. inc. in		Prairie du Chien,	
(L8)	180	Little Prairie, (K10)	100	Mindoro, (D7)	300	La Crosse.		(E9)	3,149
Hubbleton, (J9)	100	Little Suamico,		Mineral Point,(F10)	2,925	North Lake, (K9)	170	Prairie du Sac, (G9)	699
Hudson, (A6)	2,810	(L6)	300	Minnesota Junc-		North Milwaukee,		Prairie Farm, (C5)	368
Hull, (G6)	100	Little Wolf, (J7)	100	tion, (J9)	140	(L9)	1,860	Pray, (F7)	130
Humbird, (E6)	500	Livingston, (F10)	600	Minocqua, (G4)	470	Northport, (J7)	400	Preble, (L6)	180
Hurley, (F3)	2,600	Lodi, (G9)	1,044	Minong, (C3)	200	North Prairie, (J10)	200	Prentice, (F5)	606
Hustisford, (J9)	615	Loganville, (F9)	260	Misha Mokwa, (C7)	100	Norwalk, (E8)	502	Prescott, (A6)	936
Hustler, (F8)	300	Lohrville, (H7)	300	Mishicot, (L7)	450	Oakfield, (J8)	522	Preston, (F10)	100
Independence, (D7)	664	Lomira, (K8)	529	Mondovi, (C6)	1,325	Oconomowoc, (J9)	3,054	Princeton, (H8)	1,269
Indian Ford, (H10)	130	London, (J9)	180	Monico, (H4)	650	Oconto, (L6)	5,629	Prospect, (K10)	100
Ingram, (E5)	360	Lone Rock, (F9)	497	Monroe, (G10)	4,410	Oconto Falls, (K6)	1,427	Pulaski, (K6)	436
Iola, (H6)	850	Lost Creek, (B6)	100	Monroe Center,		Odanah, (E2)	2,100	Pulcifer, (K6)	100
Iron Belt, (F3)	1,000	Louisburg, (E10)	200	(G7)	150	Ogdensburg, (J7)	300	Quarry, (K6)	130
Iron Mountain, (J9)	200	Lowell, (J9)	318	Montello, (H8)	1,104	Ogema, (F5)	250	Racine, (L10)	38,002
Iron Ridge, (J9)	250	Loyal, (F6)	677	Montfort, (F10)	558	Okee, (G9)	130	Randolph, (J8)	937
Iron River, (D2)	2,500	Lublin, (E5)	100	Monticello, (G10)	671	Olivet, (B6)	150	Random Lake, (K8)	408
Ironton, (F8)	280	Luck, (B4)	383	Montreal, (F3)	500	Omro, (J7)	1,285	Rapids, (L7)	100
Itasca, (C2)	300	Luxembourg, (L7)	402	Morrison, (K7)	130	Onalaska, (D8)	1,146	Readfield, (J7)	110
Ithaca, (F9)	100	Lyndhurst, (J6)	100	Morrisonville, (H9)	200	Oneida, (K7)	2,200	Readstown, (E9)	515

WYOMING

ALASKA

HAWAII

PORTO RICO

ALASKA

HAWAII

PORTO RICO

DISTANCES BETWEEN THE LARGER CITIES OF THE UNITED STATES

The distances are by the shortest usually traveled railroad routes. Compiled from the War Department's official table of distances.

From / To	New York	Chicago	Phila-delphia	St. Louis	Boston	Baltimore	Cleveland	Buffalo	San Francisco	Pittsburg	Cincinnati	Milwaukee	New Orleans	Washington	Minneapolis
	Mls.	Mls.	Mls.	Mls.	Mls.	Mls.	Mls.	Mls.	Mls.	Mls.	Mls.	Mls.	Mls.	Mls.	Mls.
Albany	145	832	236	1,028	202	333	480	297	3,106	567	724	917	1,517	1,142	1,252
Atlanta	876	733	785	611	1,106	688	736	919	2,805	805	492	818	496	648	1,153
Baltimore	188	802	97	934	418	...	474	398	3,076	334	593	887	1,184	40	1,222
Boston	217	1,034	321	1,230	...	418	682	499	3,308	674	926	1,119	1,602	458	1,454
Buffalo	442	525	416	731	499	398	183	...	2,799	270	427	610	1,256	438	945
Chicago	912	...	821	284	1,034	802	357	525	2,572	468	298	85	912	790	420
Cincinnati	757	298	666	341	926	593	244	427	2,572	313	...	383	829	553	718
Cleveland	584	357	493	548	682	474	...	183	2,631	135	244	442	1,073	437	777
Columbus, O.	637	314	546	428	820	511	138	321	2,588	193	116	399	935	471	734
Denver	1,934	1,022	1,843	916	2,056	1,850	1,379	1,537	1,371	1,490	1,257	1,107	1,347	1,810	884
Detroit	693	272	669	488	730	649	173	251	2,546	321	263	357	1,092	655	692
Duluth	1,391	479	1,300	728	1,513	1,281	791	1,004	2,238	947	777	422	1,447	1,269	162
El Paso	2,310	1,465	2,219	1,245	2,414	2,179	1,703	1,915	1,287	1,866	1,586	1,550	1,195	2,139	1,521
Galveston	1,792	1,144	1,691	860	2,012	1,594	1,408	1,591	2,157	1,481	1,157	1,229	410	1,554	1,340
Grand Rapids, Mich.	821	178	815	462	878	796	332	379	2,452	462	308	263	1,090	764	568
Helena	2,452	1,540	2,361	1,549	2,574	2,342	1,897	2,065	1,250	2,008	1,838	1,455	2,152	2,320	1,110
Indianapolis	825	183	734	240	965	704	283	466	2,457	381	111	268	888	664	603
Jacksonville, Fla.	983	1,097	892	975	1,213	795	1,085	1,193	3,098	1,037	841	1,182	616	735	1,517
Kansas City	1,342	458	1,251	277	1,466	1,211	765	967	1,981	808	618	543	889	1,171	573
Los Angeles	3,149	2,265	3,058	2,064	3,273	3,018	2,562	2,774	475	2,705	2,425	2,350	2,007	2,978	2,301
Louisville	871	304	780	274	1,040	703	358	541	2,468	427	114	389	778	663	727
Memphis	1,157	527	1,066	311	1,287	969	738	921	2,439	807	464	612	396	920	804
Milwaukee	997	85	906	369	1,119	887	442	610	2,359	553	383	...	997	875	335
Minneapolis	1,332	420	1,241	556	1,454	1,222	777	945	2,096	888	718	335	1,285	1,210	...
Mobile	1,231	929	1,140	647	1,461	1,043	1,020	1,212	2,623	1,008	785	1,014	141	1,003	1,333
Montreal	386	841	477	1,051	330	574	623	434	3,115	704	826	920	1,655	614	1,125
Newark, N. J.	9	903	82	1,056	226	179	575	405	3,177	435	748	988	1,369	219	1,322
New Haven	76	980	167	1,141	140	264	628	445	3,254	520	833	1,065	1,448	304	1,400
New Orleans	1,372	912	1,281	699	1,602	1,184	1,073	1,256	2,482	1,142	829	997	...	1,144	1,285
New York	...	912	91	1,065	217	188	584	442	3,186	444	757	997	1,372	228	1,316
Ogden	2,496	1,404	2,315	1,414	2,528	2,296	1,851	2,019	780	1,962	1,792	1,579	1,891	2,284	1,316
Omaha	1,405	493	1,314	413	1,527	1,295	1,750	1,018	3,095	961	791	578	1,080	1,283	381
Philadelphia	91	821	...	974	321	97	493	416	2,742	353	666	906	1,281	137	1,241
Pittsburg	444	468	353	621	674	334	135	270	2,742	...	313	553	1,142	302	888
Portland, Me.	332	1,149	436	1,345	115	513	797	614	3,423	789	1,041	1,234	1,717	573	1,509
Portland, Ore.	3,204	2,292	3,113	2,312	3,326	3,094	2,649	2,817	772	2,760	2,590	2,378	2,746	3,082	2,042
Providence	190	1,034	251	1,230	45	378	682	499	3,308	654	926	1,119	1,562	418	1,454
Quebec	530	1,013	621	1,343	402	718	795	612	3,287	876	1,039	1,098	1,527	786	1,433
Richmond, Va.	342	879	252	918	573	155	552	553	3,153	417	581	964	1,046	115	1,299
Rochester, N. Y.	373	603	361	799	430	354	251	68	2,877	348	495	688	1,324	394	1,023
St. Joseph, Mo.	1,392	470	1,301	327	1,474	1,261	875	1,058	1,567	948	668	555	941	1,221	483
St. Louis	1,065	284	974	...	1,230	934	548	731	2,194	621	341	369	699	894	556
St. Paul	1,352	410	1,231	576	1,444	1,212	767	935	2,086	878	708	325	1,275	1,200	10
San Antonio	1,943	1,204	1,852	920	2,150	1,755	1,468	1,651	1,911	1,541	1,217	1,289	571	1,715	1,120
San Francisco	3,186	2,274	3,095	2,194	3,308	3,076	2,631	2,799	...	2,742	2,572	2,359	2,482	3,064	2,066
Seattle	3,151	2,239	3,060	2,232	3,273	2,941	2,596	2,764	957	2,707	2,537	2,154	2,931	3,029	1,818
Spokane	2,812	1,900	2,721	1,932	2,934	2,702	2,257	2,425	1,205	2,368	2,198	1,815	2,535	2,690	1,479
Springfield, Mass.	139	935	230	1,131	99	327	583	400	3,209	585	827	1,020	1,511	367	1,355
Tampa, Fla.	1,195	1,300	1,104	1,187	1,425	1,007	1,297	1,405	3,310	1,269	1,053	1,394	828	947	1,729
Toledo	703	244	615	437	795	595	113	296	2,518	261	203	329	1,032	595	664
Washington	228	790	137	894	458	40	437	438	3,064	302	553	875	1,144	...	1,210